Prototyping of Robotic Systems:

Applications of Design and Implementation

Tarek Sobh
University of Bridgeport, USA

Xingguo Xiong
University of Bridgeport, USA

Information Science
REFERENCE

Managing Director: Lindsay Johnston
Senior Editorial Director: Heather Probst
Book Production Manager: Sean Woznicki
Development Manager: Joel Gamon
Development Editor: Michael Killian
Acquisitions Editor: Erika Gallagher
Typesetter: Adrienne Freeland
Cover Design: Nick Newcomer, Lisandro Gonzalez

Published in the United States of America by
Information Science Reference (an imprint of IGI Global)
701 E. Chocolate Avenue
Hershey PA 17033
Tel: 717-533-8845
Fax: 717-533-8661
E-mail: cust@igi-global.com
Web site: http://www.igi-global.com

Library of Congress Cataloging-in-Publication Data

Prototyping of robotic systems: applications of design and implementation / Tarek Sobh and Xingguo Xiong, editors.
p. cm.
Includes bibliographical references and index.
Summary: "This book provides a framework for conceptual, theoretical, and applied research in robotic prototyping and its applications, covering the prototyping of various robotic systems including the complicated industrial robots, the tiny and delicate nanorobots, medical robots for disease diagnosis and treatment and simple robots for educational purposes"--Provided by publisher.
ISBN 978-1-4666-0176-5 (hardcover) -- ISBN 978-1-4666-0177-2 (ebook) -- ISBN 978-1-4666-0178-9 (print & perpetual access) 1. Robots--Design and construction. 2. Prototypes, Engineering. I. Sobh, Tarek M. II. Xiong, Xingguo, 1973-
TJ211.P77 2012
629.8'92--dc23

2011043975

British Cataloguing in Publication Data
A Cataloguing in Publication record for this book is available from the British Library.

List of Reviewers

Aarne Halme, *Aalto University, Finland*
Ahmad Taher Azar, *International Journal of System Dynamics Applications (IJSDA), USA*
Ahmed Elsayed, *University of Bridgeport, USA*
Akif Durdu, *Middle East Technical University, Turkey*
Andrew Goldenberg, *University of Toronto, Canada*
Aydan M. Erkmen, *Middle East Technical University, Turkey*
Ayssam Elkady, *University of Bridgeport, USA*
Barkan Ugurlu, *Toyota Technological Institute, Japan*
Brandon J. Stark, *Utah State University, USA*
Elif Kongar, *University of Bridgeport, USA*
Emin Faruk Kececi, *Istanbul Technical University, Turkey*
Erdem Erdemir, *Vanderbilt University, USA*
Gen'ichi Yasuda, *Nagasaki Institute of Applied Science, Japan*
Haibo Huang, *Robotics and Micro-systems Center, Soochow University, China*
Jack Toporovsky, *University of Bridgeport, USA*
Jeremy Li, *University of Bridgeport, USA*
Jianbing Hu, *Schlumberger Ltd., USA*
Jorge Manuel Miranda Dias, *University of Coimbra, Portugal*
Junling Hu, *University of Bridgeport, USA*
Kathiravelu Ganeshan, *Unitec Institute of Technology, New Zealand*
Lawrence Hmurcik, *University of Bridgeport, USA*
Linfeng Zhang, *University of Bridgeport, USA*
Madhav Patil, *University of Bridgeport, USA*
Nicola Ivan Giannoccaro, *University of Salento, Italy*
Nicolae Gari, *University of Bridgeport, USA*
Pierre Letier, *Space Applications Services, Belgium*
Qing'an Zeng, *North Carolina A&T State University, US*A
Sarosh Patel, *University of Bridgeport, USA*
Sebahattin Topal, *Middle East Technical University, Turkey*
Sedat Dogru, *Middle East Technical University, Turkey*
Srihari Yamanoor, *Stellartech Research, USA*
Tamás Haidegger, *Budapest University of Technology and Economics, Hungary*
Vicente Parra Vega, *University of Texas at Dallas, USA*
Vikas Reddy Enti, *Kiva System, Inc., USA*
Xiaojun Wu, *Data Storage Institute, A*STAR, Singapore*
Xuefu Zhou, *University of Cincinnati, USA*
YangQuan Chen, *Utah State University, USA*

Table of Contents

Detailed Table of Contents

Section 1
Robotic Prototyping: Methodologies and Design Optimizations

In this section, the general design methodologies and implementation strategies used in robotic prototyping are discussed. Several case studies are included to demonstrate the concepts. Prior to prototyping, a robotic system should be properly designed. A set of optimized design parameters needs to be decided, and the design can be verified with simulations. The modeling and design optimization strategies for some specific robotic systems are proposed. These include the modeling and simulation of discrete event robotic systems using extended Petri nets, as well as the design optimization of three-link planar manipulators using Grashof's criterion.

This chapter offers an overview of the general methodology and implementation strategy of robotic systems, supported by several case studies. Based on his practical industry experience as well as his teaching and research results as a faculty in a university, the author shares some unique views and perceptions about robotic prototyping. Three case studies are demonstrated in the chapter, which include a mobile tracker, a robot arm for internal operations in nuclear reactors, and a MRI-guided robot for prostate focal surgery. The chapter presents a general framework for robotic systems prototyping.

In this chapter, the modeling and simulation of discrete event robotic systems using extended Petri nets are introduced. Extended Petri nets are used as a prototyping tool for expressing real-time control of robotic systems. A coordination mechanism is introduced to coordinate the event activities of the distributed machine controllers through friability tests of shared global transitions. The proposed prototyping method allows a direct coding of the inter-task cooperation by robots and intelligent machines from the conceptual Petri net specification.

Space Station (ISS). The proposed robot can decrease the number of extravehicular activities of the astronauts, even for complex situations.

This chapter presents the prototyping and real-time implementation of bipedal humanoid robots based on dynamically equilibrated multimodal motion generation. The authors aim at developing a contemporary bipedal humanoid robot prototyping technology by utilizing a mathematically rigorous method to generate real-time walking, jumping and running trajectories. The main strategy is to maintain the overall dynamic equilibrium and to prevent undesired rotational actions for the purpose of smooth maneuvering capabilities while the robot is in motion. This is achieved by utilizing the Zero Moment Point criterion in spherical coordinates so that it is possible to fully exploit its properties with the help of Euler's equations of motion.

In this chapter, the prototyping of fully autonomous indoor patrolling mobile robots is proposed. The mobile robot employs a modular design strategy by using the ROS (Robot Operating System) software framework, which allows for an agile development and testing process. The primary modules—omni-directional drive system, localization, navigation, and autonomous charging—are described in detail. Special effort is put into the design of these modules to make them reliable and robust in order to achieve autonomous patrolling without human intervention. The experimental test results prove that an indoor mobile robot patrolling autonomously in a typical office environment is realizable.

In this chapter, the prototyping of a lunar excavator robotic system for participating in the 2010 NASA Lunar Excavating Competition is proposed. Remotely controlled by an operator using a computer via Wi-Fi telecommunication, the autonomous lunabotic excavator can perform the tasks of excavating regolith stimulant, collecting it in the excavator's dumpster, and depositing it into the assigned collector box. The design and implementation of the lunabotic excavator with all the functional modules are discussed. It is an interesting project, and the design strategy may offer hints leading to new and effective robotic excavators for planetary exploration.

In this chapter, two example projects about the prototyping of robotic systems in surgical procedures and automated manufacturing processes are reported. The prototyping and implementation of the robotic system is a scientific and technological integration of robotic system design, development, testing, and application. This chapter describes the recent development and applications of robotic systems to surgery procedures in biomedical engineering and automated manufacturing processes in industry. It includes the design and development, computer-aided modeling and simulation, prototype analysis, and testing of robotic systems in these two different applications.

Section 4
Prototyping of Robotic Systems for Other Applications

In this section, the prototyping and implementation of robotic systems for some other applications are discussed. Human-robot interaction (HRI) studies the dynamics of interaction between humans and robots. Research work on the use of robotic hardware and software integration to change human intentions is reported. Search and rescue (SAR) robots are used to save lives in environments which are unsafe for human first responders. The prototyping of autonomous multi-robot systems for search, rescue and reconnaissance is proposed. Finally, a heuristic approach based on a genetic algorithm for robotic disassembly sequencing optimization is reported.

This chapter proposes interesting research on the use of robotic hardware and software integration to change human intentions. Reshaping of human intention is achieved by robots moving in certain directions that have been determined a priori through observations from the interactions of humans with the objects in the scene. Being among the few studies on intention reshaping, the authors exploit spatial information by learning a Hidden Markov Model (HMM) of motion that is tailored for intelligent robotic interaction. The results obtained using the proposed approach show promising performance in reshaping the detected intentions.

Preface

The field of Robotics focuses on the study of design, simulation, implementation, and operation of robots for various applications. As programmable or remotely-controlled electromechanical machines, robots can perform certain tasks autonomously or semi-autonomously. Ever since the early eras of robotics, there has been a long history of utilizing robots to assist or replace human work. Nowadays, robotics technology has been intensively used in numerous industries including, but not limited to: manufacturing, automobile assembly, electronics, food processing, consumer goods, pharmaceuticals, health science, mining, planetary exploration, military weapons, et cetera. Robots can significantly improve the efficiency, reliability, accuracy and throughput of a traditional workforce. They are especially suitable to replace humans in performing tasks that are difficult, monotonous and tedious. They are also a good choice for delivering results in harsh environments that are too dangerous or life threatening for humans, such as workplaces with nuclear radiation, poisonous chemicals, fire, lack of oxygen or extremely high/low temperatures. Robotics, as a segment of the broader science of automation, has achieved tremendous progress in recent decades due to advances in related supporting technologies such as computing, control system, wireless communication, cameras and electronic sensing, as well as micro and nanotechnology. Many new robotic systems have been designed and implemented for various applications. For example, modern technologies in motion control, speech recognition, facial expression and human-robot interaction have led to bipedal humanoid robots that can walk, talk or even perform simple communication with human. As another example, robotic surgery utilizes computer-controlled robots to support a range of surgical procedures. Robotic surgery has enabled remote surgery, minimally invasive surgery and unmanned surgery. Compared to traditional surgery, robotic surgery has the advantages of high precision, miniaturization, smaller incisions, decreased blood loss, less pain, and quicker healing time.

Prototyping is the process of building an early sample or model of a system to prove the design concept or detect potential problems before batch fabrication. Prototyping is an important activity in engineering. Prototyping a design helps in determining system parameters, validating design concepts, debugging problems, and achieving optimized design of the system. The prototype can be used to verify design and measurements (e.g. performance, kinematics, function, architecture) and provide important information for the designer to identify which design option is better and which component may need further development and testing. Robotics is one of the industrial design fields in which prototyping is crucial for improved functionality. Prototype development is a good test for checking the viability of a proposed system. Prototyping of a robotic system is never trivial. Starting from the design specifications, a designer needs to decide the architecture of the robot best suitable for the specific application. Design optimization should be performed to decide the design parameters of the robot. During the implementation process, potential problems or challenges may be exposed. Some of the problems may be due to the design flaws and it may be necessary to further adjust or revise the design based on the information fed back from the prototyping process. Multiple iterations may be necessary before the robotic systems can be finally prototyped successfully. The prototyping of a robotic system involves many important decisions, such as deciding the architecture of the system, choosing proper materials and the electro-mechanical components, determining the design parameters and deciding the algorithmic control of the robot modules (kinematics, inverse kinematics, dynamics, trajectory planning, analog control, and digital computer control). The design for each individual module should be decided, such as the mechanical structure, power, actuation, control, telecommunication, and data-acquisition systems. Various CAD (Computer Aided Design) and CAM (Computer Aided Manufacturing) tools are available for the designers to design, visualize, and simulate a robotic system rapidly and cost-effectively. Computer simulation allows the discovery of potential problems before the robots are actually manufactured.

The objective of this book is to cover the most recent research frontiers and trends in robotic prototyping. This book discusses the design and implementation of various robotic systems and their applications, from complicated industrial robots to state-of-the-art micro and nanorobots for surgical applications, as well as robotic systems for educational purposes. Robotic systems are diverse in their structures, working principles, implementation strategies and applications. Newly emerging technologies such as computer vision, wireless communication, micro and nanotechnologies have been utilized in robots to enhance their function and performance. As a result, new robots have been proposed and prototyped for various new applications. This book aims to cover the prototyping of various robotic systems including complicated industrial robots, tiny and delicate nanorobots, medical robots for disease diagnosis and treatment, as well as simple robots for educational purposes. The design, implementation, and technical considerations in the prototyping of various robotic systems are discussed. Some case studies of robotic prototyping for industrial, medical, aerospace, and education applications are included. The applications of such robotic systems will be introduced. The new trends and most recent research frontiers in robotic prototyping and the applications are also covered.

This book is intended for researchers, industry engineers, and students working in the fields of robotics, control systems, medical electronics, computer vision, micro/nanotechnology, aerospace and other automation fields. It is also a general reference book for individuals who are interested in robotic design, prototyping and their applications. The book is designed to cover the design and implementation of a wide range of robots for different applications. Each category is accompanied with case studies.

The content of this book is arranged as follows. The book begins with an introduction about general design methodologies and implementation strategies used in robotic prototyping. Several case studies are included to illustrate the concepts. Prior to prototyping, a robotic system should be designed following the required specifications, and the design parameters should be decided. Chapter 2-3 introduces the theoretical design and optimization strategies of general robotic systems. This includes the modeling and simulation of discrete event robotic systems using extended Petri Nets in Chapter 2, and the design optimization of three-link planar manipulators using Grashof's Criterion in Chapter 3. Chapter 4 to Chapter 8 discuss the prototyping of various robotic systems for different applications, which include unmanned aerial vehicles, a portable haptic arm exoskeleton, a bipedal humanoid robot, an indoor fully autonomous patrolling mobile robot, as well as a lunabotic regolith excavator robot. The architecture, design considerations and implementation of each robot are discussed in detail. Chapters 9-12 are devoted to the topic of medical robotics. Chapter 9 provides a comprehensive overview of various medical robots being developed around the world, and future trends in this exciting field. Chapter 10 introduces the system development, assessment and clearance of surgical robotics. Chapter 11 proposes a piezoelectric driven ultrasonic cell injector that may be used as a powerful tool in medical research and disease treatment. Chapter 12 introduces two example projects about prototyping robotic systems for surgical procedures and automated manufacturing processes. Human-robot interaction (HRI) is another exciting field in robotics. It will eventually lead to robots that can directly interact with humans, which can assist people and improve human performance in daily and task-related activities. Chapter 13 discusses an interesting project addressing the use of robotics to change human intentions. Search and rescue (SAR) robots can help people to perform search and rescue tasks in environments that are unsafe or life-threatening for human rescuers, such as underwater, following an earthquake, or other natural disasters. The development of SAR robots must address challenges in robotic sensing, mobility, navigation, planning, integration and teleoperation control. Chapter 14 introduces a framework for prototyping autonomous multi-robot systems for search, rescue, and reconnaissance. Finally, robotic systems are also used for the automation of disassembly process of electronic products for recycling when they come to the end of life cycle. The disassembly process is generally very complex due to the large amount of components involved, and an optimization of the sequence is needed to improve efficiency and reduce cost. Chapter 15 proposes a heuristic approach for optimizing disassembly sequencing for robotic disassembly operations. The content of each chapter are briefly summarized below.

In Chapter 1, an overview of the general methodology for prototyping robotic systems is introduced, and some case studies are given to illustrate the whole sequence. Robotics research is the framework for studying hypotheses and conjectures, synthesizing new ideas, and discovering phenomena in the context of robotic systems. Prototypes are normally used for proof-of-principle and functional demonstration. In many cases they are part of the product design and development process. In direct relationship to successful demonstrations of new technology using prototypes and the existence or emergence of related markets, prototypes could end up being used or redesigned for commercialization. The market may even accept, albeit rarely, a demonstration prototype as the first generation product. Yet, such prototypes normally would be re-designed at a later stage as commercial prototypes. The process of prototyping is complex, costly, and risky. This chapter provides an experience-based framework of prototype development and commissioning. It introduces elements learned directly from the practice that encompass aspects of project management, technology development process, and commercialization in the context of Small and Medium Enterprises (SMEs). The contents of this chapter are based mainly on the author's practical experience of leading an SME technology developer. The author is also a faculty member working as a researcher and teacher. Because of the interrelationship between research and technology development, his views and perception of the topic may be unique, and they are personal. The chapter presents a general framework for robotic systems prototyping. Three case studies are demonstrated in the chapter, which include a mobile tracker, a robot arm for internal operations in nuclear reactors, and a MRI-guided robot for prostate focal surgery. The case studies back up the points made in the chapter and help the reader understand the outlined concepts.

The implementation of any robotic system should start from design optimization and simulation. Many different robotic systems may belong to the same category and follow similar theoretical modeling and simulation strategies. In Chapter 2, the modeling and simulation of discrete event robotic systems using extended Petri nets are introduced. This chapter deals with modeling, simulation, and implementation problems encountered in robotic manufacturing control systems. Extended Petri nets are adopted as a prototyping tool for expressing real-time control of robotic systems and a systematic method based on hierarchical Petri nets is described for their direct implementation. A coordination mechanism is introduced to coordinate the event activities of the distributed machine controllers through friability tests of shared global transitions. The proposed prototyping method allows a direct coding of the inter-task cooperation by robots and intelligent machines from the conceptual Petri net specification, so that it increases the traceability and the understanding of the control flow of a parallel application specified by a net model. This approach can be integrated with off-the-shelf real-time executives. Control software using multithreaded programming is demonstrated to show the effectiveness of the proposed method.

Design optimization is the key step to achieving a set of optimized design parameters for the robotic system. Chapter 3 introduces a novel and effective algorithm for optimal design of three-link planar manipulators, using Grashof's criterion. The design of robotic manipulators is dictated by a set of pre-determined task descriptions and performance parameters. These performance parameters are often defined in terms of workspace dexterity, manipulability, and accuracy. Many serial manipulator applications require that the manipulator have full dexterity about a work piece or a pre-defined trajectory, that is, to approach the given point within the workspace with all possible orientations about that point. Grashof's criterion defines the mobility of four-link closed chain mechanisms in relation to its link lengths. A simple assumption can convert a three-link serial manipulator into a four-link closed chain so that its mobility can be studied using Grashof's criterion. With the help of Grashof's criterion, a designer can not only predict and simulate the mobility of a manipulator during its design, but also map and identify the fully-dexterous regions within its workspace. Mapping of the dexterous workspace is helpful in efficient task placement and path planning. A simple algorithm using Grashof's criterion for determining the optimal link lengths of a three-link manipulator is proposed in order to achieve full dexterity at the desired regions of the workspace. The generated design is also tested by applying joint angle limitations.

Starting with Chapter 4, the design and prototyping of various robotic systems are introduced. In Chapter 4, a vertical take-off and landing unmanned aerial vehicle platform for personal remote sensing is proposed. Unmanned Aerial Vehicles (UAVs) for civilian applications are in a rapidly growing sector in the global aerospace industry that has only recently begun to gain traction. In this relatively immature field, there is such a steep learning curve that it can be difficult for research groups to begin development of well designed UAV systems. In this chapter, the authors present the AggieVTOL, a modular multi-rotor rotorcraft UAV prototype platform, and an overview of the prototyping phase of its development, including design parameters and the implementation of its modular subsystems. Performance results demonstrate the effectiveness of the platform. The design and implementation strategies in this project can be extended to other UAV prototyping as well.

In Chapter 5, the prototyping of a portable haptic arm exoskeleton for aerospace application is proposed. The proposed robot is a seven-degree-of-freedom force-reflective device able to produce haptic rendering of the human arm, either as master for teleoperation of a slave robot, or in interaction with a virtual reality. The project was conducted on behalf of the European Space Agency (ESA) as a prototype of the master device used for teleoperation of future anthropomorphic space robotic arms on the International Space Station (ISS). The motivation is to decrease the number of extravehicular activities of the astronauts, even for complex situations. The structure of a portable anthropomorphic exoskeleton of seven degrees of freedom was selected by ESA due to the fact that it allows a more intuitive control of anthropomorphic slave arms, and it also allows multiple contact points, offering a larger workspace (comparable to the human arm). Besides, being attached on the astronaut, the system involves only internal forces (it is self-equilibrated) and can be used in zero-gravity.

Chapter 6 presents the authors' work on prototyping and real-time implementation of bipedal humanoid robots. Dynamically equilibrated multimodal motion generation is required for the proposed bipedal humanoid robot. This chapter is aimed at describing a contemporary bipedal humanoid robot prototyping technology, accompanied with a mathematically rigorous method to generate real-time walking, jumping and running trajectories that can be applied to this type of robots. The main strategy in this method is to maintain the overall dynamic equilibrium and to prevent undesired rotational actions for the purpose of smooth maneuvering capabilities while the robot is in motion. In order to reach this goal, the Zero Moment Point criterion is utilized in spherical coordinates so that it is possible to fully exploit its properties with the help of Euler's equations of motion. Such a strategy allows the designer to characterize the rotational inertia and therefore the associated angular momentum rate change terms, so that undesired torso angle fluctuations during walking and running are well suppressed. It allows the designer to prevent backwards-hopping actions during jumping as well. The proposed approach is validated by performing simulations using a precise 3D simulator and conducting experiments on an actual bipedal robot. Results indicate that the method is superior to classical methods in terms of suppressing undesired rotational actions, such as torso angle fluctuations and backwards-hopping.

In Chapter 7, the prototyping of fully autonomous indoor patrolling mobile robots is proposed. The mobile robot employs a modular design strategy by using the ROS (Robot Operating System) software framework, which allows for an agile development and testing process. The primary modules - omni-directional drive system, localization, navigation, and autonomous charging are described in detail. Special effort is put into the design of these modules to make them reliable and robust in order to achieve autonomous patrolling without human intervention. The experimental test results prove that an indoor mobile robot patrolling autonomously in a typical office environment is realizable.

In Chapter 8, the prototyping of a lunar excavator robotic system is discussed. The lunabotic excavator was developed for participating in the 2010 NASA Lunar Excavating Competition. Being remotely controlled by an operator using a computer via Wi-Fi telecommunication, the autonomous lunabotic excavator can perform the tasks of excavating regolith stimulant, collecting it in the dumpster, and dumping it into the assigned collector box. The excavator include multiple modules including mechanical frames, front/rear wheels, excavating conveyor, steering system, dumping system, power supply and distribution system, actuation system, switch control system, data acquisition system and telecommunication system. The design and implementation of the lunabotic excavator with all the functional modules are discussed. The design concepts used in this project may offer hints leading to new and effective robotic excavators for planetary exploration.

Robotics is also finding exciting applications in the biomedical field, leading to a new interdisciplinary field of medical robotics. Medical robotics will significantly impact the health care industry, resulting in revolutionary change to the way doctors diagnose and treat diseases. For example, robotics is already being used for minimally invasive surgery (MIS), remote surgery (telesurgery), patient monitoring and stabilization, rehabilitation facilities, as well as medical training. Minimally invasive surgery based on medical robots results in smaller incisions, shorter hospital stays, improved prognoses and reduced cost. Chapter 9 to Chapter 12 are specifically devoted to this important and exciting new field. In Chapter 9, a comprehensive overview about medical robotics is proposed. As an interdisciplinary field, medical robotics focuses on developing electromechanical devices for clinical applications. The goal of this field is to enable new medical techniques by providing new capabilities to the physician or by providing assistance during surgical procedures. Medical robotics is a relatively young field, as the first recorded medical application occurred in 1985 for a brain biopsy. However, medical robotics has tremendous potential for improving the precision and capabilities of physicians when performing surgical procedures and it is believed that the field will continue to grow as improved systems become available. This chapter begins with an introduction to robotics, followed by a historical review of their use in medicine. Clinical applications in several different medical specialties are discussed. Various medical robots, ranging from commercial products from industry, to the research works of university labs are introduced. The technology challenges and areas for future research in medical robotics are also discussed.

In Chapter 10, the system development, assessment, and clearance of surgical robots are introduced. Information Technology and robotics have been integrated into interventional medicine for over 25 years. Their primarily aim has always been to provide patient benefits through increased precision, safety and minimal invasiveness. Nevertheless, robotic devices should allow for sophisticated treatment methods that are not possible by other means. Several hundreds of different surgical robot prototypes have been developed, while only a handful passed clearance procedures and were released to the market. This is mostly due to the difficulties associated with medical device development and approval, especially in those cases when some form of manipulation and automation is involved. This chapter presents major aspects of surgical robotic prototyping and current trends through the analysis of various international projects. It spans across the phases from system planning, to development, validation and clearance.

Research in medical robotics also leads to new and improved tools for medical research and disease treatment. In Chapter 11, the design and evaluation of a piezoelectric driven ultrasonic cell injector is proposed. Piezo drill, a new cell injection technology that utilizes piezo-driven pipettes with a very small mercury column for cell injection, was first invented in 1995. It was then successfully applied to intracytoplasmic sperm injection (ICSI) in a variety of mammal species. Although this technique significantly improves the survival rates of the ICSI process, shortcomings such as the damage to the cell membrane due to large lateral tip oscillations of the injector pipette, complexity of operation and toxicity of mercury immensely limit its application. In this chapter, a novel piezo-driven cell injection system for automatic batch injection of suspended cells is presented. It has a simplified operational procedure and better performance than previous piezo-driven cell injectors. Specifically, this new piezo-driven cell injector design has three advantages. First, by centralizing the piezo oscillation energy on the injector pipette, it eliminates the vibration amplitude of other parts of the micromanipulator. Second, a small piezo stack is sufficient to perform the cell injection process. Third, detrimental lateral tip oscillations of the injector pipette are attenuated to a satisfactory amount even without mercury column. The elimination of mercury enables wide applications of the proposed cell injection technology in a number of cell manipulation scenarios. Ultrasonic vibration micro-dissection (UVM) theory is utilized to analyze the piezo-driven cell injection process, and lateral oscillation of injector pipettes is investigated. Experiments on cell injection of a large amount of zebrafish embryos indicate that the injector pipette is capable of piercing through cell membranes with low injection speed and almost no deformation of the cell wall, but with a high success rate.

In Chapter 12, two example projects about the prototyping of robotic systems in surgical procedures and automated manufacturing processes are reported. The prototyping and implementation of the robotic system is a scientific and technological integration of robotic system design, development, testing and application. This chapter describes the recent development and applications of robotic systems to surgery procedures in biomedical engineering and automated manufacturing processes in industry. It includes design and development, computer-aided modeling and simulation, prototype analysis and testing of robotic systems in these two different applications.

Human-robot interaction (HRI) is another new and increasingly popular field that studies the dynamics of interaction between humans and robots. Many researchers are putting effort into developing robotic systems that are capable of performing direct, safe and effective interactions with humans. It is also interesting to look into how robotics can be used to affect human behavior and intentions. As an interdisciplinary field, HRI requires knowledge about robotics, psychology, communication, ethics and cognitive science. Chapter 13 proposes interesting research about using robotic hardware and software integration to change human intentions. Estimating and reshaping human intentions are among the most significant topics of research in the field of human-robot interaction. This chapter provides an overview of intention estimation literature on human robot interaction and introduces an approach to how robots can voluntarily reshape estimated intentions. The reshaping of the human intention is achieved by robots moving in certain directions that have been determined *a priori* through observations from the interactions of humans with the objects in the scene. Being among the few studies on intention reshaping, the authors exploit spatial information by learning a Hidden Markov Model (HMM) of motion that is tailored for intelligent robotic interaction. The algorithmic design consists of two phases. First, an approach is used to detect and track a human to estimate his/her current intention. Later, this information is used by autonomous robots that interact with the detected human to change the estimated intention. In the tracking and intention estimation phase, postures and locations of the human are monitored by applying low-level video processing methods. In the latter phase, learned HMM models are used to reshape the estimated human intention. This two-phase system is tested on video frames taken from a real human-robot environment. The results obtained using the proposed approach show promising performance in reshaping the detected intentions.

Robots have been widely used in search and rescue tasks in environments that are dangerous or life-threatening to human rescue workers. During an earthquake, nuclear accident or other disaster, rescue robots can save many lives without endangering the human rescue workers. In Chapter 14, a framework for the prototyping of autonomous multi-robot systems for search, rescue and reconnaissance is proposed. Robots consistently help humans in dangerous and complex tasks by providing information about, and executing tasks in disaster areas that are highly unstructured, uncertain, possibly hostile, and sometimes not reachable by humans directly. Prototyping autonomous multi-robot systems in disaster scenarios both as hardware platforms and software can provide the foundational infrastructure for comparing the performance of different methodologies developed for search, rescue, monitoring and reconnaissance. In this chapter, the prototyping modules of heterogeneous multi-robot networks and their design characteristics are discussed. Two different scenarios are considered in the prototyping process. One is the search and rescue in unstructured complex environments. The other is the connectivity maintenance in Sycophant wireless sensor networks, which are static ecto-parasitic clandestine sensor networks mounted incognito on mobile agents, using only the agent's mobility without intervention, and are cooperating with sparse mobile robot sensor networks.

Robots are also used for the disassembly of electronic products when they come to the end-of-life cycle. Electronic products enter the waste stream rapidly due to technological enhancements. Their parts and material recovery involve significant economic and environmental gain. To regain the value added to such products, a certain level of disassembly may be required. Disassembling electronic products is a tedious and potentially dangerous process, and robots are increasingly used for this operation. Due to large amount of components in electronic products, an efficient algorithm is needed to optimize the sequence in robotic disassembly operations. In Chapter 15, a heuristic approach for disassembly sequencing problem for robotic disassembly operations is proposed. Disassembly operations are often expensive and the complexity of determining the best disassembly sequence increases as the number of parts in a product grows. Therefore, it is necessary to develop methodologies for obtaining optimal or near optimal disassembly sequences to ensure efficient recovery process. To that end, this chapter introduces a Genetic Algorithm-based methodology to develop disassembly sequencing for end-of-life products. A numerical example is presented to provide and demonstrate better understating and functionality of the algorithm. The proposed algorithm is proven to be effective in optimizing the sequence of robotic disassembly operation and improving the efficiency of the process.

Finally, this book is a joint effort of robotic researchers and engineers from around the world. The editors hope that this book will be helpful to researchers and engineers engaged in the design and prototyping of modern robotic systems, as well as students of mechanical engineering, electrical engineering, and computer engineering who are interested in the robotics field.

Tarek Sobh
University of Bridgeport, USA

Xingguo Xiong
University of Bridgeport, USA

Acknowledgment

This book is a result of the contributions and work of many robotic researchers and engineers around the world. It is the collection of a wide range of research results from the robotics scientific research community, including scientists and engineers from universities, research institutes, and industry. The design, prototyping of various robotic systems, and their applications are explored. We hope you will enjoy the book. As editors, we would like to take this opportunity to thank all our chapter authors for their valuable contributions to this book. We are also very grateful to the editorial advisory board members for their valuable advice and suggestions to ensure the quality of the book. Special thanks to all our reviewers for their valuable time and efforts in reviewing the book chapters, as well as their constructive comments to help our authors.

Tarek Sobh
University of Bridgeport, USA

Xingguo Xiong
University of Bridgeport, USA

Section 1
Robotic Prototyping: Methodologies and Design Optimizations

Chapter 1
Prototyping Robotic Systems: Methodology and Case Studies

Andrew Goldenberg
Engineering Services Inc. (ESI), Canada & University of Toronto, Canada

ABSTRACT

This chapter provides an experience-based framework of prototypes development and commissioning. It introduces elements learned directly from the practice that encompass aspects of project management, technology development process, and commercialization in the context of Small and Medium Enterprises (SMEs). The contents of this chapter are based mainly on the author's practical experience of leading an SME technology developer. The author is also a faculty member working as a researcher and teacher. Because of the interrelationship between research and technology development, his views and perception of the topic may be unique, and they are personal. The chapter presents a general framework for robotic systems prototyping. To back up the points made in the chapter, three case studies of robotic prototyping are included to help the reader perceive the outlined concepts.

INTRODUCTION

Robotics research is the framework for studying hypotheses and conjectures, synthesizing new ideas, and discovering phenomena in the context of robotic systems. The research usually leads to peer-reviewed publications, occasional patenting, and in some cases, construction of working prototypes that may or may not have commercial value or intent. Prototypes are normally used for proof-of-principle and functional demonstration. In many cases they are part of the product design and development process. In direct relationship to successful demonstrations of new technology using prototypes, and the existence or emergence of related markets, the prototypes could end up being used or redesigned for commercialization.

DOI: 10.4018/978-1-4666-0176-5.ch001

The market may even accept, albeit rarely, a demonstration prototype as the first generation product. Yet, such prototypes normally would be re-designed at a later stage as commercial prototypes. The process of prototyping is complex, costly, and risky.

A prototype is an early sample or model of a concept, idea, or process that would allow preliminary functional and performance evaluation to be conducted through experimentation. Prototypes serve as benchmarks against which the original idea is measured in terms of representation, functionality, and even manufacturability and servicing of the related product. The early evaluation helps alleviate the concerns and uncertainty as to whether or not the new product will actually perform what and how it is expected to, as new product designs, more often than not, raise unexpected problems.

A prototype is used as part of the product design process to allow exploring design alternatives, and confirm functionality and performance prior to proceeding to the design of the commercial prototype, and much earlier than introducing the new product to the market. Prototypes are used also to confirm and verify market interest and readiness to accepting a new product; other times, prototypes are used to verify the performance or suitability of a specific product or features for executing certain tasks.

In some product development organizations, prototyping specialists are employed - individuals with specialized skills and training in general prototype development and fabrication techniques. They can help bridge between theoretical ideas, designs, and the fabrication of commercial prototypes as a prelude to products.

This chapter addresses prototyping from the perspective of new robotic product development. It is a process similar to general prototyping, but it does not benefit from the anticipated experience with earlier products as reference. Prototyping in robotics is by-and-large opportunistic, and constrained only by the original idea and its functional purpose.

BACKGROUND

There is no general agreement on what constitutes a "prototype" and the word is often used interchangeably with the word "model," which may cause confusion. In general, "prototypes" can be of four basic types:

Proof-of-Principle Prototype: In electromechanics it is sometimes called a breadboard. This type of prototype is used to test specific features of the intended design without attempting to emulate the visual appearance, required materials, or assembly process. Such prototypes are used to identify which design features may not work, and where further in-depth development and testing is necessary.

Functional Prototype: This type of prototype allows designers to explore the basic size, look and functionality of a new product. It can help assess ergonomic factors, and provides insight into industrial design of the product. These prototypes capture the intended design aesthetic and emulate the appearance of the intended product. These prototypes are intended for marketing, and are generally durable enough to be shown and use by representative users and consumers. The prototypes are suitable for use in critical design reviews and photo shoots for sales literature.

Commercial Prototype: This prototype provides the final design, aesthetics, materials and functionality of the intended product. The construction of this fully working full-scale prototype is the ultimate test of the original concept and the engineers' final check for design flaws to allow further improvements to be made before larger production runs begin.

Production Prototype: The difference between the commercial and production prototypes is expressed by three elements: (i) Material - production materials may require specific manufacturing processes involving higher capital costs than what was used for prototyping; (ii) Manufacturing - often expensive and time consuming unique tooling may be required to fabricate some custom designed parts; and (iii) Fidelity - final production design often requires extensive effort for design for manufacturing that was avoided in the earlier prototypes.

In the product development cycle, the prototype iterations may be labelled to be an "Alpha" prototype, "Beta", etc. However these distinctions are only semantically relevant, as they do not provide intrinsic meaning.

In general, functional-to-commercial prototyping costs are substantially greater than the cost of the final production prototype due mainly to extensive labor involved in the development and design. Prototypes must be amenable to revisions of the design for the purposes of refinement and improvement of functionality. It is also possible to use prototype testing to reduce the risk that a design will not perform acceptably, though prototypes generally cannot eliminate all risk. There are pragmatic and practical limitations to the ability of a prototype to match the final intended performance of the product, and some allowances and engineering judgement are often required before moving forward to the production design.

Building prototypes is also expensive in terms of material and related machine shop tasks, and software coding. As an alternative, "rapid-prototyping" techniques are used for the initial prototypes, which implement some, but not all, of the design features. This allows designers to rapidly and inexpensively test the elements of the design that are most likely to generate problems. Rapid prototyping is not addressed in this chapter.

Another approach to reducing cost of prototyping is Virtual Reality models. Virtual Reality is used to provide a software alternative to physical prototyping. The software tools are very advanced, but certain nuances of the prototype system behaviour cannot be reconciled through the VR technology available today. Such models are used more frequently at the 'proof-of-principle' stage, as they are devoid of physical properties for real life experimentation that functional-to-commercial prototypes have.

The topic of prototype development is usually addressed in the research literature as an afterthought; that is, the prototype is only used to showpiece the main intent: demonstration of a new idea, concept, or phenomenon. The actual process of prototyping is not discussed in the robotics literature because it is not considered as research. It can be found in trade literature of systems, products, and components, but there, the provided information is submerged and devoid of details in order to maintain proprietary control, and ensure enforcement and control of intellectual property.

Prototyping is a systematic *process* that can be described by the following streamlined phases: Discovery, Design, Manufacturing, and Acceptance Testing. In practice, these phases are broken down into the following steps: *information gathering, concept synthesis, development and design, manufacturing, testing, and evaluation.* These steps emphasize the streamlined aspect of the prototyping process.

Prototyping is not a mocking-up of the final product. As the prototyping work progresses, the commercial prototype becomes the product model. After building and experimenting with the model and modifications are made to it, the prototyping process as a development phase ends, leading to the production prototype. That is the model that becomes the actual product.

ISSUES IN PROTOTYPING

Expression of Idea

The method of prototyping as an expression of an idea affects the intensity of development, motivation, communication, and outcome of design. Most of the practices revolve around the creative process, but they include also the concerns with the future product cost and its servicing. Nonetheless, the practice of prototyping is mainly affected by the experience of the designers involved, in particular their previous design experience with prototyping. This is productive and counter-productive at the same time. The innovativeness is best expressed when there is experience, but not if it limits the developer from seeking "out-of-the-box" solutions for a new design because of past experience. Open-mindedness is best expressed when the experience is only used to avoid pitfalls, rather than to guide and channel the designer to already known solutions.

Cost

Frequently, the challenge is in the details of a prototype. In physically generating a working piece of robotic hardware, 80% of the total work gets finished in 20% of the total time required to finish the project. Invariably the design of the prototype will change as the project development advances. The potential customer might have suggestions; the information collected during the experimentation might raise issues of concern. Thus, variability of the internal design of the product becomes a serious concern, although the overall functionality and external aesthetics may be already satisfied. The developers must be potentially ready for re-design and modification. This implies that the prototypes must have flexible software and hardware design architectures. To reduce cost and simplify the design, at times the single goal of a prototype is to move to the next phase of the design process, whether that next step involves a more refined version or the final production prototype. A designer has to consider always that parts of the system might be reused in the next phase, thus focus any extra effort on those reusable parts. Unfortunately, this approach raises the cost of prototyping in terms of both time and materials.

Intellectual Property

Intellectual Property protection is a serious concern that has a paramount importance commercially. The exposure of the early prototypes at the inopportune time, or in an unsuitable forum, may jeopardize the chances of protection. This issue is part of the overall management of the product development.

Prototype Development

The prototype development, as part of the product development cycle, is subject to many factors that influence the outcome. If the planning and execution of the prototype development and subsequent commercial exposure are not performed adequately, even an ingenious idea with great market potential may lead to limited commercial value.

In general, product development must follow a rigorous Project Management methodology. The components of the methodology are: Study of Requirements, Determination of Feasibility, Definition of Specifications, Design and Programming, Manufacturing, Integration, Testing, Acceptance, Training of Users, and Installation. The various components usually follow each other serially. The requirements study begins after the original idea has been fully understood. Soon after it has begun, the specifications are completed. When the specifications are defined, the preliminary design begins. Shortly after the integration of the prototype the verification, acceptance testing, and the user training begin. The milestone dates of the prototype development are usually the requirements definition, preliminary design review, critical design review, and acceptance testing.

The steps described in this chapter are the generic form of Project Management methodology. These steps are generally followed for project management of any prototyping process. The only difference in robotic prototyping is that overlapping of some of the project phases listed above occurs. Actually, robot prototyping provides more of an iteration of phases instead of a non-overlapping sequence and continuation of phases.

The next sections address the issues of project management, system engineering, and quality assurance in the context of prototyping.

PROTOTYPE DEVELOPMENT METHODOLOGY

Project Management

Project Management methodologies emphasize scheduling, cost control, meeting of milestones, and interim user's signoffs. The interim user's signoffs are more important than the requirements or the final product acceptance.

The objective of project management is to direct the development, deployment, field-testing, and delivery of the prototype. The project management objectives also include the identification and mitigation of risks (technical, operational, and programmatic) associated with the development. This is done progressively as part of the Preliminary Design Review, Critical Design Review, and final acceptance with the client/user involvement up to a formal approval.

Project management is based on a management plan whose scope is to provide a structured, manageable, traceable, and accountable approach to project development. The plan includes allocation of resources, control, and management of risk. The prototype project management methodology is based on the following fundamental elements of generic project management:

- *Team:* Project Management Team (PMT) constituted of experienced managers who are also technical experts.
- *Process:* Project Management Process (PMP) based on typical four standard phases of project life cycle: Discovery, Design, Manufacture and Integration, and Acceptance Tests.
- *Plan:* Project Management Plan includes Work Break-down Structure, Work Packages, and Personnel allocation, Schedule, Managerial Risk Assessment, Milestones, and Deliverables.
- *Control:* Project Management Control includes Project Scope, Schedule, Cost, Quality, Human Resource, Communications, Integration, Risk, and Performance.

Team: The PMT is supported by a Systems Engineering Team (SET) who ensures that the product satisfies all technical requirements set forth. Due to the highly integrated nature of the development across multiple disciplines, key team members are involved in both SET and PMT activities. The intent of this overlap is to ensure balance and consistency between all technical and managerial activities.

Process: The PMP evolves in four phases. They cover all stages of project life cycle.

- *Discovery Phase:* This phase includes the derivation of system specifications, generation of conceptual designs, modeling, and system trade-offs. Through an in-depth review of the project's objectives, the PMT develops a thorough understanding of the required technologies. The PMT identifies all relevant project tasks, their interrelationships, and the resources needed to complete the project on schedule, within budget, while meeting all stakeholders' expectations. This phase consolidates the initial project's Work Breakdown Structure, Work Packages, and Schedule. As well, this phase addresses Design Concepts, Modeling, and System Trade-offs.
- *Design Phase:* The PMT and SET analyze the information gathered during the Discovery Phase, and using its findings develop, document, and validate a workable project plan that conforms to the nominal objectives. The PMT works very closely at this stage with the SET team that, in paral-

lel, leads and monitors the System Design Phase. During this phase, the SET's level of activity intensifies and the PMT operates in a "surveillance" (control) mode, tracking project progress and deliverables, intervening when necessary to steer the project back onto the right track. Subcontractors' interactions are monitored closely by the SET.

- *Manufacturing and Integration Phase:* The design is implemented, integrated, tested, and modified until all project objectives are met. Subcontractors' system integration is part of this phase of work. Payload integration is included in this phase.
- *Final Acceptance Phase:* This phase includes all activities needed for the final prototype field testing. The PMT will determine the appropriate project activities to ensure that the prototype will operate reliably for its intended use. Following the test and verifications the PMT and SET will assess whether or not further upgrades are necessary, or the project is completed.

Plan: The plan includes the four phases of system development

- **Phase 1: Discovery Phase**
  - *Objective:* Deriving system requirements from the high-level performance and technical requirements.
  - *Constraints*: Derived system requirements are subject to Project Management Team (PMT) review prior to release. The PMT is responsible for ensuring that all high-level requirements are being addressed.
  - *Benefits*: Detailed functional requirements and design constraints of each subsystem can be derived. Overall system performance measures are defined. Interface requirements are defined. Trade-offs and comparison metrics are defined. Work schedules and budget estimates are broken down by subsystem.
- **Phase 2: Design Phase**
  - **Activity 2.1: Concept Design**
    - *Objectives:* The concept of the system design is generated. Requirements and functional analysis for all levels of subsystems are generated. Planning of the design process and verification and testing stages are performed.
    - *Constraints*: Results of trade-off analyses; available technology; available resources.
    - *Benefits:* Firmed up development plan; trade-off analyses to optimize the concept; detailed subsystem and interface requirements are generated; prepare Concept Design Review.
  - **Activity 2.2: Preliminary Design**
    - *Objectives*: Specifications for all subsystems, assemblies and components are generated. Hardware and software requirements are defined in detail.
    - *Constraints*: System requirements; hardware & software test and validation plans; prototype development plan; tasks and missions to be performed.
    - *Benefits*: Definition of internal and external interfaces; preliminary design verifications; software data interfaces; breadboard models and prototypes; long-lead procurement; verification plan for each subsystem; prepare Preliminary Design Review.

- **Activity 2.3: Detailed Design**
    - *Objectives:* Hardware and software development towards production.
    - *Constraints:* Interfaces design requirements.
    - *Benefits:* Detailed design and preparation for Critical Design Review; verification plans; pre-prototype bench testing; interface design testing; production phase planning.
- **Phase 3: Manufacturing & Integration Phase**
  - *Objectives*: Manufacture and assembly of system; complete software development.
  - *Constraints:* Software verification and testing; updated test procedures; system integration planning.
  - *Benefits:* Prototype operational; field deployment operations plan; system integration test plan; payload integration test plan; perform Preliminary Acceptance Review.
- **Phase 4: Final Acceptance Phase**
  - *Objectives:* Integrated system testing.
  - *Constraints*: Integration testing plan; upgrading requirements.
  - *Benefits*: System integration testing procedures; perform Integrated System Acceptance.

Control: Management control objectives are design, schedule, cost, and programmatic

- **Design Data Control:** Design data must be controlled with high diligence by the PMT and SET due to the fact that some of the project activities bear a high technical risk, and could delay the project and increase expenditures. This control is very complex and not totally repeatable at various stages of the project. It requires a mixture of standard project management approaches, technical knowledge, and experience, to allow effective assessment of the impact of departures from, and adaptation of the plan in the technical context, and the costs involved. Design Data Control is performed by using the Configuration and Data Management Plan (CDMP) facility.
- **Schedule Control:** At the start of the project, an overall schedule based on the project Work Breakdown Structure and Work Package Definitions is prepared by the PMT. Tasks are coded numerically, estimated in terms of number of hours and projected end dates, and personnel are assigned accordingly. The schedule is updated at least once monthly. The updates are based on reported progress and related revised estimates of end dates of various project tasks. Additional tasks are added to the schedule as necessary. Major deviations from the schedule are addressed immediately by assigning backup personnel to the tasks that are behind or anticipated to fall behind schedule, or by implementing other measures of compensation for delays.
  Schedule control includes monitoring of personnel's, collaborators' and consultants' schedules. Frequent meetings or telecoms are held with the collaborators and consultants to ascertain the timing of their schedules and delivery of milestones, and face-to-face meetings are held regularly or as often as necessary. Special attention must be given to monitoring of tasks that provide risks of overruns in schedule and/or cost. These items are primarily those identified in the PMT's Management Risks list.
- **Cost Control:** Before the start of the project, Cost Control spreadsheets of estimated costs for labour, parts, materials, and

subcontractors' work are generated. The spreadsheets are updated regularly (e.g. monthly) or as often as necessary monthly to include expenditures planned, earned value, and amounts actually paid and encumbered. Balances are weighed against original estimates, and major deviations are addressed immediately, or, in case of a budget overrun, a resolution on cost reduction is implemented. In the event of an unavoidable cost overrun, the PMT will decide if: (i) specifications must be redefined; or (ii) change of scope of work is to be considered in consultation with client.

- **Programmatic: Contractual Reporting and Meetings Control:** Contractual reporting and milestone meetings are very important and must be closely monitored from the start of the project. The PMT must ensure that these obligations are met in terms of both contents and schedules.

System Engineering

System Engineering integrates human resource expertise with company's procedures, practices, and internal working methods. This approach is adapted to each specific project's objectives. The system engineering process is the key element in the prototype development. A System Engineering Team (SET) with participants from each development collaborator and discipline leads the technology development. The SET is delegated to execute its authority on technology development, and support the programmatic requirements of the PMT. Throughout the project the SET interacts with the Quality Assurance personnel, and ensures that all design activities are properly documented, and that the Configuration and Data Management (CADM) practices are followed with the intent of facilitating the storage, sharing of information, and exchanging data between various teams. The PMT and SET function in coordination at all times during the life of the project to ensure the delivery of the commitments. A good choice is to have most of the members of the SET be also members of the PMT.

System Engineering Team Functions

The main functions of the SET are to plan the technology development, to guide and monitor the design activities, to manage the integration and testing, and to support the acceptance tests. System Engineering covers:

- **System Specifications:** Determination of system specifications based on requirements
- **Concept design:** Concept formulation based of system specifications
- **System Modeling:** System level model to study feasibility of compliance to system requirements
- **System Trade-off Studies:** Performance of system trade-offs; definition of metrics for comparison; and decision criteria for each trade-off
- **System Design:** Lead, monitor, and integrate the efforts of the various design teams
- **Performance Measure and Evaluation:** Definition of system performance measures and methods of performance evaluation
- **System Validation and Verification Plan:** Generation of test plans that verify the system and subsystem requirements; monitor the execution of these plans

In addition, the SET has the following responsibilities:

- Manage system configurations, interfaces, and design data
- Support the PMT on all programmatic issues including management of technical risks

- Manage product and data documentation that will enable operation, maintenance, and updating of the prototype

In principle, the SET has the role of planning the technical work in accordance with the requirements, advising the PMT on requirements of human resources, monitoring development progress, and providing necessary support to meet the contractual requirements undertaken by the PMT. The SET tracks the technical development and reports to the PMT on either satisfaction or departure from expected results or milestones. In turn, the PMT informs the SET if the cost is under control and schedules are met, whether allocation of resources is possible, and if the client is completely satisfied with the status of the project.

The SET must operate on the basis that it has access to the most expert resources of the enterprise, and it benefits from direct link to the PMT on matters of project control, resource allocation, schedules, and liaison with the client. The SET is, in fact, the team that de-facto, and in the end, delivers the product. The PMT is the team that has the contractual responsibility for the deliverables to be on schedule and be compliant with all system requirements.

System Engineering Process

The systems engineering process is hierarchical, and consists of a series of phases, all of which are intended to take the overall system requirements through detailed specifications and verification/validation methods to the final deliverables. These phases are a combination of the activities listed in the previous sections. The SET directs the entire technical team throughout each phase, checks the results of the design activities against the overall system requirements, and takes corrective action when necessary under the supervision of the PMT.

Phase 1: Discovery - Requirements, Concept, Modeling, and Trade-offs

This initial phase involves deriving system specifications from the high-level requirements provided by the client. The input to this phase includes:

- High level system requirements
- Identified measures of performance evaluation
- Known system constraints
- Identified external interfaces

Using this input, the SET applies functional analysis techniques to map the high level requirements into system functions. These functions are then individually broken down hierarchically into sub-functions until detailed system specification are obtained. Once the set of specifications is obtained, it is checked back against the original system requirements in order to ensure that they are all met.

The system specifications are subject to PMT approval prior to release. The PMT review undertakes to confirm that the specifications capture the requirements of the system. The review also verifies that the specifications possess the following attributes:

- **Verifiable:** they are stated in a way that they can be verified using an objective analysis or testing procedure
- **Traceable:** they must be traceable to a higher level requirement to ensure that all high level requirements are flowed-down into derived specifications
- **Achievable:** they must be technically feasible within the allotted time and budget
- **Unambiguous:** they must have only one possible interpretation
- **Consistent:** they must not be in conflict with each other

- **Complete:** they must contain all of the necessary information to confirm that all high level requirements are met (in context, rationale, and terms of reference)

The output of this phase provides:

- System and subsystems design specifications
- System concept and configurations of subsystems
- Potential system trade-offs and comparison metrics
- System and subsystem performance measures
- System and subsystems interfaces definition
- Work schedules and budget estimates broken down by subsystems

Phase 2: System Design Phase

Using the output of the first phase as input, the SET launches the design phase, in which detailed concepts are generated in response to the system and subsystem specifications. The system and subsystems specifications are further refined into subassembly and components, and testing plans that verify the system and subsystem compliance with the original high level requirements are issued.

During the system design phase, the trade-offs studies are further analyzed in detail. Trade-off studies are used to support the decisions made in regards to requirements and the selection of appropriate design alternatives. The trade-off studies focus on key system performance parameters such as weight, power consumption, range and reliability of operation. Trade-off studies must also be supported by cost and schedule impact analyses. The results of these trade-off studies complement the detailed system description, and balance the requirements with other factors such as cost, manufacturability and complexity.

Long-lead items are a particular area of concern for the SET during the system design phase. Items that have long procurement schedules must be identified early in the design process. The system design must be frozen around these long-lead items so that they may be procured early in the design process, ensuring that their delivery will pose a low risk to the schedule.

The SET must monitor the progress of the design team through technical reviews and spot-checks. Technical risks are monitored and updated as necessary in order to ensure that appropriate mitigation steps are being taken. Issues and risks associated with the design are documented, and the SET conducts reviews to evaluate potential design alternatives, mitigation strategies or reversion to contingency plans. The SET provides input to potential troubleshooting steps and conduct analyses to evaluate alternatives that present high technical or programmatic risk. The output of the system design phase is:

- System and subsystem physical layouts
- Detailed design, synthesis of solutions, and analysis of compliance with the specifications
- System external and subsystem interfaces
- System performance estimates
- Detailed testing and verification plans
- Detailed system integration and implementation plans

Phase 3: Manufacturing, Assembly, Integration

This phase reflects the fact that in most high-tech environments the design team is best suited to assemble and test the overall system. The actual manufacture of parts may be done under subcontract elsewhere. The input to this phase is:

- Mechanical design drawings
- Electronics boards and circuits schematics
- Flow charts for software coding

The output of the phase is:

- The assembled system fully functional
- Bench test results
- Performance verification and validation in virtual environments
- Preparation of all hardware and software for Validation, Verification, and Acceptance Phase

Phase 4: Verification, Validation, and Acceptance

Verification is the activity that shows that the product was "built right". Validation is the activity that shows that the "right product" was built. The verification sub-phase ensures that the individual subsystems perform as per their specifications. Individual testing plans for each component or subsystem are developed during the detailed design phase, and these tests are carried out and documented during the verification sub-phase.

Deficiencies identified during verification prompt a review of the design, manufacture and integration of the subsystem that is found to be deficient. In the event that there is a previously unidentified deficiency in the design, the SET allocates the required resources to perform a detailed analysis of the design and what alternative solutions could be implemented to correct the deficiency. In the event that there is a deficiency in manufacture or integration, the SET works with the supplier or integrator to identify the problems in their process and to correct the issue.

During the validation sub-phase detailed designs and system architectures are checked against the original high level requirements. Validation activities are carried out both in the lab (bench testing, hardware in the loop simulation), as well as in the field.

In the event that validation activities yield a result indicating that the system does not meet the high-level requirements, the SET performs a review of the system concept of operations and a flow-down analysis of all of the requirements associated with the element of the system that fails the validation check. This flow down analysis will confirm that the concept responds to the mission requirements, and that the derived system requirements and specifications are adequately interrelated. In doing so, the SET isolates the source of the deficiency and takes corrective action while maintaining the integrity of the remainder of the system.

Acceptance tests include end-to-end tests and demonstrations of system functions. The tests are traceable to high-level system requirements, and upon successful demonstration, the project is considered as completed.

The SET is also responsible for the documentation of all testing and verification activities. Upon final approval, the SET prepares the final validation data package.

System Engineering Products

Specifications

System specifications are "flowed down" from high-level requirements to functional units of the project. The system specifications arise during both the Requirements Phase and the Design Phase of the product project, as a direct result of decomposition of the system into the product's constituent subsystems.

The programmatic requirements are defined by a series of documents defining the overall product system architecture. The requirements are mapped into product specifications. These are further used to define the subsystems' specifications.

System Concept

At the onset of the project, the SET guides the concept development. The conceptual design integrates the contributions of mechanical, electrical, and software development teams, as well operational constraints, interfacing requirements, and scientific user input. Selected concept invokes also the system trade-off studies. The complete system represented by a simulation model provides an end-to-end means of representing and documenting requirements, subsystems, constraints, and test plans. The model will be continually updated throughout the project as a means of tracking subsystems configurations and interfaces.

Product Database

Throughout the development, numerous designs, analyses and studies are performed to ensure conformance to system requirements. These documents indicate the progress of the design and the decisions that influenced the final configuration. These documents are managed by the SET and are archived in a single database. Designs, analyses and studies are documented according to which high level requirement they act to satisfy, in order to provide traceability from the final design up to the analysis and ultimately up to the high level requirements. By delineating which requirements are satisfied by each analysis, future reconsiderations of specific requirements can be accommodated while ensuring that all other requirements are not impacted.

Specifications, design drawings, interfaces, and analysis reports are used to document the specific baselines of the system. These documents are subjected to an approval process. Once approved, these documents are placed under configuration control and are handled according to the project Configuration and Data Management Plan.

Product Baselines

As the system life-cycle progresses, the product configuration gets defined in a series of baselines, consisting of approved documentation that is used to describe the technical specifications and interfaces of the baseline design. Each baseline identifies the mechanical, electrical and software elements of the system at various stages of system development. The general flow of baselines originates from the high-level requirements (Functional Baseline), to a system level description (Concept Baseline), to detailed design drawings and specifications (Design Baseline) and, finally, into a physical product which represents the “as built” state of the product (Product Baseline).

Quality Assurance and Safety Program

Quality Assurance

To facilitate the required level of product quality, a system of procedures has to be adopted for the regulation of company activities, from proper project requirement identification to quality inspections at different stages of prototype design and manufacturing. The responsibility for ensuring quality is shared. All employees, including top management, vendors and subcontractors are responsible for the quality of their work and of the products. This approach is provided and explained to each employee, to sub-contractors, and to vendors. The PMT retains project responsibility for the overall Quality Assurance System. The day-to-day operations related to this domain are managed by the Quality Assurance Manager. Most companies must achieve compliance with all the requirements of standard Quality Assurance Requirements as described below.

Certification to ANSI/ASQ Q 9001 (ISO 9001:2008)

Most companies have implemented, or committed to implement, an ISO 9001 Quality Management System that provides surveillance of:

- Design review, procurement, inspection, manufacturing process, inventory control and shipping
- Training, audits/surveillance, metrology/calibration, traceability
- Detection, reporting of, and corrective action arising from, discrepancies

- Failures and other indications of unsatisfactory quality, and test witnessing

SMEs usually recognize that they need professional advice and direction in order to effectively implement such a comprehensive management system, and they mostly do it through subcontracted work. Typically such subcontractors have a variety of technical staff to assign to their clients. They are expert in interpreting, applying and auditing ISO 9001 Quality Management Systems. They have experience working with engineered products in strict regulatory environments.

Quality Assurance Implementation Plan

Quality Assurance Implementation Plans (QAIPs) are typically prepared to demonstrate how an organization's Quality Management System applies to a particular project or to provide confidence to a contracting agency that all of its regulatory and contractual requirements are understood, and will be met.

A QAIP format usually follows the internationally accepted methodology for Quality Plans as defined in *ISO 10005:2005 – Quality Management Systems Guidelines for Quality Plans.* The QAIP is expected to achieve complete compliance with the Quality Assurance Requirements of each specific project.

The QAIP is managed by the project's Quality Assurance Manager. The Manager has overall responsibility for the preparation and dissemination of the QAIP. The QAIP is meant to be brief and succinct and highlight contract-specific requirements and reference existing procedures and protocols within the project Quality Management System. A comprehensive QAIP addresses at least the following areas of Quality Assurance objectives:

- Management Responsibilities
- Control of Documents and Data
- Control of Records
- Configuration Management
- Mechanical Systems Design
- Electrical Systems and Software Design
- Purchasing
- Fabrication and Assembly
- Identification and Traceability
- Final Acceptance Testing
- Customer Communication
- Infrastructure and Work Environment
- Training and Human Resources
- Technical & Program Requirements
- Reliability and Maintainability
- Safety Program
- Maintenance Program
- Control of Nonconforming Product
- Packaging and Delivery of Products
- Metrology and Instrumentation

Safety Program

The PMT must be committed to preventing the accidental loss of any of its resources, including employees and physical assets. In fulfilling this commitment to protect both people and property, management provides and maintains a safe and healthy work environment in accordance with industry standards and in compliance with legislative requirements, and strives to eliminate any foreseeable hazards which may result in property damage, accidents, or personal injury/illness.

Any company recognizes that the responsibility for health and safety are shared. All employees will be equally responsible for minimizing accidents within the company facilities and on all work sites. Accidental loss can be controlled through good management in combination with active employee involvement. Safety is the direct responsibility of all managers, supervisors, employees, and contractors. All management activities must comply with company safety requirements as they relate to planning, operation and maintenance of facilities and equipment. All employees will perform their jobs properly, in accordance with established procedures and safe work practices.

Design Failure Modes and Effects Analysis (DFMEA)

Project safety and reliability engineers work with the SET to prepare a DFMEA to properly assess and quantify the risks associated with the final design in addition to MTTR and MTBF calculations. Risks will include a series of issues:

- Risk of degraded function in the field
- Risk of loss of function in the field
- Manufacturing risks posed by the designs
- Testing risks posed by the designs
- Premature failure or reliability risk in the designs
- Safety risks to those building, operating or maintaining the vehicle.

The DFMEA is usually prepared according to widely accepted industry standards: SAE J 1739, AIAG FMEA-4, and MIL-STD-1629A. Design risks are ranked according to Severity, Probability of Occurrence, and the Strength of the mitigation Strategy. The DFMEA assists in determining where safety risks exist and what hazard-prevention measures need to be implemented. It also helps in materials selection and determination of what components and functions need redundancy or high safety-factors. An output of the DFMEA is the Design Verification Plan and Report. It defines the project's testing program in accordance with the risk profiles identified in the DFMEA.

Configuration and Data Management Plan (CDMP)

Overview

The function of the CDMP is to define a reliable, sustainable and expandable process which allows the development team to generate concurrently advanced and complex products while ensuring that the customer's objectives and requirements are always met.

Configuration Management (CM) provides a means of controlling changes to a design as it evolves from a set of objectives into a physical product operating in the field. CM requires standardized processes in which the system design is base-lined at key milestones and that changes to the design are documented and traceable.

Data Management (DM) provides a means of electronically storing design data on a file server which is capable of tracking revisions and product history information. Further, the Data Management system involves a standardized process by which design data is backed up and securely stored.

The CM plan described herein is typical and is based upon the CMII-105B Standard for Product Configuration Management by the CMII research institute (www.icmhq.com).

The usual CADM system incorporates processes that are able to:

- Identify the functional and physical characteristics of a design through all stages;
- Accommodate and track changes to a design;
- Maintain clear, concise and valid system requirements;
- Ensure that system data is accurate and secure;
- Ensure that best practices are re-used and regularly optimized;
- Ensure that the designed systems meet their requirements;
- Provide a means of assessing the total impact (schedule, cost, risk, safety) of changes to a design;
- Provide control over change audits-in-process.

All system and design data are stored in an electronic archive with a Redundant Array of Independent Disks (RAID) data storage scheme. The electronic archive is stored in a company server on the premises.

Product Baselines

As per MIL-STD-499B product configuration baselines define a set of well defined, approved and released configuration documents at various stages of the system life cycle. Current configuration documentation consists of an existing baseline plus approved changes. Four baselines are used in the system development as follows:

Functional Baseline

The Functional baseline describes the overall functional requirements of a system. It defines the key capability and performance requirements of the system and all of the necessary interfaces and operating conditions to which the system must conform. Further, the System Requirements baseline defines the verification methods necessary to demonstrate that the system meets the stated requirements.

The Functional baseline may be an extension of the System Requirements. The top level requirements are addressed with an overall system concept and plan for action, which is formally embodied in the form of the contract bid or proposal.

The Functional baseline may be extended by a System Hierarchy tree which defines the major subsystems in anticipation of the requirement to allocate high level requirements to specific subsystems.

System trade-off studies are identified and carried out after the Functional baseline has been approved. Thus, the Functional baseline represents a "safe" configuration to which the design team may revert in the event that the development baseline does not meet the system requirements or does not gain approval.

Concept Baseline

The Concept baseline is the result of a detailed Requirements Analysis and Trade-off study process. The SET decomposes the high level system requirements into derived and allocated requirements within the subsystem hierarchy. System trade-offs are performed to determine the optimal allocated design configuration. Each configuration then possesses its own set of detailed requirements and preliminary design documents (i.e. sketches, schematics, block diagrams) which address the derived high level system requirements as allocated to each configuration. These design documents are subject to approval and represent the system specifications which are required in order to move forward with a detailed design.

Detailed part selection and design proceed after the approval of the Concept baseline. Thus, the Concept baseline represents a "safe" design to which the development team may revert in the event that issues arise during the development of the Build-To baseline.

Design Baseline

The Design (or "Build-To") baseline represents the approved detailed designs of the system. The Design baseline is the result of detailed system design and analysis. It contains all approved documentation (drawings, bills of materials, software schematics) in order to fabricate or build the system.

Product Baseline

The Product (or "As-Built/As-Coded") baseline represents the approved documentation that represents a configuration as it has been build or coded. The product baseline documents the final configuration of the system as it enters its field deployment and operations phase.

Configuration Management Process

The process of moving from one baseline to another is controlled by the SET. This process develops high level requirements into detailed system descriptions and specifications, and ensures that the resultant system meets the high level requirements. Figure 1 shows the sequence of Systems Engineering phases and their associated output.

The first phase of development in systems engineering process is the Requirements Phase, in which high level system requirements are derived into detailed functional requirements and system performance measures. The output of this phase is the Functional Baseline.

The second phase of development is the System Design phase. During this phase detailed system trade-off studies and design analyses are performed in order to produce a design which is feasible and can be delivered under the project constraints (schedule, budget, etc). Major system elements and their interactions are modeled and preliminary estimates of system performance are generated from the system models. The first output of the system design phase is the Concept baseline.

Following the development of the Concept baseline, the SET tracks the detailed design and development of the system. Changes to the design are requested through a formal Change Request. Change Requests are tracked in the PDM system and are approved by the SET and, if necessary, by PMT. The SET considers each Change Request in light of its impact on the overall requirements. In the event that a Change Request impacts system performance or ability to meet requirements, the SET decides if it is appropriate to request a deviation from the system requirements or revert a specific configuration to the Concept baseline and approach the problem in a different way.

The System Design phase's next output is the Design Baseline, which provides all necessary production and ordering information to build one or more complete systems. The design baseline is subject to careful scrutiny and approval by PMT prior to hardware production and software coding. Following approval, production and software coding commences. No major changes to the design are permitted at this stage.

The final phase of development is the Validation and Verification phase, in which it is confirmed that the system that has been built and integrated satisfies the overall system requirements, and that all preliminary performance estimates are valid. Adjustments to the system are made in order to suit any variances from the requirements, and the final design is documented and approved as the Product Baseline.

Data Management Process

File Naming, Numbering and Reuse: A standardized convention for numbering and naming documents is applied to all documents, system components and configuration items such as software. All physical items are provided with unique identification (ID) numbers, as well as a name and description. All ID numbers are to be tracked through the PDM system. This numbering convention is indentified in a separate document

Figure 1. Systems engineering phases and baselines

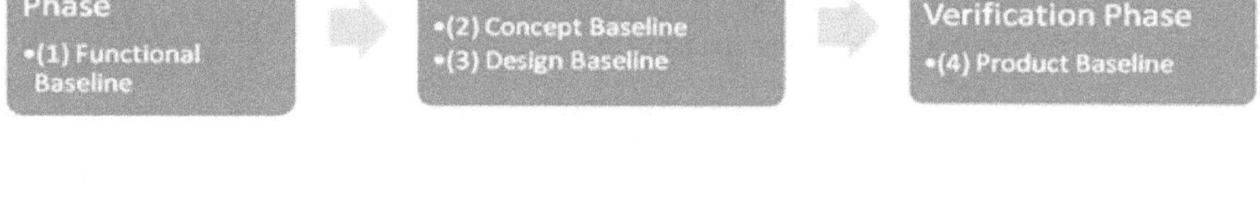

which is made available to all involved in the system design process.

Components which are identified as interchangeable (such as standard hardware) are designated the same ID number. Components which are custom fabricated from the same set of drawings are assigned the same ID number with additional information which uniquely identifies each custom component from one another.

Enabling Software: The Enterprise Product Data Management (EPDM) enables consistent tracking and approvals of system designs and baselines. The EPDM system provides shared software tools across the developer organization for:

- Product Baseline Automation
- Change Analysis
- Change Implementation Planning
- Change Tracking

The basis of EPDM software is the development of engineering "workflows" which track specific events such as approvals and changes.

CASE STUDIES

The case studies are prototypes developed by Engineering Services Inc. (n.d.). Their development cycles followed closely the framework presented in the previous sections. The presentations in this chapter are abbreviated and only focus on the new technologies that were part of the methodology described in this chapter. The study cases are also presented in Goldenberg (2007).

TRACKER – A Mobile Robot

The TRACKER is a mobile robot (Figure 2) with modular traction, modular arms, modular payload

Figure 2. TRACKER

interfaces and modular electronics, modular interfaces of payloads, open architecture controller, and field-programmable Operator Control Unit (OCU). The modular traction provides for the option of using tracks or wheels on the same platform. The modular arm and modular payload interfaces can be reconfigured for specific missions. The system is used by the military, law enforcement, and civilian security forces, as well as for personal assistance, transportation, and research purposes.

Project Management

Team: The team was assembled by first appointing a project manager, a person with experience with similar projects, who was also a very qualified person in system engineering design. That person became the delegated authority in terms of responsibility and reporting to the executive branch. Then, the project manager assembled the team consisting of: (i) APM – assistant project manager to provide for the day-to-day operations and support to the PM. That person was a junior designer with the ability to undertake project management tasks; and (ii) System Engineering team – consisting of senior project leaders and consisting at least of one representative for each of: system engineering, mechanical design, electronics, software development, production control, QA, and data base system engineering. The project leaders were technically the chief designers, while the system engineer became the chief technologist of the project.

Development Process Planning: The PM and System Engineering Team (SET) defined the critical project development path. The major milestones were: Specifications, Design, Critical Design Review, Manufacturing, and Tests. They were scheduled, and budgetary allocations were provided for each phase.

Project Control: The PM provided the tools for project management including cost control, scheduling, reporting, and delivery of that information to the executive branch.

System Engineering

Specifications

See Table 1.

Design

Base Technology

As described in the patents by Lin & Goldenberg (2009, February; 2009, October; 2010, August), the key aspect of the technology is the mobility based on a track configuration-controlling mechanism, using a pair of flippers, each being a 3-bar cam mechanism consisting of 2 moving elements and a cam fixedly mounted on the chassis. The flippers are used to lift TRACKER's pulleys over stairs or obstacles and adjust the position of its COG to avoid rolling over while ascending and descending. The variable track configuration is provided by controlling the position of the planetary wheels using the flipper (Figure 3).

The flipper has three functions: (1) to ensure the center of the planetary wheel mounted on the end of the flipper moves along an exactly elliptic path; (2) to provide the tracks with a continuously variable spring force for tensioning the tracks; and (3) to change and adjust the center of gravity of the robot system to prevent it from flipping over.

The flippers are connected by a central shaft through the center of the chassis and to a pair of special configuration cams mounted on each side of the chassis. The flippers can rotate completely around the shaft. This motion generates variable track configurations. The flippers' motion shifts the robot's center of gravity for climbing and descent of obstacles and stairs to maintain stability as shown in Figure 4. With the variable track geometry controlled by the operator, and the ability to reposition the system center of gravity, and also raise and lower the front or rear of the chassis, it is virtually impossible to get the robot stuck or rolling over on stairs. The ability to reposition the

Table 1. Specifications

Platform	Vehicle size	Configurations	1	Basic configuration - Long track with flipper Derivative configuration 1 – Short track without flipper Derivative configuration 2 – Four wheels
		Height	2	50 cm – 19.7” (planet wheel horizontal) 61 cm – 24.0” (planet wheel vertical)
		Width	3	54 cm – 21.3”
		Length	4	93 cm – 36.5” (planet wheel horizontal) 71 cm – 28.0” (planet wheel vertical)
		Clearance	5	5 cm – 2.0”
	Weight		6	50 kg – 110 lbs (with payload)
	Speed		7	On a flat ground: 2.3 m/s – 5.00 mph On a slope: 0.3 m/s – 0.62 mph On a stairway: 0.1 m/s – 0.31 mph
	Environment		8	All-weather; All-terrain; Weatherproof; Stair climbing 45 deg; Climbing 35cm obstacles; Crossing 50cm gaps; Ability to maneuver over gravel, snow, mud, sand, high grass; Sealed to operate in water at depth of 15cm – 6”
	Speed of track configuring		9	Planetary wheel: 120 deg/s
	Number of tracks		10	Two tracks
	Number of motors		11	3 + Encoders
	Electronics		12	State-of-the-art micro-controller; abundant in RAM
	Sensors & mission payload		13	Internal & external temperature sensors, GPS, inclinometer compass, mission payloads
	Payload		14	40 Kg (88.8lbs) when climbing 45 deg; 300 Kg when not climbing
	Battery		15	2.5 - 4 hours continuous mission
	Communication		16	Computer and sensor communication ports RF: Data, 2-way Video, 1-way Audio; COFDM Option
	Wi-Fi broadcast of video		17	Included
	Transportation		18	Portable by two persons
Manipulator	Gripper reach		19	Horizontal 120 cm – 47.2”; Option: 180cm with 60cm extension Vertical 170 cm – 67”and Below the chassis 60 cm – 23.6”
	DOF		20	6DOF: Turret – 300 degree Shoulder 1 – 210 degree Shoulder 2 – 210 degree Elbow – 300 degree Wrist roll – continuous Wrist pitch - continuous Gripper – 10 cm opening
	Gripper opening		21	10 cm
	Weight		22	13 Kg – 29 lb; 20 Kg – 44 lb (with extension)
	Payload lift at full-extension		23	5 kg (horizontal) 10 kg (vertical)
	Maximum payload lift		24	12.5 kg
	Payloads		524	Recoilless disrupter, X-ray, Claw, Shock tube initiators, Charge Dropper Assembly, RTVS Mount assembly, Lifting bracket with articulated movement for key opening.
	Tip Speed		26	0 to 12 cm/sec

continued on following page

Table 1. Continued

OCU	1. Weight and Size	27	Light weight 23 Kg Size 54 cm x 40 cm x 30 cm
	Variable speed joystick control	28	Included
	Daylight clarity	29	Optimal
	Battery level indicator	30	Included
	Weapons firing key	31	Removable lock out key
	Task space control of arm	32	Included
	3D display of tracks and arm configurations	33	Included

center of gravity provides a distinct advantage over competitors in tasks involving ascent and descent of stairs, slopes, obstacles, passing over ditches, and, in general, navigation over uneven terrain.

Baseline Configurations

TRACKER has three baseline configurations as follows:

- **Basic configuration:** Long track with flipper (Figure 2a-c)
- **Derivative configuration 1:** Four wheels (Figure 2d-e)
- **Derivative configuration 2:** Short track without flipper (Figure 2f)

The basic configuration has active adaptability by remote control to suit rough and unpredictable terrains, including obstacles and ditches. The active terrain adaptability of the basic configuration is performed by changing the track configuration using the flippers. The variable configuration is provided by simultaneously controlling the position of the planetary wheels whose location is controlled precisely by a rotating configuration-controlling mechanism, i.e. the flipper. For example, surmounting obstacles forward or backward can be realized by rotating the flipper and changing the position of the planetary wheel.

Based on the basic configuration, two derivative configurations without the use of flippers, the short tracks and the four-wheel configurations, respectively, have been developed. High-speed can be easily realized in the wheel mode.

The modularity provides TRACKER with high reliability, ease of in-field maintenance and re-configurability. By changing modules, the TRACKER can be switched from a tracked mobile robot to a wheeled mobile robot, and this change can be done in the field in just several minutes. Also, any malfunction can be quickly identified, and repair be done by simply replacing the malfunctioning module with a new one.

Modular Electronics

The electronics of TRACKER are highly distributed and integrated modularly within the mechanical subsystems. The electronics are composed of the following main modules (Figure 5):

- Head
- Core
- Power
- Flipper
- Flipper transmission
- Traction transmission
- Arm

Figure 3. Schematic diagram of the variable configuration track mobile robot

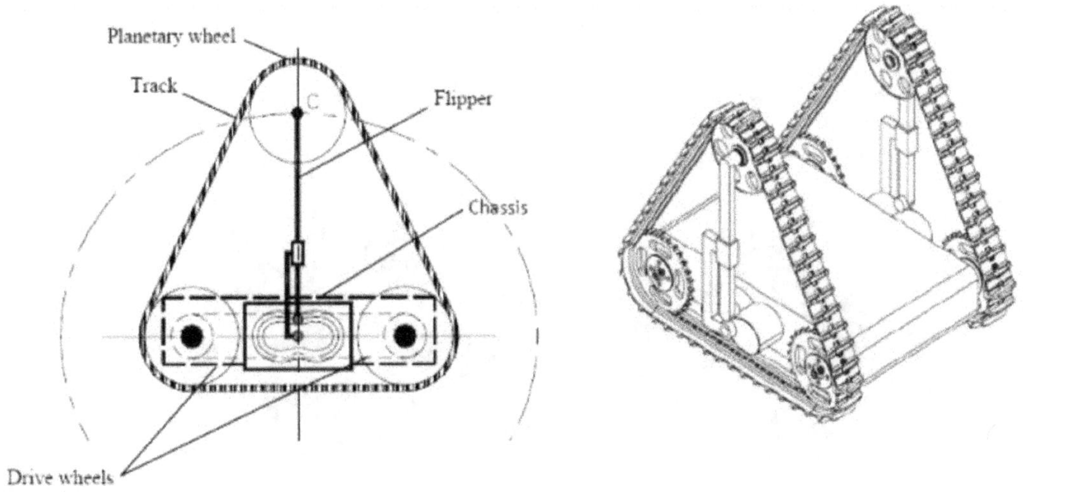

Figure 4. TRACKER climbing stairs

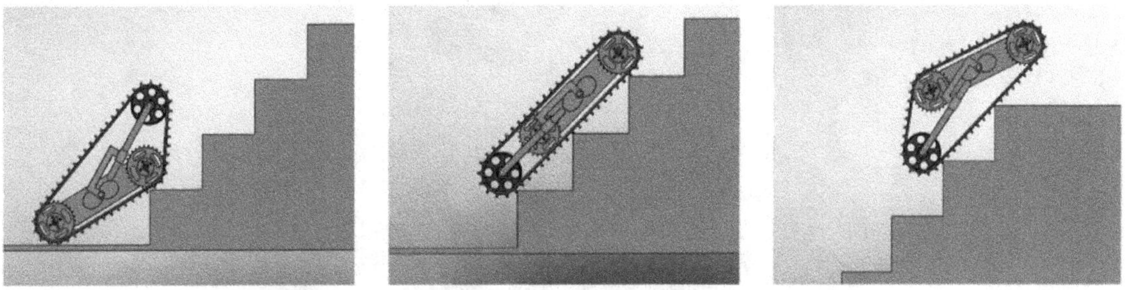

Each module is functionally complete and independent. For example, the Power Module includes battery and multiple voltage DC-DC converters, and provides all voltage and power to the entire vehicle electronics. The Core Module includes the main processor and communication system; it manages the communication with the OCU, and controls the motion of the vehicle. Some modules have the electronics and mechanical parts integrated, making them fully functional mechatronics modules. For example, the flipper module includes motor, gear head, encoder, angular position sensor, brake, servo motor driver, transmission gear pairs, cam, mechanical structure, etc. The arm module includes motors, gear heads, encoders, angular position sensors, payload interface, weapon control interface, and the mechanical structure, links, and gripper fingers integrated.

Robot Arms

The Tracker has two arms - a Modular Gripper Arm and a Modular PTZ Camera Arm (Figure 2b-c). They can be mounted and operate simultaneously.

Modular Gripper Arm: The Gripper Arm is a modular light weight manipulator. The arm, shown in Figure 6 mounted on the TRACKER, consists of a turret, shoulder pitch joint, elbow pitch joint, and specialized wrist-gripper assemblies. The baseline configuration of the Gripper Arm has 4DOF (pitch/pitch/pitch/roll). An optional

Figure 5. Modules of the TRACKER electronics

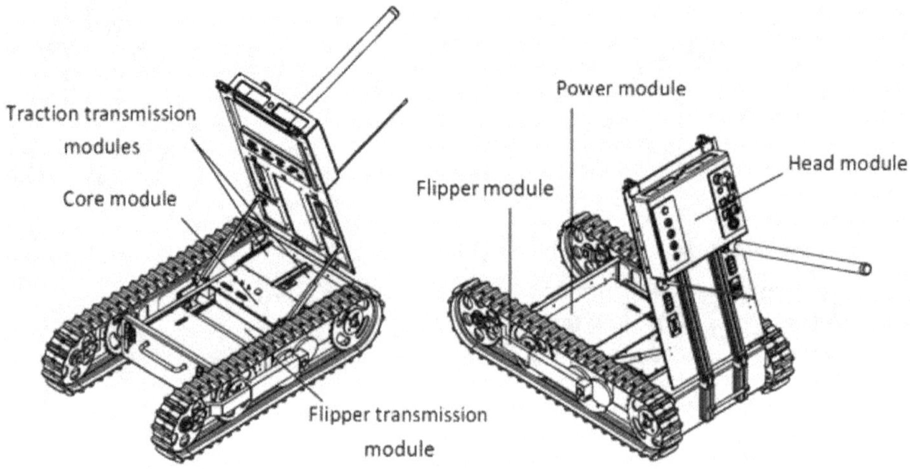

Figure 6. TRACKER with gripper arm

turret can be added on. The structure of the arm is shown in Figure 7.

The features of the Gripper Arm are:

- Wrist Roll and Pitch provide continuous rotation
- Elbow motor is placed at the base to reduce inertial weight and obtain a light-weight high payload-to-weight ratio arm
- Control electronics are placed in the turret to reduce moving parts and weight and for ease of maintenance
- All wiring is hidden and protected in the link tubes. This makes the arm look leaner and makes it safer to use
- A dedicated 32-bit controller and servo drivers control the arm motion, weapon firing, target aiming, and provide interfaces for the disruptor, relays (power output and contact output), VISCA-based zoom cam-

Figure 7. TRACKER gripper arm structure

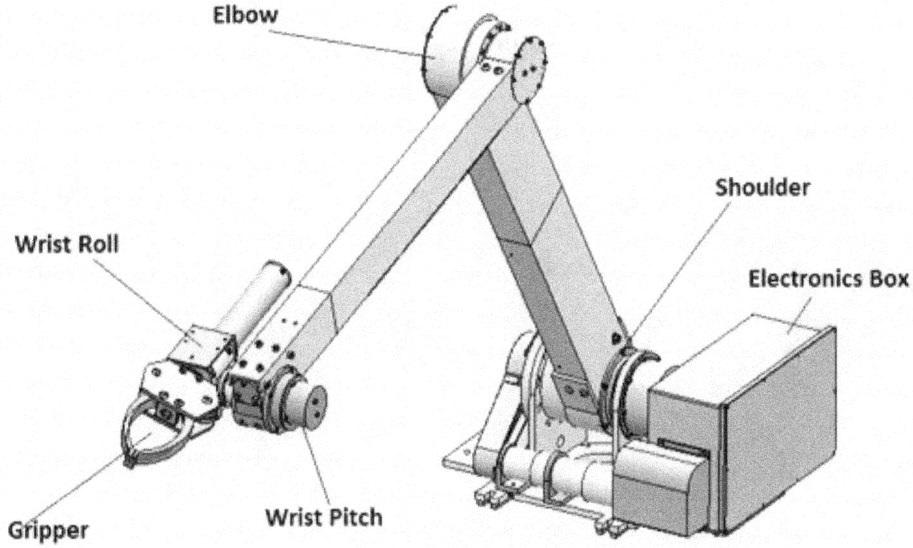

era, aiming laser, lights, re-locatable camera, etc

- Shoulder, elbow, and pitch joints have incremental encoders on the motor side and absolute angular position sensors on the output shaft.
- Coordinated linear motion of the tip by joystick control is provided.
- Pre-programmed motions are provided to facilitate the execution of various tasks. The pre-programmed tasks include homing back to stowed position, raising the arm to a vertical position, extending the arm to a horizontal position, etc.

Modular PTZ Arm: The Tracker PTZ arm is a 1-DOF light weight sub-system for mounting and supporting the PTZ camera. The PTZ arm is suitable for the Small, Medium or Large Trackers, and it consists of PTZ camera, PTZ arm electronics box, linear motor, antenna, and connectors. It can be mounted independently on a platform for surveillance, and it can be also mounted on the gripper arm for tactical tasks. The PTZ arm structure is shown in Figure 8.

Payload Interfaces

The robot provides modular interfaces, data communication and power for a range of payloads: sensors, cameras, small tools, and other task and mission accessories. On the Tracker, a typical modular payload interface includes the communication signal, video signal and power supply channels. Interfaces are provided for the firing of the disruptor, aiming the laser, activating relays (for external power supply and contacts), video cameras, the Gripper Arm Module, PTZ Arm Module, Upgrades Module, and standard computer interfaces such as USB, Ethernet, RS232, RS485, etc.

Operator Control Unit (OCU)

A user interface and supervisory controller are provided as the Operator Control Unit (OCU) (Figure 9). The OCU links can be wireless or via fiber optic cable. A Hybrid Communication Link (cable-wireless) – an extended communication range through a relay link is also available. The OCU provides control of arms (Gripper and PTZ), the flippers, and the platform, as well as control of all payloads of the robot, such as lasers, cameras,

Figure 8. (a) PTZ arm stowed (b) PTZ arm raised

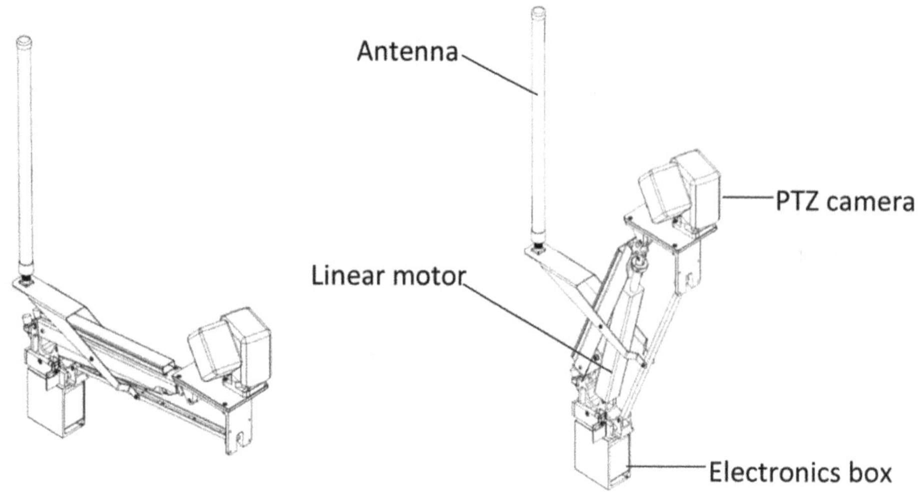

lights and disrupters. The OCU consist of two parts: (i) control/display unit that contains all user control and display interfaces, communication protocol, default configuration, and kinematics calculations; and (ii) RF unit that includes video, audio and command/data transit between the OCU and robot. The user is provided with two methods of controlling the system: (i) buttons/joysticks to control the robot and arm, firing system, relays, parking and emergency stop; and (ii) touch screen buttons/joysticks to set the configurations of joysticks, cameras, lights and 3D display.

OCU Features:

- Use of two joysticks that can be reconfigured for various control functions
- Concurrent video display from different cameras
- Display of platform and arm in 3D attitude in real-time
- Two-way audio link
- Display of battery remaining capacity of both, OCU and robot
- Can be powered by either DC or AC power, or by re-chargeable battery
- OCU fits in a single portable case

OCU Software Structure: The OCU software runs on a single board computer, with embedded Windows Operating System. The OCU software is composed of a video panel module, 3D panel module, button/joystick module, communication module, and main bus module. The OCU software package also includes developer tools, such as a debugging module, robot build tool, gripper arm testing tool, PTZ arm testing tool, and communication testing tool.

OCU Hardware Architecture: The hardware architecture is schematically shown in Figure 10.

Communication protocol between OCU and Tracker: To support the open architecture structure and offer access to new payloads through the OCU, the Tracker has a scalable main communication protocol frame between OCU and Tracker. In this protocol, information for each payload as a sub-frame is defined, and can be freely integrated into the entire communication frame.

In case the payload is controlled directly by the OCU, the Tracker communication manager does not need to know the detailed information about the content of the sub-frame. When there is a sub-frame for payload communication, the communication manager just adds it into the main

Figure 9. Tracker OCU

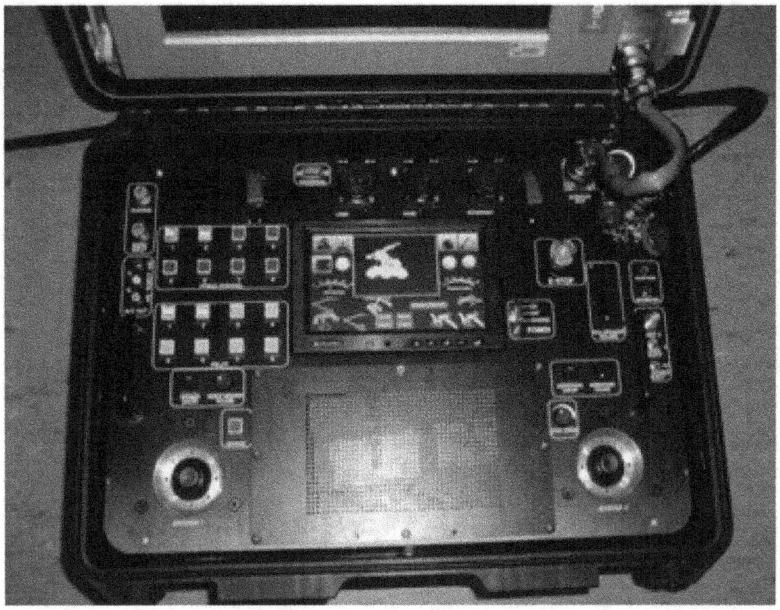

frame. Also, it sets the address and length of the sub-frame in the main frame to notify the OCU when this information appears.

For communication from the OCU to the payload, the situation is similar. If the payload is controlled directly by the OCU, the communication manager detaches the sub-frame from the main frame, and transfers it to the payload communication channel.

Open Computer Architecture (OCA) of Tracker

An OCA is necessary for modularity and reconfigurability of the system, as well as for wired or wireless links with the platform, arm, tools, and sensors. The open architecture controller consists of an open hardware interface and open software architecture. The open software architecture is composed of open protocol, open instruction sets and open language, library and Application Programming Interface (API) functions. The new generation OCA exhibits excellent scalability, extendibility and flexibility.

In the Open Architecture the electronics system provides the various standard interfaces along with necessary power supply: RS232, RS485, USB, and Ethernet. Working with the open software architecture, plug and play capability is achieved. The open software architecture and the Ethernet application program support this undertaking.

The system extendibility with additional physical modules is only limited by the vehicle platform's physical dimensions and availability of space. In the Open Architecture hardware, an Upgrading module was designed for the large Tracker version to interface with additional sensors, payloads and modules. For the Medium and Small Tracker the number of standard interfaces is limited.

The extendibility is also limited by the software driver, API and libraries. Currently, the Open Architecture and the software are capable of extending a second video RF transmitter including the Wi-Fi devices, and VISCA protocol based on SONY block camera. To be able to accept additional types of sensors and payloads, the software needs to include the related drivers in its library.

Figure 10. OCU hardware architecture

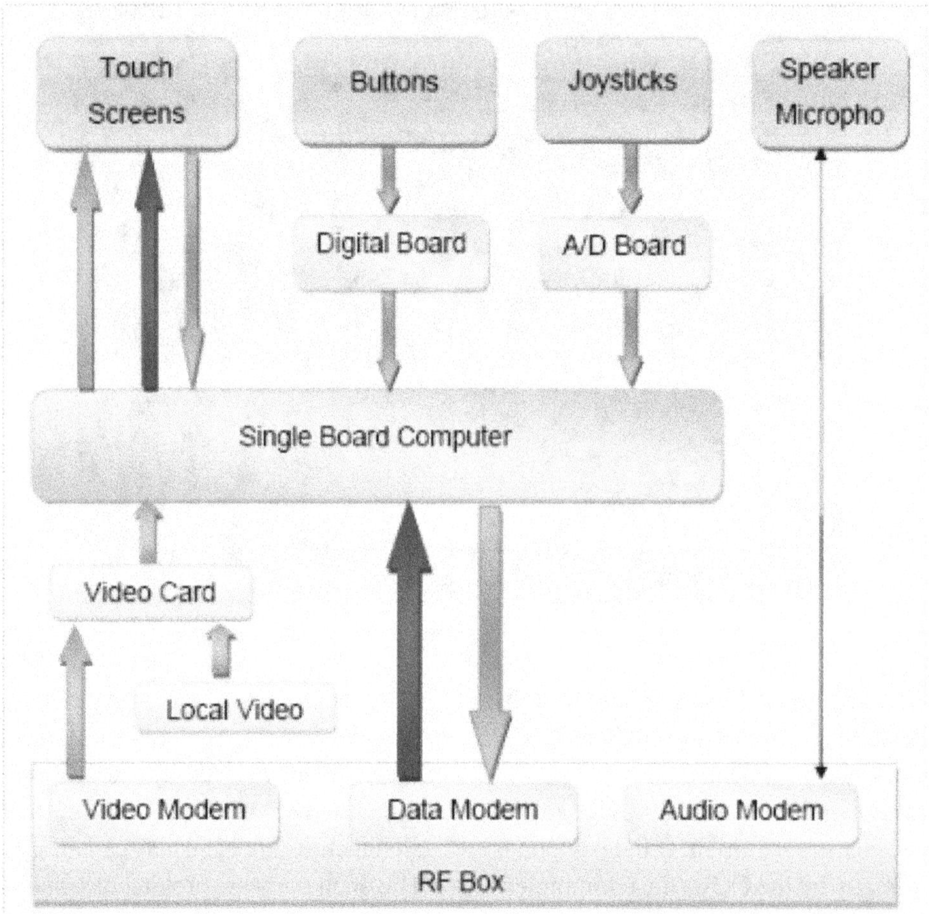

Manufacturing

Modular Assembly: The modularity can increase the productivity, by shortening the delivery time and lowering the labor cost. As the modules can be built as inventory, the assembling of the Tracker is quick. The Tracker can be assembled using an "Assembly Line" (Figure 11) manufacturing concept by serially sequencing the modules integration in order to enhance efficiency and reduce the cost of production.

Quality Assurance

Quality Assurance Implementation Plans (QAIP) were implemented following the internationally accepted methodology for Quality Plans as defined in *ISO 10005:2005 – Quality Management Systems Guidelines for Quality Plans.* The QAIP was managed by project's Quality Assurance Manager, a consultant brought into the project. The Manager had overall responsibility for the preparation and dissemination of the QAIP. In this project the QAIP addressed the following areas of Quality Assurance objectives listed in **Quality Assurance and Safety Program** section.

CMDM

Configuration Management (CM) provided a means of controlling changes to the design as it

Figure 11. Modular tracker assembly line

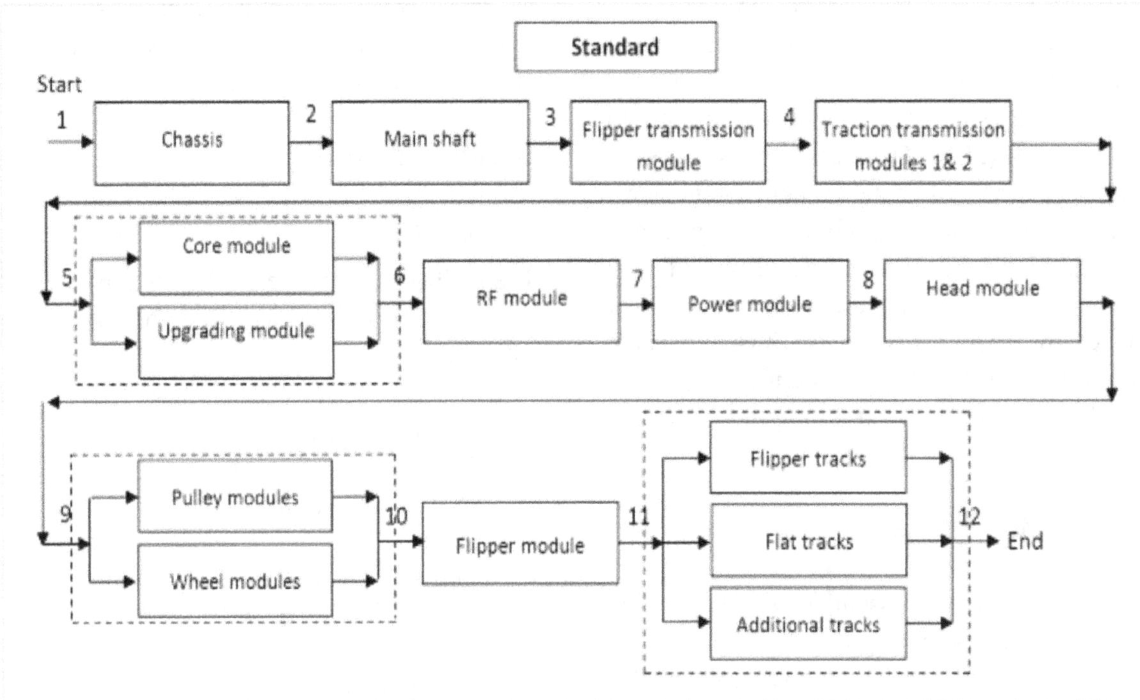

evolved from a set of objectives into a physical product operating in the field.

Data Management (DM) provided a means of electronically storing design data on a file server which is capable of tracking revisions and product history information. Further, the Data Management system involved a standardized process by which design data is backed up and securely stored.

The CM plan used described herein is typical and is based upon the CMII-105B Standard for Product Configuration Management by the CMII research institute (www.icmhq.com).

AARM - Robot Arm for Internal Operations in Nuclear Reactors

It is well known that robotic arms are used to augment and even replace human operators in hazardous environments. The AARM is a robot tool for visual-assisted remote controlled inspection and operations inside nuclear reactor cores. The robot is used in Callandria Vessels. The AARM was designed based on standard design methods of robot arms enhanced with proprietary features related to the specific environment, such as collision avoidance and radiation shielding to provide a reliable and safe tool.

Robotic manipulators have been used in nuclear reactors for various inspection and service operations as reported by Hamilton et al. (2010), Perrot et al. (2010), Caprari et al. (2010) and Zwicker et al (2010). These references reinforce the notion that designers' experience and customization are the main drivers behind the successes evident in this domain of application.

Typically, the range of motion of the robot in a reactor is limited because the base of the arm is normally fixed. Also, the robot must be provided with radiation shielding. Due to the dense network of pipes inside a typical reactor, henceforth limited access and poor visibility, there is a risk that the robot manipulator would hit internal reactor elements and cause damage, or get stuck inside the reactor. Thus, capabilities are required for collision avoidance and retrieval of the arm if power to the robot is lost.

The AARM robot by Goldenberg, Gryniewski & Campbell (2010, October) is constructed of radiation hardened material and components. The tool has built-in shielding to mitigate, and essentially eliminate, the "open channel" that is inherent when accessing the internal portions of the reactor. The AARM comes with a removable end-of-the-arm tool that can be used to pick up small debris that is found during the inspection process.

Project Management

Team: The team was assembled similarly to the previously described project.

Development Process Planning: The PM and System Engineering Team (SET) defined the critical project development path. The major milestones: Specifications, Design, Critical Design Review, Manufacturing, and Tests were scheduled and budgetary allocations were provided for each phase.

Project Control: The PM provided the tools for project management including cost control, scheduling, reporting, and delivery of that information to the executive branch.

System Engineering

Specifications

The CVI (Calandria Vessel Inspection) AARM robot system is composed of the following main subsystems:

- Boom carriage
- Boom
- End-of-robot arm
- Controller
- Operator control unit
- Pendant control panel

The maximum total weight of the system was to be 1000 lb.

Design Concept: The system operational concept is presented in Figure 12-13. Figure 12 shows schematically the AARM near the face of the reactor with the end effector inserted into

Figure 12. Schematic of the AARM inserted in the reactor

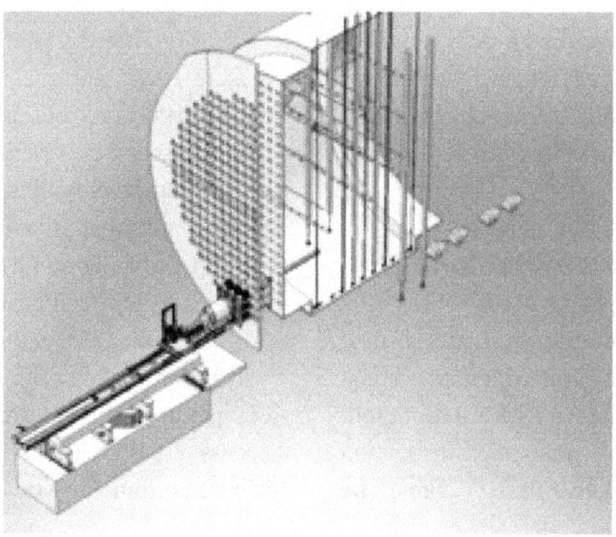

Figure 13. Schematic of the AARM inside the cavity

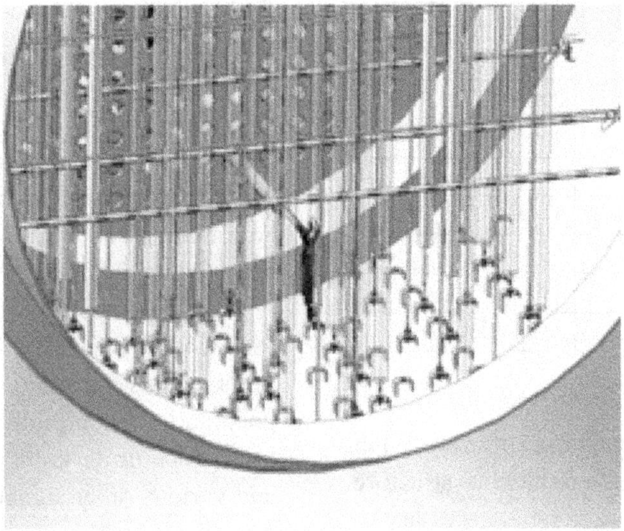

the reactor cavity, and Figure 13 shows the arm operating inside the reactor cavity.

The **AARM robot** system has 4 degrees-of-freedom: linear boom, boom roll, elbow pitch, linear extension, and gripper. The axes of motion are defined as follows:

- **Z axis** = linear horizontal motion in and out of the reactor (as per nuclear industry standard)
- **C axis** = rotation about the Z axis (standard machine tool terminology)
- **E axis** = elbow axis (end-of-robot arm pitch joint)
- **X axis** = linear motion after the elbow pitch joint

and the joint ranges and actuators are:

- **C Axis** (Boom Rotation)**:** 300° (+200°to -100°) servo motor
- **Z Axis** (Boom Insertion)**:** 245" servo motor
- **E Axis** (Elbow rotation)**:** 180° radiation hardened stepper motor
- **X Axis** (Linear extension)**:** 15.8" pneumatic with differential pressure control
- **G Axis** (Gripper fingers)**:** 180° finger rotation pneumatic with differential pressure control, 2.8" max opening, and 5lb payload

The dimensions of the AARM are:

- **Boom length:** main carriage to elbow: 224"
- **Boom diameter:** 5.118"
- **Boom end to elbow:** 36"
- **Elbow to tip (retracted/extended):** 29"/ 44.8"

Total length of extended AARM is 304.8", and the weights are presented in Table 2.

Subsystems

Boom carriage (Figure 14) mounts onto the main support frame to provide linear translation of the boom as well as rotation about a horizontal axis (roll). Using the boom carriage and the boom, the end-of-robot arm connected to the boom can be positioned at a desired depth inside the reactor, as well as be rotated to facilitate inspection and service procedures.

Table 2. Weights

Component	Weight
Main Carriage	300 lbs
Boom with End-of-Arm	300 lbs
Main Electrical Enclosure	225 lbs
Control Pendant	10 lbs
Rest Carriage	34 lbs
Front Bearing	65 lbs
Rear Bearing	55 lbs
Total	**989 lbs (450 kg)**

Boom (Figure 15) is composed of an aluminum pipe with an internal radiation shielding structure, and passage for the wires and tubes necessary for the functioning of the robot. The spacing and material of the shields have been designed to safely block the radiation escaping through the boom.

End-of-Robot Arm (Figure 16) is composed of a rotary elbow joint, a linear extension joint, a camera, a gripper, and a vacuum hose. It has been designed primarily for inspection purposes, with the added capability of picking up any debris or foreign objects that may be found. To facilitate insertion into the reactor, the arm is designed to fold up into a compact configuration. Once inside the reactor, the arm is then unfolded and controlled remotely to facilitate inspection and retrieval tasks.

Figure 16 and figure 17 show the system with the end robot arm penetrating in a mock-up vessel. The views are from the testing facilities of Automation Tooling System Ltd. (ATS) of Kitchener, Ontario.

Controller is built around Siemens S7-200 PLC 240VAC, 60 Hz, Single Phase, 15A. The input signals are analog pressure sensing of the pneumatic system, limit switches for each axis, joystick controls, collision detection of the arm, and various other signals. The output consist of commands to two servo drives for the Z and C axis, pneumatic valves for the X axis and gripper, and stepper motor drive for the E axis. User input to the system can also be through the control pendant, and Operator Control Unit with the option of remote control via Ethernet.

The servo drives operate in velocity control mode, with the velocity being proportional to the joystick movement. Joint feedback is accomplished with resolvers. The position servos operate in open loop mode only. Feedback for

Figure 14. Boom carriage frame

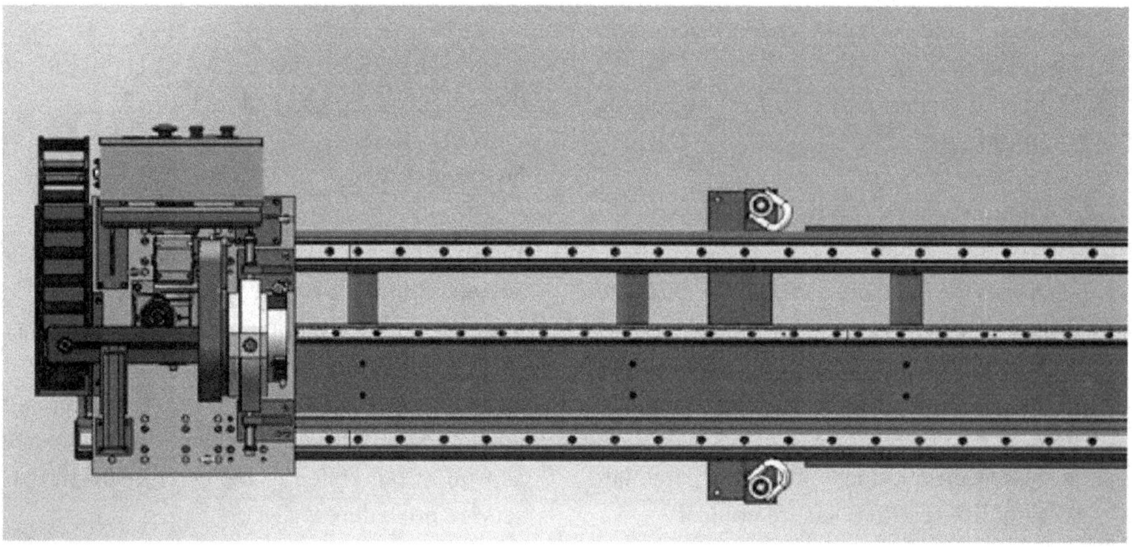

Figure 15. Boom

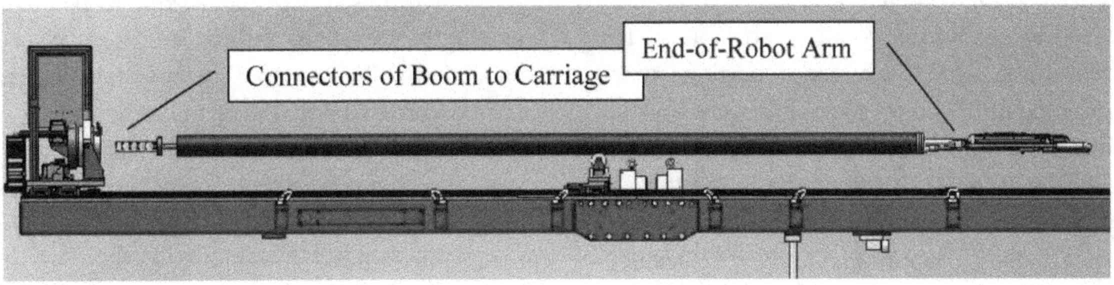

the X axis is provided by analog pressure sensors. Increased movement of the joystick in this case will increase the pressure supplied to the cylinder, thus controlling the force of the X axis. Auxiliary components that are powered but not controlled directly by the system include the IST R93 inspection camera, and the JFE SH-12 hardness tester. These devices have their own controllers. A separate control unit is specifically provided for controlling the radiation-hardened cameras.

Operator Control Unit (OCU) includes a display screen with a keypad to provide the user with information about each joint axis. Two multi-axis joysticks allow controlling each axis, either individually, or simultaneously. A "dead man" switch prevents the Z and C axis servos from responding unless it is pressed to prevent accidental movements. Figure 18 shows the OCU that includes a monitor, joysticks for all joints and E-stop.

Control pendant is shown in Figure 19. The pendant is composed of a keypad, a status screen, an E-stop button, two joysticks, and a connector for the pendant cable.

The screen on the pendant is used to display information on the position, speed, force, and limit switches for each axis. All of the axes of the system are controlled with two joysticks. The left joystick controls rotation and extension of the boom, and the right joystick controls the elbow,

Figure 16. End-of-Robot arm being inserted in a mock-up vessel (Courtesy of Atomic Energy of Canada Limited)

extension, and gripper on the arm. The left joystick has a 'dead man' switch on top, which must be pressed to enable the boom servo motors. The speeds of the Z and C axes are proportional to the displacement of the joystick. The right joystick controls the arm. The speed of the elbow joint is proportional to the displacement in the left and right direction. The pressure supplied to the extension cylinder is proportional to the displacement in the up and down direction. The gripper pressure is controlled by twisting the top knob either clockwise or counter clockwise.

Emergency stop button: The emergency stop button is located at the top left of the control pendant. When pressed, the entire system will be disabled; however, the pneumatic system will not vent the air under loss of power, so the actuators will hold their positions. In order to recover from an E-stop condition, the E-stop button must be released, and the enable button must be pressed.

Keypad Operation: The keypad on the control pendant allows the operator to enable the control pendant, view status information for each axis, and control the camera. Various status screens can be accessed with the up and down keys. The Esc button will access the main menu. When the system is first powered up, it will be in a disabled state, and to gain control the operator must press the enable button.

Auxiliary Button: The Auxiliary button allows the selection of different tools for the arm. Currently the operator can select between the gripper, hardness tester, and no tool. This function has the effect of enabling or disabling the valves that control the gripper air lines.

Collision Override: The Collision Override button allows the operator to disable the collision sensor on the robot. When pressed, the system will switch to disabled mode. The operator will then have to enable the system again to run the system without the collision sensor. This function should

Figure 17. AARM view showing the boom and controller (Courtesy of Atomic Energy of Canada Limited)

Figure 18. Operator control unit

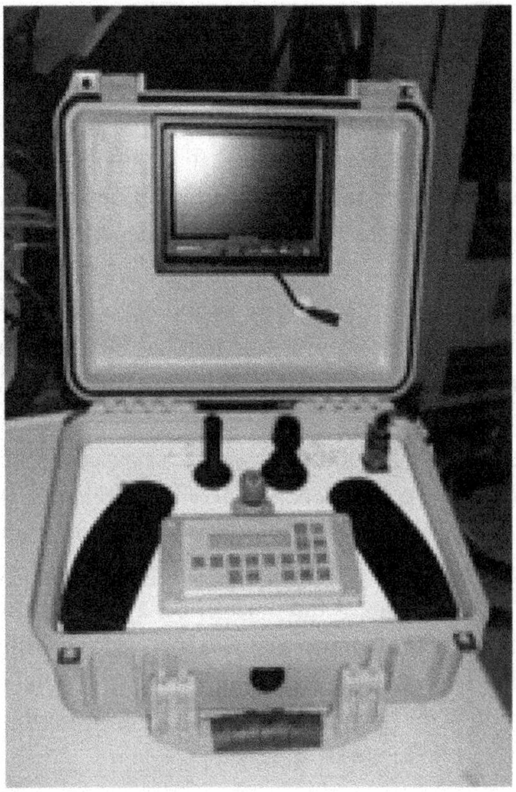

only be used under the operator's discretion. If collision sensor is ignored, damage could occur to both the arm and the reactor.

Elbow Lock: Three buttons control the various states of the elbow lock. If the lock is turned off, the elbow joint will be disengaged from the gearbox and can rotate freely. The low pressure setting is used to re-engage the lock. At start up and during all normal operations, the lock should be set to high pressure.

C axis status: The C axis (boom rotation) status is displayed in the lower left side of the screen. The speed is only shown while the axis is in motion, and is displayed to the right of the position. Each time either limit switch is activated, the position will be reset. When the system is restarted, the C axis position will show 999.9 to indicate that the axis has not been homed. The system can still be operated without homing, but position information will not be available.

Z axis status: The Z axis (boom extension) status is displayed just above the C axis information. The position and the speed are displayed. Similar to the C axis, the position is reset when either limit switch is reached, and a +/-L next to

Figure 19. Control pendant

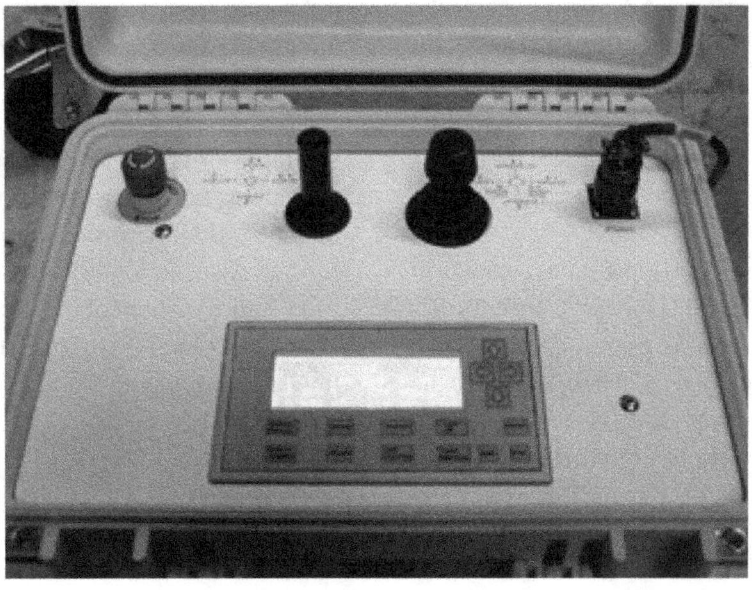

the position will indicate which limit switch is active. When the system is restarted, the Z axis position will show 999.9 to indicate that the axis has not been homed.

E axis status: The E axis (arm elbow) status is displayed in the lower right side of the screen. The position is displayed in degrees, and the speed (while the axis is in motion) is shown in degrees per second. Each time either limit switch is activated, the position will be reset, and a +/-L next to the position will indicate which limit switch is active. When the system is restarted, the E axis position will show 999.9 to indicate that the axis has not been homed. There are several things to keep in mind about the elbow joint. Since it is driven by a stepper motor, the position information is not absolute. Also, this joint can be disengaged though the lock mechanism, so the position value may not reflect the actual position of the arm. When the elbow joint has been released, it is best to home the axis after the lock is engaged again.

X axis status: The X axis (arm extension) status is displayed just above the E axis information. Unlike the other axes, the speed is not displayed, but rather the pressure, since this is a pneumatic actuator. When the X axis is in motion, the pressure is displayed as a percentage of the available system pressure. The left value is the command pressure, and the right value is the actual pressure. The X axis has only one limit switch in the retracted position. Each time the limit switch is activated, the position will be reset to zero. When the system is restarted, the X axis position will show 999.9 to indicate that the axis has not been homed.

Gripper status: The gripper status is shown in the top right of the screen. Only pressure information is available, since the gripper does not have limit switches or position feedback. Similar to the X axis, the command pressure and actual pressure are reported for the gripper, while it is in motion.

Design Requirements for Safety

Passive Mode

In the situation where control of the robot is lost, the arm has been designed to facilitate manual removal from the reactor in any orientation. All actuators of the AARM have a manual override function. On the main carriage, this involves removal of the servo motor, and installation of a manual drive adaptor and handle. On the robot arm, the pneumatics can be manually vented at the control panel, and the elbow joint lock can be released. When the elbow joint is released, the arm is free to swing open. The extension cylinder can be manually vented, allowing it to move freely. Thus, the arm can naturally straighten out when it is pulled out of the reactor, fitting through the opening through which it was inserted. The robot also has guides on it to allow the arm to slide through the opening without catching on anything during this manual removal procedure.

Conventional and Radiological Safety

Several measures have been taken to make the AARM safe to operate. E-stops are located on both the OCU and on the control pendant. Pinch points have been covered with guards. All actuators have some form of force limiting built in, either electrical or pneumatic.

The boom contains an internal radiation shielding structure. The spacing and material of the shields have been chosen to safely block the radiation within the boom at any position inside the reactor. Also, an important feature of the shields is the helical pattern of the wire and tube routing. If the wires ran through the shields in a straight line, this would allow a gap for the radiation to pass through.

Protection of Core Components

In order to prevent damage to the reactor, collision avoidance has been implemented. When the robot arm comes in contact with a metallic item, an alarm will sound. In addition to sensing, the actuators on the arm have been designed to limit the force they can apply. The elbow joint has a release mechanism in the gearbox which will disengage completely during a collision. The extension actuator is pneumatically driven. The pressure on each side of the cylinder is continuously monitored with electronic sensors, and is adjusted by the control system to maintain a constant force. The gripper operates on the same principle.

Control of Operations

The control system approach is a combination of user commands and automatic feedback control. The operations are described below.

Homing

C axis Homing: Confirmation is required based on visual feedback that the arm is either folded or has enough room to swing clear of any obstructions. Then the C axis is driven to either of limit switches. If it is already at one of the limits, is driven away slightly, and then back. The angle will be reset, depending on which limit switch was used.

Z axis Homing: Confirmation by visual feedback is required that the arm will not hit any reactor components when driven forward. The safest orientation is either completely folded up or with the C axis at 0 degrees. When outside the reactor the home is the negative limit switch (fully retracted). Z axis is driven back to the negative limit switch. If it is already at the switch, is driven forward slightly, and then back. When inside the reactor the home is the positive limit switch (fully extended). The Z position will be set to 236 inches.

E axis Homing: Confirmation is required that there is enough clearance to move the elbow joint. The E axis is open slightly, and then closed. The position will be reset to 0 degrees.

X axis Homing: When the arm is folded, the X axis cannot extend without interference. Thus, the elbow joint must be open at least 20 degrees prior to extending the X axis. Pressurization of the X axis is needed to extend it slightly past the limit switch, and then retract it. The X axis position will be set to 0 inches.

Insertion into the Reactor

The procedure is as follows: (i) Check that the Boom is straight and aligned properly. The Boom Carriage is designed to allow the Boom to swing over to the side to facilitate removal; (ii) Home the axes according to the procedures given above; (iii) Confirm that the axes display correct positions; (iv) Close up the E axis completely; and (v) Rotate the Boom 180 degrees.

The arm will now be facing up, which is the optimal orientation for insertion. Then, drive the Z axis forward slowly until the hose loop first engages with the first bearing of the vessel. Confirm that the Boom is properly aligned, and that the hose is not catching. The hose loop needs to flexible to enter the bearing. Continue driving the Z axis forward slowly until the Arm is completely though the flask and the Boom is supported on the Bearings. When the Z axis reaches 96 inches, the arm will be going through the smallest opening in the tube sheet liner. There is very little clearance and the arm will slide on the inside surface of the liner until it is through. The arm is completely inside the reactor when the Z axis reaches 141 inches.

Removal from the Reactor

The procedure is as follows: (i) Retract the X axis completely; (ii) Close the E axis; (iii) Close the gripper; (iv) Rotate the C axis 180 degrees. This is the best orientation for removal of the arm. Then:

drive the Z axis back to 141 inches; continue driving the Z axis back slowly; the arm will slide along the tube sheet liner at this point. Once 96 inches is reached, the arm is clear of the Calandria. Then slow down the Z axis once again when the base of the arm exits the bearing (52 inches). This is because the Boom will drop slightly onto the rest carriage, and part of the weight will be supported by the arm. Finally, continue driving the Z axis back until the Arm is clear of the Flask.

Picking Items up with the Gripper

The gripper is designed to be able to pick up objects weighing up to 2 lbs. To proceed, ensure the gripper is enabled by using the control pendant. The gripper pressure is proportionally controlled, so heavier objects will require higher pressure. When the gripper is fully open, only one of the jaws will be visible in the camera. Often it is helpful to be able to see both jaws. This can be done by slightly increasing the pressure until the gripper is partially closed. As the gripper jaws have been designed to operate with the vacuum nozzle in place, the relative size of the objects that can be picked up is limited. For larger objects, it may be helpful to remove the vacuum nozzle.

Using the Hardness Tester

The procedure is as follows: (i) Disconnect the gripper air lines; (ii) Use the control pendant to select hardness tester mode. This will disable the valves so the gripper. Then: (iii) Remove the gripper; (iv) Remove the vacuum nozzle; (v) Install the hardness tester using the same screws as for the gripper; and (vi) Connect the cable to the hardness tester.

Using the calibration plate, check for correct operation of the hardness tester and calibrate it if necessary, and follow the insertion procedure. Make sure the arm is fully inside the reactor (Z position of 141 inches or more). Open the E axis to approximately 20 degrees. Extend the X axis to at least 6 inches. At this position the arm can close further without any mechanical interference. The elbow joint can be closed to a minimum of 6 degrees. This will bring the hardness tester closer to the tube sheet bore.

Now, extend the X axis until it makes contact with the tube sheet wall. This will be apparent with the collision alarm or with the camera view. If the tester does not make contact when the arm is at full extension, carefully move the Z axis back until a collision is detected. Increase the pressure of the X axis to 30%, and run the hardness test.

To perform additional tests, retract the X axis slightly to clear the wall and rotate the C axis to another position. Recall that if the elbow joint is at an angle of 6 degrees, the X axis cannot retract to less than 6 inches due to mechanical interference. When all tests are complete, move the Z axis forward to clear the tester from the wall. Open the E axis to an angle of 20 degrees, fully retract the X axis, close the E axis, and follow the removal procedure.

Remove the hardness tester. The Hardness tester cable must be tucked under the air lines for the gripper and held in place with a radiation hardened cable tie. If the cable gets damaged it is very difficult to replace.

Using the Elbow Joint Lock

The elbow joint of the end–robot arm has a pneumatic lock that can disengage the gearbox and allow the arm to swing freely. Upon start up of the system, the lock should be engaged with low pressure. If the lock has been released, to re-engage it, the Elbow Lock button must be pressed on the control pendant keypad. This will allow the elbow motor to be driven until the lock engages. Then, the elbow joint motor is driven forward until the lock engages. The camera is used to confirm the locking. A high pressure is used to fully lock the Elbow joint.

Auxiliary Subsystems

Electronics Enclosure

The Main Electrical Enclosure (Figure 20) is located on the main frame of the AARM. It contains all of the drives, power supplies, controllers, and other electrical components for the system. The following operations can be controlled from the enclosure.

E-stop button: In an emergency situation, the electrical system can be disabled by pressing the E-stop button. This will remove power to all electronic actuators. In order to recover from an E-stop condition, the button must be released, and then the system has to be enabled again. In this way the actuator positions will be retained and re-homing is not required.

Push-buttons: On the front of the enclosure are three buttons. The enable button allows the operator to control the system. If the disable button is pressed, no control is allowed. The auxiliary button allows the operator to specify which end attachment is mounted to the arm. There is also a toggle-switch on the enclosure labelled Local / Remote. When set to Local, the control pendant is active, and can be used to operate the system. When set to Remote, the pendant is disabled, and control is accomplished via Ethernet.

Pneumatics Enclosure

The Pneumatics Enclosure is located on the main frame. It provides connections for the air supply and pneumatic lines, pressure gauges, vents, and an emergency stop button. There are also two connectors for cables that connect to the main electrical enclosure. Instrument grade air (approximately 6 bars) is provided to the system. The flow rate is very small, since the actuators do not operate continuously.

E-stop button: In an emergency situation the system can be disabled by pressing the E-stop button. This will close all of the valves and leave the actuators stationary. The pneumatic system is designed to trap the air rather than vent it when the valves close, so the actuators will not move. All of the emergency stop buttons disable the entire system.

Manual vent buttons: The front of the pneumatic enclosure (Figure 21) has 6 buttons for manually venting of each side of the actuators. These are provided so the system can be de-pressurized if there is loss of power and the valves are not operational. The gauges on the panel show the

Figure 20. Electronics enclosure

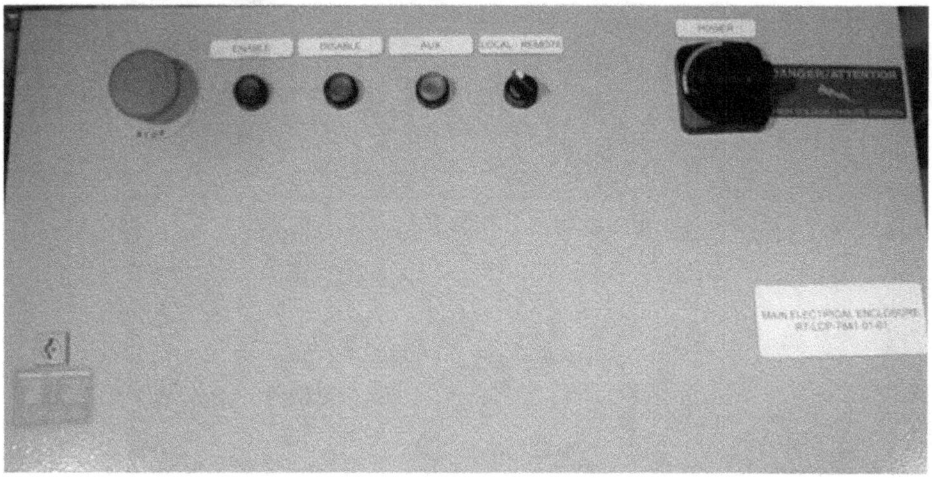

Figure 21. Pneumatic enclosure

Figure 22. Flow control valves

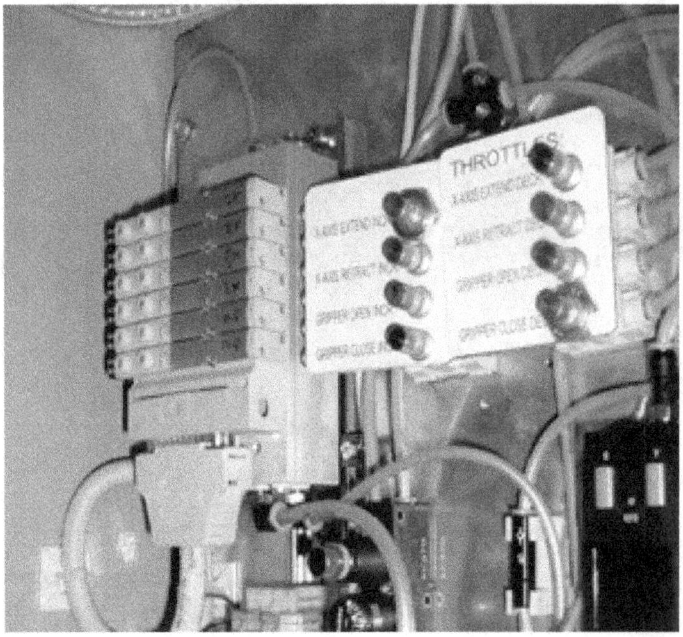

Figure 23. Compact configuration

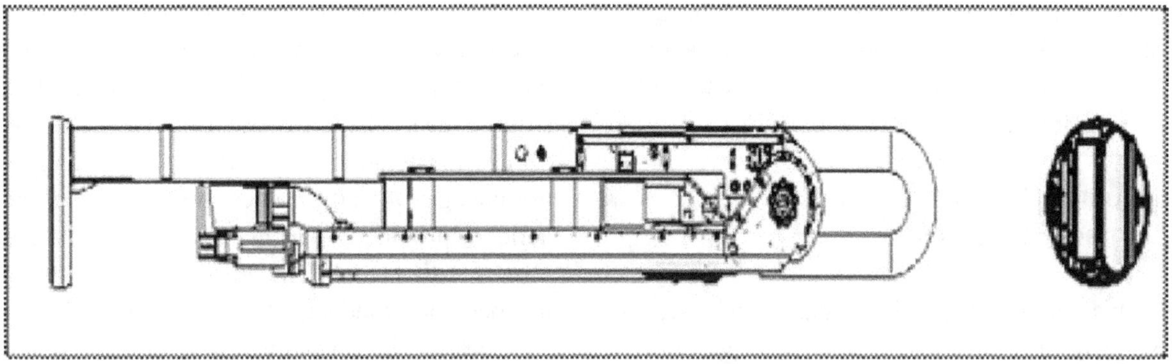

current pressure in the X axis cylinder and the gripper. The lock does not have a pressure gauge since it is not proportionally controlled.

Adjusting the flow control valves: The responsiveness of the pneumatic actuators can be adjusted with the flow control valves located inside the pneumatic enclosure (Figure 22). The flow of supply exhaust air can be adjusted for both the X axis cylinder and the gripper.

Special Features

Tool Attachments

The AARM has a removable end-of-the-arm tool that can be used to pick up small debris that is found during the inspection process. A second attachment for the end-of-the-arm tool is a vacuum nozzle that can be used for removal of smaller particles. A third attachment for the end-of-the-arm tool is a harness tester that performs hardness measurements on the Calandria Side Tube Sheet.

Compact Folding Configuration

Since most openings in the reactor are small, the end-robot arm has been designed to fold into a compact configuration (Figure 23). Once inside the reactor, the arm extends to a working configuration (Figure 24) using the elbow joint.

Contact Detection

In order to prevent damage to the reactor and robot, a collision detection sensing mechanism has been implemented. The collision detection operates on the principle of grounding the CVI structure via a contact with the reactor structure. The end-robot arm is connected to a potential of 24 V that is grounded when in contact with the infrastructure. Actuators become immobile when

Figure 24. Extended configuration

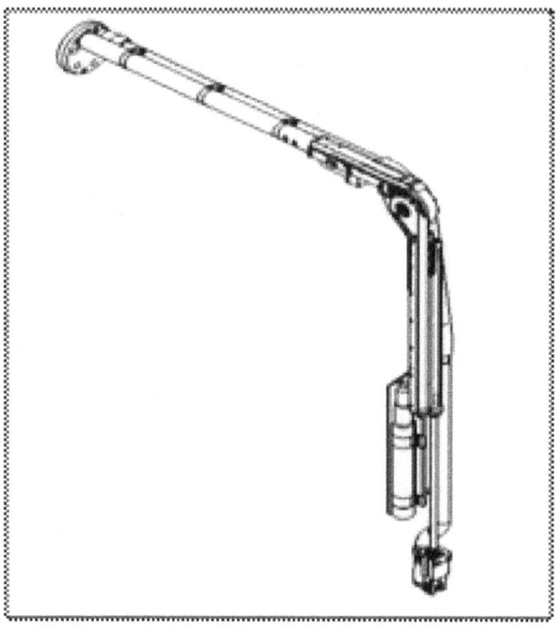

such contact occurs. The advantage with this arrangement is that the entire body of the CVI can be made sensitive to contact, and this allows the operator to decide what action to be taken to prevent any damage to the reactor.

Actuator Compliance for Contact Force Limitation

Each actuator of the end-robot arm can be set up with a limit-force threshold. When triggered, the joint actuators react as follows. The rotary elbow joint has a release mechanism in the gearbox that allows it to disengage completely during a collision to limit the impact force. The linear extension joint that is pneumatically driven has on each side of the cylinder a continuously monitored closed-loops that are adjusted to maintain a constant pressure differential across the cylinder piston. This differential limits the force that the extension joint can apply by providing compliance in the arm such that in a collision situation, the impact force is significantly reduced. The gripper is pneumatically actuated and operates on the same principle of controlled pressure differential of the piston.

Elbow Joint Release Mechanism

The elbow joint release mechanism is clutch-based using a specially designed locking system (Figure 25). The lock engages the joint that rotates the forearm of the robot. If the lock is not engaged the arm rotates freely. The locking action is provided by an air cylinder that engages a lever with a notch. The geometry of this notch has been designed such that equal force (torque on gearbox) in either direction can release the lock. If a collision or other overload occurs, the lock will release automatically before excessive force is applied by the robot arm.

The release force (torque) of the lock mechanism can be controlled by the pressure supplied to the air cylinder. It is adjustable based on the mechanism being inspected. A sufficient torque applied to the arm will force the locking pin out of its seat allowing the joint to rotate freely. The torque is proportional to the force (and pressure) in the locking cylinder.

Figure 25. Elbow locking mechanism

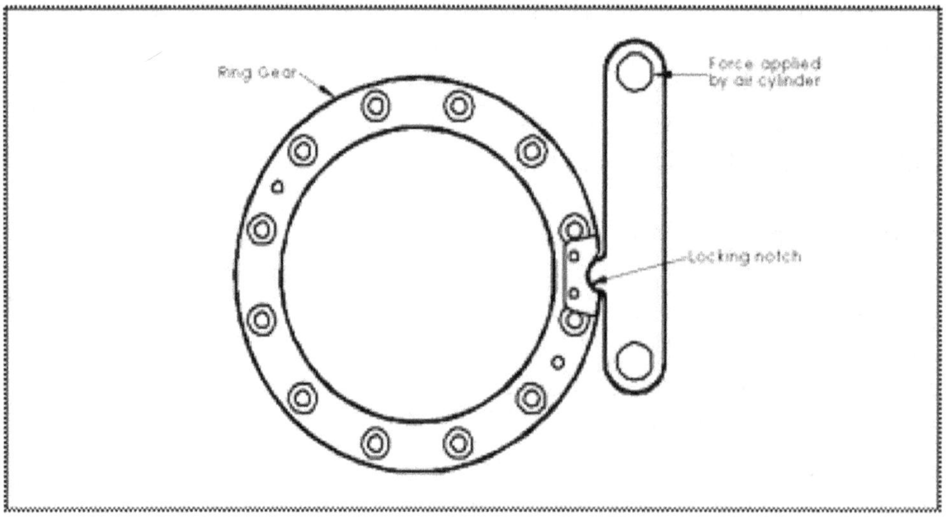

Power Loss and Safe Removal

In the situation where power of the CVI robot is lost it can be manually removed from the reactor. In this case the elbow's release mechanism is triggered manually. This leaves the arm free to swing for "fail-safe" so that the arm can straighten out when it is pulled out of the reactor from any extended configuration. The extension's cylinder can be also manually vented, allowing it to move freely. In this way, the arm can be straightened out when it is pulled out of the reactor, fitting through the opening through which it was inserted. The robot arm also has guides on it to allow the arm to slide through the opening without catching on anything during the manual removal procedure.

The entire system has been designed with a backup mode of operation for loss of power, failure of communication, or failure of actuators. The contingency operations are manual and are performed as follows:

- **Contingency for C axis and Z axis manual operations:** If the C axis or Z axis motors are not functioning correctly, manual drive units can be attached to rotate and move the Boom after the C axis or the Z axis servo motors have been removed.
- **Contingency for removal of the arm:** If control of the arm is lost, the removal is performed by the rotating the C axis to 0 degrees, either with the servo or with the manual unit; the elbow lock is released; the X axis is extended; or the Z axis is driven slowly either with the servo or the manual drive. The arm will now be hanging limp, and can be pulled out of the reactor.
- **Contingency operation of the gripper:** If an item has been picked up and control of the gripper is lost, the gripper can be released by venting it by manual control.

Radiation Shielding

When any type of tool is inserted into the nuclear reactor, it is a requirement that the tool provides shielding so that the radiation levels escaping the reactor through the tool or opening for the tool are within safe limits. This shielding is usually a solid piece of metal with enough mass to absorb the radiation. In order to function properly, the shield must be positioned at the opening to the reactor where the tool is to be inserted. This limits the use of any tool to specific fixed locations inside the nuclear reactor. If this shield were not at the opening, radiation would flow around the shield. In general, the tools in use provide a 12"slug of steel at the wall of the reaction chamber. This 12" slug is generally at a fixed point along the axis of the tool. Hence the tool has only one operational position that blocks the radiation.

The CVI takes a different approach to shielding which allows more flexibility in the positioning of the tool. In order to increase the range of motion of the robot inside the reactor, there is a series of 40mm thickness radiation shields inside the boom carriage that are spaced evenly across the length (Figure 26). The cables and hoses for the robot are routed in a helix around the shields such that there is no direct path for the radiation to escape. This allows the tool to work at any location inside the reactor. In the CVI there are 8 (steel) disks necessary for shielding, but 10-12 disks are active for shielding at all times.

MRI-P: MRI-Guided Robot For Prostate Focal Surgery

The robot reported in Goldenberg et al (2009, June), Goldenberg et al (2010, October), Goldenberg et al (2010, March), and Goldenberg et al (2008, June) was developed for minimally invasive prostate surgery (e.g. laser interstitial thermal therapy of cancerous tumors) performed in a closed bore MRI (Magnetic resonance imaging)-scanner. The robot was designed to operate with minimal electro-magnetic interference between the MRI scanner and the robot.

During the surgical procedures the patient lies on the MRI roll-in table. The robot is mounted & secured onto the table and between patient's legs (Figure 27). The robot controller and video monitor are located outside the high intensity magnetic area of the MR room in the adjacent control room where the MRI controller and video monitors are also located. The robot hand controller and the power supply of the robot and controller are located also in the adjacent room. There is a shielded waveguide through the wall between the adjacent room and MRI room through which all cables (power and communication) are passed.

The surgical tools attached to the robot are remotely manipulated using the MRI and robot images. The MR images allow identification of the tumor (target) and tool location relative to the target, and support performing suitable adjustments of the tool using the hand controller to reach the target.

After the surgical tool tip has reached the target, a sheathe covering the fiber-optic laser applicator embedded in the tool is retracted, and the ablation is performed to destroy the tumor. During the ablation process, MR image is used to display the temperature around the applicator tip to ensure that the tumor receives enough treatment, and that healthy tissue around the tumor is not harmed.

MRI-based software tools and images are used by the medical team (radiology, urology, surgery, and MRI medical physicists) to identify the tumor and select the target based on intra-operative images. The target (tumor) is defined as a point with respect to the 3D MRI scanner reference frame. The selection of the target generates automatically the coordinates of the Target in the MRI reference frame. The medical team also selects the entry (penetration) site to the perineum. The coordinates of the Entry point are also generated by the MRI system. Both the Target and Entry coordinates are transferred to the supervisory controller of the robotic operation. The display monitor connected to the robot controller provides visualization of the Target and Entry points and of the surgical tool in relation to a reference frame (of the MRI

Figure 26. Radiation shielding

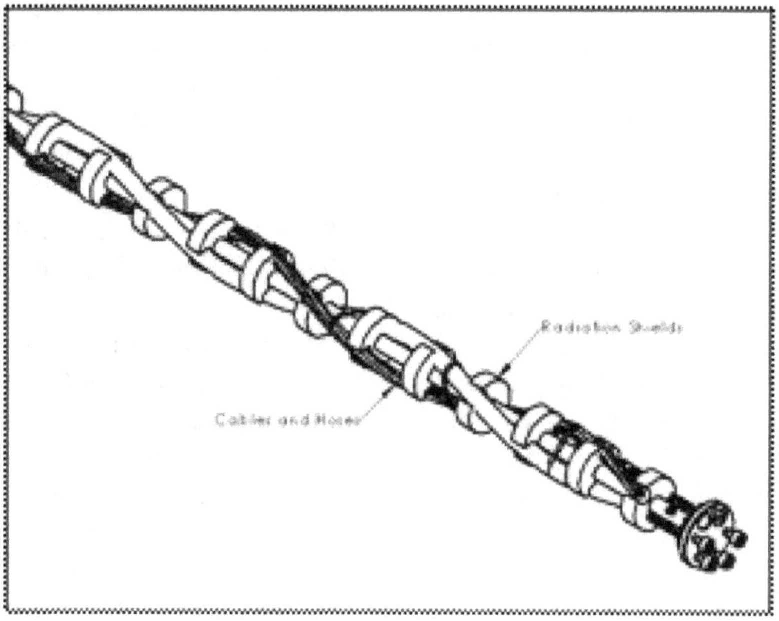

Figure 27. Schematic of a patient and MRI-P

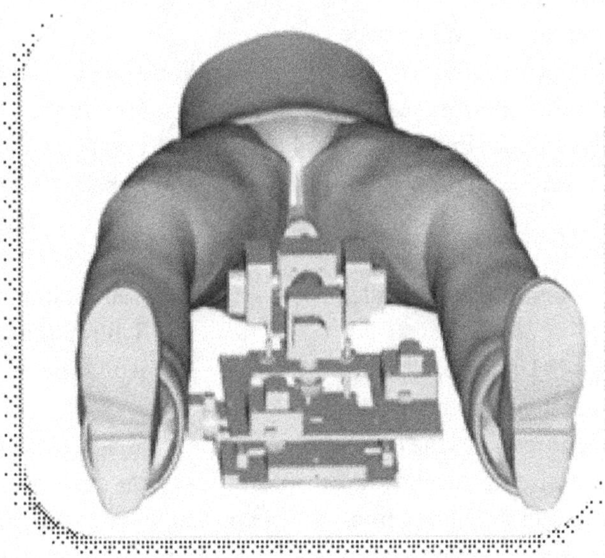

unit). In parallel, the robot kinematics model running on the robot controller generates the joint motion commands, based upon which the tool tip is automatically positioned near the entry to the perineum (at a distance selected by the medical team) and aligned with the line connecting the Target and Entry points. After the robot has been relocated to the entry location and orientation in the vicinity of the perineum, the surgical clinician manually manipulates by remote control the position and orientation of the surgical tool and proceeds with the penetration into the body of the patient towards the target. The clinician uses feedback from both the MRI and robot monitors.

Project Management

Team: The team was assembled as in the previous cases.

Development Process Planning: The PM and System Engineering Team (SET) defined the critical project development path.

The major milestones: Specifications, Design, Critical Design Review, Manufacturing, and Tests were scheduled and budgetary allocations were provided for each phase.

Project Control: The PM provided the tools for project management including cost control, scheduling, reporting, and delivery of that information to the executive branch

System Engineering

Specifications

Functional Specifications

The robot (MRI-P) is used to: (i) automatically locate the tip of the surgical tool at a desired orientation and position relative to the Target near the Entry point into the perineum; and (ii) provide the support for the tool penetration along the line connecting the Entry with the Target by remote control while images of the target and tool are projected on the MRI and robot monitors. The workflow was defined by a series of steps that provided the functional specifications:

- **Calibration:** registration of the needle in the MRI space outside the body

- **Targeting:** identified in the prostate gland and defined in the MRI space
- **Entry point:** defined in the MRI space near the perinea skin
- **Needle tip re-location:** robot moves the needle automatically to the Entry location. MRI image is taken to confirm robot location
- **Manual control:** tool penetration/retraction is performed through hand controller commands.
- **Needle is advanced** ~ 5-10 mm each time, then scanning is done, imaging the needle each time until the needle reaches the target
- **Ablation:** after the needle tip reaches the target, the sheath covering the ablation laser is retracted under user command a fixed distance (22 mm based on the type of laser used) in order to expose the laser fiber and perform the ablation. Ablation is monitored through MRI scanning.

Robot Specifications

- MRI-compatibility
    - Materials used for the robot structure must be MRI-compatible: no artifacts on the images
    - EMI (electro-magnetic interference) must be avoided: no RF noise on the images
    - SNR (signal-to-noise ratio) must be such that the images are useful to the medical staff

The design qualitative criterion was that images acquired with and without the robot in the MRI would not differ in terms of artifacts and noise. The criterion is the signal-to-noise ratio (SNR). Ultimate diagnostic decisions are based on many (inter-related) factors including SNR, resolution, and contrast. They also depend strongly on the specificity of anatomy/pathology under investigation.

- Work space
    - Horizontal translation of the base (side-by-side across the width of the bore): 40 mm
    - Vertical translation of the base: 20 mm
    - Pan rotation of the surgical tool: ±21°
    - Tilt rotation of the surgical tool: ±30°
    - Roll rotation of the surgical tool: ±25°

The work space dimensions introduced above are based on surgeons' input and their clinical experience. The work space dimensions to reach most men's prostate completely are 5 cm x 5 cm x 6 cm (left to right). The distance from the skin to the prostate is about 5 cm. The prostate is widest at the base, so one could consider it as an inverted cone. The approximate work volume would have a base consisting of a 6 cm wide circle and reach of 8-10 cm from the skin. These dimensions were used to calculate the required linear penetration and rotation of the robot tip. More precise dimensions were determined experimentally, and they will be embedded in the design of the 2nd generation robot prototype.

- Mounting of the robot

On a removable platform fitted to the MRI roll-up table that accommodates the robot and patient.

- Compact structure of the robot design to fit between the patient legs close to the perineum
- Safety of patient in the MR environment
- Image compatibility (visual and data) between MRI console and PC-supervisor displays.

Design

The functional and robot specifications were used to derive the detailed specifications of the robot system. They are listed below:

- **Dimensions (W × L × H):** 300 mm × 366 mm × 170 mm (1.8" × 14.4" × 6.7")
- **Weight:** 5 kg/11 lbs
- **Horizontal translation (H):** 40 mm
- **Horizontal speed:** 0.64 ~ 5.3mm/s
- **Vertical translation (V):** 20 mm
- **Vertical speed:** 0.36 ~ 3.0mm/s
- **Pan rotation (P):** ± 20°
- **Pan speed:** 7° ~ 45°/s
- **Tilt rotation (T):** ± 25°
- **Tilt speed:** 90° ~ 600°/s
- **Roll rotation (R):** ± 25°
- **Roll speed:** 30°~200°/s
- **Trocar penetration stroke (I):** 100mm
- **Penetration speed:** 1.6 ~ 10.6mm/s
- **Retraction to expose the fiber tip:** 24.5 mm
- **Retraction speed:** 1.6 ~ 10.6mm/s
- **Rated tissue penetration force:** 10 kg

System Structure

The system consists of five subsystems: (i) five-joint robot and trocar (tool carrier) (Figure 28); (ii) surgical tool attached to the robot; (iii) robot controller; (iv) MRI monitoring station (including temperature monitoring); (v) hand controller of the tool.

A 3D model of the MRI-P and its five degrees of freedom are illustrated in Figure 29. The robot degrees of freedom are: horizontal translation (H); vertical translation (V); pan rotation (P); tilt rotation (T); and roll rotation (R). The robot has two linear motion joints (horizontal and vertical translations) and three rotational joints (Pan, Tilt and Roll).

Artifacts and distortions in the MR image are directly related to the material used, thus the selection of materials for the robot design was a challenging process. Based on pre-testing of a series of candidate materials Aluminum 6061 was selected for most parts of the robot. A few parts are made of brass and plastic (Delrin and Nylon). The surgical tool material is titanium. ESI had initially used a 12G home-made needle made from titanium, and, more recently, they have been using custom-made 12Ga nitinol needles.

Figure 28. MRI-P

Trocar Module

Creating a modular surgical tool structure in order to perform all surgical operations (laser ablation, seed placement – brachytherapy, and biopsy) with the same robot is one of the challenges in the development of the MRI-P robot. The main requirement for modularity is not only to operate three trocar modules with the same robot, but also to facilitate the removal and mounting of each. Furthermore, the surgical tool (needle) mounted on the trocar must be sterilizable, and must have a sharp point (for tissue penetration). Titanium can be sterilized, can be sharpened, and produces very little artifact on MRI images so it was the preferred material. The needle must be long enough to reach the target, and must be as small diameter as possible to minimize tissue injury, but large enough to carry the treatment "tool" for ablation – i.e. laser fiber.

Figure 30 provides an overview of the MRI-P robot with trocar modularity such that the entire surgical robot can be considered as divided into two units: (i) robot unit with five primary degree-of-freedom (1); and (ii) trocar module unit (shown for laser ablation) (2). Other two trocar modules (for biopsy and brachytherapy) will be developed at a later stage.

In order to allow the operator to substitute easily and quickly the modular trocar units without having to make adjustments to the robot unit, simple interfaces between the base unit and the trocar module were provided (Figure 30). A positioning block (a) with two pins is attached under the shell (b) of each trocar module. By plugging the shell of the trocar module into the hollow of the support block (c) that is coupled with a large hollow bevel gear on the robot unit, and locating the positioning block against the rear side of the support block, then locking with a thumb screw (d), the trocar module is quickly mounted on the robot unit. After unlocking with the thumb screw and being pulled, the trocar module can be easily removed from the base unit.

Control System Architecture

The control system architecture comprises the following four main modules: i) PC-based supervisory controller; ii) Robot kinematics; iii) Robot embedded controller; and iv) User interface. The architecture of the modules is shown schematically in Figure 31. The supervisory controller communicates with the MRI host computer to receive data: Entry and Target points and needle tip location. As well, the controller registers the origin of the robot coordinate frame in MRI coordinates. The registration is done using a calibration phantom, as described by Goldenberg, Yang et al (2010, October). Furthermore the controller executes the kinematics module, communicates with the robot embedded controller, allows manual control of the trocar and robot joints, and provides graphical display for the user interface.

The robot kinematics calculations are done after locations Entry (E) and Target (T) are selected. Their coordinates are transferred automatically to the robot supervisory controller and displayed on the robot monitor. The robot kinematics resolves T & E into a space line TE (path of tool penetration). Kinematics calculations also generate the joint commands to locate the tool tip very near the Entry point, and orient the tool parallel to the TE line.

Robot Controller

The robot embedded closed-loop controller provides for aligning automatically the position and orientation of the tool with the TE line. The tip of surgical tool moves automatically toward E near the perineum along the TE line, and stops short of the perineum (as selected by the medical staff).

The robot uses ultrasonic motors as actuators. The ultrasonic motor has important advantages over other types of actuators:

Figure 29. 3D model of the robot

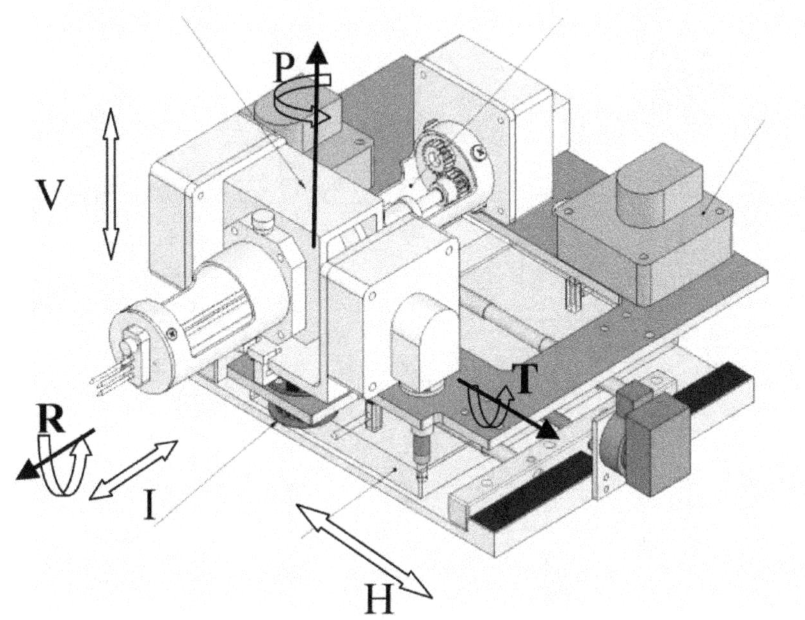

Figure 30. MRI-P robot and modular trocar

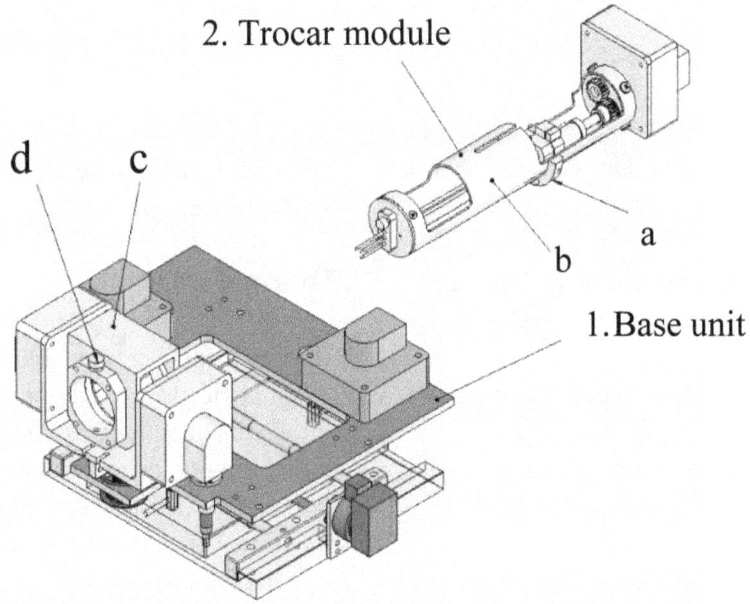

Figure 31. Control system architecture

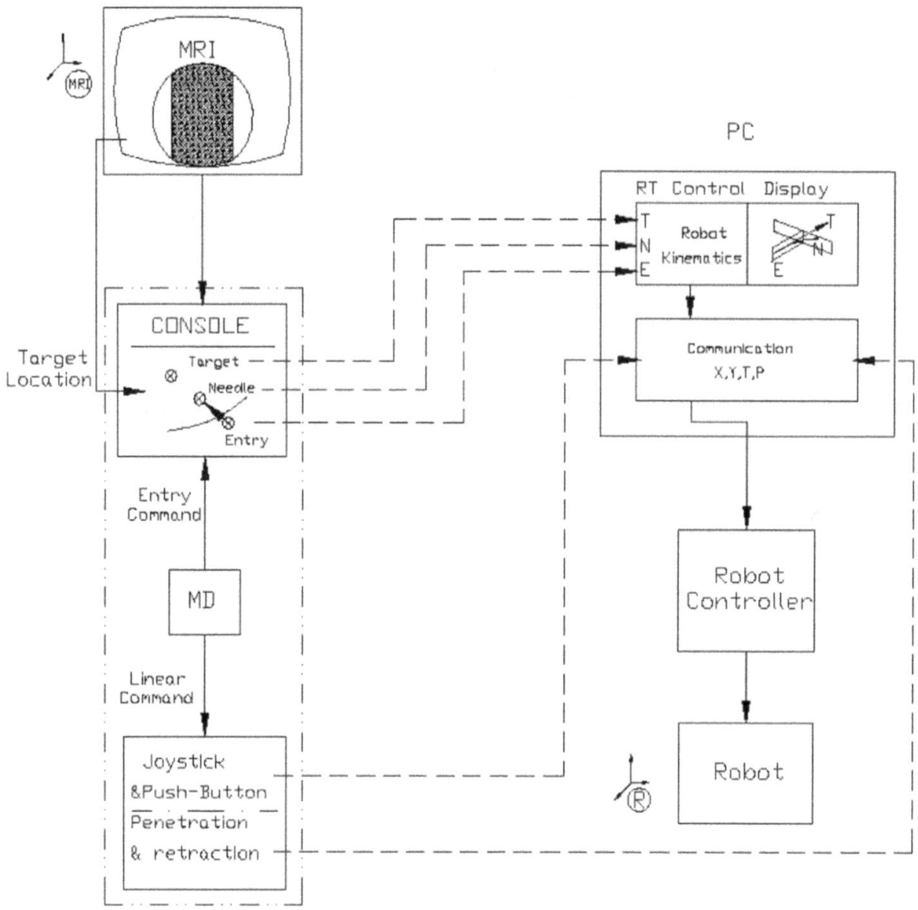

- Can be positioned on the robot such that it does not interfere with the MRI magnetic field
- Allows for direct drive (no gear) as it generates high torque at a low speed
- It is bidirectional
- It is retentive, thus a brake might not be necessary
- It is compact and light as has a simple structure.

Based on these advantages ultrasonic motors were selected for the MRI-P robot. As a result of tests of ultrasonic motors that were conducted in the MR environment, one USR30-E3N and four USR60-E3N ultrasonic motors, all made by SHINSEI Corporation Japan, were selected.

Thermal Imaging

The treatment fiber is capable of creating a thermal lesion in the gland resulting in cell death without untoward patient effects. The robot is capable of delivering one or more treatment fiber(s) to an image-identified target(s) in the prostate gland under real-time visual feedback via fast MR imaging methods.

The effect of delivery of focal ablation therapy is addressed through imaging to minimize the risks of damaging healthy tissue of other organs (in particular the urethra and rectum). Visualization provides the medical staff with the feedback that is necessary to ensure the safety of the patient during the procedure. A 3D-image-based evaluation of the treatment-effects delivered by the treatment fiber is performed in real-time while the treatment is being delivered.

FUTURE OF PROTOTYPING

Prototyping - techniques, cost, and modularity – is gaining importance because of the proliferation of robotics in every day's aspect of life. From automotive to military and law enforcement and on to medical and service tasks, the robots are close to becoming personal, much like the PC and wireless communication tools of today. This raises the need, expectations, and role of prototypes in the process of product development. It is a trend that is already prevalent, and is rapidly gaining strength.

The process of prototyping must be not only judicious technically but also must be linked to the commercial endeavors onto which the building of the prototype is related. The industry of robotics is dominated by research-minded people, and prototyping is the link to the applications and markets. This undertaking is at least as important as the generation of the original ideas and related research in order to develop the knowledge that would propel the industry further.

The trends in robotics markets are clearly indicating that the traditional markets – automotive, military and law enforcement – cannot sustain the growth of the industry to propel it to the forefront of the high technology boom. Personal robots for ever day life, from home-care for the elderly and disabled to domestic helpers and further on to entertainment, are becoming the wave of the future. In fact they are already present.

Prototyping for this "tsunami of robotics" is a major concern. The cost of prototyping today is high. Methods of rapid and low cost prototyping while expressing the functionality to allow adequate examination of the representations of ideas are required. This need may be satisfied by developing a theory that supports preliminary design of a reduced scale prototype with similar functionality. The closest parallel would be wind tunnels on earth that simulate aerodynamic effects in the upper atmosphere.

CONCLUSION

The domain of prototyping must be addressed as a fundamental element of product development. It must be provided with a theoretical foundation that would allow developing and comparing prototypes in terms of quality, reliability, and cost. Prototyping is the key element in the quest for commercial success.

The chapter has presented a tutorial of prototyping methodology with the intent to support such activities and highlight the need for further work in the development of prototyping methodology for adequate and inexpensive representation of new ideas with scientific or commercial promise.

REFERENCES

Caprari, G., Breitenmoser, A., Fischer, W., Hurzeler, C., Tache, F., & Siegwart, R. ... Moser, R. (2010, October). *Highly Compact Robots for Inspection of Power Plants*, Paper presented at *2010 1st International Conference on Applied Robotics for the Power Industry (CARPI)*, (pp. 1-6), Montreal, QC.

Engineering Services Inc. (n.d.). Retrieved September 2, 2011, from http://www.esit.com

Goldenberg, A. (2007). Robotics, mechatronics, and intelligent automation. In Avallone, E. A., Baumeister, T., & Sadegh, A. (Eds.), *Marks' standard handbook for mechanical engineers*. New York, NY: McGraw Hill.

Goldenberg, A., Grinewsky, M., & Campbell, T. (2010, October). AARM: A robot for internal operations in nuclear reactors. *2010 1st International Conference on Applied Robotics for the Power Industry (CARPI)*, (pp. 1-6). Montreal, QC.

Goldenberg, A., Trachtenberg, J., Kucharczyk, W., Yang, Y., Haider, M., & Ma, L. (2008, June). Robotic system for closed-bore MRI-guided prostatic interventions. *IEEE/ASME Transactions on Mechatronics, 13*(3), 374–379. doi:10.1109/TMECH.2008.924122

Goldenberg, A., Trachtenberg, J., Kucharczyk, W., Yang, Y., Haider, M., Sussman, M., ... Weersink, R. (2010, March). Robot-assisted MRI-guided prostatic interventions. *Robotica – Special Issue on Surgical Robotics, 28*(2), 215-234.

Goldenberg, A., Yang, Y., Ma, L., Trachtenberg, J., Kucharczyk, W., & Haider, M. ... Weersink, R., (2009, June). *MRI compatible robot with calibration phantom & phantom.* US Patent Application No. 12/457,707, filed.

Goldenberg, A., Yang, Y., Ma, L., Trachtenberg, J., Kucharczyk, W., & Haider, M. ... Weersink, R. (2010, October). *MRI compatible robot with calibration phantom & phantom*, PCT Patent Application No. PCT/CA2009/000846, filed in Jun 2009, 35 claims accepted Oct 2010.

Hamilton, A., Burany, S. J., Peralta, S. B., & Greenland, L. (2010, October). Robotic removal of high-activity debris from a nuclear primary heat transfer system. *2010 1st International Conference on Applied Robotics for the Power Industry (CARPI)*, (pp. 1-6). Montreal, QC.

Lin, J., & Goldenberg, A. (2009, February). *Variable configuration articulated tracked vehicle.* U.S. Patent #7,493,976.

Lin, J., & Goldenberg, A. (2009, October). Variable configuration articulated tracked vehicle. *US CIP Patent*, *7*(600), 592.

Lin, J., & Goldenberg, A. (2010, August). *Variable configuration articulated tracked vehicle.* Canadian Patent #2,555,455.

Perrot, Y., Gargiulo, L., Houry, M., Kammerer, N., Keller, D., & Measson, Y. ... Verney, A. (2010, October). Long reach articulated robots for inspection in hazardous environments, recent developments on robotics and embedded diagnostics. *2010 1st International Conference on Applied Robotics for the Power Industry (CARPI)*, (pp. 1-5). Montreal, QC.

Zwicker, E., Zesch, W., & Moser, R. (2010, October). A modular inspection robot platform for power plant applications. *2010 1st International Conference on Applied Robotics for the Power Industry (CARPI)*, (pp. 1-6). Montreal, QC.

Chapter 2
Modeling and Simulation of Discrete Event Robotic Systems Using Extended Petri Nets

Gen'ichi Yasuda
Nagasaki Institute of Applied Science, Japan

ABSTRACT

This chapter deals with modeling, simulation, and implementation problems encountered in robotic manufacturing control systems. Extended Petri nets are adopted as a prototyping tool for expressing real-time control of robotic systems and a systematic method based on hierarchical Petri nets is described for their direct implementation. A coordination mechanism is introduced to coordinate the event activities of the distributed machine controllers through friability tests of shared global transitions. The proposed prototyping method allows a direct coding of the inter-task cooperation by robots and intelligent machines from the conceptual Petri net specification, so that it increases the traceability and the understanding of the control flow of a parallel application specified by a net model. This approach can be integrated with off-the-shelf real-time executives. Control software using multithreaded programming is demonstrated to show the effectiveness of the proposed method.

1. INTRODUCTION

Complex robotic systems such as flexible manufacturing systems require sophisticated distributed real-time control systems. A major problem concerns the definition of the user tasks and the cooperation between the subsystems, especially since the intelligence is distributed at a low level (machine level). Controlling such systems generally requires a hierarchy of control units corresponding to several abstraction levels. At the bottom of the hierarchy, i.e. machine control level, are the programmable logic controllers

DOI: 10.4018/978-1-4666-0176-5.ch002

(PLC). The next level performs coordination of the PLCs. The third level implements scheduling, that is, the real-time assignments of workpieces and tools to machines. At the machine level, PLCs perform local logical control operations of flexible, modular, high-speed machines through the use of multiple independent drives (Holding, et al. 1992). Implementation languages can be based on ladder diagrams or more recently state machines (Silva, et al. 1982). However, when the local control is of greater complexity, the above kinds of languages may not be well adapted. The development of industrial techniques makes a sequential control system for robotic manufacturing systems larger and more complicated one, in which some subsystems operate concurrently and cooperatively (Neumann, 2007).

In the area of real-time control of discrete event robotic systems the main problems that the system designer has to deal with are concurrency, synchronization, and resource sharing problems. Presently, the implementation of such control systems makes a large use of microcomputers. Real-time executives are available with complete sets of synchronization and communication primitives. However, coding the specifications is a hazardous work and debugging the implementation is particularly difficult when the concurrency is important. It is important to have a formal tool powerful enough and allowing developing validation procedures before implementation. Conventional specification languages do not allow an analytical validation. Consequently, the only way to validate is via simulation and step by step debugging. On the other hand, a specification method based on a mathematical tool may be more restricting, but analytical procedures can strongly reduce the simulation step. Rapid prototyping is an economical and crucial way to experiment, debug, and improve specifications of parallel applications. Increasing complexity of synchronizing mechanisms involved in concurrent system design makes necessary a prototyping step starting from a formal, already verified model. Petri nets allow to validate and to evaluate a model before its implementation. The formalism allowing a validation of the main properties of the Petri net control structure (liveness, boundedness, etc.) guarantees that the control system will not fall immediately in a deadlocked situation. In the field of flexible manufacturing cells, the last aspect is essential because the sequences of control are complex and change very often.

When using Petri nets, events are associated with transitions. Activities are associated to the firing of transitions and to the marking of places which represents the state of the system. The network model can describe the execution order of sequential and parallel tasks directly without ambiguity (Silva, 1990). Pure synchronization between tasks, choices between alternatives and rendezvous can be naturally represented. Moreover at the machine level Petri nets have been successfully used to represent the sequencing of elementary operations (Yasuda, et al. 1992). In addition to its graphic representation differentiating events and states, Petri nets allow the possibility of progressive modeling by using stepwise refinements or modular composition. Libraries of well-tested subnets allow components reusability leading to significant reductions in the modeling effort. The possibility of progressive modeling is absolutely necessary for large and complex systems, because the refinement mechanism allows the building of hierarchically structured net models. Furthermore, a real-time implementation of the Petri net specification by software called a token player can avoid implementation errors. This is because the specification is directly executed by the token player and the implementation of these control sequences preserves the properties of the model. In this approach, the Petri net model is stored in a database and the token player updates the state of the database according to the operation rules of the model. For control purposes, this solution is very well suited to the need of flexibility, because, when the control sequences change, only the database needs to be changed.

Some techniques derived from Petri nets have been successfully introduced as an effective tool for describing control specifications and realizing the control in a uniform manner (Murata, et al. 1986). However, in the field of flexible manufacturing cells, the network model becomes complicated and it lacks the readability and comprehensibility (David, et al. 1992). Therefore, the flexibility and expandability are not satisfactory in order to deal with the specification change of the control system. Despite the advantages offered by Petri nets, the synthesis, correction, updating, etc. of the system model and programming of the controllers are not simple tasks (Zhou, et al. 1993), (Desrochers, et al. 1995), (Lee, et al. 2006). Some Petri net implementation methods have already been proposed for simulation purposes or for application prototyping (Butler, 1991), (Garcia, 1998), (Piedrafita, et al. 2008). However the implementation method for hierarchical and distributed control of complex robotic systems has not been established sufficiently so far (Breant, et al. 1992), (Girault, et al. 2003), (Zhou, et al. 1999). If it can be implemented using Petri nets, the modeling, simulation and control can be realized consistently.

This chapter describes a Petri net based prototyping method for real-time control of complex robotic systems. The presented method, based on the author's previous works (Yasuda, et al. 2010), (Yasuda, 2010), involves three major steps, and progressively gathers all information needed by the control system design and the code generation for simulation experiments. The first step consists in specifying the conceptual net model for overall system control. The second step consists in transforming the net model into the detailed net model. Based on the hierarchical and distributed structure of the system, the specification procedure is a top-down approach from the conceptual level to the detailed level. The third step consists in decomposing the detailed net into local net models for machine control and the coordination model. The coordination algorithms are simplified since the robots and machines in the system are separately controlled using dedicated task execution programs. In order to deal with complex models, a hierarchical approach is adopted for the coordination model design. In this way, the macro representation of the system is broken down to generate the detailed nets at the local machine control level. Finally, the C++ code generation using multithreaded programming is described for the prototype hierarchical and distributed control system.

2. MODELING OF DISCRETE EVENT SYSTEMS USING EXTENDED PETRI NETS

A Petri net is a directed graph whose nodes are places shown by circles and transitions shown by bars. Directed arcs connect places to transitions and transitions to places. Formally, a Petri net is a bipartite graph represented by the 4-tuple $G=\{P,T,I,O\}$ such that:

$P=\{p_1,p_2,\ldots,p_n\}$ is a finite, not empty, set of places;

$T=\{t_1,t_2,\ldots,t_m\}$ is a finite, not empty, set of transitions;

$P\cap T=\phi$, i.e. the sets P and T are disjointed;

I: $T\rightarrow P^\infty$ is an input function, a mapping from transitions to bags of places;

O: $T\rightarrow P^\infty$ is an output function, a mapping from transitions to bags of places.

The input function I maps from a transition t_j to a collection of places $I(t_j)$, known as input places of a transition. The output function O maps from a transition t_j to a collection of places $O(t_j)$, known as input places of a transition. The pre-incidence matrix of a Petri net is $C^-=[c_{ij}^-]$ where $c_{ij}^-=1$ $(p_i\in I(t_j))$, $c_{ij}^-=0$ $(p_i\notin I(t_j))$; the post–incidence matrix is $C^+=[c_{ij}^+]$ where $c_{ij}^+=1$ $(p_i\in O(t_j))$, $c_{ij}^+=0$ $(p_i\notin O(t_j))$, then the incidence matrix of the Petri net $C = C^+ - C^-$. Each place contains integer (positive or zero) marks or tokens. The number of tokens in each place is defined by the marked vector or marking $M = (m_1, m_2, ..., m_n)^T$. The number of tokens in one place p_i is simply indicated by $M(p_i)$. The firing of a transition will change the token distribution (marking) in the net according to the transition firing rule. In the basic Petri net, a transition t_j is enabled if, $\forall p_i \in I(t_j)$, $M_k(p_i) \geq w(p_i, t_j)$, where the current marking is M_k and $w(p_i,t_j)$ is the weight of the arc from p_i to t_j. A sequence of firings will result in a sequence of markings. A marking M_n is said to be reachable from a marking M_0 if there exists a sequence of firings that transforms M_0 to M_n. The set of all possible markings reachable from M_0 is denoted as $R(M_0)$. A Petri net is said to be k-bounded or simply bounded if the number of tokens in each place does not exceed a finite number for any marking reachable from M_0, i.e., $\forall p_i \in P$, $\forall M \in R(M_0)$, $M(p_i) \leq k$ (Reisig, 1985), (Murata, 1989).

In the basic Petri net, bumping occurs when despite the holding of a condition, the preceding event occurs. This can result in the multiple holding of that condition. From the viewpoint of discrete event process control, bumping phenomena should be excluded. So, the firing rule was modified so that the system is free of this phenomenon. Because the modified Petri net must be 1-bounded, for each place p_i, $m_i = 0$ or 1, and the weight of every arc is 1. A Petri net is said to be ordinary if all of its arc weights are 1's. Thus the axioms of the modified Petri net are as follows.

1. A transition t_j is enabled if for each place $p_k \in I(t_j)$, $m_k = 1$ and for each place $p_l \in O(t_j)$, $m_l = 0$.
2. When an enabled transition t_j fires, the marking M is changed to M', where for each place $p_k \in I(t_j)$, $m'_k = 0$ and for each place $p_l \in O(t_j)$, $m'_l = 1$.
3. In any initial marking, there must not exist more than one token in each place.

A transition without any input place is called a source transition, and one without any output place is called a sink transition. A source transition is unconditionally enabled, and the firing of a sink transition consumes a token in each input place, but does not produce any. According to these axioms, the number of tokens in a place never exceeds one, thus the modified Petri net is essentially 1-bounded and said to be a safe graph.

Besides the guarantee of safeness, considering not only the modeling but also the actual system control of robotic systems, the additional capabilities of input and output interfaces which connect the net to its environment are required. The extended Petri net adopts the following two elements: 1) Gate arc, 2) Output signal arc (Hasegawa, et al. 1984). A gate arc connects a transition with a signal source, and depending on the signal, it either permits or inhibits the occurrence of the event which corresponds to the connected transition. Gate arcs are classified as permissive or inhibitive, and internal or external. An output signal arc sends a command request signal from a place to an external machine. The interfaces are a set of transitions which represent the communication activities of the net with its environment. Thus the firing rule of a transition is described as follows: an enabled transition may fire when it does not have any internal permissive arc signaling 0 nor any internal inhibitive arc signaling 1 and it does not have any external permissive arc signaling 0 nor any external inhibitive arc signaling 1.

A robotic action is modeled by two transitions and one condition as shown in Figure 1. At the "Start" transition the command associated with the transition is sent to the corresponding robot or machine. At the "End" transition the status report is received. When a token is present in the

Figure 1. Extended Petri net model of robotic action with external permissive and inhibitive gate arcs

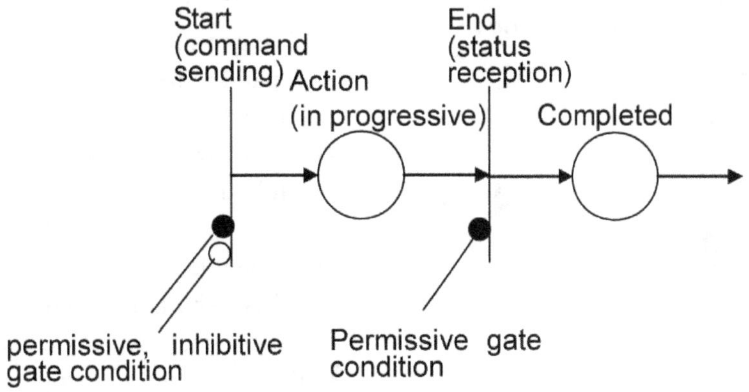

"Action" place, the action is in progressive. The "Completed" place can be omitted, and then the "End" transition is fused with the "Start" transition of the next action. Activities can be assigned an amount of time units to monitor them in time for real performance evaluation.

The firing of a transition is indivisible; the firing of a transition has duration of zero. Extended Petri nets to consider timing conditions where each activity is assigned an amount of time units can also be used as shown in Figure 2. Through the simulation steps, the transition vector table is efficiently used to extract enabled or fired transitions.

The flow chart of simulation and evaluation procedure is shown in Figure 3. At a step of simulation of robotic task, the configuration of the robots can be seen with graphic simulation. The data structure of the extended Petri net simulator is made up of several tables corresponding to the structural information of the net specifying the robotic task. These tables are the following ones:

1. The table of the labels of the input and output places for each transition,
2. The table of the transitions which are likely to be arbitrated for each conflict place,
3. The table of the gate arcs which are internal or external, permissive or inhibitive, for each transition,
4. The table of marking which indicates the current marking for each place,
5. The table of places to tasks mapping which points out the tasks that have to be put into the ready state when the corresponding place receives a token,
6. The table of the "end of task" transitions, which associates with each of task the set of transitions with external gate arcs switched each time an "end of task" message is received by the simulator. The "end of task" transitions are only fired on the reception of an "end of task" message.

3. PETRI NET MODELS OF COOPERATIVE CONTROL OF CONCURRENT TASKS

A robotic task consists of several subtasks or operations, and a subtask consists of several actions. Conceptually, robotic processes are represented as sequential constructs or state machines, where each transition has exactly one incoming arc and exactly one outgoing arc. The structure of a place having two or more output transitions is referred to

Figure 2. Examples of representation of timing conditions: external timer with output signal arc and external gate arc, (b) timed transition

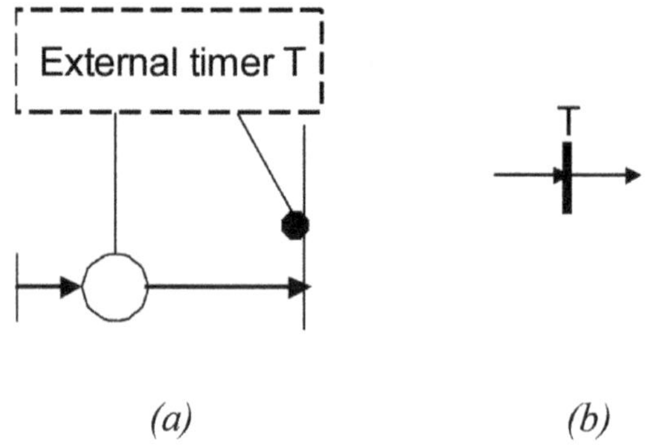

as a conflict, decision, or choice, depending on applications. State machines allow the representation of decisions, but not the synchronization of parallel activities. In a net model of robotic task, the set of places can be classified into three groups: idle, operation, and resource places. A token in an idle place indicates that the robot is ready to work and waiting for a specified signal from another robot or its environment. An operation place represents an operation to be processed for a workpiece or part in a manufacturing sequence and initially it has no token. Resource places represent resources (robots and machines) and their initial tokens represent the number of available resource units.

A task executed by a robot or machine can be seen as some connection of more detailed subtasks. For example, transferring an object from a start position to a goal position is a sequence of the following subtasks: moving the hand to the start position, grasping the object, moving to the goal position, and putting it on the specified place. Figure 4 shows the net representation of a robotic task: pick and place operation with the input and output conveyors. While the place "Robot" in Figure 4(a) indicates that the state of the robot is "ready" when the token is in the place, in Figure 4(b) it indicates that the state of the robot is "operating". The place also indicates the macro representation of the pick and place operation. The parallel net in Figure 4(b) is equivalent to the cyclic net in Figure 4(a) with respect to the enabling conditions of all the transitions in the net. The parallel net assures that the robot can load or unload only one workpiece at a time. Figure 4(c) shows the possible evolution of dynamic behavior of the net.

Furthermore the subtasks are translated into more detailed actions. A hierarchical approach consists in building a model by stepwise refinements. At each step some parts of the model are replaced by a more complex model. A modular approach by model composition affords to build complex validated Petri nets. Figure 5 shows the view of a hierarchical net representation by the graphic net simulator.

A specification procedure for discrete event robotic systems based on Petri nets is as follows. First, the conceptual level activities of the system are defined through a net model considering the task specification corresponding to the aggregate discrete event process. The places which represent the subtasks indicated as the task specification are connected by arcs via transitions in the specified order corresponding to the flow of subtasks and a workpiece. The places representing robots and machines used for the subtasks are also added to connect transitions which correspond to the beginning and ending of their subtasks. Then, the places describing the subtasks are substituted by a subnet based on activity specification and required control strategies in a manner which maintains the structural properties (Yasuda, et al. 2010).

For concurrent control in the conceptual net model, two implementation methods of synchronous interaction between two tasks which are executed respectively by one robot are shown in Figure 6(a), (b) (Yasuda, 2000), while an implementation with asynchronous communication based on the well-known signal/wait (semaphore) concept is shown in Figure 6(c).

A coordination mechanism is introduced to coordinate concurrent systems with separate ro-

Figure 3. Flow chart of simulation and evaluation procedure

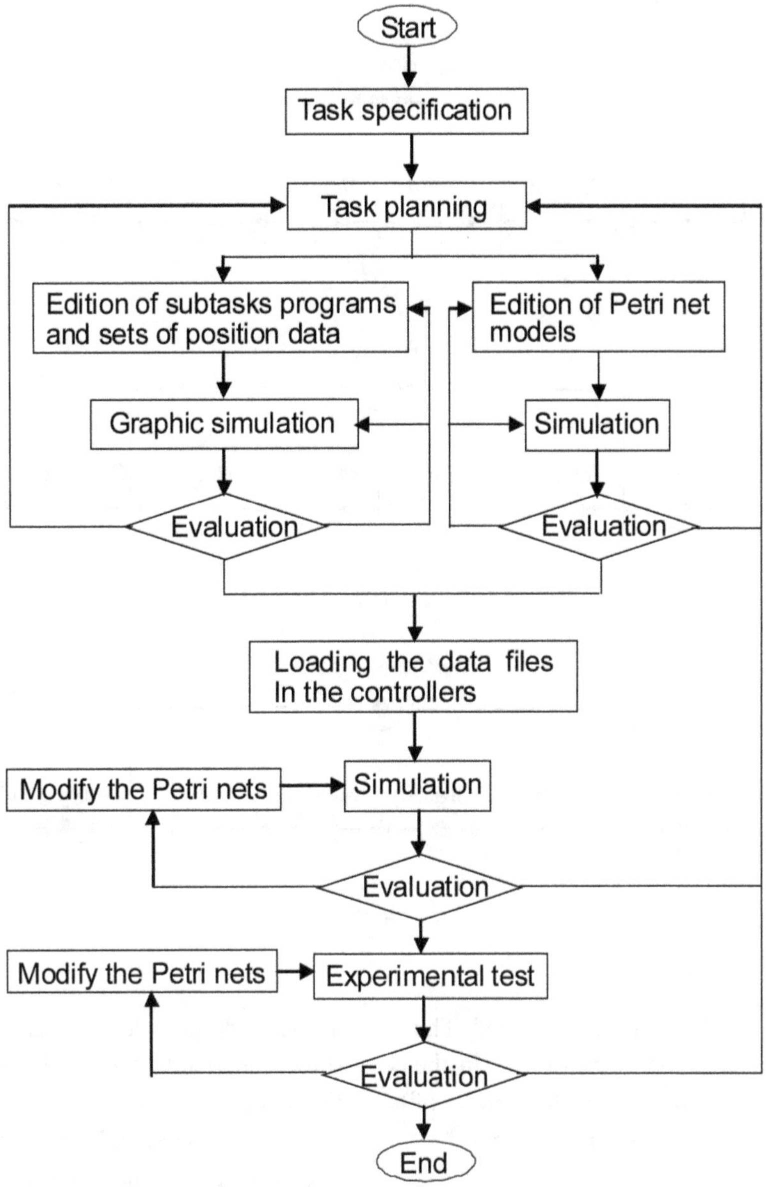

bots which require interaction between each other, such as synchronization and resource conflict. Figure 7 shows the net model of a coordination mechanism to conduct synchronous interaction by means of synchronous communication between two robots. This is also the detailed representation of the shared transition. A shared transition for synchronous interaction by separate robots is said to be a global transition, while a transition for independent action by a single robot is said to be a local transition. For synchronous interaction, the coordination algorithm is formally expressed using logical variables such that the global transition is fired if all of associated

Figure 4. Net representation of robotic task: (a) cyclic net model of robot, (b) equivalent parallel net model, (c) possible state of net model

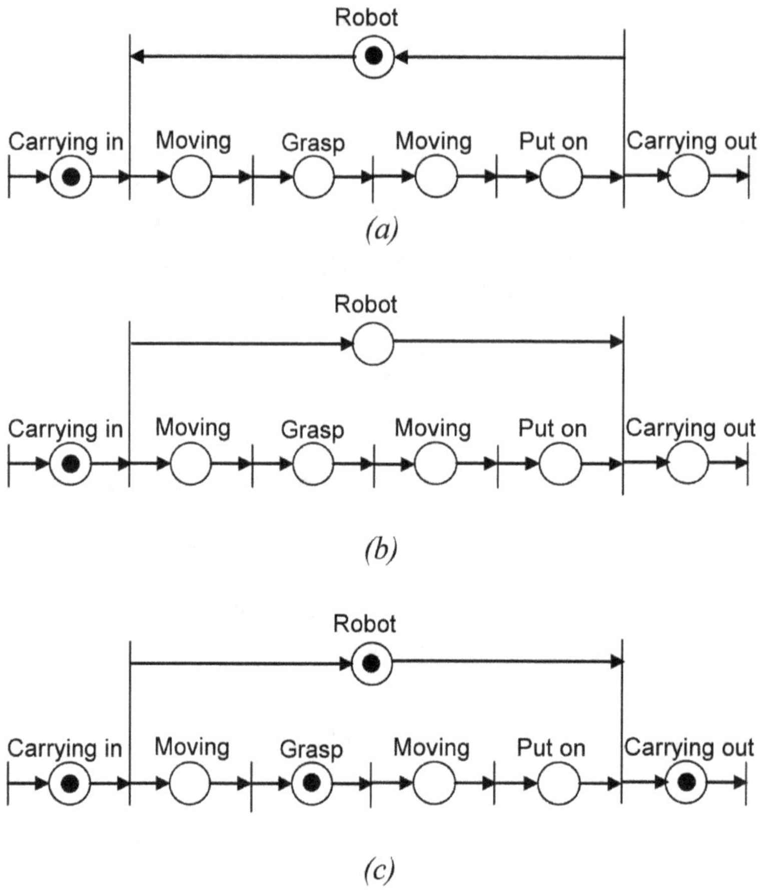

transitions in the local net models are fired. The firing condition of a global transition G_j in the conceptual net which represents the event of synchronous action by S robots is written as

$$G_j = \bigcap_{sub=1}^{S} t_{jsub} \tag{1}$$

where the corresponding event of action by each robot is represented by t_{jsub} ($sub = 1 \cdots S$), and $\bigcap$ denotes logical product operation. Figure 8 shows a net model of a coordination mechanism to execute selective control by means of synchronous communication, where the decision place executes any arbitration rule such as order or priority, to select independent action by a robot or cooperative action by two robots.

A hierarchical and distributed control system is composed of one system controller and several machine controllers. The coordination mechanism as well as the conceptual net model of the system is implemented in the system controller and detailed net models are allocated to machine controllers. The coordination program is substantially the firing test program of global transitions using the firing rule and a set of relational data tables of global transitions with their input and output places and associated gate arcs. The coordination procedure through communica-

Figure 5. Hierarchical net representation of robotic task (loading) on graphic net simulator

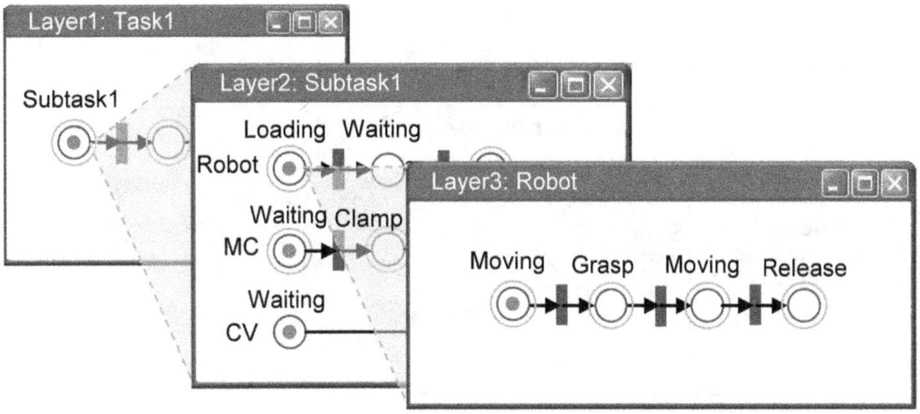

Figure 6. Net representation of synchronous interaction between two concurrent tasks: (a) synchronous communication with a shared transition, (b) interlock with mutual gate arcs,(c) asynchronous communication (signal/wait)

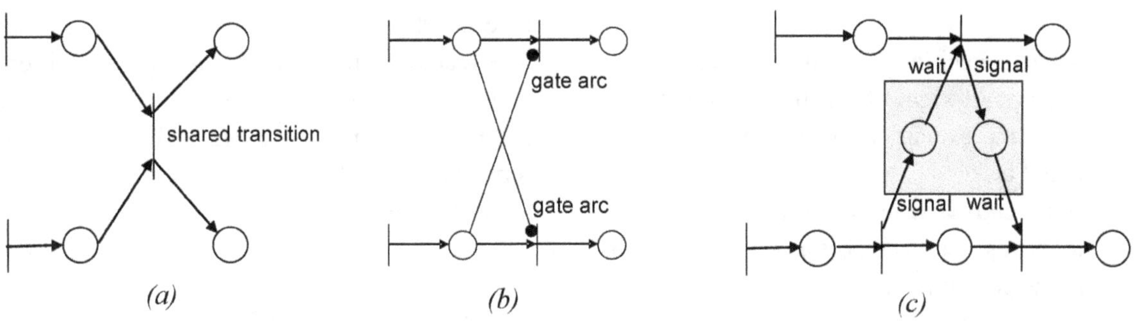

Figure 7. Net model of coordination mechanism for synchronous interaction by means of synchronous communication with a shared transition

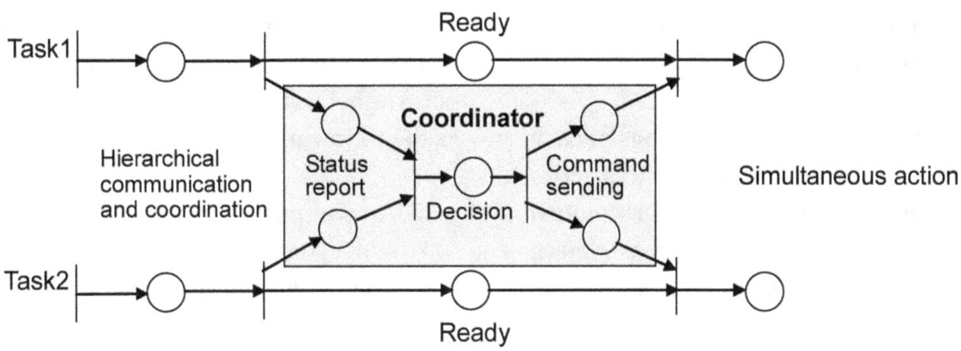

Figure 8. Net model of coordination mechanism for selective control

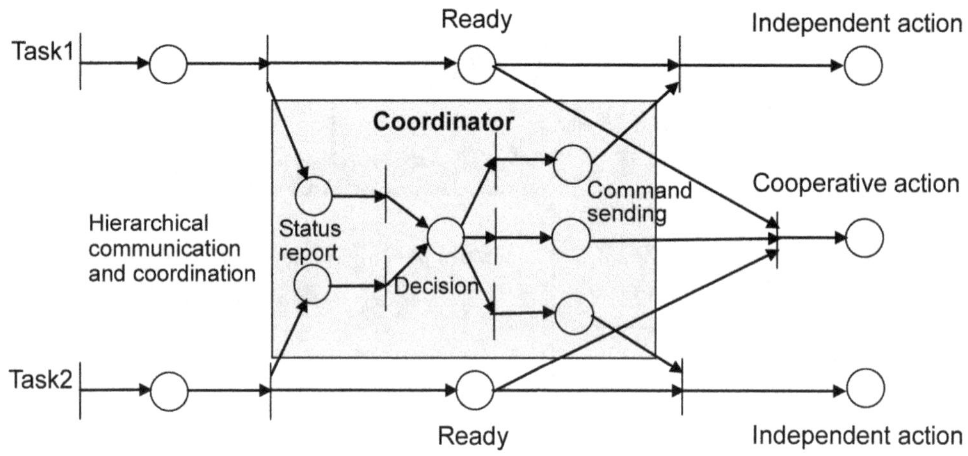

tion between the coordinator and machine controllers is shown as follows:

1. When a machine controller receives a command start signal from the coordinator, it starts the execution of the requested command task.
2. At the end of execution, the machine controller sends a status report as a gate condition to the coordinator.
3. When the coordinator receives the new gate condition, it updates the net model and tests the firing condition associated with the gate condition in the net. If a new transition is fired, then a command associated with the transition is sent to the corresponding machine controller.

For the actual control, the operations of each machine or robot are broken down into a series of unit motions, which is represented by mutual connection between places and transitions. A place means a concrete unit motion of a machine. From these places, output signal arcs are connected to the external machines, and external gate arcs from the machines are connected to the transitions of the net when needed, for example, to synchronize and coordinate operations. When a token enters a place that represents a subtask, the machine defined by the machine code is informed to execute a specified subtask with positional and control data; these code and data are defined as the place parameters. Figure 9 shows the net representation of real-time execution control of a robotic unit action.

4. IMPLEMENTATION OF DISTRIBUTED CONTROL WITH MULTITHREADS

The example robotic system has one robot, one machining center, and two conveyors, where one is for carrying in and the other one is for carrying out. The main execution of the system is indicated as the following task specification:

1. A workpiece is carried in by the input conveyor.
2. The workpiece is loaded to the machining center by the robot.
3. The workpiece is processed by the machining center.

Figure 9. Net representation of execution control of robotic action using output signal arc and external permissive gate arc

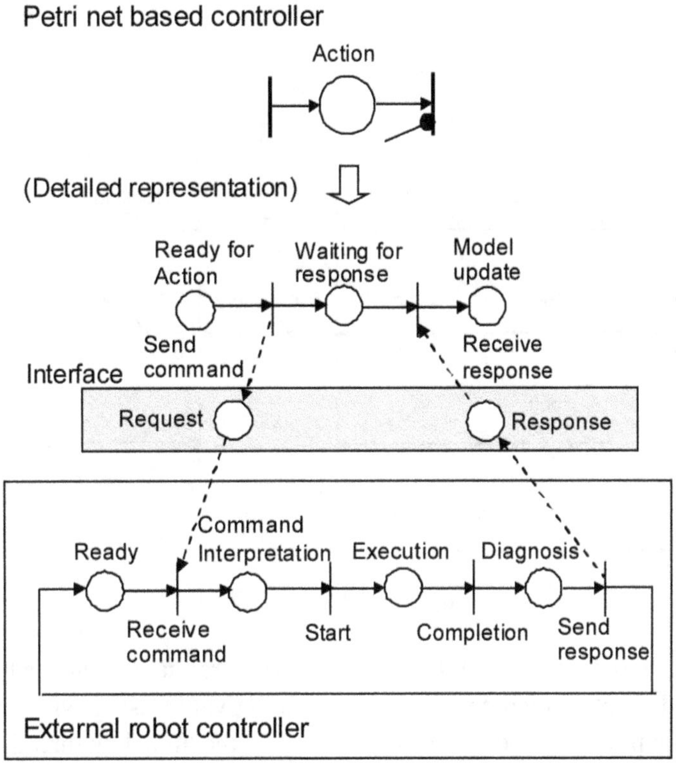

4. The workpiece is unloaded from the machining center by the robot.
5. The workpiece is carried out by the output conveyor.

The robotic system works in the following way: A workpiece comes on the input conveyor up to the take up position. The robot waits in the waiting position in front of the conveyor, and, on stopping, approaches in the take up position, grips the object and returns to the waiting position. Then it turns, goes into the working space of the machine center and there it leaves the workpiece. After automatic gripping of the object the robot draws back and it waits for the machining center to complete object processing. After object processing, the robot goes to the machining center, takes the workpiece from an opened vice and carries it over to a free position on the output conveyor.

The discrete event processes of the robotic system are represented at the conceptual level as shown in Figure 10. All the transitions are used to coordinate the overall system. Shared transitions between the robot and a machine, represent synchronous interaction and are coordinated as global transitions to be fired simultaneously. In this step, if necessary, control conditions such as the capacity of the system between the respective subtasks must be connected to regulate the execution of the manufacturing process. Next, each place representing a subtask at the conceptual level is translated into a detailed subnet, which is used for local control of each robot or machine.

The prototyping method was applied to produce semi-automatically C++ program from the net models on a general PC using multithreaded programming (Grehan, et al. 1998). Then, by executing the coordination program and net based

Figure 10. Net representation of robotic task at conceptual level

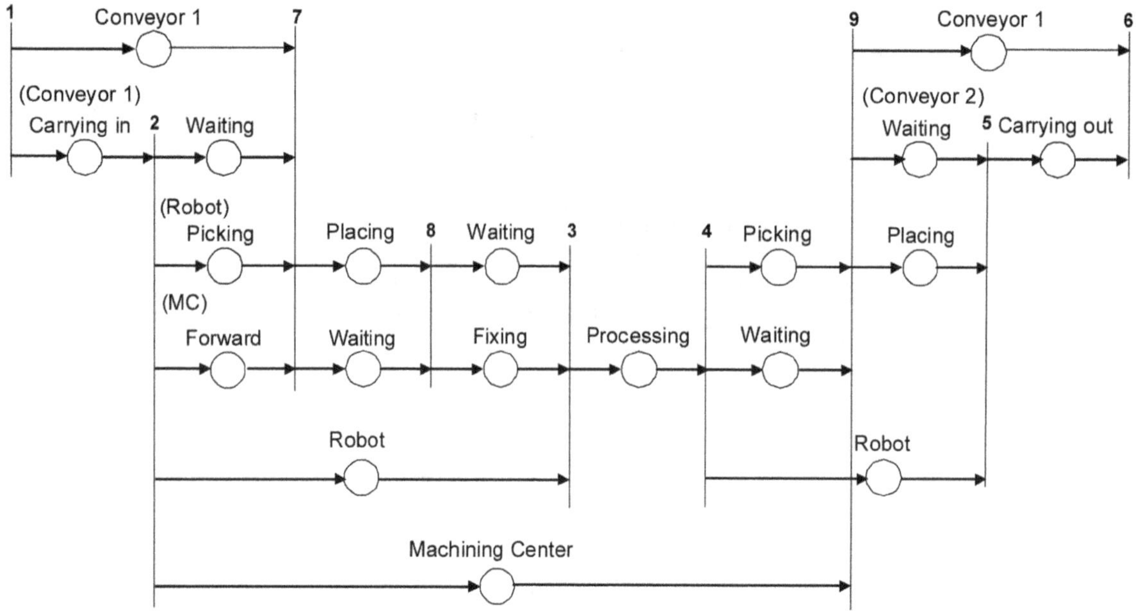

controllers algorithms based on loaded information on a network of dedicated microcomputers, experiments on the final test can be performed (Yasuda, 2010). Multithreads control software composed of one modeling thread, one simulation thread and several task execution threads, were designed and written in Microsoft Visual C# under OS Windows XP SP3. The simulation thread executes the coordination program and the conceptual net based controller algorithm, while the task execution threads execute local net based controller algorithms, which control robots and machines through serial interfaces using the command/response concept. An example diagram of two-level net based concurrent real-time control of two external machines using one simulation and two task execution threads is shown in Figure 11.

The modeling thread, which is the main thread, executes the event driven net modeling, drawing and modification based on task specification using windows button clicks and mouse operations, as shown in Figure 12. When the transformation of graphic data of the net model into internal structural data is finished, the simulation thread is activated using window buttons by the user from the modeling thread. The simulation thread executes the enabling and firing test using gate conditions as shown in Figure 13, and when a transition is fired, the simulation thread activates the task execution thread and initiates the execution of a subtask by sending commands attached to the fired transition. When all the subtasks in the system are in progressive, the simulation thread waits for the turning on of any gate condition repeatedly. If a subtask is completed, the gate condition is turned on, then the simulation thread receives the gate signal through the external gate arc and updates the table of gate conditions.

The program structure and the main C# code of the task execution thread are illustrated in Figure 14 and Listing 1. One task execution thread is allocated in each machine. When a task execution thread is activated, it sends the direct commands with specified positional and control data through the serial interface to the dedicated robot or machine. Then the thread searches the target gate condition and waits for any status report

Figure 11. Two-level net based concurrent real-time control of two external machines

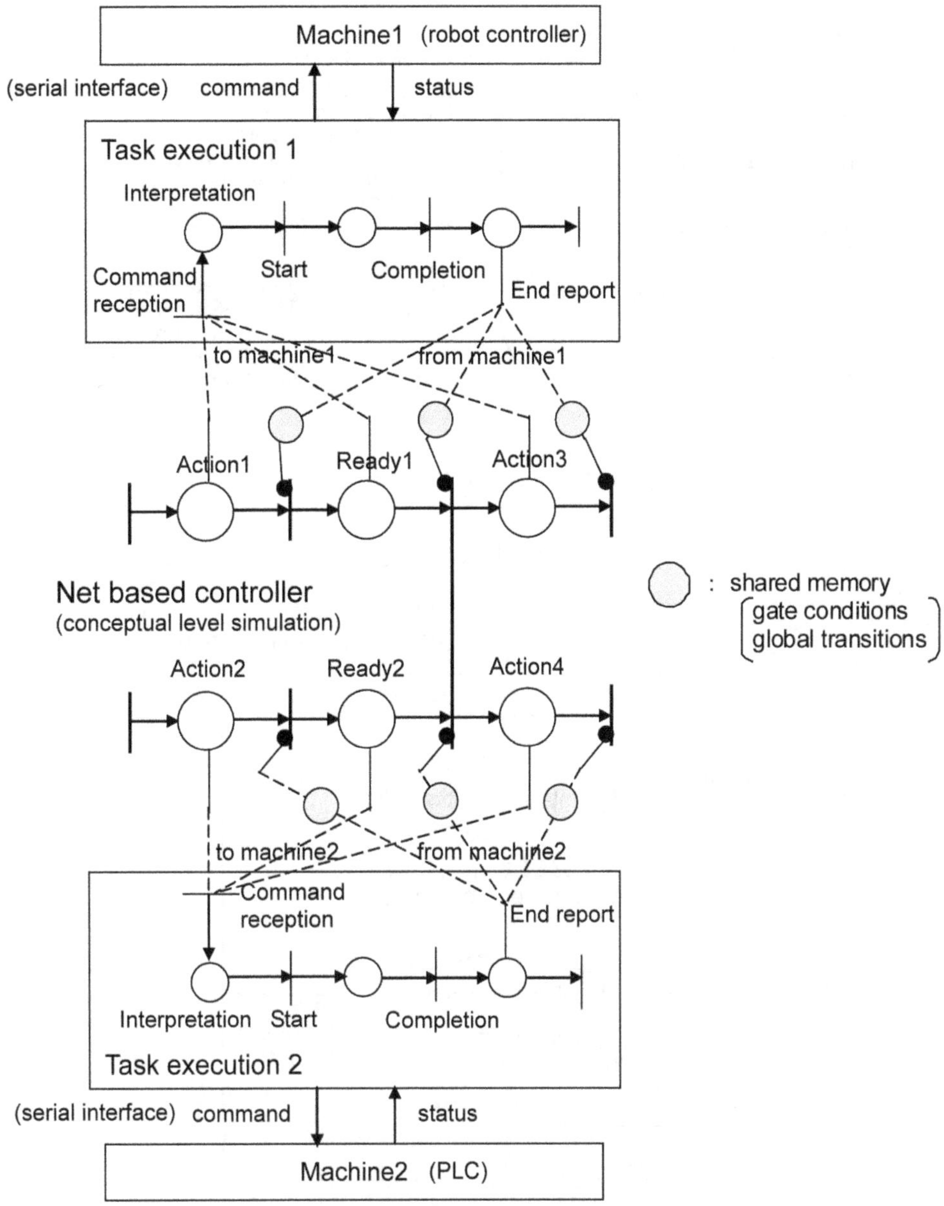

through the interface repeatedly. When the subtask ends its activity normally, the thread receives the normal end report. Then, the thread turns the target gate condition on and the current gate condition off, so that the simulation thread can proceed with the next activations through the external gate arc. During simulation and task execution, it is decided by the simulator that whether it is in a deadlocked situation or not, and the user can stop the task execution through the simulator at any time.

Figure 12. Program structure and main C# code of modeling thread

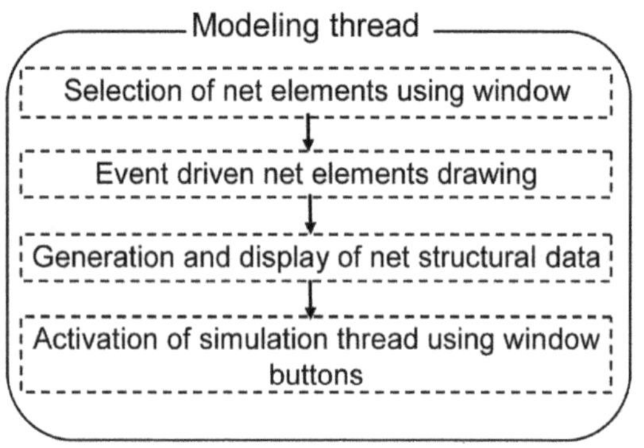


```
button1_Click()
MouseDown_Drawing()
MouseUp_Drawing()
MouseMove_Drawing()
pictureBox1_MouseDown()
pictureBox1_MouseUp()
pictureBox1_Move()
button14_Click()
button9_Click()
Thread trd0 = new Thread(new
                ThreadStart(ThreadTask0));
                trd0.IsBackground = true;
                trd0.Priority =ThreadPriority.Lowest;
                trd0.Start();
```

Figure 13. Program structure of simulation thread

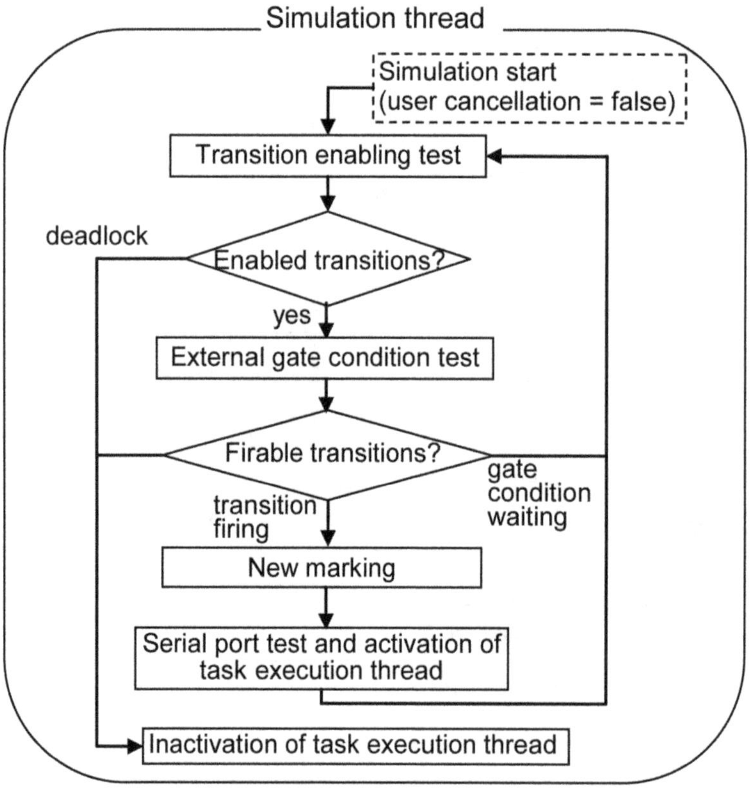

In the real-time control both the simulation thread and the task execution threads access to the external gate variables as shared variables; task execution threads write the values of the gate variables after the completion of subtasks and the simulation thread reads them for the friability test

Figure 14. Program structure of task execution thread

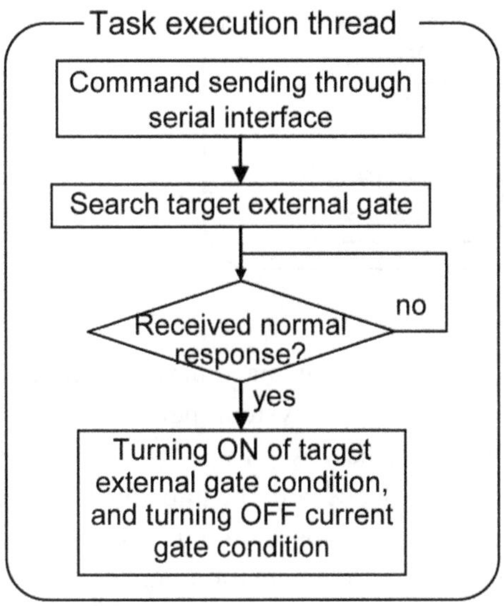

considering gate conditions. Mutual exclusive access control and the C# code were implemented as shown in Figure 15 and Listing 2 respectively. The method function uses a "lock" statement of C# for the mutual exclusive access to shared variables so that, while one thread calls the function, the other thread can not call it. Using the method function call, the simulator thread waits for the external gate signal, and after a task execution thread writes the permissive value of the target gate variable, the simulator thread reads it.

Experimental results of multithreads scheduling for one and two task execution threads are shown in Figure 16(a), (b), respectively. In case of two threads, one thread takes charge of a robot controller, and another takes charge of a PLC for sequence control of a conveyor respectively through serial interface. Here, the time slice of the OS is about 15ms, and the timer program is inserted in the reference position of each thread to capture the time the method function is called.

Numerous experiments show that the gate condition is transferred through the shared memory from the task execution thread to the simulation thread as quickly as possible. Experiments using a real robot and conveyors show that

Listing 1. Task execution thread

```
private void ThreadTask1()
{
     serialPort1.WriteLine(CC1[TH1 - 1] + "\r\n");          // command of fired transition no.TH1
     for (int j = 0; j < TP_R; j++){
          if (TPC[TH1 - 1, j] > 0){                           // search output place of fired transition
               PO = TPC[TH1 - 1, j];
               PH[PO - 1] = 1;  }}                            // output place no.PO
               for (int i = 0; i < T_No; i++){                // search output transition of place no.PO
                    for (int j = 0; j < TP_R; j++){
                         if (TPC[i, j] == -(PO + 1)){
                              for (int m = 0; m < EG_R; m++){
                                   if (TGE[i, m] > 0)  GC1 = TGE[i, m];    // gate no.GC1
                              }}}
                         do{
                              // method call
                         }
                         while (THN == 1);                    // gate condition write mode on
               }
     if (THN == 0) receiveData1 = "EG-WAIT";
}
```

Figure 15. Program structure of method function for mutual exclusive access

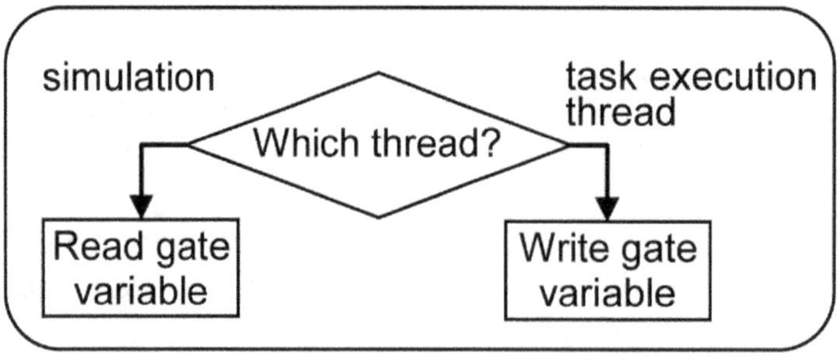

the simulation thread and the task execution threads are concurrently executed with even priority, and the values of external gate variables are changed successfully in the conceptual net model. The global transitions fire simultaneously with the transition of the conceptual net of the whole system task. The robot cooperated with the conveyors and the machining center, and the example robotic system performed the task specification successfully.

In accordance with the implementation using multithreaded programming, hierarchical and distributed implementation under a real-time operating system on a network of microcomput-

Listing 2. Method function for mutual exclusive access

```
private void EG_SET()
{
    lock (this)
    {
    if (EG_f == 1){                                  // simulation thread
        for (int i = 0; i < T_No; i++){              // gate condition test
        if (TI[i] == 1){                             // enabled transition
            for (int j = 0; j < EG_R; j++){
                GC1 = TGE[i, j];                     // gate no.j associated with transition no.i
                if (GC1 > 0){                        // permissive gate
                    if (EG[(GC1 - 1)] == 0) TI[i] = 0;          // read permissive gate signal off
                }
                else if (GC1 < 0) {                  // inhibitive gate
                    if (EG[Math.Abs(GC1 + 1)] == 1) TI[i] = 0;  // read inhibitive gate signal on
                    }}}}}

    else if (EG_f == 2){                             // task execution thread
        if (recieveData1 == "RDY"){                  // received status report 'ready'
            GC2 = TGE[TH1-1, j];                     // write gate no.GC2 off
            if (GC2 > 0) EG[GC2 - 1] = 0;
            else if(GC2 < 0) EG[Math.Abs(GC2 + 1)] = 0;
            EG[GC1 - 1] = 1;                         // write gate no.GC1 on
            TH[TH1 - 1] = 0;                         // fired teansition no. initialize
            PH[k] = 0;                               // working place no. initialize
            THN = 0;                                 // gate condition write mode off
        }
        Application.DoEvents();                      // serial communication message buffer initialize
    }}
```

Figure 16. Experimental results of multithreads scheduling: one task execution thread, (b) two task execution threads

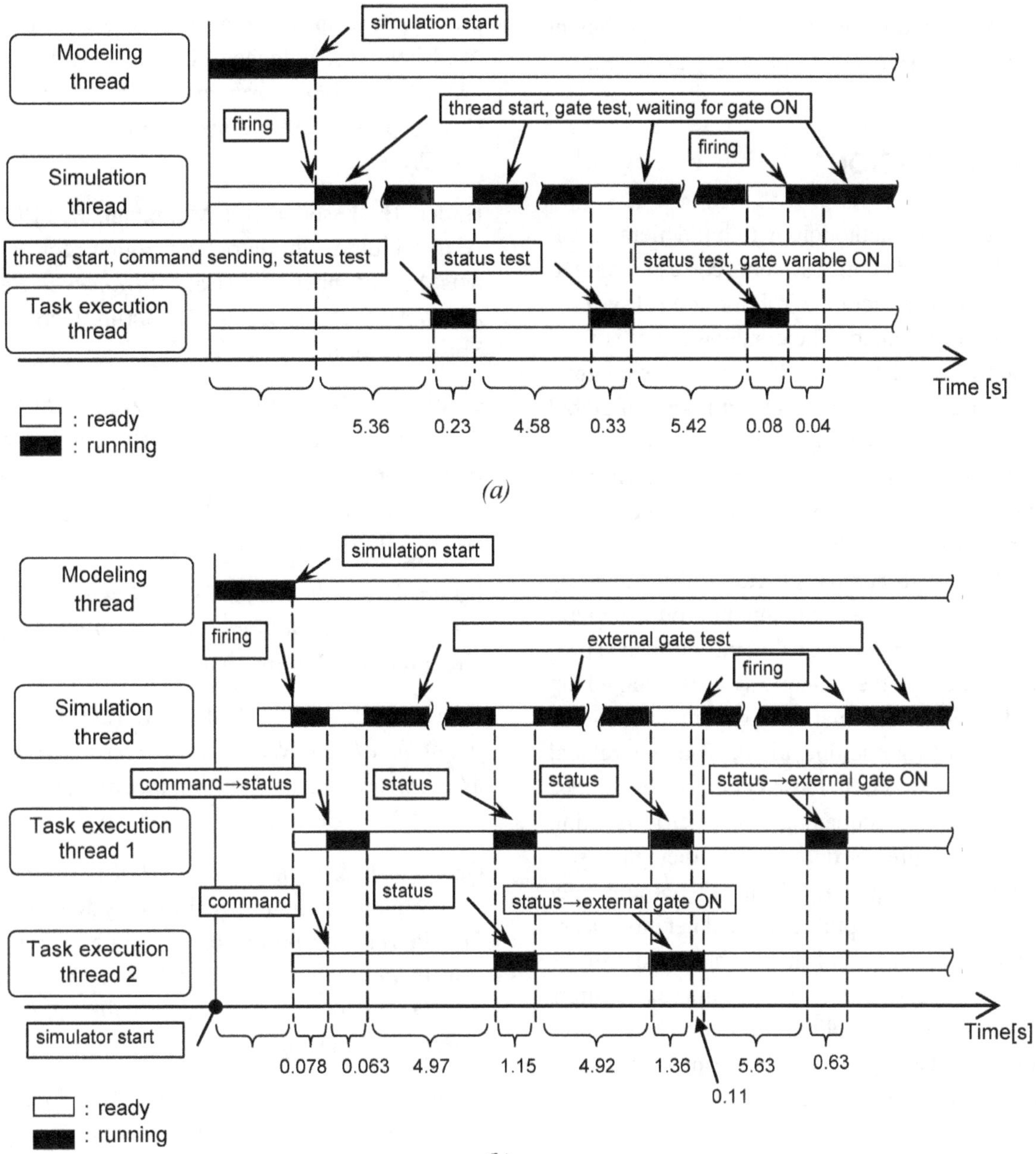

ers connected via a shared serial bus is also possible, where each microcomputer is dedicated to the local net model of a subsystem in the overall robotic system. After the arrival of a request, the response carrying the status information crosses the communication network and gets to the input buffer of the network board of the controller and written in the cache memory. The control data issued by the controller is written in the cache memory shared with the network board, sent to its target microcomputer, and there causes the new controlled status. By a shared communication network replacing traditional directly wired systems, a more flexible, reliable and efficient control performance can be surely expected.

5. CONCLUSION

A prototyping methodology to build hierarchical and distributed control systems corresponding to the hardware structure of robotic control systems has been presented. The conceptual net is used to coordinate distributed local machine controllers using the decomposed information of global transitions representing cooperative interaction between machines: the coordination mechanism can be implemented in each layer of the control hierarchy of the system repeatedly.

The overall control structure of the example robotic system was implemented on a general PC with serial interface using multithreaded programming. For the example system, detailed net models can be automatically generated using the database of robotic operations. The hierarchical approach allows us to reuse validated net models, such as loading, unloading, and specific handling operations, already defined for other purposes, that is an efficient way to deal with complex net models. The conceptual and local net models are not so large that all of the net based controllers can be implemented on general microcomputers or PLCs. Thus, modeling, simulation and control of large and complex manufacturing systems can be performed consistently using Petri nets.

REFERENCES

Breant, F., & Paviot-Adet, E. (1992). OCCAM prototyping from hierarchical Petri nets. In Becker, M., Litzler, L., & Trehel, M. (Eds.), *Transputers '92 Advanced Research and Industrial Applications* (pp. 189–209). Amsterdam, Netherlands: IOS Press.

Butler, B., Esser, R., & Mattmann, R. (1991). A distributed simulator for high order Petri nets. In Rozenberg, G. (Ed.), *Advances in Petri Nets LNCS 483* (pp. 47–63). Berlin, Germany: Springer-Verlag.

Caloini, A., Magnani, G., & Pezze, M. (1998). A technique for designing robotic control systems based on Petri nets. *IEEE Transactions on Control Systems Technology*, *6*(1), 72–87. doi:10.1109/87.654878

David, R., & Alla, H. (1992). *Petri Nets and Grafcet: Tools for modelling discrete events systems*. UK: Prentice-Hall International.

Desrochers, A. D., & Al-Jaar, R. Y. (1995). *Applications of Petri Nets in manufacturing systems: Modeling, control and performance analysis*. New York, NY: IEEE Press.

Garcia, F. J., & Villarroel, J. L. (1998). Decentralized implementation of real-time systems using time Petri nets. Application to mobile robot control. In *Proceedings of the 5th IFAC Workshop on Algorithms and Architectures for Real-time Control*, (pp. 11-16).

Girault, C., & Valk, R. (2003). *Petri Nets for systems engineering*. Berlin, Germany: Springer-Verlag.

Grehan, R., Moote, R., & Cyliax, I. (1998). *Real-time programming: A guide to 32-bit embedded development*. Reading, MA: Addison-Wesley.

Hasegawa, K., Takahashi, K., Masuda, R., & Ohno, H. (1984). Proposal of mark flow graph for discrete system control. *Transactions of SICE, 20*(2), 122–129.

Holding, D. J., & Sagoo, J. S. (1992). A formal approach to the software control of high-speed machinery. In G. W. Irwin & P. J. Fleming (Eds.), *Transputers in real-time control,* (pp. 239-282). Taunton, Somerset, UK: Research Studies Press.

Lee, E. J., Togueni, A., & Dangoumau, N. (2006). A Petri net based decentralized synthesis approach for the control of flexible manufacturing systems. In *Proceedings of the 2006 IMACS Multiconference Computational Engineering in Systems Applications*.

Murata, T. (1989). Petri nets: Properties, analysis and applications. *Proceedings of the IEEE, 77*(4), 541–580. doi:10.1109/5.24143

Murata, T., Komoda, N., Matsumoto, K., & Haruna, K. (1986). A Petri net based controller for flexible and maintainable sequence control and its application in factory automation. *IEEE Transactions on Industrial Electronics, 33*(1), 1–8. doi:10.1109/TIE.1986.351700

Neumann, P. (2007). Communication in industrial automation – What is going on? *Control Engineering Practice, 15*(11), 1332–1347. doi:10.1016/j.conengprac.2006.10.004

Piedrafita, R., Tardioli, D., & Villarroel, J. L. (2008). Distributed implementation of discrete event control systems based on Petri nets. In *Proceedings of the 2008 IEEE International Symposium on Industrial Electronics,* (pp. 1738-1745).

Reisig, W. (1985). *Petri Nets*. Berlin, Germany: Springer-Verlag.

Silva, M. (1990). Petri nets and flexible manufacturing. In Rozenberg, G. (Ed.), *Advances in Petri Nets LNCS 424* (pp. 374–417). Berlin, Germany: Springer-Verlag.

Silva, M., & Velilla, S. (1982). Programmable logic controllers and Petri nets: A comparative study. In Ferrate, G., & Puente, E. A. (Eds.), *In IFAC Software for Computer Control 1982* (pp. 83–88). Oxford, UK: Pergamon.

Yasuda, G. (2000). A distributed control and communication structure for multiple cooperating robot agents. In *IFAC Artificial Intelligence in Real Time Control 2000* (pp. 283–288). Oxford, UK: Pergamon.

Yasuda, G. (2010). Distributed cooperative control of industrial robotic systems using Petri net based multitask processing. In H. Liu, H. Ding, Z. Xiong, & X. Zhu (Eds.), *Proceedings of the 3rd International Conference on Intelligent Robotics and Applications LNAI 6425*, (pp. 32-43). Berlin, Germany: Springer-Verlag.

Yasuda, G., & Ge, B. (2010). Petri net model based specification and distributed control of robotic manufacturing systems. In *Proceedings of the 2010 IEEE International Conference on Information and Automation*, (pp. 699-705).

Yasuda, G., & Tachibana, K. (1992). Implementation of real-time control schemes on a parallel processing architecture using transputers. In *Proceedings of IEEE Singapore International Conference on Intelligent Control and Instrumentation*, (pp. 760-765).

Zhou, M., DiCesare, F., & Desrochers, A. A. (1993). *Petri Net synthesis for discrete event control of manufacturing systems*. London, UK: Kluwer.

Zhou, M., & Venkatesh, K. (1999). *Modeling, simulation, and control of flexible manufacturing systems: A Petri Net approach.* Singapore: World Scientific.

Chapter 3
Optimal Design of Three-Link Planar Manipulators Using Grashof's Criterion

Sarosh H. Patel
RISC Laboratory, University of Bridgeport, USA

Tarek Sobh
RISC Laboratory, University of Bridgeport, USA

ABSTRACT

The design of robotic manipulators is dictated by a set of pre-determined task descriptions and performance parameters. These performance parameters are often defined in terms of workspace dexterity, manipulability, and accuracy. Many serial manipulator applications require that the manipulator have full dexterity about a work piece or a pre-defined trajectory, that is, to approach the given point within the workspace with all possible orientations about that point. Grashof's criterion defines the mobility of four-link closed chain mechanisms in relation to its link lengths. A simple assumption can convert a three-link serial manipulator into a four-link closed chain so that its mobility can be studied using Grashof's criterion. With the help of Grashof's criterion, it is possible not only to predict and simulate the mobility of a manipulator during its design, but also to map and identify the fully-dexterous regions within its workspace. Mapping of the dexterous workspace is helpful in efficient task placement and path planning. Next, the authors propose a simple algorithm using Grashof's criterion for determining the optimal link lengths of a three-link manipulator, in order to achieve full dexterity at the desired regions of the workspace. Finally, the authors test the generated design by applying joint angle limitations.

NOMENCLATURE

- *D*: Dexterity index of the manipulator at a point.
- D_{Mean}: Mean dexterity index over a region or trajectory.
- *N*: Number of points along the trajectory.
- d_x, d_y, d_z: Dexterity indices about the X, Y and Z axis.
- α, β, γ: Yaw, pitch and roll angels of the end-effector.
- *a,b,c,d*: Link lengths of the four-bar kinematic chain.

DOI: 10.4018/978-1-4666-0176-5.ch003

l_1,l_2,l_3: Link lengths of the three-link planar manipulator.
$\theta_1, \theta_2, \theta_3$: Joint angles of the three-link planar manipulator.
d: Distance between a task-point and base of the manipulator
d_{min}: Minimum distance between a task-point and base of the manipulator.
d_{max}: Maximum distance between a task-point and base of the manipulator.

1. INTRODUCTION

The problem of designing an optimal manipulator configuration is very complex, as the equations governing the motion of the end-effector in the workspace are both non-linear and complex, often having no closed solutions. Prototyping methods such as kinematic synthesis and numerical optimization are complex and very time consuming. The inherent complexity of kinematic synthesis has helped to make a strong case for rapid prototyping methods in which manipulators are designed with very specific performance requirements or tasks point specifications. Rapid prototyping allows designers to spend more time on design, simulation and evaluation of different manipulator configurations instead of solving mathematical models describing kinematics chains.

The study of mobility of closed chain mechanisms has interested researchers for a very long time. Understanding the mobility of chain mechanisms in relation to their link lengths can help us to design better and highly dexterous manipulators. In 1833, Grashof first introduced a simple rule to understand the mobility of four-link mechanisms [6]. This rule, commonly known as the Grashof's theorem, helps analyze the rotatability of links in a closed four-bar mechanism. This was further extended by Paul (1979), who introduced an inequality into the Grashof's theorem and proved that Grashof's criterion is both a necessary and sufficient condition for the existence of a crank in the four-bar mechanism (Chang, Lin, & Wu, 2005).

Researchers have applied Grashof's criterion to understand and study the workspace mobility of both closed and open chain planar mechanisms. Barker (1985), using Grasshof's criterion, classified four-bar planar mechanisms based on their mobility. Grashof's criterion was applied to the study of three-link planar mechanism by Li and Dai (2009). Furthermore, they developed equations for the orientation angle and presented a simple program to analyze the orientation angle for a manipulator, given the link parameters. The mobility and orientation of open chain mechanisms can also be analyzed using Grashof's criterion. Dai and Shah (2002, 2003) studied the mobility of serial manipulators by introducing a virtual ground link between the end-effector and the base so as to form a virtual closed chain. In (Li, & Dai, 2009; Dai, & Shah, 2003), the authors proposed workspace decomposition based on the orientation capability of the manipulator.

Grashof's Theorem has been extended to include more than four-bar chain mechanisms. Grashof's criterion for five bar chain was proposed by Ting (1986). Ting and Liu (1991) extended this work to evaluate the mobility of N-bar chain mechanisms. Nokleby and Podhorodeski (2001) applied Grashof's criterion for the optimized synthesis for five-bar mechanisms.

In this work we present a simple algorithm for the optimal design of a three-link planar manipulator, using Grashof's criterion. We begin by adding a virtual link to the three-link planar manipulator in order to make it a closed four-bar chain mechanism, so that Grashof's criterion can be applied. We evaluate the generated manipulator designs using dexterity index as a performance measure. Our proposed optimization algorithm generates the required link lengths such that the manipulator has maximum dexterity in the region specified by the user. This region of interest can either be a set of task points or a trajectory. Furthermore, we have also demonstrated, with the help of simulations, the influence of the serial chain link ordering on the dexterous workspace. Finally, we simulate our design under practical conditions such as joint angle limitations.

2. DEXTERITY INDEX AS A PERFORMANCE MEASURE

The performance evaluation of a robotic manipulator in its workspace is central to the design process of the manipulator. Both a bottom-up approach – designing a manipulator to meet a certain performance standards (Gosselin, 1992; Sobh, & Toundykov, 2004; Paden, & Sastry, 1988) and a top-down approach – optimal task placement in workspace of the manipulator (Santos, Steffen, & Saramago, 2010), have been thoroughly investigated. In either case, it is essential to define proper performance parameters and study the variation of these parameters in the workspace.

In this process, numerous parameters have been proposed to quantify and measure the performance of the manipulator in its workspace, such as dexterity index, manipulability index, condition number, minimum singular value, etc. (Tanev, & Stoyanov, 2000). Two parameters stand out in quantifying the manipulator's performance in its workspace: dexterity and manipulability. A manipulator has to be highly dexterous or manipulable in its workspace to meet high performance standards. Many applications of robotic manipulators require the manipulator to be fully dexterous to perform specified tasks.

The dexterity of a manipulator is defined as the number of ways in which any given point in the workspace can be approached by the end-effector. The dexterity index of a manipulator at a point in the workspace can be defined as a measure of a manipulator to achieve varying orientations at that point (Tanev, & Stoyanov, 2000).

In their work, Kumar and Waldron (1981) introduced the parameter dexterity index as another measure for manipulator performance. They defined dexterous workspace as *'the volume within which every point can be reached by the manipulator end-effector with any desired orientation'* (Tanev, & Stoyanov, 2000, p. 1). While the manipulability index is a function of both the manipulator configuration and joint angles (the manipulability index can be different at the same point if the manipulator has multiple inverse kinematic solutions), the dexterity index is only a function of the manipulator configuration.

The orientation at any given point in the workspace can be represented in terms of the yaw (α), pitch (β) and roll (γ) angles (Spong, & Vidyasagar, 1989) as:

$$R_{xyz} = R_{x,\gamma} R_{y,\beta} R_{z,\alpha} \tag{1}$$

All three of the angles have a range 0 - 2π to provide all possible orientations. The dexterity index can be defined as the summation of the dexterity indices about each of the axes (Tanev, & Stoyanov, 2000) given by Equation (2):

$$D = \frac{1}{3}(d_x + d_y + d_z) \tag{2}$$

$$D = \frac{1}{3}(\frac{\Delta\gamma}{2\pi} + \frac{\Delta\beta}{2\pi} + \frac{\Delta\alpha}{2\pi}) \tag{3}$$

where d_x, d_y, and d_z are X, Y and Z dexterity indices. Δα, Δβ and Δγ are the range of possible yaw, pitch and roll angles about a point.

As seen in Equation (3), the dexterity index can vary between a minimum of 0 to a maximum of 1, hence is a well bounded parameter. Therefore, points in the workspace with multiple inverse kinematic solutions will have a higher dexterity index when compared to points with unique solutions.

The mean dexterity index of a manipulator for a region of the workspace or trajectory can be defined as (Tanev, & Stoyanov, 2000):

$$DMean = \frac{\sum D}{N} \tag{4}$$

The manipulator is said to be completely dexterous at a given point if the dexterity at that point is equal to unity. Similarly, an area or point in the workspace can be said to be completely X-dexterous or Y-dexterous if d_x or d_y is equal to unity. In the case of a planar manipulator operating in the XY – plane

$$d_x = d_y = 0 \tag{5}$$

A simple algorithm for calculating the dexterity index of a workspace was proposed in (Tanev, & Stoyanov, 2000).

3. GRASHOF'S CRITERION

The study of mobility of a closed chain mechanism has interested researchers for a very long time. In 1833, Grashof first introduced a simple rule to understand the mobility of four-link mechanisms (Kumar, & Waldron, 1981). This rule, commonly known as Grashof's theorem, helps judge the rotatability of links in a four bar mechanism. This was further extended by Paul (1979), who introduced an inequality into Grashof's theorem and proved that Grashof's criterion is both a necessary and sufficient condition for the existence of a crank in the four-bar mechanism.

Consider a four-link kinematic chain consisting of four links a, b, c, and d, as shown in Figure 1. Let a be the longest link and d be the short link in the chain, such that $a>b\geq c>d$. According to Grashof's criterion, there exists at least one link that can fully revolve with respect to the other links if:

$$a+d\leq b+c \tag{6}$$

Figure 1. Four-link kinematic chain

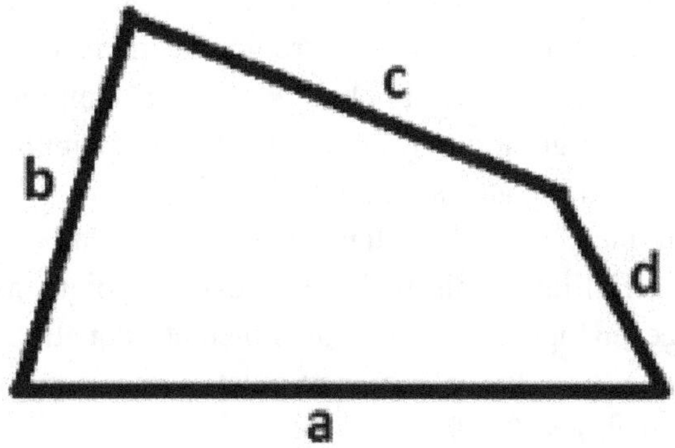

i.e. the sum of the longest and the shortest link should be less than or equal to the sum of the other two links. And none of the links are fully revolute if:

$$a+d>b+c \tag{7}$$

Paul (1979) proved that this criterion was both a necessary and sufficient condition for the existence of a fully rotatable link in the chain. Such a mechanism is also known as a Grashof linkage (Chang, Lin, & Wu, 2005; Li, & Dai, 2009). In a Grashof linkage, the shortest link in the chain is the always fully revolvable with respect to the other links (Li, & Dai, 2009).

The complete classification and behavior of four-bar linkages is explained in (Chang, Lin, & Wu, 2005; Barker, 1985).

4. DESIGN OPTIMIZATION

It is impossible for a manipulator to be equally dexterous or highly dexterous at all points in the workspace. Some regions of the workspace have high dexterity while other regions can only be attained by a unique set of joint angles. Therefore, it is important that the manipulator be designed in such a way that it has maximum dexterity in its region of operation.

By design optimization we mean adjusting the link lengths in such a way that the manipulator has maximum dexterity in the region of interest. As the dexterity index is a function of the manipulator configuration, by optimizing the link lengths, the manipulator's dexterous workspace can be engineered. A three-link planar manipulator is a serial link chain with three revolute joints, as seen in Figure 2. The forward kinematic equations for a three-link planar manipulator are given by Equations (8), (9)

$$x = l_1 \cos(\theta_1) + l_2 \cos(\theta_1 + \theta_2) + l_3 \cos(\theta_1 + \theta_2 + \theta_3) \tag{8}$$

$$y = l_1 \sin(\theta_1) + l_2 \sin(\theta_1 + \theta_2) + l_3 \sin(\theta_1 + \theta_2 + \theta_3) \tag{9}$$

In order to apply Grashof's criterion, essentially meant for four-bar linkages, to design three-link planar manipulators, we assume the distance between the base of the manipulator and the center of the end-effector as the imaginary fourth link of the chain. The length of this imaginary fourth link is not constant and depends on the joint angles, which in turn determine the position of the end-effector in the workspace. Hence, the links lengths should be selected such that Grashof's criterion is satisfied at all points in the region of operation as d varies from d_{min} to d_{max}.

Consider a three-link planar manipulator with links l_1, l_2 and l_3, as shown in Figure 2. Let d be the variable radial distance between the base and the end-effector shown in Figure 2. In order to optimize the link lengths to achieve maximum dexterity in the area of interest or trajectory, we propose the following algorithm.

1. Let d_{max} be the maximum distance from the base and d_{min} minimum distance while following a trajectory such that $d_{min} \leq d \leq d_{max}$
2. Let l_3 be the shortest link in the manipulator such that $l_3 < d_{min}$. The length of l_3 is equal to the minimum link length practically possible this is determined by the other factors such as the size of the motors and loading on the manipulator.

Next, we calculate l_1 and l_2 as follows:

$$l_1 = \left(\frac{d_{\max} + l_3}{2} \right) + 1 \tag{10}$$

Figure 2. 3-Link planar manipulator

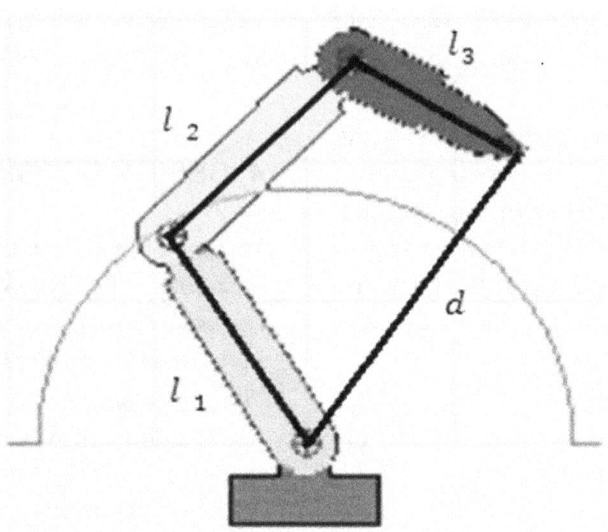

$$l_2 = \left(\frac{d_{\max} + l_3}{2} \right) \tag{11}$$

Since d_{max} and l_3 are finite and known l_1 and l_2 can easily be determined.

4. The dexterity index plot of the complete workspace is generated to check if the manipulator has maximum dexterity in the area of interest.

The link lengths are chosen such that the manipulator behaves as a Grashof's linkage as the radial distance from the base to the task point varies from d_{min} to d_{max}. l_1 and l_2 are chosen to be the larger links in the chain so that the last link or the end-effecter l3 is imparted full rotatability at all points within this range. The fourth link d is an imaginary link whose length depends on the position of the task point. Equations (10) and (11) give the minimal link lengths required to satisfy Grashof's criterion such that:

At $d=d_{max}$ we have $l_1>l_2>d>l_3$ and $l_1+l_3>l_2+d$

At $d=d_{min}$ we have $l_1>l_2>d>l_3$ and $l_1+l_3>l_2+d$

Section 5.2 discusses a condition with changed link order where the second link is the shortest link in the chain.

As long as Grashof's criterion is satisfied, the shortest link (end-effector) will be completely revolute, as seen in Figures 3 and 4. Figure 3 shows few of the infinite orientations that the manipulator can attain about a given point when Grashof's criterion is satisfied. The manipulator will obviously not behave as a Grashof's linkage at all points in the workspace since the fourth link in the chain d is not constant. Optimization ensures that the manipulator will behave as a Grashof's linkage in the region $d_{min}\leq d\leq d_{max}$, thereby making the end-effector completely revolute and hence, providing maximum dexterity to manipulator in the region.

As the task point keeps moving in the workspace, the radial distance between the base of the manipulator and the task point, the length of virtual fourth link 'd' keeps changing. Accordingly, depending of the position of the task point the shortest link in the chain will be as shown in Table 1.

It is important to mention here that Grashof's criterion only sets a condition for one of the links in the chain to be fully revolute with respect to the other links, but does not comment on the order of the links in the chain. Grashof's criterion helps only in deciding the link lengths, however the dexterity of the manipulator depends not only on the link lengths but also on their relative positioning in the kinematic chain.

5. EXAMPLE 1: TRAJECTORY

Let us consider a design problem where we need to design an optimal manipulator configuration while following a cubic trajectory joining the task points A (14, 14), B (12, 8), C (10, -3), D (9,-7), and E (8,-8), shown in Figure 5. The manipulator follows a cubic trajectory from one task point to another, given by Equation (12)

$$y = -0.2292x^3+7.625x^2-78.83x+252 \tag{12}$$

Following the steps in the proposed algorithm, we have:

1. The minimum and maximum distances from the base while following the trajectory are 10 and 20, such that $10\leq d\leq 20$, assuming the manipulator will be based at the origin O (0,0).
2. We chose $l_3 = 8$ such that $l_3 < d_{min}$.
3. From Equation (12), (13) we have $l_1 = 15$ and $l_2 = 14$.

Using the above generated link lengths, we calculate the Dexterity Index (D) of the manipulator at all points on its workspace. Assuming that the manipulator operates in the XY-plane,

Figure 3. Different orientations of the manipulator about a set task point

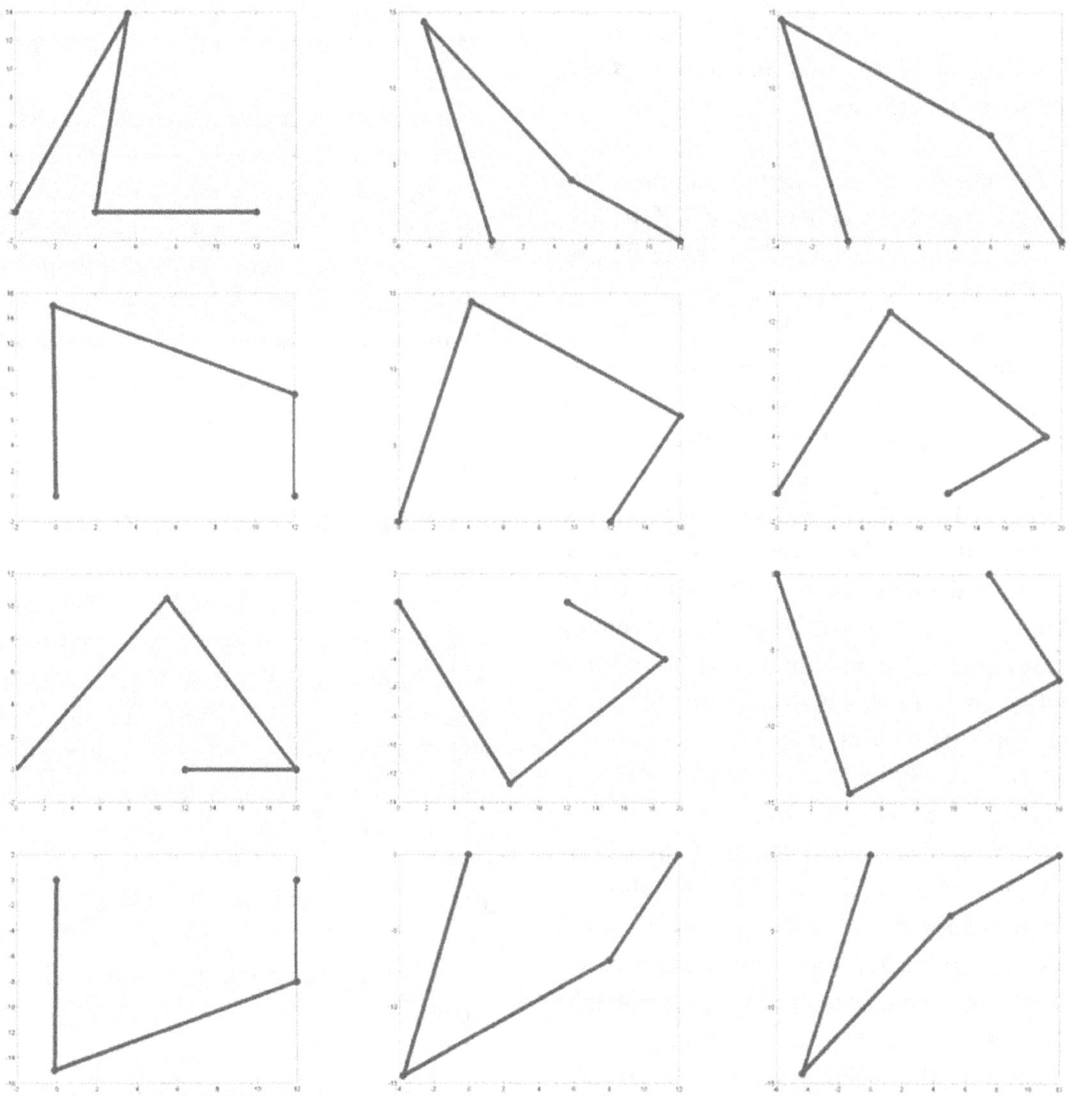

the maximum value of the Dexterity index (D) = (1/3) and the maximum value of the $d_z = 1$ while $d_x = d_y = 0$.

Case 1: Ideal Case

We simulate the manipulator to generate the dexterity plot in the ideal case. We assume that all three joints are completely revolute and the joint angles are not limited. The joint angles are only limited such that the links do not overlap each other, should the angle between the links be close to zero.

In this case, the manipulator has a very symmetric dexterous workspace spread over the four quadrants, as seen in Figure 6. At points in the

Table 1. Grashof's criterion with varying d

Link Lengths	Grashof's Criterion	Shortest Link
$l_1>l_2>l_3>d$	$l_1+d\leq l_2+l_3$	d
$l_1>l_2>d>l_3$	$l_1+l_3\leq d+l_2$	l_3
$l_1>d>l_2>l_3$	$l_1+l_3\leq d+l_2$	l_3
$d>l_1>l_2>l_3$	$d+l_3\leq l_1+l_2$	l_3

Figure 4. Manipulator demonstrating full dexterity about a given task point

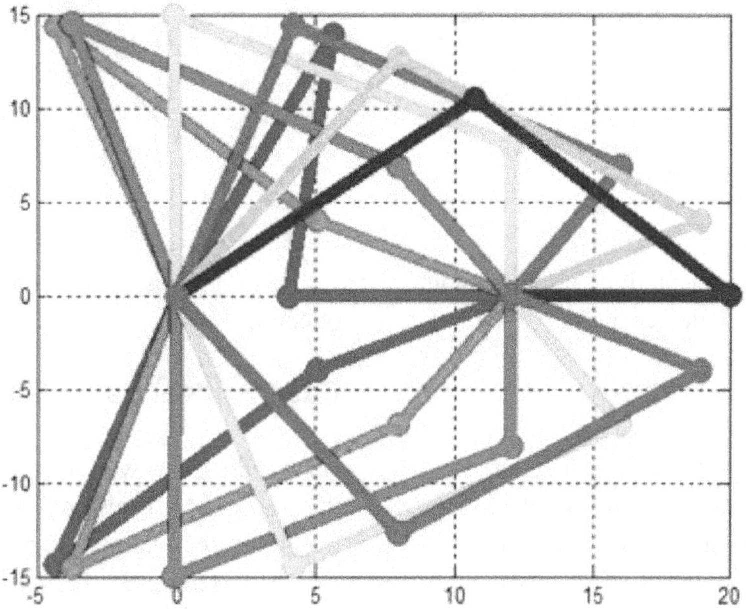

workspace where Grashof's criterion is satisfied, the shortest link in the chain is completely revolute, therefore the end-effector can have infinite orientations about the point. As seen in Figures 4 and 5, the manipulator has maximum Dexterity Index (D) = 0.33 or d_z =1, when $10\leq d\leq 20$, which was the design criteria. The variance of dexterity index in the workspace can be seen more clearly in the sectional view, as seen in Figure 7.

The behavior of the manipulator within its dexterous workspace can be better analyzed by dividing the plot into four regions, as shown in Figure 8:

Region 1: $0\leq d\leq l_3$

Within this region, the shortest link in the chain is not l_3 but d. In this case, when Grashof's criterion is satisfied for a point the link, will be fully revolute. However, since is the distance between the end-effector and base, and is not a physical link, it is fixed and non-revolute. The points within this region will have limited inverse kinematic solutions and hence, low manipulability.

Region 2: $l_3<d\leq l_1+l_2-l_3$

In this region, Grashof's criterion is satisfied for all values of. For all points within this region the shortest link in the chain (end-effector) is completely revolute for all values of , thereby

Figure 5. Desired manipulator trajectory

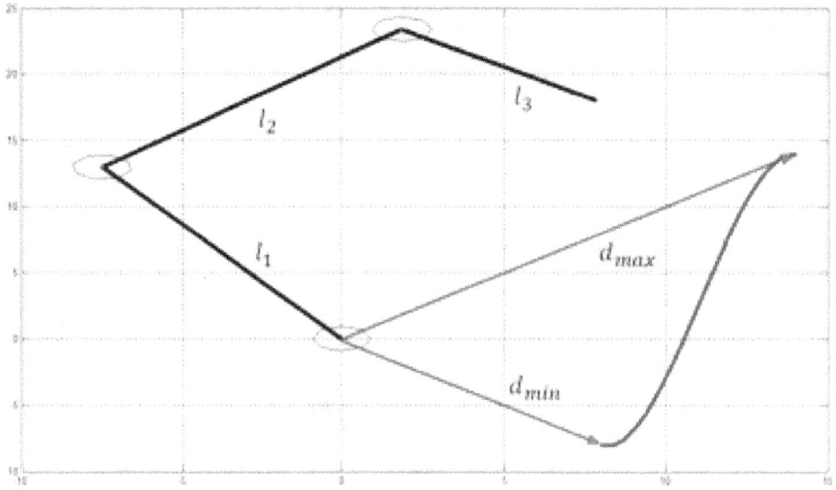

able to have infinite orientations about the point and hence, maximum dexterity.

Region 3: $l_1+l_2-l_3 < d \leq l_1+l_2+l_3$

Here, the value of d is so large that the sum of the largest link (d) and the shortest link (l_3) is no longer less than the sum of the other two links. In this region, the manipulator behaves as a non-Grashof linkage. As $d \rightarrow l_1+l_2+l_3$, the boundaries of the workspace, the dexterity decreases because there are very few inverse kinematic solutions for points close to edges of the workspace.

Region 3: $d > l_1+l_2+l_3$

This region is beyond the limits of the end-effector.

Case 2: Ideal Case; Changed Order

As noted earlier, we know that the dexterity of the manipulator is also a function of the relative positioning of the links. To prove this, in this case we change the order of the links by swapping the lengths of links l_2 and l_3; ($l_2 = 8$ and $l_3 = 14$). The new link order for this case is $l_1 = 15$; $l_2 = 8$ and $l_3 = 14$.

Unlike the previous case, the high dexterity band is very thin, as seen in Figure 9. This is so because, when Grashof's criterion is met, the shortest link in the chain becomes fully revolute,

Figure 6. 3D plot of the dexterity index in the workspace

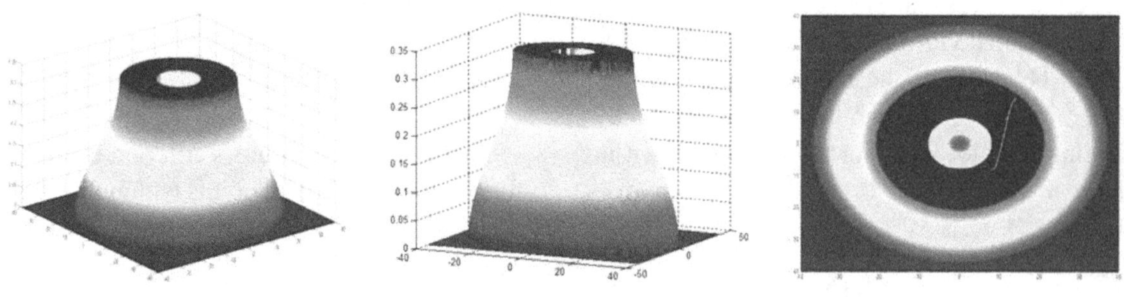

Figure 7. Sectional view of the dexterity index (d)

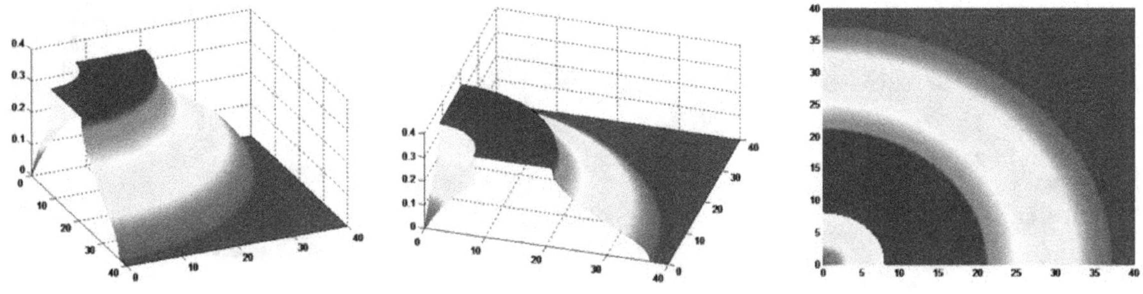

Figure 8. Different regions in the dexterity plot

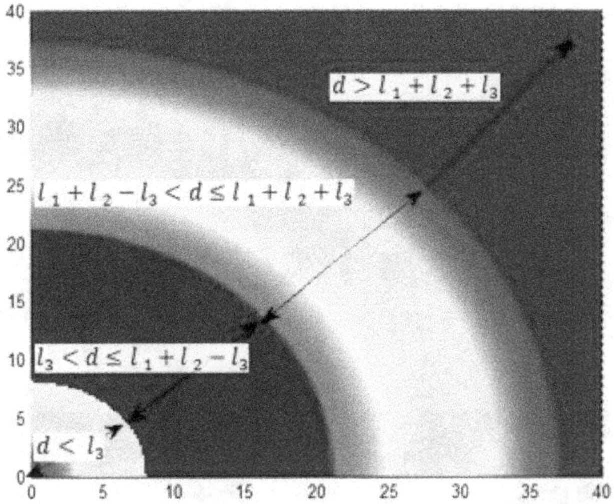

which in this case is the second link (l_2) and not the last link (l_3). The high dexterity band (Region 2) is given by:

$$l_3 < d \leq l_1 + l_2 - l_3 \tag{13}$$

which in this case, will be $8 < d \leq 9$

Figure 10 gives a comparative view of the dexterity plots in Case 1 and Case 2. As seen in Figure 10, with the positioning of the links changed, the boundaries of the workspace remain the same but the maximum dexterity region is significantly diminished in Case 2. This simulation reinforces the fact that the dexterity index of a manipulator is also a function of the relative positioning of the links and not just the link lengths.

Case 3: Limited Joint Angles

Grashof's criterion assumes that the joints are completely flexible, i.e., able to attain any angle between 0 and 2π. However, this is not possible in practice. The rotation of joints is often limited due to various mechanical and/or workspace constraints. In this case, we consider a practical scenario where the range of the joint angles assumes practical values.

Figure 9. Dexterity plot with the changed order of the links

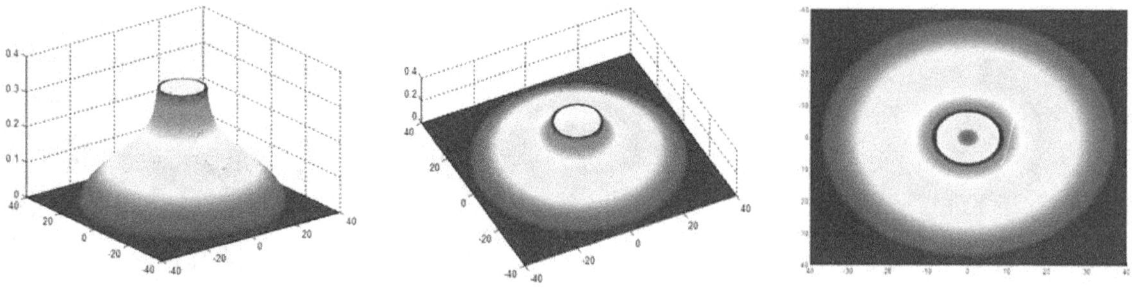

Figure 10. Comparative views of workspace dexterity (case 1 vs. case 2)

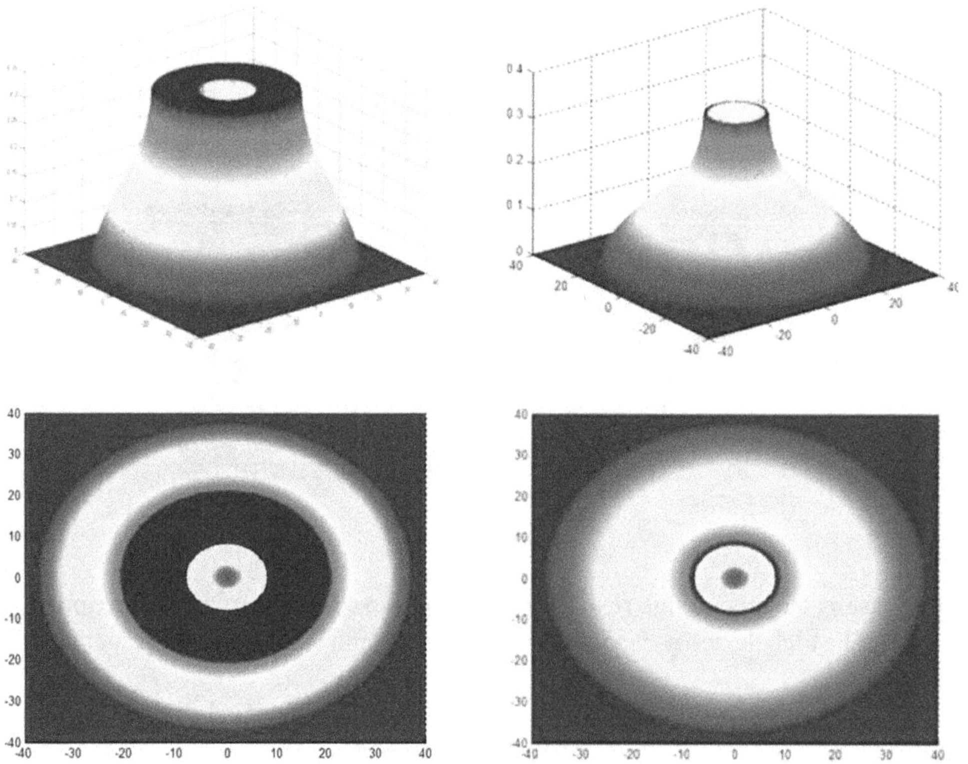

$-150\leq\theta_1\leq150$
$-150\leq\theta_2\leq150$ and
$-175\leq\theta_3\leq175$

As seen in Figure 11, the dexterity plot of the left half is distorted. This is due to the limitations of the first two joint angles. With limited joint angles, there will be fewer orientations possible about the point in the workspace when compared to Case 1.

EXAMPLE 2: INDEPENDENT TASK POINTS

Let us consider another design example where we need to design an optimal manipulator configuration to have maximum dexterity while following the independent task points A(16,19), B(18,0), C(-8,-25), D(-15,11), E(-10,16), F(20,-21), and G(-7,27). We assume that the manipulator is at the origin of coordinate frame O (0, 0). Following the steps in the proposed algorithm, we have:

1. The minimum and maximum distances of the points from the manipulator base are 29 (Point F) and 18 (Point B), such that $18 \leq d \leq 29$
2. We chose $l_3 = 15$ such that $l_3 < d_{min}$
3. From Equation (12), (13) we have $l_1 = 23$ and $l_2 = 22$

The simulation plots for the manipulator under all three cases can be seen in Figure 12. In Case 1 where ideal joints are assumed, we see that all the task points are within the most dexterous band. The design ensures that the manipulator will be able to attain every possible orientation about each of the task points.

In Case 2 the lengths l_2 and l_3 are interchanged. As expected, the most dexterous band is greatly reduced and is very thin in this case. All the task points lie outside the most dexterous band. This simulation once again proves that in order to achieve maximum dexterous area, the shortest link in the manipulator chain should be the end-effector or the last link.

In Case 3 we assume that the joints are with constraint, i.e. the joints can only attain a range of joint angles. As seen in Figure 12 the dexterity regions are very unsymmetrical when compared to the previous two plots, but the high dexterity bands

Figure 11. Workspace dexterity plot under joint angle constraints

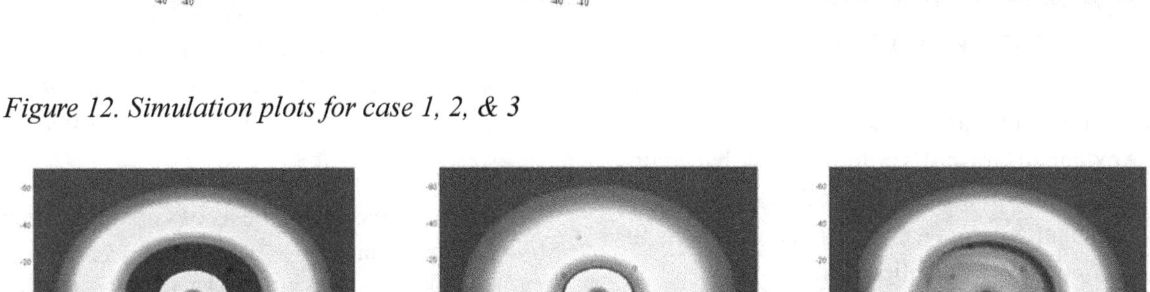

Figure 12. Simulation plots for case 1, 2, & 3

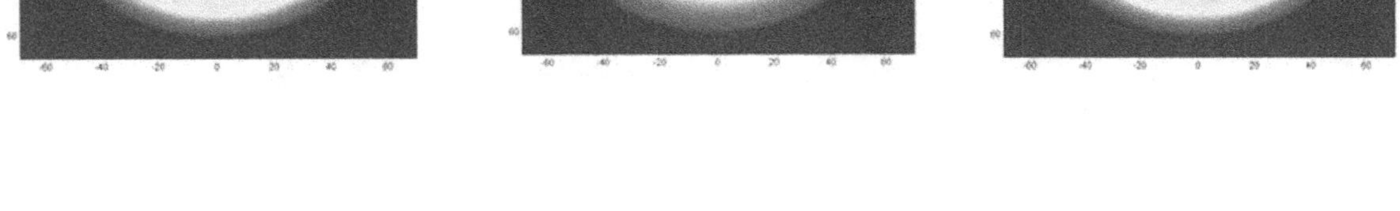

are similar to the bands in Case 1. The dexterity bands are distorted to the left because the point in this region will have limited solutions due the joint angle constraints put in place. In spite of the joint constraint, three task points lie in the most dexterous region while the rest are located in high dexterity band, as in Figure 12.

6. CONCLUSION

In this chapter, we have proposed a simple algorithm for generating optimal link lengths for a three-link planar manipulator. By adding a virtual stationary link in the chain, the three-link manipulator is converted into a four-link closed chain. Using Grashof's criterion, we optimize the link lengths to achieve maximum dexterity in the desired regions of the workspace. Dexterity index plots of the workspace generated by simulating the optimal manipulator configurations meet our design. Simulations with the link order changed (Case 2) have shown that the manipulator has a maximum high dexterity area when the shortest link is also the last link (end-effector) of the manipulator. The dexterity index plots of the workspace generated by applying Grashof's criterion help in better task placement and trajectory planning, especially when the manipulators mobility is limited due to joint angle limitations (Case 3).

7. FUTURE RESEARCH

In our future research, we would focus on applying the extended Grashof's criterion for N-bar chains for the prototyping of minimum degree of freedom (DoF) manipulators with optimal link lengths for accomplishing a specified task.

REFERENCES

Barker, C. R. (1985). A complete classification of planar four-bar linkages. *Mechanism and Machine Theory*, *20*(6), 535–554. doi:10.1016/0094-114X(85)90071-0

Chang, W., Lin, C., & Lin, L. (2005). A note on Grashof's theorem. *Journal of Marine Science and Technology*, *13*(4), 239–248.

Dai, J. S., & Shah, P. (2002). Orientation capability of planar serial manipulators using rotatability analysis based on workspace decomposition. *Proceedings of the Institution of Mechanical Engineers. Part C, Journal of Mechanical Engineering Science*, *216*(C4), 275–288. doi:10.1243/0954406021524990

Dai, J. S., & Shah, P. (2003). Orientation capability of planar manipulators using virtual joint angle analysis. *Mechanism and Machine Theory*, *28*(3), 241–252. doi:10.1016/S0094-114X(02)00118-0

Gosselin, C. (1992). The optimum design of robotic manipulators using dexterity indices. *Robotics and Autonomous Systems*, *9*(4), 213–226. doi:10.1016/0921-8890(92)90039-2

Kumar, A., & Waldron, K. (1981). The workspace of mechanical manipulator. *Journal of Mechanical Design. Transactions of the ASME*, *103*, 665–672. doi:10.1115/1.3254968

Li, R. Q., & Dai, J. S. (2009). Orientation angle workspace of planar serial three-link manipulators. *Science in China Series E: Technological Sciences*, *52*(4), 975–985. doi:10.1007/s11431-009-0083-7

Nokleby, S. B., & Podhorodeski, R. P. (2001). Optimization-based synthesis of Grashof geared five-bar mechanisms. *Journal of Mechanical Design*, *123*(4), 529–534. doi:10.1115/1.1401736

Paden, B., & Sastry, S. (1988). Optimal kinematic design of 6R manipulators. *The International Journal of Robotics Research*, *7*(2), 43–61. doi:10.1177/027836498800700204

Paul, B. (1979). A reassessment of Grashof's criterion. *Journal of Mechanical Design. Transactions of the ASME*, *101*(3), 515–518. doi:10.1115/1.3454089

Santos, R. R., Steffen, V., & Saramago, S. (2010). Optimal task placement of a serial robot manipulator for manipulability and mechanical power optimization. *Intelligent Information Management*, *2*(9), 512–525. doi:10.4236/iim.2010.29061

Sobh, T., & Toundykov, D. (2004). Kinematic synthesis of robotic manipulators from task descriptions. *IEEE Magazine on Robotics and Automation*, *11*(2), 78–85. doi:10.1109/MRA.2004.1310944

Spong, M. W., & Vidyasagar, M. (1989). *Robot dynamics and control*. New York, NY: John Wiley & Sons.

Tanev, T., & Stoyanov, B. (2000). On the performance indexes for robot manipulators. In *Problems of Engineering* (pp. 64–70). Cybernetics and Robotics.

Ting, K. L. (1986). Five-bar Grashof criteria. *Journal of Mechanisms. Transmissions and Automation in Design*, *108*(4), 533–537. doi:10.1115/1.3258765

Ting, K. L., & Liu, Y. W. (1991). Rotatability laws for N-bar kinematic chains and their proofs. *ASME Journal of Mechanical Design*, *113*(3), 32–39. doi:10.1115/1.2912747

Section 2

Implementation of Robotic Systems and their Applications

Chapter 4
AggieVTOL:
A Vertical Take Off and Landing Unmanned Aerial Vehicle Platform for Personal Remote Sensing

Brandon J. Stark
Center for Self-Organizing & Intelligent Systems (CSOIS), Utah State University, USA

YangQuan Chen
Center for Self-Organizing & Intelligent Systems (CSOIS), Utah State University, USA

Mac McKee
The Utah Water Research Laboratory, Utah State University, USA

ABSTRACT

Unmanned Aerial Vehicles (UAVs) for civilian applications are in a rapidly growing sector in the global aerospace industry that has only recently begun to gain traction. In this relatively immature field, there is such a steep learning curve that it can be difficult for research groups to begin development of well designed UAV systems. In this chapter, the authors present the AggieVTOL, a modular multi-rotor rotorcraft UAV prototype platform, and an overview of prototyping phase of its development, including design parameters and the implementation of its modular subsystems. Performance results demonstrate the effectiveness of platform.

1. INTRODUCTION

In recent years, the development of Unmanned Aerial Vehicles (UAVs) has increased dramatically. Largely spearheaded by military applications, UAVs have grown from impractical toys to highly sophisticated tools. While most of the attention has been on military applications, UAVs are expected to play a major role in redefining remote sensing in civilian applications such as traffic control, border patrol, fire fighting management, agricultural monitoring, and natural resource management. However, as with many robotic platforms, UAV development is often slow and expensive. While there is such a demand for

DOI: 10.4018/978-1-4666-0176-5.ch004

UAV systems, each system is often tailored to a specific purpose or a specific architecture which slows down development in the event of application or architecture changes. In this immature field, a prototyping environment is absolutely critical to achieve substantial progress.

The Center for Self-Organizing and Intelligent Systems (CSOIS) at Utah State University has been developing UAVs for the specific emerging field of low cost personal remote sensing. By equipping low cost autonomous UAV systems with scientifically calibrated sensor arrays, end users are able to gather valuable data easily without requiring the hiring of manned aircraft or years of platform specific training. As the field matures, these platforms will be available at such low cost that they will be affordable even to small farm owners wanting to monitor the growth of their crops and water usage on every part of their farm. The goal at the CSOIS is to develop these UAVs in a Cyber-Physical System (CPS) context (Chao 2010, Tricaud 2010); implementing them as a mobile actuator/sensor network to identify, estimate and control a distributed parameter system (DPS), a system where the states evolve over both time and space.

CSOIS's success with fixed-wing UAVs has led to the development of the AggieAir Flying Circus ("AggieAir Flying Circus," 2011) service center: a remote sensing operation that specializes in high-resolution, multi-spectral aerial imagery for monitoring soil moisture, riparian and wetland mapping, and fish and wildlife tracking. The low cost platform allows for these applications at a much higher resolution, both temporally and spatially than conventional solutions. The UAV platform, named "AggieAir", has the capability of carrying multiple cameras that capture imagery in the visual and near-infrared bands at a spatial resolution of 2.5 to 25 *cm*, depending on the altitude of flight. In Figure 1, the imagery was collected and geo-rectified within 72 hours and given to the Rushing Rivers Institute to be annotated. This type of high resolution imagery is invaluable to researchers and quite often exceeds their expectations.

While research continues in the fixed-wing UAV research, development has begun on AggieVTOL; the Vertical Take-Off and Landing (VTOL) rotary wing platform for remote sensing. While rotary wing platforms lack the range of fixed-wing, they have the advantage in maneuverability, from hovering in place with accurate orientation and positioning as well as precise and steady translational movement. The VTOL platform selected for AggieVTOL is known as a Quadrotor, a mechanically simple vehicle with a higher payload capacity compared to other rotorcraft platforms. Specializing in low-altitude, high-resolution imagery of specific targets, AggieVTOL applications includes on-demand spatial data collection and target inspection from multiple angles as well as some of the standard AggieAir fixed-wing UAV applications.

One specific application of AggieVTOL is analyzing of the spread of *Phragmites Autralis*, an invasive species in the wetlands in North America. In previous work with the AggieAir platform, aerial imagery in the red, green, blue, and near-infrared spectra was obtained for approximately 50 square miles of the Bear River Migratory Bird Refuge (BRMBR) and used to produce a high-resolution base map of the entire area wherein the invasive weed *Phragmites* is a problem. These data were used in combination with on-ground field inspections to identify specific areas or patches, totaling approximately 12 square miles, where UAVs were used to conduct aerial sampling at three other points in time. The imagery that was acquired was used to identify the distribution of *Phragmites* within that 12-square-mile area and to measure the spread of *Phragmites* throughout year. Each of these four high-elevation UAV flights (i.e., one 50 square mile flight and three flights of 12 square miles each) produced mosaiced and georectified images that have a resolution of approximately 25 *cm* (see Figures 2a and 2b).

Figure 1. Annotated mosaic a few days after flight with annotation by Rushing Rivers Institute ("AggieAir Flying Circus," 2011)

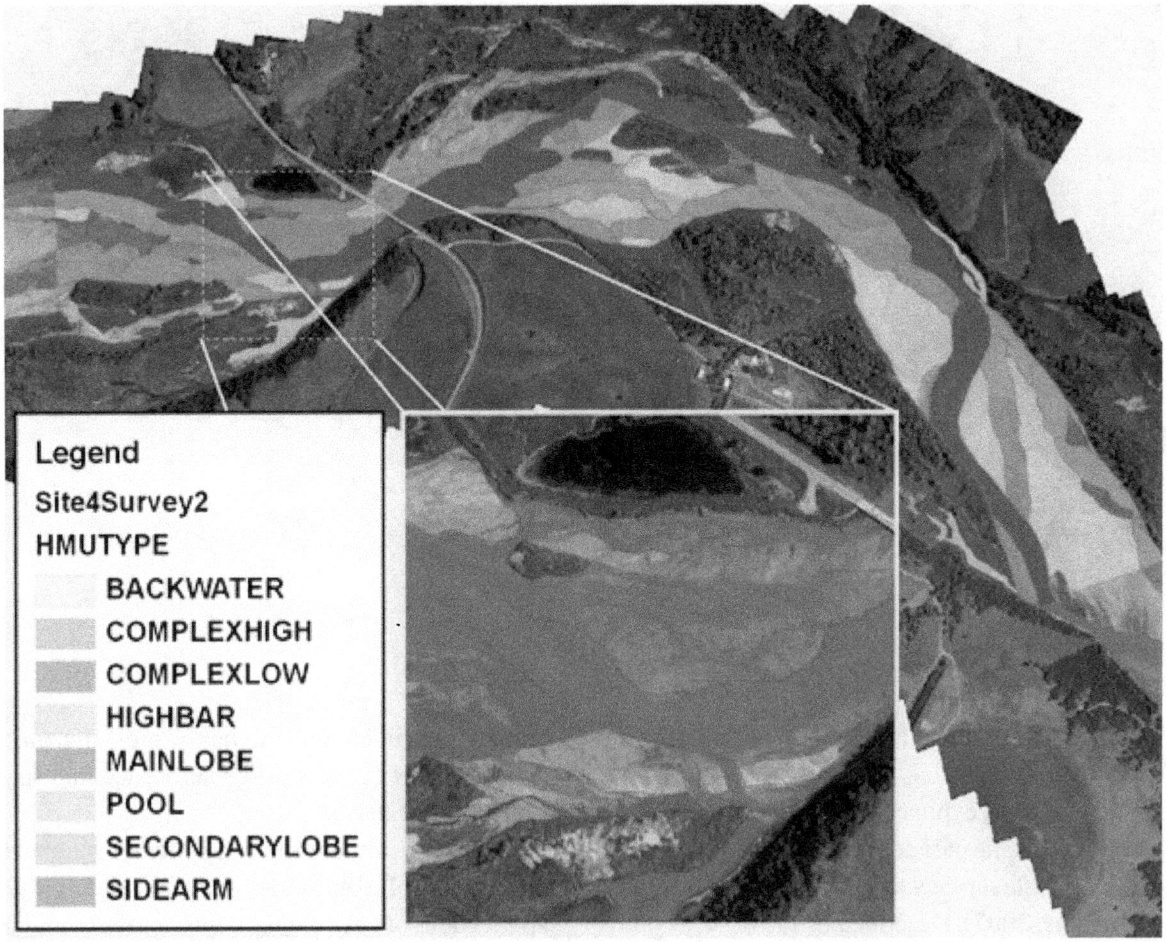

However, before the 25 *cm* resolution images can be used to recognize areas where *Phragmites* and other species are growing, it is necessary to intensively sample approximately 100 very small sites, each of only a few square meters, to establish the base data for use in training the pattern recognition algorithms to detect different wetlands species and other cover types (e.g., open water, roads, etc.). These sites must be sampled with either on-ground field activities or high-resolution photographs taken by UAVs flying only a few feet off the ground, wherein AggieVTOL is the perfect solution. The data obtained can be used to train and test a learning machine for classification of wetlands species using the spectral data from the 25 *cm* resolution images.

The development of this platform has been an involved process and is the motivation for this chapter. This chapter presents the AggieVTOL system (Figure 3), focusing on the prototype phase of development including evaluation of design parameters, the modularization of subsystems, and the implementation of navigation algorithms and control systems. The ultimate contribution is to shorten the learning curve for interested readers based on the reproducible development efforts summarized. Experimental results are included to demonstrate the effectiveness of the prototyping platform.

Figure 2. (a) Visual spectrum RGB, (b) NIR imagery (false grey scale), (c) RVM classified (Phragmites in white) (Zaman et al., 2011)

2. UAV DEVELOPMENT

There has been exponential growth in UAV research within the past decade due in no small part to the valuable services these vehicles can provide (Valavanis, 2007). Described by the IEEE as one of the past decade's top 11 new technologies (Ross, 2011), UAVs in the military have paved the way for UAVs to become accepted into the public consciousness. No longer are they science fiction, but a very real and tangible tool for a multitude of purposes, and nothing depicts this more than the dramatic increase in research funding supplied by the U.S. Department of Defense. In the past decade, the annual funding towards UAV research has increased from a mere \$284 million in 2000 to \$3.113 billion (Department of Defense, 2005). While the U.S. represents the largest market for UAVs, the global UAV market is expected to reach \$80 billion by 2020 according to a recent independent survey by the Teal Group (Teal Group, 2011). This same survey predicts that UAVs will continue to be the most dynamic growth sector of the world's aerospace industry (Teal Group, 2011).

However, while military applications remain the most public and most well funded, civilian applications of UAV technology has seen a similar growth (Valavanis, 2007). UAVs have a breadth of applications, with surveillance a common theme. Civilian UAVs can monitor traffic congestion, track the growth of invasive weeds, provide real-time surveillance on large plots of land, investigate hazardous environments, inspect infrastructure of bridges, walls, power lines and buildings, and even search and rescue operations. Armed with a variety of sensors, these vehicles have the potential to dramatically change the field of remote sensing. In applications covering large areas of land, current methods of remote sensing are often inadequate; satellite imagery often suffers from low resolution and low update rates while ground probes can only cover a small area with adequate resolution before cost becomes a significant issue.

Figure 3. AggieVTOL Quadrotors, 2011

Using manned craft, or sending research crews can be expensive and can be a risk to the personnel involved. On the other hand, low cost UAVs are uniquely equipped for the task of remote sensing. Specially equipped UAVs can be a great boon for water management and irrigation control, where the target area can be immense but still require a high spatial resolution (Chao et al., 2008). This kind of sensing can easily be accomplished using geo-referenced multi-spectral imagery obtained with even low cost UAVs (Jensen, Baumann, & Chen, 2009).

Beyond the collection of multi-spectral imagery on large plots of land, the goal of specific target reconnaissance requires a different skill set. Remote sensing over fields can be best accomplished with large fixed-wing UAVs; large, long range vehicles moving rapidly over their target area. But to examine the infrastructure of a remote bridge, precisely targeted imagery is required. VTOL UAVs, with their ability to hover during data collection, is a rapidly growing section of the UAV market (Valavanis, 2007). These vehicles are capable of deftly navigating through enclosed environments, enabling a whole new set of applications. Flying low and close to their targets, VTOL UAVs are a valuable tool for detailed inspections and surveillance (Herisse, Oustrieres, Hamel, Mahony, & Russotto, 2010).

Rotary-Wing UAV Development

However, unlike the fixed-wing UAVs, VTOL rotary-wing UAVs are a newer topic of research. Only recently, with the development of small scale sensor solutions and advances in microprocessors have VTOL vehicles been possible. These vehicles have taken a number of forms. While the traditional helicopter design remains very common due to its accessibility from the RC hobbyist community, many other novel designs have been developed. The Quadrotor design in specific as a UAV began to gain traction in the early 2000's (Pounds, Mahony, Hynes, & Roberts, 2002). Ironically, the Quadrotor was the one of the first VTOL designs and predates the modern helicopter design by several decades, but was untethered flight was impossible due to instability (Leishman, 2002). While the necessary active control system did not exist in 1907, modern electronics have allowed this design to reach fruition. The Quadrotor design boasts several advantages over other VTOL vehicles (Bouabdallah, Becker, & Siegwart, 2007). While the platform suffers from high power consumption, it is a mechanically simple vehicle, capable of stationary flight with a higher payload capacity than rotorcraft of equivalent size. Unlike other rotorcraft, the Quadrotor does not require the complex mechanical control linkages for rotor pitch control, since all vehicle attitude control is accomplished by differential thrust on fixed-pitch rotors (Hoffmann, Huang, Waslander, & Tomlin, 2007). In addition, the thrust is generated by four smaller rotors which store less kinetic energy than a single rotor; the danger of a large source of kinetic energy is one of the biggest detractors of a traditional helicopter design (Pounds et al., 2002). Other variants of VTOL vehicles include six or eight blade multi-rotor systems, or a Quad-Tilt-Wing (QTW) that combines the Quadrotor design with tilting rotors to achieve a fixed-wing like flight (Nonami, Kendoul, Suzuki, Wang & Nakazawa, 2010). Larger Quadrotor designs have also been examined specifically to increase the payload capacity of the vehicle (Pounds, Mahony & Corke, 2010).

With the recent advances in microprocessors and MEMS sensors, Quadrotors are more commonly found as small radio-controlled (RC) toys, such as the DraganFlyer ("Draganfly.com - Industrial Aerial Video Systems & UAVs," n.d.), and the iPhone controlled A.R. Drone ("AR.Drone.com USA - Parrot Wi-Fi quadricopter," n.d.). With some effort, these toys may be slightly modified for research purposes. As a prototyping platform, these pre-built systems can be quite useful, removing many of the obstacles of developing a VTOL craft such as frame development and motor controller development. However, these systems lack sufficient payload capacity for many of the previously described applications.

In many research groups, Quadrotor systems are built from off-the-shelf components to a great deal of success. These vehicles serve as test beds for control systems, remote sensing, multi-vehicle cooperative controls. Over the past decade, a number of groups have demonstrated quite a number of advances. Some of the earliest works demonstrated not only the feasibility and controllability of the design, but also that it can be made from off-the-shelf components (Pounds et al., 2002; Bouabdallah, Murrieri, & Siegwart, 2004; Hoffmann et al., 2005). It has been demonstrated that a simple rotational rate or attitude control system is sufficient for control and has been proven to be stable (Bouabdallah, Noth, & Siegwart, 2004). While simple PID control algorithms are the most common, a variety of additional control schemes have been demonstrated (Tayebi & McGilvray, 2005; Bouabdallah & Siegwart, 2005).

Control systems are not the only topic of research with Quadrotor platforms. Given the need for accurate attitude information, inertial navigation systems (INS) have been heavily researched. The majority of simple autonomous systems purely rely on inertial measurement units (IMUs) for their attitude estimates which has led to many investigations into improving low cost IMUs (Tournier, Valenti, How, & Feron, 2006; Pounds, Hamel, & Mahony, 2007). Commonly, many groups augment IMU by adding a GPS module for position tracking outdoors (Hoffmann et al., 2007), but other attempts to improve the performance have experimented with visual feedback (Altug, Ostrowski, & Mahony, 2002), optical flow (Herisse et al., 2010) or laser range-finders (He, Prentice, & Roy, 2008). At the University of Pennsylvania, the visual feedback method using an external camera system has led to some impressive demonstrations by the GRASP Lab of cooperative control and aggressive maneuvers (Mellinger, Michael, & Kumar, 2010).

Another important topic is the modeling of the dynamics of these vehicles. The OS4 simulator, developed by Bouabdallah (Bouabdallah & Siegwart, 2007b) has proven to be a valuable tool for Quadrotor simulations, providing accurate modeling of the aerodynamic effects. Using the simulator, more algorithms can be evaluated such as a fractional order flight controller (Luo, Di, Han, Chao, & Chen, 2010). Further system identification was researched in (Pounds, Mahony, & Corke, 2006), detailing a model for the rotor actuator system and blade flapping effects on the propeller. More aerodynamic effects were investigated by the Stanford STARMAC group for aggressive maneuvering (Huang, Hoffmann, Waslander, & Tomlin, 2009). New finely tuned designs for actuator systems have been developed based on these important modeling advances (Pounds et al, 2009).

But despite all these advances, very little remote sensing has been accomplished using the Quadrotor platform. Outside of research groups, several commercial companies market Quadrotor vehicles for surveillance, equipping their vehicle platforms with high-definition video. While well-developed, these commercial systems are expensive and proprietary, limiting the size of their market. As a surveillance device, these platforms are satisfactory. However, these platforms are not equipped to provide precision imagery needed to do high quality image mosaics. This is the difference between knowing that there is a tree in a large field and knowing the tree's precise location. Given the wealth of knowledge on Quadrotor platforms, it is feasible to develop an inexpensive working prototype for remote sensing using off-the-shelf components.

QUADROTOR MODEL

The Quadrotor is a mechanically simple device, using only four fixed-pitch rotors to achieve complete flight control. Attitude control of the vehicle is accomplished by differential thrust across the desired rotation axis. In the free-body diagram in Figure 4, the rolling motion of the vehicle is accomplished by differencing the thrust of actuators T_2 and T_4. Adjustment of the heading of the vehicle, or a yawing motion, is accomplished by differential thrust between the two pairs of counter-rotating rotors.

Quadrotor Dynamics

Quadrotor dynamics has been well documented. Including many of the additional aerodynamic effects described by Pounds et al. (2006) and Huang et al. (2009), this model can be quite complex. However, in this chapter, we make the following assumptions for a remote sensing platform:

- No aggressive maneuvers,
- Steady hovering,
- Limit target velocity to slow steady movements.

Given these assumptions, factors such as blade flapping and compensating for aggressive maneuvers can be safely ignored and a simplified derivation can be presented. The following derivation is performed in the North-East-Down (NED) inertial and body fixed coordinates and follows the derivation by Hoffman et al., (2007). The nonlinear dynamics of the vehicle is that of a rigid body with external forces applied to the center of mass. Expressed in the inertial frame, this can be described as

$$\begin{pmatrix} \mathbf{F} \\ \mathbf{M} \end{pmatrix} = \begin{pmatrix} m & 0 \\ 0 & \mathbf{I} \end{pmatrix} \begin{pmatrix} \ddot{\mathbf{r}} \\ \dot{\omega}_B \end{pmatrix} + \begin{pmatrix} 0 \\ \omega_B \times \mathbf{I}\omega_B \end{pmatrix} \tag{1}$$

Figure 4. Quadrotor body diagram

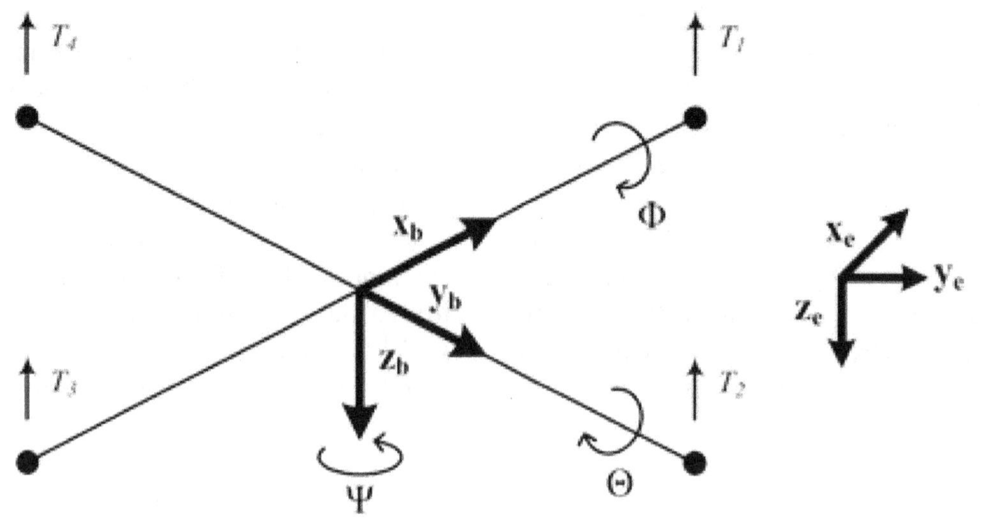

where, **F** is the total force, **M** is the total moment, **I** is the inertia matrix, **r** is the position vector of the vehicle from the origin of the inertial system to the vehicle center of gravity (CoG), ω_B is the body angular velocity, and *m* is the system mass.

Let the Earth-fixed frame be represented as $\{\mathbf{x}_e, \mathbf{y}_e, \mathbf{z}_e\}$ and the body-fixed frame as $\{\mathbf{x}_b, \mathbf{y}_b, \mathbf{z}_b\}$. The Euler angles of the body axis are denoted as $\{\Phi, \Theta, \Psi\}$ with respect to the body-fixed frame. The actuator system lies along the body axes $\{\mathbf{x}_b, -\mathbf{x}_b, \mathbf{y}_b, -\mathbf{y}_b\}$ at a distance *l*. The rotor plane of the actuator system ideally lies parallel with the body frame, however this is not always the case and thus let $\{\mathbf{x}_r, \mathbf{y}_r, \mathbf{z}_r\}$ represent the unit vectors associated with the rotor plane for the ith rotor. Each rotor *i* produces both a thrust, T_i, that is perpendicular to the rotor plane along the $\mathbf{z}_r$ axis and an aerodynamic moment, M_i, which is comprised of aerodynamic effects such as the body gyro effect, propeller gyro effect, hub force and a rolling moment effect (Bouabdallah & Siegwart, 2007a).

The total force and moment can be summed as

$$\mathbf{F} = m\mathbf{g}\mathbf{z}_d + \sum_{i=1}^{4}\left(-T_i\mathbf{R}_{i,I}\mathbf{z}_{R,i}\right) \quad (2)$$

$$\mathbf{M} = \sum_{i=1}^{4}(M_i + l \times (-T_i\mathbf{R}_{i,B}\mathbf{z}_{R,i})) \quad (3)$$

where **g** is the gravity vector, $\mathbf{R}_{i,I}$ is the rotation matrix from the plane of the rotor *i* to inertial coordinates, and $\mathbf{R}_{i,B}$ is the rotation matrix from the plane of the rotor *i* to body coordinates.

Actuator Dynamics

The thrust at each actuator system is produced by a rotor spun by a brushless DC motor. The torque generated by the motor can be described as follows (Hoffmann et al., 2007)

$$Q = K_q I \quad (4)$$

$$V = R_a I + K_e \omega \quad (5)$$

where *Q* is the torque generated by the motor, *V* is the voltage across the motor, *I* is the current through the motor, and ω is the angular rate at which the motor is spinning. The constants, K_q, K_e and R_a are motor specific, representing current to torque ratio, motor speed to back EMF ratio and the total armature resistance of the motor, respectively. The steady state power consumption can be given by

$$P = IV = \frac{Q}{K_q} V = \frac{\kappa_t T}{K_q} V \tag{6}$$

where κ_t is a constant ratio that relates the torque of the motor to the thrust produced and is unique depending on the rotor blade geometry. The nominal power at hovering, P_h can be described as the thrust, T, times the speed it is applied at, or

$$P_h = T v_h \tag{7}$$

where the induced velocity at hovering, v_h, is the change in air speed induced by the rotor blades with respect to the free stream velocity v_∞. The velocity of the air column can be derived by using momentum theory (Hoffmann et al., 2007) as

$$v_h = \sqrt{\frac{T}{2\rho A}} \tag{8}$$

where T is the thrust generated by the rotor to remain in hovering, A is the area swept by the rotor, ρ is the density of air. For a Quadrotor, the thrust T to remain in hovering is equal to 0.25 m, where m is the mass of the vehicle. The relation between applied voltage and the thrust is found by equating the power produced with the ideal power consumed at hovering

$$\frac{\kappa_t T}{K_q} V = \frac{T^{\frac{3}{2}}}{\sqrt{2\rho A}} \tag{9}$$

Solving for T,

$$T = \frac{2\rho A \kappa_t^2}{K_q^2} V^2 \tag{10}$$

It is apparent that the thrust produced by the rotors is proportional to the square of the voltage across the motor. However, brushless motors are dissimilar to regular brushed DC motors. Rather than being driven by a constant voltage, they are driven by a tri-phase high current signal generated by a special controller. It is then more accurate to replace V with V_{RMS} for the average voltage over several sinusoidal cycles.

While this actuator model is commonly used, it is valid only at slow velocities and hovering states. More advanced aerodynamics effects have been modeled with great success by Hoffman et al. (2007), Pounds et al. (2006), and Bouabdallah & Siegwart (2007a).

QUADROTOR DESIGN PRINCIPLES

A prototyping platform for a VTOL UAV can be challenging to develop. While it is worthwhile to maintain a modular system to allow for a true prototyping setup, in practice this is rarely done due to the overall complexity of the system design. More so than many robotic platforms, a fully modular VTOL prototype requires careful advanced planning to ensure a robust design. This section details several design principles towards achieving a well designed platform.

Frame Design

The construction of the physical platform is often overlooked by electrical engineers. A common misconception is that due to the simplicity of the mechanical system, a good control system can make anything fly as long as it has four motors. However, there is a significant difference between lifting off the ground and controllable flight.

Frame designs range from simple motors mounted on top of four PVC pipes in a cross formation to intricate designs carefully machined from carbon fiber. A list of desirable attributes for a good VTOL frame follows:

- **Low cost:** For a prototyping platform, damage is unavoidable. Even if the target goal is a high end system, the initial prototyping platforms should be cheap and easily obtainable.
- **Repairable:** Some impressive looking designs are often machined from a single piece. These can be well designed, but an unlucky gust of wind can set back development until a new frame can be made. The best design is one that can be repaired in the field and cheaply.
- **Sufficient protection of expensive electronics or payloads:** Even the most expensive frames are less expensive compared to the cost of the electronics. Poor designs will leave these valuable components vulnerable in the event of a hard landing or crash.
- **Lightweight:** As with most air vehicles, the frame of a VTOL must be lightweight enough to allow for flight and allow for a payload capacity.

Material Construction

Given the previous design considerations, the next step is to evaluate which materials will be used. Metals can be good due to strength, availability and cost, but be mindful of weight, especially when using metals other than lightweight aluminum. Plastics such as ABS or Polycarbonate (PC) can provide an adequate strength and are lightweight as well as low cost, however, they often must be machined before use and can add costly time during rebuilding after each unsuccessful landing. Carbon fiber is quickly becoming a favorite among the hobbyist community for its light weight and strength, but is more costly and difficult to machine. It has also been known to act as an RF shield, which can interfere with radio signals from a transmitter or modem. Fiberglass can be used as well; lightweight, and easily cut. Wood is generally avoided due to its weight, or in the case of balsa wood, lack of strength, though it can be used effectively as part of a composite.

Modular Design

The key to an easily repairable frame is a modular design with replaceable parts. Detachable motor mounts and arms allow for quick repairs as well as transportability. This however does add special requirements on the selection of connectors, necessitating a balance of strength and weight.

Frame Optimization

Compared to fixed-wing UAVs, Quadrotor frame designs are not as aerodynamically designed. Since the rotor system provides all of the lift and an active control system maintains flight, designers normally do not spend much analyzing the actual physical construction. However, there are several key parameters that should be kept in mind during the design process.

Center of Gravity (CoG)

The center of gravity has a profound effect on stability of the craft (Pounds, Mahony, Gresham, Corke, & Roberts, 2004). While placing the center of gravity below the rotor plane is the most common set up, this makes the vehicle inherently oscillatory. Placing the CoG above the rotor plane causes the vehicle to be unstable but non-oscillatory. If the CoG and the rotor plane are coplanar, the system is marginally stable, but this is difficult to achieve.

Motor Placement

Since the thrust is generated by the rotors, it is best to ensure that the motors are far away enough from any vehicle structure such that the rotor swash is not obstructed. On vehicles with thin and light arms, the arms may be subject to vibration from the motors which can reduce motor efficiency and overall performance. Adding a cross brace may be advantageous to improve rigidity, as long as the rotor swash is not disturbed (Hoffmann et al., 2007). Landing gear beneath the arms may provide an excellent rigid, yet modular solution, but can also disrupt the airflow causing the thrust vector to no longer be normal to the rotor plane. Moving the motors farther out improves the roll and pitch actuator force as the arms act as a rigid lever, but combined with a slow actuator response or an improperly tuned controller may cause unstable oscillations. Similarly, moving the motors too close inwards reduces the actuator force and may also cause disturbances between the rotor swashes decreasing performance.

Rotor Shrouds

Rotor shrouds, such as the one found on the A.R. Drone, should be designed carefully if used. The shrouds can disrupt airflow causing random variations in torque which can reduce the effectiveness of yaw tracking (Hoffmann et al., 2007). On the other hand, rotor shrouds are an effective method to prevent damage to the propellers and motors in the event of a crash.

Rotor Plane Height from Ground

An effect known as the In Ground Effect (IGE) can affect the controllability of a small Quadrotor during takeoff and landing. For a rotorcraft, the IGE describes the increase in rotor efficiency when in close proximity to the ground. This causes the vehicle to bounce or hover close to the ground prior to producing enough thrust to achieve flight. However, IGE can be mitigated by increasing the height of the rotor plane to above one rotor diameter above the ground (Bouabdallah & Siegwart, 2007b).

Weight Distribution

Given the symmetric design of the Quadrotor, the distribution of weight should be equally symmetric as much as possible. This includes the deployment of the payload and the battery, neither of which is commonly square shaped. Discrepancies may lead to undesired movement or decrease the efficiency of the control system.

Actuator System

The actuator system can be realized through a variety of means. While it is feasible to make everything in-house, off-the-shelf-parts are also commonly used with a great deal of success. Many of these components may be found in Hobby Stores as RC Helicopter parts.

Motors

The efficiency of the motor is a critical factor in deciding on a motor solution. Though typically more expensive, brushless DC motors are the most commonly used motors. Whereas brushed DC motors suffer from rapid wear out, contact bounce, and maintenance requirements, brushless DC motors have permanent fixed magnets that allow for more reliable performance. However, since the polarities of the motor cannot be switched as the rotor crosses a commutation plane, this must be done with electrical switching by a dedicated motor controller. However, this does add another advantage of the brushless DC motor. Because the rotation is related to the electronic switching and the field strength of the permanent magnet is independent of the current through the coil, the torque-speed of brushless DC motors is more linear than their brushed counterparts, allowing for better open-loop control, a necessity for a vehicle that utilizes differential thrust (Silva, 2005).

Propellers

Unlike many other larger rotary-wing platforms, the propellers on a Quadrotor are commonly fixed-pitch. While fixed-pitch propellers do not require any additional complex mechanics, they are reliant on the motor system for changes in thrust. Ideally, the rotor radius should be as large as the system can support; the power required for flight is inversely proportional to rotor radius (Pounds et al., 2010). However, too large propellers can have an adverse on response time due to increased inertia. The selection of choosing the right propeller is a balance of response time and efficiency. One other constraint is the use of counter rotating propellers which limits the off-the-shelf options for propellers which are only available in a limited number of sizes and shapes. One possible work-around if an adequate solution is not found is to substitute the counter rotating pair by using four identical propellers all spinning in the same direction, and pitching a pair of opposite rotors by 15 degrees to provide the counter rotational torque.

Electronic Speed Controllers

Brushless motors require a special drive controller known as an Electronic Speed Controller (ESC) to generate the three-phase control signal. Since many Quadrotor projects began with the hobbyist community, off-the-shelf hobby aircraft ESCs are commonly used. Cheap and low cost, they have been shown to be capable, though they suffer from a number of drawbacks. Generally, these solutions are not capable of update rates above 50 *Hz* and are interfaced with an imprecise PWM signal. More specialized motor controllers with more microprocessor friendly interfaces such as I^2C have been made available recently, while many other groups opt to design their own (Pounds et al., 2010). While these solutions are much improved over older ESCs, they do suffer from drawbacks including the necessity to multiplex instructions to motors, which may be a timing issue in platforms with large numbers of motors. In all however, the ideal ESCs have a high update rate and can support high current draw.

Autopilot System

While the autopilot system may be one of the most critical components of a Quadrotor UAV, it is a flexible choice. The autopilot system is responsible for several key objectives (Chao, Cao, & Chen, 2010). In addition to attitude stabilization, it can be utilized for altitude hold, position hold, autonomous take-off and landing and waypoint navigation. Autopilot systems have been developed in such a wide variety of processing devices, ranging from PICs (Hoffmann et al., 2005) to real-time Linux processors and everything in between. A high degree of computational power is desired, as is a variety of interface ports.

There are a variety of off-the-shelf UAV autopilots available, as well as open-source projects. Many commercial autopilots are comprehensive packages, including a full suite of sensors and telemetry built-in. Some of the autopilots available include the Procerus Kestral autopilot ("Procerus Technologies," n.d.), the Cloud Cap Piccolo autopilot ("Cloud Cap Technology," n.d.) and the MicroPilot autopilot ("MicroPilot," n.d.). Open-source autopilot systems are also a viable solution for many groups, especially those interested in working on control systems and navigation algorithms which may not be adjustable in commercial systems. For rotary-wing platforms, some open-source projects include Paparazzi, and ArduCopter ("Arducopter," n.d.).

It is also useful that the autopilot system incorporates a user interface on the ground through data telemetry communication. Connected to a ground control station, this grants the user to view relevant navigational data as well as allow the user high level navigation control such as changing waypoints or activating payload systems. A data telemetry system must be a reliable communication link with long range capability

Navigation Sensors

A navigation system for a UAV is often composed of two components; an Attitude and Heading Reference System (AHRS) and an Integrated Navigation System (INS). The AHRS is used to determine the rotations of the platform and may be accomplished with an Inertial Measurement Unit (IMU). An Integrated Navigation System is a comprehensive set of inertial sensors for position and velocity calculations and may be augmented with additional sensors.

Inertial Measurement Unit

The AHRS may be accomplished with commercial IMU modules or developed in-house. These IMU systems are used to measure the inertial movement of a vehicle, most commonly with a total of 6 Degrees of Freedom (DoF). Commercial IMU modules can be divided into several categories, depending on the performance and characteristics (Chao, Coopmans, Long, & Chen, 2010). A brief summarization of the four types of IMUs can be found in Table 1.

For low cost systems such as the industrial and hobbyist grade systems, the IMUs are commonly composed of Microelectromechanical Systems (MEMS) accelerometers and gyroscopes. These on-chip MEMS sensors are also often the primary choice for in-house IMU designs. They are small, cheap and consume little power compared to other solutions. While they are typically not as accurate as the larger sensors found in navigation or tactical grade IMUs, they have shown to be effective in many applications.

MEMS Sensors

A MEMS accelerometer measures the inertial force generated by movement, which can be used to estimate the acceleration. This is done by measuring the displacement of a proof mass inside the accelerometer and converting it to an electrical signal. This can be done by a variety of methods; the most common uses either piezoresistors for inexpensive accelerometers or differential capacitive sensors in higher end accelerometers.

MEMS gyroscopes measure the rotational motion of a body, however not through acceleration as with the accelerometers. These sensors measure the rotational rate of a body around an axis by measuring the Coriolis force generated by rotational movement.

MEMS Accelerometer and Gyroscope Models

In order to use these MEMS sensors accurately, several considerations must be examined. While older mechanical sensors are larger and more expensive, they do not suffer from as many error sources as the smaller MEMS sensors do. Special care must be taken to ensure a high level of ac-

Table 1. IMU categories (Chao et al., 2010)

IMU Type	Navigation Grade	Tactical Grade	Industrial Grade	Hobbyist Grade
Cost ($)	> 50 k	10-20 k	0.5-3 k	< 500
Weight	> 2 kg	~ .5 kg	< 50 g	
Gyro Bias	< 0.1 °/h	0.1-10 °/h	≤ 1 °/s	> 1 °/s
Gyro Random Walk Error	< 0.005 °/$\sqrt{h}$	0.2-0.5 °/$\sqrt{h}$		
Accel Bias	5-10 μg	0.02-0.04 mg		
Example	Honeywell HG9848	Honeywell HG1900	Microstrain GX2	ArduIMU

curacy. The following section describes several major sources of errors and calibration techniques to minimize these errors.

Error Sources

Manufacturing variabilities form the majority of the error sources in MEMS devices. These include variation in bias, scale factor, nonlinearities, temperature variations and nonorthogonalities.

- **Bias:** As an electronic sensor, the majority of MEMS sensors are voltage ratiometric, in that the output is proportional to the source voltage. This means that the supply voltage must stable, and that any change to the power supply system may affect the calibration. The bias value, when the sensor is at rest, is generally at half of the operating voltage, but may vary due to manufacturing imperfections.
- **Scale Factor:** The scale factor of a MEMS sensor is the ratio of change in output to the change in the true value. Ideally, this scale factor is a linear relationship over all measurements of the sensor; however this is not always the case. Proper calibration is necessary for accurate performance. As with the bias, this error source is also ratiometric and should be evaluated with the actual power system used.
- **Temperature Variations:** Temperature variations can affect both bias and scale factor and can be detrimental in UAV platforms that are exposed to a large range of temperatures.
- **Nonorthogonalities:** Nonorthogonalities, or misalignment errors are often very small, but occur due to imperfections in manufacturing or in mounting. These errors cause undesired measurements on stationary axes.

MEMS Magnetometer

While not an inertial measurement sensor, magnetometers are often found used with IMUs to augment heading calculations. Similar to the other MEMS sensors, magnetometers also suffer from a variety of errors and should be calibrated. However, magnetometers function very differently than accelerometers and gyroscopes.

In a 2D example, the heading of a magnetometer can be determined by a mapping of the magnetic field in two orthogonal directions on a unit circle. The heading angle, Ψ, is calculated as the angle of the vector to the coordinates of the two axes of the magnetic measurements. However, in the presence of a constant additive in the magnetic field, the measurements of the magnetic field may be off-center. This is known as a hard-iron distortion and can be easily be calibrated out by determining the bias offset of the unit circle. Soft-iron distortions are caused by a local variation in the magnetic field that varies with the orientation of the sensor (Konvalin, 2009). This manifests as a distortion in the unit circle to an elliptical shape and/or rotation.

Proper calibration can improve the performance of the magnetometer in the presence of these hard- and soft-iron distortions. However, it is important to note that these distortions are caused by external sources. For a UAV that will be flying outdoors away from large iron deposits, magnetic calibration must be done in a similar environment and not in the safety of a lab, where magnetic sources and iron deposits are abundant. In addition, since motors are significant sources of magnetic field distortion, proper calibration should be done with the motors engaged.

Inertial Measurement Sensor Calibration

Given the set of error sources in MEMS sensors, a comprehensive set of calibrations are necessary. This is also a significant area of research. While

INS solutions and other high-end systems are factory calibrated using specialized equipment, the majority of UAV research groups use low-end IMUs that need certain calibration efforts with limited resources. A bare minimum calibration technique is known as the 6-Point Calibration Technique (Aggarwal, Syed, Noureldin, & El-Sheimy, 2010).

6 Point Calibration

In a simplified model, the output of a MEMS accelerometer or gyroscope has the following form,

$$x_{true} = m_x x_{meas} + b_x \tag{11}$$

where x_{true} is the true measurement in m/s (or $°/s$ for gyroscope), x_{meas} is the discrete value as read by an analog-to-digital converter (ADC), and m_x and b_x are the scale factor and bias associated with measurement axis x.

The 6 point calibration test involves mounting the IMU on a level surface measuring the acceleration on each axis for positive and negative gravity vectors. The values for m_x and b_x can then be found by

$$b_x = \frac{x_{meas,+} + x_{meas,-}}{2} \tag{12}$$

$$m_x = \frac{2g}{x_{meas,+} - x_{meas,-}} \tag{13}$$

where $x_{meas,+}$ and $x_{meas,-}$ are the measured values for the positive and negative gravity vectors, and g is the known ideal gravity vector.

This method may also be used to calibrate a gyroscope axis if it can be rotated at a known rotational speed. This is a simple method; however a higher accuracy may be achieved if a significant number of samples are collected as in the Multiple Point Calibration method.

Multiple Point Calibration

Given a set of known true measurements x_t and the sensor outputs x_m, a linear model can formed with tall matrices in the form

$$\mathbf{X}_t = \mathbf{X}_m \mathbf{B} \tag{14}$$

or,

$$\begin{pmatrix} x_{t,1} \\ x_{t,2} \\ \vdots \\ x_{t,n} \end{pmatrix} = \begin{pmatrix} x_{m,1} & 1 \\ x_{m,2} & 1 \\ \vdots & 1 \\ x_{m,n} & 1 \end{pmatrix} \begin{pmatrix} m_x \\ b \end{pmatrix} \tag{15}$$

This linear equation can be solved easily in the least squares sense. The coefficients **B** can be solved as,

$$\mathbf{B} = \left(\mathbf{X}_m^{\ T} \mathbf{X}_m\right)^{-1} \mathbf{X}_m^{\ T} \mathbf{X}_t \tag{16}$$

Although this method is simple, it is effective for most applications. Further accuracy may be obtained if the model is expanded to include nonorthogonalities and temperature variations.

Integrated Navigation System Sensors

Whereas the IMU is used to calculate the Euler angles of a vehicle, the INS is used to calculate the position and velocity of the vehicle in three dimensions. While it may use only inertial measurements, it is often augmented with other sensors such as Global Positioning System's (GPS) (Hoffmann et al., 2007), barometers (Bouabdallah & Siegwart, 2007a), visual systems (Mellinger et al., 2010) or sonar systems (Hoffmann et al., 2007). The addition of these extra sensors improves position tracking significantly.

Addition of a GPS module to a Quadrotor INS is nearly ubiquitous because of its advantage of drift-free position sensoring, even considering its limitations. While GPS data does not suffer from drift over time, it has a low resolution, especially for a vehicle that is often less than 1 *m* in diameter. In addition, a slow update rate, which can be as slow as 1 *Hz* limits its effectiveness as a standalone solution (Aggarwal et al., 2010). Beyond the standard GPS system, Differential GPS (DGPS) has been shown to provide a more accurate positioning, though this requires a stationary secondary GPS receiver. In DGPS systems, the stationary receiver calculates a set of offsets of each satellite relative to the vehicle GPS. This set of offsets is then used with GPS positioning calculations to improve position resolution to 1-2 *cm* as opposed to ±1 *m* with traditional GPS positioning (Hoffmann et al., 2007).

For altitude measurements, several alternative sensors are available. GPS positioning may be used, however with its low update rate, other sensors may be better suited if available. One alternative is the addition of a barometer which measures pressure and is proportional to altitude. While barometric sensors also suffer from a low resolution and a significant amount of noise, like GPS, they can provide useful information when used in conjunction with integration algorithms such as a Kalman Filter (KF) (Tanigawa, Luinge, Schipper, & Slycke, 2008). Sonar systems measuring the distance to the ground can also be used to augment altitude measurements; however these systems are often limited by range and the environment in which the UAV is deployed. Sonar systems require that the system remains within the range of the sonar, which may be as low as 5 *m*, and that the ground remains level and uniform. However, its use in autonomous take-off and landing procedures has been well documented (Iwakura, Wang, Nonami & Haley, 2011).

Visual (camera) systems have also been used in a variety of capacities. Attitude estimation using a single camera system has been demonstrated as well as for autonomous take-off and landing procedures (Tournier et al., 2006). Optical flow systems have also been demonstrated for attitude estimation (Herisse, Russotto, Hamel, & Mahony, 2008) and furthered to including terrain following (Herisse et al., 2010). Advanced work with visual systems has successfully used a stereo camera to assist with visual odometry (Achtelik, Bachrach, He, Prentice, & Roy, 2009). However, despite these advances, many of these systems are as of yet impractical for truly autonomous flight as they often require known markers in a known environment. One exception is the practical implementation of visual systems to aid in controlled maneuvers such as take-offs and landings (Herisse et al., 2008).

More advanced navigation systems may include a scanning laser range-finder. The addition of a laser range-finder may be used for velocity estimations for positioning calculations (Achtelik et al., 2009) or simultaneous localization and mapping (SLAM) purposes (Grzonka, Grisetti, & Burgard, 2009).

Payload System

The primary application of the AggieVTOL platform is for remote sensing and as such payload development is important. A variety of payloads are available, most notably camera systems. These may range from low cost web cameras to high-definition video systems. However, during development, special consideration should be taken if live wireless video is desired. Many recent off-the-shelf video systems transmit on the 2.4 *GHz* band, as do many wireless modems and RC transmitter/receivers. Unlike the digital communication systems, analog 2.4 *GHz* video can cause communication disruptions.

While video systems are much more common, regular digital cameras can provide superior image quality. Digital cameras with 10 *MP* resolution are easily affordable and can easily be interfaced with an autopilot or payload controller. Combined with accurate attitude and position information, images from these cameras can easily be geo-rectified and formed into a mosaic to provide large high resolution maps.

SOFTWARE ARCHITECTURE

With the major components selected, a system architecture can be developed. An overview of the on-board system architecture used in AggieVTOL can be seen in Figure 5. In this system, four sensor sources are used to calculate the attitude and the position. The 6 DoF IMU and the magnetometer are fused together by the Attitude and Heading Reference System (AHRS) module. The AHRS provides this data to the Integrated Navigation System (INS) along with positioning data from a GPS module and a barometer. These data are then processed to form an estimation of the UAV's true position and altitude. The attitude and position information is used by the controller, along with any control information from either an RC transmitter for manual control or the Ground Control Station (GCS) data telemetry link. The actuators for the UAV are controlled by specialized ESCs. A payload can be controlled autonomously by the controller or manually through the GCS. The GCS provides the operator with real-time status information as well as the ability to form navigational paths for the UAV to follow.

It is common to find Quadrotor platforms in which many of the previously mentioned modules are interconnected and coupled (Mahony, Cha, & Hamel, 2006). While this may result in a well-designed end result, it lacks the modularity of a good prototyping system. The modular architecture described in this section allows for a variety of algorithms to be evaluated independently of each other. In the description of the modules that follow, the algorithms in use on the AggieVTOL platform are described in detail. Other possible algorithms are mentioned along with respective references for the readers to explore.

Attitude and Heading Reference System Module

The Attitude and Heading Reference System (AHRS) is a specialized module that calculates the estimated attitude and heading of a system. In general parlance, this is known as the pitch, roll and yaw of the vehicle. The development of an accurate AHRS using low cost sensors is a well-researched topic and many sources are available for the reader to peruse. While many solutions exist, the simplest solution is the use of a Complementary Filter (CF) and is most commonly found within low cost systems where computational load is carefully monitored (Hamel & Mahony, 2006). The complementary filter is a simple combination of low and high pass filtered data streams (Higgins, 1975). Other common solutions use a Kalman filter or one of its variants such as Extended Kalman Filter (EKF) or Unscented Kalman Filter (UKF).

Complementary Filter

The CF for attitude estimation is not a new concept and has been in use for many decades. Computationally simple and efficient, it provides an adequate performance for systems with low processing power. The basic filter is designed around two measurements, $\mathbf{x}$, and $\mathbf{y}$ of some signal $\mathbf{z}$, $\hat{\mathbf{z}}$ as the estimate of $\mathbf{z}$. If the measurements x and y are corrupted by noise ($\mathbf{x}$ by high frequency noise and $\mathbf{y}$ by low frequency noise), a low pass filter, $\mathbf{G}(s)$, can be designed to remove the noise of x. The complement $[1\text{-}\mathbf{G}(s)]$ can then be used as a high pass filter for $\mathbf{y}$. The summation of the two filtered signals then form the estimate $\hat{\mathbf{z}}$, which ideally is noise free.

Given the measurement sources available in the attitude estimation problem, the complementary

Figure 5. AggieVTOL architecture

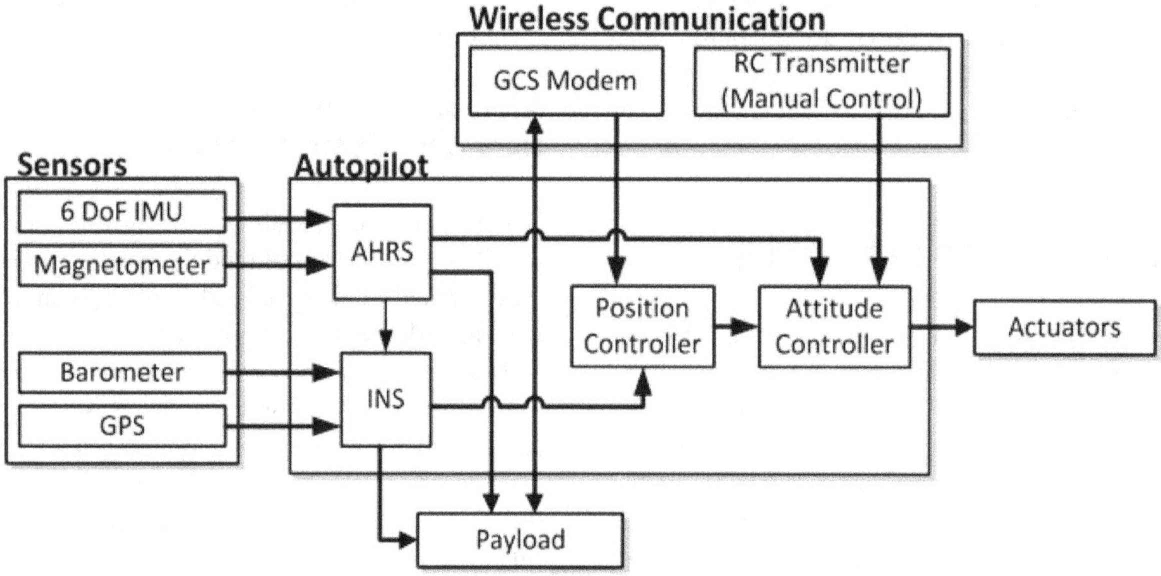

filter is quite adequate. From the IMU, the set of acceleration sensors can be used to find the gravity vector. Calculating the roll and pitch from the gravity vector requires only simple trigonometry. The gyroscopes that measure the angular rate on the roll and pitch axis need only be integrated over time to also calculate the roll and pitch angles. However, both of these sources of measurements are corrupted by noise. The complementary filter can be designed to low-pass the accelerometer data and high pass the integrated gyroscope data. The block diagram of this complementary filter implementation can be found in Figure 6.

However, this implementation has some drawbacks. The primary assumption that the accelerometers can be used to calculate the gravity vector requires that the vehicle is stationary. In a VTOL vehicle, this assumption can be met by the prior assumption that the vehicle is to be a remote sensing platform and thus will only be moving slowly and will do more hovering in place than aggressive maneuvering. The complementary filter also does not consider the statistics of the noise on the measurement signals (Higgins, 1975) nor does it take into account the drift error associated with gyroscopes. In situations where those issues may affect the performance, other solutions may be used. On the other hand, nonlinear complementary filters have also been researched and have demonstrated excellent performance (Hamel & Mahony, 2006).

Other Solutions

In addition to the CF, there are a variety of other solutions. Of the alternative solutions, the Kalman filter and its variants are commonly used. The KF is an adaptive filter that can not only estimate the attitude of the vehicle, but the gyro bias as well. An explanation of the KF can be found in the following section on Integrated Navigation Systems as it is used predominantly there.

As a final note, in this system, the roll, pitch and yaw dynamics are all assumed to be decoupled from each other and thus separate filters exist for each axis. However, attitude calculations can also be accomplished by representing the attitude of a vehicle in a quaternion and using the KF or EKF on the quaternion (Kingston & Beard, 2004).

Figure 6. Complementary filter for Φ

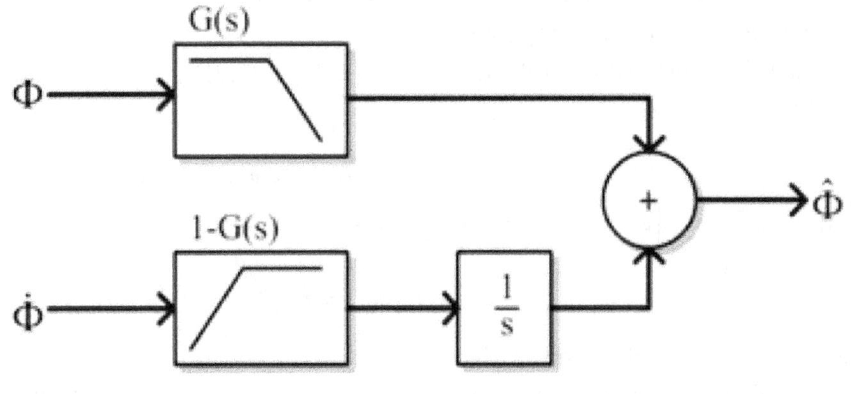

Integrated Navigation System Module

While inertial measurements can be used to calculate position, the accuracy degrades with time from the errors associated with the sensors. However, if the data was fused together with a separate data source, the position estimate errors can be significantly reduced over time. In essence, this problem can be viewed as the estimation of the states of a stochastic process from noisy measurement sources. While many solutions exist, the most common solution is the Kalman filter (Aggarwal et al., 2010). This robust filter for state estimation is effective and computationally efficient, making it an ideal solution for airborne INSs.

Kalman Filter

While both inertial measurements and GPS data can be used for positioning, they suffer from a number of drawbacks. IMU-based navigation suffers from a high degree of drift over time, especially in a mobile platform. GPS measurements may not suffer from drift errors; however, they are subject to a low resolution and a low update frequency. However, combined together they can form seamless position updates.

There are two common forms of integration: Loosely coupled and tightly coupled (Aggarwal et al., 2010). In a loosely coupled mode of integration, the GPS data is used as an update for the navigation KF, which is often decoupled along each axis, under the assumption that the vehicle can travel in any direction independently. While this may not be completely true for all UAVs, the advantage of three small KF over one large one helps reduce the airborne computational load. On the other hand, a tightly coupled mode of integration often includes the aerodynamic model of the vehicle as well as a GPS model for additional information regarding the positional errors in the GPS. Either implementation of integration is utilized effectively, though for low cost systems where computational load is a concern, the loosely coupled form is often more commonly seen (Aggarwal et al., 2010).

A short introduction to the Kalman Filter and a simple implementation of a loosely coupled mode of integration follows. While a detailed derivation of the Kalman Filter and its other variants are beyond the scope of this chapter, the reader is encouraged to browse the Additional Reading Section for additional information.

Given a single axis, the dynamic model can be expressed as

$$\mathbf{x}_t = \mathbf{F}\mathbf{x}_{t-1} + \mathbf{B}u_t + \mathbf{Q} \tag{17}$$

where $\mathbf{x}_t$ represents the state $\mathbf{x}$ at time t, $\mathbf{F}$ is the state matrix, $\mathbf{B}$ is the matrix for the system input u_t and $\mathbf{Q}$ is the noise associated with the input. Applying this to the vehicle estimation problem, the equation can be expressed as below

$$\begin{pmatrix} x_t \\ \dot{x}_t \end{pmatrix} = \begin{pmatrix} 1 & \Delta t \\ 0 & 1 \end{pmatrix} \begin{pmatrix} x_{t-1} \\ \dot{x}_{t-1} \end{pmatrix} + \begin{pmatrix} \frac{\Delta t^2}{2} \\ \Delta t \end{pmatrix} \ddot{x}_{t-1} \tag{18}$$

where Δt is the step size or sampling period, the position x can be found through the integration of the system input, the acceleration, $\ddot{x}$. Whereas the dynamic model expresses the change of state as a function of the input, the measurement model expresses the relationship between the states and the available measurement in the form

$$\mathbf{z}_t = \mathbf{H}\mathbf{x}_t + \mathbf{R} \tag{19}$$

where $\mathbf{z}_t$ is the measurement at time t, $\mathbf{H}$ is the measurement matrix, and $\mathbf{R}$ is the noise associated with the measurement.

The KF algorithm can be described in two stages: A Prediction Stage and an Update Stage. In the prediction stage, the filter predicts the next state, $\hat{\mathbf{x}}_{t|t-1}$ and the a posteriori error covariance matrix, $\mathbf{P}_{t|t-1}$ which represents an estimate of the accuracy of the state estimation,

$$\hat{x}_{t|t-1} = F\hat{x}_{t-1|t-1} + Bu_t \tag{20}$$

$$\mathbf{P}_{t|t-1} = \mathbf{F}\mathbf{P}_{t-1|t-1}\mathbf{F}^T + \mathbf{Q} \tag{21}$$

The update stage provides the blending of the state estimation with the measurement data. This is accomplished through the Kalman gain which gives weight to either the state estimation or the measurement depending on the uncertainty of either source. The Kalman gain, $\mathbf{K}$, can be expressed by

$$\mathbf{K} = \mathbf{P}_{t|t-1}\mathbf{H}^T(\mathbf{H}\mathbf{P}_{t|t-1}\mathbf{H}^T + \mathbf{R})^{-1} \tag{22}$$

A more accurate estimate of the state $\hat{\mathbf{x}}_{t|t}$ can now be calculated

$$\hat{x}_{t|t} = \hat{x}_{t|t-1} + K(z_t - H\hat{x}_{t|t-1}) \tag{23}$$

where $\mathbf{z}_t$ is the position measurement from the GPS. And finally, the error covariance $\mathbf{P}$ can be calculated and be available for use in the next prediction stage of the KF as

$$\mathbf{P}_{t|t} = (\mathbf{I} - \mathbf{K}\mathbf{H})\mathbf{P}_{t|t-1} \tag{24}$$

Other INS Solutions

The linear KF has been demonstrated to be a feasible solution, however many groups have shown much improved results using other variants of the Kalman filter such as the Extended Kalman Filter (Li, Wang, Rizos, Mumford, & Ding, 2006) or the Unscented Kalman Filter (Zhang, Gu, Milios, & Huynh, 2006). If another sensor is available for altitude positioning, such as a barometer, it can be used as part of the measurement update.

Control System Module

The control system is the most common topic researched on Quadrotor vehicles and for good reason; since they are inherently unstable vehicles they require an active stabilization system to stay aloft. However, all research on these platforms agree on one particular statement: A simple PID controller is adequate for attitude control (Bouabdallah et al., 2004). Position control as well can be adequately accomplished with a simple PID controller (Hoffmann et al., 2007). However, as expected, control algorithms that are based on a system model have shown superior performance. A large variety of control algorithms have been tried for both attitude and position control: Double Lead Compensator (Pounds et al., 2002), Backstepping (Altug et al., 2002), LQR (Bouabdallah et al., 2004), Sliding Mode (Hoffmann et al., 2005), PD (Tayebi & McGilvray, 2006), Integral Backstepping (Bouabdallah & Siegwart, 2007a), and PID with aerodynamic compensation (Huang et al., 2009) to name a few.

PID Control

The computationally simple and intuitive proportional-integral-derivative (PID) controller is the most commonly used control algorithm for both attitude and position controllers. While it is well documented, a quick overview of PID control follows.

In a PID controller, the control action is generated by a summation of a set of three feedback components: proportional error, integrated error, and derivative error (Levine, 2010). The control law can be described as

$$u(t) = K_p e(t) + K_i \int_0^t e(s)ds + K_d \frac{de(t)}{dt} \tag{25}$$

where $u(t)$ is the control action, K_p, K_i, K_d are the tunable gains for the controller and $e(t)$ is the error of the system given by $e(t) = y_{sp}(t)-y(t)$ where $y_{sp}(t)$ is the set point command. The most difficult aspect of a PID controller implementation is the selection of the gains. The proportional part of the PID is a simple feedback. The integral part is advantageous for removal of steady-state errors. The derivative part anticipates action and can be useful in improving the response time of a system.

For attitude control, because each axis can be rotated independently, each Euler angle can be controlled with its own PID controller and gains. While a Quadrotor is designed to be symmetric, often the weight distribution is not. A rectangular battery is a common culprit though minor construction flaws also may play a role. Allowing for tuning control gains individually on each attitude controller greatly improves flight performance. The final control action is used to dictate the difference in thrust between the pairs of rotors, or in the case of yawing motion, the difference between the two sets of rotors.

While many control systems may opt to use a single controller for position and attitude, a dual control loop may be implemented (See Figure 7). This control scheme allows for the attitude controller to be separate from the position controller. During the development phase, it is often advisable to implement critical components sequentially rather than all at once, ensuring that one system is fully satisfactory before moving to the next system. In the simplest case, the position control may also use a simple PID controller. As with the attitude controller, the position controller may be decoupled along each axis of movement. The horizontal position dynamics differ greatly from the altitude dynamics and should be developed individually. Furthermore, horizontal control gains may differ individually based on weight distribution and other imperfections.

Other Control Systems

While attitude and position control systems can provide full control of the Quadrotor, many systems also incorporate separate control systems for autonomous take-offs and landings. These systems often incorporate visual systems (Herisse et al., 2008) or sonar systems (Iwakura et al., 2011). Multi-UAV systems may incorporate additional controllers to ensure optimal swarm and collision avoidance behavior (Nonami et al, 2010).

Figure 7. Dual control loop implementation

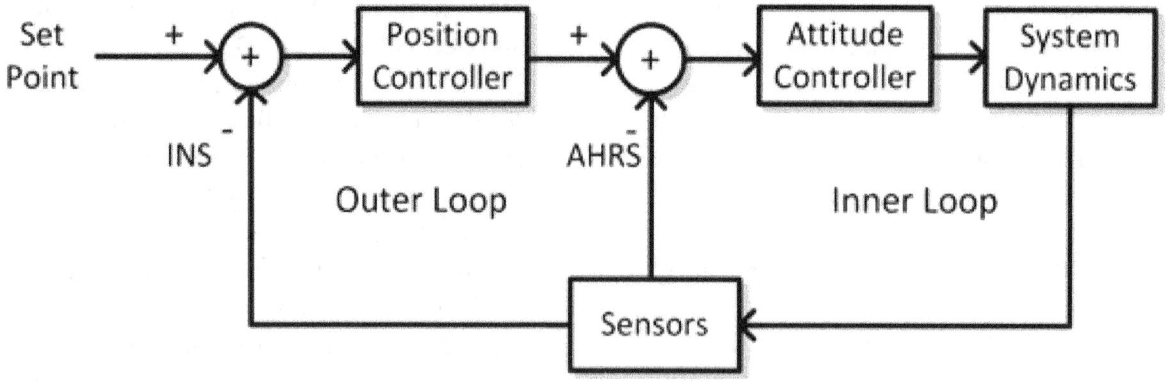

3. AGGIEVTOL PROTOTYPE PLATFORM

The AggieVTOL prototype platform is a constantly evolving platform, undergoing iterations of testing and performance evaluation. The basis of this prototyping platform is the open-source hardware and software UAV system known as Paparazzi ("Paparazzi," n.d.). The modular architecture has allowed for the platform to evolve. In this section, the AggieVTOL platform is described in detail (Figure 8).

Platform Overview

The AggieVTOL is comprised of several subsystems: Paparazzi Autopilot, Sensor System, Actuator system, Power System, Communication System, and Payload System. The distribution of weight and component costs can be found in Table 2. While many Quadrotor platforms are under 1 *kg*, AggieVTOL performs comfortably even at a weight of 2.3 *kg*, ensuring its performance with a payload of up to 500 *g*. It is capable of altitudes above 200 *m*, travels at a nominal speed of 5 *m/s* and has a flight time of 20 minutes. Safety beacon LEDs on the arms were added to aid in visually determining the heading of the vehicle at large distances and at low light situations.

Paparazzi Autopilot

The Paparazzi hardware main board for rotorcraft is known as the Lost Illusions Serendipitous Autopilot (LISA). This main board is driven by a STM32 processor, a 32-bit ARM Cortex MCU that operates at 72MHz. While a separate IMU is necessary, several interfaces are available including SPI, UART and I^2C for communications. Onboard features include power management, two barometric sensors and a connector for a separate GUMSTIX Overo single board computer which features a 600 MHz OMAP3530 processor and USB 2.0 connectivity. In the close up image in Figure 9, the main board can be identified by the

Table 2. Weight and cost distribution

Component	Weight	Cost
Autopilot	237 *g*	$350
Battery	640 *g*	$325
Frame	440 *g*	$20
Ground Communication	108 *g*	$225
GPS	76 g	$90
IMU	52 *g*	$550
Motor/Prop x 4	352 *g*	$360
Payload Camera	275 *g*	$210
Payload Control/Communication	165 *g*	$130
Total	**2345 *g***	**$2260**

Figure 8. AggieVTOL in action

two barometric sensors and USB port. The two boards above are the IMU and GPS boards, while the boards below are ESCs.

Sensor System

On a platform such as AggieVTOL, the selection of sensors is a careful balance of performance and cost. While none of the selected sensors are beyond commercial grade quality, they have suited the platform well. The sensor array on the AggieVTOL platform consists of 3 ADXL320 single-axis accelerometers, 2 ADXRS610 dual-axis gyroscopes, a MS2100 tri-axis magnetometer, an MPXx6115 absolute pressure sensor and an uBlox LEA-5h GPS module. However, with the use of low cost inertial measurement sensors, it is important that they are calibrated properly.

Actuator System

The actuator system consists of three parts: Propeller blade, Motor, and ESC. Without relying on in-house production, the selection of the propeller blade was limited. Currently the AggieVTOL vehicle uses MaxxProducts APC12038 Slow-Fly counter-rotating pair propellers, though other sizes have been explored. High efficiency, high torque motors are ideal for the AggieVTOL platform and as such the motors selected are AXI 2217-20. With a rated efficiency >75% and a current draw of up to 18 A, these motors are well suited for a Quadrotor platform. The selection of a good ESC depends on the maximum current capacity and the update rate. While standard hobby aircraft ESCs are usable, an ESC specifically designed for Quadrotors is superior. The Mikrokopter project ("Mikrokopter," n.d.) ESCs have a maximum current capacity of 35 A and use an I^2C interface reducing the interconnectivity of the autopilot to motor control a triviality.

Figure 9. Close up of the LISA board with IMU, GPS and ESC

Power System

The power system is a critical component, however in terms of design considerations, there are not a significant number of concerns. An evaluation of the power consumption of the UAV shows that the motors consume the greatest amount of power, the power consumption of the rest of the system can be considered negligible (See Table 3). Therefore the battery should be able to supply a sufficient amount of current to the motors for as long as possible. The AggieVTOL platforms typically use 8 *Ahr* batteries with a discharge rate of 20 *C* and weigh 640 *g*, providing AggieVTOL with up to 20 minutes of flight time.

Communication System

A reliable communication system is required for any UAV, and the AggieVTOL system is no different. Digi Xtend 900 *Mhz* serial modems provide the major communications between the UAV and the GCS, providing reliable transmission to up to 64 *km* in an unobstructed outdoor environment. Manual control of the Quadrotor can be accomplished through the use of a Radio Control (RC) system. This system also provides a convenient kill-switch for safety considerations. A Spektrum 2.4 *GHz* RC system is used on AggieVTOL due to its simple and robust performance.

Table 3. Power distribution of AggieVTOL

Component	Power
Actuator System (at nominal)	295.4 *W*
Autopilot	0.62 *W*
Camera Payload	3.2 *W*
Communication (900Mhz, Wifi, RC receiver)	7.25 *W*
IMU	0.21 *W*
Safety Beacon LEDs	4.0 *W*
Total	**310.7 *W***

AggieCAP Payload System

One of the payload systems that AggieVTOL is capable of carrying is a sophisticated imaging system called AggieCAP for high resolution imagery (Figure 10). This system uses a GUMSTIX Overo, a Cannon SX110 camera and a 5 *GHz* Ubuiti Networks Bullet M WiFi Radio. AggieCAP receives position and orientation information from the autopilot and remotely captures imagery using the camera. Both the image and the position and orientation information are streamed down to the Ground Station through the 5 *GHz* WiFi link. Images from the payload system (Figure 11) can be georectified and mosaiced to form large high resolution mapping

4. RESULTS

The testing phase of VTOL development can be a long process. While the algorithms presented are easy to implement, the tuning of the gains poses a challenge that should not be underestimated. As with any mobile robotic platform, safety is a major concern. Special caution should be exercised, especially during control system tuning as the vehicle may perform unexpectedly.

Attitude Tracking

Attitude tracking is the most important aspect of a VTOL vehicle. Since the VTOL requires active stabilization to remain aloft, this phase of testing is best done on a tethered test bench. Uncontrolled oscillations involving sharp propeller blades spin-

Figure 10. Sample imagery from AggieVTOL flying over a field

Figure 11. Sample imagery from AggieVTOL from 100 m altitude

Figure 12. Simple tethered test bench

Figure 13. Attitude tracking performances during flight

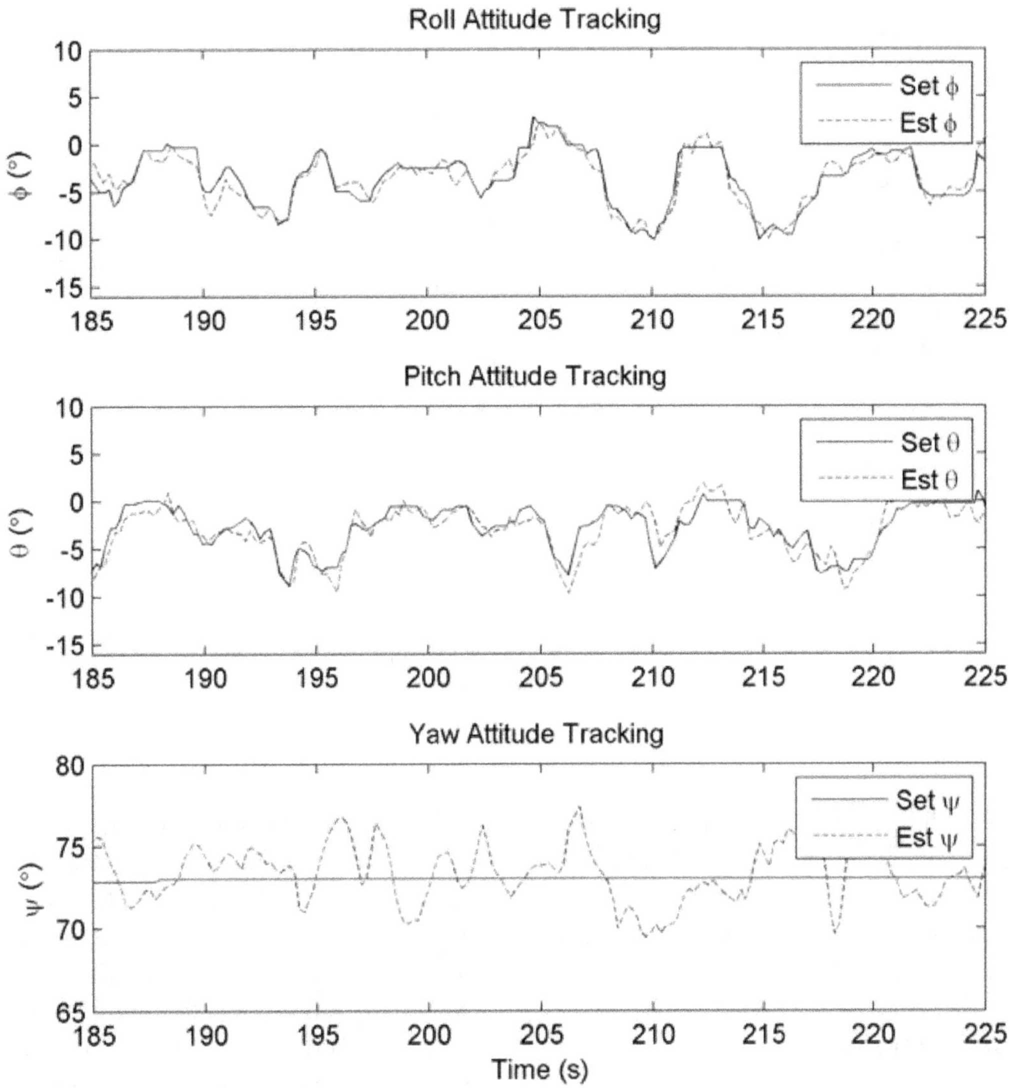

ning at upwards of 7000 RPMs are dangerous. In a system, such as AggieVTOL, where attitude tracking is accomplished by a PID controller, tuning the gains can be difficult. Since each axis of rotation is decoupled, the roll, pitch and yaw controller gains can be tested independently of each other on a specially design test bed. The desired attitude tracking response should have no steady state error, be free of overshoot and oscillations, with a quick response being a tertiary goal. During the AggieVTOL testing phase, a simple test bed was constructed of PVC pipes to hold two arms of the vehicle in place (See Figure 12). This test bed limits the vehicle movement to only a single rotation axis at a time. This safely allows for online gain tuning without risk of damage to the researchers or the vehicle.

The roll and pitch tracking plots in Figure 13 demonstrate the effectiveness of a well tuned PID controller for attitude tracking during an outdoor flight with a light breeze (sustained wind speed of 5-7 knots). Roll and pitch attitude tracking

Figure 14. Altitude tracking performance during flight

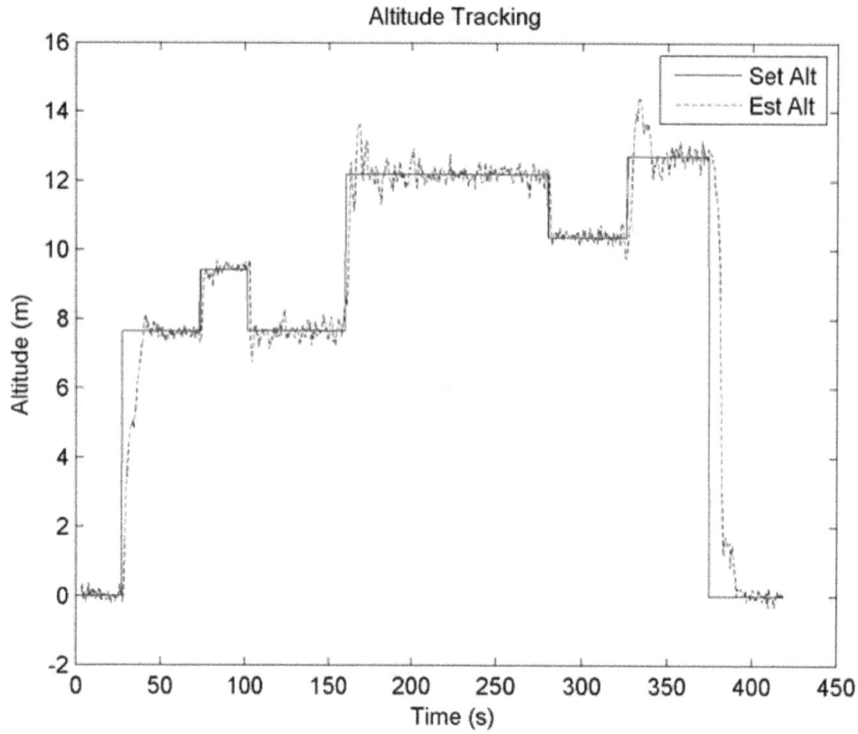

controllers exhibited an absolute tracking error of ±2° and a variance of roughly 1°. The yaw tracking could be improved, though in this untuned controller, it exhibited an absolute error of ±5° and a variance of 3°.

Position Tracking

Once attitude tracking is satisfactory and manually controlled flight is achieved, the position control systems can be implemented. The greatest difficulty of implementing a good positioning controller lies not in the control scheme but rather in ensuring the INS is accurate. Once this is achieved, altitude and positioning may be controlled accurately.

In Figure 14, a sample of the step response of the altitude controller performance is shown. The performance of a PID controller, even one that has not been rigorously tuned is adequate. Overshoot is a common problem, but the steady state response can be easily controlled. In tuning this controller, the primary concern is in the stability of the controller, and a low steady state error. Response time, while useful, is not critical to system performance. Wind disturbances are particularly problematic for altitude control performance. Gusts of wind may cause temporary shifts in barometric pressure sensors (causing incorrect altitude readings) and will also change the air velocities above the propeller that affect the actuator efficiency.

Position control, likewise, has a similar performance though atmospheric disturbances are common in outdoor environments and can degrade performance. Figure 15 depicts AggieVTOL traveling in a loop from waypoint to waypoints. This performance did suffer from some overshoot and wind disturbance but otherwise was successful.

While the performance of a simple PID based attitude and position controller has been demon-

Figure 15. Waypoint navigation performance during flight

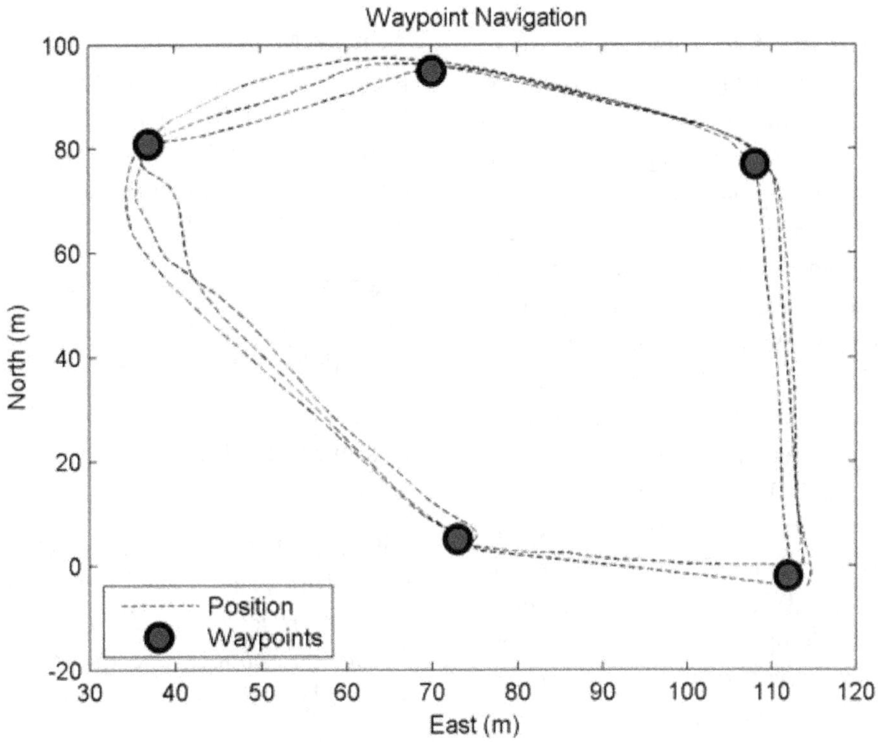

strated to be adequate, more robust controllers are suggested for better performance against wind disturbances. Improvements in attitude and position estimation can be accomplished by more accurate sensors or with more robust filter algorithms. However these improvements may bring the cost of the platform beyond the low cost goal.

5. FUTURE RESEARCH DIRECTIONS

The start of the 2000's sparked a new wave of research into UAVs and the Quadrotor platform, but as the decade has come to a close, the rate of progress has slowed down. Rather than a call for alarm, this is a sign of the maturation of the field. The next frontier of publications will not be on control systems or the technical aspects of VTOL implementations, but rather on applications. As a mobile remote sensing platform, it is capable of limitless potential. Applications such as personal remote sensing, cooperative control, search and rescue operations, airborne robotic manipulators and resource monitoring are just starting to be developed. Currently many systems are simplistic or are based on toy platforms with limited capability, but as the field matures further, the multi-rotor UAV platform will prove to be a valuable tool in "*remote sensing*" as well as "*remote actuation*" applications. While applications will be a major front for future research directions, much research will still be needed. The Quadrotor platform performs well in limited environment situations, but soon applications will demand a more robust performance. Atmosphere disturbance rejection, improved flight endurance, increased payload capacity, fault tolerant flight controls and intelligent collision avoidance are topics that are ripe for future research endeavors.

6. CONCLUSION

The development of a fully autonomous VTOL platform for remote sensing applications can be a difficult process without a proper prototyping platform. While we mature the AggieVTOL platform, we figured out in our initial work how to ensure a modular system towards a strong experimental/prototyping platform for components and algorithms to be evaluated without the need for repeated system overhaul. Based on the platform prototyping efforts described within this chapter, the real-world mission ready AggieVTOL personal remote sensing system can be expected in near future (Stark et al. 2012). As a final note, it is a pleasure to update that AggieVTOL was sent to the 9th AUVSI (Association of Unmanned Vehicle Systems International) SUAS (Student Unmanned Aerial Systems) Competition in June 2011 and AggieVTOL is the ever first VTOL UAV that can complete fully autonomous take-off and waypoint navigation in the history of this competition.

ACKNOWLEDGMENT

We sincerely thank our research sponsors: Utah Water Research Laboratory (UWRL) MLF Seed Grants (2006-2011) and National Science Foundation grant #0851709 (NSF REU: Research Experience for Undergraduates 2006-2012). We also thank Professor Raymond L. Cartee, Research Farms Director of Utah State University for providing the 600 acre research farm at Cache Junction, Utah as our flight test field. Last but not least, we thank the Paparazzi community of which we feel proud of being a part!

REFERENCES

Achtelik, M., Bachrach, A., He, R., Prentice, S., & Roy, N. (2009). Stereo vision and laser odometry for autonomous helicopters in GPS-denied indoor environments. In *Proceedings of SPIE Conference on Unmanned Systems Technology XI*, Orlando, FL, USA. doi:10.1117/12.819082

Aggarwal, P., Syed, Z., Noureldin, A., & El-Sheimy, N. (2010). *MEMS-based integrated navigation*. Artech House.

AggieAir Flying Circus. (2011, March 3). *Home*. Retrieved March 25, 2011, from http://aggieair.usu.edu/

Altug, E., Ostrowski, J. P., & Mahony, R. (2002). Control of a quadrotor helicopter using visual feedback. In *Proceedings of IEEE International Conference on Robotics and Automation 2002,* (vol. 1, 72–77). doi:10.1109/ROBOT.2003.1242264

Arducopter. (n.d.). *Project hosting on Google Code*. Retrieved March 11, 2011, from http://code.google.com/p/arducopter/

Bouabdallah, S., Becker, M., & Siegwart, R. (2007). Autonomous miniature flying robots: Coming soon! Research, development, and results. *IEEE Robotics & Automation Magazine*, *14*(3), 88–98. doi:10.1109/MRA.2007.901323

Bouabdallah, S., Murrieri, P., & Siegwart, R. (2004). Design and control of an indoor micro quadrotor. In *Proceedings of IEEE International Conference on Robotics and Automation, 2004,* (vol. 5, 4393-4398). doi:10.1109/ROBOT.2004.1302409

Bouabdallah, S., Noth, A., & Siegwart, R. (2004). PID vs LQ control techniques applied to an indoor micro quadrotor. In *Proceedings of IEEE/RSJ International Conference on Intelligent Robots and Systems 2004*, (vol. 3, pp. 2451-2456). doi:10.1109/IROS.2004.1389776

Bouabdallah, S., & Siegwart, R. (2005). Backstepping and sliding-mode techniques applied to an indoor micro quadrotor. In *Proceedings of IEEE International Conference on Robotics and Automation, 2005*, 2247–2252. doi:10.1109/ROBOT.2005.1570447

Bouabdallah, S., & Siegwart, R. (2007a). Full control of a quadrotor. In *Proceedings of IEEE/RSJ International Conference on Intelligent Robots and Systems, 2007,* (pp. 153-158). doi:10.1109/IROS.2007.4399042

Bouabdallah, S., & Siegwart, R. (2007b). Design and control of a miniature quadrotor. In Valavanis, K. (Ed.), *Design and control of a miniature quadrotor*. Springer.

Chao, H. (2010). *Cooperative remote sensing and actuation using networked unmanned vehicles*. Doctoral Dissertation, Utah State University. Retrieved from http://digitalcommons. usu.edu/ctd/597

Chao, H., Baumann, M., Jensen, A., Chen, Y. Q., Cao, Y., Ren, W., & McKee, M. (2008). Band-reconfigurable multi-UAV-based cooperative remote sensing for real-time water management and distributed irrigation control. In *Proceedings of International Federation of Automatic Control World Congress*. Seoul, Korea.

Chao, H., Cao, Y., & Chen, Y. Q. (2010). Autopilots for small unmanned aerial vehicles: A survey. *International Journal of Control Automation and Systems*, *8*(1), 36–44. doi:10.1007/s12555-010-0105-z

Chao, H., Coopmans, C., Long, D., & Chen, Y. Q. (2010). A comparative evaluation of low-cost IMUs for unmanned autonomous systems. In *Proceedings of IEEE Conference on Multisensor Fusion and Integration for Intelligent Systems, 2010*, 211–216. doi:.doi:10.1109/MFI.2010.5604460

Cloud Cap Technology. (n.d.). *Piccolo II Autopilot for unmanned systems*. Retrieved March 11, 2011, from http://www.cloudcaptech.com/piccolo_II.shtm

CSOIS - Center for Self-Organizing Intelligent Systems. (n.d.). Retrieved February 21, 2011, from http://www.csois.usu.edu/

Department of Defense. (2005). *Unmanned aircraft systems (UAS) roadmap, 2005-2030*. Washington, DC: Office of the Secretary of Defense.

Draganfly.com. (n.d.). *Industrial aerial video systems & UAVs*. Retrieved February 21, 2011, from http://www.draganfly.com/

Drone, A. R. com USA. (n.d.). *Parrot Wi-Fi quadricopter- Augmented reality games on iPhone, iPod touch and iPad.* Retrieved February 21, 2011, from http://ardrone.parrot.com/parrot-ar-drone/usa/

Grzonka, S., Grisetti, G., & Burgard, W. (2009). Towards a navigation system for autonomous indoor flying. In *Proceedings of IEEE International Conference on Robotics and Automation, 2009*, 2878–2883. doi:.doi:10.1109/ROBOT.2009.5152446

Hamel, T., & Mahony, R. (2006). Attitude estimation on SO[3] based on direct inertial measurements. In *Proceedings of IEEE International Conference on Robotics and Automation, 2006*, 2170–2175. doi:.doi:10.1109/ROBOT.2006.1642025

He, R., Prentice, S., & Roy, N. (2008). Planning in information space for a quadrotor helicopter in a GPS-denied environment. In *Proceedings of IEEE International Conference on Robotics and Automation, 2008*, 1814–1820. doi:.doi:10.1109/ROBOT.2008.4543471

Herisse, B., Oustrieres, S., Hamel, T., Mahony, R., & Russotto, F. (2010). A general optical flow based terrain-following strategy for a VTOL UAV using multiple views. In *Proceedings of IEEE International Conference on Robotics and Automation, 2010*, 3341–3348. doi:.doi:10.1109/ROBOT.2010.5509536

Herisse, B., Russotto, F. X., Hamel, T., & Mahony, R. (2008). Hovering flight and vertical landing control of a VTOL unmanned aerial vehicle using optical flow. In *Proceedings of IEEE/RSJ International Conference on Intelligent Robots and Systems, 2008,* (pp. 801–806.) doi:10.1109/IROS.2008.4650731

Higgins, W. T. (1975). A comparison of complementary and Kalman filtering. *IEEE Transactions on Aerospace and Electronic Systems, 3*, 321–325. doi:10.1109/TAES.1975.308081

Hoffmann, G., Huang, H., Waslander, S. L., & Tomlin, C. J. (2007). Quadrotor helicopter flight dynamics and control: Theory and experiment. In *Proceedings of American Institute of Aeronautics and Astronautics Guidance, Navigation, and Control Conference 2007*. Hilton Head, South Carolina.

Hoffmann, G., Rajnarayan, D. G., Waslander, S. L., Dostal, D., Jang, J. S., & Tomlin, C. J. (2005). The Stanford testbed of autonomous rotorcraft for multi agent control (STARMAC). In *Proceedings of Digital Avionics Systems Conference, 2004.* Salt Lake City, Utah.

Huang, H., Hoffmann, G. M., Waslander, S. L., & Tomlin, C. J. (2009). Aerodynamics and control of autonomous quadrotor helicopters in aggressive maneuvering. In *Proceedings of IEEE International Conference on Robotics and Automation, 2009*, 3277–3282. doi:.doi:10.1109/ROBOT.2009.5152561

Iwakura, D., Wang, W., Nonami, K., & Haley, M. (2011). Movable range-finding sensor system and precise automated landing of quad-rotor MAV. *Journal of System Design and Dynamics*, *5*(1), 17–29. doi:10.1299/jsdd.5.17

Jensen, A. M., Baumann, M., & Chen, Y. Q. (2009). Low-cost multispectral aerial imaging using autonomous runway-free small flying wing vehicles. In *Proceedings of IEEE International Geoscience and Remote Sensing Symposium, 2008,* (vol. 5, pp. 506-509). doi:10.1109/IGARSS.2008.4780140

Kingston, D. B., & Beard, R. W. (2004). Real-time attitude and position estimation for small UAVs using low-cost sensors. In *Proceedings of American Institute of Aeronautics and Astronautics 3rd Unmanned Unlimited Systems Conference and Workshop*. Chicago, Illinois.

Konvalin, C. (2009). Compensating for tilt, hard-iron, and soft-iron effects. *Sensors Magazine*. Retrieved February 22, 2011, from http://www.sensorsmag.com/sensors/motion-velocity-displacement/compensating-tilt-hard-iron-and-soft-iron-effects-6475

Leishman, J. G. (2002). The breguet-richet quadrotor helicopter of 1907. *Vertiflite*, *47*(3), 58–60.

Levine, W. S. (2010). *The control handbook: Control system fundamentals* (2nd ed.). Taylor & Francis US.

Li, Y., Wang, J., Rizos, C., Mumford, P., & Ding, W. (2006). Low-cost tightly coupled GPS/INS integration based on a nonlinear Kalman filtering design. In *Proceedings of ION National Technical Meeting*, (pp. 18–20).

Luo, Y., Di, L., Han, J., Chao, H., & Chen, Y. Q. (2010). VTOL UAV altitude flight control using fractional order controllers. In *Proceedings of 4th IFAC Workshop Fractional Differentiation and its Applications*. Badajoz, Spain.

Mahony, R., Cha, S., & Hamel, T. (2006). A coupled estimation and control analysis for attitude stabilisation of mini aerial vehicles. In *Proceedings of the Australasian Conference on Robotics and Automation*. Auckland, New Zealand.

Mellinger, D., Michael, N., & Kumar, V. (2010). Trajectory generation and control for precise aggressive maneuvers with quadrotors. In *Proceedings of International Symposium on Experimental Robotics 2010*. Delhi, India.

MicroPilot. (n.d.). *World leader in small UAV autopilots*. Retrieved March 11, 2011, from http://www.micropilot.com/

MikroKopter. (n.d.). *Wiki*. Retrieved March 15, 2011, from http://www.mikrokopter.de/

Nonami, K., Kendoul, F., Suzuki, S., Wang, W., & Nakazami, D. (2010). *Autonomous flying robots*. Springer. doi:10.1007/978-4-431-53856-1

Paparazzi. (n.d.). *Main page*. Retrieved March 11, 2011, from http://paparazzi.enac.fr/wiki/Main_Page

Pounds, P., Hamel, T., & Mahony, R. (2007). Attitude control of rigid body dynamics from biased IMU measurements. In *Proceedings of IEEE Conference on Decision and Control, 2007*, 4620–4625. doi:.doi:10.1109/CDC.2007.4434173

Pounds, P., Mahony, R., & Corke, P. (2006). Modelling and control of a quad-rotor robot. In *Proceedings of Australasian Conference on Robotics and Automation*. Auckland, New Zealand.

Pounds, P., Mahony, R., & Corke, P. (2009). Design of a static thruster for microair vehicle rotorcraft. *Journal of Aerospace Engineering*, *22*(1), 85–94. doi:10.1061/(ASCE)0893-1321(2009)22:1(85)

Pounds, P., Mahony, R., & Corke, P. (2010). Modelling and control of a large quadrotor robot. *Control Engineering Practice*, *18*(7), 691–699. doi:10.1016/j.conengprac.2010.02.008

Pounds, P., Mahony, R., Gresham, J., Corke, P., & Roberts, J. (2004). *Towards dynamically-favourable quad-rotor aerial robots*. In Australasian Conference on Robotics and Automation. Canberra, Australia.

Pounds, P., Mahony, R., Hynes, P., & Roberts, J. F. (2002). Design of a four-rotor aerial robot. In *Proceedings of Australasian Conference on Robotics and Automation*. Auckland, New Zealand.

Procerus Technologies. (n.d.). *Fly light with world's smallest UAV autopilot*. Retrieved March 11, 2011, from http://www.procerusuav.com/index.php

Ross, P. (2011). Top 11 technologies of the decade. *Spectrum, IEEE*, *48*(1), 27-63. doi:10.1109/MSPEC.2011.5676379

Silva, C. W. D. (2005). *Mechatronics: An integrated approach*. CRC Press.

Stark, B., Jensen, A., Chen, Y. Q., & McKee, M. (2012) A personal remote sensing system with VTOL UAV for natural resource management applications. (submitted) In *Proceedings of IEEE Aerospace Conference 2012*. Big Sky, MT, USA.

Tanigawa, M., Luinge, H., Schipper, L., & Slycke, P. (2008). Drift-free dynamic height sensor using MEMS IMU aided by MEMS pressure sensor. In *Proceedings of 5th Workshop on Positioning, Navigation and Communication, 2008*, (pp. 191–196).

Tayebi, A., & McGilvray, S. (2004). Attitude stabilization of a four-rotor aerial robot. In *Proceedings of IEEE Conference on Decision and Control, 2004*, (vol. 2, pp. 1216–1221). doi:10.1109/CDC.2004.1430207

Tayebi, A., & McGilvray, S. (2006). Attitude stabilization of a VTOL quadrotor aircraft. *IEEE Transactions on Control Systems Technology*, *14*(3), 562–571. doi:10.1109/TCST.2006.872519

Teal Group. (2011). *Teal Group predicts worldwide UAV market will total over $80 billion in its just released 2010 UAV market profile and forecast*. Retrieved February 21, 2011, from www.tealgroup.com

Tournier, G. P., Valenti, M., How, J. P., & Feron, E. (2006). Estimation and control of a quadrotor vehicle using monocular vision and Moire patterns. In *Proceedings of American Institute of Aeronautics and Astronautics Guidance, Navigation and Control Conference and Exhibit,* (pp. 21–24). Keystone, Colorado.

Tricaud, C. (2010). *Optimal sensing & actuation policies for networked mobile agents in a class of cyber-physical systems*. Doctoral Dissertation, Utah State University. Retrieved from http://www.digitalcommons.usu.edu/ctd/673

Valavanis, K. (2007). *Advances in unmanned aerial vehicles: State of the art and the road to autonomy*. Springer. doi:10.1007/978-1-4020-6114-1

Zaman, B., McKee, M., & Jenson, A. (2011) Use of high-resolution multispectral imagery acquired with an autonomous unmanned aerial vehicle to quantify the spread of an invasive wetland species (to appear) In *Proceedings of IEEE International Geoscience and Remote Sensing Symposium 2011*.

Zhang, P., Gu, J., Milios, E. E., & Huynh, P. (2006). Navigation with IMU/GPS/digital compass with unscented Kalman filter. In *Proceedings of IEEE International Conference Mechatronics and Automation, 2005,* (vol. 3, pp. 1497–1502). doi: 10.1109/ICMA.2005.1626777

ADDITIONAL READING

Aggarwal, P., Syed, Z., Niu, X., & El-Sheimy, N. (2008). A Standard Testing and Calibration Procedure for Low Cost MEMS Inertial Sensors and Units. *Journal of Navigation, 61*, 323–336. doi:10.1017/S0373463307004560

Brown, A., & Lu, Y. (2004). Performance Test Results of an Integrated GPS/MEMS Inertial Navigation Package. In *Proceedings of Institute of Navigation Global Navigation Satellite System Conference, 200*, 1251–1256.

Conte, G., & Doherty, P. (2008). An Integrated UAV Navigation System Based on Aerial Image Matching. In *Proceedings of IEEE Aerospace Conference, 2008*. doi:.doi:10.1109/AERO.2008.4526556

Doherty, P., Haslum, F., Heintz, F., Merz, T., Nyblom, P., Persson, T., & Wingman, B. (2004) A Distributed Architecture for Autonomous Unmanned Aerial Vehicle Exploration. In *Proceedings of the 7th International Symposium on Distributed Autonomous Robotic Systems, 2004.* Toulouse, France.

Dorf, R., & Bishop, R. (2008). *Modern Control Systems* (11th ed.). Upper Saddle River, N.J.: Pearson Education, Inc.

Eisenbeiss, H. (2004) A Mini Unmanned Aerial Vehicle (UAV): System Overview and Image Acquisition. In *Proceedings of International Workshop on Processing and Visualization Using High Resolution Imagery*. Pitsanulok, Thailand.

Grewel, M., & Andrews, A. (2008). *Kalman Filtering Theory & Practice using MATLAB* (3rd ed.). Hoboken, N.J.: Wiley-IEEE Press.

Grewel, M., Weill, L., & Andrews, A. (2007). *Global Positioning Systems, Inertial Navigation and Integration* (2nd ed.). Hoboken, N.J.: John Wiley & Sons, Inc.doi:10.1002/0470099720

Guenard, N., Hamel, T., & Mahony, R. (2008). A Practical Visual Servo Control for an Unmanned Aerial Vehicle. *IEEE Transactions on Robotics, 24*(2), 331–340. doi:10.1109/TRO.2008.916666

Gurdan, D., Stumpf, J., Achtelik, M., Doth, K. M., Hirzinger, G., & Rus, D. (2007) Energy-Efficient Autonomous Four-rotor Flying Robot Controlled at 1Khz. In *Proceedings of IEEE International Conference on Robotics and Automation*, 361-366. doi:10.1109/ROBOT.2007.363813

Heintz, F., Rudol, P., & Doherty, P. (2007) From Images to Traffic Behavior - A UAV Tracking and Monitoring Application. In *Proceedings of 20th International Conference on Information Fusion, 2007,* 1-8. doi: 10.1109/ICIF.2007.4408103

Herwitz, S. R., Johnson, L. F., Dunagan, S. E., Higgins, R. G., Sullivan, D. V., & Zheng, J. (2004). Imaging from an Unmanned Aerial Vehicle: Agricultural Surveillance and Decision Support. *Computers and Electronics in Agriculture, 44*(1), 49–61.doi:10.1016/j.compag.2004.02.006

Mokhtari, A., & Benallegue, A. (2004). Dynamic Feedback Controller or Euler Angles and Wind Parameters Estimation for a Quadrotor Unmanned Aerial Vehicle. In *Proceedings of IEEE Conference on Robotics and Automation, 3*, 2359–2366. doi:.doi:10.1109/ROBOT.2004.1307414

Pettersson, P. O., & Doherty, P. (2004) Probabilistic Roadmap Based Path Planning for Autonomous Unmanned Aerial Vehicles. In *Proceedings of the Workshop on Connecting Planning and Theory with Practice at 14th International Conference on Automated Planning and Scheduling*. Whistler, British Columbia, Canada.

Pounds, P., & Mahony, R. (2009). Design principles of large quadrotors for practical applications. In *Proceedings of IEEE International Conference on Robotics and Automation, 2009*, 3265–3270. doi:.doi:10.1109/ROBOT.2009.5152309

Ryan, A., Zennaro, M., Howell, A., Sengupta, R., & Hedrick, J. K. (2004). An Overview of Emerging Results in Cooperative UAV Control. In *Proceedings of IEEE Conference on Decision and Control, 1*, 602–607.

Saripalli, S., Montgomery, J. F., & Sukhatme, G. S. (2003). Visually Guided Landing of an Unmanned Aerial Vehicle. *IEEE Transactions on Robotics and Automation, 19*(3), 371–380.doi:10.1109/TRA.2003.810239

Shakernia, O., Vidal, R., Sharp, C. S., Ma, Y., & Sastry, S. (2002). Multiple View Motion Estimation and Control for Landing an Unmanned Aerial Vehicle. In *Proceedings of IEEE International Conference on Robotics and Automation, 2002*, 2793–2798. doi:.doi:10.1109/ROBOT.2002.1013655

Sinopoli, B., Micheli, M., Donato, G., & Koo, T. J. (2001). Vision Based Navigation for an Unmanned Aerial Vehicle. In *Proceedings of IEEE International Conference on Robotics and Automation, 2*, 1757–1764. doi:.doi:10.1109/ROBOT.2001.932864

Skog, I., & Handel, P. (2006). Calibration of a MEMS inertial measurement unit. In *Proc. XVII IMEKO World Congress,* Rio de Janeiro, Brazil.

Sukkarieh, S., Gibbens, P., Grocholsky, B., Willis, K., & Durrant-Whyte, H. F. (2000). A Low-Cost, Redundant Inertial Measurement Unit for Unmanned Air Vehicles. *The International Journal of Robotics Research, 19*(11), 1089–1103. doi:10.1177/02783640022067995

Titterdon, D., & Weston, J. (2004). *Strapdown Inertial Navigation Technology* (2nd ed.). Bodmin, U.K.: American Institute of Aeronautics and Astronomy (AIAA).

Wan, E., & Van Der Merwe, R. (2000) The Unscented Kalman Filter for Nonlinear Estimation. In *Proceedings of Adaptive Systems for Signal Processing, Communications and Control Symposium*. Lake Louise, AB, Canada.

Wendel, J., Meister, O., Schlaile, C., & Trommer, G. (2006). An Integrated GPS/MEMS-IMU Navigation System for an Autonomous Helicopter. *Aerospace Science and Technology*, *10*(6), 527–533.doi:10.1016/j.ast.2006.04.002

Wu, A. D., Johnson, E. N., & Proctor, A. (2005) Vision-Aided Inertial Navigation for Flight Control. In *Proceedings of American Institute of Aeronautics and Astronautics Guidance, Navigation and Control Conference and Exhibit, 2005.* San Francisco, California.

KEY TERMS AND DEFINITIONS

AHRS: Attitude and Heading Reference System. A navigation system that calculates the Euler angles of an aircraft and is often comprised of an IMU

Autopilot: A system that allows for augmented control or fully autonomous control of an aircraft. This system requires the use of an AHRS and INS for accurate performance.

IMU: Inertial Measurement Unit. An array of inertial measurement sensors, often used for the calculation of attitude estimation.

INS: Integrated Navigation System. An array of sensors used for the calculation of position estimation. Often composed of inertial measurement sensors augmented with other non-inertial sensors such as GPS or sonar systems.

Personal Remote Sensing: Low cost data collection of remote target locations for individual use.

UAV: Unmanned Aerial Vehicle. An aircraft without an onboard human pilot, that can be controlled remotely or autonomously

VTOL aircraft: Vertical Take Off and Landing aircraft. An aircraft, usually with rotary wings, that is able to launch and land in a vertical manner, removing the need for a runway.

APPENDIX: LIST OF ABBREVIATIONS

ABS: Acrylonitrile Butadiene Styrene Plastic
ADC: Analog-to-Digital Converter
Ahr: Ampere hours
AHRS: Attitude and Heading Reference System
CF: Complementary Filter
CoG: Center of Gravity
CPS: Cyber Physical System
CSOIS: Center for Self-Organizing and Intelligent Systems
DC: Direct Current
DGPS: Differential Global Positioning System
DoF: Degree of Freedom
DPS: Distributed Parameter System
EKF: Extended Kalman Filter
EMF: Electromagnetic Field
ESC: Electronic Speed Controller
GCS: Ground Control Station
GPS: Global Positioning System
I^2C: Inter-Integrated Circuit
IGE: In Ground Effect
IMU: Inertial Measurement Unit
INS: Inertial Navigation System
KF: Kalman Filter
LISA: Lost Illusions Serendipitous Autopilot
MEMS: Microelectromechanical Systems
NED: North, East, Down
PC: Polycarbonate
PID: Proportional-Integral-Derivative
PVC: Polyvinyl Chloride Plastic
RC: Radio Control
RF: Radio Frequency
RPM: Revolution per Minute
SLAM: Simultaneous Localization and Mapping
SPI: Serial Peripheral Interface Bus
UART: Universal Asynchronous Receiver/Transmitter
UAV: Unmanned Aerial Vehicle
UKF: Unscented Kalman Filter
USB: Universal Serial Bus
VTOL: Vertical Take Off and Landing

Chapter 5
Portable Haptic Arm Exoskeleton

Pierre Letier
Space Applications Services N.V./S.A., Belgium

André Preumont
Université Libre de Bruxelles (ULB), Belgium

ABSTRACT

This chapter describes a seven degree of freedom force-reflective device able to produce haptic rendering on the human arm, either as master for teleoperation of a slave robot, or in interaction with a virtual reality. This project was conducted on behalf of the European Space Agency (ESA) as a prototype of the master device used for teleoperation of future anthropomorphic space robotic arms on the International Space Station (ISS). The motivation is to decrease the number of extravehicular activities of the astronauts, even for complex situations. The structure of portable anthropomorphic exoskeleton of 7 degrees of freedom has been selected by ESA because it allows a more intuitive control of anthropomorphic slave arms; it also allows multiple contact points, offers a larger workspace (comparable to the human arm). Besides, being attached on the astronaut, the system involves only internal forces (it is self-equilibrated) and can be used in zero-gravity.

1. INTRODUCTION

Manned Extra-Vehicular Activities (EVA) have been a core part of the space exploration either in Earth orbit (space stations, satellites servicing) and planetary exploration. Due to the complex nature, the risks and the costs associated with such tasks, Space Agencies have implemented a number of robotic devices to support or even perform EVAs. The motivation is to decrease the number of spacewalks for astronauts. The ERA and the Canadarm 2 are examples of large robotic arms currently in use on the International Space Station (ISS) to support handling of heavy payloads and modules. Recently, new interest has grown in the development of humanoid robots with

DOI: 10.4018/978-1-4666-0176-5.ch005

Figure 1. Concept of teleoperation on-board the ISS

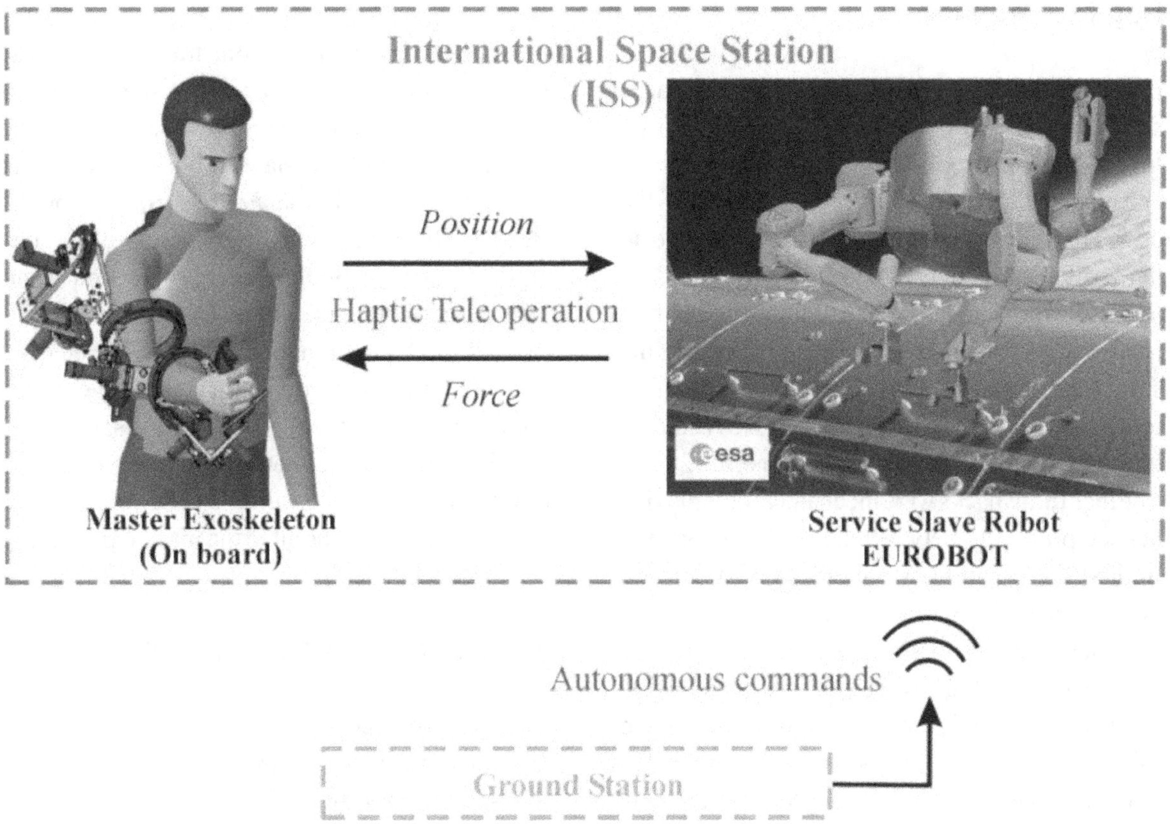

advanced dexterous manipulation capabilities. Eurobot (ESA) and Robonaut (NASA), launched on-board the ISS in 2011 for preliminary testing, are examples of on-going research developments. They are equipped with anthropomorphic arms, kinematically equivalent to the human arm. These robots will either be controlled autonomously or through manual teleoperation by an astronaut staying inside the space station.

The device presented in this chapter was designed at the initiative of the European Space Agency as a prototype of the master device used for teleoperation of these future anthropomorphic space robotic arms on the ISS (Figure 1). Called "SAM", for *Sensoric Arm Master*, it consists of a seven degrees of freedom force-reflective interface, able to produce haptic rendering on the human arm, either as a master for teleoperation of a slave robot, or in interaction with a virtual reality. Compared to simple joysticks controllers, the structure of a portable anthropomorphic exoskeleton with seven degrees of freedom has been selected because it allows a more intuitive control of this family of slave robots with a larger workspace (comparable to the human arm). Furthermore, the addition of force information to the operator, with multi-point contacts capabilities, improves the immersion and the quality of the manipulation. Besides, being attached on the astronaut, the system involves only internal forces (self-equilibrated) and can be used in zero-gravity.

After a short review of the principle of force-feedback and the concept of exoskeleton, this chapter discusses the various steps of the design of the SAM project: kinematics, mechatronics and control.

1.1. Introduction to Teleoperation Haptic Feedback

Force-feedback (or kinesthetic) haptic interfaces are used to increase the user's performances during tasks. They can produce realistic haptic feedback and improve the sensation of immersion of the user. Historically, their developments started in the 1950's in the context of master/slave manipulators for the nuclear industry. Initially, the two parts were mechanically connected allowing the slave robot arm (located in the irradiated room) to replicate the motion of the master device (hold by the operator in the safe place) (Bicker et al., 2004). The main drawbacks were mechanical complexity, needed proximity between the manipulators and limited lift capabilities. Goertz and Thompson developed the first haptic teleoperation system, by incorporating actuators on both sides and replacing the mechanical link by an electrical connection to exchange position and force information. In this way, the slave arm is remotely controlled while the operator can feel the efforts exerted on the slave. Since this first development, the use of haptic interfaces for remote teleoperation has been extended to other fields of applications like robotic arm teleoperation in harsh environment (deep-sea, space), mini-invasive surgery (Okamura, 2009) or nanomanipulation (Sitti & Hashimoto, 2003) (Bolopion et al., 2010). Since the early 1990's, with the increase of the computer processing power, virtual haptic applications have emerged in parallel to the teleoperation developments. The slave device is then replaced by a simulated virtual environment. The haptic interface allows controlling and physically interacting with the virtual objects. This opened haptic technologies to new applications like, among others, virtual learning and training, virtual assembly and manufacturing (automotive and aerospace industry) and games.

1.2. Operating Principle

The operating principle is illustrated in Figure 2; it is based on the exchange of position and force information between the master device, held by the operator, and the slave robot that performs the task. In the simple *Force-Position* control strategy illustrated here, the position of the operator θ_m is measured by the master encoder and sent as set point to the slave. The difference with the measured position of the slave θ_s is then used by the position controller to compute the motor command of the slave robot τ_s. At the same time, the force interaction between the slave and its environment τ_e is measured by the force sensor and sent back as set point to the master, producing the force feedback τ_m.

1.3. Performance and Stability

The performance of a teleoperation system is measured by the *transparency*, which means that the operator has the feeling to perform the task directly in place of the slave. Ideal transparency can be defined by the capacity of the system to insure continuously position and force tracking (Yokokohji & Yoshikawa, 1994),

$$\forall t \begin{cases} \theta_s(t) = \theta_m(t) \\ \tau_h(t) = -\tau_e(t) \end{cases}$$

Another way to quantify transparency and the quality of the teleoperation system is to compare the impedance felt by the user to that of the environment. Impedance is defined by the relation between the force and the position (or velocity). The friction in free motion (when the slave has no interaction) and the stiffness in contact are the two main impedances used to analyze the performance of the system. Impedance in free motion has to be minimized to ensure that the operator is not disturbed by, for instance, device friction; the felt contact impedance must be as close as possible

Figure 2. Principle of a haptic teleoperation system (Force-Position control method)

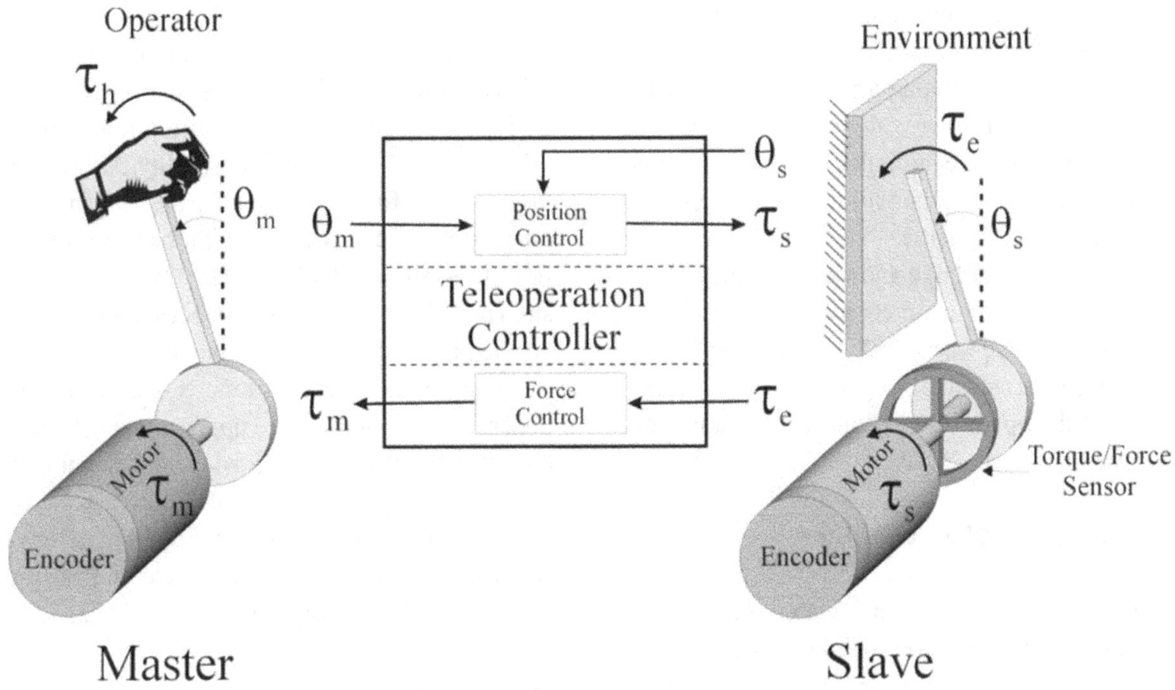

of the contact stiffness (or damping) of the slave on the environment.

The stability of the system is also essential to guarantee the quality and the safety of haptic feedback. However, increasing controller gains to improve tracking performances and allow higher potential stiffness rendering, have a detrimental effect on stability, and a compromise must be found. The sampling frequency of the haptic loop plays an important role on stability (a minimum value of 1Khz is generally used to ensure stability). Transmission delays have also a detrimental impact on stability.

1.4. Control Strategies

Beside the simple *Force-Position* method illustrated in Figure 2, other control strategies can be implemented:

The *Position-Position* control method is historically the first strategy implemented for teleoperation systems; it transmits only position information between the master and the slave. The main advantages are its simplicity of hardware integration, with the absence of force/torque sensor, and a better stability. However, performances, in free motion are degraded.

A variant of the *Force-Position* method is to add local force control on the master side, with the purpose to decrease the felt friction and inertia in free motion and improve the force tracking in contact. This can be done either through feedforward model-based compensation, local feedback loop (need of a master force sensor) or a combination of the two with the hybrid method (Bernstein et al., 2005), as it will be detailed later in this chapter.

In order to overcome performances limitation of the previous two channels control methods, a *four-channel* scheme has been proposed in (Lawrence, 1993). It is based on the exchange of the four positions and forces information; it has the potential to produce an ideally transparent system.

The *admittance* method is another approach. The forces of the operator, τ_h, and from the environment τ_e are measured and applied on a dynamic virtual entity. The position of this entity is then used as position set-point both for the master and the slave. The main purpose is to replace the dynamic of the master with the one of the virtual entity, which can be minimized (keeping in mind stability). It is interesting for devices that present high friction and inertia.

In presence of large and varying time delays, more advanced control methods can be considered based on "Wave Variables" or passivity (Ryu et al., 2010) that can take also into account the variations of the operator or environment parameters (Passenberg et al., 2010).

1.5. Exoskeleton Devices

An exoskeleton is an articulated mechanical structure made of several links and joints that follow the anatomy of the operator. By opposition to prosthesis, the device is attached externally to the body working in parallel with the user's limbs (legs, arms or hands). There can be many applications:

Human Power Amplification

The device helps the user during tiring activities. The actuators increase the power of the operator so that he is able, to carry heavy loads during a long time or on a long distance. They are characterized by high power output (almost beyond human capabilities) provided, as for example in the Sarcos XOS Exoskeleton, by internal-combustion engine that provide power to hydraulic actuators (Guizzo & Goldstein, 2005). They are however characterized by a reduced dexterity (limited number of degrees of freedom, d.o.f.). For instance, HAL-5, developed in a first time by the University of Tsukuba and commercialized now by the Cyberdine Company, is a full-body exoskeleton that has only two active articulations on the arm (shoulder and elbow), compared to the seven basic d.o.f. of the human arm (Guizzo & Goldstein, 2005).

Force (Haptic) Feedback

At the opposite of amplifiers, the main design criterion for haptic feedback is to transmit, with the highest fidelity, the force information from the teleoperated side or the virtual reality. Although a high level of force can improve the realism, it is not required to provide most of the haptic information (O'Malley & Goldfarb, 2002). However, it shall be able to follow and interact smoothly with all the human motions. As they are mainly used for manipulation tasks, haptic exoskeletons are generally designed for the upper limbs (arm and hand). Examples of arm exoskeletons developed initially for haptic virtual applications are the L-Exos from the Scuola Superiore Sant'Anna (Italy) (Frisoli et al., 2009) and the exoskeleton from the Washington University (Perry et al., 2007).

Rehabilitation

A patient performs medical exercises inside the exoskeleton. The device is used to produce repetitive task-oriented activities either by guiding the motion of the user or producing adaptable resistance. Typically, they have output force capabilities greater than the haptic needs, but with more focus on a specific set of joints. Examples of arm exoskeletons dedicated primary to rehabilitation are the MGA from the Georgetown University (Carigan & Liszka, 2007), the Armin from the University of Zurich (Nef et al., 2009) or the PMA from the Salford University (Tsagarakis & Caldwall, 2003). The differences between haptic and rehabilitation devices can be very small and some of them can be used in both types of applications. Another kind of rehabilitation consists to use arm exoskeletons to filter and suppress non-voluntary tremor (Rocon & Pons, 2011).

Desktop haptic interfaces represent the great majority of haptic devices; they interact with the user through one end-effector point. Arm exoskeleton devices present several advantages over desktop haptic interfaces: (i) Their anthropomorphic structure allows reaching a larger workspace, comparable to that of the human arm. (ii) By similitude to the human kinematics, they offer a more intuitive way of controlling a redundant anthropomorphic slave robot arm. (iii) Multi-point contacts can be generated through the exoskeleton fixations (arm, forearm and hand) for better immersion and control.

A distinction can be made between grounded exoskeletons that are fixed on an external support and fully portable ones which are carried by the operator. Although the entire weight is carried by the operator, the latter are easier to transport to crisis sites (nuclear, earthquake) or to enclosed environment (submarine, spacecrafts); they allow free user displacement on a larger surface (e.g. large virtual training environment). Besides, all interaction forces are internal forces to the system (human + exoskeleton) and there is no net external force reaction on the operator when contacting surfaces under microgravity conditions, which facilitates operations in the ISS.

Due to the design constraints of a fully portable device, very few examples can be found in the literature. The Arm-Master was the first one, developed during the 90's by the EXOS Company (which has since disappeared). It used small sized electrical motors in association with water cooling to allow a sufficient high level of forces (Bin et al., 1995). The Chinese ZJUESA Arm-Exoskeleton addressed this problem by using pneumatic actuators that present a better power to weight ratio than electrical technologies (Ying et al., 2007). However, these two solutions needed an external annex system that limits the capacity of displacement with the exoskeleton. The interface designed at the SAIT (Samsung Advanced Institute of Technology) use electrical brakes to produce force feedback (Kim et al., 2005). Although the use of brakes decreases the total weight of the system, it also limits the range of haptic sensations that can be rendered, like spring repulsion.

2. DESIGN

The objectives of the *Sensoric Arm Master* (SAM) project were to develop a portable system easy to carry, to calibrate and to maintain, and providing comfortable and full haptic rendering. The following requirements were taken into account in the specification of the project: (1) The workspace should be as close as possible to that of the human arm. (2) The power to weight ratio of the device has to be optimized in order to procure sufficient haptic forces while limiting the weight and so the tiredness of the user. (3) Minimize friction, weight and inertia which limit the transparency (either by proper selection of the actuation technology, or by control). (4) Good ergonomics with stable and comfortable fixations which can be adjusted to the body of the operator. (5) Safety of operation should be addressed, by hardware and by software, in order to protect the user. (6) On board power electronics and communication to reduce the external link to the minimum.

2.1. Kinematics

The human arm has 7 basic degrees of freedom (d.o.f.) of rotation, Figure 3. Most of the existing arm exoskeletons address 5 to 7 of these joints, depending mainly on the presence of the last two wrist motions. The kinematic structure can be defined in two ways. In the anthropomorphic design used in most of the exoskeletons, there is a one to one correspondence between the anatomic joints and the exoskeleton joints. This approach requires a good alignment of the joint axes to avoid any constraints during the arm motion. The other possibility is to differentiate the kinematics of the exoskeleton from that of the arm. There is no need for alignment, but the design may require an increased number of d.o.f., associated with more complex mechanical design and control; the EXARM exoskeleton is an example of this approach (Schiele, 2008). In the SAM project, given the anthropomorphic nature of the slave robot, an anthropomorphic design was selected, with 7 active d.o.f and 6 passive sliders to adapt to the user morphology (fixed during operations).

In an anthropomorphic design, the shoulder and the wrist may be represented by spherical joints, with an additional rotation joint at the elbow. In the initial design, the two spherical joints were materialized by three orthogonal rotation joints (Figure 4). However, when joint 2 is at 90°, joint 1 and joint 3 are aligned and the system loses one d.o.f. The grey scale in the right side of Figure 4 illustrates the value of the condition number of the Jacobian within the workspace: the dark zone, when joint 2 approaches 90°, corresponds to large values of *c(J)*, meaning that the system is ill-conditioned.

In order to move this singularity from the centre to the frontier of the workspace, the joint configuration of the shoulder has been modified by two successive rotations, respectively α around *Y* and β around *Z* of the inertial reference frame, to obtain the modified configuration illustrated in Figure 5. This solution has been implemented in several other existing exoskeletons (Perry et al., 2007), (Carrigan et al., 2007), (Frisoli et al, 2009). The values of these two rotations have been selected based on iterations between a numerical kinematic model analysis of the exoskeleton and a real kinematic prototype with modifiable angles. The singularity still happens when joint 2 is at ±90, but it is removed from the useful workspace, as can be seen from the condition number in the right side of the figure. For the wrist, due to the limited anatomic range of motion of the wrist joints, the succession of the exoskeleton joint has been modified such as the singular configuration happens outside this range. Nevertheless an angle γ has been inserted to avoid mechanical interactions between the various parts of the exoskeleton. The table given in 3 compares the anatomical range of the arm with those of the exoskeleton. The limitations are mainly due to mechanical interactions between the exoskeleton parts and with the user's body. The global work-

Figure 3. The 7 basic motions of the human arm and the comparison between anatomical (Kapandji, 1997) and SAM joints range of motion

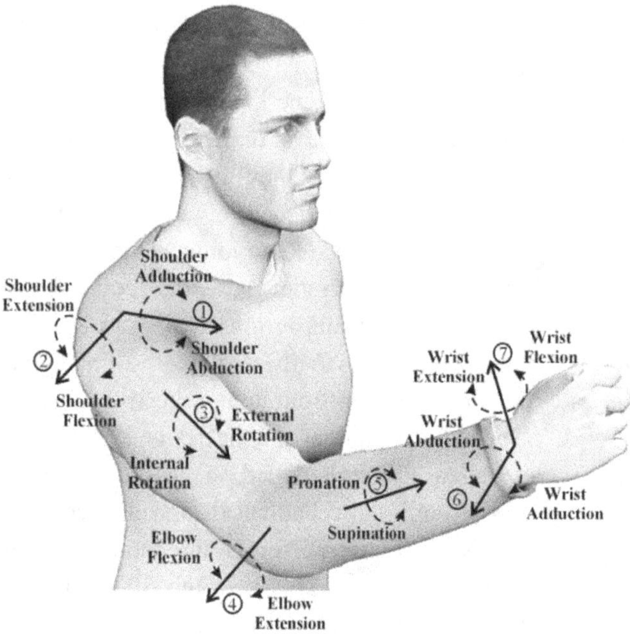

Joint	Anatomic Angles	SAM Angles
1	-50/180 °	-10/180°
2	-45/180°	-20/180°
3	-80/90°	-50/90°
4	0/145°	0/110°
5	-90/85°	-90/85°
6	-15/30°	-15/30°
7	-70/80°	-30/80°

Figure 4. Initial kinematic configuration of the exoskeleton and representation of the condition number c(J) in the workspace, associated with the degree of singularity

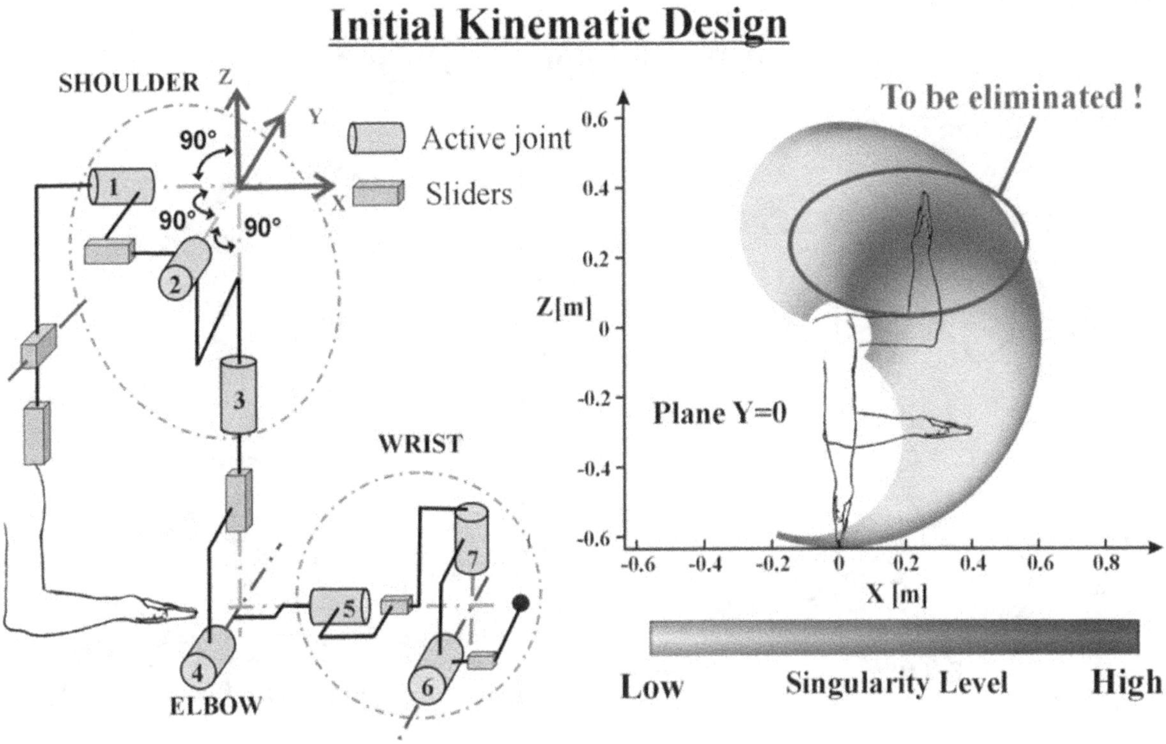

space has been experimentally measured and it reaches 73% of a theoretical estimation of the natural human workspace adapted from (Klopcar et al., 2007).

2.2. Mechanical Structure

The *Sensoric Arm Master* SAM is shown in Figure 6. The system is built mostly using aluminium material. The links have a U shape to increase the strength to weight ratio, with reinforcements in the straight angles. The links support on-board electronics boards that are responsible for the sensors reading, the current amplification and for the communication with the main controller. These local boards limit the number of wires along the structure.

To ensure a good ergonomics, fixations have to be stable and comfortable. They should allow prolonged use of the exoskeleton without pain or constraints on the bloodstream, and a good transmission of the efforts. The exoskeleton is attached to the user body by four points at the back, the upper and lower arm, and the hand. The ergonomic dorsal backplate is made from hard plastic and implements the body fastening system with large straps for a good distribution of the weight of the exoskeleton. On the arm, shell polypropylene orthosis, with internal foam layers, are used. They allow comfortable fixation and adaptation to different arm diameters; they are place inside large diameter roller bearings of joints 3 and 5 (Figure 7). The upper part of a joystick is used as end-effector, implementing high quality buttons interfaces to control the system (dead-man button, slave end-effector...).

For safety of the user and the exoskeleton itself, mechanical stops are added on each joint to protect the device and the user. The total weight of the exoskeleton is 7.4kg, including the mechani-

Figure 5. Modified and final kinematic configuration of the exoskeleton and representation of the condition number c(J) in the workspace, associated with the degree of singularity. The singularity is now at the limit of the workspace

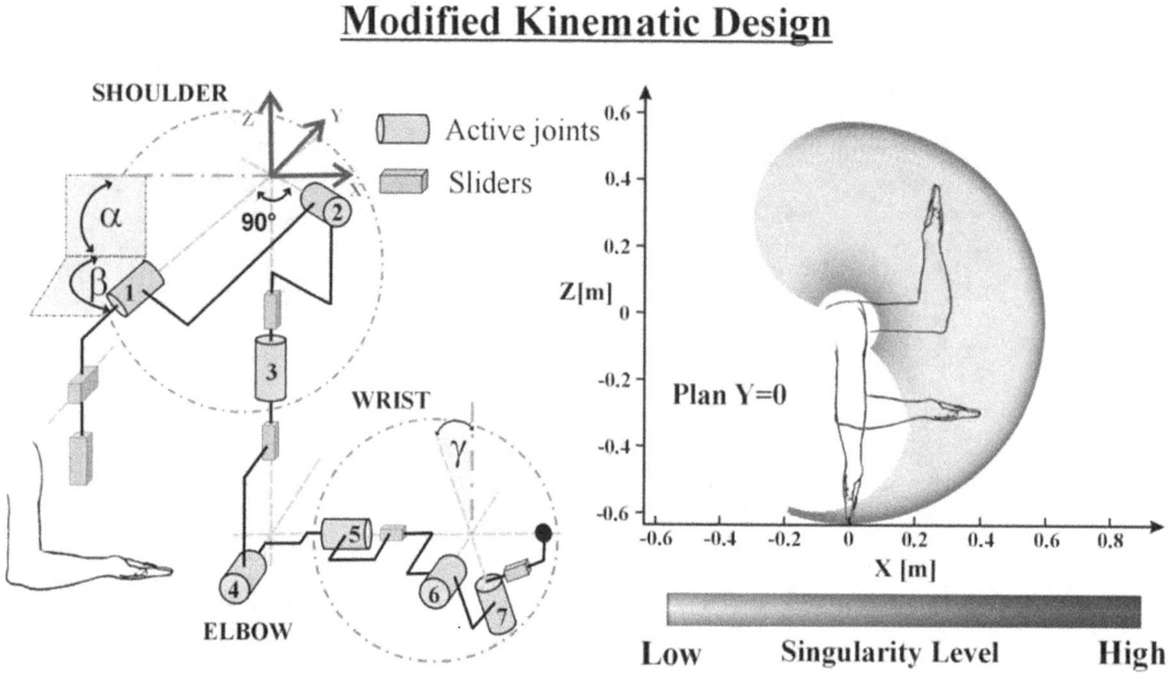

cal structure (3.9kg) and the actuators (3kg). 4.5kg are carried by the arm, the rest being supported by the backplate. This allows a correct use of the system, but the internal gravity compensation improves undoubtedly the comfort, even if 100% is not required.

2.3. Actuation Technology

The choice of the actuation technology has a great influence on the quality of the interface, on the haptic performances and the user comfort. Indeed, it will affect the output power capabilities, the weight and the dynamics (inertia, friction) of the device. In this project, the selection of the actuation technology was done on the basis of a preliminary analysis on a 1 d.o.f. setup, in which several technologies were compared, including magneto-rheological (MR) brakes and cable transmission (Letier et al., 2006). The study showed that cable transmission deteriorates the quality of the torque signal due to added and unpredictable friction effects. MR technology was not selected in this project (in spite of its good torque to weight ratio) because of its inability to represent spring repulsion, but was used for a rehabilitation device in another project (Avraam et al, 2011).

On-joint actuation has been preferred to cable transmission because of a smoother transmission and less friction; the main drawback of the on-joint actuation is the added mass located on the exoskeleton parts carried by the operator, but this can be overcome by partial gravity compensation (this, however, has to be taken into account when sizing the actuators). A compact system composed of a brushed DC motor, a capstan reducer and a gearbox has been selected for each joint

Figure 6. Haptic arm exoskeleton SAM and its main specifications

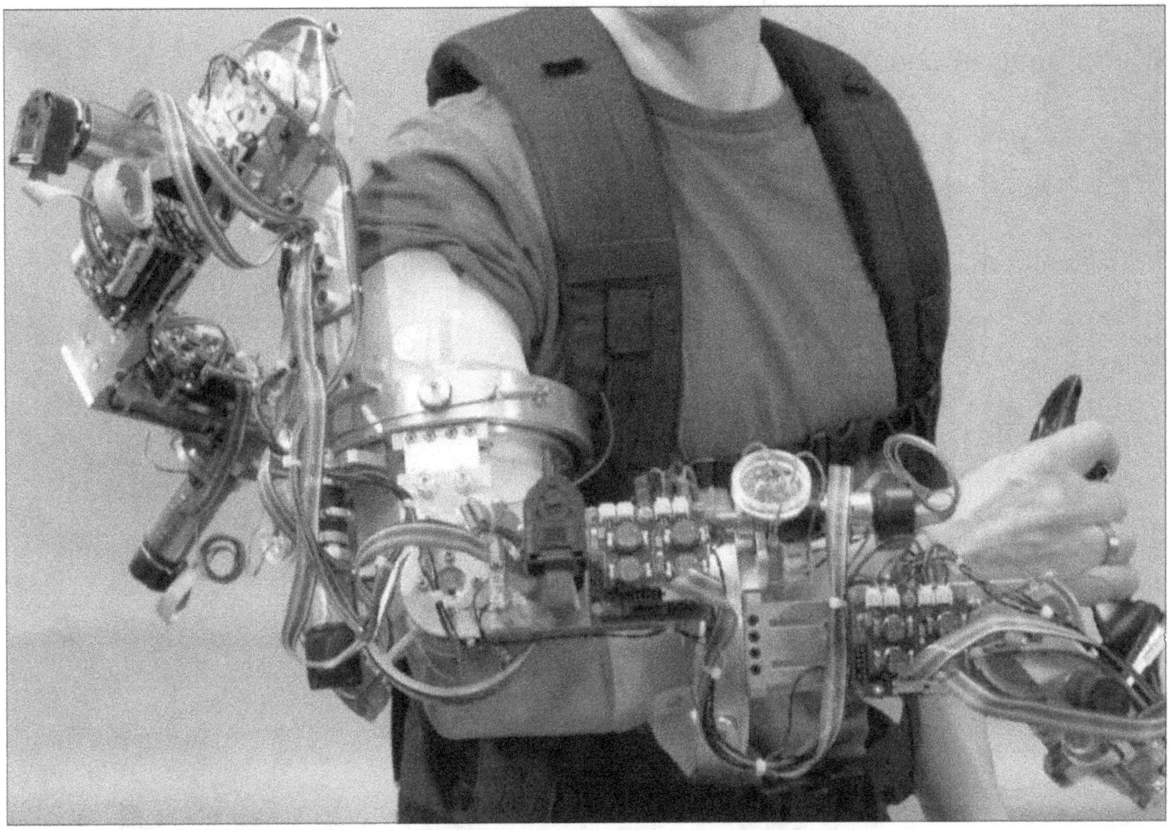

- 7 active DOF, 6 passive adjustments
- Aluminum based structure
- Compact joint design with local DC actuation and integrated sensors (position and torque)
- Permanent output torque : 10Nm to 1Nm from the shoulder to the wrist
- Workspace: 73% of the human
- Ergonomic fixations
- On-board power, conditioning and communication electronic
- Weight: 7.4 kg

(Figure 8). Table 1 summarizes the selection of the components.

The DC technology has been chosen for its ease of implementation and its good controllability with a direct relation between the current and the output torque. However, it presents intrinsically a very poor torque to mass ratio. It is then required to combine it with a reducer that will increase the output torque while keeping acceptable volume and mass. The capstan type reducer, often used in haptic devices, is composed of a wheel, wrapped by a cable on the motor shaft. It allows zero-backlash transmission as well as low friction at the expense of a low torque to volume ratio. The reduction ratio is proportional to the diameter of the wheel (it has been limited to 10 in this project). At the opposite, planetary gearboxes present a good compactness but with a higher level of friction (mainly influenced by the number of stages) and some backlash. In this project, the two types of reducers have been combined, to achieve a sufficient torque combined

Figure 7. Large diameter roller bearings of joints 3 and 5. The outer part of the bearing is used as capstan and the inner part supports the orthosis

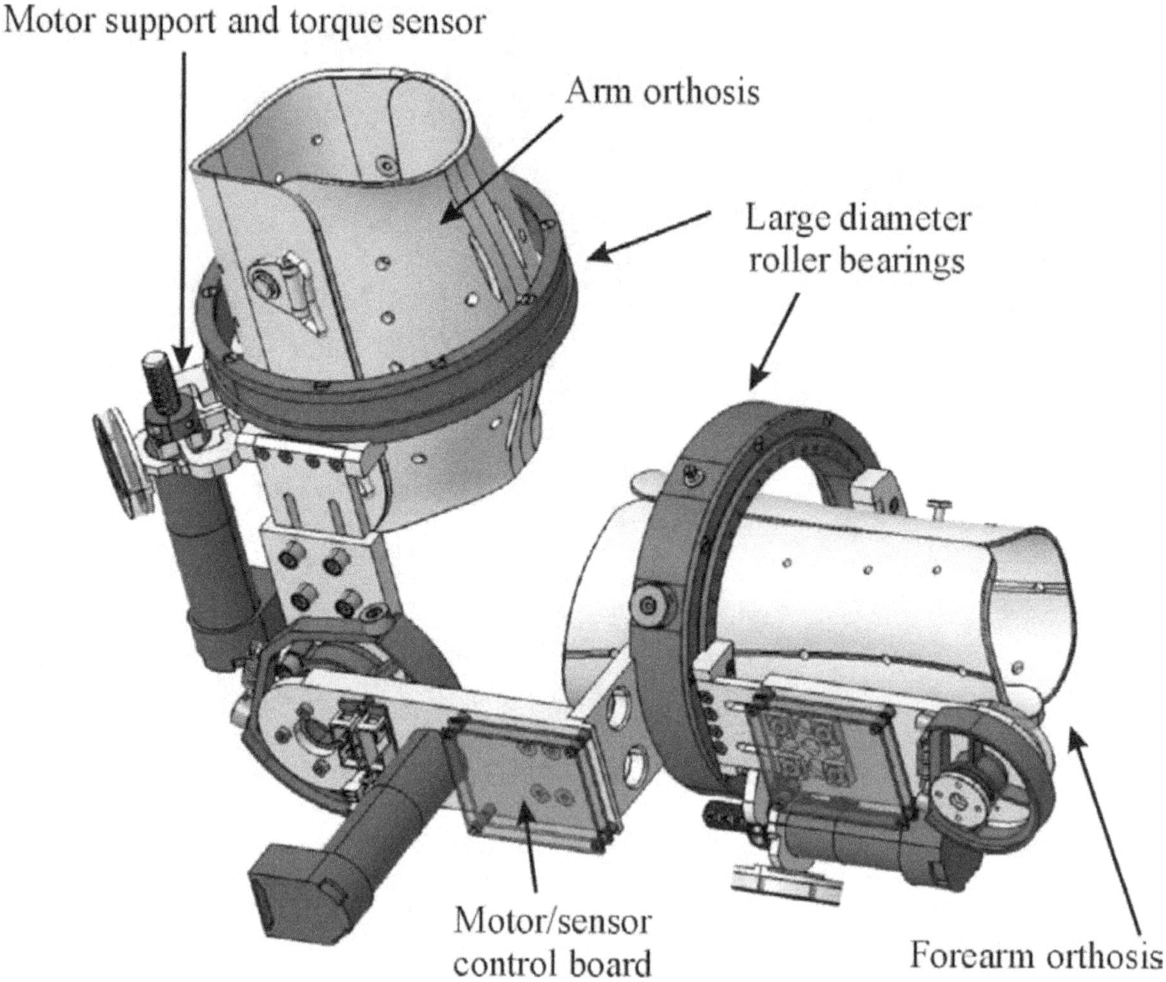

with high compactness, low friction and low backlash transmission. The various reduction ratios are summarized in Table 1.

As mentioned earlier, high level of forces are not mandatory to produce haptic information. In order to achieve a compromise between weight and quality of the haptic feedback, the motors have been design to produce continuous torque of 5% of the maximum human capability. This is comparable to commercially available desktop interfaces. Note that a continuous control of the force around 15% of the maximum capacity will induce fatigue to the user.

The sizing of the motors is based on the fact that the maximum current required to represent the contact forces is necessary only during a fraction of the operation of the device (during contact), and that the current necessary for the gravity compensation is needed continuously. The relationship between the maximum permanent

Figure 8. Typical joint (#2 in the shoulder) including a brushed DC motor, gearbox and capstan reducer with integrated strain gage for torque measurement and conditioning electronics

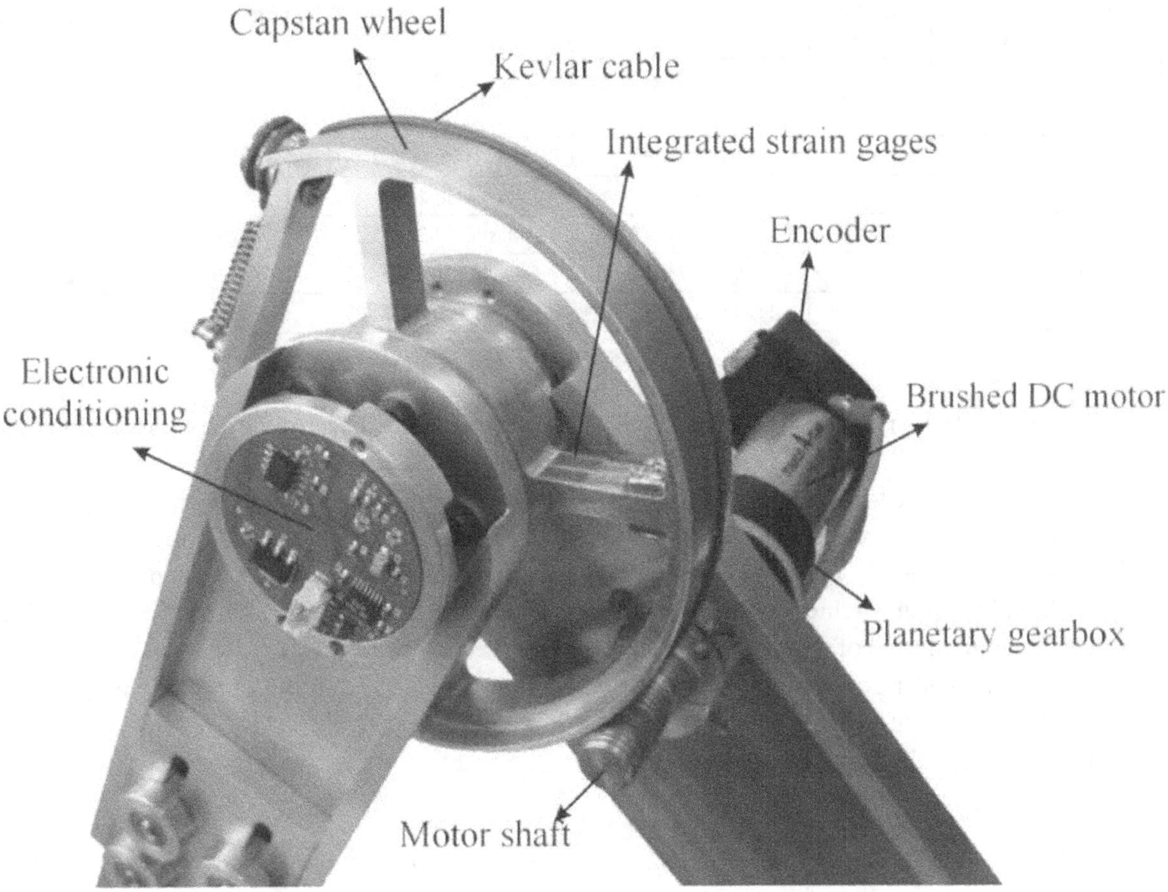

current I_{perm} (which is a motor characteristics) and the current I_{on} allowed during a fraction of time T_{on} (T_{on}=1 represents a continuous use) without deteriorating the motor is given by

$$I_{on} = I_{perm}\sqrt{\frac{1}{T_{on}}}$$

This is the maximum current available, from which the current necessary for gravity compensation, I_{grav}, must be subtracted; what remains is available to represent contact forces, I_{wall}.

$$I_{wall} = (I_{perm} - I_{grav})\sqrt{\frac{1}{T_{on}}}$$

Since the current is directly proportional to the torque, the torque available for contact, τ_{wall}, may be evaluated as follows,

$$\tau_{wall} = (I_{perm} C K_T \eta - \tau_{grav})\sqrt{\frac{1}{T_{on}}}$$

with τ_{grav} being the torque needed for gravity compensation, C the total reduction ratio (capstan + gearbox), K_T the torque constant of the motor and η the efficiency of the actuation chain, assumed to

Table 1. Joint actuation data information. (1) Adapted from (Carigan et al. 2005)(Rosen et al., 2005)

Joint	Max Human Torque [Nm] (1)	Torque in daily Life [Nm] (1)	Motor	Gear	Total Reduction Ratio	Maximum Permanent Torque [Nm]
1	134	10	Maxon A-max 32	Maxon GP-32C (1:23)	264	9.3
2	115	10	Maxon A-max 32	Maxon GP-32C (1:23)	264	9.3
3	60	3.1	Maxon A-max 26	Maxon GP-26B (1:19)	342	4.8
4	72	3.8	Maxon A-max 26	Maxon GP-26B (1:19)	218	3.1
5	9	0.4	Maxon A-max 22	Maxon GP-26A (1:19)	299	1.3
6	21	0.38	Faulhaber 2232-024	Faulhaber GP-20/1 (1:3.7)	33	0.6
7	21	0.25	Faulhaber 2232-024	Faulhaber GP-20/1 (1:3.7)	33	0.6

be 0.8. As an illustration, Figure 9 represents the acceptable torque as a function of T_{on} for selected components of joints 2 and 4. When in operation, the user is not always in contact and not always with the maximum torque. For this reason, we selected, for all joints, the lighter system that can produce the required level of torque during 20% of the cycle time under 100% of gravity, while ensuring 100% without gravity compensation. Note however, that these requirements cannot be met exactly, because of the limited choice of motors and gearboxes. This explains why joint 4 is slightly undersized with no gravity compensation. Finally, joint 3 and 5 have a specific design, because their axis of rotation is parallel to the human arm. The arm is inserted inside large size thin section bearings (Kaydon) allowing the arm motion inside and at the same time playing the role of capstan.

Figure 9. Maximum output torque on joints 2 and 4, related to the relative duration of contact (1 = permanent contact), for several levels of gravity. The saturation is due to the current saturation of the amplifier

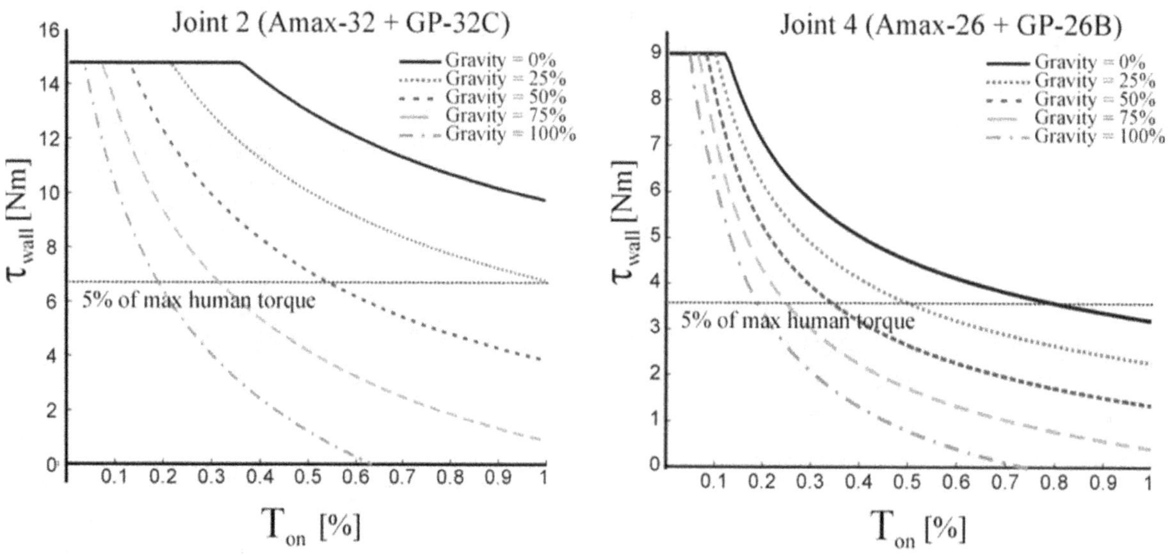

2.4. Sensors Technology

In order to compare various control strategies, position and force/torque are measured on the exoskeleton. Incremental encoders, located on each actuator, measure the angular position of the joints. A value of 512 pulses per revolution, associated with the reduction ratio, provides a high resolution with good speed measurement capabilities (derived from digital position signal). Each joint is also equipped with a local torque sensor, based on strain gages technology (Figure 8); it measures the deformation of the ribs inside the capstan reducer, through a full Wheatstone bridge. The rib dimensions define the sensitivity of the torque sensor. They were chosen to resist a torque two times larger than the maximum deliverable torque of the actuators. A local programmable electronic conditioning board amplifies the torque signal and sends it to the controller through a current loop to decrease the noise sensitivity.

3. CONTROL EXPERIMENTS

This section discusses the various control strategies which have been implemented on the exoskeleton, and compares the experimental results obtained in the two main operating modes: virtual reality and teleoperation.

3.1. Virtual Reality Mode

In the virtual reality (VR) mode, the exoskeleton is directly connected to a VR world. It controls the motion of a proxy, a massless virtual entity, which interacts with the VR environment. The position of the proxy corresponds to a specific point of the exoskeleton (e.g. end-effector, elbow) and several proxies can be simultaneously implemented to build, for example, a virtual manikin. This approach differs from the teleoperation where a slave robot is driven, with its own mass and dynamic (see next section).

For the VR experiments, the control of the exoskeleton was managed by a dSpace DSP robotic board (ds1103) associated to Matlab/Simulink, at a sampling frequency of 1 kHz. External linear current amplifiers were used to drive the motors. The virtual reality environment as been built around the haptic software library CHAI 3D, that manages the physical engine, the communications with the device, and the graphical rendering.

In the following, we describe the implementation of the two main control schemes used with exoskeleton in haptic and rehabilitation applications: Force-Position and admittance control. We will highlight the limitations of the simple open-loop Force-Position scheme and the improvements brought by the hybrid local torque control.

3.2. Force-Position Control

The Force-Position algorithm is the most used as it allows to produce haptic effects, in a simple manner, without the need of force sensors or computation of inverse kinematics and dynamics. Examples of previously cited exoskeletons using this method are (Perryet al., 2007), (Tsagarakis & Caldwall, 2003), (Frisoli et al., 2009) and (Bin et al., 1995).

Figure 10-(a) provides the architecture of the Force-Position scheme, when linked to a VR environment. The Cartesian position of the end-effector X_m, obtained by the forward kinematics computation on the vector of joint angle measurements Θ_m, is sent to control the proxy in the VR. In case of interaction with the environment, the contact force F_e is computed,

$$F_e = -Z_e X_m$$

with Z_e the impedance of the environment (stiffness and damping). This force is then converted in motor control command,

$$\tau_m = \tau_{md} = J_m^T F_e + G(\Theta_m)$$

Figure 10. Control schemes of the exoskeleton when linked to a virtual environment. Dashed lines correspond to optional

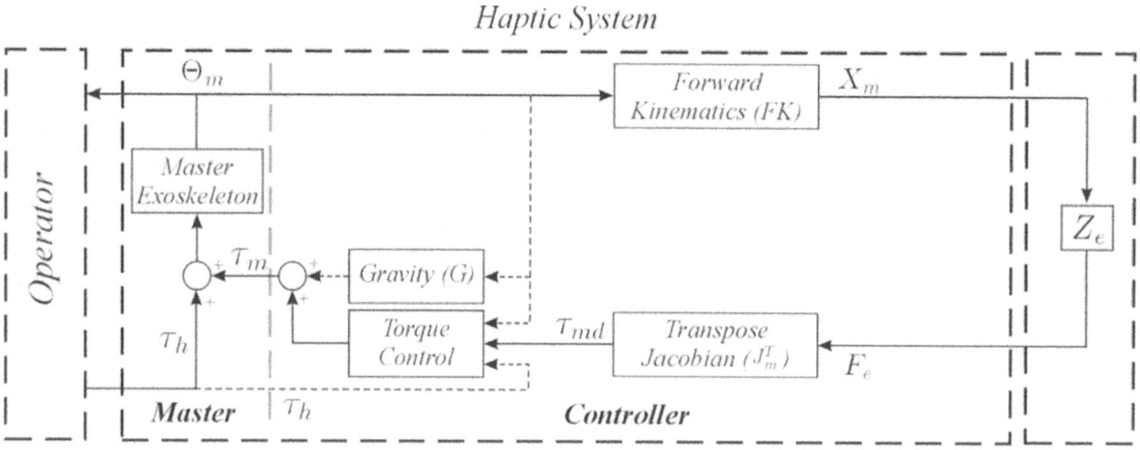

(a) Force-Position control with virtual reality

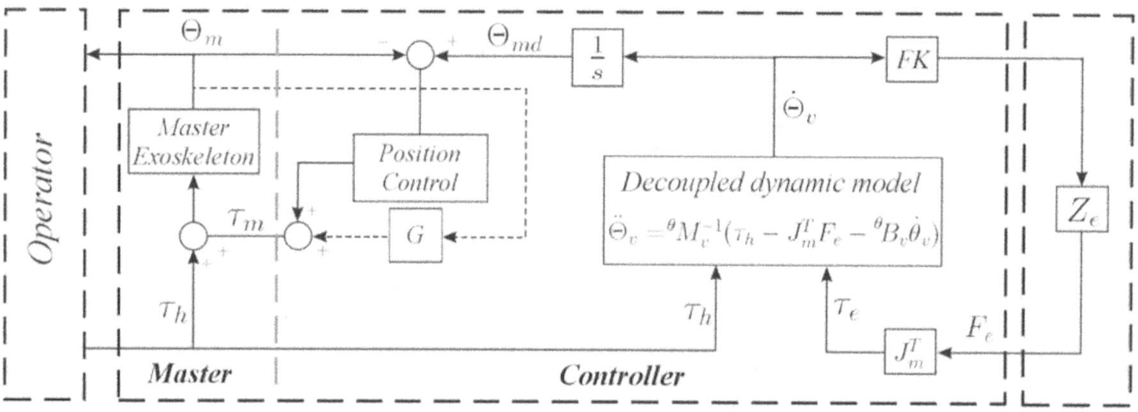

(b) Admittance Control (joint space) with virtual reality

with J_m^T the exoskeleton Jacobian transpose that does the transformation between the cartesian space and the joint space, and the potential gravity compensation $G(\Theta_m)$ to reduce the tiredness of the user and the perception of the interface weight. Figure 11 illustrates an experiment of shape following. The good matching shows the capacity for the exoskeleton to provide haptic effect and to control correctly the direction of the Cartesian force.

However, this simple scheme has limitations. Figure 12-(I) provides end-effector cartesian force and stiffness tracking for an experiment of soft contact (100 N/m) of a virtual plane $Z = cst$. We can observe bad tracking performances both in free motion and in contact. In free motion, the user is sensitive to the natural friction of the exoskeleton, although the experienced level is here acceptable (the user can move the arm). In contact, we have observed variations of the joints friction, related to the output torque. This effect can be explained by the increasing Coulomb friction between the teeth of the gearbox under load (Gogoussis & Donath, 1987). When penetrating the surface, this results in a noticeable increase of the felt stiffness and sometimes to the blocking of the device, which is unacceptable. The same problem

Figure 11. Experiment of virtual shape following with the exoskeleton, under the Force-Position control method

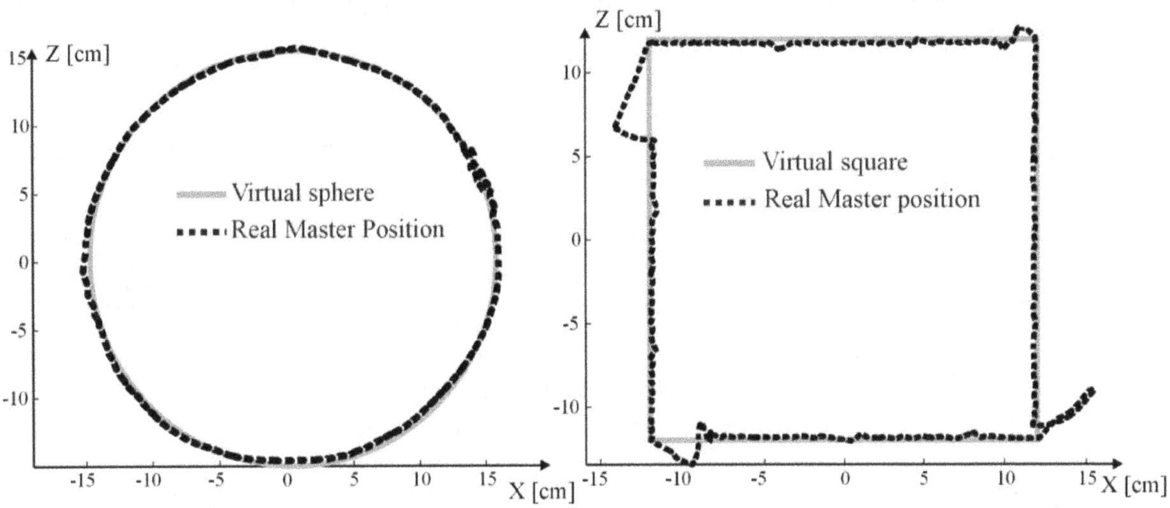

arises when high internal gravity compensation is applied.

3.3. Hybrid Local Torque Control

To improve tracking performances, the previous open loop control method can be upgraded, on each joint, with a hybrid local torque controller, inspired from (Bernstein et al., 2005). It consists of the coupling between a torque feedback regulator and a feedforward compensator. The latter is based on the friction model,

$$\hat{Z}_{mi} = \tau_{c_i} sign(\dot{\Theta}_{m_i}) + b_{c_i} \dot{\Theta}_{m_i}$$

with τ_{c_i} and b_{c_i} being respectively the Coulomb friction and the viscous coefficient estimated for joint *i*. Due to the discontinuity of the model around null speed, instabilities or sticking sensations can appear. For this reason, when the velocity of the joint approaches zero, the controller switches to the torque regulator,

$$H = \tau_{md_i} + K_{f_i}(\tau_{md_i} - \tau_{h_i})$$

with τ_{md_i} the joint torque set-point from the VR, τ_{h_i} the measured torque and K_{f_i} the feedback gain for joint *i*. Stability considerations limit the feedback gains and thus the level of compensation that can be achieved. By merging both methods, the Hybrid controller can attenuate their drawbacks. The use of the regulator at low speed avoids any troubles with the discontinuity of the friction model. At the same time, going to the model at higher speed reduce the authority of the regulator, allowing a higher gain keeping the same level of stability. For a smooth transition between the two methods, the output signal of each of them are multiplied by an adaptive gain related to the joint velocity, which gives the total joint motor control command,

$$\tau_{m_i} = \tau_{md_i} + K_{f_i}(\tau_{md_i} - \tau_{h_i})\frac{\omega_{a_i}}{\omega_{a_i} + \left|\dot{\Theta}_{m_i}\right|} + \hat{Z}_{m_i}\frac{\left|\dot{\Theta}_{m_i}\right|}{\omega_{b_i} + \left|\dot{\Theta}_{m_i}\right|} + G_i(\Theta_m)$$

Figure 12. Contact and stiffness rendering in virtual reality [free motion, followed by a contact in Z direction, 100 N/m (soft)]; Comparison of the force and stiffness tracking for different control strategies. The end-effector cartesian force has been estimated from joint measurements

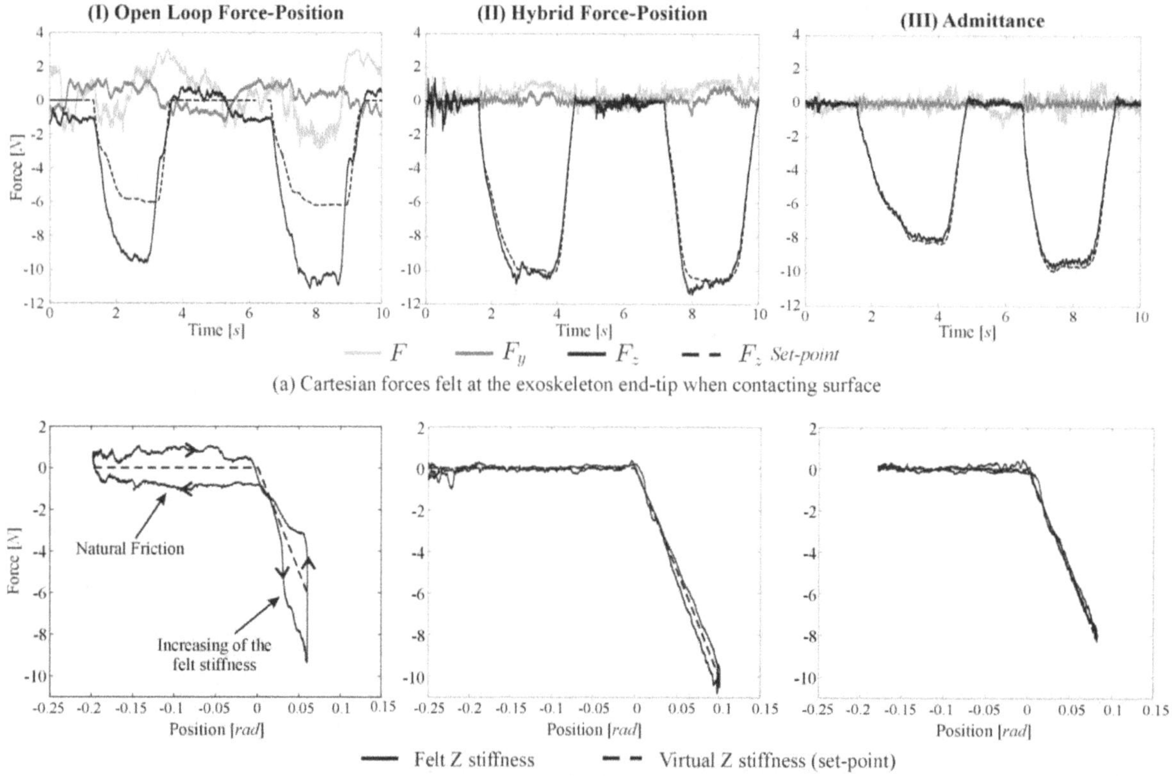

(a) Cartesian forces felt at the exoskeleton end-tip when contacting surface

(b) Stiffness in Z direction

with ω_{a_i} and ω_{b_i} that fix the range of velocity on which each individual strategy is effective.

This method has been tested with the soft-contact experiment, Figure 12-(II). Compared to the open-loop result, force and stiffness tracking are improved both in free motion and in contact. It is now possible to really feel the soft penetration inside the surface. However, a compromise has to be found between the improvement of the performances and the stability, especially for hard contact.

3.4. Admittance Control

The presence of torque sensors on the exoskeleton joints allows the implementation of the admittance control, Figure 10-(b). From the user side, the exchange of position and force information is performed in the opposite direction, as compared to the Force-Position strategy. The master joints are measured and send to a virtual entity, whose position is then used as set-point for the master and the slave or the VR proxy. This allows keeping the same virtual reality software.

Through the use of high gains in the master position controller, the idea is to replace the exoskeleton dynamics by that of the virtual entity. This method fits well haptic interfaces with high friction and inertia. It has been implemented for instance on the MGA that use harmonic drives with a very high level of friction (Carigan et al., 2007). As the MGA is equipped with a 6 d.o.f. force sensor at the end-tip, the admittance control is performed in the cartesian space with the virtual entity consisting of a mass-dashpot system. However, SAM is equipped with local torque sensors on each joint and the control is done in the joint space. The virtual entity is defined by a decoupled dynamic model with the same kinematics as the master exoskeleton. The local torque measurements τ_h and the cartesian forces from the VR (end-effector, elbow…) are applied on the entity, giving the dynamic behavior,

$$\ddot{\Theta}_v = {}^{\theta}M_v^{-1}(\tau_h - J_m^T F_e - {}^{\theta}B_v\dot{\Theta}_v)$$

with ${}^{\theta}M_v$ and ${}^{\theta}B_v$ the joint inertia and damping diagonal matrices of the virtual entity. Thanks to the decoupling, each joint is controlled individually. The joint position Θ_v is then used as set-point for the virtual end-effector proxy and the master exoskeleton,

$$\tau_m = {}^{\theta}K_p(\Theta_v - \Theta_m) + {}^{\theta}K_d(\dot{\Theta}_v - \dot{\Theta}_m) + G(\Theta_m)$$

with ${}^{\theta}K_p$ and ${}^{\theta}K_d$ the matrices of proportional and derivative gains of the master position controller. The inertia and the damping of the virtual entity have to be minimized to reduce the dynamic effects felt by the operator, while ensuring the stability in free motion. At the same time, the position controller gain must be large enough so that the natural dynamics of the master is erased and replaced by that of the virtual entity.

As illustrated in Figure 12-(III), the admittance control presents a very good behaviour both in free motion and in contact, even better than the hybrid method. Moreover, higher stiffness could be presented stably to the user. The main drawback of the admittance control is the complexity of the implementation, with the computation of the virtual entity dynamic and the absolute need of force sensors.

3.5. Teleoperation Mode

In a second step, the SAM exoskeleton has been tested in teleoperation, in the frame of the EXOSTATION project. EXOSTATION (Figure 13) is an ESA program aiming at building a complete haptic control station, as a demonstrator for future haptic teleoperation on-board the ISS or for planetary exploration. The project was driven by a Belgian consortium composed of the Active Structures Laboratory (ULB) and the companies MicroMega Dynamics and Space Application Services. The final objectives of the project were:

- The development of a new portable arm exoskeleton with force-feedback capabilities.
- The integration of this exoskeleton in a demonstrator of master/slave teleoperation, based on the simulation of a slave robot, for concept validation.

The haptic station is composed of four main components:

- The **SAM exoskeleton**, described above, is the master device in the haptic loop. It measures the user position and, at the same time, provides him with force-feedback from the slave.
- The **ECO controller** is built around a desktop computer and on-board electronic boards. The computer implements the haptic loop and control strategies on a QNX real time operating system and manages the exchange of data and the synchronisation between all the system components. The haptic frequency loop has been fixed here at 500Hz due to limitation in the computation of the physics simulator. Each on-board electronic board manages two joints. They interface the actuators and sensors, providing PWM power amplification and position/torque sensors conditioning. All

Figure 13. Components of the haptic station EXOSTATION

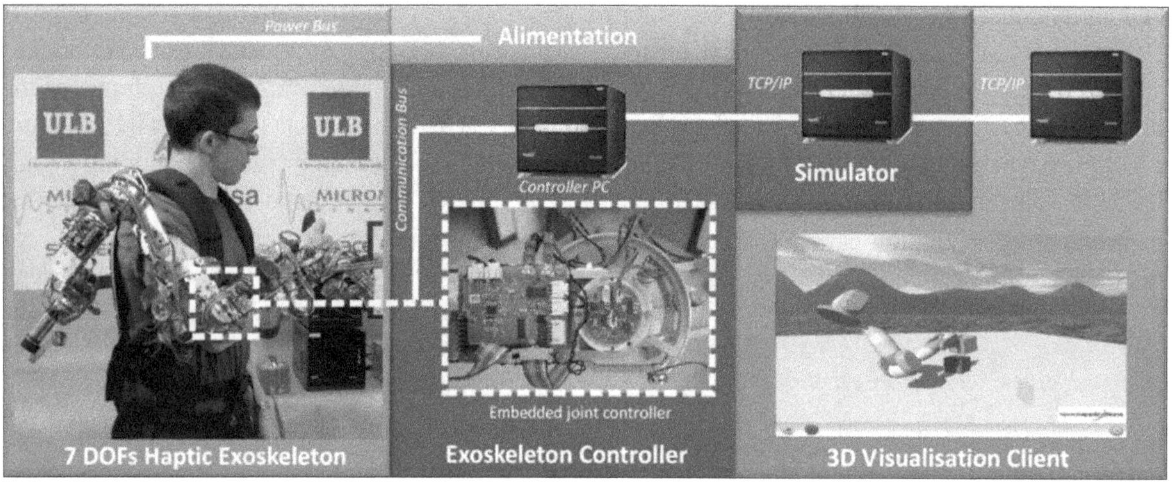

these components are linked by a serial data communication bus, with a limited number of wires on the exoskeleton.

- The **Simulator** is a program that manages the dynamic simulation of a slave anthropomorphic robot (PA-10-Mitsubishi) and its environment on the basis of the ODE dynamic engine. A multi-task architecture allows separating simulation and communication aspects from less crucial processes like telemetry and logging. Python scripts facilitate the definition and the modification of the scenarios. The communication with ECO is based on a TCP/IP protocol.
- The **3D client** allows visualizing the simulation and provides graphical user interfaces to control the parameters of the simulation. It is running slower than the haptic loop (around 40Hz).The graphical rendering is based on OpenGl 2.1 with multi-clients and stereoscopic vision support.

The main difference with the virtual reality mode is the presence of the slave robot. Its position control can either be done in the joint space or the cartesian space. The latter is based on the exchange of position information in the cartesian space and allows to control a robot with a dissimilar kinematics. However, it requires the computation of the inverse kinematics, which can be complex for redundant robotic arms. Since in this project the slave robot is anthropomorphic, it was decided (for this first setup prototype) to control the slave robot in the joint space with a joint to joint correspondence (Figure 14),

$$\tau_s = {}^{\theta}K_p(\Theta_{sd} - \Theta_s) + {}^{\theta}K_d(\dot{\Theta}_{sd} - \dot{\Theta}_s)$$

The main drawback of this basic control is that each joint is controlled individually, without taking into account the global configuration of the robot and the couplings. It is thus impossible to select optimal gains for all configurations.

Three haptic control methods have been tested on the teleoperation setup (Figure 14): the classical Force-Position control, with and without hybrid upgrade, and a 3 channels strategy.

Figure 15 gives the master/slave joint tracking during soft contact experiment. Only the main joints involved in the contact are provided (i.e. joints 1, 2 and 4).

As above, the performances of the open-loop Force-Position strategy are limited in terms of tracking and transparency (friction in free motion).

Figure 14. Control schemes of the exoskeleton in the teleoperation mode (dashed lines are optional). FK stands for the Forward Kinematics.

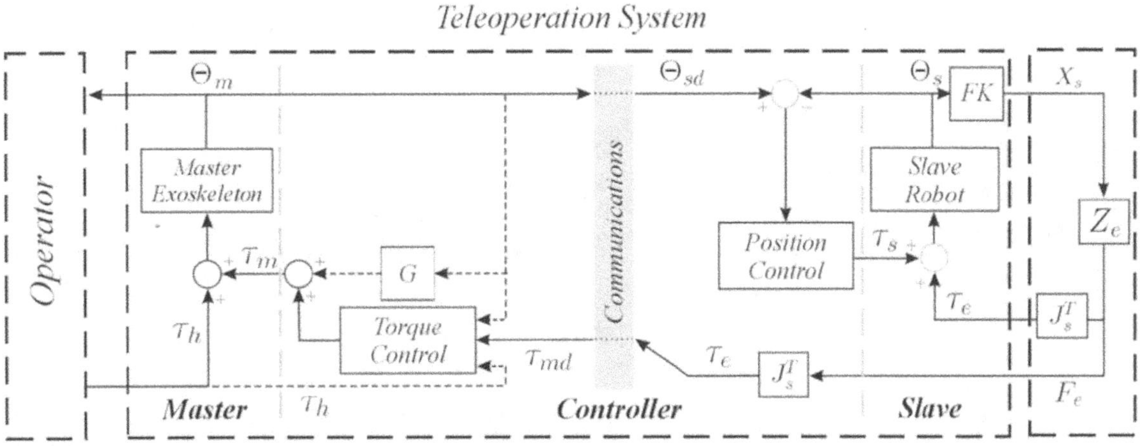

(a) Force-Position Control in the joint space for a teleoperation system

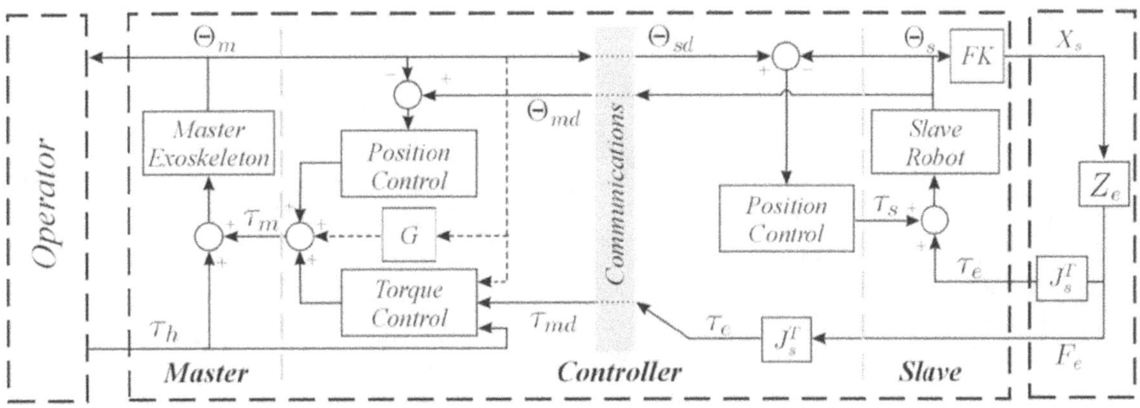

(b) 3 channel control in the joint space for a teleoperation system

The addition of the hybrid controller improves the behaviour. However, due to the relative low frequency (500 Hz) of the haptic loop, the reduction of the friction had very bad consequences on the maximum stable stiffness rendering (around 1200 N/m), preventing the possibility to have in the same scenario soft and rigid contacts. To solve this problem, a 3 channel strategy has been proposed (Figure 14-(b)). It is based on the Hybrid Force-Position controller with the addition of the slave to master position channel. The main advantage of a 2 channel position/position controller is its stability. However, it increases the friction felt by the operator. By coupling the position/position approach with the hybrid control, the purpose is to preserve this better stability while limiting the additional friction by adjusting the parameters of the hybrid controller. As shown in the experimental results (Figure 15), the 3 Channel strategy keeps good tracking and low friction in free motion (although slightly worse than pure hybrid control) while increasing the maximal stiffness rendering to 2500 N/m, which is sufficient to stimulate hard contact. The adjustment of the hybrid controller offers a possibility to play between haptic transparency and stability.

Figure 15. Master and slave joint torques measured during an experiment of soft contact, for the different control schemes implemented in teleoperation (Z=cst, K=100N/m)

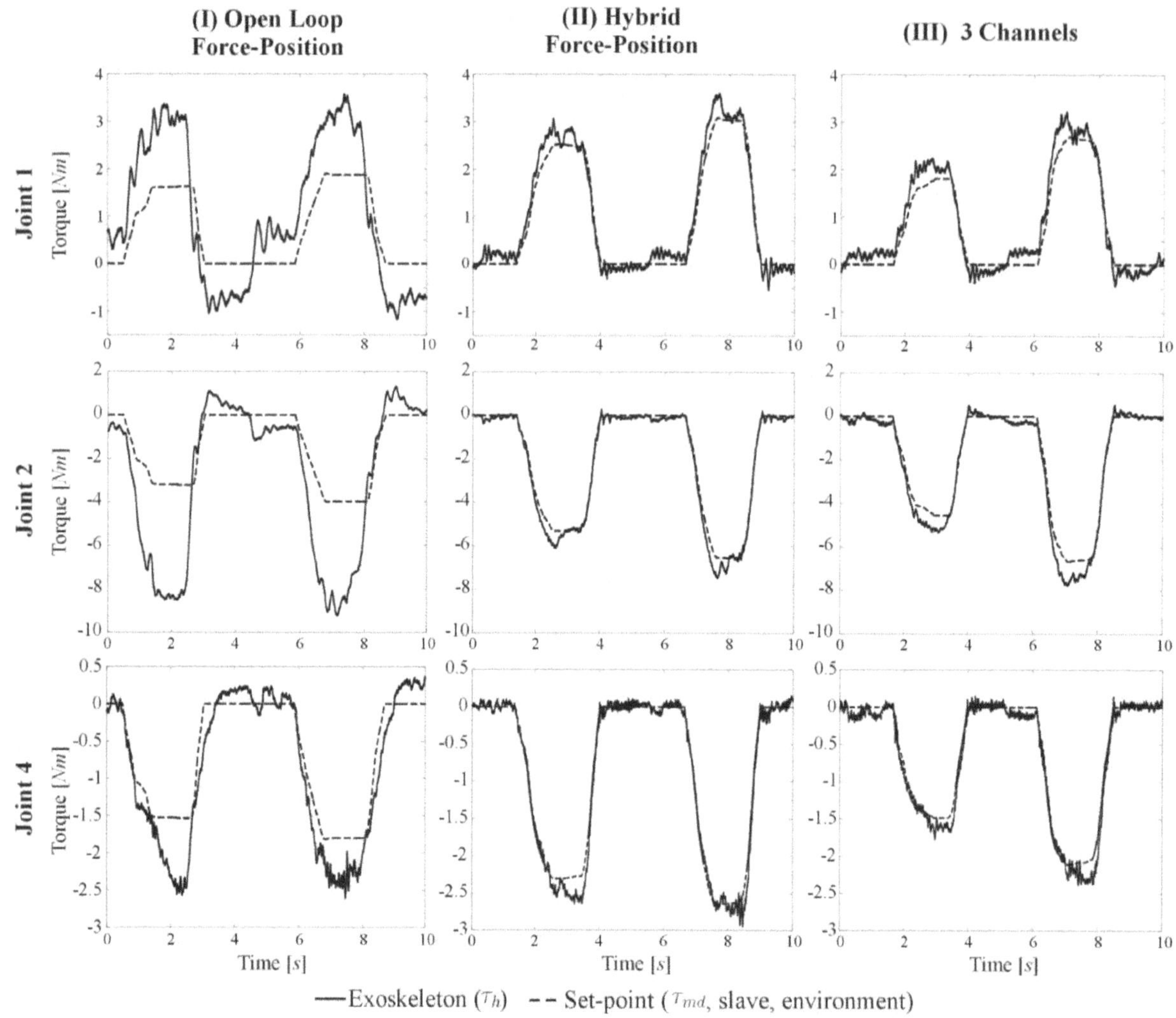

4. CONCLUSION

This chapter has provided the main aspects of the design and implementation of a 7 d.o.f. portable haptic arm exoskeleton. The kinematics structure and the selection of the actuators and sensors have been described. Compared to previous exoskeletons developments, the main characteristics of this device are:

- Portable system with 7 active d.o.f. to drive each motion of the human arm from the shoulder to the wrist;
- Compact joint design with integrated actuator, position and torque sensor;
- Optimization of the actuation chain to insure at the same time sufficient haptic feedback and portability, with the possibility of internal gravity compensation;
- On-board conditioning and power electronics that limit the number of wires along the structure and out of the exoskeleton.

Only four wires are needed to connect to the external computer;
- Hardware components based mostly on usual parts to simplify integration and reduce costs.

Several control strategies have then been compared with experiments in virtual reality and teleoperation. Thanks to the natural back drivability of the exoskeleton, the classical open loop Force-Position method can offer basic haptic rendering (only hard contact and without gravity compensation) in an easy and safe way. It is this strategy that was most of the time used during SAM demonstrations. In terms of pure performances and in case of more complex rendering or gravity compensation, although it requires the use of the torque sensors, best results, in terms of transparency and torque tracking, have been obtained with a hybrid Force-Position (coupling of model-based friction compensation and local torque feedback) and admittance control. In teleoperation, a 3 channels strategy has also been successfully implemented, improving the stability of the hybrid control in hard contact. In the frame of the EXOSTATION project, the system succeeded to simulate real life applications as contacts, manipulation of objects, peg in the hole, screwing,... through the use of the anthropomorphic slave arm robot. Users who tested the system were impressed by the ease of operation with the exoskeleton and recognized the advantages of the addition of force feedback information.

Although most of the used desktop haptic interfaces are now commercial products, haptic exoskeletons are still at the level of research developments. This is mainly due to the complexity of their design and control. An industrialization phase is still needed to enhance their global ergonomics (weight, fixations) and robustness (hardware and software). These improvements should open new application fields and a new ways to control robots. Exoskeletons are particularly adapted to the control of anthropomorphic arms. Several terrestrial applications are asking for remote teleoperation actions with human dexterity and intuitiveness. Beside space activities, examples of future applications are CBRNE (chemical, biological, radiological, nuclear and explosive) crisis management, bomb disposal, nuclear dismantling and submarine manipulations. The coupling between a slave robot arm with an intuitive and immersive controller in the form of a haptic exoskeleton would improve manipulation performances, reactivity and ability in such situations, avoiding risking life of operators.

ACKNOWLEDGMENT

This research was conducted, in the frame of the project EXOSTATION, jointly by *Space Application Services*, *Micromega Dynamics* and *ULB* on behalf of the *European Space Agency* within the GSTP program. Elvina Motard, Michel Ilzkovitz, Jean-Philippe Verschueren, Serge Cattoul, Sam Veillerette, Mihaita Horodinca, Iulian Romanescu and More Avraam contributed to the success of this project.

REFERENCES

Avraam, M., Horodinca, M., Romanescu, I., & Preumont, A. (2010). Computer controlled rotational MR-brake for wrist rehabilitation device. *Journal of Intelligent Material Systems and Structures*, *21*(15), 1543–1557. doi:10.1177/1045389X10362274

Bernstein, N. L., Lawrence, D. A., & Pao, L. Y. (2005). Friction modeling and compensation for haptic interfaces. *First IEEE Joint Eurohaptics Conference and Symposium on Haptic Interfaces for Virtual Environment and Teleoperator Systems,* (pp. 290-295).

Bicker, R., Burn, K., Hu, Z., Pongean, W., & Bashir, A. (2004). The early development of remote tele-manipulation systems. *International Symposium on History of Machines and Mechanims*, (pp. 391-404).

Bin, A., Massie, T., & Vayner, V. (1995). *Sensory feedback exoskeleton Armmaster.* (US Patent 1995/006376).

Bolopion, A., Cagneau, B., Redon, S., & Régnier, S. (2010). Haptic teleoperation for 3D microassembly of spherical objects. *IEEE Transaction on Mechatronics*, *99*, 1–12.

Carigan, C., Liszka, M., & Roderick, S. (2005). Design of an arm exoskeleton with scapula motion for shoulder rehabilitation. *International Conference on Advanced Robotics ICAR05* (pp. 524-531).

Carigan, C., & Liszka, M. S. (2007). *Portable arm exoskeleton for shoulder rehabilitation*. US Patent 2007/0225620.

Carigan, C., Tang, J., Roderick, S., & Naylor, M. (2007) A configuration-sapce approach to controlling a rehabilitation arm exoskeleton. *IEEE 10th International Conference on Rehabilitation Robotics,* (pp. 179-187).

Frisoli, A., Montagner, A., Borelli, L., Salsedo, F., & Bergamasco, M. (2009). A force-feedback exoskeleton for upper limb rehabilitation in Virtual Reality. *Applied Bionics and Biomechanics*, *6*(2), 115–126. doi:10.1080/11762320902959250

Gogoussis, A., & Donath, M. (1987). Coulomb friction joint and drive effects in robot mechanisms. *IEEE International Conference on Robotics and Automation*, (vol. 4, pp. 828-836).

Guizzo, E., & Goldstein, H. (2005). The rise of the body bots. *IEEE Spectrum*, *42*(10), 50–56. doi:10.1109/MSPEC.2005.1515961

Kapandji, I. A. (1997). *Physiologie articulaire: Schemas commentés de mécanique humaine (5ème edition), Tome 1: Members supérieurs*. Maloine.

Kim, Y., Lee, J., Lee, S., & Kim, M. (2005). A force reflected exoskeleton-type masterarm for human-robot interaction. *IEEE Transactions on Systems, Man and Cybernetics. Part A*, *35*(2), 198–212.

Klopcar, N., Tomsic, M., & Lenarcic, J. (2007). A kinematic model of the shoulder complex to evaluate the arm-reachable workspace. *Journal of Biomechanics*, *40*(1), 86–91. doi:10.1016/j.jbiomech.2005.11.010

Lawrence, D. A. (1993). Stability and Transparency in bilateral teleoperation. *IEEE Transactions on Robotics and Automation*, *9*(5), 624–637. doi:10.1109/70.258054

Letier, P. (2010). *Bras Exosquelette Haptique: Conception et Contrôle*. PhD Thesis, Active Structures Laboratory, Université Libre de Bruxelles, July 2010.

Letier, P., Avraam, M., Horodinca, M., Schiele, A., & Preumont, A. (2006). Survey of actuation technologies for body-grounded exoskeletons. *Eurohaptics Conference*, (pp. 497-500).

Nef, T., Guidali, M., & Riener, R. (2009). ARMin III - Arm therapy exoskeleton with an ergonomic shoulder actuation. *Applied Bionics and Biomechanics*, *6*(2), 127. doi:10.1080/11762320902840179

O'Malley, M., & Goldfarb, M. (2002). The effect of force saturation on the haptic perception of detail. *IEEE/ASME Transactions on Mechatronics*, *7*(3), 280–288. doi:10.1109/TMECH.2002.802725

Okamura, A. M. (2009). Haptic feedback in robot-assisted minimally invasive surgery. *Current Opinion in Urology*, *19*(1), 102–107. doi:10.1097/MOU.0b013e32831a478c

Passenberg, C., Peer, A., & Buss, M. (2010). A survey of environment-, operator-, and task-adapted controllers for teleoperation systems. *Mechatronics*, *20*(7), 787–801. doi:10.1016/j.mechatronics.2010.04.005

Perry, J. C., Rosen, J., & Burns, S. (2007). Upper-limb powered exoskeleton design. *IEEE/ASME Transactions on Mechatronics*, *12*(4), 408–417. doi:10.1109/TMECH.2007.901934

Rocon, E., & Pons, J. L. (2011). *Exoskeletons in rehabilitation robotics: Tremor suppression*. Springer Tracts in Advanced Robotics.

Rosen, J., Perry, J. C., Manning, N., Burns, S., & Hannaford, B. (2005). The human arm kinematics and dynamics during daily activities – Toward a 7 DOF upper limb powered exoskeleton. *International Conference on Advanced Robotics* (pp. 532-539).

Ryu, J. H., Artigas, J., & Preusche, C. (2010). A passive bilateral control scheme for a teleoperator with time-varying communication delay. *Mechatronics*, *20*(7), 812–823. doi:10.1016/j.mechatronics.2010.07.006

Schiele, A. (2008). Case study: The ergonomic EXARM exoskeleton. In *Wearable robots: Biomechatronic exoskeleton* (pp. 248–255). John Wiley & Sons Ltd.

Sitti, M., & Hashimoto, H. (2003). Teleoperated touch feedback from the surfaces at the nanoscale: Modeling and experiments. *IEEE/ASME Transactions on Mechatronics*, *8*(2), 287–298. doi:10.1109/TMECH.2003.812828

Tsagarakis, N., & Caldwell, D. (2003). Development and control of a "soft-actuated" exoskeleton for use in physiotherapy and training. *Autonomous Robots*, *15*, 21–33. doi:10.1023/A:1024484615192

Yinga, C., Jia-Fan, Z., Can-Jun, Y., & Bin, N. (2007). Design and hybrid control of the pneumatic force-feedback systems for Arm-Exoskeleton by using on/off valve. *Mechatronics*, *17*(6), 325–335. doi:10.1016/j.mechatronics.2007.04.001

Yokokohji, Y., & Yoshikawa, T. (1994). Bilateral control of master-slave manipulators for ideal kinesthetic coupling - Formulation and experiment. *IEEE Transactions on Robotics and Automation*, *10*(5), 605–620. doi:10.1109/70.326566

KEY TERMS AND DEFINITIONS

Anthropomorphic: Parallel and similar to the human kinematics

Exoskeleton: Articulated mechanical structure made of several links and joints that follow the anatomy of the operator and is attached to him

Force-feedback Device: Apparatus able to provide force/torque information to the user

Haptic: Linked to the sense of touch

Teleoperation: Remote control of a robotic device

Virtual Reality: Computer based simulation of a physical and/or graphical world

Chapter 6
Prototyping and Real-Time Implementation of Bipedal Humanoid Robots:
Dynamically Equilibrated Multimodal Motion Generation

Barkan Ugurlu
Toyota Technological Institute, Japan

Atsuo Kawamura
Yokohama National University, Japan

ABSTRACT

This chapter is aimed at describing a contemporary bipedal humanoid robot prototyping technology, accompanied with a mathematically rigorous method to generate real-time walking, jumping, and running trajectories that can be applied to this type of robots. The main strategy in this method is to maintain the overall dynamic equilibrium and to prevent undesired rotational actions for the purpose of smooth maneuvering capabilities while the robot is in motion. In order to reach this goal, Zero Moment Point criterion is utilized in spherical coordinates, so that it is possible to fully exploit its properties by the help of Euler's equations of motions. Such a strategy allows for characterization of the rotational inertia and therefore the associated angular momentum rate change terms, so that undesired torso angle fluctuations during walking and running are well suppressed. It enables prevention of backwards-hopping actions during jumping as well. To validate the proposed approach, the authors performed simulations using a precise 3D simulator and conducted experiments on an actual bipedal robot. Results indicated that the method is superior to classical methods in terms of suppressing undesired rotational actions, such as torso angle fluctuations and backwards-hopping.

DOI: 10.4018/978-1-4666-0176-5.ch006

1. INTRODUCTION

Starting from 1970s, humanoid robots have evolved in many aspects and they are still expected to be adapted within the dynamic human environment. Considering such an environment, people may walk, run, jump and interact with others in various cases. Therefore, a well-adapted bipedal humanoid robot should function this kind of human-like dexterity for the purpose of providing compatible human-humanoid interaction capabilities. Some futuristic aspects regarding this matter are discussed by Kajita & Sugihara (2009), within an extensive manner.

In humanoid robotics technology, bipedal trajectory generation is known to be one of the most challenging issues. Formally speaking, humanoids interact with the outer environment through floating base points, i.e., through their feet and hands. These unilateral contacts make them underactuated and thus nonlinear. Besides this structural property, yet another nonlinearity arouses from the motion interference between sagittal and lateral planes through intrinsic angular momentum (Ugurlu, 2010) rate change terms. All together with other unmentioned nonlinearities, bipedal trajectory generation may be regarded as a complicated task.

In general, bipedal motion planning algorithms bifurcate into two main approaches: a) Multi-mass based approach; b) Abstracted dynamics approach. The first approach is known to be accurate and usually utilized to build dynamic and precise simulation environments (Fujimoto, & Kawamura, 1998; Kanehiro et al., 2004). On the other hand, its real life applications on actual robots may suffer from various deficiencies. For instance, inaccurate CAD (Computer Aided Design) data of mechanical hardware and unmodelled dynamics -such as joint frictions and inherent compliance- can be listed in this matter. In addition to these difficulties, multi mass approach is considered to be computationally expensive and it has inherent drawbacks during the online implementation on built-in computers, which are usually placed in humanoid robots.

In the second approach, the overall humanoid dynamics are usually abstracted by employing different modeling tools, such as, planar and spatial inverted pendulum (Kajita, &Tani, 1991; Kajita et al. 2001), spring-loaded pendulum (Poulakakis, &Grizzle, 2009), variable impedance (Jafari et al. 2010) pendulum and so on. The basic idea is to capture the essence of dominant humanoid dynamic characteristics, so that researchers can obtain applicable motion generation and control techniques. Even though there may be no clear and quantitative way to express the actual accuracy of these methods in representing the dynamics, it is possible to quantify their effectiveness through experimental studies. What is more, dynamic error compensators (Takenaka et al., 2009a) might allow us to regulate some of the unmodelled aspects in these approaches.

ZMP (Zero Moment Point) (Vukobratovic, &Borovac, 2004; Kajita, &Espiau, 2008) appears to be a de facto standard while indicating the dynamic balance of a humanoid robot. Some other ground reference points can be examined in (Popovic et al., 2005) and (Hirukawa et al., 2006) as well. By the definition, the ZMP is a point on the floor, in which the total moment acting on the robot due to gravity, centroidal torques and inertial forces equals to zero. In order to ensure the dynamic balance, the ZMP response must be within the convex hull of the support polygon. If the actual ZMP reaches to the edge of the support polygon, the robot is considered to be losing its dynamic equilibrium. Three different scenarios are illustrated in Figure 1.

In comparing ZMP equations with inverted pendulum's mathematical model, one may easily think that they resemble to each other. Oftentimes, ZMP's mathematical model is considered to be forced pendulum equations. In other words, ZMP equations become equivalent to inverted pendulum equations in case of zero ZMP values, both for x-axis and y-axis (the homogenous case). Gener-

Figure 1.Three distinct scenarios of Zero Moment Point, illustrated in the case of a one-legged hopper (Ugurlu, & Kawamura, 2010a). a) The actual ZMP is within the boundaries of the support foot in the lateral plane, so it is dynamically balanced on this plane. Same principle is hold valid for the sagittal plane as well's) A dynamically unbalanced scenario in the sagittal plane. The ZMP is at the edge of the support polygon, thus the robot may be about to tip over. c) ZMP is undefined when the robot is in the air, however, if the ZMP reaches the edge of the support foot at the moment of touch-down, it may not recover from tipping over.

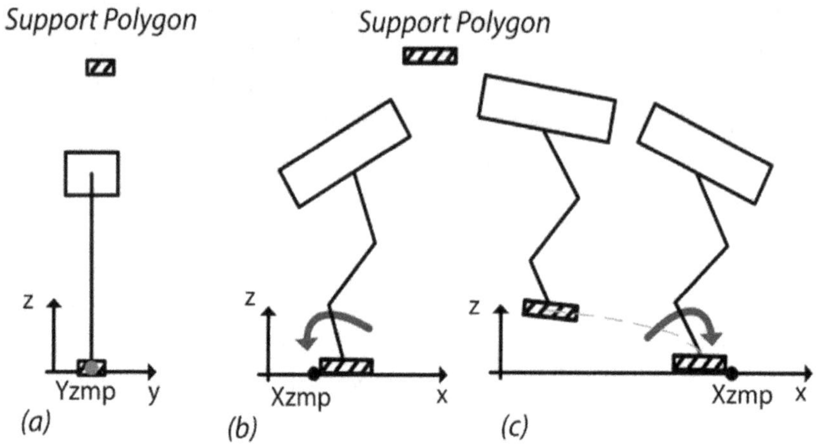

ally speaking, the main difference in these approaches is that the pivot is only a point in inverted pendulum model while it constitutes a certain area in x-y plane in ZMP model. This area is simply the support polygon.

Having captured an imported essence of humanoid dynamics, ZMP is widely used for trajectory generation and control algorithms. As we discuss later in detail, ZMP model includes inertial forces terms, gravity term and intrinsic angular momentum (Featherstone, 2008) rate change (IAMR) terms. Conventionally, IAMR terms are often omitted due to the hardship of melting these terms in simplified humanoid dynamics approach. On the other hand, these terms are associated with the rotational inertia, an important characteristic of humanoid dynamics. Ignoring IAMR terms, we lose rich information concerning the rotational behavior of the robot. Furthermore, ZMP equations become mass and inertia independent without IAMR terms. This fact may be interpreted as "*dynamic balance conditions for an 80 [kg] human-size humanoid and a 7.5 [kg] small-scale soccer humanoid are the same.*" Of course, we can heuristically claim that they have different characteristics in representing their own dynamic balance conditions. This simple example indicates the necessity of IAMR terms.

Some of the main drawbacks of ignoring IAMR terms can be summarized as follows:

- Actual stepping motion could be unusual. The body and legs perform twisting (Kajita et al., 2007) to cancel the certain portion of angular momentum as it is omitted during the motion generation stage.
- It may create unnatural torso angle fluctuations (Ugurlu, & Kawamura, 2010a; Tajima et al., 2009). To give a brief idea, some experimental results on a one-legged robot are displayed in Figure 2.
- The motion interference between sagittal and lateral plane vanishes. (Ugurlu, 2010).

Figure 2. Forward jumping experiments from a one-legged robot, (Ugurlu, & Kawamura, 2010a). a) IAMR terms are included in motion equations, so that torso angle variations are measured to be about 5 degrees at most. b) IAMR terms are omitted. Torso angle varied between 20 to –17 degrees and subsequently the robot tipped over.

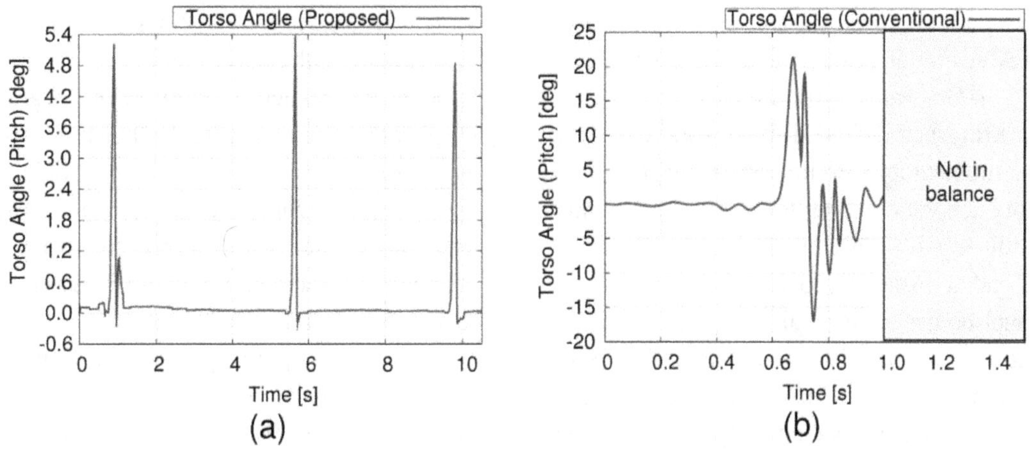

- ZMP becomes independent of inertial parameters (mass, rotational inertia) and loses rich information.

In the light of facts highlighted above, we defend that we are in need of bipedal trajectory generation techniques that fully exploit ZMP modeling by injecting IAMR terms into the pattern generation algorithm. To this end, we composed a method to generate ZMP-based feasible bipedal motion patterns in which IAMR terms are smoothly included. It is applicable to various motion types; bipedal walking, jumping and running while remedying some of the problems that may be accompanied with conventional approaches.

This chapter is organized as follows. Firstly, we provide literature review in bipedal motion planning. Afterwards, our experimental hardware and simulation environment are introduced with a special emphasis on the mechatronic prototyping technology of our robot. Subsequently, we disclose the mathematical description of our proposed bipedal motion generation techniques for walking, jumping and running, respectively. Related simulation and experimental results are also accompanied with each section. Following that, we discuss future research directions and then conclude the chapter. Some of the key mathematical derivations are presented in appendices.

2. BACKGROUND

As stated previously, we need to inject certain amounts of torques to CoM (Center of Mass) through IAMR terms during the motion generation task. On this matter, Kajita et al. proposed Resolved Momentum Control (RMC) (Kajita et al., 2007) in which joint rates can be computed by referential momenta inputs. In a similar manner, Orin, & Goswami (2008) derived centroidal momentum matrix of a general humanoid. Furthermore, Sugihara, & Nakamura (2007) developed a technique for 3D hopping motion of bipedal robots. In their ZMP-based method, IAMR terms are firstly ignored then included in the joint motion computation stage. The common ground in both these methods is that the angular momentum information is related with joint motions by employing different properties of multi-mass dynamics. Divergently, we defend that angular momentum information could be directly included in the CoM trajectory generation stage. In this manner, we can enrich abstracted bipedal motion generation techniques, generate more feasible trajectories and compute joint motions by using simple inverse kinematics. A similar strategy for this purpose is also pursued in (Goswami, 2008) as well.

Due to the facts defended above, we aim to inject angular momentum information during the CoM trajectory generation. Therefore, IAMR terms must be combined with other terms in ZMP equations. To this end, a misled approach is to sum up all the individual bodies' centroidal torques (Park, 2001; Popovic et al., 2005; Hofmann et al., 2009). Such an addition could be performed for inertial forces, however, individual links' torques are not additive unless they are projected to the CoM (Orin, &Goswami, 2008). Another problem in computing IAMR terms also appears when duplicating the planar ZMP equation to express ZMP equations in 3D (Huang et al., 2001). Strictly speaking, this is not acceptable for the 3D case, as motions in different planes mutually have influence on each other. In addition, IAMR terms associated with different axes have different formulations. Likewise, it may be argued that the techniques developed for planar cases (Mitobe et al., 2004; Pratt et al., 2006) cannot be fully applied to 3D bipedal motion planning.

Flywheel based methods also enable researchers to approximate angular momentum computations for bipedal motion planning (Pratt et al., 2006; Takenaka et al., 2009b). However, such methods can represent only one single rotation of the body, usually about the sagittal plane. On the other hand, basic straight walking and/or running motions include at least two simultaneous rotations, namely about sagittal and lateral planes. Figure 3 illustrates rotational actions both in flywheel models and our approach. Taking into account this fact, we may claim that they cannot represent the rotational motion in the lateral plane. The same fact also makes them impotent to indicate the motion interference between sagittal and lateral planes. Moreover, such methods treat rotational inertia as a single constant element. On the contrary, the real rotational inertia of a 3D bipedal is defined with tensors and its elements are functions of joint motions (Balafoutis & Patel, 1991a).

Considering the above discussions, we propose a method to generate ZMP-based feasible bipedal motion patterns that may ensure the dynamic balance. In our proposed method, ZMP equations are treated in spherical coordinates, so that IAMR terms can be combined with inertial forces by using Euler's equations of motion. Since the intrinsic angular momentum rate change is computed by using Euler's equations of motion, we name it Eulerian ZMP Resolution (EZR).

Authors group firstly developed this method for a planar case to generate jumping motions on a one-legged robot (Ugurlu, & Kawamura, 2010a). Subsequently, the method is enhanced for a 3D scenario in order to generate running motions for bipedal robots by considering unsymmetrical robot geometry (Ugurlu, & Kawamura, 2009). Afterwards, EZR is employed for bipedal walking pattern generation (Ugurlu, & Kawamura, 2010b).

This book chapter is intended to be the complementary report about Eulerian ZMP Resolution method, through the helpful discussions and improvements. Being composite rigid bodies, bipedal robot's rotational inertia tensor elements are functions of joint motions. This chapter provides an improved and detailed mathematical description on this matter. What is more, we disclose an efficient algorithm to implement classical ZMP-based bipedal motion generation task. Latest simulation and experimental results are also included together with detailed comparison studies to demonstrate our method's efficiency clearly.

In addition, this chapter also provides detailed explanation about development procedures of the bipedal robot MARI-3, with a special emphasis on its unique features. Real-time implementation challenges are also addressed.

Figure 3.a) Illustration of flywheel model based rotational motion planning (Takenaka et al. 2009b). As it may be observed in the figure, this model can only represent a single rotation of the robot body, usually in the sagittal plane. b) Rotational motions in our approach (Eulerian ZMP Resolution). In this method, the actual robot body is considered to be an equivalent composite rigid body that is oriented identically to the CoM. In contrast to flywheel models, it can represent two successive rotations, both in sagittal and lateral planes.

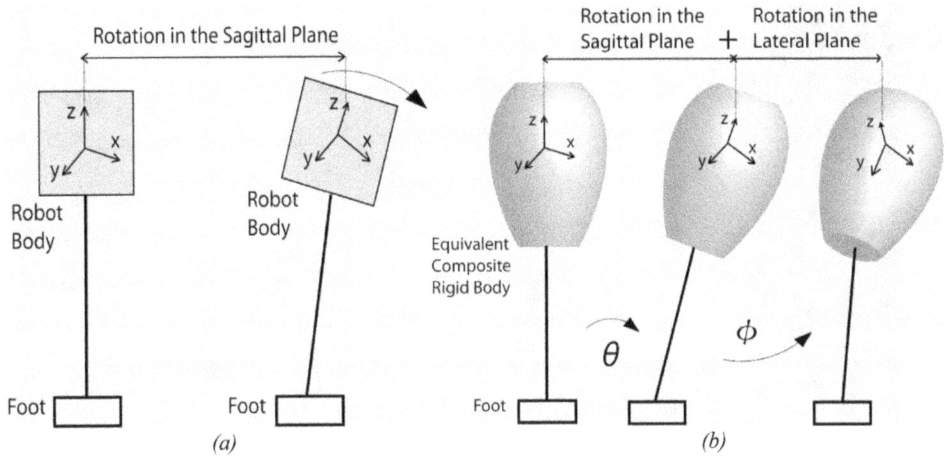

3. PROTOTYPING OF BIPEDAL ROBOTS

Bipedal Robot MARI-3: General Specifications

In order to explore bipedal motion dynamics, we developed a bipedal robot called MARI-3 (Zhu, & Kawamura, 2010). Table 1 summarizes its mechanical specifications. MARI-3 has a total of 13 DOF(degrees of freedom); 6 DOF in each leg and 1 DOF at the waist to be able to compensate the undesired yaw moment. Furthermore, it is equipped with 6-axis Force/Torque (F/T) sensors at the ankles. In addition, MARI-3 has a triaxial rate gyro sensor and an accelerometer located at the pelvis. In its electronic hardware structure, the main controller is a 32 bit built-in Renesas SH-4 microprocessor. Data communication is performed via RON (RObot Network) bus, which is a serial communication bus system, developed in our laboratory. Figure 4 shows the actual robot, joint frames and its ROCOS (Fujimoto, & Kawamura, 1998) simulation model. ROCOS is a 3D dynamic simulator, also developed in our laboratory.

Power/Weight Ratio

In order to achieve dynamic and fast maneuvering, the robot has to be as light as possible whilst it is able to exert high power. On this matter, knee joints are of primary importance. Therefore, we utilized two identical motors in parallel to obtain high power output. Figure 5 (a) illustrates this design approach.

Table 1. Mechanical specifications of bipedal robot MARI-3

Size	Upper Leg Length: 300 [mm]
	Lower Leg Length: 300 [mm]
	Ankle-Sole Length: 110 [mm]
Weight	Each Leg: 13.47 [kg]
	Upper Body: 6.683 [kg]
	Total: 33.64 [kg]

Figure 4. a) Bipedal robot MARI-3. b) Joint frames. c) MARI-3's simulation model in ROCOS.

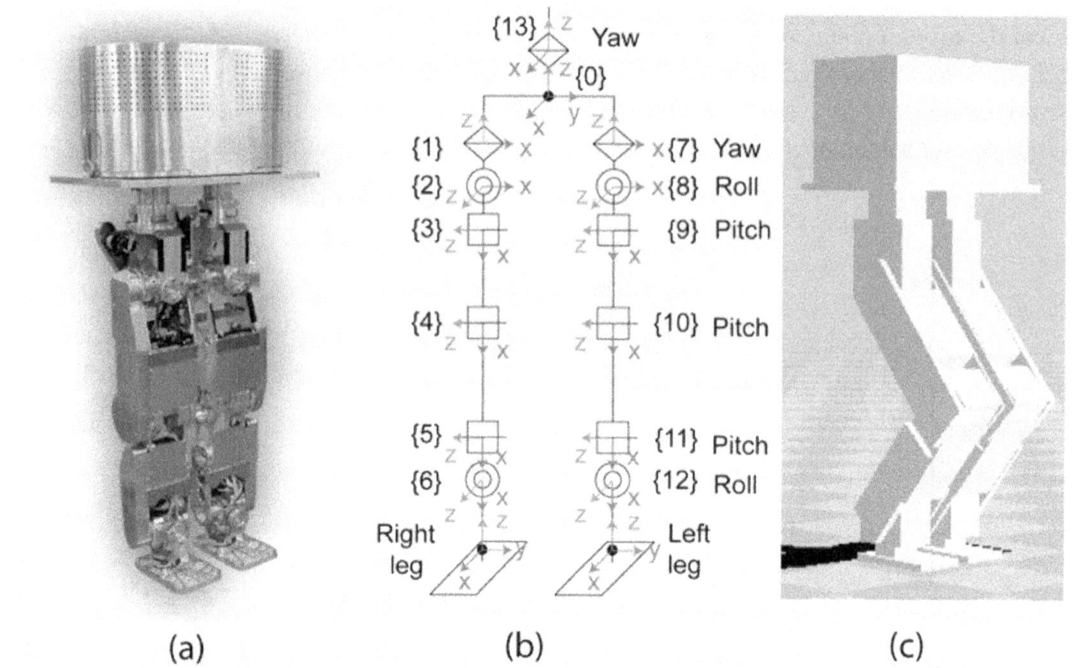

The main structural material used in MARI-3 is aluminum alloy A7075, which is light but has high strength and good machinability. While further lightening the weight without compromising the strength and rigidness, various measures are taken. To name a few, hollow block structures are used for hip and ankle joints, parallel-plate beams enhanced with hollow rods are utilized in link structures, titanium bolts are employed to connect mechanical parts. Together with other design efforts, power/weight ratio is maximized to achieve dynamically fast maneuvering.

Joint Ranges

One of the main objectives in our design is to have wide range of motion capabilities in the joint space. For this purpose, we put a special emphasis on the mechanical design and maximized joint ranges. Figure 5-(b) displays an extreme situation to prove the motion capabilities of MARI-3, especially its pitch axis joints'. Some of the key specifications of MARI-3's joints are listed in Table 2.

On the other hand, some of the control and communication boards are distributed within thigh and shin links as shown in Figure 5 (c). Hence, the structure of links and joints are elaborately designated to prevent potential interference between the electrical components and the mechanical structure. Yet another problem may occur when a joint is driven to rotate beyond its range because of an electronic malfunction and/or software failure. This may cause it to harm the robot beyond the maintenance. To avoid this situation, we placed mechanical stoppers in each joint to make sure that the joints cannot exceed their designated ranges. Figure 5 (d) depicts one of these mechanical stoppers, located at the hip yaw joint.

Electrical Design

The overall electronic control system is summarized in Figure 6. The software layer can be grouped into two branches, namely hardware interface and controller layer. The hardware in-

Figure 5. a) Two identical motors are connected in parallel in knee joints to amplify power output. b) Wide joint ranges of MARI-3 enable it to have great motion flexibility. c) Each RON unit and servo amplifier is distributed in a close proximity to the corresponding joint. d) Mechanical stoppers are placed to prevent accidental extreme motions for providing safety while experimenting.

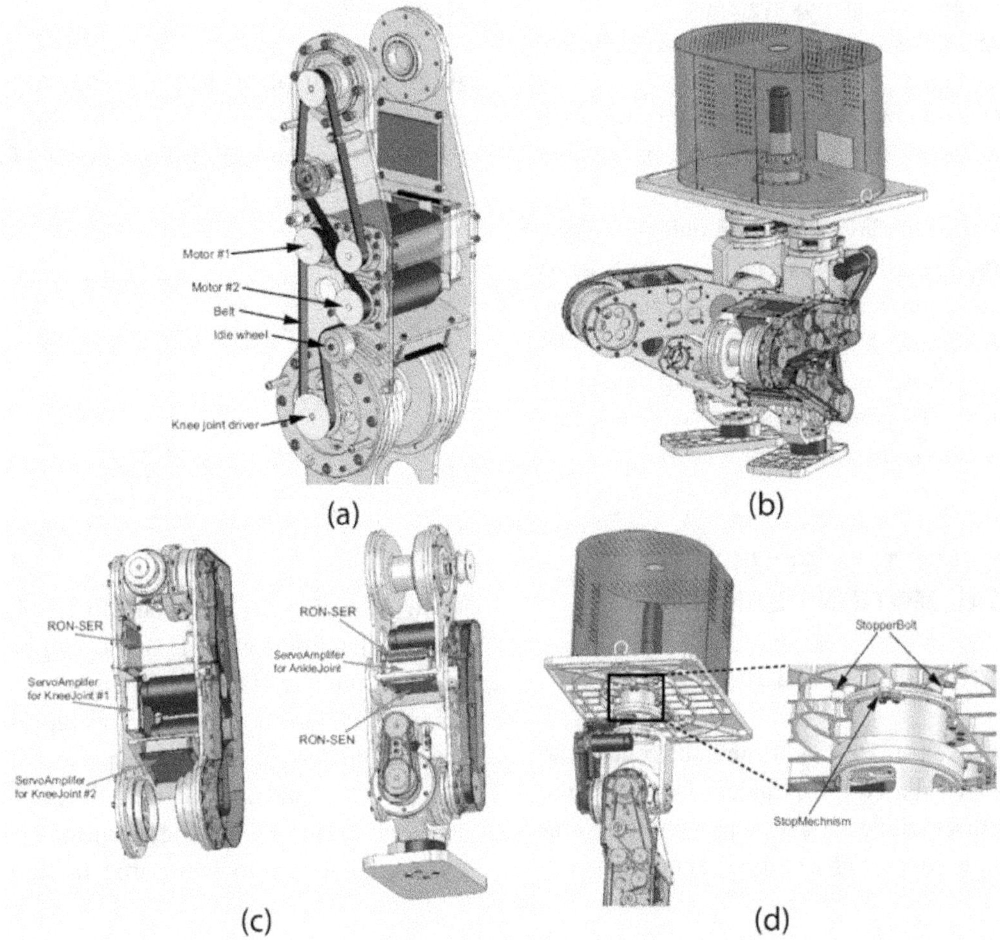

Table 2. Joint specifications of MARI-3. Note that corresponding joints in two legs have identical specifications.

Joint	Axis	Motor Power [W]	Conveyor-Belt Ratio	Gear Ratio	Movable Range [Deg]	Max. Torque [Nm]
Waist	Yaw	150	-	50×4.3	+15 ~ -15	260
Hip	Yaw	60	1.6	100	+45 ~ -45	86
Hip	Roll	90	1.6	100	+45 ~ -45	100
Hip	Pitch	150	1.2	100	+120 ~ -45	145
Knee	Pitch	150×2	1.2	100	+70 ~ -150	290
Ankle	Pitch	150	1.6	100	+100 ~ -80	97
Ankle	Roll	90	1.6	100	+30 ~ -30	100

terface can further be divided into three groups as follows.

1. **RON (RObot Network) interface:** Communication bus system.
2. **Data Transform layer:** Transforms binary data to real values.
3. **I/O layer:** Manages the communication with the host PC (Personal Computer).

The controller layer reads the state of the system from the hardware interface layer and sends the torque commands to each joint depending on the application. As it may be understood, RON bus is a key factor in providing an efficient communication between all the subsystems. Its main configuration is displayed in Figure 7.

4. DYNAMICALLY EQUILIBRATED BIPEDAL MOTION PLANNING

Eulerian ZMP Resolution (EZR)

In a basic sense, EZR is a method to generate feasible ZMP-based CoM trajectories for bipedal motions that includes single support phases. Naturally, single support phases are the common both in walking, jumping and running motions. As for double support phases during walking and flight phases during jumping/running, we define proper boundary conditions to connect these phases seamlessly.

A. Single Support Motion Planning

Firstly, let us analyze ZMP equations (Tajima et al., 2009):

$$X_{zmp} = \overbrace{\frac{x(\ddot{z}+g)-z\ddot{x}}{(\ddot{z}+g)}}^{1^{st}} - \overbrace{\frac{\dot{L}_y}{m(\ddot{z}+g)}}^{2^{nd}} \qquad (1)$$

$$Y_{zmp} = \overbrace{\frac{y(\ddot{z}+g)-z\ddot{y}}{(\ddot{z}+g)}}^{1^{st}} + \overbrace{\frac{\dot{L}_x}{m(\ddot{z}+g)}}^{2^{nd}} \qquad (2)$$

In (1) and (2), x, y and z stand for CoM position in Cartesian frame while L_x and L_y symbolize roll axis and pitch axis intrinsic angular momentum about CoM. One dot and two dots represent first and second derivatives with respect to time. Further, m is the total mass and g is the gravitational acceleration. In EZR method, we are going to group ZMP equations into two parts: Inertial Forces

Figure 6. Hierarchical control system

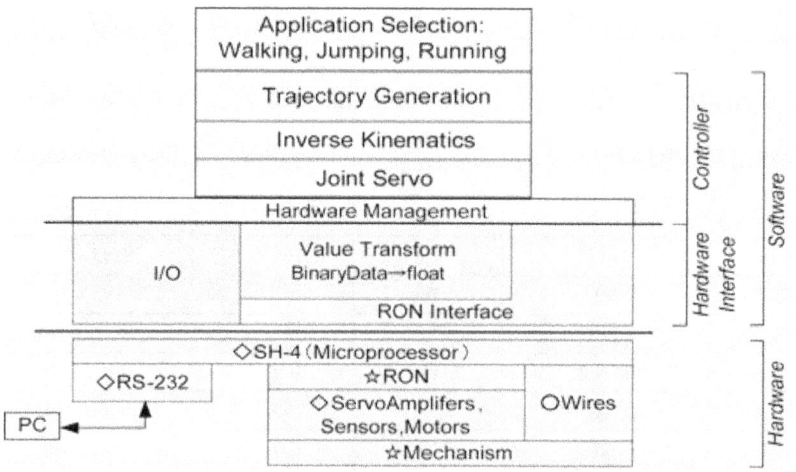

Figure 7. System configuration in RON bus communication system

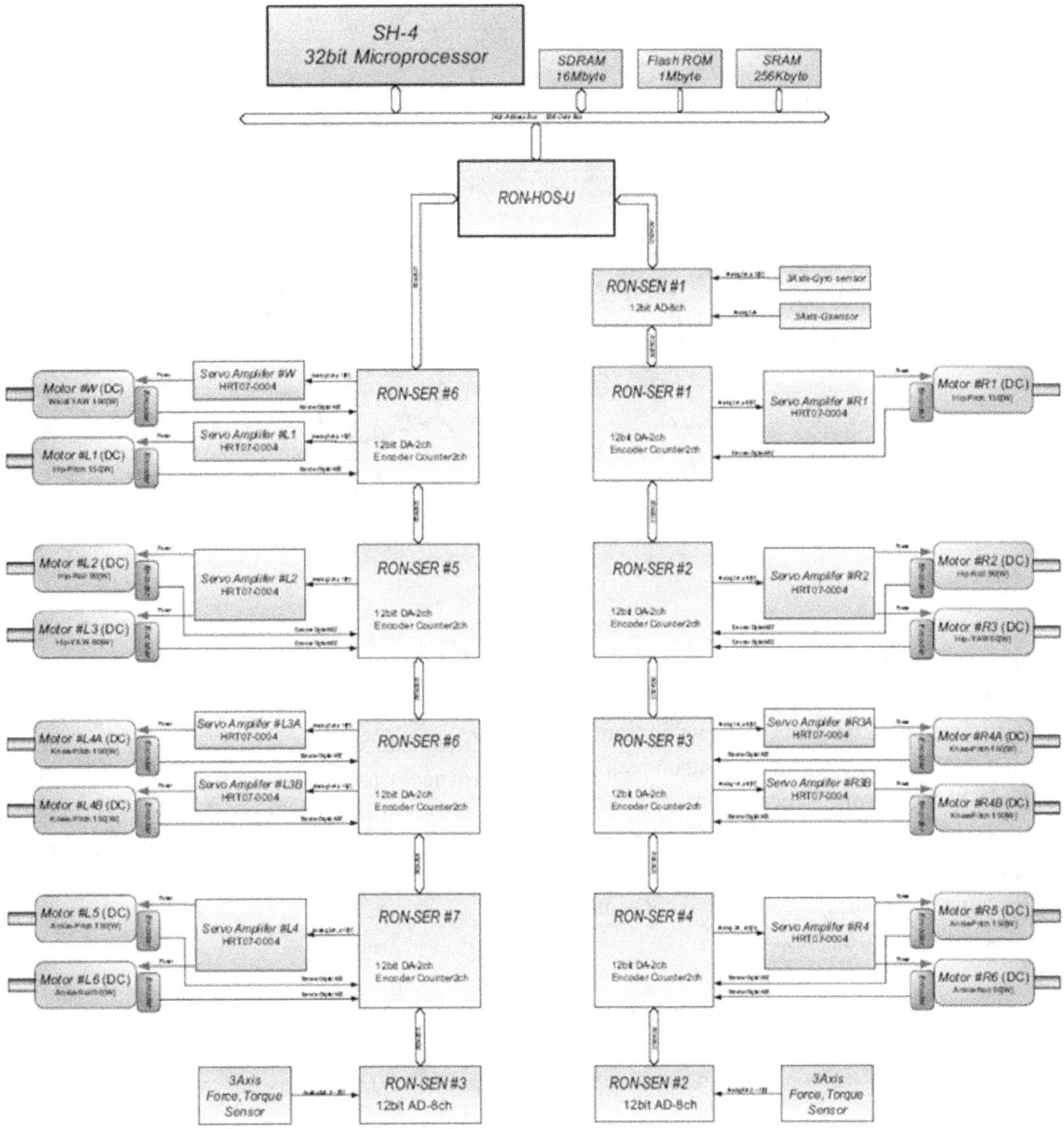

Terms and Rate Change of Angular Momentum Terms, which are the first and second terms in (1) and (2) respectively.

Throughout single support phases, the robot is considered as a composite rigid body, rotating about a fixed point (pivot frame's origin) and it is in contact with the floor through a rectangular foot as illustrated in Figure 8. For simplicity, we assume that the frame that is attached to the CoM is oriented in an identical manner to the equivalent rigid body. A similar strategy is also used in Timoshenko's (Timoshenko, & Young, 1948) pendulum as well. In this model, the CoM configuration is defined in the spherical coordinate frame by using the parameters r (CoM length), θ angle and ϕ angle.

Figure 8. Composite rigid body model and spherical coordinates on MARI-3. The robot is considered as a composite rigid body with variant rotational inertia. Its CoM is assumed to be oriented identically. The origin of Cartesian frame and pivot frame are in fact coincided. The offset is left for illustrative purposes.

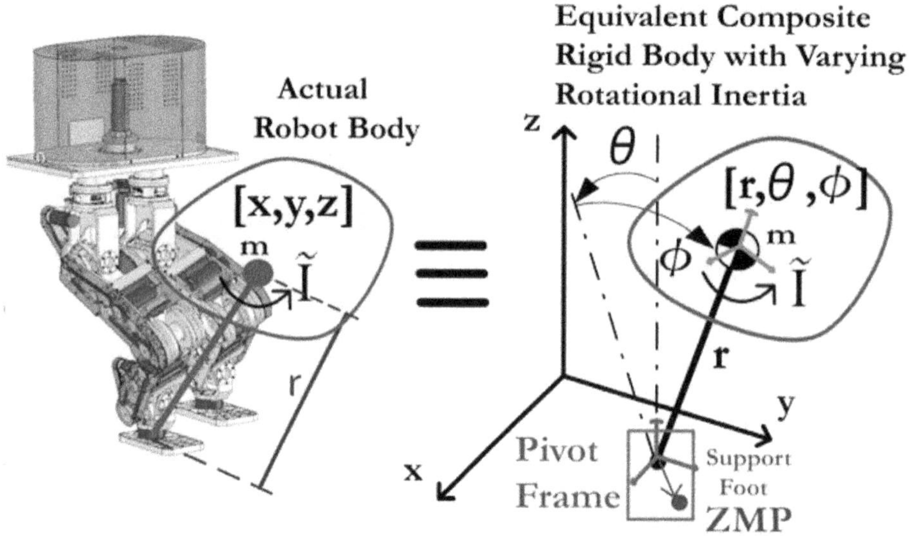

The main reason for utilizing spherical coordinates is to express equivalent rigid body's angular velocity in compact forms, using θ and ϕ parameters. Hereby, it becomes quite convenient to obtain IAMR terms, which are often omitted in classical methods. We later discuss the way we express this angular velocity vector.

For our convenience, the spherical coordinate frame is defined as follows:

1. Go r units along the +z axis (translational).
2. Rotate the frame through θ about +y axis (pitch, the rotational action in sagittal plane).
3. Rotate the frame through ϕ about +x axis (roll, the rotational action in lateral plane).

Employing such a model, we are able to combine IAMR terms with inertial force terms that appear in ZMP and yaw moment equations. Henceforward, we are going to express these terms in the spherical coordinate frame that is defined above.

A.1. Inertial Forces Terms

In order to express inertial forces terms by using the spherical coordinate system's parameters, we perform necessary coordinate transformation for CoM position. Subsequently, we differentiate position expression two times and obtain CoM acceleration.

$$x = r\sin\theta \tag{3}$$

$$y = r\cos\theta\sin\phi \tag{4}$$

$$z = r\cos\theta\cos\phi \tag{5}$$

$$\ddot{x} = \ddot{r}\sin\theta + 2\dot{r}\dot{\theta}\cos\theta + r\ddot{\theta}\cos\theta - r\dot{\theta}^2\sin\theta \tag{6}$$

$$\begin{aligned}\ddot{y} =& \ddot{r}\cos\theta\sin\phi + r\ddot{\phi}\cos\theta\cos\phi - r\ddot{\theta}\sin\theta\sin\phi \\ &+2\dot{r}\dot{\phi}\cos\theta\cos\phi - (r\dot{\theta}^2 + r\dot{\phi}^2)\cos\theta\sin\phi \\ &-2r\dot{\theta}\dot{\phi}\sin\theta\cos\phi - 2\dot{r}\dot{\theta}\sin\theta\sin\phi\end{aligned} \tag{7}$$

$$\ddot{z} = \ddot{r}\cos\theta\cos\phi - r\ddot{\phi}\cos\theta\sin\phi - r\ddot{\theta}\sin\theta\cos\phi$$
$$-2\dot{r}\dot{\phi}\cos\theta\sin\phi - (r\dot{\theta}^2 + r\dot{\phi}^2)\cos\theta\cos\phi$$
$$-2r\dot{\theta}\dot{\phi}\sin\theta\sin\phi - 2\dot{r}\dot{\theta}\sin\theta\cos\phi \tag{8}$$

A.2. Intrinsic Angular Momentum Rate Change Terms

Supposing that the friction between the floor and the foot sole is sufficient and there is no foot rotation, we can consider the support phase motion as a rotation of a rigid body about a fixed point as previously stated. For this case, Euler's equations of motion for composite rigid bodies might give us insights,

$$\dot{L}_x = I_{xx}\dot{\omega}_x - (I_{yy} - I_{zz})\omega_y\omega_z + \dot{I}_{xx}\omega_x \tag{9}$$

$$\dot{L}_y = I_{yy}\dot{\omega}_y - (I_{zz} - I_{xx})\omega_z\omega_x + \dot{I}_{yy}\omega_y \tag{10}$$

$$\dot{L}_z = I_{zz}\dot{\omega}_z - (I_{xx} - I_{yy})\omega_x\omega_y + \dot{I}_{zz}\omega_z \tag{11}$$

in which I_{xx}, I_{yy} and I_{zz} are moments of inertia at composite CoM about principle axes, ω_x, ω_y and ω_z are angular velocities about principle axes and $\dot{L}_x$, $\dot{L}_y$ and $\dot{L}_z$ are IAMR terms about roll pitch and yaw axes. Usually, Euler's equations are defined for rigid bodies with constant rotational inertia elements. In this study, we presented a modification to this equation set in a way that it includes rotational inertia elements' time derivatives, as they are not constant in bipedal robots. Necessary derivation can be examined in the Appendix A. In addition, we omit products of inertia due to the fact that they are 50 to 300 times smaller than diagonal elements for MARI-3 while it performs walking and running.

As it may be understood from (9), (10) and (11), we primarily need to express angular velocity, $\vec{\omega}$, and angular acceleration, $\vec{\dot{\omega}}$. For this purpose, let us write down equivalent rigid body's CoM orientation by considering 2nd and 3rd rotational actions while defining the spherical coordinate frame.

$$Q_s = \begin{bmatrix} \cos\theta & 0 & \sin\theta \\ -\sin\theta\sin\phi & \cos\phi & \cos\theta\sin\phi \\ -\sin\theta\cos\phi & -\sin\phi & \cos\theta\cos\phi \end{bmatrix} \tag{12}$$

Utilizing a tensorial approach (Balafoutis, & Patel, 1991a), CoM angular velocity can be expressed in skew symmetric form ($\omega^\dagger$).

$$\omega^\dagger = \dot{Q}_s Q_s^T = \begin{bmatrix} 0 & \dot{\theta}\sin\phi & \dot{\theta}\cos\phi \\ -\dot{\theta}\sin\phi & 0 & \dot{\phi} \\ -\dot{\theta}\cos\phi & -\dot{\phi} & 0 \end{bmatrix} \tag{13}$$

CoM angular velocity can simply be extracted from (13) as shown in (14).

$$\omega_x = -\dot{\phi}$$
$$\omega_y = \dot{\theta}\cos\phi \tag{14}$$
$$\omega_z = -\dot{\theta}\sin\phi$$

Differentiating CoM angular velocity vector, we can obtain CoM angular acceleration vector as well.

$$\dot{\omega}_x = -\ddot{\phi}$$
$$\dot{\omega}_y = \ddot{\theta}\cos\phi - \dot{\theta}\dot{\phi}\sin\phi \tag{15}$$
$$\dot{\omega}_z = -\ddot{\theta}\sin\phi - \dot{\theta}\dot{\phi}\cos\phi$$

Inserting (14) and (15) into (9), (10) and (11), we can obtain $\dot{L}_x$, $\dot{L}_y$ and $\dot{L}_z$ terms.

$$\dot{L}_x = (I_{yy} - I_{zz})\dot{\theta}^2\sin\phi\cos\phi - Ixx\ddot{\phi} - \dot{I}_{xx}\dot{\phi} \tag{16}$$

$$\dot{L}_y = I_{yy}(\ddot{\theta}\cos\phi - \dot{\theta}\dot{\phi}\sin\phi)$$
$$+\dot{I}_{yy}\dot{\theta}\cos\phi + (I_{xx} - I_{zz})\dot{\theta}\dot{\phi}\sin\phi \tag{17}$$

$$\begin{aligned}\dot{L}_z &= -I_{zz}(\ddot{\theta}\sin\phi + \dot{\theta}\dot{\phi}\cos\phi) \\ &-\dot{I}_{zz}\dot{\theta}\sin\phi + (I_{xx} - I_{yy})\dot{\theta}\dot{\phi}\cos\phi\end{aligned} \tag{18}$$

In above computations, rotational inertia elements and their time derivatives are computed with respect to joint motions by using the algorithm presented in (Balafoutis, & Patel, 1991b). We later discuss this issue in the next subsections.

A.3. Solving ZMP Equations for θ and ϕ Trajectories

Having expressed both inertial forces terms and IAMR terms by using (r, θ, ϕ) parameters, we are going to combine them in ZMP equations and then solve to obtain θ and ϕ trajectories. Firstly, let us define Tx and Ty which come out from cross multiplication in (1) and (2) to eliminate fractional expressions.

$$\begin{aligned}T_x &= \overbrace{mX_{zmp}(\ddot{z}+g)}^{LHS} = \overbrace{mx(\ddot{z}+g) - mz\ddot{x} - \dot{L}_y}^{RHS} \\ T_y &= mY_{zmp}(\ddot{z}+g) = my(\ddot{z}+g) - mz\ddot{y} + \dot{L}_x\end{aligned} \tag{19}$$

While solving ZMP equations, z-axis CoM trajectory (z) is usually predetermined. Mostly, it is defined by polynomials (Ugurlu, & Kawamura, 2010a) for jumping and running planning, in order to meet the vertical motion constraints. During walking, it may be kept constant (Ugurlu et al., 2009) or given in advance for energy efficiency (Kormushev et al., 2011). Since it is predetermined, we know its value as well as its derivatives' values for all times. In this case, we need to express r and $\dot{r}$ in terms of z, θ, ϕ and their time derivatives.

$$\begin{aligned}r &= z\sec\theta\sec\phi \\ \dot{r} &= \dot{z}\sec\theta\sec\phi + r(\dot{\theta}\tan\theta + \dot{\phi}\tan\phi)\end{aligned} \tag{20}$$

Having designated feasible X_{zmp} and Y_{zmp} referential inputs, and using the predetermined z-axis CoM trajectory, left hand sides (LHS) of (19) can be calculated. Subsequently, if we insert (3), (4), (5), (6), (7), (8), (16), (17), (18) and (20), into right hand sides of (RHS) (19), we obtain following motion equations.

$$\begin{aligned}T_x &= -0.5\sin 2\theta\sin\phi(J_a(\ddot{\phi}+\dot{\phi}^2+\dot{J}_a\dot{\phi}) \\ &-\ddot{\theta}(J_a\cos\phi - I_{yy}\cos\phi) \\ &-\dot{\theta}(\dot{I}_{yy}\cos\phi + \dot{J}_a\cos\phi) \\ &+\dot{\theta}\dot{\phi}\sin\phi(J_a\sin^2\theta - I_n) + mgr\sin\theta\end{aligned} \tag{21}$$

$$\begin{aligned}T_y &= -\ddot{\phi}(I_{xx}J_a\cos^2\theta) + \dot{\theta}\dot{\phi}J_a\sin 2\theta \\ &+I_t\dot{\theta}^2 + mgr\cos\theta\sin\phi\end{aligned} \tag{22}$$

In (21) and (22), we employed the repeating sub-expressions expressed below, for the purpose of easing our computations.

$$\begin{aligned}J_a &= mr^2 \\ J_a &= 2mr\dot{r} \\ I_n &= I_{xx} - I_{yy} - I_{zz} \\ I_t &= 0.5(I_{yy} - I_{zz})\sin 2\phi\end{aligned} \tag{23}$$

Equations (21) and (22) describe a pair of coupled 2nd order differential equations. These equations can be solved numerically by using Runge-Kutta method (Press et al., 1992) when following parameters are provided.

- **z-axis CoM trajectory:** z
- **Referential ZMP inputs:** X_{zmp} and Y_{zmp}
- **Initial values:** $\theta_{[0]}, \dot{\theta}_{[0]}, \phi_{[0]}, \dot{\phi}_{[0]}$.
- **Support Phase Time Interval:** (t_{start}, $t_{start}+t_{width}$)
- **Swing Leg Parameters:** Stride, Foot Height

Having computed θ and ϕ, we may calculate x-axis and y-axis CoM trajectories simply by performing a conversion from spherical coordinates to Cartesian coordinates. As next step, we apply an explicit CoM-based inverse kinematics algorithm (Ugurlu. & Kawamura, 2010b) to obtain referential joint angles. These referential joint angles are treated in local servo blocks to compute joint torques in which position control, velocity control and disturbance observers are employed (Zhu, & Kawamura, 2010). When needed, necessary control algorithms (Ugurlu et al., 2009) could be utilized to ensure stability.

A.4. Inertia Update Process

Figure 9 summarizes θ and ϕ trajectories' computation diagram, along with the inertia update process in a basic manner. In this figure, subscripts denote discrete steps, (0, 1, 2…). At the beginning of the first cycle (i=0), inertia tensor elements are already known. Further, their derivatives are zero at this time. Inserting these parameters all together with initial conditions and z-axis CoM trajectory, we can solve (21) and (22) for the next cycle's (θ,ϕ) values with their derivatives. This (θ,ϕ) couple is used to compute joint motions, which are associated with the next step. Since we obtain the new joint motions, inertia tensor elements and their derivatives can be computed by using the algorithm described in (Balafoutis, & Patel, 1991b). These new inertia tensor elements and their derivatives are used for the subsequent cycle's (θ,ϕ) calculation. This process is recursively repeated throughout the single support phase in each control cycle.

B. Double Support Phase Planning for Walking Motion

Double support phase and swing leg trajectories are planned in a similar fashion as proposed in (Ugurlu et al., 2009). They are generated using 5th and 6th order polynomials (Press et al., 1992) to create a seamless connection between switching single support phases, both in position, velocity and acceleration levels.

C. Flight Phase Planning for Jumping and Running Motion

In the flight phase, a well-designed foot trajectory ensures a safe landing and stable initial conditions for the next support phase. Moreover, adjusting z-axis of foot trajectory may extend the flight time as well. Figure 10 depicts a flight phase where light and dark dashed lines indicate CoM and foot trajectories respectively. ψ_x is the angle between z and x axes while ψ_y is the angle between z and y axes when the robot is in the air. Subscripts lo and td symbolize lift-off and touch-down. As we mentioned previously, we cannot use ZMP here. Thus, we employ projectile motion dynamics in this phase.

Firstly, z-axis CoM trajectory follows a parabola simply formulated as below.

Figure 9. Inertia elements and their derivatives are updated concurrently in each control cycle

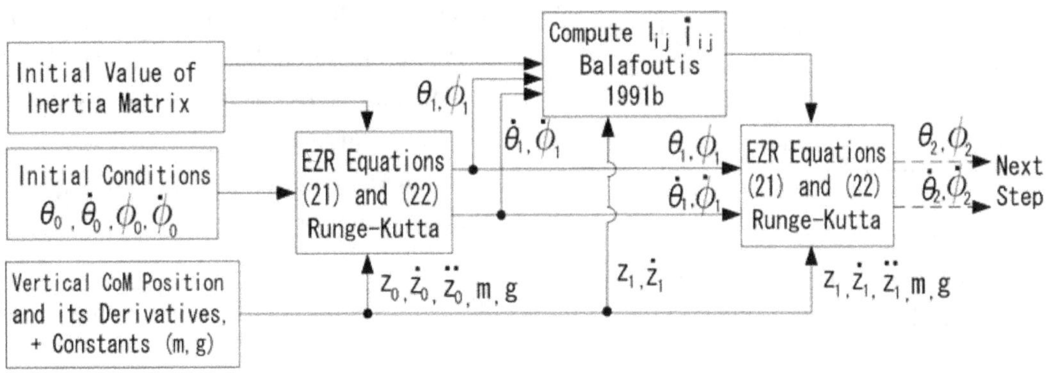

Figure 10. Illustration of a flight phase

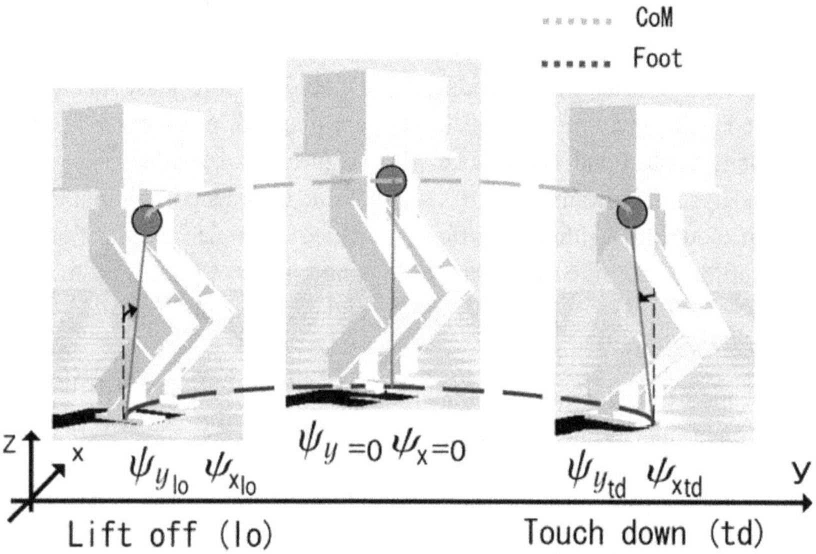

$$z_{CoM} = z_{lo} + \nu_{zlo}t - \frac{1}{2}gt^2 \tag{24}$$

In (24), z_{CoM}, z_{lo}, ν_{zlo} and t stand for z-axis CoM trajectory, z-axis lift-off position, z-axis lift-off velocity and time variable. Furthermore, z-axis foot trajectory (z_{foot}) can be calculated using polynomials as done in double support phase planning. Below, we explain the flight phase motion planning for the lateral plane. An identical approach is executed for the sagittal plane as well.

For y-axis CoM trajectory, we use the translational momentum conservation (Kajita et al., 2007) since there is no force acting on the robot along this axis. In other words, y-axis CoM acceleration is zero; $\ddot{y}_{CoM} = 0$. Following, we discretize this equation using the sample time of Δt.

$$\ddot{y}_{CoM} = \frac{y_{CoM[k+1]} - 2y_{CoM[k]} + y_{CoM[k-1]}}{\Delta t^2} \tag{25}$$

Above, k denotes the discrete step. Adding one delay into (25), we may solve it for $y_{CoM[k]}$.

$$y_{CoM[k]} = 2y_{CoM[k-1]} - y_{CoM[k-2]} \tag{26}$$

Utilizing (26), we can obtain y-axis CoM trajectory during the flight phase when $(y_{CoM[0]}, y_{CoM[1]})$ pair is given.

What is more, ψ_y angular trajectory can be calculated by polynomials between ψ_{lo} and ψ_{td} angles. Mostly, ψ_{lo} is determined during single support phase by considering the single support period, referential ZMP trajectory and mechanical capabilities of the given robot. In MARI-3's case, this angle could be set between 4 to 12 degrees. As for ψ_{td}, we simply equalize it to ψ_{lo}. In other words, if the robot leaves the ground with 7 degrees, it returns with the same angle in a cyclic manner. However, we think that if ψ_{td} angle is chosen differently, the robot can accelerate or decelerate.

As we calculated y_{CoM}, z_{CoM}, z_{foot} and ψ_y, foot location planning for y-axis can be expressed by the following formula.

$$y_{foot} = y_{CoM} - (z_{CoM} - z_{foot})\tan\psi_y \tag{27}$$

Since we follow identical approach for flight phase motion planning in x-z plane, x-axis CoM and foot trajectories can be shown as in the following.

$$x_{CoM[k]} = 2x_{CoM[k-1]} - x_{CoM[k-2]} \quad (28)$$

$$x_{foot} = x_{CoM} - (z_{CoM} - z_{foot})\tan\psi_x \quad (29)$$

In jumping and running motion, another important issue is to connect support and flight phases in a continuous way. For that purpose, Sugihara and Nakamura (2007) assigned a variable with weight function to make the seamless shift between support and flight phases. As another example, Kajita et al. (2007) overcame this difficulty by assigning proper boundary conditions. Comparing these methods, we followed a similar technique as Kajita et al. (2007) proposed.

In Figure 11, boundary condition assignment is illustrated for three main transition cases, namely C1, C2, and C3. In this figure, necessary initial conditions are indicated for each transition cases.

- **Case 1:** In this situation, we initiate a support phase from a stationary case, e.g., at the beginning of an experiment. Initial θ and ϕ angles $(\theta_{[0]}, \phi_{[0]})$ can be calculated by initial joint angles. Moreover $(\dot{\theta}_{[0]}, \dot{\phi}_{[0]})$ are set to zero as the robot is initially motionless.

Case 2: At this moment, the robot lifts off and enters the flight phase. This is also the final moment of the support phase, namely lift-off. Initial values $(x_{CoM[0]}, x_{CoM[1]}, y_{CoM[0]}, y_{CoM[1]})$ can be obtained by using lift-off values of vertical CoM position (z_{lo}), θ angle and ϕ angle.

$$\begin{aligned}
x_{CoM[0]} &= z_{lo}\tan\theta_{lo}\cos\phi_{lo} \\
y_{CoM[0]} &= z_{lo}\tan\theta_{lo}\sin\phi_{lo} \\
\dot{x}_{CoM[0]} &= \dot{z}_{lo}\tan\theta_{lo}\cos\phi_{lo} + z_{lo}\dot{\theta}_{lo}(1+\tan^2\theta_{lo})\cos\phi_{lo} \\
&\quad - z\dot{\phi}_{lo}\tan\theta_{lo}\sin\phi_{lo} \\
\dot{y}_{CoM[0]} &= \dot{z}_{lo}\tan\theta_{lo}\sin\phi_{lo} + z_{lo}\dot{\theta}_{lo}(1+\tan^2\theta_{lo})\sin\phi_{lo} \\
&\quad + z\dot{\phi}_{lo}\tan\theta_{lo}\cos\phi_{lo} \\
x_{CoM[1]} &= x_{CoM[0]} + \Delta t\dot{x}_{CoM[0]} \\
y_{CoM[1]} &= y_{CoM[0]} + \Delta t\dot{y}_{CoM[0]}
\end{aligned} \quad (30)$$

Case 3: In this transition case, the robot touches down to the ground and enters the support phase. In other words, it is the initial moment of a support phase. Initial values $(\theta_{[0]}, \dot{\theta}_{[0]}, \phi_{[0]}, \dot{\phi}_{[0]})$ can be computed by using touch down values of CoM position.

$$\begin{aligned}
r_{td} &= \sqrt{x_{td}^2 + y_{td}^2 + z_{td}^2} \\
\dot{r}_{td} &= \frac{x_{td}\dot{x}_{td} + y_{td}\dot{y}_{td} + z_{td}\dot{z}_{td}}{r_{td}} \\
\theta_{[0]} &= \cos^{-1}\frac{z_{td}}{r_{td}} \\
\phi_{[0]} &= \tan^{-1}\frac{y_{td}}{x_{td}} \\
\dot{\theta}_{[0]} &= \frac{\dot{z}_{td}r_{td} - \dot{r}_{td}z_{td}}{r_{td}\sqrt{x_{td}^2 + y_{td}^2}} \\
\dot{\phi}_{[0]} &= \frac{\dot{y}_{td}x_{td} - x_{td}y_{td}}{x_{td}^2 + y_{td}^2}
\end{aligned} \quad (31)$$

D. A Brief Road Map to Apply Eulerian ZMP Resolution

In this section, we summarize Eulerian ZMP resolution and list the subsequent steps to be taken.

1. Decide the type of application, i.e., walking, jumping or running.
2. Determine the timing profile. In the case of walking motion, timing profile refers to double support and single support periods. In jumping and running motion, it refers to flight and support periods.

3. Designate feasible referential ZMP inputs.
4. Using Runge-Kutta method (Press et al., 1992), numerically solve (21) and (22) to obtain CoM trajectories.
5. If the application is decided as jumping and running, utilize (24), (26),(27),(28) and(29) to obtain CoM trajectories during the flight phase. Connect subsequent support and flight phases using (30) and (31).

5. SIMULATION AND EXPERIMENTAL RESULTS

Bipedal Walking Simulations

In order to validate our proposed method, we firstly simulated bipedal walking motion on a 3D dynamic simulator (Fujimoto, & Kawamura, 1998) using the realistic model of our bipedal robot. In these simulations, single and double support phases are planned as 0.5 [s] and 0.1 [s]. Referential ZMP inputs are constants and they are substituted as 0.05 [m] and ±0.02 [m] for x-axis and y-axis, respectively. Results can be examined in Figure 12 and Figure 13.

In Figure 12, ZMP responses of right foot are illustrated. In these figures, solid lines stand for ZMP responses while dotted lines indicate support polygons. Furthermore, ZMP responses are only plotted when the robot is in the right leg's single support phase. Other periods are blocked with mat areas. Since ZMP responses for both axes are always within the support polygon, we obtained dynamically equilibrated walking cycles. Note that, ZMP responses of the left foot are identical to these results and therefore not plotted.

Pitch and torso angle variations are plotted in Figure 13. Dotted lines show pitch and roll angle variations while IAMR terms are ignored. Contrarily, solid blue and solid red lines indicate pitch and roll angle variations while IAMR terms are included. Comparing these two cases, it is obvious that IAMR terms are very crucial in suppressing undesired torso angles.

CoM trajectories in bipedal walking simulations are identical to experimental results as we used the same walking parameters in both cases. Hence, we did not include CoM trajectories in simulation results.

Bipedal Walking Experiments on MARI-3

Having obtained satisfying simulation results, we conduct bipedal walking experiments on the actual robot MARI-3. Walking parameters are designed in the same way as performed in simulations. Results can be seen from Figure 14 to Figure 17.

X-axis and y-axis ZMP responses are illustrated in Figure 14. Solid lines indicate ZMP responses

Figure 11. Boundary condition assignment for jumping and running motions

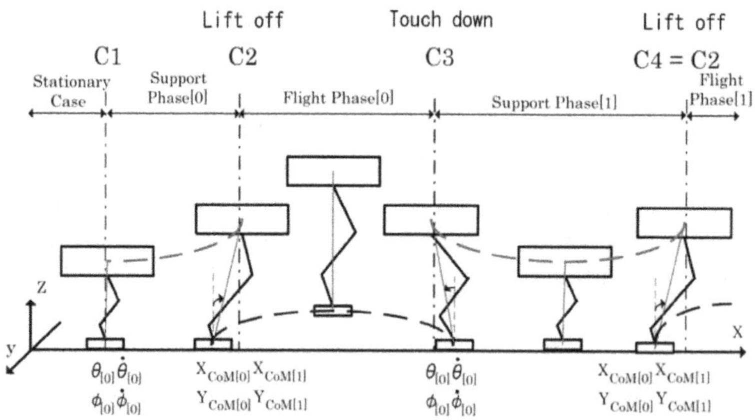

while dotted lines stand for support polygons. Based on these figures, we may claim that we obtained dynamically equilibrated walking cycles since ZMP responses are always within support polygons. What is more, ZMP responses are undefined while the right foot is swinging and these periods are symbolized with mat areas. In addition, left foot's ZMP responses are not included as they are identical with these results.

CoM trajectories for x-axis and y-axis can be observed in Figure 15. In these figures, solid and dotted lines show response and referential values, respectively. These results indicate that trajectory responses follow their references well. Nonetheless, left/right single support phases and double support phases are tied seamlessly.

Pitch axis and roll axis torso angle variations are depicted in Figure 16. In these figures, dashed lines show pitch and roll axis torso angle responses when IAMR terms are not included. In this case, pitch axis angle is fluctuated between ±15 degrees while roll axis angle is fluctuated between ± 10 degrees. On the contrary, solid lines stand for pitch and roll axis torso angle responses when IAMR terms are in-

Figure 12. Measured ZMP responses of the right foot for x-axis and y-axis, respectively (simulation)

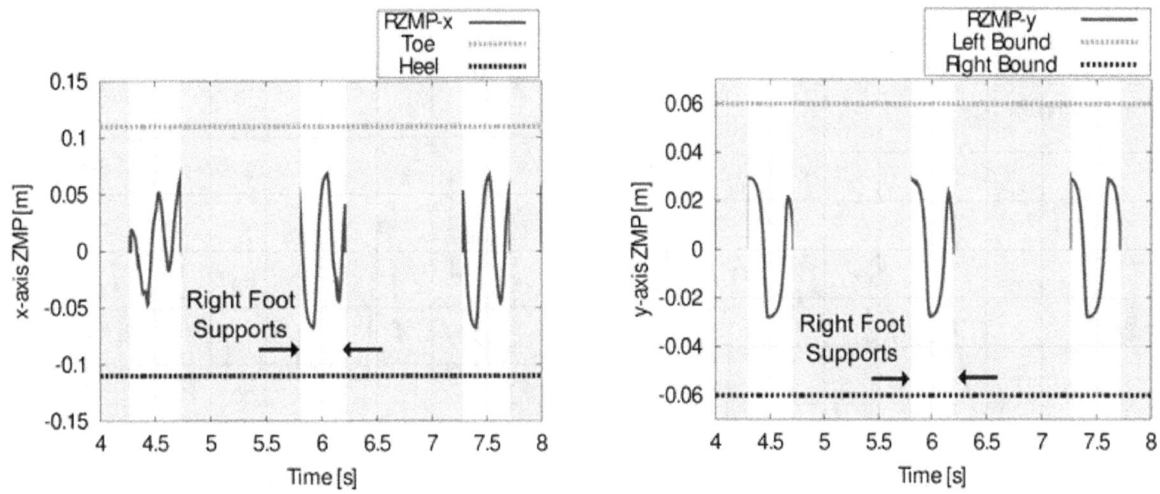

Figure 13. Torso angle variations for pitch axis and roll axis, respectively (simulation)

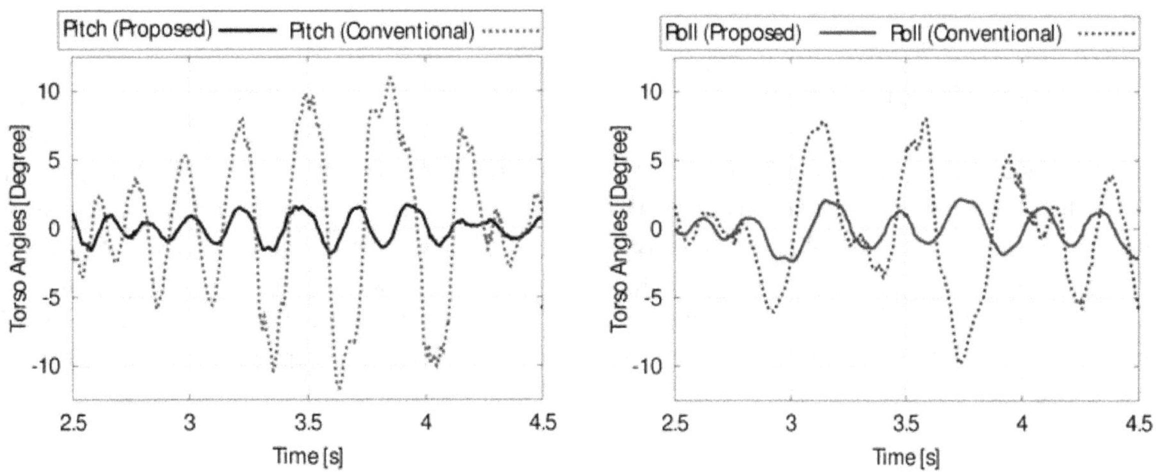

Figure 14. Measured ZMP Responses of the right foot for x-axis and y-axis, respectively (experiment)

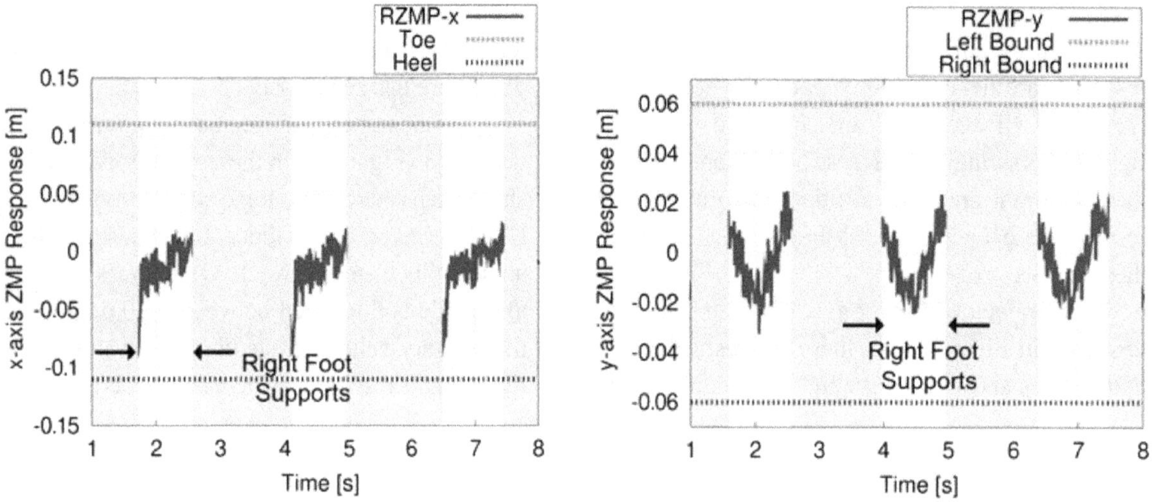

Figure 15. Response and referential variations of the CoM trajectories for x-axis and y-axis, respectively (experiment)

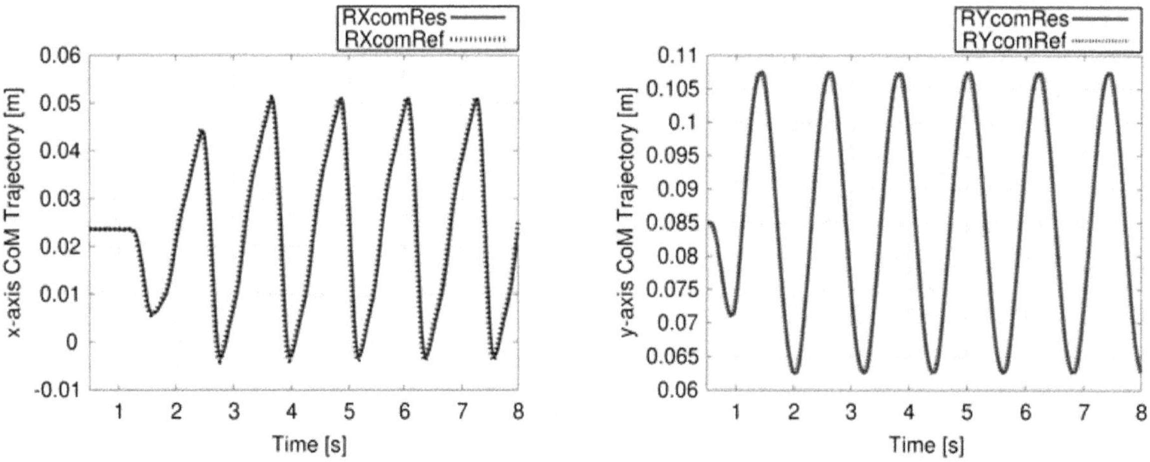

cluded. In opposition to the first case, these angles' variations are less than 5 degrees. All in all, we are able to reduce peak-to-peak torso angles about 75% for pitch axis and about 66% for roll axis. Considering this result, we can say that IAMR term is a very important index in bipedal motion planning. To this end, EZR can provide dynamically equilibrated walking cycles in which undesired torso angles are well suppressed.

Finally, Figure 17 illustrates some significant moments from the bipedal walking experiment on MARI-3. As partially displayed in this figure, the robot was able to walk in a dynamically equilibrated manner, throughout the whole experimentation period. In these snapshots, we can also roughly observe that torso orientation does not vary so much and the body tray is kept parallel to the floor as the robot performs walking on a flat surface.

Figure 16. Torso angle variations for pitch axis and roll axis, respectively (experiment)

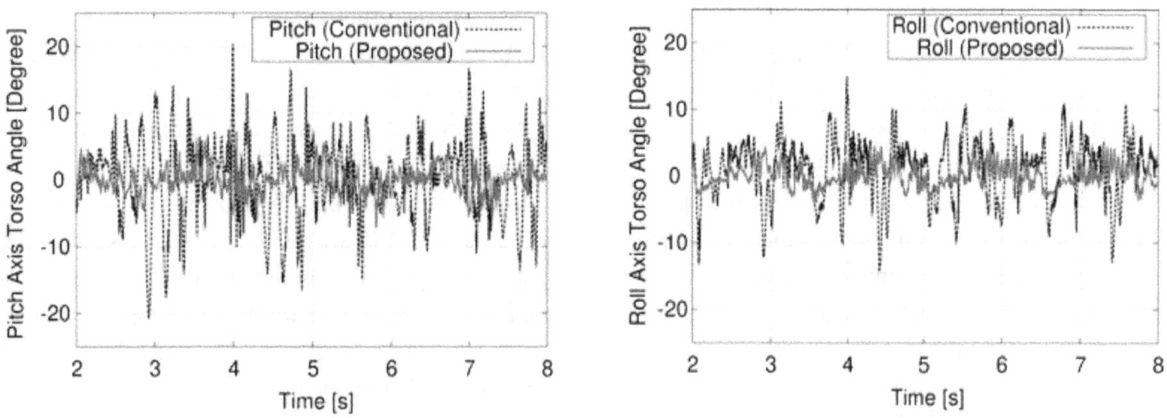

Figure 17. Significant moments from the bipedal walking experiment, a couple of steps. The robot is able to walk continuously within a dynamically equilibrated manner without undesired rotational actions. Note that the robot walks with no support from the paramount lift, which is used just in case of emergency situations. Although snapshots display a couple of steps, the robot performed dynamically balanced bipedal walking throughout the whole experimentation period.

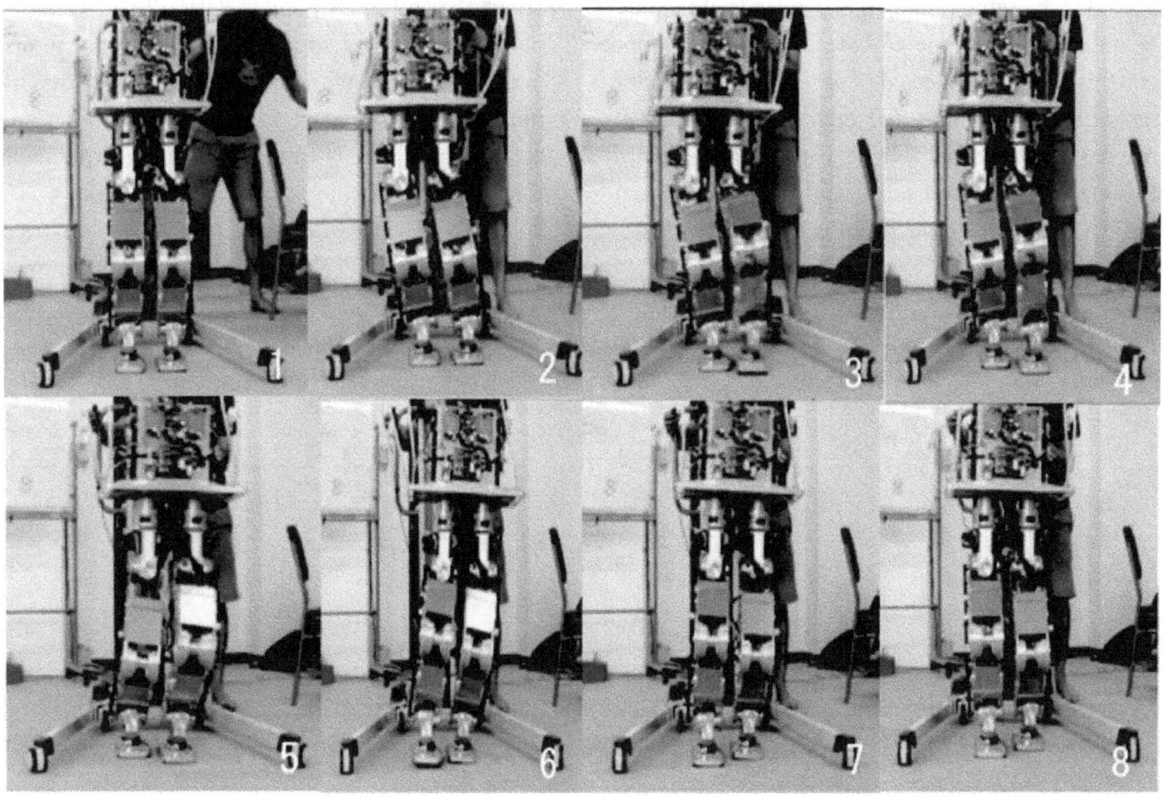

Jumping Experiments

Being the previous step of succeeding running motion, jumping pattern generation is of importance. In this sense, we conducted single leg jumping experiments on bipedal robot MARI-3. Primarily, we decided to conduct vertical hopping, so that x-axis and y-axis ZMP inputs are chosen as close to zero. Flight time and support time are decided as 0.12 [s] and 0.75 [s], respectively. As the result, we obtain repetitive, dynamically balanced and successful jumping cycles within these values. Results might be observed from Figure 18 to Figure 20.

Ground reaction force (GRF) response is depicted in Figure 18, left hand side. In this figure, zero GRF period indicates a successful flight phase. Additionally, touch-down impact can also be examined. Z-axis CoM trajectory is also displayed in the same figure, right hand side. As it may be observed, we firstly pull the CoM at the bottom and this action is indicated as pre-jump phase. After that, CoM is manipulated upwards and when it reaches the maximum value it performs a jump. Having successfully landed on the floor, the robot turns back to the bottom position for the next cycle.

Measured x-axis and y-axis ZMP response values are shown in Figure 19. As stated previously, we set ZMP inputs close to zero for the purpose of vertical jumping. Having observed the response values, we may claim that response values almost follow their references by the help of ZMP feedback (Ugurlu & Kawamura, 2010a). Only during touch-down we observe a peak. However, they are always within the support polygon throughout the whole period, indicating a dynamically equilibrated jumping motion.

Another prominent result of EZR is that it solves the backward jumping problem which is encountered in (Zhu, & Kawamura, 2010) and (Maeda et al., 2008). In (Zhu, & Kawamura, 2010), MARI-3 performed a single jump and developed an undesired rotational action during the flight phase, then landed on the paramount lift and tipped over. As an improvement, Maeda et al., performed dynamically balanced experiments, however, MARI-3 jumped backward about 15 [cm] in each cycle. Using EZR, we definitely remedied these problems. The robot jumps and lands stably without any backwards hopping motion. In this matter, Figure 20 provides some significant moments from the jumping experiment for further observation.

Figure 18. Ground reaction force response at the right foot and z-axis CoM trajectory (experiment)

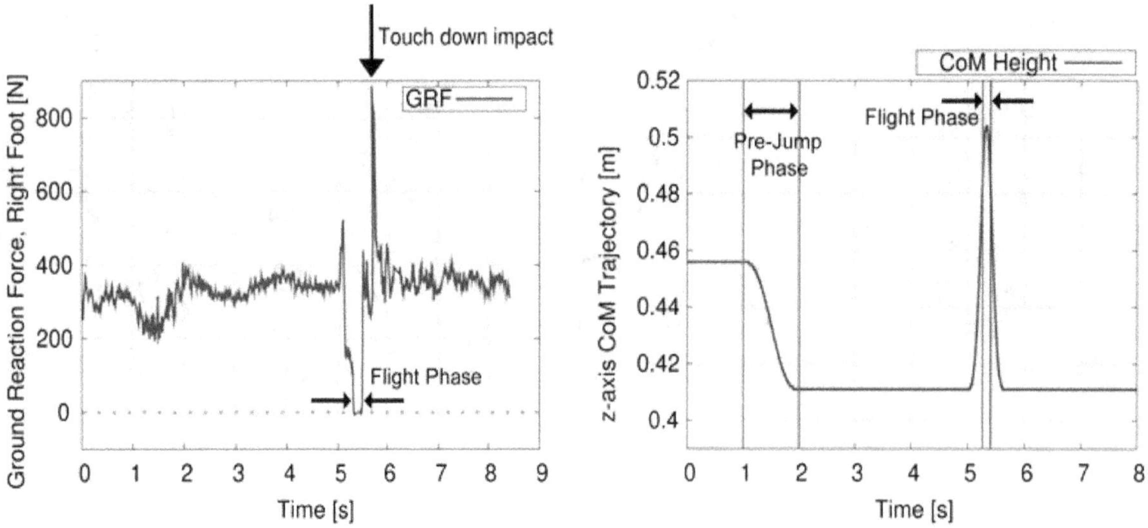

Figure 19. Measured ZMP responses of the right foot for x-axis and y-axis, respectively (experiment)

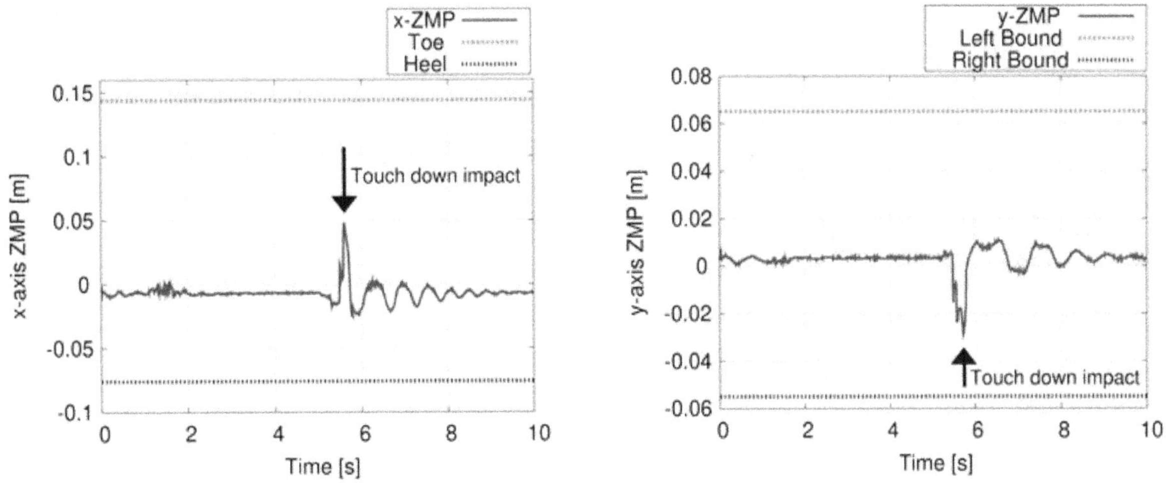

Running Simulations

Having obtained successful and repetitive jumping cycles on MARI-3, we performed running motion simulations on ROCOS. Results may be seen from Figure 21 to Figure 23.

In Figure 21, left hand side, GRF responses are plotted both for right and left legs. In this figure, it is possible to see successful flight phases when both of the legs' GRF responses become zero. What is more, touch down impact peaks are indicated with circles. These peaks are changing between 900 [N] to 1200 [N], due to the landing orientation of the foot. Since the orientation does not change too much, touch-down peaks always stay within these values and do not influence the overall balance.

In Figure 21, left hand side, GRF responses are plotted both for right (solid line) and left (dotted line) legs. In this figure, it is possible to see successful flight phases when both of the legs' GRF responses become zero. What is more, touch down impact peaks are indicated with circles. These peaks are changing between 900 [N] to 1200 [N], due to the landing orientation of the foot. Since the orientation does not change too much, touch-

Figure 20. Significant moments from the one-legged jumping experiment on bipedal robot MARI-3: 1) Rest, 2) Lift-off, 3) Apex, 4) Touch-down. Jumping height is estimated to be about3 [cm] which is pretty enough for running

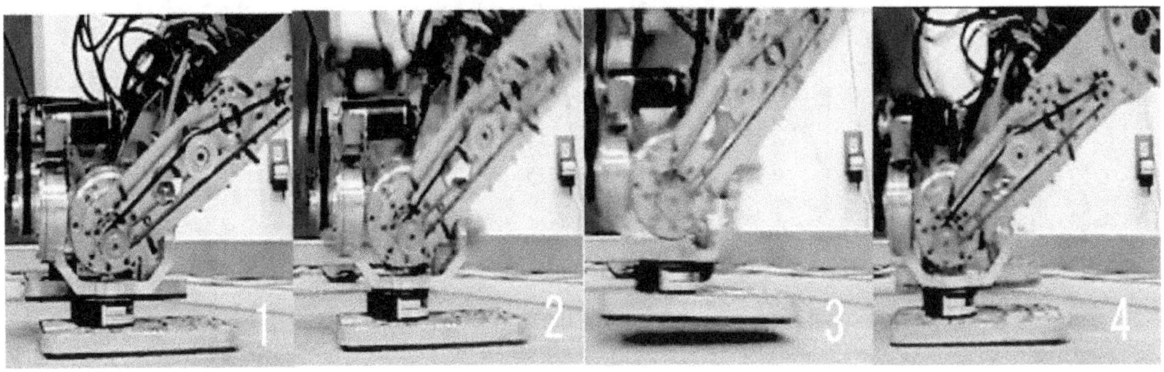

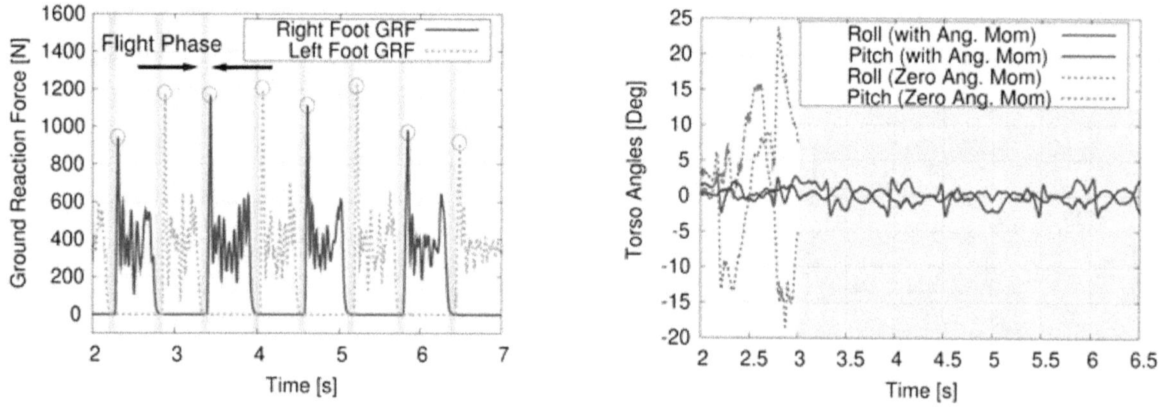

Figure 21. Ground reaction force and torso angle variations (simulation)

down peaks always stay within this values and do not influence the overall balance.

Undesired pitch and roll axes torso angle variations are also depicted in the same figure, right hand side. In this plot, solid lines indicate torso angle variations when IAMR terms are included. Since the angular momentum is not forced to be zero, certain amount of torso angle fluctuation is decreased. For a comparison, we ignored IAMR terms in our motion equations and repeated the simulation. Dotted lines show this case's results. As it may be observed, torso angles fluctuated about 20 degrees and the robot tipped over after the 3rd second.

CoM trajectories for y-axis and z-axis are illustrated in Figure 22. In these plots, dotted lines indicate referential trajectories while solid lines indicate response trajectories. Trajectory responses follow their responses well. Additionally, z-axis CoM trajectory is plotted with respect to left foot sole center. Same trajectory with respect to right foot sole center is identical to these results. We both use these z-axis trajectories since single support phases are switched between two legs.

ZMP response and reference values for x-axis and y-axis are depicted in Figure 23. Solid lines symbolize measured response values while dotted lines indicate referential inputs. Support polygon

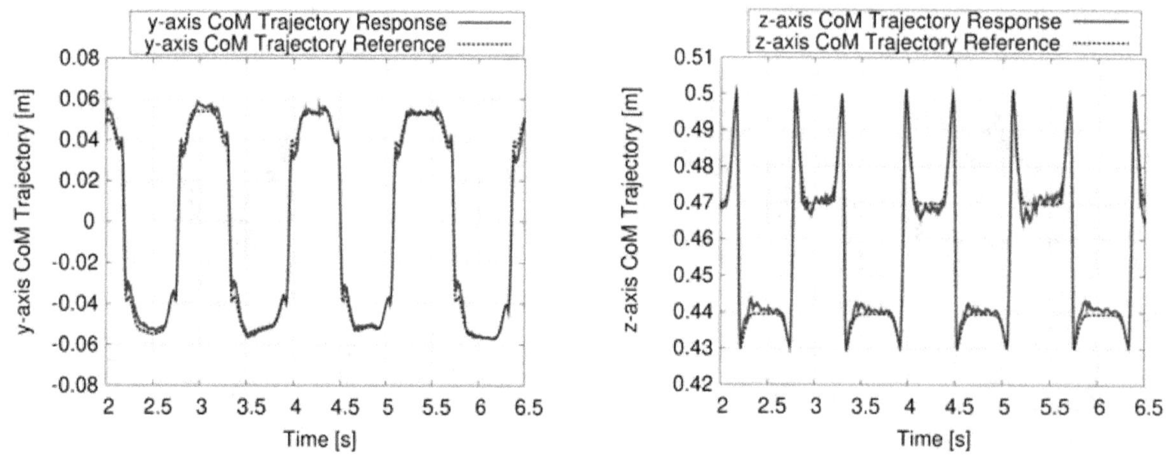

Figure 22. Center of mass trajectories for y and z axes, reference and response values (simulation)

Figure 23. Measured ZMP response values for x-axis and y-axis, respectively (simulation)

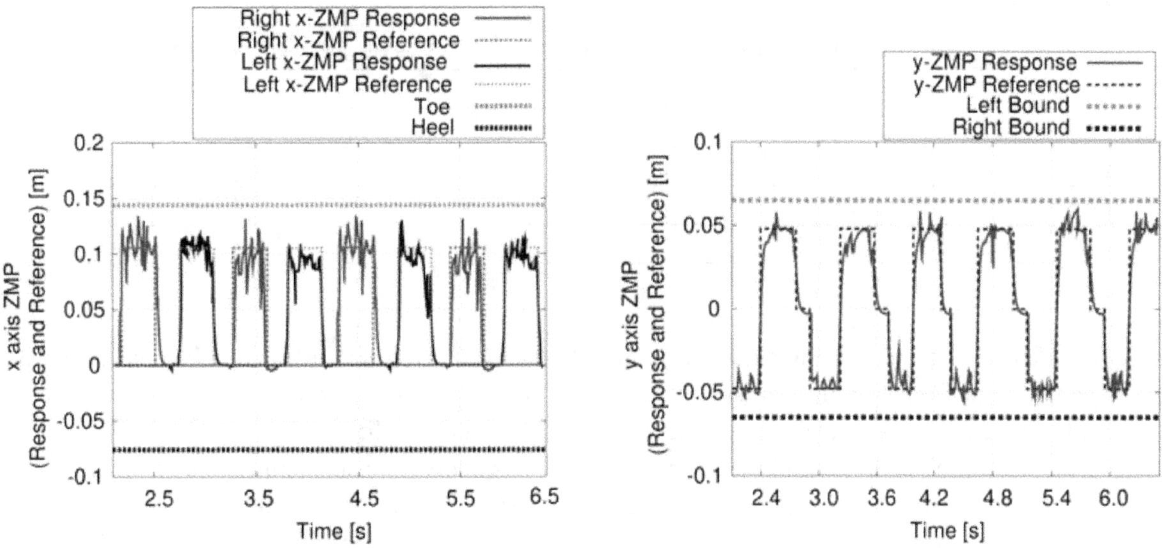

boundaries are also displayed with dotted lines. Observing these plots, ZMP responses are always within the support polygon and follow their references well. Note that ZMP is undefined during flight phases and we substituted zero value when the robot's feet have no contact with the floor.

Comparison with Other ZMP-Based Methods

In this subsection, we compare the proposed method with other recently published ZMP-based trajectory generation methods. Considering our particular purposes, we define the comparison criteria as angular momentum inclusion, computation time, mechanism independence and initial condition determination. Table 3 provides the resulting comparison. In this table, x, Δ, and ○ symbolize the unfavorable, likely unfavorable and favorable, respectively.

Kajita et al. (2006) proposed a bipedal walking pattern generation by using a preview control of ZMP. Erbatur and Kurt (2009) treated ZMP equations using Fourier series approximation. As different approaches, Motoi et al. (2009) proposed virtual inverted pendulum mode, and Tajima et al. (2009) implemented a discrete domain approach.

In first three methods, angular momentum information is either ignored or zero-referenced. As previously stated, forcing angular momentum around zero causes undesired rotational actions and torso angle fluctuations. Only Tajima et al. implemented an approach to include angular momentum using translational CoM velocity and empirically tuned rotational inertia element. On the other hand, this approach does not provide a rigorous and substantial solution since angular momentum is related with CoM angular velocity instead of its translational counterpart. What is more, the rotational inertia of a bipedal robot is not constant and it is expressed with tensors whose elements are functions of joint motions (Balafoutis, & Patel, 1991b). From the point of view of angular momentum inclusion, the proposed method has clear advantage over these methods.

Considering the computation time, the second and third methods consist of hyperbolic functions which are unlikely to be computed real-time, using built-in microprocessors. On the contrary, first two methods and the proposed method are already

Table 3. The comparison chart includes some recently proposed ZMP-based trajectory generation methods

	Angular Momentum	Computation Time	Mechanism Independency	Initial Condition Determination
1-(Kajita et al., 2006)	x	○	○	○
2-(Erbatur, & Kurt, 2009)	x	Δ	○	○
3-(Motoi et al., 2009)	x	Δ	○	○
4-(Tajima et al. 2009)	Δ	○	Δ	x
5-Proposed Method	○	○	○	x

implemented in real systems in online, so that their sufficient computational efficiencies are proved.

While focusing on the mechanism independency, the first three methods and the proposed method can be applied to any type of bipedal robots. However, Tajima et al., implemented their technique on a humanoid robot with toe joint, which is advantageous for torque optimization but creates redundancy.

As stated previously, ZMP-based trajectory generation requires solving two second order differential equations, namely ZMP equations. For this purpose, one has to define initial position and velocity values to obtain the bipedal motion patterns. Properly defined initial conditions play an important role in generating desired movements. To this end, analytical methods (Kajita et al., 2006, Erbatur, & Kurt, 2009, Motoi et al., 2009) possess clear advantage over numerical methods (Tajima et al., 2009). Numerical solution is also required in the proposed method and this disadvantage remains as an open problem. Generally speaking, ZMP equations cannot be solve analytically when angular momentum terms are included. Thus, there is a trade-off between angular momentum inclusion and initial condition determination.

Real-Time Implementation Challenges

In prototyping and manufacturing MARI-3, we prioritized power/weight ratio for the purpose versatile bipedal motion generation tasks. As the result, we obtained a suitable and useful experimental bipedal robot platform with sufficient motion capabilities.

Although we put our utmost effort in designing the robot, some of the challenges were encountered during the real-time implementation of the motion generation algorithm. Main issues can be listed as follows.

- MARI-3's computation system may not be the best solution for this type of robots. Even though we did not encounter any problems during the online implementation of the proposed method, real-time Linux based solution are experienced to be superior to those based on microprocessors (Ugurlu et al., 2011). A bipedal humanoid robot has many degrees of freedom and it is equipped with several sensors. When considering data recording, signal conditioning and other related issues, the computation system should be able to handle all these tasks in real-time, in addition to online pattern generation.
- Communication system is another important factor during the real-time implementation of humanoid robots. RON bus was able to handle this task; however, real-time Ethernet (Ugurlu et al., 2011) appears to be a faster and more flexible solution.
- To increase power/weight ratio, we focused on pitch axis joints. Intuitively speaking, pitch axis joints are the dominant ones dur-

ing bipedal walking, running and jumping. On the other hand, lateral motion has a substantial influence on the overall movement of the robot. Especially, a recent study proved that roll axis hip joints consumes the highest amount of energy during bipedal walking (Kormushev et al., 2011). As we placed 60 [W] motors for these joints (see Table 2), there were serious temperature problems in these motors. What is more, they could not output the necessary torques, required for running motions. Because of this fact, we were not able to conduct running experiments and only performed simulations.

- Home positioning of joints appear to be another hardware issue at the beginning of each experiment. As we have incremental encoders, metrics are placed to handle this issue. Figure 24 shows our solution. Regrettably, hand-tuning of home positioning is quite difficult and cannot be done precisely. A better solution for this problem may be the utilization of extra potentiometers in each joint.

Considering all these hardware issues, we are one step closer to the "perfect humanoid prototyping" goal, since bitter past experiences enables researchers to develop more solid solutions.

6. FUTURE RESEARCH DIRECTIONS

A different impact of Eulerian ZMP Resolution might be accompanied with the definition of support polygon. Generally speaking, such polygons are defined when Center of Pressure in mind. Center of Pressure equations are argued to be mathematical equivalence of conventional ZMP equations without IAMR terms (Hyon, & Cheng, 2007). This actually makes sense, since Center of Pressure indicates the point where the resultant GRF is acting. Naturally, GRF has to act on some point on the robot foot sole, as long as there is contact with the floor.

On the other hand, IAMR terms introduce additional terms in ZMP equations, creating a difference from Center of Pressure equations. Thus, one may defend that the real support polygon of a humanoid robot may be wider or narrower than the usual one. What is more, it may not be a static quantity. Instead, the support polygon area should be a function of the given robot's inertial parameters. On this matter, Eulerian ZMP Resolution may enable researchers to provide a new definition regarding the dynamic support polygon.

Heretofore, Eulerian ZMP Resolution is mostly applied to bipedal motion generation task. It may also provide insights about control implications concerning the real-time stabilization of bipedal humanoid robots and other legged mechanisms.

Considering the issues of environmental interaction, terrain adaptability and safety, compliant mechanisms are of importance in humanoid robotic technology. At this juncture, it is experimentally verified that compliant humanoids also inherently more stable in comparison with rigid robots (Ugurlu et al. 2011). However, a precise and extensive mathematical description that connects compliance and dynamic balance is still lacking. A mathematical tool for that purpose certainly enables researchers to create better motion generation and control algorithms.

7. CONCLUSION

In summary, we proposed a systematic way of generating ZMP-based bipedal motion generation that is on the basis of combining inertial forces and intrinsic angular momentum rate changes. We may claim that the technique can successfully characterize the rotational inertia, a crucial parameter of humanoid dynamics. Hence, it enriches simplified motion generation algorithms and control implications.

Moreover, we validated our approach by performing bipedal walking, jumping, running simula-

Figure 24. In MARI-3, metric measurements are placed in each joint for home positioning. On the other hand, this solution does not provide a solid basis for home-positioning problem

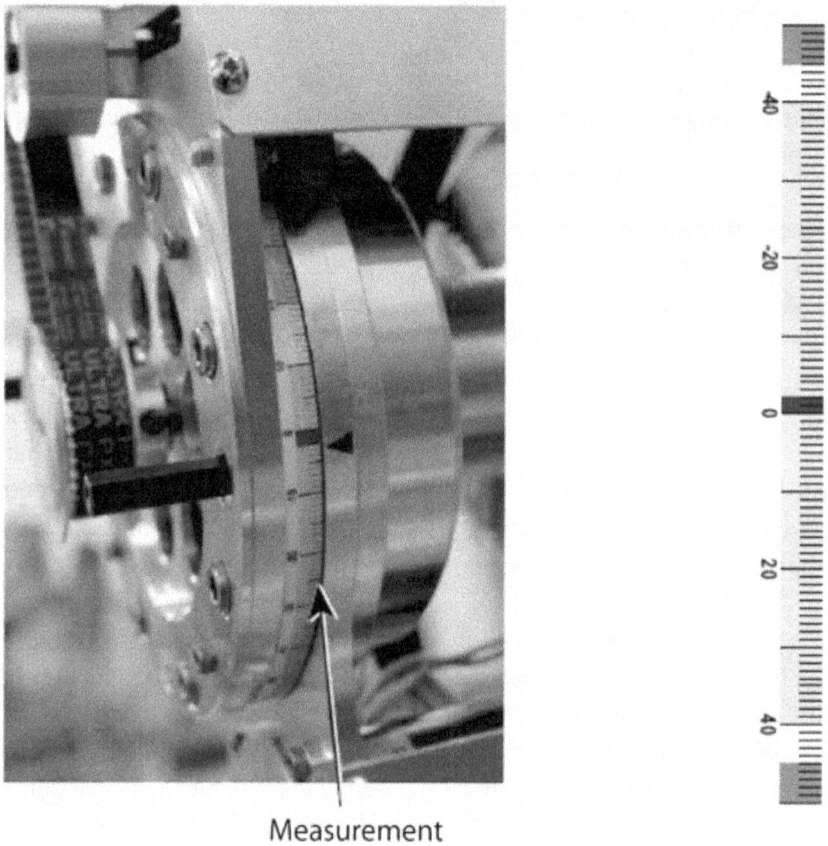

tions and experiments. Based on these results, we examine that Eulerian ZMP Resolution is superior to conventional ZMP-based methods in suppressing undesired rotational actions such as torso angle fluctuations and backwards-hopping. This property of Eulerian ZMP Resolution definitely enables us to obtain more feasible bipedal motion patterns.

Recently, we also implemented Eulerian ZMP Resolution for regulating undesired yaw moment that is induced about the support foot sole (Ugurlu et al., 2012). In doing so, bipedal robots may walk on low frictional surfaces and/or their performance might be enhanced in terms of faster motion capabilities. Therefore, Eulerian ZMP Resolution is fully utilized to exploit the motion generation capabilities of ZMP-based algorithms within a complete manner.

REFERENCES

Balafoutis, C. A., & Patel, R. V. (1991a). *Robot dynamic analysis of robot manipulators: A Cartesian tensor approach*. Dordrecht, The Netherlands: Kluwer Academic Publishers Group.

Balafoutis, C. A., & Patel, R. V. (1991b). Efficient computation of manipulator inertia matrices and direct dynamics problem. *IEEE Transactions on Systems, Man, and Cybernetics*, *19*(5), 1313–1320. doi:10.1109/21.44054

Erbatur, K., & Kurt, O. (2009). Natural ZMP trajectories for biped robot reference generation. *IEEE Transactions on Industrial Electronics*, *56*(3), 835–845. doi:10.1109/TIE.2008.2005150

Featherstone, R. (2008). *Rigid body dynamics algorithms*. New York, NY: Springer Publishing. doi:10.1007/978-0-387-74315-8

Fujimoto, Y., & Kawamura, A. (1998). Simulation of an autonomous biped walking robot including environmental force interaction. *IEEE Robotics & Automation Magazine*, *5*(2), 33–42. doi:10.1109/100.692339

Goswami, A. (2008, December). Kinematic and dynamic analogies between planar biped robots and the reaction mass pendulum (RMP) model. In *IEEE International Conference on Humanoid Robotics* (pp. 182-188).

Greenwood, D. T. (2003). *Advanced dynamics*. New York, NY: Cambridge University Press.

Hirukawa, H., Hattori, S., Harada, K., Kajita, S., Kaneko, K., & Kanehiro, K. … Morisawa, M. (2006, October). A universal stability criterion of the foot contact of legged robots – Adios ZMP. In *IEEE International Conference on Robotics and Automation* (pp. 1976-1983).

Hofmann, A., Popovic, M., & Herr, H. (2009, April). Exploiting angular momentum to enhance bipedal center-of-mass control. In *IEEE International Conference on Robotics and Automation* (pp. 4423-4429).

Huang, Q., Yokoi, K., Kajita, S., Kaneko, K., Arai, H., Koyachi, N., & Tanie, K. (2001). Planning walking patterns for a biped robot. *IEEE Transactions on Robotics and Automation*, *17*(3), 280–289. doi:10.1109/70.938385

Hyon, S., & Cheng, G. (2007, April). Disturbance rejection for biped humanoid robots. In *IEEE International Conference on Robotics and Automation* (pp. 2668-2673).

Jafari, A., Tsagarakis, N. G., Vanderborght, B., & Caldwell, D. G. (2010, October). A novel actuator with adjustable stiffness (AwAS). In *IEEE International Conference on Intelligent Robots and Systems* (pp. 4210-4215).

Kajita, S., & Espiau, B. (2008). Legged robots. In Siciliano, B., & Khatib, O. (Eds.), *Springer handbook of robotics*. Berlin, Germany: Springer Publishing. doi:10.1007/978-3-540-30301-5_17

Kajita, S., Kanehiro, F., Kaneko, K., Yokoi, K., & Hirukawa, H. (2001, October). The 3D linear inverted pendulum mode: A simple modeling for a biped walking pattern generation. In *IEEE International Conference on Intelligent Robots and Systems* (pp. 239-246).

Kajita, S., Kaneko, K., Morisawa, M., Nakaoka, S., & Hirukawa, H. (2007, April). ZMP-based biped running enhanced by toe springs. In *IEEE International Conference on Robotics and Automation* (pp. 3963-3969).

Kajita, S., Morisawa, M., Harada, K., Kaneko, K., Kanehiro, F., Fujiwara, K., & Hirukawa, H. (2006, October). Biped walking generator allowing auxiliary ZMP control. In *IEEE International Conference on Intelligent Robots and Systems* (pp. 2993-2999).

Kajita, S., & Sugihara, T. (2009). Humanoid robots in the future. *Advanced Robotics*, *23*(11), 1527–1531. doi:10.1163/016918609X12469692711804

Kajita, S., & Tani, K. (1991, April). Study of dynamic biped locomotion on rugged terrain. In *IEEE International Conference on Robotics and Automation* (pp. 1405-1410).

Kanehiro, F., Hirukawa, H., & Kajita, S. (2004). OpenHRP: Open architecture humanoid robotics platform. *The International Journal of Robotics Research*, *23*(2), 155–165. doi:10.1177/0278364904041324

Kormushev, P., Ugurlu, B., Calinon, S., Tsagarakis, N. G., & Caldwell, D. G. (September, 2011). Bipedal walking energy minimization by reinforcement learning with evolving policy parameterization. *Proceedings of the IEEE International Conference on Intelligent Robots and Systems.*

Maeda, T., Ugurlu, B., & Kawamura, A. (2008, March). Two legged jumping simulation and experiment on biped robot MARI-3. In *IEEE International Workshop on Advanced Motion Control* (pp. 301-305).

Mitobe, K., Capi, G., & Nasu, Y. (2004). A new control method for walking robots based on angular momentum. *Mechatronics*, *18*(2), 163–174. doi:10.1016/S0957-4158(03)00028-X

Motoi, N., Suzuki, T., & Ohnishi, K. (2009). A bipedal locomotion planning based on virtual linear inverted pendulum mode. *IEEE Transactions on Industrial Electronics*, *56*(1), 835–845. doi:10.1109/TIE.2008.2004663

Orin, D. E., & Goswami, A. (2008, October). Centroidal momentum matrix of a humanoid robot: Structure and properties. In *IEEE International Conference on Intelligent Robots and Systems* (pp. 653-659).

Park, J. H. (2001). Impedance control for biped robot locomotion. *IEEE Transactions on Robotics and Automation*, *17*(6), 870–882. doi:10.1109/70.976014

Popovic, M., Goswami, A., & Herr, H. (2005). Ground reference points in legged locomotion: Definitions, biological trajectories and control implications. *The International Journal of Robotics Research*, *24*(12), 1013–1032. doi:10.1177/0278364905058363

Poulakakis, I., & Grizzle, J. W. (2009). The spring loaded inverted pendulum as the hybrid zero dynamics of an asymmetric hopper. *IEEE Transactions on Automatic Control*, *54*(8), 1779–1793. doi:10.1109/TAC.2009.2024565

Pratt, J., Carff, J., Drakunov, S., & Goswami, A. (2006, December). Capture point: A step toward humanoid push recovery. In *IEEE International Conference on Humanoid Robotics* (pp. 200-207).

Press, W. H., Tekolsky, S. A., Vetterling, W. T., & Flannery, B. P. (1992). *Numerical recipes in C: The art of scientific computing* (2nd ed.). New York, NY: Cambridge University Press.

Strikwerda, J. C. (1989). *Finite difference schemes and partial differential equations*. Wadsworth & Brooks/Cole Advanced Books & Software.

Sugihara, T., & Nakamura, Y. (2007, October). Enhancement of boundary condition relaxation method for 3D hopping motion of biped robots. In *IEEE International Conference on Intelligent Robots and Systems* (pp. 444-449).

Tajima, R., Honda, D., & Suga, K. (2009, April). Fast running experiments involving a humanoid robot. In *IEEE International Conference on Robotics and Automation* (pp. 1571-1576).

Takenaka, T., Matsumoto, T., & Yoshiike, T. (2009a, October). Real time motion generation and control for biped robot – 3rd report: Dynamics error compensation. In *IEEE International Conference on Intelligent Robots and Systems.* (pp. 1594-1600).

Takenaka, T., Matsumoto, T., Yoshiike, T., & Shirokura, S. (2009b, October). Real time motion generation and control for biped robot –2nd report: Running gait pattern generation. In *IEEE International Conference on Intelligent Robots and Systems* (pp. 1092-1099).

Timoshenko, S., & Young, D. H. (1948). *Advanced dynamics*. US: McGraw-Hill Book Company.

Ugurlu, B. (2010). *Bipedal motion planning based on composite rigid body angular momentum resolution.* Doctoral dissertation, Yokohama National University, Yokohama, Japan.

Ugurlu, B., Hirabayashi, T., & Kawamura, A. (2009, November). A unified control frame for stable bipedal walking. In *IEEE International Conference on Industrial Electronics and Control* (pp. 4162-4172).

Ugurlu, B., & Kawamura, A. (2009, October). Real-time running and jumping pattern generation for bipedal robots based on ZMP and Euler's equations. In *IEEE International Conference on Intelligent Robots and Systems.*(pp. 1100-1105).

Ugurlu, B., & Kawamura, A. (2010a). ZMP-based online jumping pattern generation for a one-legged robot. *IEEE Transactions on Industrial Electronics*, *57*(5), 1701–1709. doi:10.1109/TIE.2009.2032439

Ugurlu, B., & Kawamura, A. (2010b, May). Eulerian ZMP resolution based bipedal walking: Discussions on the intrinsic angular momentum rate change about center of mass. In *IEEE International Conference on Robotics and Automation* (pp. 4218-4223).

Ugurlu, B., & Kawamura, A. (2010c, December). Bipedal walking trajectory generation based on ZMP and Euler's equations. In *IEEE International Conference on Humanoid Robotics* (pp. 468-473).

Ugurlu, B., Saglia, J. A., Tsagarakis, N. G., & Caldwell, D. G. (2012, May). Yaw moment compensation for bipedal walking robots based on Euler's equations. In *IEEE International Conference on Robotics and Automation* (Under Review).

Ugurlu, B., Tsagarakis, N. G., Spyrakos-Papastravridis, E., & Caldwell, D. G. (2011, April). Compliant joint modification and real-time dynamic walking implementation on bipedal robot cCub. In *IEEE International Conference on Mechatronics* (to be presented).

Vukobratovic, M., & Borovac, B. (2004). Zero moment point – Thirty five years of its life. *International Journal of Humanoid Robotics*, *1*(1), 157–173. doi:10.1142/S0219843604000083

Zhu, C., & Kawamura, A. (2010). Development of biped robot MARI-3 for jumping. *Advanced Robotics*, *24*(11), 1661–1675. doi:10.1163/016918610X512640

ADDITIONAL READING

Atkeson, C. G., & Stephens, B. J. (2007, December). Multiple balance strategies from one optimization criterion. In *IEEE International Conference on Humanoid Robotics.* (pp. 924-929).

Balafoutis, C. A., & Patel, R. V. (1991). Efficient computation of manipulator inertia matrices and direct dynamics problem. In *IEEE Transactions on Systems, Man, and Cybernetics*, *19*(5), 1313–1320. doi:10.1109/21.44054

Erbatur, K., & Kurt, O. (2009). Natural ZMP trajectories for biped robot reference generation. *IEEE Transactions on Industrial Electronics*, *56*(3), 835–845. doi:10.1109/TIE.2008.2005150

Fujimoto, Y. (2004, April). Trajectory generation of biped running robot with minimum energy consumption. In *IEEE International Conference on Robotics and Automation.* (pp. 3803-3808).

Fujimoto, Y., & Kawamura, A. (1998). Simulation of an autonomous biped walking robot including environmental force interaction. *IEEE Robotics & Automation Magazine*, *5*(2), 33–42. doi:10.1109/100.692339

Fujimoto, Y., Obata, S., & Kawamura, A. (1998, April). Robust bipedal walking with active interaction control between foot and ground. In*IEEE International Conference on Robotics and Automation,* (pp. 2030-2035).

Goswami, A. (2008, December). Kinematic and dynamic analogies between planar biped robots and the reaction mass pendulum (RMP) model. In *IEEE International Conference on Humanoid Robotics.*(pp. 182-188).

Hirukawa, H., Hattori, S., Harada, K., Kajita, S., Kaneko, K., Kanehiro, K., et al. (2006, October). A universal stability criterion of the foot contact of legged robots – Adios ZMP. In *IEEE International Conference on Robotics and Automation.*(pp. 1976-1983).

Hyon, S. (2009). Compliant terrain adaptation for biped humanoids without measuring ground surface and contact forces. *IEEE Transactions on Robotics, 25*(2), 171–178. doi:10.1109/TRO.2008.2006870

Hyon, S., & Cheng, G. (2007, April). Disturbance rejection for biped humanoid robots. In *IEEE International Conference on Robotics and Automation.*(pp. 2668-2673)

Hyon, S., Hale, J. G., & Cheng, G. (2007). Full-body compliant human-humanoid interaction: Balancing in the presence of unknown external forces. *IEEE Transactions on Robotics, 23*(5), 884–898. doi:10.1109/TRO.2007.904896

Kajita, S., Kanehiro, F., Kaneko, K., Fujiwara, K., Harada, K., Yokoi, K., & Hirukawa, H. (2003, September). Biped walking pattern generation by using preview control of zero-moment-point. In *IEEE International Conference on Robotics and Automation.*(pp. 1620-1626).

Kajita, S., Kanehiro, F., Kaneko, K., Fujiwara, K., Harada, K., Yokoi, K., & Hirukawa, H. (2003, October). Resolved momentum control: Humanoid motion planning based on the linear and angular momentum. In *IEEE International Conference on Intelligent Robots and Systems.*(pp. 1644-1650).

Kajita, S., Kanehiro, F., Kaneko, K., Yokoi, K., & Hirukawa, H. (2001, October). The 3D linear inverted pendulum mode: A simple modeling for a biped walking pattern generation. In *IEEE International Conference on Intelligent Robots and Systems.*(pp. 239-246).

Kajita, S., Kaneko, K., Morisawa, M., Nakaoka, S., & Hirukawa, H. (2007, April). ZMP-based biped running enhanced by toe springs. In *IEEE International Conference on Robotics and Automation.* (pp. 3963-3969).

Kajita, S., Morisawa, M., Harada, K., Kaneko, K., Kanehiro, F., Fujiwara, K., & Hirukawa, H. (2006, October). Biped walking generator allowing auxiliary ZMP control. In *IEEE International Conference on Intelligent Robots and Systems.*(pp. 2993-2999).

Kajita, S., & Sugihara, T. (2009). Humanoid robots in the future. *Advanced Robotics, 23*(11), 1527–1531. doi:10.1163/016918609X12469692711804

Kajita, S., & Tani, K. (1991, April). Study of dynamic biped locomotion on rugged terrain. In *IEEE International Conference on Robotics and Automation.* (pp. 1405-1410).

Kanehiro, F., Hirukawa, H., & Kajita, S. (2004). OpenHRP: Open architecture humanoid robotics platform. *The International Journal of Robotics Research, 23*(2), 155–165. doi:10.1177/0278364904041324

Khatib, O. (2004). Inertial properties in robotic manipulation: An object-level framework. *The International Journal of Robotics Research, 13*(1), 19–36.

Kormushev, P., Ugurlu, B., Calinon, S., Tsagarakis, N. G., & Caldwell, D. G. (2011). Bipedal walking energy minimization by reinforcement learning with evolving policy parameterization. In *IEEE International Conference on Intelligent Robots and Systems.*(to be presented).

Li, Q., Takanishi, A., & Kato, I. (1991, October). A biped walking robot having a ZMP measurement system using universal force-moment sensors. In *IEEE International Conference on Intelligent Robots and Systems.*(pp. 1568-1573).

Orin, D. E., & Goswami, A. (2008, October). Centroidal momentum matrix of a humanoid robot: Structure and properties. In *IEEE International Conference on Intelligent Robots and Systems.*(pp. 653-659).

Park, J. H. (2001). Impedance control for biped robot locomotion. *IEEE Transactions on Robotics and Automation, 17*(6), 870–882. doi:10.1109/70.976014

Poulakakis, I., & Grizzle, J. W. (2009). The spring loaded inverted pendulum as the hybrid zero dynamics of an asymmetric hopper. *IEEE Transactions on Automatic Control, 54*(8), 1779–1793. doi:10.1109/TAC.2009.2024565

Pratt, J., Carff, J., Drakunov, S., & Goswami, A. (2006, December). Capture point: A step toward humanoid push recovery. In *IEEE International Conference on Humanoid Robotics.*(pp. 200-207).

Sentis, L., Park, J., & Khatib, O. (2010). Compliant control of multicontact and Center-of-Mass behaviors in humanoid robots. *IEEE Transactions on Robotics, 26*(3), 483–501. doi:10.1109/TRO.2010.2043757

Sugihara, T. (2009, December). Solvability-unconcerned inverse kinematics based on Levenberg-Marquardt method. In *IEEE International Conference on Humanoid Robotics.* (pp. 555-560).

Sugihara, T. (2009, October). Dynamics morphing from regulator to oscillator on bipedal control. In *IEEE International Conference on Intelligent Robots and Systems.*(pp. 2940-2945).

Sugihara, T. (2009, May). Stabilizability and stepping maneuver in planar bipedalism on the best COM-ZMP regulator. *IEEE International Conference on Robotics and Automation.*(pp. 1966-1971).

Sugihara, T. (2010, May). Consistent biped step control with COM-ZMP Oscillation based on successive phase estimation in dynamics morphing. In *IEEE International Conference on Robotics and Automation.*(pp. 4224-4229).

Sugihara, T., & Nakamura, Y. (2007, October). Enhancement of boundary condition relaxation method for 3D hopping motion of biped robots. In *IEEE International Conference on Intelligent Robots and Systems.*(pp. 444-449).

Sugihara, T., & Nakamura, Y. (2009). Boundary condition relaxation method for stepwise pedipulation planning of biped robots. *IEEE Transactions on Robotics, 25*(3), 658–669. doi:10.1109/TRO.2008.2012336

Tajima, R., Honda, D., & Suga, K. (2009, April). Fast running experiments involving a humanoid robot. In *IEEE International Conference on Robotics and Automation.*(pp. 1571-1576).

Takenaka, T., Matsumoto, T., & Yoshiike, T. (2009, October). Real time motion generation and control for biped robot – 1st report: Walking gait pattern generation. In *IEEE International Conference on Intelligent Robots and Systems.*(pp. 1084-1091).

Takenaka, T., Matsumoto, T., & Yoshiike, T. (2009, October). Real time motion generation and control for biped robot – 3rd report: Dynamics error compensation. In *IEEE International Conference on Intelligent Robots and Systems.*(pp. 1594-1600).

Takenaka, T., Matsumoto, T., Yoshiike, T., Hasegawa, T., Shirokura, S., Kaneko, H., & Orita, A. (2009, October). Real time motion generation and control for biped robot – 4threport: Integrated balance control. In *IEEE International Conference on Intelligent Robots and Systems.*(pp. 1601-1608).

Takenaka, T., Matsumoto, T., Yoshiike, T., & Shirokura, S. (2009, October). Real time motion generation and control for biped robot – 2ndreport: Running gait pattern generation. In *IEEE International Conference on Intelligent Robots and Systems.*(pp. 1092-1099).

Ugurlu, B. (2010). *Bipedal motion planning based on composite rigid body angular momentum resolution.* Doctoral dissertation, Yokohama National University, Yokohama, Japan.

Ugurlu, B., Hirabayashi, T., & Kawamura, A. (2009, November). A unified control frame for stable bipedal walking. In *IEEE International Conference on Industrial Electronics and Control.* (pp. 4162-4172).

Ugurlu, B., & Kawamura, A. (2009, October). Real-time running and jumping pattern generation for bipedal robots based on ZMP and Euler's equations. In *IEEE International Conference on Intelligent Robots and Systems*.(pp. 1100-1105).

Ugurlu, B., & Kawamura, A. (2010). ZMP-based online jumping pattern generation for a one-legged robot. *IEEE Transactions on Industrial Electronics*, *57*(5), 1701–1709. doi:10.1109/TIE.2009.2032439

Ugurlu, B., & Kawamura, A. (2010, May). Eulerian ZMP Resolution based bipedal walking: Discussions on the intrinsic angular momentum rate change about center of mass. In*IEEE International Conference on Robotics and Automation*. (pp. 4218-4223).

Ugurlu, B., Tsagarakis, N. G., Spyrakos-Papastravridis, E., & Caldwell, D. G. (2011, April). Compliant joint modification and real-time dynamic walking implementation on bipedal robot cCub. In *IEEE International Conference on Mechatronics*.(to be presented).

Vukobratovic, M., & Borovac, B. (2004). Zero Moment Point – Thirty five years of its life. *International Journal of Humanoid Robotics*, *1*(1), 157–173. doi:10.1142/S0219843604000083

Zhu, C., & Kawamura, A. (2010). Development of biped robot MARI-3 for jumping. *Advanced Robotics*, *24*(11), 1661–1675. doi:10.1163/016918610X512640

KEY TERMS AND DEFINITIONS

Angular Momentum: A vector quantity of mass or system of masses turning about an inertial frame, which is conserved in the absence of the action of external forces

Bipedal Motion Generator: An algorithm that produces feasible patterns to move a bipedal robot.

Bipedal Running: A type of bipedal motion which includes flight phases.

Bipedal Walking: A type of bipedal motion in which the robot is always in contact with the floor, either on both feet (double support phase) or on a single foot (single support phase).

Euler's Equations: Three coupled equations that describe the rotational behavior of a 3D body. It is obtained through the time differentiation of intrinsic angular momentum.

Humanoid Robot: A robot with anthropomorphic appearance and motion capability.

Intrinsic Angular Momentum: Angular momentum that is associated with the rotational inertia

Jumping Robot: A robot with sufficient jumping ability.

Rotational Inertia: A tensorial quantity of a 3D object's resistance to variations to its rotational acceleration.

Spherical Coordinate System: A coordinate system for 3D space where the position is specified by three quantities; the radial distance from the origin, the polar angle and the azimuth angle.

Zero Moment Point: A point on the floor, in which the total moment acting on the robot due to gravity, centroidal torques and inertial forces equals to zero.

APPENDIX A. RATE CHANGE OF INTRINSIC ANGULAR MOMENTUM

In this section, we slightly modify Euler's equations of motion, which are actually obtained by differentiating intrinsic angular momentum. The word *intrinsic* is proposed by Featherstone (2008), in order to distinguish the angular momentum that is associated with the rotational inertia. In other words, intrinsic angular momentum is the multiplication of rotational inertia tensor and angular velocity vector of a rigid body. Referring (Featherstone, 2008), we may express intrinsic angular momentum as in the following.

$$\vec{L} = \tilde{I}\vec{\omega} \tag{A.1}$$

In (A.1), $\vec{L}$, $\tilde{I}$ and $\vec{\omega}$ are intrinsic angular momentum vector, rotational inertia tensor and angular velocity vector, respectively. We may differentiate (A.1) as below.

$$\dot{\vec{L}} = \dot{\tilde{I}}\vec{\omega} + \tilde{I}\dot{\vec{\omega}} \tag{A.2}$$

In order to differentiate $\tilde{I}$ tensor, we employ dyadic notation (Greenwood, 2003).

$$\tilde{I} = \sum_{i=1}^{3}\sum_{j=1}^{3} I_{ij}\vec{e}_i\vec{e}_j \tag{A.3}$$

$$\dot{\tilde{I}} = \sum_{i=1}^{3}\sum_{j=1}^{3}\left[\dot{I}_{ij}\vec{e}_i\vec{e}_j + I_{ij}\dot{\vec{e}}_i\vec{e}_j + I_{ij}\vec{e}_i\dot{\vec{e}}_j\right] \tag{A.4}$$

In (A.4), $\vec{e}_i$ and $\vec{e}_j$ are unit vectors. For rigid bodies, rate change of rotational inertia tensor elements, $\dot{I}_{ij}$, are zero. Since our interest is composite rigid bodies, inertia elements' derivatives may not be assigned to zero, $\dot{I}_{ij} \neq 0$. Diverging from the solution explained in (Greenwood, 2003), we define $\dot{I}^*$, element-wise time derivative of inertia matrix.

$$\dot{I}^* = \begin{bmatrix} \dot{I}_{xx} & \dot{I}_{xy} & \dot{I}_{xz} \\ \dot{I}_{xy} & \dot{I}_{yy} & \dot{I}_{yz} \\ \dot{I}_{xz} & \dot{I}_{yz} & \dot{I}_{zz} \end{bmatrix} \tag{A.5}$$

Furthermore, unit vectors are also rotating as they are attached onto it, shown in Figure 5. In this case, following equalities hold valid (Greenwood, 2003).

$$\begin{aligned} \dot{\vec{e}}_i &= \vec{\omega} \times \vec{e}_i \\ \dot{\vec{e}}_j &= -\vec{e}_j \times \vec{\omega} \end{aligned} \tag{A.6}$$

Combining (A.5) and (A.6) in (A.4), $\dot{\tilde{I}}$ is obtained.

$$\dot{\tilde{I}} = \dot{\tilde{I}}^{*} + \vec{\omega} \times \tilde{I} - \tilde{I} \times \vec{\omega} \tag{A.7}$$

$$\dot{\vec{L}} = \tilde{I} \cdot \dot{\vec{\omega}} + (\dot{\tilde{I}}^{*} + \vec{\omega} \times \tilde{I} - \tilde{I} \times \vec{\omega}) \cdot \vec{\omega} \tag{A.8}$$

$$\dot{\vec{L}} = \tilde{I} \cdot \dot{\vec{\omega}} + (\dot{\tilde{I}}^{*} + \vec{\omega} \times \tilde{I}) \cdot \vec{\omega} \tag{A.9}$$

If we expand (A.9) row-by-row and omit non-diagonal elements, we obtain (9), (10) and (11). This equation set appears to be a generalized version of Euler's equations of motion for composite rigid bodies, having rotational inertia elements' time derivatives.

APPENDIX B. BIPEDAL MOTION GENERATION BASED ON CONVENTIONAL ZMP APPROACH

In this section, we provide an easy-to apply algorithm (Ugurlu et al., 2009) to generate bipedal motion generation, based on conventional ZMP approach in which IAMR terms are omitted. Ignoring these terms in (1) and (2), we obtain the following formulae.

$$X_{zmp} = \frac{x(\ddot{z}+g) - z\ddot{x}}{(\ddot{z}+g)} \tag{B.1}$$

$$Y_{zmp} = \frac{y(\ddot{z}+g) - z\ddot{y}}{(\ddot{z}+g)} \tag{B.2}$$

As it may be observed, all inertial parameters including the total mass vanished. Moreover, ZMP equations for x-axis and y-axis also became identical. Thus, we continue using only (B.1) for x-axis motion generation as an identical approach can be executed for y-axis.

Keeping in mind that z-axis CoM trajectory is predetermined; we may obtain its and its second derivatives' values for all times. Therefore, we discretize (B.1) for x parameter as follows.

$$\ddot{x} = \frac{x[i+1] - 2x[i] + x[i-1]}{\Delta t^2} \tag{B.3}$$

In (B.3), Δt is the sampling period, i is the discrete event. i starts from 0 to n, which is the total number of events. Inserting (B.3) into (B.1), following expressions are expressed.

$$x(i+1) = -\frac{b[i]}{c[i]}x[i] - x[i-1] + \frac{X_{zmp}[i]}{c[i]}$$
$$c[i] = -\frac{z[i]}{(\ddot{z}[i]+g)\Delta t^2} \tag{B.4}$$
$$b[i] = 1 - 2c[i]$$

In order to solve this tridiagonal efficiently, we employ Thomas algorithm (Strikwerda 1989). In this algorithm, we need to define initial and terminal position of x-axis CoM position. Therefore, for a given set of ZMP trajectory, initial position and terminal position, we are able to obtain x-axis CoM trajectory. For this purpose, we re-arrange the tridiagonal equations as follows:

$$x[i] = e[i+1]x[i+1] + f[i+1] \tag{B.5}$$

In (B.5), e[i+1] and f[i+1] can be defined as below.

$$e[i+1] = \frac{c[i]}{c[i]e[i] + b[i]} \tag{B.6}$$

$$f[i+1] = \frac{X_{zmp}[i] - c[i]f[i]}{c[i]e[i] + b[i]} \tag{B.7}$$

Combining (B.5), (B.6) and (B.7), we obtain the final motion equation in discrete domain.

$$x[i] = -\frac{c[i]}{c[i]e[i] + b[i]}x[i+1] + \frac{X_{zmp}[i] - c[i]f[i]}{c[i]e[i] + b[i]}$$

Having obtained initial and terminal x-axis CoM position values previously $(x[0] = x_0 \quad x[n] = x_n)$; e(1) and f(1) are determined as 0 and x_0,respectively.

Utilizing the Thomas algorithm for the solution of this tridiagonal equation, we can obtain the CoM trajectory's x-axis component. If an identical approach is also executed for y-axis CoM position, we could derive all the components of CoM trajectory in real-time.

Chapter 7
Prototyping of Fully Autonomous Indoor Patrolling Mobile Robots

Xiaojun Wu
*Data Storage Institute, A*STAR, Singapore*

Bingbing Liu
*Institute for Infocomm Research, A*STAR, Singapore*

Jun-Hong Lee
Dyson Operations, Inc. Singapore

Vikas Reddy
Kiva Systems, Inc. USA

Xi Zheng
Thinking Dots, Inc. Singapore

ABSTRACT

In this chapter, the design and prototyping of an indoor patrolling mobile robot is presented. This robot employs a modular design strategy by using the ROS (Robot Operating System) software framework, which allows for an agile development and testing process. The primary modules - omni-directional drive system, localization, navigation, and autonomous charging - are described in detail. Special efforts have been put into the design of these modules to make them reliable and robust in order to achieve autonomous patrolling without human intervention. With experimental tests, the authors show that an indoor mobile robot patrolling autonomously in a typical office environment is realizable.

1. INTRODUCTION

Robotics, especially mobile robotics, has attracted intensive research efforts in the past decades. A real autonomous mobile robot, i.e. a robot that can run by itself without any human intervention, can function as a reliable platform and transform into many practical and useful service robot forms, and subsequently benefit humans in their daily life. For example, an indoor service robot can act as a patrolling security guard, a house servant, a tour guide, an entertainer and a tutor.

The first indoor mobile robot that was really deployed for practical applications probably is

DOI: 10.4018/978-1-4666-0176-5.ch007

RHINO, a museum robot guide developed by Burgard *et al.* (1998). As pioneering work in this field, this robot guide was deployed mainly for experimental purposes. During its 6-day deployment and tens of hours of operation, it had successfully demonstrated its localization and navigation, and functioned well to fulfill tour requests with very few exceptional failures. According to the authors, this robot achieved "overall success rate of 99.75%" and thus concluded that "time is ripe for developing AI-based commercial service robots that assist people in everyday life". It has been more than twelve years since this promising prediction about service robots was made by these expert researchers in this field. However, the development and practical application of service robots, even in the relatively simple indoor environments, have not reached the optimistic level as predicted. A fully autonomous robot platform that requires no human intervention seems like "mission impossible".

The reason for slow progress lies in the challenges that the robot has to deal with in typical people inhabited indoor environments. Let us assume that an indoor mobile robot needs to perform a simple task, e.g. patrolling, inside a typical, medium size office (in a scale of 50×50 meters). Inside the office are cubicles, normal office furniture and obviously, a lot of people, with normal working behaviors. The robot has to function in a fully autonomous mode (i.e. 24 hours in a day with 7 days a week) without any human operation and it has to be able to localize itself accurately and reliably while planning a drivable path (loop) to patrol the environment. It has to avoid any obstacles en-route by maneuvering itself in any direction to pass narrow passages safely, and return to the docking station for power recharging whenever the onboard energy is lower than a threshold.

The above challenges are similar to those faced by an indoor patrolling robot. A robot is a complex system comprising multiple electronic and mechanical sub-systems. Thus, it is important to design the robot such that each sub-system can be built and tested separately. This requires the robot to be highly modular and modularity is a key element in prototyping robotic systems. The concept of modular design is best illustrated with an example. Let us consider our robot *Rio*, which is an indoor holonomic, patrolling robot. Rio's mechanical system consists of static structures such as the frames and dynamic structures such as the drive train. Rio's electronic systems consist of Li-Ion battery packs, voltage regulators, charging and discharging circuitry, power distribution circuits, motor and emergency stop circuits. Rio's sensors include a single axis Gyro, a 2D Laser Radar (LADAR), a 3D scanning LADAR, optical wheel encoders, and a docking station for autonomous recharging. Given the large number of components present in Rio, it is important to follow a systematic approach in the design where the components are grouped into modules as shown in Figure 1.

Out of all the modules listed inside Figure 1, in this chapter, we will mainly discuss those key modules that handle the challenges defined for a patrolling robot, i.e. the omni-directional drive module, the localization module, the path planning and obstacle avoidance module, and the auto-docking module for power recharging. The rest of the chapter is arranged as follows. We will provide the background information in the next section. After that, each of these key modules we developed for the indoor patrolling mobile robot will be discussed, each in one individual section, i.e., omni-directional drive module, localization module, path planning and obstacle avoidance, and auto-docking module. After these sections, an integrated testing case of our developed patrolling robot will be given. Finally, conclusions and future research directions will be presented.

2. BACKGROUND

In the literature, each of the challenges we defined for the patrolling robot has been addressed individually in certain degree.

Figure 1. Structure and modular diagram of Rio, the indoor patrolling robot

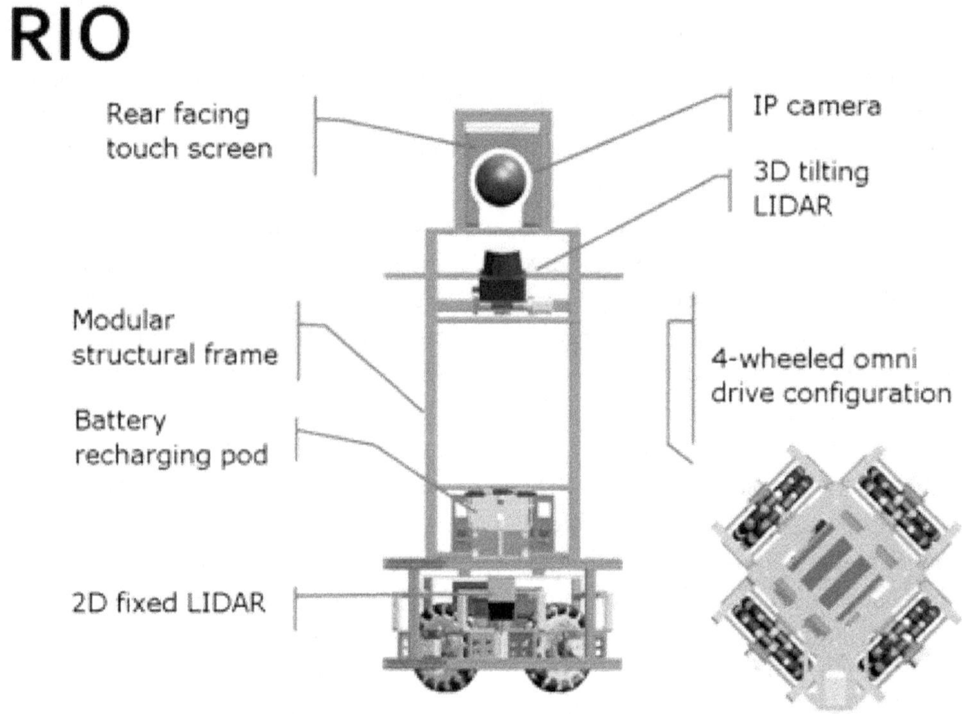

In order to move effectively in narrow areas and to avoid static or dynamic obstacles, an omni-directional drive is preferred for patrolling mobile robot compared to traditional differential drive, in terms of agility and maneuverability. For this purpose, there are a few existing mobile robot designs adopting omni-directional drive. For instances, the Uranus robot from CMU by Muir *et al.* (1987) employs Mecanum wheels to achieve omni-directional mobility, the latest PR2 robot from Willow Garage (2011a) adopts steerable castors, and a similar mobile robot design using powered caster wheels was reported by Lim *et al.* (2004).

Thrun *et al.* (2005) provide a comprehensive introduction into the field of probabilistic robotics, especially on mobile robot localization in a book. In this book, from preliminary mathematics related to robotics and localization, e.g. recursive state estimation theories, robot motion and perception models, to complicated and state-of-the-art mobile robot localization algorithms, e.g. Markov and Gaussian based localizations, grid based and Monte Carlo localizations, are introduced clearly and in very good details. Of course, with respect to mobile robot localization, we must mention simultaneous localization and mapping, i.e., the SLAM problem. Regarded as the "holy grail" in the mobile robotics community as discussed by Dissanayake *et al.* (2001), SLAM is about using a robot to build a map of the unknown environment from an unknown starting point, and use the map being built to localize itself simultaneously. As a main advantage, SLAM eliminates the necessity for artificial infrastructure or a prior map of its environment. Since the concept was first consolidated by Smith *et al.* (1990), the SLAM problem has been studied by numerous researchers. A detailed survey on the prevailing algorithms and state-of-the-art work of SLAM can be found in the papers of Durrant-Whyte *et al.* (2006) & Bailey *et al.* (2006).

With respect to the topic of path planning and obstacle avoidance, a dedicated book is written by LaValle (2006). This book is self contained, covering introductory materials, motion planning in continuous space, planning under various kinds of constrains and planning with uncertainty. The planning problem it covers is not only concerned in the robotics, but also in the field of artificial intelligence and control theory. All of these three areas share a common ground for the planning algorithms addressed inside the book. Certainly we are more interested in the motion planning algorithms for mobile robots, which were thoroughly introduced in a survey paper by Meyer *et al.* (2003). In a brief review by Lavalle (2005), the graph searching algorithms are widely employed for global path planning, such as *Breadth first*, *Depth first*, *Dijkstra*, *A** and variants, *D** and variants, where the environment is often represented in a discrete format, and path planning is converted into solving a least cost problem between two states on a graph. For obstacle avoidance, an effective approach need consider factors like the overall goal, the actual speed and kinematics of the robot, the on-board sensors, and the actual and future risk of collision etc. Algorithms implemented on actual robots for this purpose include *Bug algorithm* by Lumelski *et al.* (1986), *Vector Field Histogram (VFH)* by Borenstein (1991), *Dynamic Window Approach (DWA)* by Fox *et al.* (1997) etc. Furthermore, other than objects with regular shape, such as walls, doors, cabinets, etc., indoor environments quite often involve other objects with complex three-dimensional structure such as tables and chairs. Because of the different cross-section along the height direction, the shape features of such objects cannot be fully detected by a fixed laser scanner, which can lead to collisions during the robot movement. Thus, the robot demands the capability of three-dimensional obstacle sensing. This can be done with the help of a 3D LADAR. A recent research effort by Navarro-Serment *et al.* (2010) has been reported on using 3D LADAR to detect and track pedestrians. The proposed algorithm is to fit 3D laser points to different object models using statistical pattern recognition approaches. By extracting and using geometric and motion features, the motion characteristics of human being can be recognized and thus pedestrians can be detected.

Existing recharging approaches for mobile robots have also been reported in literature. To name a few, the *Roomba* vacuum cleaning robot from the iRobot Inc. has a 360 degree infrared (IR) turret integrated at the front to receive docking signals from the docking base, the Pioneer robot from the MobileRobots (2011) uses a laser scanner to recognize the shape of the docking station, Silverman *et al.* (2002) in Willow Garage used vision to assist docking of a Pioneer 2-DX robot and a Texas robot by Willow Garage (2011b) respectively.

Thus, according to the literature, solutions to each of these challenges, i.e., omni-directional drive, localization, path planning and obstacle avoidance and auto-docking have been individually accessed in various degrees. However, to the knowledge of the authors, there is scarcely integrated system that can complete the patrolling task mentioned above. Looked back, the earlier mobile robots include, for example, the indoor mobile robot by Crowley (1985) tested for navigation using a rotating ultrasonic range sensor, the HERMIES-IIB robot developed by Burks *et al.* (1987) for visual perception, goal recognition and navigation in unknown environments, and the omni-directional wheeled robot Uranus developed by Muir *et al.* (1987). Slightly later, the CARMEL robot, developed in the University of Michigan by Borenstein *et al.* (1989), was used to test obstacle avoidance approaches at a fast running speed. A 'Robuter' mobile robot, equipped with 6 static sonar transducers was developed by Leonard *et al.* (1991) and utilized in the implementation of a beacon based localization algorithm. In 1995, a pair of mobile robots connected by a compliant linkage was reported by Borenstein (1995) as an innovative in the control and kinematic design. The first indoor mobile robot deployed for *practical* applications, according to literature, is RHINO developed by Burgard *et al.* (1998), a museum robot guide, although deployed mainly for experimental purposes. More recently, there are a few robots found in the literature that have potential to fulfill some requirements of the patrolling robot we defined. Michaud *et al.* (2005) reported a socially interactive robot with modular design, called *Spartacus*. Although Spartacus (and its previous version, called U2S robot) has a humanoid shape, its mobility function is actually based on a differential-drive mobile base. A 2D Sick LADAR is used on the mobile base for localization and navigation purposes. Spartacus is also equipped with a pan-tilt-zoom color camera, an array of microphones and a touch-screen, which are all used to implement the robot's main design features, i.e., the human-robot interactions. Although Spartacus is aiming for the human-robot interaction aspect, it is not capable of being a fully autonomous patrolling robot, for the lack of capabilities for reliable navigation and dynamic obstacle avoidance. In addition, it needs manual connection to an electric outlet for power recharging. In the year of 2010, Intel's personal robotics group released their Home Exploring Robotic Butler robot or HERB in short, as reported by Srinivasa *et al.* (2010). HERB can fetch drinks, get a pack of chips, and sort dishes, thanks to a pair of dexterous arms that are driven by cables. HERB sits on a particularly modified Segway base, which is basically differential driven. Although it looks big with the arms extended, the developer Siddhartha Srinivasa claimed that "it will fit through most doorways". A 2D LADAR on the top of HERB helps generate 3D data so the robot has superb sensing ability in this way. There's also a camera to help HERB "see." With the development focused on object handling, HERB is not an appropriate candidate for our patrolling task either. Although it does have good 3D laser-based sensing capability, it lacks the auto-charging function. Additionally, it is over bulky due to its arms and is not omni-directional driven, thus not perfect for our patrolling robot application.

Besides these robots developed in universities and institutes discussed above, a lot of off-the-shelf robots also seem to be able to handle the challenges we raised earlier for the patrolling robot. One group of robots from MobileRobots Inc., i.e., *PeopleBot*, *Research PatrolBot* and *Pioneer 3-DX*, in our point of view, have the potential to compete for our patrolling robot application. As both the PeopleBot and Research PatrolBot are based on differential drive base similar to the Pioneer 3-DX, here we only discuss the latter. Pioneer 3-DX robot is probably the world's most popular research robot base. Its simple but rugged design guarantees high reliability. The control of the robot is easy and its odometry data is very accurate when it runs on normal indoor surfaces. Pioneer 3-DX robot is readily to integrate additional hardware to expand its ability. For example, with a 2D LADAR, a Laser Mapping and Navigation System and a docking system, a Pioneer 3-DX robot is able to reliably patrol in a pre-explored indoor environment in an autonomous mode. The problem with Pioneer 3-DX robot could be, Mobile Robot Inc. does not provide a suitable 3D LADAR solution for it. Its current SICK laser based 3D solution is too bulky for Pioneer 3-DX. Additionally, Pioneer 3-DX has a differential type of drive mechanism, which is not the best choice for dense environment. Segway Inc. provides a mobile robot platform with Mecanum type of omni-drive wheels, *RMP 50 Omni*, which can move sideways. However, this platform is designed for relatively heavy duty, and thus has a relatively large footprint, which actually hinders its application as an indoor patrolling robot to travel through narrow passages. Gostai Inc. has a patrolling robot model, called *Jazz Security*. With a camera equipped, Jazz Security is basically a tele-presence robot, similar to the *QB* tele-presence robot from *Anybots*. However, Gostai Inc. plans to mount a laser ranger finder on Jazz Security to localize itself and navigate autonomously in a given indoor environment. Furthermore, an auto-recharging module is also provided. Jazz Security could be a good patrolling robot candidate, although, again, it lacks 3D sensing and omni-directional mobility. Kiva Systems Inc. develops an automation system for warehouse, where a batch of mobile robots perform localization and navigation task under high accuracy. By reading positioning information encoded in 2D barcodes on the floor, Kiva robots can track their own position and orientation very accurately. However, the constraint of this localization mechanism also stands out obviously, i.e. the robots have to follow pre-defined routes, usually straight lines, to read the barcodes and achieve the positioning accuracy.

A most promising mobile robot that can handle all the challenges we require for a patrolling robot, could be probably the *PR2* robot developed by Willow Garage Inc. In terms of hardware, the PR2 robot is more than qualified to act as the patrolling robot. For the localization and navigation purposes (and more than these), it has two cameras on its head, one is a high resolution color camera and the other is a stereo camera, both with Ethernet interface. Besides the cameras, it is equipped with two LADARs, one of them fixed on the base and the other located above the shoulder, with a tilting mechanism to make it a 3D scanning sensor. For indoor maneuvering, PR2 robot has an Omni-directional base with 4 individually steered and driven casters. For auto-charging, PR2 robot does not have a dedicated charging station. Yet, with its onboard powerful computational resources (2×Quad-core i7 Xeon processors) and vision sensors, it can correctly recognize electric outlets on the wall. With its dexterous arms and hands, it can accurately plug into an electric outlet and charge the onboard batteries by itself. In a word, PR2 has been provided more capabilities than those we defined for a patrolling robot. Compared to the hardware and its modular design, what equally impressive is the software developed for PR2, called ROS (Robot Operating System). In brief, ROS contains low-level drivers and provides synchronized communications between different nodes (could be a sensor or another robot), but leaves robot developers to focus on coding high-level algorithms to make the robot that uses ROS more intelligent. Another good thing about ROS is its nature of open source, as implied by another name, Robot Open Source. The only reason we do not recommend PR2 to act for the patrolling task we defined is that, obviously, this robot is too complicated and expensive, since it is designed mainly as a service robot, for tasks such as object manipulation, human-robot interaction etc. That is why we have developed our own patrolling robot, which is also modular, able to fulfill all the challenges, but much simpler and less costly, thus much more suitable as a prototyping patrolling robot. In the following sections, we will discuss the key modules among our prototyping of the indoor patrolling robot, in order to address the challenges we have defined for the patrolling robot.

3. OMNI-DIRECTIONAL DRIVE MODULE

3.1 Background

3.1.1. Omni-Directional Drive versus Differential Drive

Indoor environments usually involve various objects and human being activities. Thus, a patrolling robot must have the capability to move in tight areas and to avoid static or dynamic obstacles. For such scenarios, an omni-directional drive is preferred for patrolling mobile robot compared to others such as traditional differential drive, in terms of agility and maneuverability.

A differential drive refers to a drive mechanism for mobile robots that two fixed standard wheels are placed on either side of the robot body and these two wheels are separately driven. The main feature of the differential drive is that the robot turning speed is determined by the speed difference between the wheels, i.e. "differential". For fixed standard wheels, there exists a sliding constraint for motion so that there is no motion in the direction perpendicular to the wheel's forward velocity, otherwise the wheel will skid. Based on this wheel constraint and its configuration feature, a robot using the differential drive can instantly move forward and back, but cannot move to the right or left without rotating beforehand. This is also called non-holonomic constraint.

In comparison, an omni-directional drive is able to realize sideway motion without rotating. This gives it advantages in confined or congested spaces, such as offices, workshops, warehouses, hospitals etc. Thus the omni-directional drive is chosen for our patrolling mobile robot.

3.1.2. Omni Wheel Design

There are three most popular types of wheel designs for omni-directional drive.

1. Universal Omni-wheel with passive rollers

This type of omni-wheel is built using small rollers attached along the outer periphery of the main wheel. This wheel can move in the direction perpendicular to the motor axis, similar to a fixed standard wheel, and can also slide along the motor axis direction through the small rollers. Figure 2 shows an example 6 inch omni-wheel design using passive rollers available on the market, produced by AndyMark Inc.

On one hand, universal omni-wheel is compact with simple mechanical design and low weight, easy to install and control, and operates well in flat, indoor environments. On the other hand, this wheel type is not suitable for heavy loading, and the wheel slippage might seriously affect the dead-reckoning accuracy. Currently on the market, there have been various options provided for universal omni-wheel to cater for user requirements, such as different wheel sizes, wheel configurations (single, double or even triple layers), body material (aluminum or plastic), roller hardness (polyurethane or rubber etc), bearing and mount options.

2. Mecanum wheel

Similar to the universal omni-wheels, the Mecanum wheel consists of a hub carrying a number of free moving rollers, but angled at 45° about

Figure 2. A 6" aluminum omni wheel from AndyMark Inc. (Image courtesy of AndyMark Inc.)

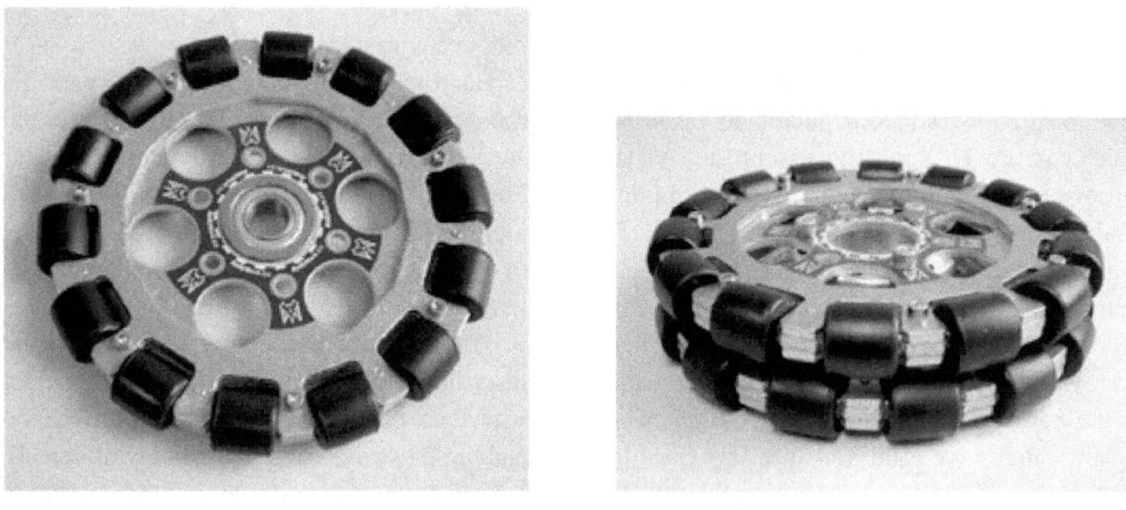

Figure 3. 6" Mecanum wheels from AndyMark Inc. (Image courtesy of AndyMark Inc.)

the hub's circumference. Figure 3 shows a typical Mecanum wheel design from AndyMark Inc.

The most distinct property of this wheel type is the high load capacity. In overall, the Mecanum wheel has more complex mechanical design and its wheel slippage is also obvious when rotating on the planar ground. A typical example employing Mecanum wheel is the Uranus robot from CMU developed by Muir *et al.* (1987).

3. Powered steering or castor wheel

These designs have a wheel that can be steered by an orthogonal steering actuator, as shown in Figure 4. Using steerable wheels can achieve near omni-directional motion, i.e. the resulting vehicle kinematics are subject to non-holonomic constraints, and sometimes will undergo wheel slip and/or scrubbing to change directions, as discussed by Udengaard (2007).

The powered steering can employ standard wheels, and the contact between wheels and ground is continuous, which makes the wheels

Figure 4. Concept of powered steering wheel

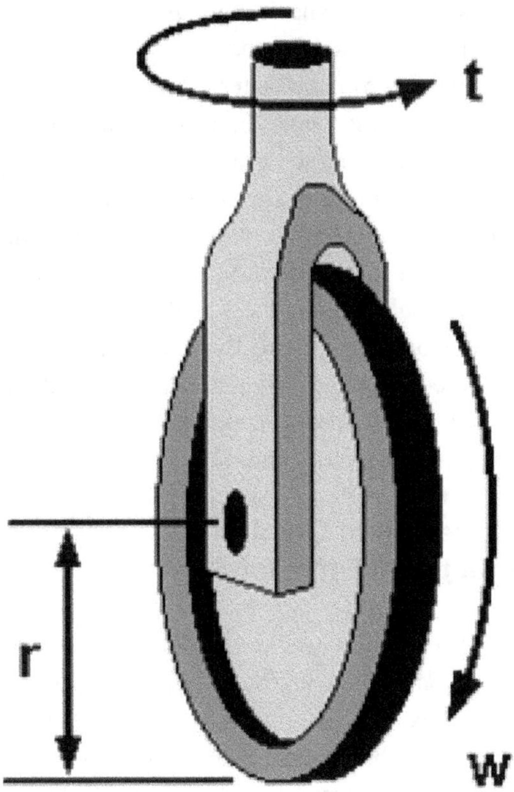

robust to floor conditions. The mechanical and control of the steerable wheels are more complex compared to other omni-wheels using rollers. A typical mobile robot using steerable castors is the latest PR2 robot from Willow Garage (2011a).

As described above, these three wheel types own different advantages and disadvantages. The selection of omni-wheel type depends on the actual requirements of the robot to be designed. For design of an indoor patrolling mobile robot, taking into consideration the complexity for mechanical design and control, also the availability on the market, the universal omni-wheel with passive rollers is chosen for the robot prototyping.

3.1.3. Three-Wheel Design versus Four-Wheel Design

The most common mechanical configurations for omni-directional robots are based on three or four wheels. According to Oliveira (2009), three wheeled systems are mechanically simpler but robots with four wheels have more acceleration with the same kind of motors. Four wheeled robots are expected to have better effective floor traction, that is, less wheel slippage – assuming that all wheels are pressed against the floor equally. Four wheeled robots also have higher manufacturing cost, increased energy consumption and may require some kind of suspension to distribute forces equally among the wheels. Mainly based upon its property of effective floor traction, i.e., less wheel slippage, and better stability and load capacity, the four-wheel configuration is chosen for our robot prototyping.

3.2 Design of Omni-Directional Drive for Rio

Based on the above review and analysis, an omni-directional drive with four-wheel configuration, using universal omni-wheels is chosen for our patrolling robot. Figure 5 shows the kinematics diagram of the omni-directional drive base.

We can derive the kinematic model of the omni-directional drive base, with respect to the linear velocity (v_x,v_y) and angular velocity $\dot{\omega}$ of the robot base, as

$$\begin{bmatrix} v_x \\ v_y \\ \dot{\omega} \end{bmatrix} = \begin{bmatrix} \cos\theta & \sin\theta & -\cos\theta & -\sin\theta \\ \sin\theta & -\cos\theta & -\sin\theta & \cos\theta \\ \frac{1}{4R} & \frac{1}{4R} & \frac{1}{4R} & \frac{1}{4R} \end{bmatrix} \begin{bmatrix} v_1 \\ v_2 \\ v_3 \\ v_4 \end{bmatrix}.$$

To be noticed, the above equation is based on the assumption that the mass of the robot base is evenly distributed and thus the constant *1/4R* appears as a convention in this case. Based on the

Figure 5. The diagram of the omni-directional drive base. In the diagram, R is the radius of the robot base and r is the radius of each of the four wheels. (x_r, y_r) is the body frame coordinate of the robot base, while (X,Y) is the global coordinate. $v_{1...4}$ denotes the linear velocity of the four wheels respectively. ω denotes the angular velocity of the robot base. θ defines the heading angle of the robot, which is the angle from positive X to the axel of the fourth drive wheel.

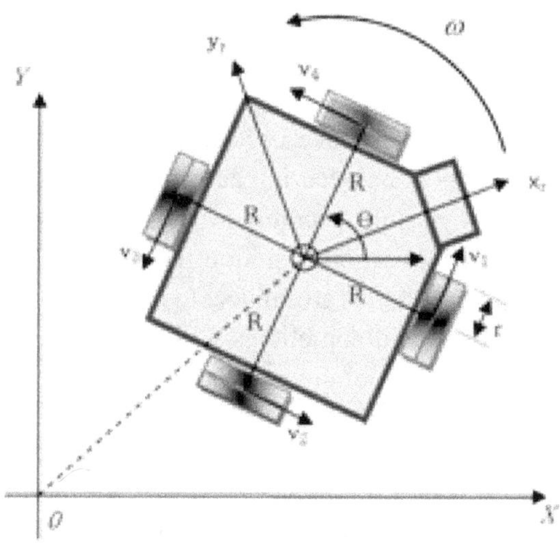

kinematic model, the inverse kinematic model for the control purpose can be derived as,

$$\begin{bmatrix} v_1 \\ v_2 \\ v_3 \\ v_4 \end{bmatrix} = \begin{bmatrix} \frac{1}{\cos\theta} & \frac{1}{\sin\theta} & 4R \\ \frac{1}{\sin\theta} & \frac{-1}{\cos\theta} & 4R \\ \frac{-1}{\cos\theta} & \frac{-1}{\sin\theta} & 4R \\ \frac{-1}{\sin\theta} & \frac{1}{\cos\theta} & 4R \end{bmatrix} \begin{bmatrix} v_x \\ v_y \\ \dot{\omega} \end{bmatrix}.$$

To be noticed that, this kinematic model is for the control input computation under certain circumstance (constant heading angles), thus singularity can be safely avoided.

The detailed development of the robot base (see Figure 6 (b)) includes the mechanical design for omni-directional wheel assembly and robot chassis, the electronics design for the motor control and the interfacing microcontroller module.

3.2.1 Mechanical Design

For the part of mechanical design, four universal omni-wheels of 6" diameter from AndyMark Inc. are adopted. The wheel consists of two layers of aluminum body plus 28 rollers made of black SBR rubber, as shown in Figure 2, and its load capacity is 120 pounds. Each of the four wheels is driven by a separate Dynamixel 106+ DC brush servo motor from Robotis Inc. The four omni-directional wheels are positioned at 90° apart to achieve omni-directional mobility, i.e., to allow the robot to move in *any direction* along the ground plane at any time regardless of the orientation of the robot around its vertical axis.

The proposed design leads to a symmetric design which will result in high stability for the robot base. A stiff suspension mechanism together with the interface microcontroller module, the bi-directional motor controller board, DC brush motor and optical shaft encoder are built into a complete set of wheel assembly module (see Figure 6 (a)). At the centre of the robot base, the Lithium-Ion battery packs, power management system module and onboard computer are hosted.

3.2.2 Electronics and Control

The electronics of Rio drive train consists of a Fit-PC2 computer for basic control, an intelligent battery & power system (IBPS) from Ocean Server Technology Inc., DC brush geared motors, optical shaft encoders, motor controllers and interface microcontroller modules. All components used to

Figure 6. (a) The omni-directional wheel assembly module, and (b) the robot base

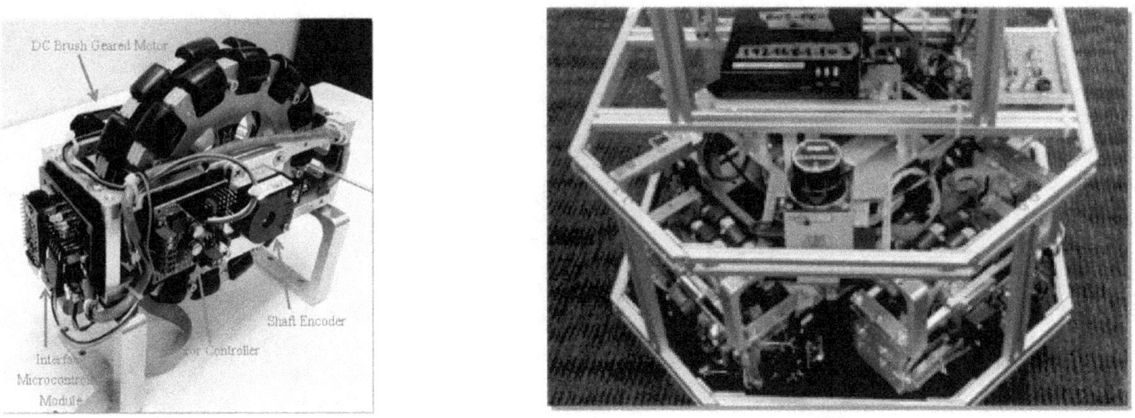

(A) ***(B)***

construct the robot base are chosen from conventional, commercially available hardware, thus to reduce the manufacturing and maintenance cost of the robot. Especially, the selected DC brush servo motor, Dynamixel EX106+ from Robotis Inc., is a very compact actuator which integrates reduction gear, controller, driver and network function together. This motor can provide holding torque up to 106 kg•cm. The adoption of this actuator greatly reduces the design complexity of control system for the patrolling robot.

The wheel position is sensed by separate optical shaft encoder. The position and velocity of each wheel are thus available to the onboard control computer from the interfacing 8-bit microcontroller module through I2C communication bus. The interfacing microcontroller module also handles the PID control of the DC brush motor through the motor controller. The desired robot trajectory and velocity of each motor are computed by the onboard computer. The diagram of the whole electronics and control system of the robot base is shown in Figure 7.

3.2.3 Handling Large Odometry Error via Adapting Motion Model

The Rio robot has a holonomic drive system, providing greater degree of freedom in the motion of the robot. However, the universal omni-wheels with passive rollers are prone to slippages and this adversely affects the accuracy of the resulting odometry readings. We did a simple test in our office, where the ground is covered by carpet, to

Figure 7. Diagram of the hardware control system of the robot base

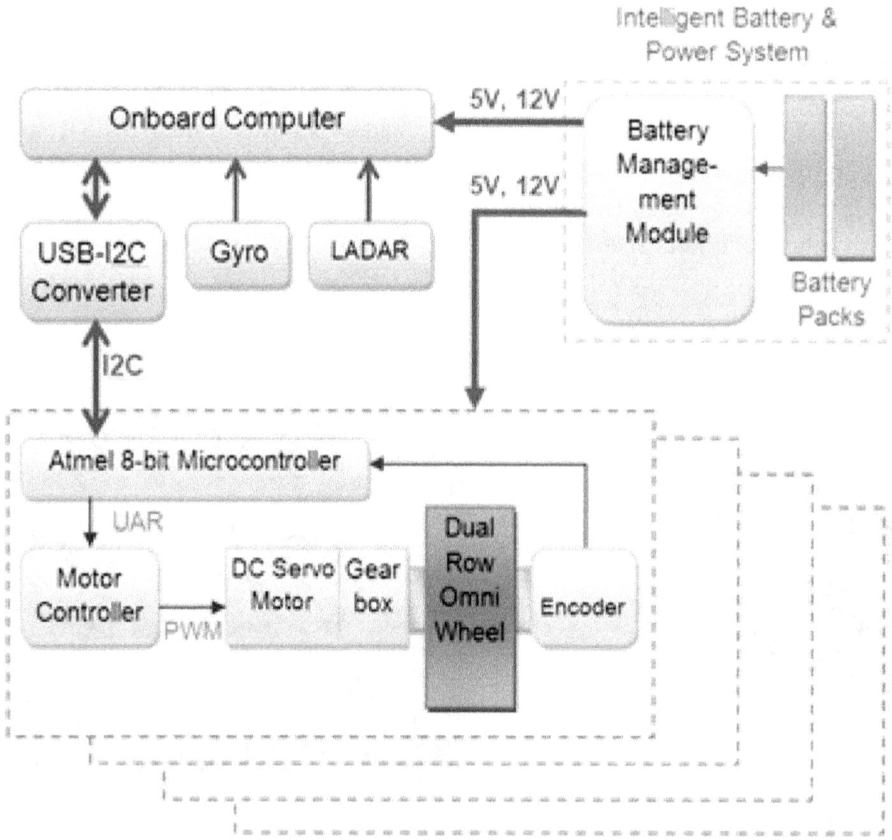

check the odometry accuracy of Rio. During the test, a forward moving command keeps being issued to Rio until the robot odometry reports the travel distance reaches 2 meters. The actual distance that the robot traverses is checked using measuring tape, which averages around 1.8 meters. In this case, the robot odometry suffers an error rate of about 10%. This large odometry error also causes the Monte-Carlo localization (MCL) algorithm (the core of our localization system, described in next Section) to fail in the navigation test, where particles are observed to accumulatively diverge from the actual pose of the robot. After a few iterations, it becomes evident that the particles are unable to accurately track the motion of the robot as it moves around the room, with none surviving near the actual pose of the robot.

This slippage error is difficult to correct since it relates to various factors such as the material of both the ground and the wheels, the actual robot loading and speed etc. In order to address this issue, we choose to take into consideration the odometry error in localization module. The MCL approach uses robot-specific noise parameters to represent motion error, as discussed by Thrun *et al.* (2005). Thus we first tried to inflate these noise parameters of the robot to compensate for the noise in the odometry readings. This strategy reflects the cumulative nature of the drift in the odometry readings, in other words, the amount of noise injected is in proportional to the motion of the robot. As the time step gets smaller, smaller motions are realized and thus less noise is injected into the model. However, in actual test, the particles still accumulatively diverges from the actual pose of the robot despite the above inflation of noise parameters. This could be due to the fact that other than the uncertainty item in proportional to robot motion, there are constant uncertainties arising from limited map and laser resolution etc., which is not properly modeled by the noise parameter inflation. Thus, besides the proportional noise injection, alterations were also made to include a consistent amount of Gaussian noise into the robot motion. The predicted pose is first calculated based on the readings obtained from the odometry. A consistent amount of Gaussian noise is then added to the predicted pose to compensate for the odometry inaccuracies and noise in the system. Taking into consideration the re-sampling process in the previous time step which works to cluster the particles, the addition of Gaussian noise produces a scattering effect, increasing the particle diversity and thus the number of effective particles. The weighing and re-sampling process would help to keep the scattering in check so that the particles are always distributed in close proximity to the pose of the actual robot.

The benefit of this method lies in its simplicity and effectiveness. It is straightforward to implement and the amount of noise to be injected into the system can be easily adapted to meet different needs. The scattering effect increases the particle diversity and allows the system to be tolerant to noise. However, as more noise is injected into the system, the particles would deviate from the actual pose more rapidly. Therefore weighing and re-sampling has to be carried out more frequently to keep the scattering effect in check.

3.3 Section Summary

In this section, we presented the module of omni-directional drive for the patrolling robot. As different from the traditional differential drive model used for indoor mobile robots, we discussed the advantages of using the omni-directional drive. Three different kinds of omni-wheels were reviewed and the reason that the universal omni-wheel with passive rollers was chosen for the patrolling robot was explained. Three wheeled configuration was also compared with four wheeled one, and the later was chosen for the properties of less wheel slippage, and better stability and load capacity. The direct and inverse kinematic models of the omni-directional drive were derived. The omni-directional drive module is a perfect example to show the concept of modular design. For example, each of its four wheels is a complete set of assembly module that consists of different components, such as interface microcontroller module, bi-directional motor controller board, DC brush motor and optical shaft encoder. This design concept renders great convenience for testing, debugging, repairing and replacing each functional module and part. In order to handle the problem of large odometry error of the omni-directional drive, we adopt an approach of adapting the error within MCL's motion model, thanks to the fact that the motion model of MCL uses the odometry reading as input.

4. LOCALIZATION MODULE

4.1 Overviews

Mobile robot localization has been a highly heated research topic for robotics community, for the simple reason that localization ability is a fundamental function for mobile robots to perform any further tasks. In order to achieve a robust and accurate localization solution for the patrolling robot, we design a 3-tiered hierarchical localization system, which is illustrated by the framework in Figure 8. The core of the localization system is the 2nd tier, quasi-Monte Carlo localization (QMCL), which is a variant of the popularly utilized MCL algorithm, as discussed by Thrun *et al.* (2005), but using a more efficiently initialized random particle set. As being a probabilistic model, QMCL can handle slightly dynamic noise in the environment although the input, pre-explored map only consists of static objects. However, as the dynamic noise level increases significantly (i.e., larger than a threshold), e.g. when too many human being hanging around the robot which causes the wrong scan matching results, QMCL will fail to produce correct pose information for the robot. In this case, the localization system will resort to the 3rd tier, which is a *dead reckoning* method, consisting of odometry information for translational measurements and a single-axis gyroscope for orientation. In order to achieve reliable localization ability after the robot's long term running, its pose is reset to *0* when it returns to the docking point for recharging. This simple *zero positioning* is set with top priority, i.e. the 1st tier of the localization system, because we are fully certain that back at the docking point, which is set as the origin of the global coordinate for the robot, the pose of the robot is *0*. And this again re-initializes the QMCL and dead reckoning methods to restart pose computing for the robot. In this subsection, we will mainly discuss QMCL.

4.2 Basic Monte Carlo Localization

Compared to the SLAM problem, which is a mobile robot to explore an unknown environment

Figure 8. Framework of the 3-tiered localization system

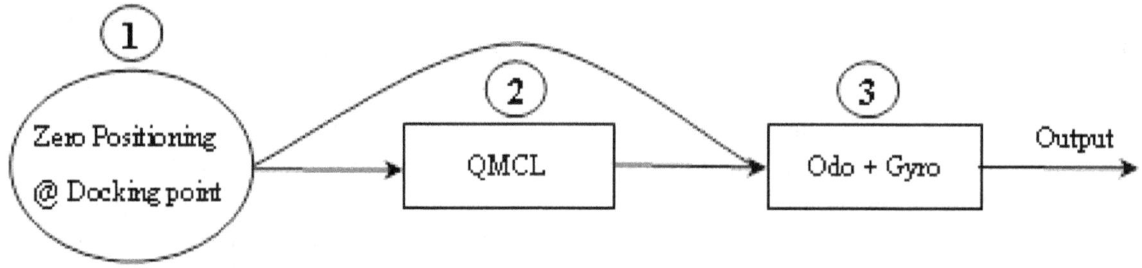

from an unknown starting point and then to incrementally build a map of the environment while using this map to compute the robot's location simultaneously, a practical problem confronted more frequently in the mobile robot applications, is to estimate the robot's global position, with an *a priori* available map. This is because, as a quick argument, for applications such as path planning and obstacle avoidance, the robot needs to know its positioning information online while the map of the environment could easily be explored and built offline, in advance. This problem, known as global position estimation for mobile robots with existing maps, has also been studied thoroughly in the past decade. The challenge with global position estimation is that the robot's starting point within the map is totally unknown and must be found quickly and accurately before the position can be tracked. The most well known solution to the global position problem is probably the family of Monte Carlo Localization (MCL) methods.

The basic MCL method was proposed by Dellaert *et al.* (1999). In this method, a set of samples or particles are randomly drawn to represent the probability density function (PDF) of the robot's pose to be estimated, instead of describing the PDF by its 1st and 2nd order moments as a Kalman filter (KF) does. The samples are maintained to update the PDF over time by using Monte Carlo methods. Two major advantages have been obtained by using this sampling representation based localization over the common KF based localization methods. The first one is that the sampling essence of MCL enables it to represent multi-modal distributions and thus to achieve global localization and re-localization (solving the "kidnapped robot problem") for a robot, which a KF technique cannot. The second advantage is MCL breaks the strong assumption of Gaussian distribution that a KF technique requires the prediction and observation noises to be. The strong theoretical foundation, which is the essential duality between a sample and the probability density from which it is drawn, and the relatively easy implementation enable MCL a successful localization method for mobile robots and continuously attract researchers to study and extend its applications, such as Kwon *et al.* (2010) & Lee *et al.* (2010).

The basic MCL actually takes the format of a Bayes filter to estimate the PDF of the robot pose, based on motion and perception sensors' data. It is a version of sampling/importance resampling method. The prime idea is to recursively estimate the posterior distribution of the robot poses, which is called "belief" and represented by a set of weighted, random samples. The belief is denoted here as,

$$Bel(\boldsymbol{x}_t) = p(\boldsymbol{x}_t \mid \boldsymbol{S}_{0\to t}),$$

where, x_t denotes the robot pose at the time instant t, which in indoor environments, consisting of the robot position in a 2D coordinate frame and the heading angle. $\boldsymbol{S}_{0\to t}$ denotes the sensor data (motion and perception) from time instant 0 till time t. Commonly, an odometer is used as motion sensor and a laser range finder is used as perception sensor, so is in this work. With applying the Bayes conditional probability rule, the Markov theorem and other theorems, a series of deductions by Thrun *et al.* (2005) lead to

$$Bel(\boldsymbol{x}_t) = \\ kp(z_t \mid \boldsymbol{x}_t)\int p(\boldsymbol{x}_t \mid \boldsymbol{x}_{t-1}, \boldsymbol{u}_{t-1})Bel(\boldsymbol{x}_{t-1})d\boldsymbol{x}_{t-1}.$$

where the probability $p(\boldsymbol{x}_t \mid \boldsymbol{x}_{t-1}, \boldsymbol{u}_{t-1})$ is actually the *motion model* (***u*** denotes odometry data), and the probability $p(z_t \mid \boldsymbol{x}_t)$ is the *observation model* (*z* denotes laser data). The belief $Bel(x_t)$ described in the format of above equation defines a recursive, posterior estimator for the basic MCL algorithm and is the basis of various MCL algorithms proposed by Dellaert *et al.* (1999), Kwon *et al.* (2010) & Lee *et al.* (2010), as well as our QMCL here. In order to start the estimator to estimate the robot pose, an initial density is needed. In the case of global localization, $p(x_0)$is usually modeled as a *uniform* distribution by Monte Carlo methods, represented by multiple samples drawn from it, to cover all free spaces. However, the uniform distribution by Monte Carlo method is not an efficient way to generate random sequences and in order to avoid a large sample set size from the very initialization period; a better random sequence generator is needed.

4.3 Quasi Monte Carlo Localization with Modified Motion Model

Although being a mature localization solution, there do exist some shortcomings of MCL left untouched, for example, the inefficient uniform random sequence generated by Monte Carlo methods and hence the unnecessary large sample set used for the initialization of MCL. In order to address this problem, we present a novel approach called quasi Monte Carlo localization (QMCL). Quasi Monte Carlo method reported by Caflisch (1998) will be applied to generate a smaller number, but more uniform samples, to cover the whole free space sufficiently, during initialization of localization and re-localization. QMCL also improves solving the kidnapped robot problem, which needs to generate random samples to cover all allowable space again. A new motion model will also be developed by injecting noise to the motion model more efficiently and thus reduces the convergence time of MCL and improves its efficiency.

The aim of initialization is to generate sufficient random samples to cover every possible pose for the robot. For an environment with a pre-explored map, an intuitive way is to increase the sample set size to ensure there will be a sample at every possible pose. However, the sample set size determines the computational overhead and for an online robot localization algorithm, the size is bounded to a certain number. With a fixed number of sample set, an efficient way is desired to generate the samples to cover the free space on the map as evenly as possible. In the previous work by Dellaert et al. (1999) & Thrun *et al.* (2005), Monte Carlo method was used to generate pseudo-random sequences to cover all allowable space during initialization. Pseudo-random sequences are software-generating sequences of numbers that approximates the properties of random numbers. Pseudo-random sequences typically exhibit statistical randomness and have the benefit of being able to produce repeatable numbers that are useful for algorithm testing. Compared to these pseudo-random sequences that mimic the properties of random sequences, the random sequences generated by quasi Monte Carlo can provide better *uniformity*, as reported by Caflisch (1998). Uniformity of a sequence is measured in terms of its *discrepancy* and quasi random sequences are also called low-discrepancy sequences. Generally speaking, the discrepancy of a sequence is low if the number of points in the sequence falling into an arbitrary set is close to proportional to the measure of that set. The precise definition is presented here. For a sequence of N points x_N in the unit cube I^d, its discrepancy D_N is defined as,

$$D_N = \sup_{J \in E} \left| \frac{1}{N} \#\{\boldsymbol{x}_n \in \boldsymbol{J}\} - \lambda_d(\boldsymbol{J}) \right|$$

for any subset $\boldsymbol{J}$ of I^d. Usually $\boldsymbol{J}$ is restricted to be a rectangular set or "box" of the form

$$\prod_{i=1}^{d} [a_i, b_i) = \{x \in \boldsymbol{I}^d : a_i \le x_i < b_i\},$$

where, $0 \le a_i < b_i \le 1$.

The above equations tell the discrepancy is, for any subset J in the d-dimensional unit cube I^d, the supremum of the error between the numbers of points in the sequence that falls into subset $\boldsymbol{J}$ and measuring of the volume of $\boldsymbol{J}$ ($\lambda_d(\boldsymbol{J})$ is the d-dimensional Lebesgue measure). The smaller the discrepancy of a sequence is the better uniformity it has. For an extreme case, an infinite sequence $\{x_n\}$ is uniformly distributed, if

$$\lim_{N \to \infty} D_N = 0.$$

For quasi random sequences,

$$D_N \leq c(\log N)^d N^{-1},$$

where c is a constant and d is the dimension. Compared with $O(N^{-1/2})$, which is the discrepancy of Monte Carlo sequences, quasi random sequences have a considerably smaller discrepancy for small dimensions and thus have a better uniformity. For localization application of indoor mobile robots with dimension not higher than 3, this argument is valid.

In MCL, a motion model is required to predict the propagation of the particles. The motion of the robot is modeled using linear piecewise equations. Without losing generality, a motion model based on a differential drive model is used to illustrate how to modify the model to generate more robust localization solution with faster converge rate. This differential driver based motion model firstly needs to generalize a rotation (Δa_1) for the robot to face the destination, followed by a translation (Δl) to the destination and another rotation (Δa_2) to match the final heading of the robot, as illustrated in the follows,

$$\begin{aligned}\Delta a_1 &= \arctan(y_0^t - y_0^{t-1}, x_0^t - x_0^{t-1}) - \theta_0^{t-1} \\ \Delta l &= \sqrt{(y_0^t - y_0^{t-1})^2 + (x_0^t - x_0^{t-1})^2} \\ \Delta a_2 &= (\theta_0^t - \theta_0^{t-1}) - \Delta a_1,\end{aligned}$$

where x_0, y_0 and θ_0 are x, y position and heading angle read from odometry, with t and t-1 denote two consecutive time instants.

The motion of the robot is derived from the odometry readings of the robot. However there is uncertainty in the readings due to factors such as wheel slippage, wear and tear and uneven surface which would distort the readings. The uncertainty is modeled by injecting noise into the motion of the robot. In the context of MCL, sufficient noise has to be injected otherwise the particles would not be able to track the motion of the robot, which would lead to divergence of the particles, causing localization failures. The conventional approach injects noise based on statistical model of the motion of the robot. Noise is added at each of the rotation and translation of the motion. This addition of noise would distribute the particles at possible poses of the robot, as illustrated with the follows,

$$\begin{aligned}\theta_i^{t-} &= \theta_i^{t-1} + \Delta a_1 + \omega_{\Delta a_1} \\ x_i^t &= x_i^{t-1} + (\Delta l + \nu_{\Delta l})\cos\theta_i^{t-} \\ y_i^t &= x_i^{t-1} + (\Delta l + \nu_{\Delta l})\sin\theta_i^{t-} \\ \theta_i^{t+} &= \theta_i^{t-} + \Delta a_2 + \omega_{\Delta a_2},\end{aligned}$$

where, $i=1...N$ with N is the total particle number. $\omega_{\Delta a_1}$, $\nu_{\Delta l}$ and $\omega_{\Delta a_2}$ are zero mean, white Gaussian noise associated with Δa_1, Δl and Δa_2 respectively.

The conventional method is to inject noise based on the possible motion of the robot, which is a sound method, provided that an accurate starting pose is given. If sufficient noise is added, all possible motions of the robot would be covered. However during global localization, the starting pose is unknown and the particles are distributed randomly throughout the configuration space. Although there probably are particles in the proximity of the actual pose, seldom do any of them coincide perfectly with the actual pose. The fewer particles used, the larger probability this non-coincidence will be. Therefore by using the conventional method of injecting noise, there might cause situations whereby the actual pose will not be predicted by the current set of particles. The algorithm would thus take longer to converge to the actual pose.

In order to illustrate this point more clearly, consider the two scenarios below:

1. The nearest particle has a heading parallel to that of the actual pose: Most of the indoor mobile robots are non-holonomic and cannot traverse sideways. Therefore by using the conventional method, it would be unable to converge to the actual pose within the next iteration. It would need to undergo a few more iteration till the actual pose is within the range of motions allowed by the motion model. Although our patrolling robot is omni-directional and able to traverse sideways, we would like to make the motion model be as general as to adapt to most robots.
2. The robot undergoes pure rotation: As the robot moves and gain more data, it should be able to better deduce its location. By performing a full rotation it should have acquired sufficient data from its sensors to pinpoint its location. However, by using the conventional method, only rotational noise will be injected and there would not be any improvement in the pose of the particles.

A different approach to inject noise into the motion model is thus proposed in order to better respond to sensor measurements and decrease the convergence time. In this new approach, the linear piecewise equation is used to model the motion of the robot, without any noise injected into the rotation and translations. Instead the noise is added to the pose of the robot directly, as shown in the following equation. Although the errors in odometry readings grow exponentially due to cumulative effect, the additional error introduced at each step is small enough to be modeled by Gaussian noise.

$$x_i^t = x_i^{t-1} + \nu_x$$
$$y_i^t = y_i^{t-1} + \nu_y$$
$$\theta_i^t = \theta_i^t + \omega_\theta,$$

where, v_x, v_y and ω_θ are Gaussian noise to be added on *x, y* and θ respectively.

The advantages of this approach are at least in two aspects. Firstly, the estimated pose is calculated without the addition of noise, to obtain the best prediction with the current set of data available. Subsequently Gaussian noise is added to the estimated pose of each particle directly, which would disperse the particles in a small radius around the calculated pose, covering all the possible poses in the region. The proposed model is thus not limited by the range of motion of the robot. The particle filter would complement the method by eliminating particles that have drifted too far from the actual pose. This would concentrate particles at possible poses and keep the dispersion in check.

4.4 Experiment and Simulation Results

In order to evaluate the proposed QMCL method, experiments were carried out with our robot equipped with a 2D LADAR. The testing site was a typical indoor office environment cluttered with walls, cubicles, and cabinets etc. There was a corridor which formulated a loop with a length of approximate *100* meters in the office. An accurate map was acquired using a SLAM based map generating function, with careful hand calibration using real site measurements and an original layout of the office from the building's constructor, as shown in Figure 10 (a).

Figure 9 (a) shows the comparison of the initialization between QMCL and MCL. As clearly seen, the 2500 particles generated by QMCL for initialization is more evenly distributed in the configurable space of the office map, which will increase the probability of matching

Figure 9. Comparison of 2500 particles initialized by QMCL and MCL respectively. The random particles are more uniformly distributed in the configurable space by QMCL.

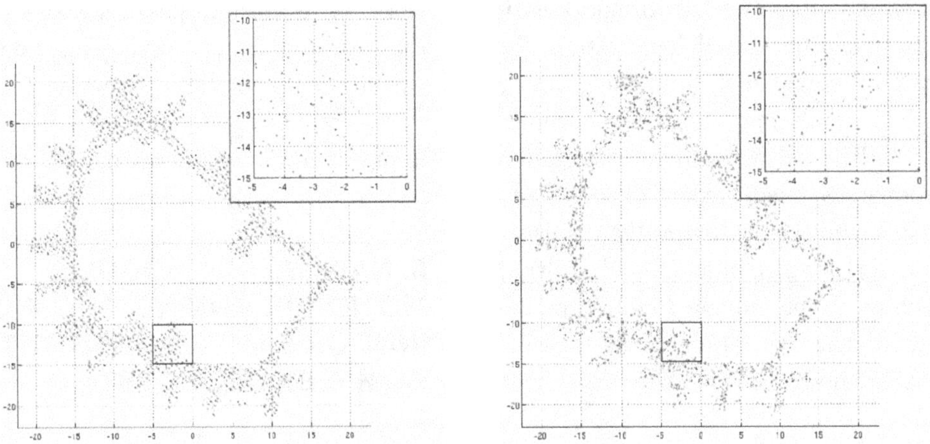

(a) QMCL initialization *(b) Normal MCL initialization*

a particle with the actual robot's position and thus enhance the efficiency of localization. Together with the modified motion model, QMCL is able to provide an accurate, efficient and reliable localization solution. As shown in Figure 10 (a), the robot was driven to run five continuous loops along the corridor of the office with a total length around 500 meters while QMCL can localize the robot's pose well all the time, with a perfect fit into the generated map. A video clip by Bingbing Liu (2011) can be viewed to see the whole process of how the particles track the robot pose. The comparison of the trajectories generated by QMCL and pure odometry respectively in Figure 10 (b) can also show the advantage of the proposed QMCL method. In Figure 11 the performances of

Figure 10. (a) The QMCL trajectory fit into the pre-explored map; (b) Comparison of QMCL trajectory with pure odometry

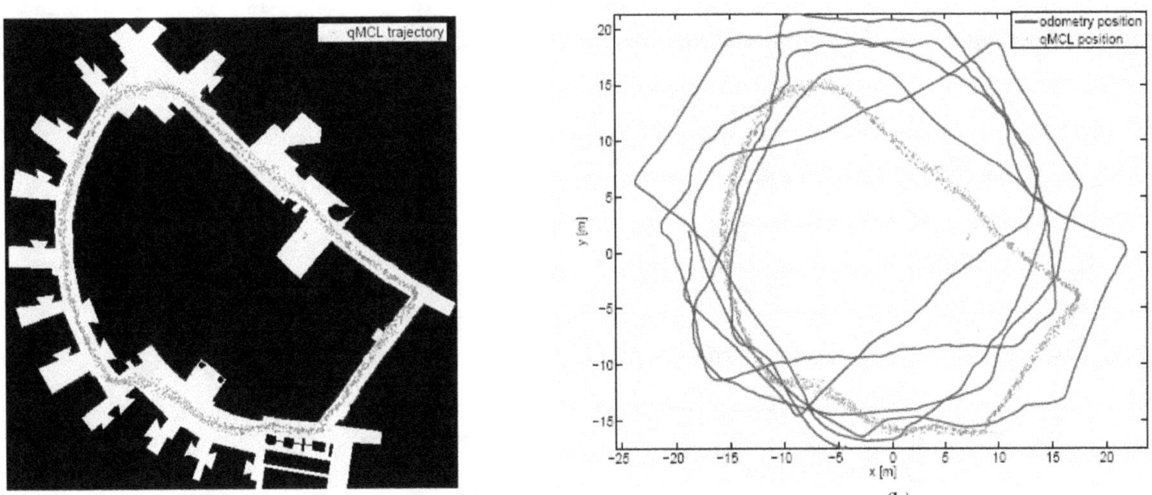

(a) *(b)*

the modified motion model and conventional model are compared, in two different aspects. In Figure 11 (a), as an important performance indicator for particle filter based localization, the highest particle weight plot from the propose motion model is much more smooth and becomes stable much earlier, which means the localization solution converges much faster. This is even confirmed in Figure 11 (b), where the displacement error (using prediction as ground truth) of the approach using the proposed motion model shows a much faster convergence rate than the conventional method.

4.5 Section Summary

In this section, we presented our approach to address the localization problem for the patrolling robot. A 3-tirered localization system was developed to achieve reliable and accurate positioning result for the robot, in a lightly dynamic indoor environment. The main component of this localization system is QMCL, which has adopted Quasi Monte Carlo for initialization and the advantage of QMC lies in its ability to generate more uniform random particles than a normal MC approach. Thus fewer particles can be used for the particle based localization method and less computation load is induced. A modified motion model has been proposed to replace the conventional one for QMCL. With the new motion model, the particles can cover almost all possible robot motions and thus lead to a faster convergence rate for the localization. By combining QMC and the modified motion model, QMCL becomes more efficient in term of particle number and convergence rate.

5. NAVIGATION MODULE: MOTION PLANNING IN TYPICAL INDOOR ENVIRONMENT WITH COMPLEX OBJECTS

5.1 Introduction

For a fully autonomous patrolling robot, the capability of auto navigation is crucial. Since the patrolling robot we are studying is constrained to drive on flat ground, its task of auto patrolling can be defined as: with a pre-recorded map of the environment, the robot autonomously plans the path to move around, and more specifically, to avoid obstacles which might be static furniture or human hanging around etc.

Although the above navigation problem for a patrolling robot typically involves only three degree of freedom *(x, y, θ)*, it has to deal with difficulties arising from several aspects. First, the

Figure 11. Comparison of the proposed motion model with conventional model

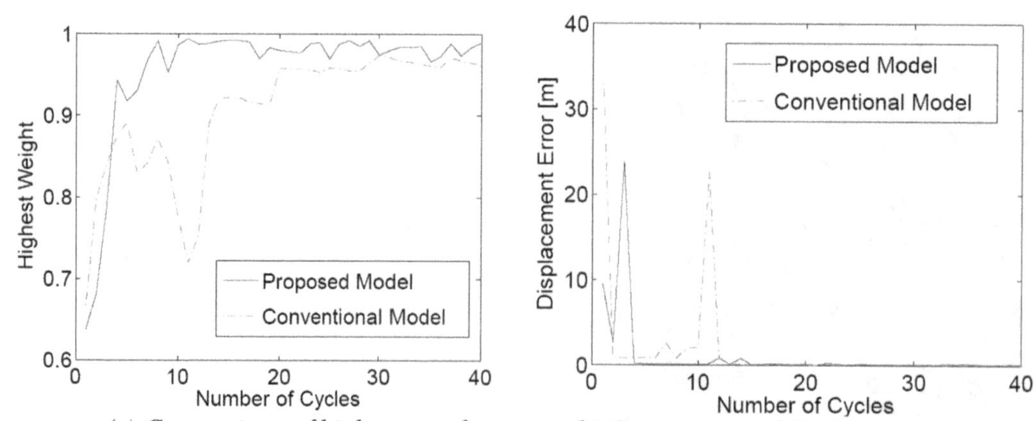

(a) Comparison of highest weight *(b) Comparison of displacement error*

navigation problem is more than pure geometric path planning, since a geometrically collision-free trajectory cannot ensure that the robot can execute it, due to the kino-dymanic constraints of the robot. Therefore, the navigation problem has to deal with not only path planning, but also the robot kinematics and dynamics constraints. Secondly, the real-world environment might change as time elapses, thus the navigation problem is dynamic. Supposing that at time moment *t*, the robot receives a target position at *G*, and a path can be planned out immediately starting from the robot current place to *G*, then the robot can start off right away. But during its motion execution, the environment changes as the robot moves forward, therefore the robot must collect new information about the environment changes from its sensors and incorporate it to adjust its path or re-plan its movement. Furthermore, indoor environments typically involve a variety of objects, among which those have regular shape, such as walls, doors, cabinets, etc., can be reliably detected by a fixed laser scanner. However, there are many other objects with complex three-dimensional structure such as tables and chairs. Because of the different cross-section along the height direction, the shape features of such objects cannot be fully detected by a fixed laser scanner, which can lead to collisions during the robot movement. Thus, in order for a patrolling robot to navigate safely, the robot demands the capability of three-dimensional obstacle sensing.

In this section, we will briefly review some existing navigation methods for mobile robots, including those for global path planning and obstacle avoidance, followed by the description of our implementation for the Rio patrolling robot.

5.1.1 Popular Methods for Mobile Robot Global Path Planning

Potential Field Path Planning: Robot is treated as a point under the influence of an artificial potential field, where the goal generates an attractive force, and obstacles generate repulsive forces, as discussed by Khatib (1986). As a main advantage, this method can produce both a plan and the corresponding control. The disadvantage is that the method suffers from local minima problem.

Graph Search algorithm: The environment is often represented in a discrete format (*visibility graph*, *Voronoi diagram*, and *cell decomposition etc.*) using graph structure, and path planning is converted into solving a least cost problem between two states on a (directed) graph. *Breadth first*, *Depth first*, *Dijkstra* and variants (such as *wavefront* propagation algorithm by Mitchell (1996)), *A** and variants, *D** and variants are the most popular algorithms under this category, as discussed by Lavalle (2005). The graph searching methods are deterministic, i.e., they can ensure to find a collision-free path if it exists. The main limitation of graph search methods is that the feasibility of paths is often not inherently encoded because the robot's kino-dymanic constraints are not taken into consideration.

5.1.2 Popular Methods for Local Obstacle Avoidance

The goal of obstacle avoidance algorithms is to ensure safety during robot motion, and it can be regarded as planning based on a local map. The obstacle avoidance should be optimal with respect to a few aspects, including the overall goal, the actual speed and kinematics of the robot, the on-boards sensors, and the actual and future risk of collision, according to Siegwart (2004). Typical algorithms for local obstacle avoidance are listed as follows.

Bug algorithm: The main idea of the bug algorithm family is to follow along the obstacle to avoid it, as suggested by Lumelski *et al.* (1986). Thus it owns such advantages that no global map is required, and completeness is guaranteed; as for disadvantage, the solution is frequently highly suboptimal.

Vector Field Histogram (VFH): The environment is represented in a grid, and cell values are equivalent to the probability that there is an obstacle, and the opening with the lowest cost function for the robot to pass is selected as the robot's steering direction, according to Borenstein (1991). As a limitation that has been observed, local minima sometimes cannot be avoided, and the method does not perform well if narrow areas have to be passed through.

Dynamic Window Approach (DWA): The main idea is to consider the kinematics of the robot via search in velocity space and only circular trajectories are considered, as discussed by Fox *et al.* (1997). The dynamic window restricts the admissible velocities to those that can be reached within a short time interval given the limited accelerations of the robot. An objective function is computed incorporating a criteria of target heading, distance to the closest obstacle and forward velocity of the robot.

5.2 Navigation Implementation for Rio

The navigation module under the open-source Robot Operating System (ROS) is for navigation of wheeled mobile robots in planar space. It has been verified as safe and robust. For example, in an endurance test, a PR2 robot with ROS was left running unattended in Willow Garage Inc.'s office, and the robot successfully completed 42.2 km (26.2 miles) of travelling, as reported by Marder-Eppstein *et al.* (2010). The whole ROS software is available under open-source license. For prototyping of our patrolling robot, we choose ROS as the software tool to implement the navigation functionality. Figure 12 gives the architecture of the implemented planners, including its interfacing to the robot drive train.

Figure 12. Framework of the motion planners and controller system of the robot

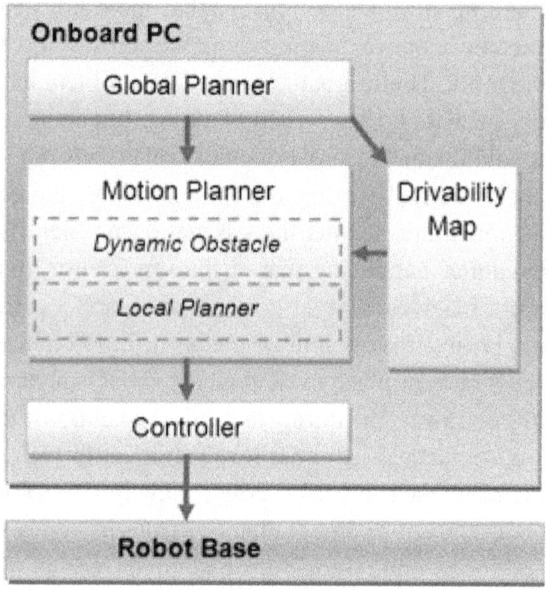

In the rest of this section, the concept of costmap employed in ROS and how the ROS navigation is applied on our patrolling robot will be described in details.

5.2.1 Costmap

ROS navigation employs costmap to represent the environment, in format of 2D occupancy grid. Figure 13 shows such an example costmap captured during the testing of Rio robot.

In Figure 13, the dark cells represent obstacles in the costmap, which are assigned a lethal cost. The polygon represents the footprint of the robot. For the robot to avoid collision, the footprint of the robot should never intersect a dark cell. The grey cells represent obstacles inflated by a user-specified inflation radius. The inflation is performed in two dimensions, applying an exponential decay function, to propagate costs from obstacles cells out to the inflation radius.

The costmap is initialized with a static environmental map (if available), and updates as new sensor data come in to maintain an up-to-date view of the robot's local and global environment.

5.2.2 2D Navigation with a Fixed LADAR

A Hokuyo URG-*04LX* laser scanner is mounted on Rio's base, about 20cm above the ground. This fixed laser scanner detects obstacles along a horizontal slice of the world, therefore is suitable for detecting vertical structures that dominate typical indoor environments. Due to its short response time and high accuracy, the fixed laser is also effective for detecting the position change of moving objects and human beings.

Based on data collected from the fixed laser scanner, the motion planning is executed at two levels, i.e., global and local level. At the global level, with the obstacle information contained in the environment map, the position information from the robot's localization system and a goal, the planner creates a high-level guide to reach the goal location for the robot to follow. In ROS, an interface named *nav_core::BaseGlobalPlanner* is provided to call the global planner. ROS also supports user to write alternative global planning algorithm in format of *plugin*, which can be seamlessly integrated into ROS navigation stack. The default global planning algorithm adopted is the classical A* searching algorithm. At the local level, the planner is responsible for generating velocity commands for the mobile base. In ROS, similar to the global planner, an interface named *nav_core::BaseLocalPlanner* is provided for calling the local planner. Currently, user can choose from a few algorithms, including *Trajectory Rollout* and *Dynamic Window Approach (DWA)*, for local planning. Of course, alternative local planning algorithm is supported via plug-

Figure 13. Concept of costmap

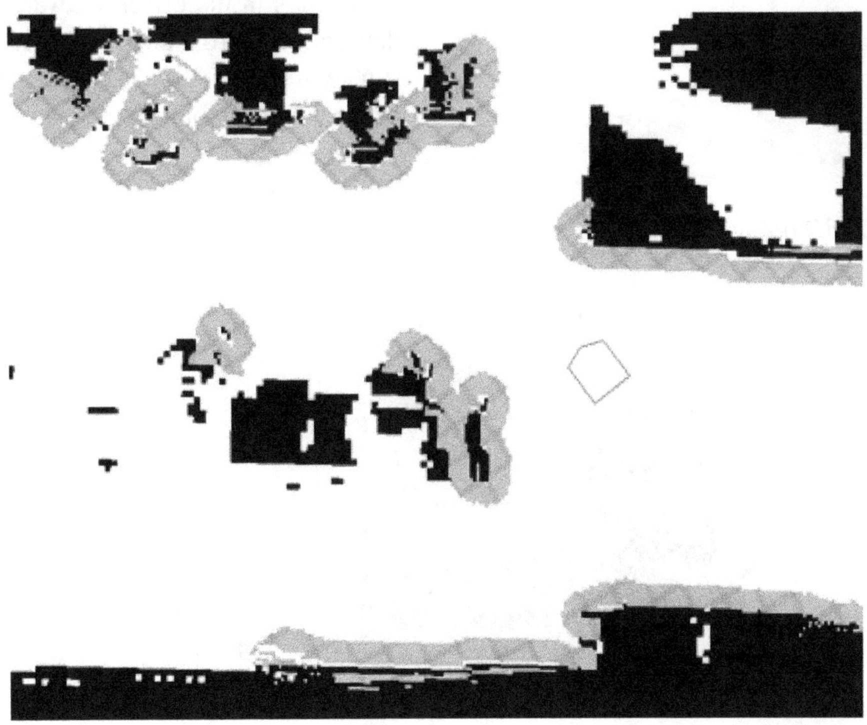

gings. The main steps of the local planner inside the navigation procedure are summarizes below:

- Sample the robot control space discretely ($dx, dy, d\theta$)
- For each sampled velocity, perform forward simulation from the robot's current state to predict the new robot state if the sampled velocity has been applied for some time.
- Score each trajectory resulting from the forward simulation using such metrics: proximity to obstacles, proximity to the goal, proximity to the global path, and speed. Discard those illegal trajectories that collide with obstacles.
- Choose the highest-scoring trajectory and send the associated velocity to the mobile base.
- Repeat the above steps.

More details about the global planner and local planner in ROS are discussed by Marder-Eppstein *et al.* (2010).

5.2.3 Handling Complex Objects with a 3D Tilting LADAR

As discussed before, for a patrolling robot to navigate safely in indoor environments, the capability of three-dimensional obstacle sensing is necessary. For this purpose, we designed a compact 3D LADAR for the patrolling robot. Applying a tilting mechanism to a normal 2D LADAR enables it to continuously scan the space from different angles and thus generates a three-dimensional description of the nearby environment.

As shown in Figure 14, the tilting mechanism design consists of a servo motor (AX12+ servo from Robotis Inc.), its companion controller, an USB2Dynamixel adaptor, the mechanical frame and rotating parts. Commands are sent from the control PC to the servo thus the laser scanner can be tilted periodically.

A ROS driver for Robotis Dynamixel AX12+ servo controller has also been written in c++ language so that the tilting device works properly under ROS environment. The rotation speed and the minimum/maximum angles are adjustable parameters in the driver software. The original data collected from the tilting LADAR is further synchronized with the instantaneous joint status of the AX12+ motor to generate 3-dimensional

Figure 14. The 3D tilting LADAR

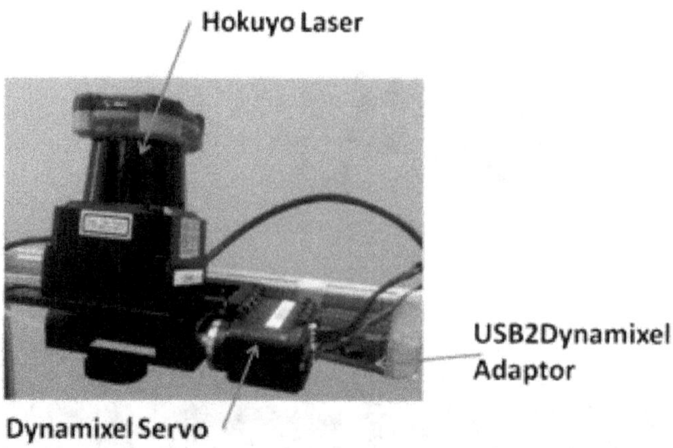

point cloud. The Random Sample Consensus (RANSAC) algorithm is further applied to detect the ground plane and remove it from the point cloud. An example office environment and the associated laser data generated from this 3D scanner are shown in Figure 15.

Especially, the 3D LADAR is able to detect presence of human beings. Figure 16 shows another typical scan result with a human standing in front of the robot.

In our testing of the tilting laser, we further observed that when a human passes in front of the tilting laser rapidly, the scanned human body displays in a distorted shape. This is because the human movement is continuously captured while the laser is being tilted so that within a single tilting cycle, the human position at the end is different from that at the beginning. The tilting can be actuated under different speed. A slow tilting speed will give dense obstacle information but the time between two consecutive updates for a specific region of the environment will increase, and vice versa. Thus the actual appropriate titling speed has to be determined depending on the

Figure 15. An example environment and the associated 3D laser data (a) The actual environment (b) The 3D laser data overlapped with a pre-explored 2D map, where red dots are the 3D laser points, while white color denotes free space and black lines denote objects and grey denotes uncertain area

(a)

(b)

Figure 16. The 3D laser data of the same example environment with a human being standing in front of the robot

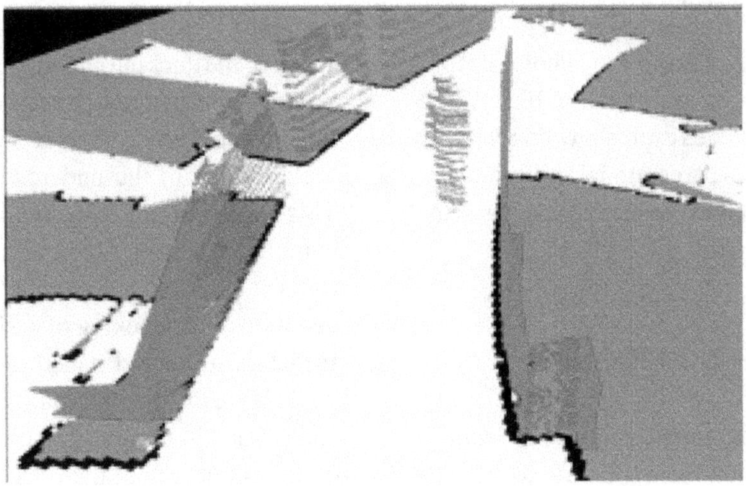

actual robot moving speed and the accuracy requirement for 3D obstacle sensing.

The 3D scanning data of objects from the tilting laser are represented by *voxel* grid, which are three dimensional occupancy grid of the space. Each 3D cell in the grid has one of three states: occupied, free, or unknown. In order for the motion planner to utilize the occupancy information contained in the 3D voxel grid, the 3D voxel columns are further projected down onto the planar ground where they are assigned with costs. For example, columns with occupied cells are assigned a lethal cost. In this way, the data from the 3D tilting laser are fused into the costmap. With the fused costmap, the global and local planning described above can then be executed to plan the robot motion. The testing result of the Rio robot patrolling using both 2D and 3D lasers will be described later in the Section of the integrated test.

5.3 Section Summary

In this section, we introduced our navigation module which can navigate the robot in typical indoor environments with complex objects. We briefly reviewed the existing algorithms for both global path planning and obstacle avoidance. And we also presented our implementation of navigation for Rio robot based upon the open source ROS software platform, including motion planning with regular objects sensing using a 2D fixed LADAR. In order to handle complex objects in such an environment, a 3D LADAR was developed and the result show that it highly enhanced the robot's perceptional ability and thus could handle obstacle avoidance problem in actual environments robustly.

6. AUTO-DOCKING MODULE

The auto recharging function can help to achieve long-term robotic autonomy. In this section, the development of a general auto recharging solution (GARS) employing modulated infrared as sensing signal for mobile robot is presented. The developed GARS is generally applicable to various mobile robots with different dimension and battery types, and the design is simple and cost effective. The system has been tested on a Pioneer 3-DX robot equipped with a differential drive and lead-acid battery, and on Rio equipped with an omni-directional drive and lithium ion battery, and the testing result is optimistic.

6.1 Introduction

For service robots working indoors, the auto-recharging function can refill their battery when necessary and thus enables them to execute tasks on a fully autonomous basis and completely exempt human intervention. In literature, there are a few existing recharging approaches for mobile robots. The *Roomba* vacuum cleaning robot from the iRobot Inc. has a 360 degree infrared (IR) turret integrated at the front to receive docking signals from the docking base. Two charging plates are located at the bottom of Roomba which keeps contact with two charging spots at the docking base. This recharging mechanism works safely and reliably. However, this mechanism is designed to suit the dimension and electrical property of Roomba, thus it cannot be used on other robots. The charging system of the Pioneer robot from the MobileRobots (2011) uses a laser scanner to recognize the shape of the docking station. A plate placed on the ground with upward protruding spring contacts for the robot to drive over and mating contacts are mounted to the undercarriage of the robot. This charging mechanism requires additional complex electronics to be installed internally on robot at a high cost. Silverman *et al.* (2002) used vision to assist docking of a Pioneer 2-DX robot. In detail, a colored paper was mounted on the wall above the docking station as a vision target, and a pan-tilt-zoom (PTZ) camera was mounted on the robot to find the docking station. A laser range finder on the Pioneer robot scanned for a laser beacon mounted above the docking station. However, the vision-based approaches required constant lighting so that the color patters could be reliably detected. From the literature, it can be summarized that the existing recharging solutions for mobile robots are either designed for special robot type or require special environmental conditions to ensure reliable operations, which leads to such a result that the same design is neither not applicable for other robots or is only applicable in limited environments.

In this section, aiming at designing a general auto recharging solution (GARS) which can adapt to different mobile robots and work reliably in indoor environments, we propose the solution employing modulated infrared (IR) as communication signal.

6.2 System Design

6.2.1 The Main Design Concept

Throughout our development of GARS, the top design target is that it should be generally applicable to various mobile robots. Mobile robots have different battery type and charging specs on voltage and current *etc.* This issue is addressed by designing GARS as a plug-in device for an existing robot. Almost every mobile robot on the market has been provided a companion charger which requires a human operator to manually plug it into a power socket for recharging, and this charger can surely work fine with the robot battery. GARS is designed in such a way that for an existing robot, GARS module will connect to its companion charger in serial and work together for recharging. Therefore, the GARS module does not need to worry about what type of battery the robot is actually using, and the system complexity is also kept to minimum. Moreover, mobile robots may have different shapes and dimensions. In order to adapt to this variety, a special mechanical structure has been adopted for GARS so that the vertical position of the charging block on the docking base can be adjusted to match the height of the robot.

The second design target is the GARS system should work reliably in typical indoor environments. For this purpose, we choose modulated infrared as sensing signal between the robot and docking base. Modulated infrared is not only easy to configure and use, but also reliably in indoor environments, rejecting those interferences and noises like the sun light, light from bulbs and heating system *etc*. Additionally, via modulation, the IR beam can carry binary data stream for communication. No heavy computation is required for IR transmission and receiving. Furthermore, low power modulated IR is both safe and suitable for homes with pets and/or children, and in busy office environments, compared to higher power systems such as scanning laser beams, sonic beacons etc. The developed GARS includes two parts. One part is mounted onto the robot, called charging mate (CM) and the other part is docking base (DB), standing next to an AC power socket. The system architecture is shown in Figure 17. The main features of GARS are further explained in the next subsections.

6.2.2 Signal Sensing and Robot Alignment Technology for Docking

In this subsection, the principle of robot alignment during docking operation, together with the infrared modulation scheme are explained.

Robot Position Alignment

In order for the robot to locate itself relative to the docking base, area information must be provided. Figure 18 (a) illustrates the theory used for this purpose. As shown in the figure, the IR transmitting unit consists of duo IR transmitting LEDs with a separator and the two IR transmitters send out different IR signals to provide area information to the detector so that the robot can

Figure 17. The GARS system architecture

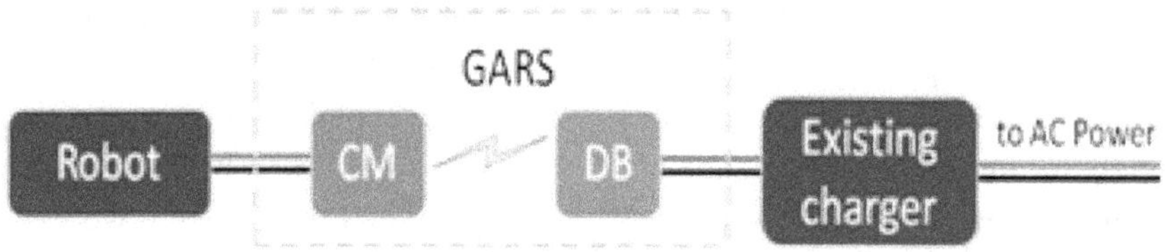

approach the docking station. Only in Area III, the robot can detect IR signal from both transmitters, while in Area I or II, the robot can only sense the signals from either the left or the right transmitter respectively. Thus the sensing result from the IR receiver provides an input for the robot position alignment.

Robot Orientation Alignment

Other than position alignment, the robot also needs to adjust its own orientation in respective to the docking station. In our design, this is achieved by using two adjacent IR receiving sensors and using the difference of signal strength from them as a clue for relative orientation. This idea is inspired by Schur (2011) in his on-line tutorial on "Docking Logic". As illustrated in Figure 18 (b), when the vane is directly aiming at the IR source, both sensors are exposed in IR beam and will receive equal signal. When the vane deviates from this aiming, the signal strength from two sensors will be different and this difference can be further used to align the robot towards the IR source.

Infrared Modulation

Using the technology of infrared modulation, the signal from two transmitting IR LEDs can be recognized by a single IR receiver.. In principle, the control signal is attached to a carrier with higher frequency such as 38 kHz, and the IR receiver is tuned to that carrier frequency and ignore anything else, thus make signal stand out above noises like from sun lights, light bulbs, heating system *etc*. There are many different infrared protocols proposed by industrial parties for remote control of respective appliances. However, we choose to design a customized modulation scheme for robot docking so that the robot can reject possible infrared interferences coming from other

Figure 18. (a) Illustration of IR transmitter unit with separator (b) Illustration of IR comparator (Image courtesy of Chris Schur)

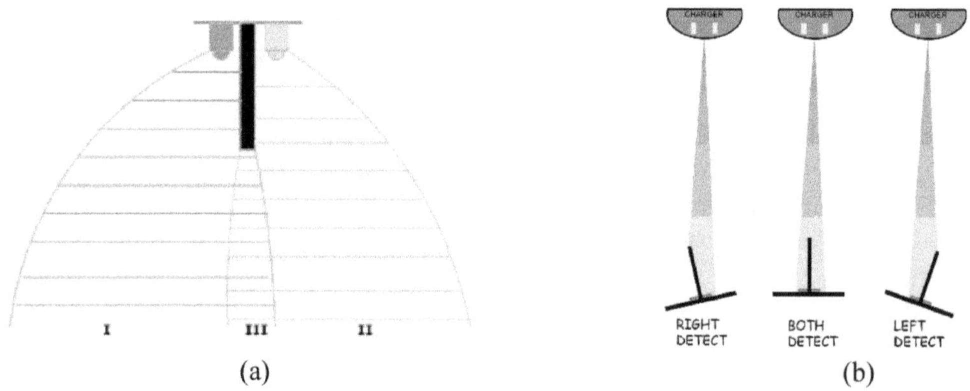

electronic devices or other robots operating in the same environment.

The designed modulation scheme is of phase distance type. The basic representation of binary 1 and 0 of the proposed modulation is depicted in Figure 19 (a). As regards the stream data format, each IR signal includes four segments, i.e., a start bit, address bits, command bits and an end bit. An example of such a data stream format is shown in Figure 19 (b), which represents the binary string of 0b1001 100100.

Using the above IR modulation scheme, the two IR transmitting LEDs on the docking base will send out different modulated value thus the robot could differentiate that it is located in area I, II or III, as shown in Figure 18 (a).

6.2.3 Mechanical Design

Mechanical design for GARS involves structure design of the charging mate (CM) on the robot and the stand-alone docking base (DB). Especially, the design of charging contacts in both CM and DB is important for reliable charging. To ensure enough contact area between the two parts, the type of surface contact is obviously preferred compared to either point or line contact. A spring-loaded charging contact is designed for the CM, and two charging plates are adopted on the DB to match the CM charging contact. These charging plates connect to the positive and negative pole of the robot companion charger respectively. Figure 20 shows the CM prototype in (a), the DB prototype in (b), and the Solidworks models of DB and CM in well docked position in (c), respectively. Furthermore, the DB is adjustable at the vertical direction to suit the installation requirement on different robots.

6.2.4 Other Design Considerations

For safety reason, the charging plates on the DB are only energized when a robot is present at the docking base. This is achieved by applying a +5v voltage at the charging contact on the CM. Only when the robot is correctly docked, the +5v voltage can be detected by the DB charging plate and the circuit on DB will activate a relay to let the charging plate switch to the charger. It is possible that during the docking process, the robot may not well align with the docking base, and adjustment of robot orientation is necessary in this case. For this purpose, Micro-switches have been added to the charging mate, as shown in Figure 20(a).

Furthermore, when the robot gets contact with the base at the last stage of docking, impact may generate at horizontal direction. To reduce possible damage from such impact, existing solutions address this issue from a hardware aspect, e.g., adding a mechanical spring to further damp the robot, as reported by Silverman *et al.* (2002). We handle it from a software perspective, i.e., regulating the robot speed at the last stage of docking based on received signal strength. In Area III, when the robot is far away from the docking base, the signal sensed by the IR receivers on the CM is relatively weak, so the robot can approach the base in relatively fast speed. While the robot gets closer to the base and the received IR signal gets stronger, the robot can move with a lower

Figure 19. (a) Binary representation scheme (b) An example of modulated data stream

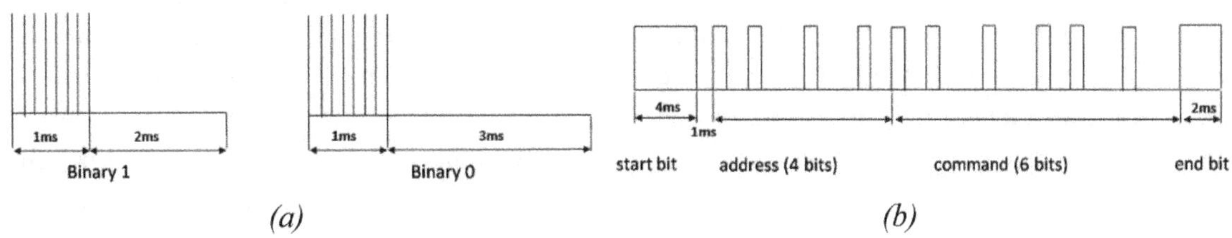

Figure 20. (a) The CM prototype (b) The DB prototype (c) Cross-section view of the DB and CM in well docked position (in Solidworks Software)

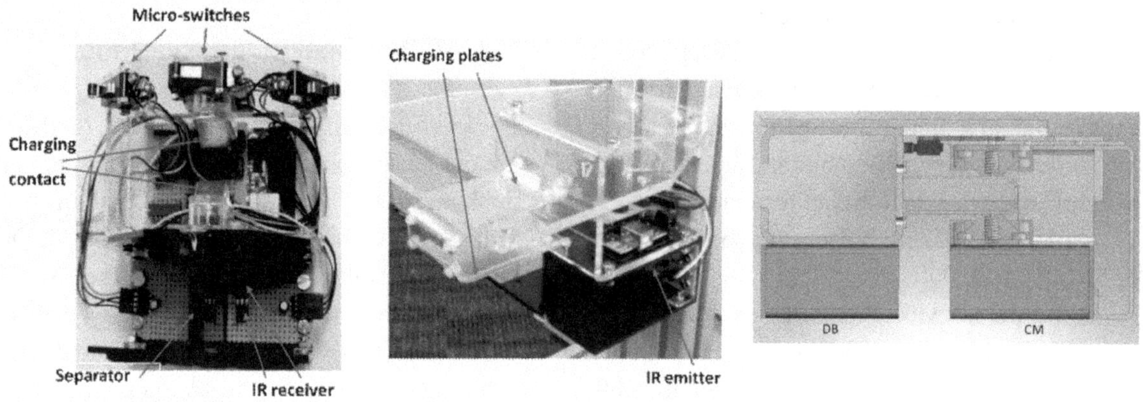

(a) *(b)* *(c)*

speed. In other words, a reverse relationship between the IR signal strength and the distance from robot to docking base is assumed here. This technique has been verified useful to reduce the robot impact. Of course, the exact signal strength from a dedicated IR receiver is also affected by many other factors, such as the specifications of both the transmitting IR LED and receiving sensor, the relative angle between IR transmitter and receiver, environment conditions such as whether IR reflections are present etc. There are some references discussing this issue, such as the research work by Roufas *et al.* (2001). Yet till now there is no general equation to model all these factors. Thus, here we simplified the situation based on the fact that when the robot is within Area III, both the IR transmitting LED and receiving sensor are pointing to each other.

6.3 Software Design

The recharging process starts when the battery level drops below a preset threshold and the main procedures include five steps in sequence to complete a re-charging loop, which are:

1. **Scan:** the robot rotates on the spot to detect whether an IR source is available in the nearby. If an IR source is detected, the robot further rotates back and forth to find the two boundary angular values for IR signal. The robot then rotates to the average of these two angles, which is deemed as the center of the IR source.
2. **Align:** the robot starts approaching to the IR source and rotate simultaneously, trying to reach Area III, as explained in Figure 18 (a).
3. **Dock:** at this stage, the robot is already in Area III. It further approaches the docking base with a regulated speed until completely docks.
4. **Charge:** when the robot is docked at the base, power from the charger flows through the charging contacts to fill the battery. A current sensor is used on the docking base side to monitor the charging status.
5. **Escape:** when charging is done, the robot is ready to move off the docking base.

To implement the above auto recharging function, three ROS packages have been written. The first package is in charge of IR signal sensing, the second one monitors the battery condition and reports its status in certain interval, and the third package takes care of the robot docking and charging. The most obvious advantage of writing our recharging management software under ROS is that the same program can easily be adapted to a different mobile robot if only that robot has a corresponding ROS driver.

6.4 System Verification

Our GARS system has been verified on different robots equipped with different battery and chargers. Figure 21 shows the odometry traces of the Rio robot when it completed docking trials successfully, for a few times. It can be clearly seen from the figure that, at the initial period of each docking trial, the robot tried to find the IR signal and then aligned itself with Area III (Figure 18 (a)). After aligning, the robot approached the docking base until fully docked, as can be seen that all the docking trails end at location *(0, 0)*.

6.5 Section Summary

In this section, a general auto recharging solution (GARS) employing modulated infrared as sensing signal for mobile robots has been presented. The developed GARS is generally applicable to various robot types, with a simple and cost effective design, and works reliably and safely in indoor environments. The system has been verified on different robots including the Rio robot equipped with an omni-directional drive and lithium ion battery. In summary, GARS enables a mobile robot to autonomously recharge, which makes the long-term autonomy possible.

Figure 21. Odometry traces of six docking testing trials ((0, 0) is the location where the docking base is located)

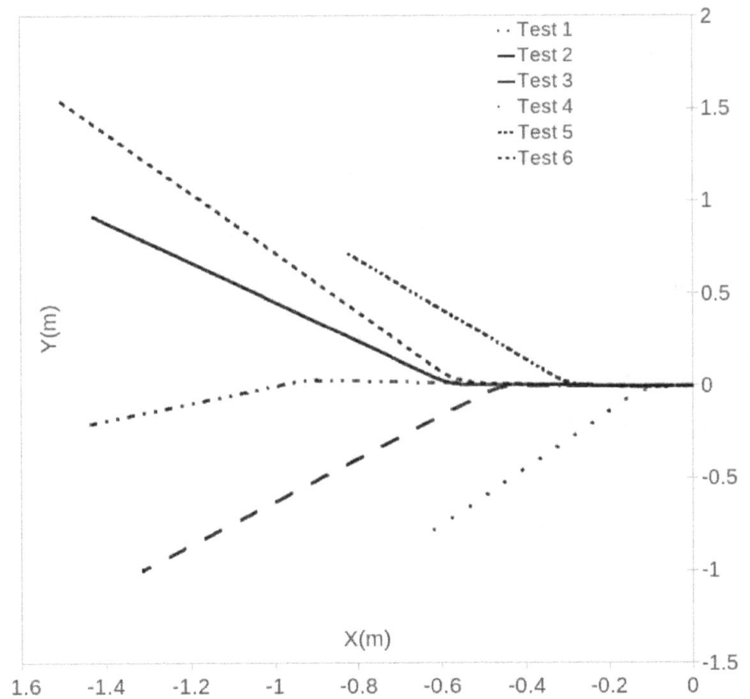

7. AN INTEGRATED TEST: ROBOT AUTO PATROLLING IN A TYPICAL INDOOR ENVIRONMENT WITH AUTO-CHARGING

An integrated auto-patrolling test has been executed in our office located at Lever 14, Fusionopolis Building, Singapore. All the key modules described in this chapter are integrated to realize the autonomous patrolling. Figure 22 shows the map of the testing scenario.

During its patrolling, the robot periodically moves between via points ①, ② and ③. On its journey, the robot moves along the office passage, corridor and also enters a laboratory room, where not only objects with a consistent shape at the vertical dimension, but also complex objects such as chairs and tables are cluttered. During the robot patrolling, once the battery monitor detects that the battery level is below the pre-set threshold, it will trigger a docking event out to the system. After receiving the docking message, the robot will approach to the neighborhood of the docking bay, and then try to dock and recharge. Figure 23 shows an example snapshot captured during the test.

8. CONCLUSION AND FUTURE RESEARCH DIRECTIONS

In this chapter, we have presented the design of the major modules contained in a fully autonomous patrolling robot, Rio, which performs a continuously patrolling task inside a pre-explored indoor environment with no or

Figure 22. The map of testing environment (The locations marked with ①②③ are via points for the robot patrolling, and location D is the place where the docking station stands)

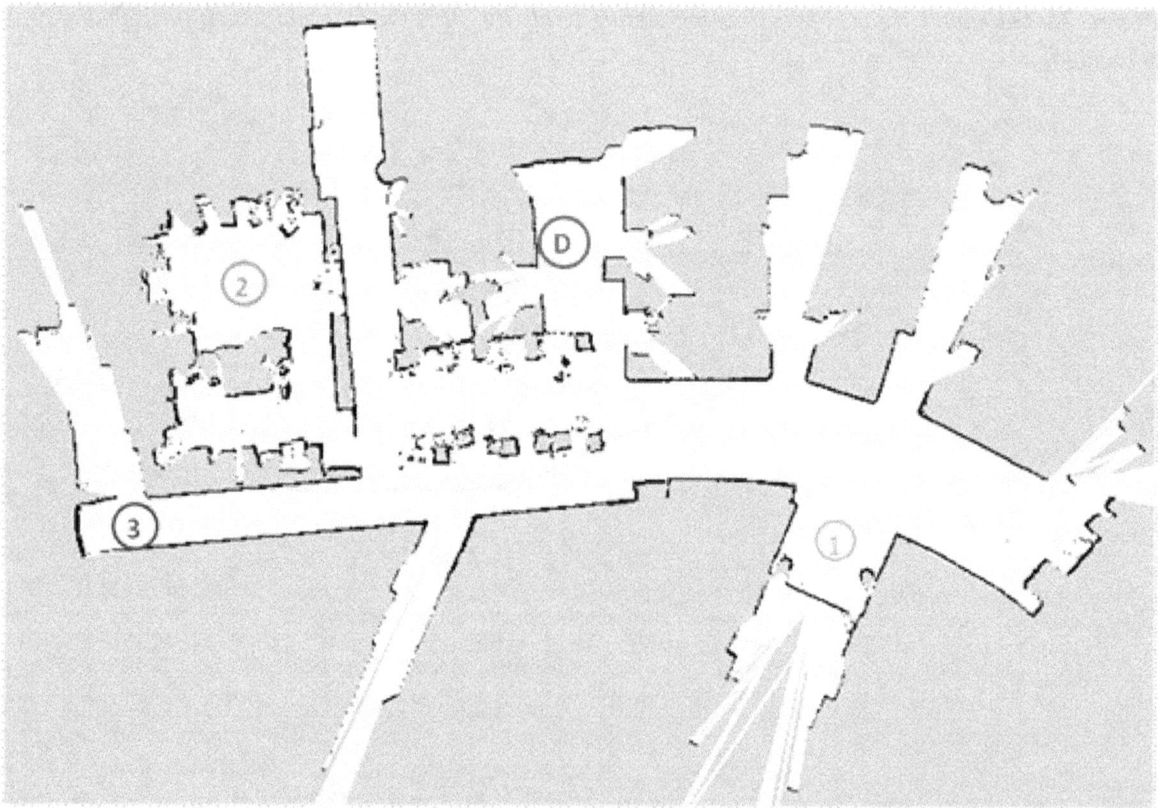

Figure 23. Example snapshot of robot patrolling: (a) The costmap shown in ROS Visualizer (b) The robot in action

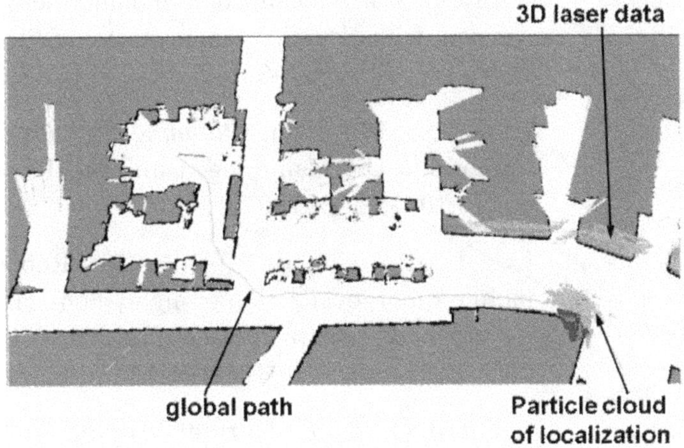

(a)

(b)

bare need of human operation or intervention. The challenges raised from this patrolling task require the robot design to be reliable and robust. The design of each module we presented here has shown our deep concern on this requirement. By experiment verifications, we have shown that these modules can achieve this design goal.

It has been more than one decade since the first indoor mobile robot that was deployed for practical applications. Compared with intensively utilized industrial robots, home or service robots face a booming period. The patrolling robot presented in this chapter actually forms a reliable platform. Its modular design enables easy transformation into many other practical and useful service robot forms, and subsequently may benefit human beings in daily life in many aspects. Definitely, in order for a patrolling robot to provide some further services, a lot of other functions can be added, e.g. object grasping and human-robot interaction. Concurrent research is going on all over the world to develop these functions and in the near future, we can foresee service robots become indispensable utilities in human being daily lives.

REFERENCES

Bailey, T., & Durrant-Whyte, H. F. (2006). Simultaneous localisation and mapping (SLAM): Part II state of the art. *Robotics and Automation Magazine*, *13*(3), 108–117. doi:10.1109/MRA.2006.1678144

Borenstein, J. (1995). Control and kinematic design for multi-degree-of-freedom mobile robots with compliant linkage. *IEEE Transactions on Robotics and Automation*, *11*(1), 21–35. doi:10.1109/70.345935

Borenstein, J., & Koren, Y. (1989). Real-time obstacle avoidance for fast mobile robots. *IEEE Transactions on Systems, Man, and Cybernetics*, *19*(5), 1179–1187. doi:10.1109/21.44033

Borenstein, J., & Koren, Y. (1991). The vector field histogram – Fast obstacle avoidance for mobile robots. *IEEE Journal on Robotics and Automation*, *7*(3), 278–288. doi:10.1109/70.88137

Burgard, W., Cremers, A. B., Fox, D., Hähnel, D., Lakemeyer, G., & Schulz, D. … Thrun, S. (1998). The interactive museum tour-guide robot. In *Proceedings of the AAAI Fifteenth National Conference on Artificial Intelligence*.

Burks, B. L., de Saussure, G., Weisbin, C. R., Jones, J. P., & Hamel, W. R. (1987). Autonomous navigation, exploration, and recognition using the HERMIES-IIB robot. *IEEE Expert*, *2*(4), 18–27. doi:10.1109/MEX.1987.5006527

Caflisch, R. E. (1998). Monte Carlo and quasi-Monte Carlo methods. *Acta Numerica*, *7*, 1–49. doi:10.1017/S0962492900002804

Crowley, J. (1985). Navigation for an intelligent mobile robot. *IEEE Journal on Robotics and Automation*, *1*(1), 31–41. doi:10.1109/JRA.1985.1087002

Dautenhahn, K., Woods, S., Kaouri, C., Walters, M. L., Koay, K. L., & Werry, I. (2005). What is a robot companion – Friend, assistant or butler? In *Proceedings of IEEE/RSJ International Conference on Intelligent Robots and Systems (IROS)*, (pp. 1488-1493).

Dellaert, F., Fox, D., Burgard, W., & Thrun, S. (1999). Monte Carlo localization for mobile robots. In *Proceedings of IEEE International Conference on Robotics and Automation (ICRA)*, (pp. 1322-1328).

Dissanayake, G., Newman, P., Durrant-Whyte, H. F., Clark, S., & Csobra, M. (2001). A solution to the simultaneous localisation and map building (SLAM) problem. *IEEE Transactions on Robotics and Automation*, *17*(3), 229–241. doi:10.1109/70.938381

Durrant-Whyte, H. F., & Bailey, T. (2006). Simultaneous localisation and mapping (SLAM): Part I the essential algorithms. *Robotics and Automation Magazine*, *13*(2), 99–110. doi:10.1109/MRA.2006.1638022

Fox, D., Burgard, W., & Thrun, S. (1997). The dynamic window approach to collision avoidance. *IEEE Robotics & Automation Magazine*, *4*(1), 23–33. doi:10.1109/100.580977

Khatib, O. (1986). Real-time obstacle avoidance for manipulators and mobile robots. *The International Journal of Robotics Research*, *5*(1), 90–98. doi:10.1177/027836498600500106

Kwon, T. B., Song, J. B., & Joo, J. H. (2010). Elevation moment of inertia: A new feature for Monte Carlo localization in outdoor environment with elevation map. *Journal of Field Robotics*, *27*(3), 371–386.

LaValle, S. M. (2006). *Planning algorithms*. Cambridge, UK: Cambridge University Press. doi:10.1017/CBO9780511546877

Lee, J.-S., & Chung, W. K. (2010). Robust mobile robot localization in highly non-static environments. *Autonomous Robots*, *29*, 1–16. doi:10.1007/s10514-010-9184-1

Leonard, J. J., & Durrant-Whyte, H. F. (1991). Mobile robot localization by tracking geometric beacons. *IEEE Transactions on Robotics and Automation*, *7*(3), 376–382. doi:10.1109/70.88147

Lim, C. W., & Li, Y. P. (2004). *Torque distribution and slip minimisation in an Omni-directional mobile base*. Technical Report, Singapore Institute of Manufacturing Technology.

Liu, B. (2011). *5 actual rounds localization by QMCL*. Retrieved August 10th, 2011, from http://www.youtube.com/watch?v=-pCzy9oFZL8

Lumelski, V. J., & Stepanov, A. A. (1986). Dynamic path planning for a mobile automaton with limited information on the environment. *IEEE Transactions on Automatic Control*, *31*(11).

Marder-Eppstein, E., Berger, E., Foote, T., Gerkey, B., & Konolige, K. (2010). The office marathon: Robust navigation in an indoor office environment. In *Proceedings of IEEE International Conference on Robotics and Automation (ICRA)*.

Meyer, J.-A., & Filliat, D. (2003). Map-based navigation in mobile robotics: II. A review of map-learning and path-planning strategies. *Cognitive Systems Research*, *4*(4), 283–317. doi:10.1016/S1389-0417(03)00007-X

Michaud, F., Brosseau, Y., Cote, C., Letourneau, D., Moisan, P., & Ponchon, A. ... Kabanza, F. (2005). Modularity and integration in the design of a socially interactive robot. In *Proceedings of IEEE International Workshop on Robot and Human Interactive Communication (ROMAN)*, (pp. 172-177).

Mitchell, J. S. B. (1996). Shortest paths among obstacles in the plane. *International Journal of Computational Geometry & Applications*, *6*(3), 309–332. doi:10.1142/S0218195996000216

MobileRobots. (2011). *MobileRobots Pioneer docking station.* Retrieved March 30th, 2011 from http://mobilerobots.com/ResearchRobots/Accessories/ChargingSolutions.aspx

Muir, P., & Neuman, C. (1987). Kinematic modeling for feedback control of an omnidirectional wheeled mobile robot. In *Proceedings of IEEE International Conference on Robotics and Automation (ICRA)*, (pp. 1772-1778).

Navarro-Serment, L. E., Mertz, C., & Hebert, M. (2010). Pedestrian detection and tracking using three-dimensional LADAR data. *The International Journal of Robotics Research*, *29*, 516–1528. doi:10.1177/0278364910370216

Oliveira, H. P., Sousa, A. J., Moreira, A. P., & Costa, P. J. (2009). Modeling and assessing of Omni-directional robots with three and four wheels. In Rodić, A. D. (Ed.), *Contemporary robotics - Challenges and solutions*. InTech.

Oliveira, H. P., Sousa, A. J., Moreira, P., & Costa, P. J. (2008). Dynamical models for Omni directional robots with 3 and 4 wheels. In *Proceedings of International Conference on Informatics in Control, Automation and Robotics.*

Roufas, K. D., Zhang, Y., Duff, D. G., & Yim, M. H. (2001). Six degree of freedom sensing for docking using IR LED emitters and receivers. *Experimental Robotics*, *VII*, 91–200.

Schur, C. (2011). *Chris & Dawn Schur's docking logic.* Retrieved March 30th, 2011 from http://www.schursastrophotography.com/robotics/dockinglogic.html

Siegwart, R., & Nourbakhsh, I. R. (2004). *Introduction to autonomous mobile robots*. The MIT Press.

Silverman, M., Nies, D. M., Jung, B., & Sukhatme, G. S. (2002). Stay alive: A docking station for autonomous robot recharging. In *Proceedings of IEEE International Conference on Robotics and Automation (ICRA).*

Smith, R., Self, M., & Cheeseman, P. (1990). Estimating uncertain spatial relationships in robotics. In Cox, I. J., & Wilfon, G. T. (Eds.), *Autonomous robot vehicles* (pp. 167–193). Springer-Verlag. doi:10.1007/978-1-4613-8997-2_14

Song, J.-B., & Byun, K.-S. (2004). Design and control of a four-wheeled Omnidirectional mobile robot with steerable Omnidirectional wheels. *Journal of Robotic Systems*, *21*, 193–208. doi:10.1002/rob.20009

Srinivasa, S. S., Ferguson, D., Helfrich, C. J., Berenson, D., Collet, A., & Diankov, R. (2010). HERB: A home exploring robotic butler. *Autonomous Robots*, *28*(1), 5–20. doi:10.1007/s10514-009-9160-9

Thrun, S., Burgard, W., & Fox, D. (2005). *Probabilistic robotics.* Cambridge, MA: MIT Press.

Tsai, C.-C., Wu, Z.-R., Wang, Z.-C., & Hisu, M.-F. (2010). Adaptive dynamic motion controller design for a four-wheeled omnidirectional mobile robot. In *Proceedings of International Conference on System Science and Engineering.*

Udengaard, M., & Iagnemma, K. (2007). Kinematic analysis and control of an omnidirectional mobile robot in rough terrain. In *Proceedings of IEEE/RSJ International Conference on Intelligent Robots and Systems (IROS)*, (pp. 795-800).

Willow Garage. (2011a). *PR2 users manual*. Retrieved August 10th, 2011, from http://pr2support.willowgarage.com/wiki/PR2%20Manual

Willow Garage. (2011b). *Docking station and autonomous parking for the Texas robot*. Retrieved August 10th, 2011 from http://www.willowgarage.com/blog/2010/01/26/docking-station-and-autonomous-parking-texas

KEY TERMS AND DEFINITIONS

3D Tilting LADAR: A sensing device for the patrolling robot that employs a tilting mechanism to a normal 2D laser scanner, so that the latter can continuously scan the space from different angles and generate a three-dimensional description of the nearby environment.

Localization: A basic function for the patrolling robot to know where it is and which direction it heads to in the environment, at any time of performing a patrolling task.

Modular Design: A design strategy that subdivides a system into smaller parts (modules) that can be independently created, tested and replaced easily.

Navigation: A basic function for the patrolling robot to plan an optimal route to reach a given goal from the current position it stands.

Omni-directional Drive: A robot driving mechanism that uses omni-directional wheels and enables the robot to move in any direction, at any angle, without rotating beforehand.

Patrolling Mobile Robot: An autonomous mobile robot used for patrolling inside a typical indoor office environment, which basically have the functions of localization, navigation, omni-directional drive and auto-docking.

Robot Auto-Recharging: A robot function that can autonomously recharge itself without human intervention.

Chapter 8
Prototyping of Lunabotic Excavator Robotic System

Nicolae Gari
University of Bridgeport, USA

Xingguo Xiong
University of Bridgeport, USA

ABSTRACT

In this chapter, the design and prototyping of a lunar excavator robotic system is proposed. The lunabotic excavator was developed for participating in 2010 NASA Lunar Excavating Competition. Being remotely controlled by operator using a computer via Wi-Fi telecommunication, the autonomous lunabotic excavator can perform the tasks of excavating regolith stimulant, collecting it in the dumpster, and dumping it into the assigned collector box. The design and prototyping of the lunabotic excavator robotic system are discussed. The excavator includes multiple modules including mechanical frames, front/rear wheels, excavating conveyor, steering system, dumping system, power supply and distribution system, actuation system, switch control system, data acquisition system, and telecommunication system. The lunabotic excavator robotic system has been implemented. The excavator can perform the expected functions as designed. The lunabotic excavator designs in NASA Lunar Excavating Competition may lead to new and effective robotic excavators for planetary exploration. The design and prototyping strategies used in the lunabotic excavator may be extended to other ground wheeled vehicles or remotely-controlled robotic systems as well.

1. INTRODUTION

Robotic systems have been widely used for aerospace applications for a long time. They have been proven to be very useful tools to help explore the outer space and extend the human presence in other planets. The outer space generally has harsh conditions, such as extremely high or low temperatures, high-energy radiations, different gravity levels, lack of oxygen and water, etc. These harsh conditions pose potential hazards for astronauts to perform the expected tasks in the outer space directly. As a result, robotic systems are widely used in aerospace exploration to expand the ca-

DOI: 10.4018/978-1-4666-0176-5.ch008

pacity of astronauts in space operation tasks, or to help explore the planets which are not suitable for human landing. Robotic systems can perform various jobs such as terrain exploration, planet soil sample collection, taking images/videos, maintaining/repairing satellites, etc. Robotic systems have accomplished many outer space missions, such as Apollo moon landing, Skylab, Mir, the Space Shuttle, and the International Space Station. For example, as described by NASA Lyndon B. Johnson Space Center (1973), NASA (National Aeronautics and Space Administration) has accomplished both unmanned and manned moon landing with its Apollo missions ever since 1960s. Various robotics systems have been used in the missions of moon landing. In 2003, NASA has started its Mars Exploration Rover (MER) mission to launch two robotic rovers, Spirit and Opportunity, to explore the terrain and geology in Mars. As described by Crisp et al. (2003), the rovers are six-wheeled, solar-powered robots equipped with navigation system and cameras. The robotic rovers have sent back high resolution panoramic images and other valuable information from Mars. Robotics has also been used in International Space Station (ISS). As a habitable artificial satellite, ISS travels around the Earth in low Earth orbit. It allows scientists to do experiments in various fields (e.g. biology, physics, astronomy) in a microgravity environment. As described by Stieber et al. (1997), robotic systems play a critical role in the on-orbit assembly, external maintenance and operations of the International Space Station.

With increased interest in the colonization of the Moon, NASA is planning to construct a lunar outpost, a permanent human-inhabited facility on the surface of the Moon between 2019 and 2024, as described in NASA Lunar Outpost (n.d.). The ambitious plan requires special tools and equipment adapted for the environment on the Moon, which may be very different from that on the Earth. Some easy tasks on the Earth such as digging or excavating may become serious challenges on the Moon, due to different physical properties of regolith, reduced gravity level, the absence of air and water on the Moon. In order to attract more students into the robotics field for aerospace applications, NASA wishes to extend their robotic research to the undergraduate students in US universities. Since 2010, NASA is conducting annual Lunabotic Mining Competition (n.d.) for students enrolled in US schools and universities. The goal of the competition is to promote interest among students in the robotics field, and attract more students to become aerospace robotic engineers. The competition also helps promote students' interest in space activities and STEM (Science, Technology, Engineering, and Mathematics) fields. The competition focuses on regolith excavation, which is the necessary first step towards moon resources mining and building human-habitable bases on the Moon. The unique physical/chemical properties of lunar regolith, reduced 1/6th gravity and vacuum environment make excavation much more challenging than on the Earth. The ideas and innovations developed in the competition may potentially benefit NASA in its future planetary exploration missions. In this chapter, we proposed our work on design and prototyping of "UB (University of Bridgeport) Excavator", a lunabotic excavator for participating in NASA's 2010 annual Lunabotic Mining Competition. Being remotely-controlled by operator via Wi-Fi telecommunication, the excavator can perform the tasks of digging regolith stimulant, collecting it in the dumpster, and dumping it into the collector box. The design and prototyping strategies used in the lunabotic excavator may be extended to other ground wheeled vehicles or similar robotic systems as well.

Some research works about the design and prototyping of robotic systems for planetary exploration have been reported. Muff et al. (2004) proposed a bucket wheel excavator which has the capability of continuous excavation for digging regolith on the Moon, Mars and Phobos. A small bucket wheel excavator with similar size as the rovers carried to Mars on the Mars Exploration

Rover Mission was prototyped. One of the advantages of the bucket wheel excavator was that the excavation forces are horizontal rather than vertical. This minimized the need for excavator mass and it is suitable for operations in reduced gravity field. Zacny et al. (2007) reported a pneumatic drill and excavator for planetary exploration. They investigated a method for penetrating into the regolith and mining of regolith using a gas powered (pneumatic) drill and "jet-lift" method respectively. The proposed architecture would allow subsurface access and mining without requiring solar illumination, potentially large nuclear power system, or complicated distributed power systems. Maniscalco et al. (2007) discussed the design concept for a robotic system to harvest lunar regolith and support a lunar base. The proposed semi-autonomous, robotic system consisted of modular, low mass, low power equipment that excavates, handles, and transports lunar regolith for site preparation, construction activities and in-situ resource utilization. The motivation for regolith harvesting, lunar environmental design factors, robotic system tasks and constraints were discussed. Thangavelautham et al. (2008) presented autonomous multi-robot lunar excavation and in-situ resource utilization using artificial-neural-tissue (ANT) controllers. ANT controllers required much less human supervision and pre-programmed human expertise than other techniques. With lunar in-situ resource utilization (ISRU) efforts in mind, ANT controllers was tested on a multi-robot excavation task in which teams of robots with no explicit supervision could avoid obstacles, interpret excavation blueprints, perform layered digging, avoid burying or trapping other robots and clear/maintain digging routes. Craft et al. (2009 & 2010) presented a percussive digging system for robotic exploration and excavation of planetary and lunar regolith. Experimental results showed the percussive digging system leads to a 15× reduction in the downforce needed to penetrate the compacted regolith soil, thus reducing the required reaction loads and robot mass. Cooper et al. (2010) reported their work in design and tele-operation of a lunar excavator for attending the 2008 NASA Regolith Excavation Challenge. The design was based on a scrapper excavator from Miskin Scraper Works (Miskin Scraper Works, n.d.). The excavator was implemented by UBC-Thunderbird Robotics Team (TREAD), a University of British Columbia student club. All these work on robotic system design for planetary exploration may offer some helpful hints to our lunabotic excavator design. On the other hand, our excavator design follows a set of rules announced by NASA for 2010 Lunabotic Mining Competition. As a result, our lunabotic excavator design may use strategies different from the previous works to meet the challenges in the contest.

2. NASA LUNABOTIC MINING COMPTITION

The first annual Lunabotic Mining Competition was conducted by NASA at U.S. Astronaut Hall of Fame, Kennedy Space Center, Florida on May 25-28, 2010. The logo of NASA Lunabotic Mining Competition (n.d.) is shown in Figure 1. Undergraduate and graduate student teams enrolled in a

Figure 1. NASA 2010 Lunabotics mining competition (courtesy of NASA)

U.S. college or university are eligible to enter the competition. Each team participating in the contest should develop their own lunabotic excavator model following the competition rules announced by NASA.

During the competition, a sandbox filled with compacted lunar regolith simulant supplied by NASA was used, as shown in Figure 2. As described by NASA Lunabotic Mining Competition rules (2010), the regolith simulant matches as closely as possible to the lunar regolith described by Heiken, Vaniman & French (1991). The sandbox allows two teams to perform the competition at the same time. The size of the sandbox for each team is 7.38m (length) × 3.88m (width). It includes an obstacle area (3.88m × 2.94m) and a mining area (3.88m × 2.94m). A collector box is placed adjacent to the sand box sidewall with height of 1m from the top surface of the regolith simulant. The size of the collector top opening is 1.65m (length) × 0.48m (width). It can hold a maximum soil weight of 250 kilograms (kg). Three obstacles are randomly placed on top of the compressed lunar regolith simulant surface in the obstacle zone. Each obstacle has a diameter of approximately 20~30cm and an approximate mass of 7~10kg. There are also two craters of varying depth and width, being no wider or deeper than 30cm. During the 15 minutes' mining competition, the operator of each team needs to remotely control their lunabotic excavator to perform multiple tasks. Starting from the start position beside the collector box, the operator needs to drive the excavator through the obstacle area to the mining area, avoiding the obstacles and craters placed on the way. In the mining area, the excavator needs to excavate the regolith, collect it and then carry it back through the obstacle area to the collector box, avoiding again the obstacles and craters on the way. The regolith is then dumped into the collector box, and the excavator can continue the next round of excavating process, till it comes to the end of the competition time. Each team has 15 minutes to perform as many excavating tasks as they can during this time period. A minimum excavated mass of 10 kg must be met in order for a team to qualify to win the competition. The team whichever excavates and collects the most regolith (more than 10kg) in the collector box will win the contest.

The 2010 Lunabotic Mining Competition is governed by NASA Lunabotic Mining Competition rules, (2010). Some of the rules are summarized as below.

- Weight limit: The total mass of the excavator hardware shall not exceed 80kg;
- Size limit: The excavator hardware must be contained within 1.5m (width) × 0.75m (length) × 2m (height);
- Time limit: The excavation hardware operates during the 15-minute time limit of the competition attempt.

Figure 2. Sandbox with lunar regolith stimulant supplied by NASA (courtesy of NASA)

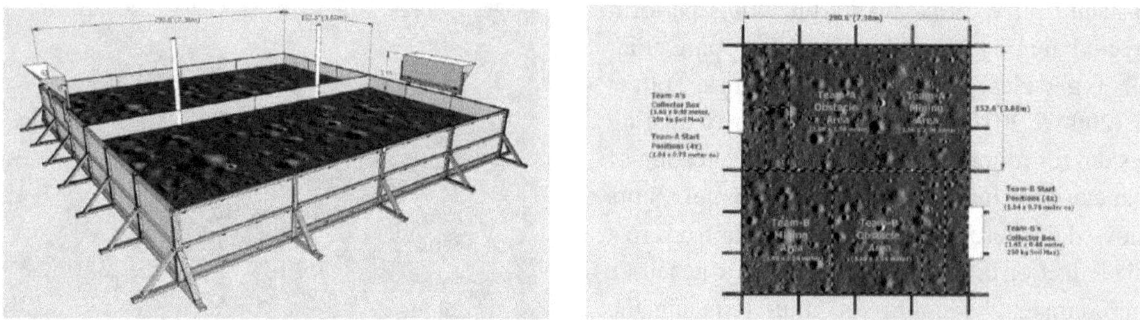

(a) side view *(b) top view*

- During the competition, excavator is limited to autonomous and telerobotic operations only. No physical access to the excavation hardware is allowed. Operators are only allowed to use data and video originating from the excavator. During the competition attempt, the telerobotic operators control the excavator in the mission control room (~60 meters from the sandbox), which has visual and auditory isolation from the excavator.
- Communication with and control of the excavator during competition must be performed solely through the Wi-Fi infrastructure supplied by NASA. No other extra Wi-Fi, RF (radio-frequency), or IR (infrared) device is allowed.
- Bandwidth limit: The communications link used for telerobotic operations is required to have a total bandwidth of no more than 5.0 megabits/second in Wi-Fi communication.
- The excavation hardware cannot employ any fundamental physical processes (e.g., suction or water cooling), gases, fluids or consumables that would not work in the lunar environment;
- The excavator must be equipped with an easily accessible red emergency stop button ("kill" switch) of minimum diameter 5cm. The emergency stop button should be able to stop excavator motion and disable all power to the excavator with one push motion on the button.
- 10 kg is the minimum excavated mass which must be met in order to qualify to win the competition.

As described by Heiken, Vaniman & French (1991), the lunar regolith has many different physical/chemical properties from the sand on the Earth. The environment on the Moon also differs from that on the Earth significantly. For example, compared to the sand on the Earth, lunar regolith particles are very fine and abrasive with sharp and angular shapes. The grain size of regolith ranges in about 5~45 microns. There is no moisture on the Moon. This affects the manner in which the regolith clumps together. The gravity on the Moon is $1.62m/s^2$, which is only 1/6 as that on the earth ($g=9.8m/s^2$). The temperature fluctuation on the Moon is very large, ranging from -233°C to 123°C. The density of the regolith varies between $1.3g/cm^3$ on the surface to $1.9g/cm^3$ at depth greater than 30cm. There are strong radiations on the Moon due to cosmic rays, solar flare events and solar winds. There is no air on the Moon, thus there is no convection on the Moon. The vacuum on the Moon varies between 10^{-12}torr (in daytime) and 10^{-14}torr (at night). The surface of the Moon is covered by many small craters due to micrometeoroids bombardment.

During the lunabotic mining competition, the challenges include the abrasive characteristics of the lunar simulant, the weight and size limitations of the lunabot, and the ability to control the lunabot from a remote control center. The abundant presence of air and fluids on the Earth allows forces created by pressurized fluids, known as hydraulic forces to be used for excavation. However, the absence of air and any type of fluid on the Moon makes it impossible to use hydraulic forces for digging and excavating. As a result, pure electromechanical forces (e.g. DC motors, linear actuators) should be used to drive the excavator. The reduced $1/6^{th}$ gravity on the Moon makes the digging and excavating process even more complicated. The stability of the excavator model is significantly degraded during the excavation task due to the lower gravity level. All these technical challenges should be carefully considered in the lunabotic excavator design and operation.

3. ROBOTIC SYSTEM DESIGN OF UB EXCAVATOR

"UB (University of Bridgeport) Excavator" is the name of the lunabotic excavator we developed for participating the 2010 NASA Lunabotic Mining Competition. It is the result of the joint effort of an interdisciplinary team (UB Excavator team) from University of Bridgeport. It consists of students and faculty advisors from Electrical Engineering, Mechanical Engineering, as well as Computer Science and Engineering. The collaboration of students and faculty advisors from different engineering backgrounds is vital to the design and prototyping of the UB Excavator lunabotic robotic system. The UB Excavator consists of multiple modules, including mechanical frames, excavating conveyor, steering system, front/rear wheels, power supply and distribution system, actuation system (DC motors, linear actuators/sliders), switch control system, data acquisition system, telecommunication system, etc. The design considerations of the UB Excavator robotic system are discussed in details in the following sections.

3.1 UB Excavator Design

UB Excavator is a battery-powered autonomous robotic system remotely controlled by operator using a computer via Wi-Fi telecommunication. It can perform the functions of excavating, storing, and transporting lunar regolith under human remote control. The implementation of the lunabotic excavator follows the set of rules and regulations issued by NASA for 2010 Lunabotic Mining Competition. The UB Excavator design with all the components marked is shown in Figure 3. The model is designed using Pro/Engineer software by PTC Inc. (n.d.). As shown in the figure, the components are marked as follow: (1). excavator main frame; (2). excavating conveyor; (3).electrical linear slider; (4). front/rear wheels; (5). steering system; (6). dumpster; (7). electrical linear actuators; (8). power supply and distribution system; (9). telecommunication and control system; (10). front web camera; (11). back web camera; (12). weight sensor monitoring regolith weight in the dumpster; (13). distance sensor monitoring distance to the fence; (14). rear wheel driving motors; (15). excavating conveyor motor; (16). suspension springs; (17). regolith sliding plate.

3.2. The Working Modes of the UB Excavator System

According to the rule of NASA 2010 Lunabotic Mining Competition, each participating team should operate their lunabotic excavator to excavate, collect and dump as much regolith simulant as possible during the 15 minutes' time limit. To accomplish this goal, the UB Excavator is designed to work in three modes: traveling

Figure 3. UB Excavator system design with the components marked

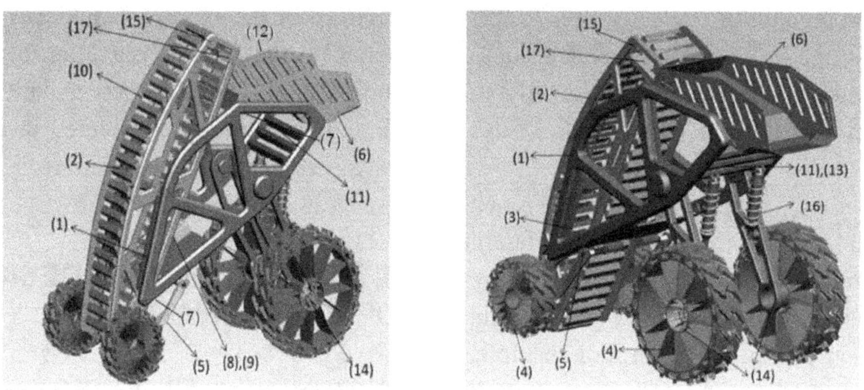

(a) side view (excavating conveyor is lifted up) *(b) back view (excavating conveyor is pulled down)*

mode, excavation mode, and dumping mode. At the beginning of the competition, the lunabotic excavator is set to high speed traveling mode to quickly travel from the start position to the mining area, passing through the obstacle area. The video cameras installed on the excavator monitor the road condition and send back the video signal to the operator via Wi-Fi telecommunication. The operator can turn the excavator left or right as needed via the front wheel steering system to avoid the obstacles and craters on the way. When it comes to the proper mining area, the excavator will stop traveling and switch to excavating mode. It starts excavating regolith simulant using its grabs, and passes the collected regolith into its dumpster via the excavating conveyor. The excavator stays in the excavating spot and excavates till the regolith simulant beneath it is exhausted. After that, the excavator needs to move around (in slow speed) to another nearby spot to continue the excavation. A weight sensor is installed at the bottom of the dumpster to monitor the weight of the regolith collected in the dumpster. Once the excavator has collected enough regolith simulant in its dumpster, it again turns into traveling mode to move from the mining area to the collect box, passing through the obstacle area. Proper steering of the excavator is needed to avoid the obstacles and craters on the way. When the excavator comes near the collect box, it switches to dumping mode. The operator controls the excavator to turn around and move back toward the collect box. A distance sensor installed on the back side of the excavator will monitor the distance between the back of the excavator and the sand field fence. When the excavator backs off to the collect box, the electrical linear actuator lifts up the dumpster, and the regolith simulant in the dumpster is dumped into the collect box. The excavator then continue to the next round of excavating, collecting and dumping process till the competition time is over. The design considerations for each mode are discussed as below.

3.2.1. The Traveling Mode

Traveling mode allows the excavator to quickly travel from one location to another. At the beginning of the excavating competition, the lunabotic excavator is set to be in traveling mode (high speed) to quickly move from the start position to the mining area, passing through the obstacle area in between. To save time, the excavator is set to high speed mode at this time. In order to avoid the obstacles and craters on its way, the excavator needs to be properly steered via the front wheel steering system. After the excavator arrives at proper mining area, it turns to excavating mode and start digging and collecting regolith simulant into its dumpster. Again, after the excavator collects enough regolith simulant in its dumpster, the excavator turns back to traveling mode to move in high speed toward the collect box carrying the load (regolith). The collected regolith should be dumped into the collector box as required by the competition. It will pass the obstacle area again and proper steering of the excavator is needed to avoid the obstacles and craters on the way. For stability consideration, rear wheel driving is used for the lunabotic excavator, and the two back driving wheels have larger diameter than the two front driven wheels. Two 24V DC motors are attached to the rear wheels. By using a set of switches to switch the direction of current flowing through the DC motors, the two rear wheels can be driven to move either forward or backward. By introducing four voltage-dividing resistors, the two DC motors can be driven to rotate (either forward or backward) in two different (high and low) speed levels. If a small serial resistor is connected, the actual voltage drop on the DC motors is larger, hence resulting in a high speed of the motors. On the contrast, if a large serial resistor is connected, the actual voltage drop on the DC motors is smaller, hence leading to a lower speed of the motors. The rotation speed of both wheels should be precisely matched so that they move either forward or backward with the same speed to ensure the stability of the excavator during traveling. The high speed mode is used for quickly moving the excavator from one location to another (e.g. from the start position to the mining area or from the mining area toward the collector box, etc.). The low speed mode is used for moving the excavator from one excavating spot to another during the excavating mode, turning the excavator around toward the collector box before the dumping mode, or steering the excavator left or right to avoid the obstacles and craters while passing through the obstacle area.

While the excavator is passing through the obstacle area, it needs to be properly steered to avoid the obstacles and craters in its way. During the excavating mode, the excavator may need to move around to choose the proper mining spot. Before the dumping mode, the excavator also needs to turn around toward the collect box to be ready to dump the collected regolith simulant. This is achieved by the steering system acting on the two front wheels. The steering system is used to change the moving direction of the excavator (i.e. turn it left or right) as needed. It is the key component used to control the pose of the excavator. The front wheel steering system is manipulated by a 12V DC electrical linear actuator.

3.2.2. The Excavating Mode

The excavating mode allows the excavator to dig and collect the regolith with its excavating conveyor. Once the excavator travels to the proper spot in the mining area, it switches from traveling mode into excavating mode. In this mode, the excavating conveyor is lowered down to the ground level to touch the surface of the regolith simulant. This ensures better contact between the excavating grabs and the regolith simulant for effective excavation. On the same time, the DC motors attached to the rear wheels are stopped and the excavator stays in the spot for mining. The excavating conveyor is now driven by a 24V DC motor to rotate at constant speed. The excavating conveyor is equipped with 36 aluminum excavating grabs. The excavating grabs are cut from aluminum sheet and folded into long and rectangular cups. The shallow cups of the excavating grabs allow the regolith simulant to be temporarily held inside them. Each aluminum grab has sharp teeth along its outer edge to ensure effective digging on the compacted regolith simulant. As the excavating conveyor is rotating with constant speed, the regolith simulant collected in the excavating grabs are continuously passed to the top of the excavating conveyor and dumped into the dumpster via a regolith sliding plate. The regolith coming out of the excavating grabs is collected into the dumpster. To prevent the dumpster from touching the grabs when the excavating conveyor is rotating, there is a gap between the excavating conveyor and the dumpster in the design. The regolith sliding plate acts as a connector between the excavating conveyor and the dumpster in this gap. The design of the regolith sliding plate is shown in Figure 4. It is a smooth aluminum plate with two hollow dumping channels along its both sides. The middle empty space of the regolith sliding plate is reserved for the excavating conveyor and its chain. The regolith sliding plate collects the regolith simulant dispersed from the excavating grabs and directs it into the dumpster via its side dumping channels. After an excavating grab moves to the top of the excavating conveyor, its starts to turn upside-down and the regolith held in its cup falls out into the regolith sliding plate. Due to gravity, the regolith slides down along the plate and falls into the dumpster along both dumping channels. The lunabotic excavator stays in the spot to perform the excavation task till the regolith beneath the excavator is exhausted. It will then move to another nearby spot in the mining area and continue the excavating task.

In order to monitor the amount of the regolith accumulated in the dumpster in real time, a weight sensor is placed in the bottom of the dumpster. When the amount of the regolith reaches the desired value, the excavating conveyor is shut off

Figure 4. Sliding plate used to pass the regolith from the running grabs to the dumpster

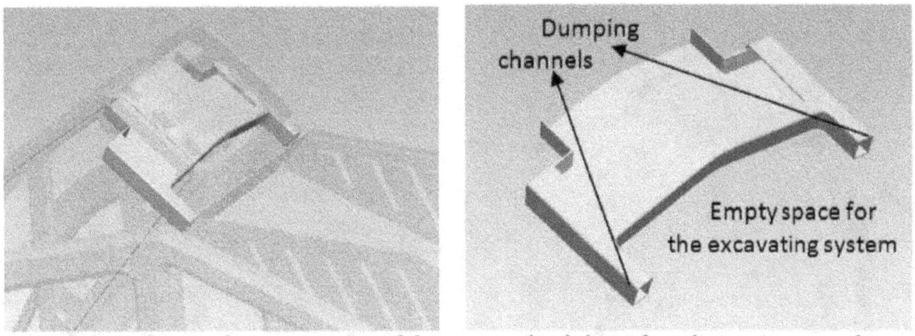

(a) sliding plate in the excavator model *(b) sliding plate design (separated)*

manually to stop the excavation. The excavating conveyor is then lifted up by an electrical linear slider so that it will not touch the ground any more. The excavator then switches to traveling mode, and moves toward the collect box to dump the collected regolith. The excavating conveyor system is connected to the main mechanical frames of the excavator via a sliding bar. The excavating conveyor system can be lifted up or pulled down by a 24V DC electrical linear slider attached to the main frame of the excavator. When the excavator is in excavating mode, the excavating conveyor needs to be pulled down so that the bottom excavating grab touches the ground. This allows the grabs to contact the regolith simulant surface for effective excavation. In the traveling mode, the excavating conveyor needs to be lifted up so that it will not scratch the ground and prevent smooth movement of the excavator. In the dumping mode, the excavating conveyor should remain to be lifted up as well.

3.2.3. The Dumping Mode

Once the excavator collects enough regolith in its dumpster in the excavating mode, it stops excavation and switches to traveling mode to move toward the collector box. Once it comes near the collector box, it turns around (via its front wheel steering system) to have its back facing the collector box. The excavator then switches to dumping mode to dump the collected regolith to the collector box. In dumping mode, the excavator moves backward toward the collector box, because the dumpster is designed to dump the regolith to its back side. The backwards movement is achieved by reversing the polarity of the DC motors at the rear wheels, and thus rotating the wheels in the opposite direction. The DC motors run at low speed to allow the excavator to adjust itself to the proper position beside the collector box. The back view web camera helps the operator to monitor the back side of the excavator. The distance sensor installed on the back of the excavator monitors the distance between the excavator back and the fence of the sand field to prevent the excavator from hitting the fence when it backs off. These are important measures to ensure proper "parking" of the robot by the collector box. Once the excavator is settled at the proper position for the dumping task, the DC motors at rear wheels stop running, and the brakes associated with the back-wheel DC motors are turned ON to ensure the stability of the robot during the dumping process. Next, the dumpster is driven by a 12V DC linear actuator to rotate around a plastic rod support placed underneath the dumpster, as shown in Figure 5. When the front end of the dumpster is lifted up enough by the linear actuator, the back door of the dumpster slides open and the collected regolith in the dumpster is dumped out to the collector box. After that, the linear actuator brings the dumpster

Figure 5. The excavator system in the dumping mode

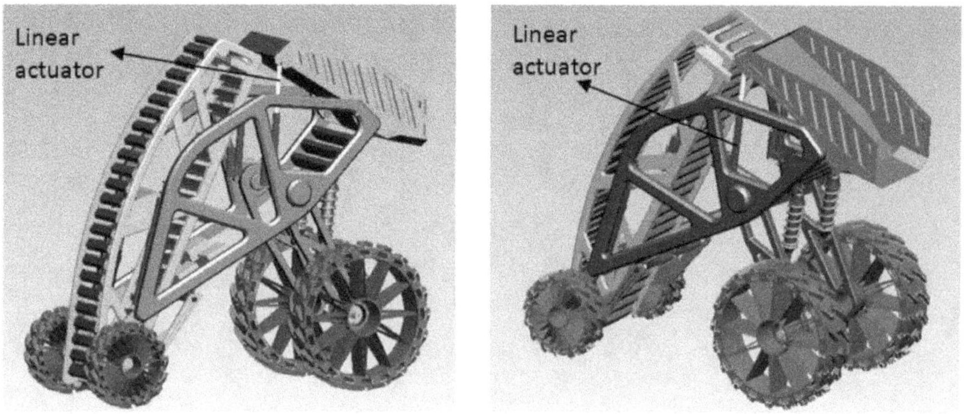

(a) side view *(b) back view*

back to its initial flat position. The brakes of the rear wheels are released, and the excavator turns into traveling mode again to move to the proper mining area for the next cycle of excavation. In the dumping mode, the excavating conveyor is in the same (lifted-up) position as it was in the traveling mode, and it stays in the same position until the excavator turns again into the excavating mode.

4. DESIGN OF UB EXCAVATOR LUNABOTIC SYSTEM

The lunabotic excavator consists of multiple modules, including mechanical frames, front/rear wheels, excavating conveyor, steering system, dumping system, power supply and distribution system, actuation system (DC motors, linear actuators/sliders), switch control system, data acquisition system, telecommunication system. The mechanical structure of the UB Excavator model is designed using Pro/Engineer software by PTC Inc. (n.d.). Pro/Engineer is powerful 3D CAD modeling software for mechanical designs with the capability of stability simulation of the model. According to the rules of NASA Lunabotic Mining Competition, the maximum weight limit of the entire design including all the equipment is 80kg, and the size limit for the excavator design is 1.5m (width) × 0.75m (length) × 2m (height). After trying many different design options, finally we achieve our design as 1.48m (width) × 0.749m (length) × 1.5m (height), and estimated weight of 71kg. The stability of the mechanical structure with all the forces involved is simulated with Pro/Engineer. The final design of our UB Excavator is shown in Figure 6. The design considerations of all the functional modules of the excavator system are discussed in the following sections.

Figure 6. Design of the UB Excavator lunabotic system in Pro/Engineer

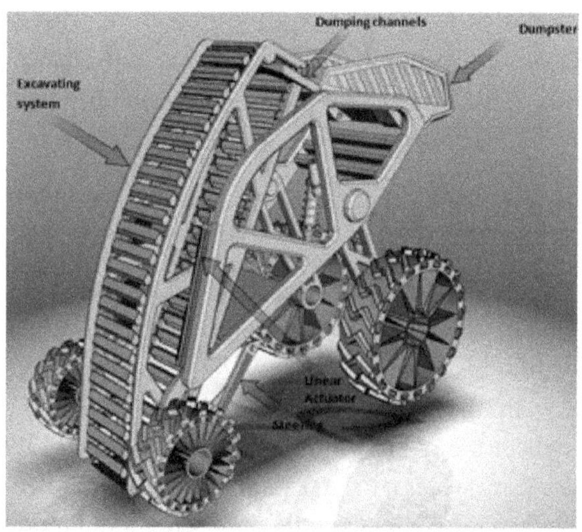

4.1. The Excavator Frame

The mechanical frame is the backbone of the entire excavator. It supplies the platform for the excavator and all the functional components are attached to it. Stability (both in the static mode and working mode) is the most important consideration in designing the excavator frame. The excavator frame design of the excavator is highlighted in Figure 7. It consists of two frames (left and right) connected with joints. Considering the weight limit, the frames are designed as hollow structure. Wood is used as the structure material for the frames due to its low cost and easy machinability. The mechanical frames support the overall weight of all the functional components, hence it should be very robust and tightly connected. The mass center of the mechanical frame structure should be maintained low for better stability. The stability of the mechanical frame structure has been carefully simulated using Pro/Engineer software. The design ensures that the excavator will maintain stable not only in its traveling mode, but also will not lose balance and tip over when it is making left/right turn or dumping the full load of regolith into the collector box.

Figure 7. The Excavator frame

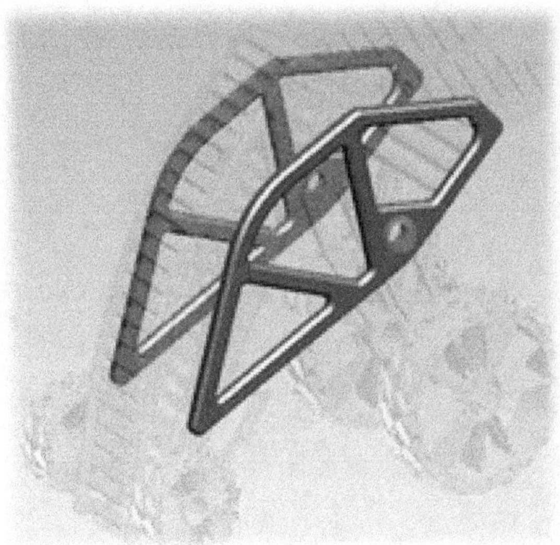

4.2. The Wheels

The excavator relies on wheels to move. The wheels also support the load of the entire excavator and they are important for the stability of the system. Our lunabotic excavator is equipped with four wheels: two back driving wheels and two front driven wheels. The diameter of the front wheels is 355mm, and the diameter of the rear wheels is 590mm. The rear wheels are designed to be bigger for more efficient driving because larger diameter of wheels allows the excavator to move longer distance when driven by the motors. The front wheels are designed to be small for better maneuverability when steering left or right. Since there is no air on the Moon, regular wheels with air-filled tube won't work on the Moon. Further, plastic wheels are not deformable and they are less capable to overcome obstacles on the way. Thus we select rubber tubeless wheels for our excavator. The excavator uses rear wheel driving for better stability. Each rear wheels are powered by a 24V DC motor with estimated speed of 30RPM (revolutions-per-minute). A set of switches and four voltage-dividing resistors are connected to the back-wheel DC motors so that the excavators can be driven forward and backward as needed, with high or low speed levels.

4.3. The Excavating Conveyor with Grabs

The excavating conveyor and grabs perform the key function of excavating and grabbing the lunar regolith from the ground and passing them into the dumpster. As highlighted in Figure 8, it consists of two excavating conveyor frames, two steel chains, and 36 excavating grabs (cups) connected to the chains. Each grab has a line of teeth along its outer edge for effective "digging" on the compacted regolith surface. Two steel chains powered by a 24V DC motor are used to drive the excavating conveyor with grabs. The excavating conveyor runs at constant speed during excavat-

Figure 8. The excavating conveyor and the grabs

(a) Excavating conveyor

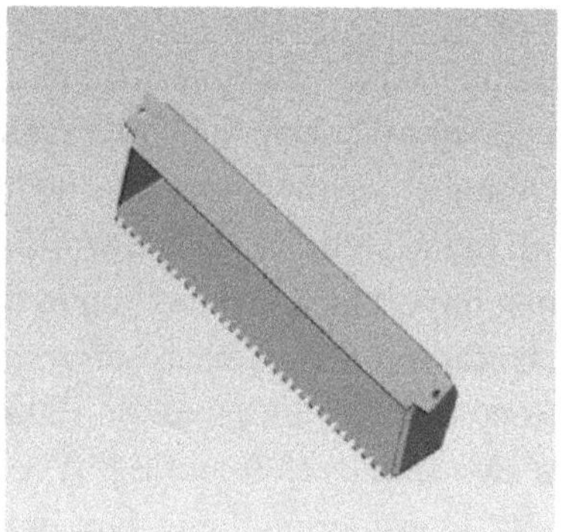

(b) Aluminum grab

ing mode. The total driving torque from the DC motor should be able to overcome the torque required for an excavating grab to penetrate the compacted regolith surface, plus the total torque due to the weight of all the excavating grabs filled with regolith on the rotating conveyor.

When the excavator is in the excavating mode, the excavating conveyor should be pulled down to touch the ground level to "dig" the regolith. However, in traveling mode and dumping mode, the excavating conveyor should be lifted up so that it will not scratch the ground and block the movement of the excavator. This is achieved by connecting the conveyor frame to the main excavator frame through a sliding bar, as shown in Figure 9. In this way, the excavating conveyor can be lifted up or pulled down by a 24V electrical linear slider connecting between the conveyor frame and the main excavator frame. When the linear slider extends and retraces, the excavating conveyor can be lifted up or pulled down along the sliding bar as needed.

4.4. The Steering System

When the excavator is in traveling mode, it may need to turn left or right to avoid obstacles on the way. Before the excavator dumps its load, the excavator also needs to turn around and back off toward the collector box. The adjustment of moving direction is achieved by the steering system attached to the two front wheels. The steering

Figure 9. The sliding bar for excavating conveyor

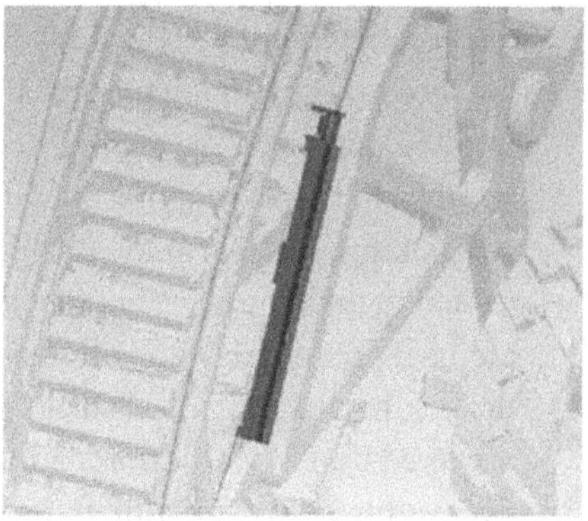

(a) sliding bar located in the excavator

(b) conveyor sliding bar

system consists of two steering arms (one for each front wheel) connected by a steering shaft (plastic rod). Each steering arm consists of two joint segments: one segment connected to the steering shaft, and another segment connected to the front wheel. As an example, the steering arm for left front wheel is shown in Figure 10. A 12V DC linear actuator is used to activate the steering shaft to move left or right. As a result, both front wheels are steered to turn left or right at the same time.

4.5. The Dumping System

In dumping mode, the excavator moves backward to the collector box, and dumps the collected regolith in the dumpster into the collector box. The dumpster, along with the dumping actuator is shown in Figure 11(a). The dumpster can hold up to 40kg of regolith. However, if the dumpster holds too much regolith, the excavator may lose balance and fall over while dumping the regolith to the collector box. This may lead to hazardous situations during the contest. In order to ensure the stability of the excavator in the dumping process, the excavator may start dumping process when the accumulated regolith in the dumpster approaches 20~25kg. The dumpster is connected to the main frame through a cylindrical pin. The front side of the dumpster is connected to a vertical dumping actuator (a 12V DC electrical linear actuator). The amount of regolith collected in the dumpster is monitored by a mass sensor installed at the bottom of the dumpster. When it is ready to dump the collected regolith, the linear actuator is powered by 12V DC battery to extend, and the front end of the dumpster is lifted up. The dumpster gets inclined to the back end, and the collected regolith is dumped into the collector box, as shown in Figure 11(b). The back door of the dumpster is top-hinged. As a result, it flips open to allow the regolith to be easily dumped out when the dumpster is lifted up. With careful design, the backend of the dumpster reaches the height of 1m, which coincides with the collector's height of 1m as described in NASA's rules. This allows a smooth dumping of the regolith into the collector box without spilling. After the regolith is dumped, the linear actuator retreats, and the dumpster comes back to its initial flat position. During the dumping process, the regolith in the dumpster may not be thoroughly dumped out. There may be still some residual regolith in the dumpster after dumping due to the stiction of the regolith. But this won't be a problem for the excavator to continue its next cycle of excavation effort. The excavator is most likely to lose balance and fall off during dumping process. Thus it should be handled with care to avoid falling-off of the excavator.

Figure 10. The steering arm for left front wheel

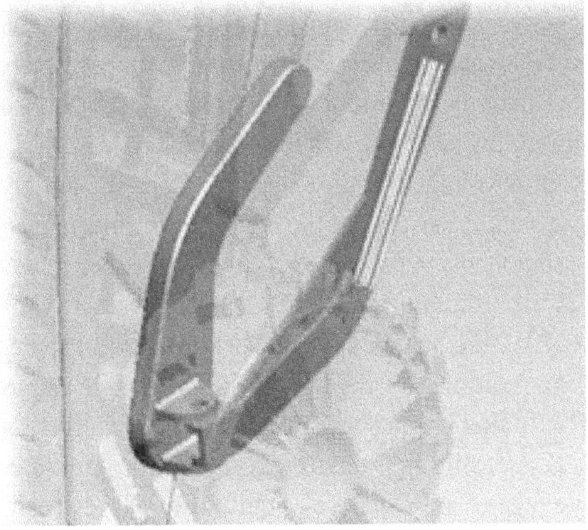

4.6. Power Supply and Distribution System

Power supply and distribution system supplies the necessary energy to drive the excavator to perform the expected functions during regolith mining. Rechargeable DC batteries are used as the power supply for the lunabotic excavator. In order to drive the lunabotic excavator with weight of 71kg to work continuously for 15 minutes during the

Figure 11. The dumpster and the dumping actuator

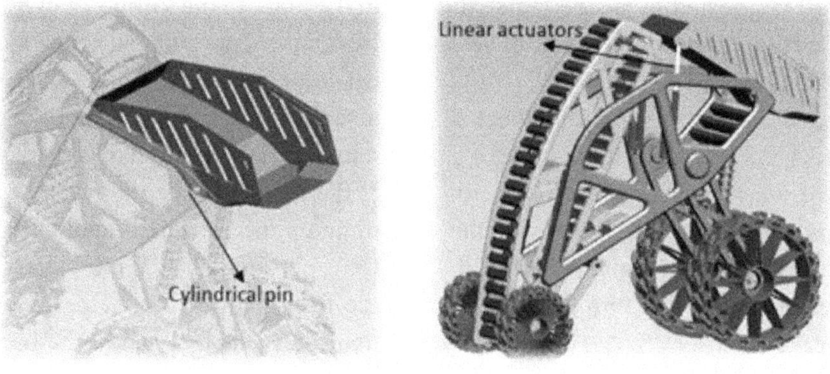

(a) dumpster in rest *(b) dumpster being lifted up*

regolith mining competition, high capacity batteries (with large mAh values) are preferred. We don't want the battery to run out of power prematurely during the regolith mining competition. On the other hand, large capacity batteries also lead to increased battery weight and size. Considering the strict weight and size limitation of the excavator design in the competition, the batteries should be carefully selected. Various batteries are available in the market, such as lead acid, nickel cadmium (NiCd), Sodium Sulphur (NaS), Lithium Ion, Vanadium Redox (VRB), Zinc Bromine, Metal air, Fuel cells, etc. For midsize robotic design like our lunabotic excavator, we use three 12V DC lead acid batteries for a balanced weight/size and performance. Among them, one battery is used to offer 12V DC voltage output, and the other two batteries are connected in series to offer 24V DC voltage output. Both voltage outputs are used to power the various electromechanical components of the excavator.

The power supply should be able to generate enough electrical current for all the electrical loads (e.g. DC motors, linear actuators, routers, switches, web cameras, etc.). An electromechanical component from the market generally has its rated power (*P*) and the working voltage (*V*) labeled. Thus its rated current (I) can be easily derived ($I=P/V$). By summing up the rated currents of all the electromechanical components involved in a certain working mode of the excavator, we can estimate the worst-case maximum load current of the excavator. The electrical load calculations help us to determine the right amperage for the batteries. After some careful calculations, we find the maximum current required for the excavator in worst-case estimation is 12.4A. This happens when the excavator works in the excavating mode when the excavating conveyor motor and all the switches/router/sensors/web-cameras are powered on. This guides us in selecting the proper batteries for the excavator.

The power distribution system delivers power from the power supply to all the electromechanical components in the excavator. Different electromechanical components require different working voltages, and we need to deliver the power in a reasonable way so that the overall load current will not exceed the maximum current of each battery set. We use three 12V DC batteries to offer both 12V and 24V supply voltages for the excavator. Correspondingly, we should also choose the proper electromechanical components with rated DC voltage of either 12V or 24V. Furthermore, each power supply voltage (12V or 24V) is connected to a power hub to be distributed into multiple identical outputs, as shown in Figure 12. In this way, each power supply can deliver the voltage to multiple devices working at the same rated voltage. A regular five-outlet power stripes easily

Figure 12. Power hub devices for distributing 12V and 24V power supplies

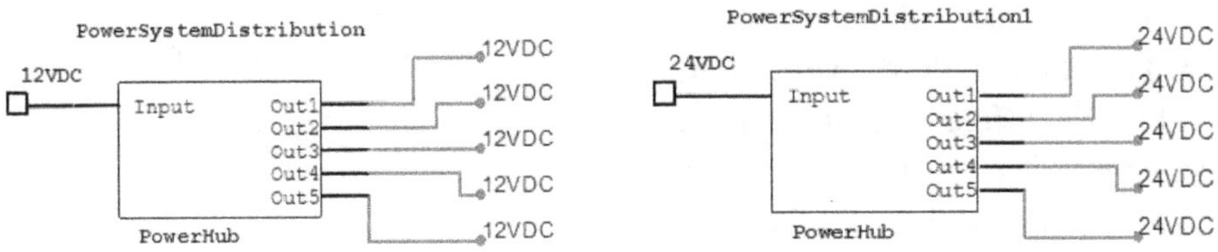

bought from an office supply store can be used as the power hub device for the excavator.

The wires used to distribute the power should follow the national electric code published by National Fire Protection Association (2011). High amperage usage of a wire may cause hazardous situation such as fire or wire damage. Proper wires should be selected to match the current it is expected to deliver to the electromechanical components. To ensure safety, fuses are also used for short circuit and overload protection of the power supplies and electromechanical components.

According to NASA's rules, it is mandatory for all the excavators to have a red emergency switch to disconnect the power of the entire excavator in case of emergency. Figure 13 shows an example of the power distribution system with protection fuses and emergency switch, assuming an electromechanical device is connected to 12V DC power supply. In the figure, fuse1 (maximum current: 15A) is used as protection for the power supply, and fuse2 (maximum current: 5A) is used as protection for the load. Although the figure shows a single emergency switch is being connected to the 12V power supply, in reality the emergency switch is a double-pole double-throw switch which is connected to both 12V and 24V power supplies. In this way, when the emergency switch is pressed, it can disconnect the robotic system entirely from both 12V and 24V power supplies.

4.7. Actuation System

The actuation system activates the electromechanical components of the excavator to perform the expected tasks for regolith excavation. It includes six actuators: two 24V DC motors for driving the rear wheels, one 24V DC motor to drive the excavating conveyor, one 24V DC linear slider to drive the excavating conveyor lifter, one 12V DC

Figure 13. The emergency switch and fuses for overload protection

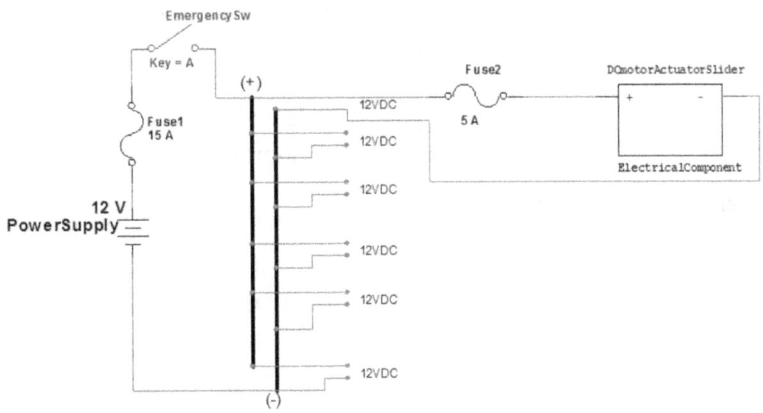

linear actuator for the steering system, and one 12V DC linear actuator for the dumping system.

As described by Hughes (2005), various motors are available for different applications. In order to select appropriate DC motors from the market to drive the rear wheels of the lunabotic excavator, we need to decide the RPM (Revolutions per minute), the torque, and the power of the motors. As described by Ehsani, et al. (2004), consider a vehicle moving along a slope with acceleration *a*, as shown in Figure 14. Assume the mass of the vehicle is M_v (unit: kg), the speed of the vehicle along the slope is v (unit: m/s). The density of air is ρ (unit: kg/m^3). The air density at 20°C under 1atm pressure is ρ=1.2kg/m^3. Assume the aerodynamic drag coefficient is C_d (unitless), the vehicle front surface area is A_v (unit: m^2), and the rolling friction coefficient between the vehicle tire and the ground is f_{rr} (unitless). For car tires moving on concrete or asphalt floor, f_{rr}=0.01. Assume the length and height of the slope is L and H separately. Thus: $sin\alpha=H/L$. Consider the direction along the slope, the vehicle experiences driving force ("traction effort") F_v from the vehicle propulsion, the aerodynamic drag force F_d from the air, rolling friction F_r from ground, and gravity force component along slope direction (F_g), which can be expressed in following equations separately:

$$F_d = \frac{1}{2}\rho C_d A_v v^2 \tag{1}$$

$$F_r = f_{rr} M_v g \cos\alpha \tag{2}$$

$$F_g = M_v g \sin\alpha \tag{3}$$

In order to accelerate the vehicle with acceleration *a (a=dv/dt)*, we have

$$F_v - (F_d + F_r + F_g) = M_v \cdot a \tag{4}$$

According to Ehsani, et al. (2004), the total driving force F_v required from the propulsion system is

$$\begin{aligned} F_v &= (F_d + F_r + F_g) + M_v a \\ &= \frac{1}{2}\rho C_d A_v v^2 + f_{rr} M_v g \cos\alpha \\ &+ M_v g \sin\alpha + M_v \frac{dv}{dt} \end{aligned} \tag{5}$$

For the sand box in NASA lunabotic excavation, the regolith simulant surface is approximately flat. Thus, we assume the lunabotic excavator is moving along a horizontal surface, i.e., $\alpha\approx0$. Further, we assume the lunabotic model moves with almost constant speed, i.e., acceleration $a=0$. As a result, the required total driving force F_v from the propulsion system is

$$F_v = \frac{1}{2}\rho C_d A_v v^2 + f_{rr} M_v g \tag{6}$$

The required output torque from the motor driving the rear wheels of the excavator is

$$T = F_v \times r = \frac{1}{2}\rho C_d A_v v^2 r + f_{rr} M_v g r \tag{7}$$

where r is the radius of the rear wheels. In our design, $r=295mm$.

The power P_v required from the DC motor to drive the excavator to move with velocity v is:

$$P_v = F_v v = \frac{1}{2}\rho C_d A_v v^3 + f_{rr} M_v g v \tag{8}$$

The above estimations can help us to select the appropriate DC motors for driving the rear wheels of the excavator. In our design, we used 24V DC motor with 30 RPM rotation frequency.

Similarly, we can also decide the required torque of the DC motor for driving the excavating conveyor. The output torque from the motor driving the excavating conveyor needs to overcome the torque due to the weight of the excavator conveyor with collected regolith in the grabs, as well as torque

Figure 14. Vehicle moves along a slope discussed in Ehsani, et al. (2004)

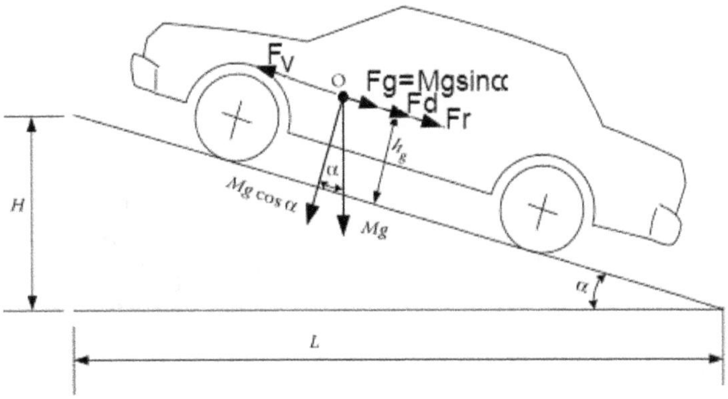

due to the force required for an excavating grab to penetrate the regolith during the excavation process. The friction in the sprocket (diameter Φ=100mm) which connects the excavating conveyor chains and the DC motor should also be considered. After some careful estimation, we decide the torque required from the DC motor for driving the excavating conveyor to be approximately 25N·m. A 24V DC motor with 30 RPM rotation frequency is used to drive the excavating conveyor.

The lifter for excavating conveyor is driven by a DC linear slider. By estimating the weight of the excavating conveyor with full load in its grabs (including the weight of the excavating conveyor frames, chains, grabs and the regolith held in the cups), we can roughly derive the required toque of the linear actuator to lift up the excavating conveyor. After calculation, a 24V DC linear slider is used for the excavating conveyor lifter.

The steering system of the lunabotic excavator is driven by an electrical linear actuator. We refer to the steering system design by Volkswagen of America, Inc. (2004) to decide the steering torque needed for the excavator. Based on the weight and the traveling speed of the lunabotic excavator, we estimate the torque required to drive the steering system to be approximately 12 lbs-ft (i.e. 16.25N·m). A 12V DC linear actuator is used to drive the steering system.

In the dumping mode, the dumping system uses an electrical linear actuator to lift the dumpster to dump the collected regolith into the collector box. Since there is no water on the Moon, hydraulic actuators involving the usage of any type of fluid is not practical for lunar mining. Instead, electrical linear actuator is a good choice for this purpose. Various electrical linear actuators are available in the market. They are generally defined by the parameters such as the type and the level of the working voltage, the stroke length, the stroke speed, maximum overcoming weight, etc. In our excavator design, the electrical linear actuator for the dumping system has to be able to overcome the weight of the dumpster with the collected regolith load in it. The electrical linear actuator used in our design has a mechanical worm system which extends the stroke length from its regular size in the form of a screw. A 12V DC linear actuator is used for the dumping system of the excavator.

4.8. Switch Control System

According to NASA's rules, the robotic excavator should be remotely controlled by operators using a computer via Wi-Fi telecommunication. A set of simple switch networks are used to control the robotic excavator to work in different modes. The USB switch control board is connected to the Wi-Fi infrastructure provided by NASA during

Figure 15. The switch network controlling the left rear wheel motors

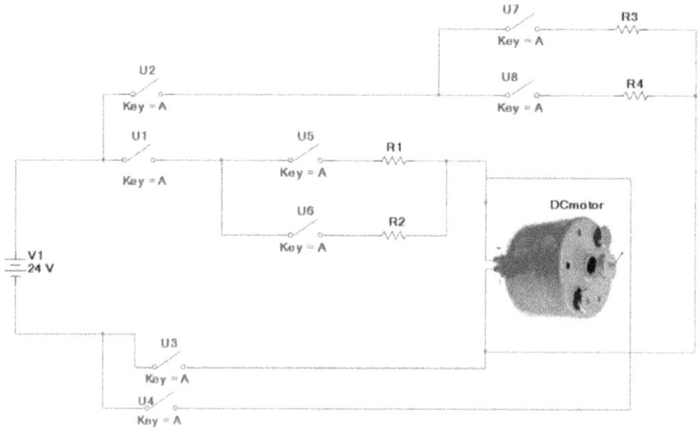

the contest via a serial server. The switches for each electromechanical component are discussed as below.

In the traveling mode, the lunabotic excavator needs to move forward or backward with different speed levels (high or low) as needed. The lunabotic excavator is driven to move forward or backward through two 24V DC motors, one attached to each rear wheel. The two motors are controlled by a set of switches which act on both rear wheel DC motors simultaneously. As we know, a motor shaft changes its rotation into reverse direction when the polarity of the voltage applied to the DC motor is reversed. This helps us to drive the robotic excavator to move backwards when necessary. Taking the left rear wheel as an example, the switch network used is shown in Figure 15.

The DC motor speed is proportional to the applied voltage. In the switch network, four voltage dividing resistors (R1~R4) are connected to the DC motor in series to adjust the actual voltage drop across the DC motor, hence controlling its speed. Assume resistors R1=R3 (small resistance), R2=R4 (large resistance). By default, all the switches are turned off. When switches U1, U5 and U3 are turned ON, R1 (small resistance) is connected to the DC motor. The DC motor is driven by a large positive voltage, hence the excavator moves forward with a high speed. If the excavator needs to travel backward with high speed, switches U2, U7 and U4 are turned ON (and all other switches are turned OFF), resistor R3 (small resistance) is connected to the DC motor, and the DC motor is driven by a large reversed voltage. As a result, the excavator moves backward with a high speed. Similarly, when switches U1, U6, U3 are turned ON (and all other switches remain OFF), the excavator moves forward with low speed. When switches U2, U8, U4 are turned ON (and all other switches remain OFF), the excavator moves backward with low speed.

An identical switch network is used to control the right rear wheel motor. However, since the left and right rear wheels are mirror images to each other, in order to coordinate both wheels to move forward or backward simultaneously, the left and right rear wheels needs to be driven with reverse rotations. That is, if left rear wheel rotates clockwise, the right rear wheel should rotate counter-clockwise. Thus the left and right rear wheel motors need to be driven by DC voltages with reverse polarities. The status of all the switches for left and right rear wheels of the excavator in different working modes is shown in Table 1. In the table, the subscripts "L" and "R" for each switch indicate the corresponding switch is used to control the left and right back wheels respectively. During the operation of the excavator, the operator needs

Table 1. The switch status controlling the left/right rear wheels in different modes (L: left, R: right)

Excavator Traveling Modes		Switches turned ON for left rear wheel motor	Switches turned ON for right rear wheel motor
move forward	high speed	U_{1L}, U_{5L}, U_{3L}	U_{2R}, U_{7R}, U_{4R}
	low speed	U_{1L}, U_{6L}, U_{3L}	U_{2R}, U_{8R}, U_{4R}
move backward	high speed	U_{2L}, U_{7L}, U_{4L}	U_{1R}, U_{5R}, U_{3R}
	low speed	U_{2L}, U_{8L}, U_{4L}	U_{1R}, U_{6R}, U_{3R}
Stop		-	-

to be very careful in controlling the status of the switches. When the excavator is changing from one traveling mode to another, all the switches currently in "ON" state should be turned off first before the set of switches for the next traveling mode can be turned on. This can avoid signal confusion and possible large current if the motor is by-passed by switches mistakenly turned on.

Since the excavating conveyor only rotate in one direction with constant speed, we just need one switch to turn it ON or OFF. Thus the control of the excavating conveyor is much easier.

In addition to DC motors, electrical linear actuators are used in the dumping system, steering system, as well as the lifter of excavating conveyor. The linear actuators need to be controlled to extend or retreat as needed. In order to retreat a linear actuator/slider, we just need to reverse the voltage polarity applied to it. This is achieved by using a switch network as shown in Figure 16. By default, all the switches are turned OFF. If switches U17 and U18 are turned ON, a positive voltage is applied to the linear actuator/slider so that it extends. If switches U19 and U20 are turned ON, a reverse voltage is applied to the linear actuator/slider, hence it retreats.

4.9. Data Acquisition System

Data acquisition system is vital to the proper operation of the excavator. The success in the NASA lunabotic competition relies on not only the design but also the proper operation of the excavator. Data acquisition system collects real time information about the status of the excavator and the surrounding environment. The information is sent to the operator via Wi-Fi telecommunication, so that the operator can make correct judgement and send appropriate commands to control the excavator to accomplish the necessary tasks during the lunabotic mining

Figure 16. The switch network for extending/retreating the electrical linear actuator/slider

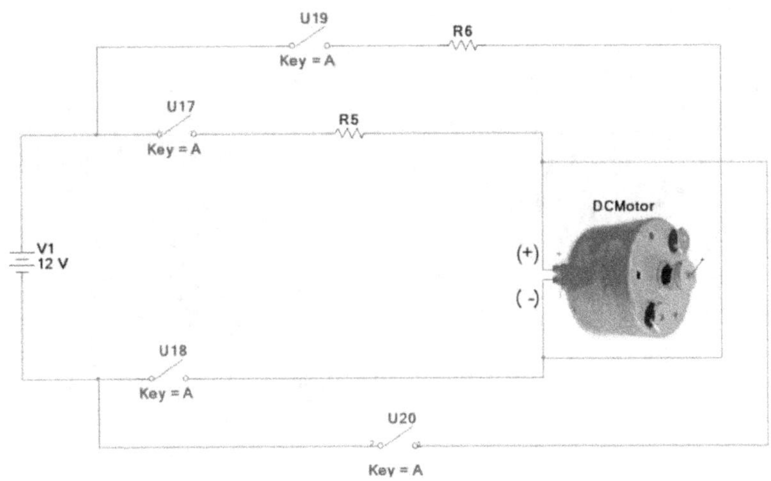

competition. Data acquisition system includes one weight sensor, one distance sensor and two web cameras. The weight sensor is installed on the bottom of the dumpster to sense the weight of regolith collected in the dumpster, as shown in Figure 17(a). This helps the operator to estimate the amount of collected regolith in the dumpster in real time, and decide when the excavator should stop excavating and dump the collected regolith into the designated collector box. The distance sensor is installed on the back of the excavator to sense the distance of the back side of the excavator from the fence of the sand field, as shown in Figure 17(b). In the dumping mode, the excavator moves backward to the collector box. According to NASA's rules, the operators sit inside a room far away from the sand field and cannot directly see the excavator. NASA does not allow the excavator to touch the fence of the sand field any time during the mining competition, even during the dumping process. The distance sensor helps the operator to monitor the distance between the back of the excavator and the fence. This allows the operator to move the excavator as close as possible to the collector box, without touching the fence during the dumping process. Two web cameras are attached to the excavator (one in the front, and one in the back) to collect the video signals about the surrounding conditions in real time. The web cameras help the operator to see the road conditions and surroundings around the excavator, so that the excavator can be controlled to avoid obstacles in the sand field during traveling, find the proper mining area for excavation, and locate the sand box to dump the collected regolith.

4.10. Telecommunication System

According to NASA's rules, communication with and control of the excavator during competition must be performed solely through the Wi-Fi infrastructure supplied by NASA. No other extra Wi-Fi, RF (radio-frequency), or IR (infrared) device is allowed. Furthermore, the total bandwidth of wireless communication should be no more than 5.0 megabits/second on all the data and video sent to and received from the excavator hardware. This bandwidth limit should be considered when we select the proper server, distance sensor, mass sensor and web cameras. The telecommunication system consists of an 802.11b wireless serial server, as shown in Figure 18. It converts all the information from the data acquisition system, which is in the form of USB or RS232 data into Wi-Fi signals, and sends it to the operator via Wi-Fi network. It also receives the commands from the operator to control the switch network to set the excavator in the appropriate working modes.

Figure 17. The sensors used in data acquisition system

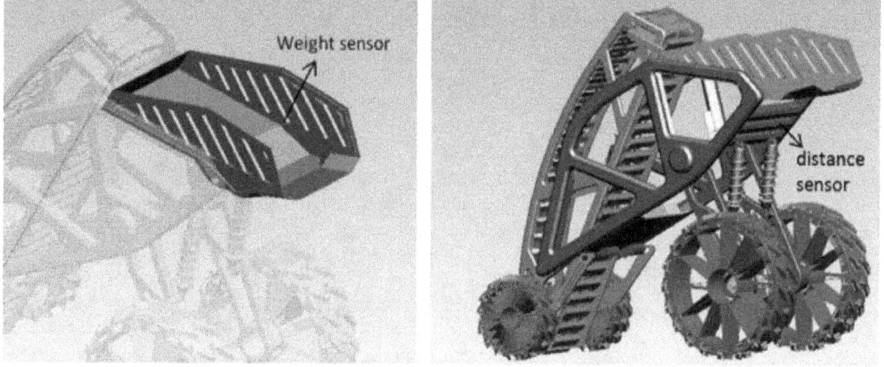

(a) weight sensor *(b) distance sensor*

5. PROTOTYPING OF UB EXCAVATOR ROBOTIC SYSTEM

The prototyping of UB Excavator robotic system involves machining of mechanical frames, selection of proper electromechanical components from the market, integration of all the parts together, as well as testing and debugging of the implemented robotic system. To save time and reduce the cost, most of the materials and components of the excavator are directly bought from local stores or online. For example, the plywood for the mechanical frames, aluminum sheets to make the excavating grabs, electrical wires/connectors, tubeless wheels are bought from local home improvement stores (Home Depot, Lowe's). DC motors, linear actuators, serial server, switch boards and web cameras are bought online. It is rare that a robotic system can be prototyped in the very first try. During the implementation, some problems in the design phase may be exposed. The design may be dynamically adjusted based on the problems found in the prototyping process. Some critical thinking and problem-solving skills are necessary for the prototyping of the lunabotic excavator. The prototyping of various modules of the UB Excavator is discussed as below.

5.1. Prototyping of Mechanical Frames

The mechanical frames of the excavator include the main excavator frame, excavating conveyor frame, rear wheel support frame. Mechanical frames offer the platform for all the other electromechanical components. The frames should be stable and maintain proper balance during the excavation. They also should be light-weight considering the strict weight limit imposed by NASA. As a student project, cost is also an important consideration in the prototyping. After some careful consideration, we chose wood as the structure material for the mechanical frames of the excavator due to its low cost, light weight and easy machining. We start from the plywood bought from local home improvement stores. The mechanical frames are designed in Pro/Engineer and printed to paper with exact 1:1 full scale. The prints are then used as the patterns to cut the plywood into the desired shapes. Carpenter tools such as saws, drills, chisel, milling machine are used to pattern the plywood.

The main excavator frame includes two symmetrical pieces (left and right) connected by shafts. Each part has a hollow structure to further reduce the weight. The model design, printout and the final implementation of the main excavator frame are shown in Figure 19.

Figure 18. Block diagram of telecommunication system design

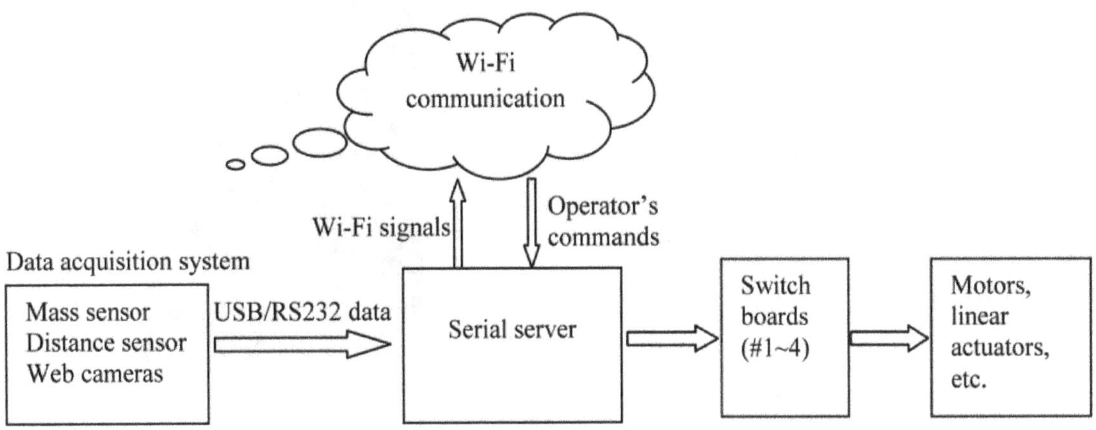

Figure 19. The main excavator frame: design, printout and implementation

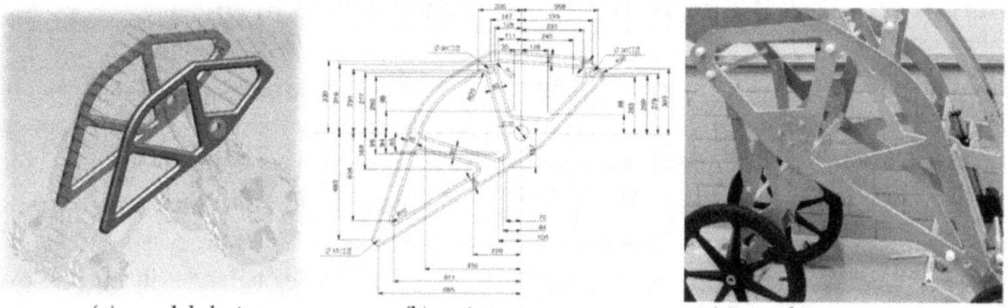

(a) model design *(b) printout* *(c) implementation*

The excavating conveyor frame is used to support the excavating conveyor and its chains. It is connected to the main excavator frame via two sliding bars. This allows the excavating conveyor to be lifted up (during traveling mode) and pulled down (during excavating mode) as needed. It consists of two symmetrical pieces (left and right) connected by shafts. The model design, printout and implementation of the excavating conveyor frame are shown in Figure 20.

The two rear wheels are connected to the main excavator frame through rear wheel support frame. It includes two identical mirrored parts, manufactured by wood using the same approach as mechanical frame. The model design, drawing and implementation of the rear wheel support frame are shown in Figure 21.

5.2. Prototyping of Excavating Conveyor System

The excavating conveyor system includes two chains, two gears and 36 excavating grabs. The chains are driven by a 24V DC motor to rotate at constant speed. The excavating grabs connected to the chains move together to "dig" and hold the regolith. The excavating conveyor keeps rotating so that the regolith held in the grabs can be passed to the top of the conveyor and then dumped into the dumpster via the regolith sliding plate. We use 7075 aluminum sheet to make

Figure 20. The excavating conveyor frame: design, drawing and implementation

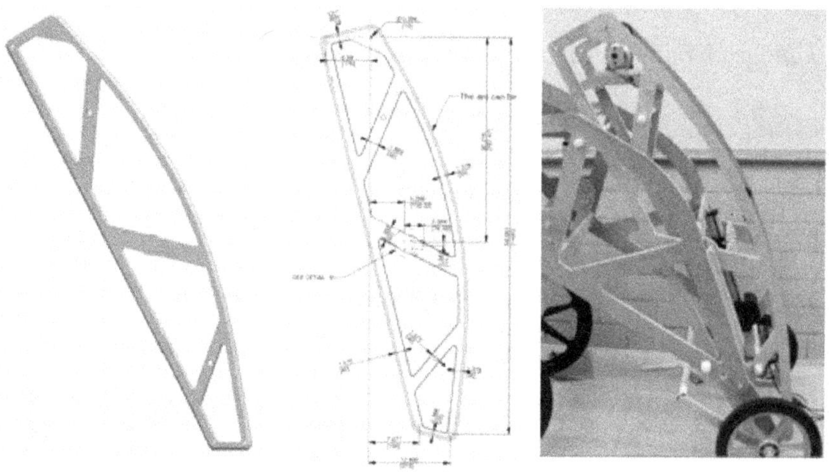

(a) model design *(b) drawing* *(c) implementation*

Figure 21. The rear wheels support frame – model design, drawing and implementation

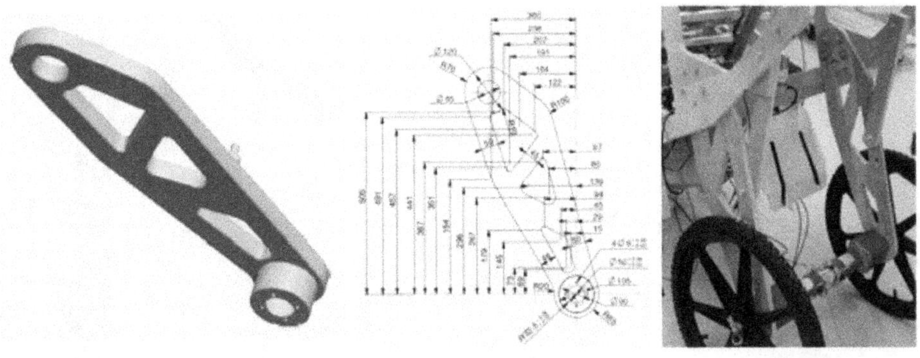

(a) model design *(b) drawing* *(c) implementation*

the excavating grabs due to its excellent strength and light weight. The design of a grab is printed out onto paper with exact 1:1 scale. The printout is then used to cut the aluminum sheet into the desired shape with strong scissors. After that, the 2D pattern was wrapped into an aluminum grab. Figure 22 shows the design, drawing and the implementation of an aluminum grab. Each grab has repeated teeth structure along its outer edge for more efficient digging on compacted regolith simulant.

5.3. Prototyping of Steering System

Steering system attached to both front wheels is used to turn the front wheels left or right. It includes steering arms installed on both front wheels and a steering shaft powered by a 12V DC linear actuator. Taking the steering arm for the left front wheel as an example, it has two pieces: a rigid front wheel support and a movable part activated by an electrical actuator to turn the front wheel support left or right. The model design, drawing and the implementation result of the steering system for left front wheel are shown in Figure 23.

Figure 24 shows the implementation of the steering system. A 12V DC linear actuator is installed on an aluminum bracket support, and its expanding end is directly connected to the steering arm of the left front wheel. The steering arms for both left and right front wheels are connected with a steering shaft – a PVC (polyvinyl chloride) pipe. As a result, if the steering arm of left front wheel is driven by the linear actuator to turn left or right, the steering arm of right front wheel will turn left or right together. That is, the steering shaft synchronizes the steering of left and right front wheels by moving them toward

Figure 22. The excavating grabs: design, drawing and practical implementation

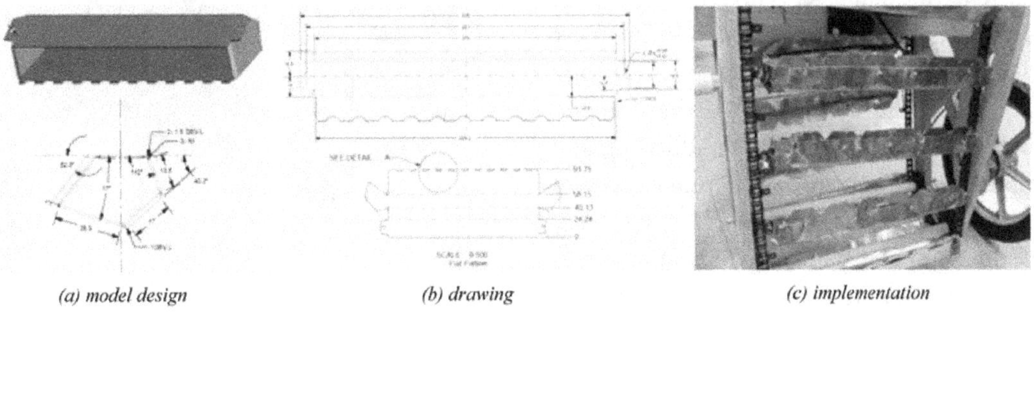

(a) model design *(b) drawing* *(c) implementation*

Figure 23. The steering arm for left front wheel: design, drawing and implementation

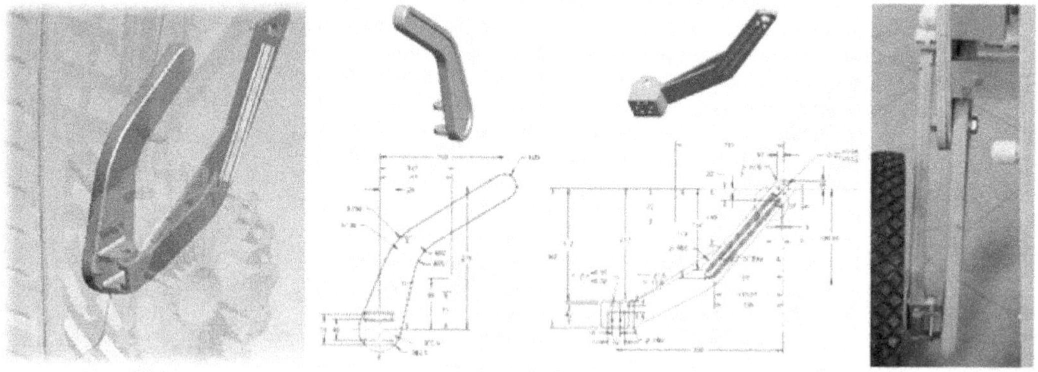

(a) model design *(b) drawings* *(c) implementation*

the same direction simultaneously. In this way, we just need to use one linear actuator (instead of two) to steer both front wheels at the same time.

5.4. Prototyping of the Dumping System

The dumping system consists of a dumpster and a linear actuator used to lift the dumpster during dumping process. The dumpster is used to collect the regolith from grabs during excavating mode, and dump the regolith into the collector box in the dumping mode. The dumpster is made by wood as well. It is connected to main excavator frame via a PVC pipe in the bottom acting as a rotation axis. The front bottom side is connected to a 12V DC linear actuator. In the dumping mode, the linear actuator extends to push the dumpster up so that it rotates around the axis. The back door of the dumpster is top-hinged and can slide open from the bottom. When the dumpster is in rest, its back door is closed due to gravity force. When the dumpster is pushed up, its back door slides open and the regolith inside the dumpster will be dumped into the collector box. The design, drawing and the implementation result of the dumpster are shown in Figure 25. In Figure 25(c), the back door of the dumpster is intentionally left open to show how the regolith can be dumped out from the back door. The dumpster is designed to be able to hold up to 40kg of regolith. However, to ensure stability of the excavator, it is recommended that the excavator starts dumping the regolith when the accumulated regolith in it approaches 20~25kg.

Figure 26 shows the electrical linear actuator associated with the dumping system. As shown in the figure, the actuator bracket is mounted on a horizontal wood support bar connected to excavator frames. The other expandable end is connected to the front side of the dumpster from the bottom. When the linear actuator extends, the dumpster rotates around the cylindrical axis sup-

Figure 24. Implementation of the steering system

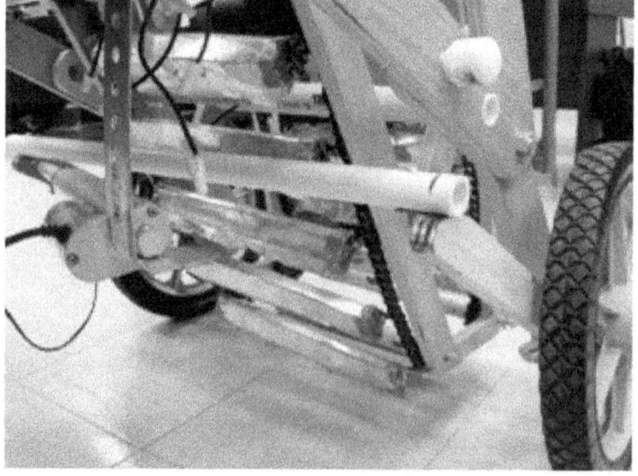

Figure 25. The dumpster: design, drawing and implementation

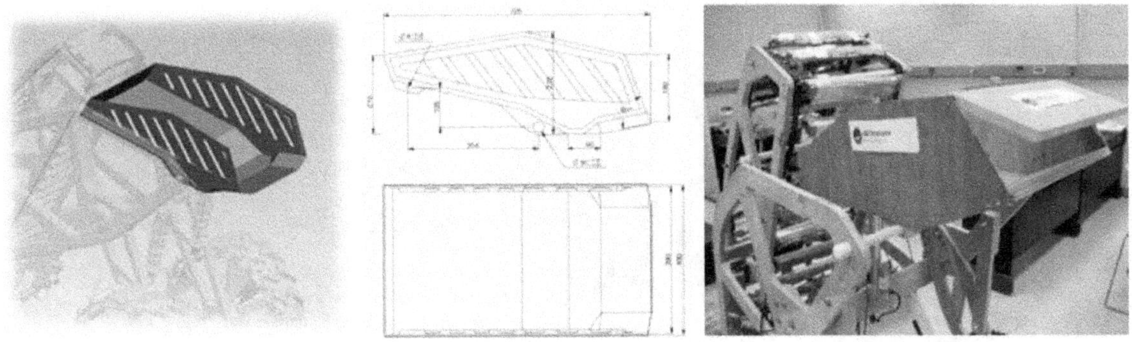

(a) design (b) drawing (c) implementation

port. In this way the back door of the dumpster inclines and the regolith accumulated in the dumpster is dumped out into the collector box.

5.5. Prototyping of Actuation System

The actuation system activates the electromechanical components of the excavator to perform the desired tasks as needed. It includes two 24V DC motors to drive both back wheels, one 24V DC motor to drive the excavating conveyor, one 24V linear slider for lifter of the excavating conveyor, one 12V linear actuator for front wheel steering system and one 12V linear actuator for dumping system. Based on the power estimation of the electromechanical components, proper DC motors and electrical linear actuators/sliders should be carefully selected.

The rear wheels are driven with a pair of identical 24V DC motors (model: PM6015-PL7390) made by Groschopp (n.d.). The motors run at rated current of 3.3A, delivering power of 57W. The maximum rotation speed is 26.7 RPM (revolutions per minute) with delivered torque of 104.5 in-lbs. Figure 27 shows both DC motors associated with the rear wheels of the lunabotic excavator. Two DC motors are mounted back to back and tied with each other by four long screws through the mounting plates. The other ends of both motors are mounted on the rear wheel support frames. The rotation shafts of both motors are connected to the axes of rear wheels via an interface, so that back wheels can be driven by the motors accordingly. The DC motors are equipped with electromagnetic brakes. By default, the brakes are always on unless an external driving voltage

Figure 26. The linear actuator associated with the dumping system

Figure 27. DC motors mounted on rear wheels of the excavator

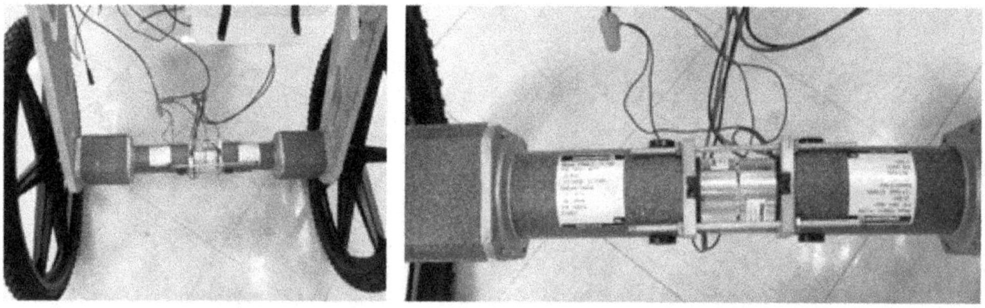

(a) DC motors mounted to rear wheels *(b) connection between both motors*

is applied to the motors. When driving voltage is applied, the electromagnetic coil associated with the brakes will release the brakes, hence the rear wheels can be activated.

The excavating conveyor is driven by a Groschopp PM6015-PS1960 DC parallel shaft gear motor. It has rated electrical current of 3.3A, output power of 38W, and delivers output torque of 80.8lb-in.The excavating conveyor runs at constant speed, and it works only during excavating mode. The rotation of excavating conveyor allows the attached aluminum grabs to dig the regolith, and passes the regolith held in the cups of the grabs into the dumpster through the regolith sliding plate. The excavating conveyor motor is mounted on a pre-designed platform connected to the main excavator frame, as shown in Figure 28(a). A timing belt is used to couple the rotation of the motor shaft to a metal shaft with chain sprockets attached to both ends. In the excavating mode when the DC gear motor starts running, the metal shaft with the pair of sprockets is coupled to rotate together, hence driving both chains of the excavating conveyor to rotate. The excavating grabs mounted on the excavating conveyor start to dig the ground and collect the regolith. The excavating conveyor chains with mounted grabs are shown in Figure 28(b).

In the excavating mode the lunabotic excavating conveyor is pulled down to touch the ground level to dig the regolith. When the excavation process ends, the excavator switches to traveling mode and dumping mode. The excavating conveyor needs to be lifted up in order to avoid scratching the ground while moving. This is achieved by driving the lifter for the excavating conveyor via a 24V DC electrical linear slider (model: FA-200-TR-24-10") by Firgelli Automation Inc. The bottom of the electrical slider is mounted to the main excavator frame, and the extendable end of the slider is mounted to the excavating conveyor frame via a wood board, as shown in Figure 29. The electrical slider has 200 lbs full load capacity with a stroke value of 10 inches. As shown in Figure 29, the bottom side of the slider is mounted to the wood support connected to the excavator main frame. The top extension side of the linear slider is connected to the frame of the excavating conveyor. When the electrical linear slider expands, the excavating conveyor is lifted up from the ground level to allow smooth traveling of the excavator without scratching the ground. Although the linear slider can extend for up to 10 inches, the actual height which the excavating conveyor can be lifted up for is limited by the maximum allowed sliding distance of the sliding bar, and it may be smaller than 10 inches.

The lunabotic excavator uses one electrical linear actuator for the front wheel steering system and another one for the dumping system. Both of them are 12V DC linear actuators (model: FA-240-S-12-9) made by Firgelli Automation Inc.,

Figure 28. Excavating conveyor motor of the lunabotic excavator

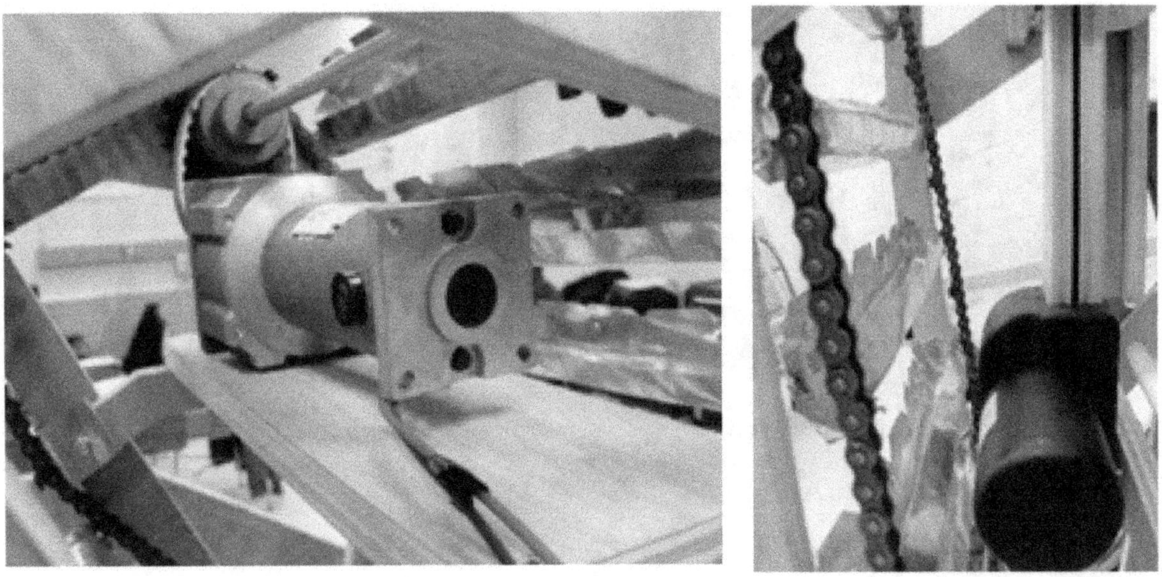

(a) Excavating conveyor motor *(b) Excavating conveyor chains*

with full load of 240lbs and stroke of 9 inches. They are shown in Figure 24 and 26 as discussed before.

The power supply and distribution system is shown in Figure 30. The power supply of the excavator consists in three 12VDC rechargeable sealed lead acid batteries (model: SLA1116) made by Interstate Inc. One battery is used to supply 12V DC voltage output and the other two batteries are connected in series to supply 24V DC voltage output. The capacity of each battery is 18Ah. The three batteries are mounted on the wood electrical panel connected to the main excavator frame, as shown in Figure 30(a). Each power supply (12V or 24V) is distributed to multiple identical voltage outputs via a power hub, as shown in Figure 30(b). Two multi-outlet power strips bought from office supply stores are used the power hub devices. The power supplies are also connected to proper fuses before connecting to the electromechanical components to protect the circuit from overload or short circuit.

Figure 29. Electrical linear slider used as lifter of the excavating conveyor

Figure 30. The power supply and distribution system

(a) DC batteries

(b) power hubs and fuses

5.6. Prototyping of the Data Acquisition System

The success in lunabotic mining competition not only depends on the design but also the proper control of the excavator. The data acquisition system collects the information about the excavator and the surrounding environment, and sends it back to the operator via Wi-Fi telecommunication. This is very important for the operator to make correct decisions and control the excavator to perform the required tasks in regolith mining competition. The data acquisition system includes a weight sensor, a distance sensor and two web cameras. The weight sensor is installed on the bottom of the dumpster to sense the amount of regolith collected in the dumpster. Once the excavator collects enough amount of regolith in its dumpster, the excavator stops excavation and moves to the collector box to dump the regolith. The distance sensor is installed on the back side of the excavator to monitor the distance of the excavator back from the fence of the sand field. This helps the excavator avoid hitting the fence of the sand field while it is backing off toward the collector box to dump the regolith. Two web cameras are installed on the front and back side of the excavator. During the mining competition, the operators sit in a control room far away from the excavator. Thus the operators cannot directly see the excavator. The operators rely solely on the video signals captured by the web cameras to judge the road conditions, find proper mining area, and locate the collector box for dumping the regolith. The signals from the mass sensor, the distance sensor and the web cameras are sent to the operator via Wi-Fi communication. Considering NASA's bandwidth limit (5.0 MB/sec) for Wi-Fi communication, we need to limit the bandwidth of each component of the data acquisition system. The web cameras used in the excavator are two D-link DCS 920 cameras with 9600bps data rate, as shown in Figure 31. The wireless web cameras directly send the video signals to the computer via Wi-Fi telecommunication without going through the serial server. Both the mass sensor and the distance sensor should either directly output Wi-Fi signals, or output RS232 data to be converted into Wi-Fi signals via a serial server, so that they can be sent to the operator via Wi-Fi communication.

5.7. Prototyping of Telecommunication System

Telecommunication system transfer data between the operator and the excavator via a Wi-Fi infrastructure supplied by NASA. It sends the signals from data acquisition system to the operator, and sends the operator's commands to the switch control boards to control all the electromechanical components. According to NASA's rules, the participants at the contest should use an autonomous robotic system controlled via a local Wi-Fi infrastructure supplied by NASA. No direct RF (Radio Frequency) or infra-red (IR) communication is allowed. As shown in Figure 32, the telecommunication system consists of a router, which is a 4-port 802.11 b/g wireless serial server (model: Connect Port TS W) by Digi International Inc. (n.d.). The operator can see the serial server from the virtual COM port of computer via Wi-Fi communication. The switch (relay) boards are connected in series to the router. The operator can control all the relays connected to the serial server using the Pro-XR software supplied by the manufacturer, hence controlling the working mode of the excavator. The mass sensor and the distance sensor are connected to the port #2 and port #3 of the serial server, so that they can be seen and controlled from the additional COM ports (e.g. COM2 and COM3) of the operator's computer.

Figure 31. D-link DCS920 web camera used in data acquisition system

The electromechanical components (e.g. DC motors, electrical linear actuators/sliders) in the lunabotic excavator are controlled by switches, so that the excavator can be driven to perform the required tasks as needed. The switches used are high power relays with rated current of 10A or 20A. The relay boards used for lunabotic excavator are RS232 serial relay systems made by National Control Devices, LLC. (n.d.). Considering the number of electromechanical components and their working modes, we use totally four relay boards. The serial server is connected to one main relay board, which is in turn connected to three extension boards in series. The main relay board (model: ADR410PROXR) is shown in Figure 33(a). It has four 10A relays, and an expansion port. The RS232 port is marked with red color. The four relays are on the top side, and 12V DC power supply connections are on the right side. The extension port is shown on the top left corner. The main relay board is connected to two exten-

Figure 32. The wireless router (serial server) for telecommunication system

sion relay boards with eight 20A relays on each board, and another extension relay board with four 20A relays. They are connected in series. Thus totally we have 24 relays, which is good enough to control the excavator to work in different modes. One extension board with eight 20A relays is shown in Figure 33(b).

Figure 34 shows the connection of the relay boards. The expansion boards are connected in series, which in turn are connected to the main relay board via a 9-pin SR232 adapter. Serial cable further connects the main relay board to the router. The commands from the wireless serial server will control the status of these switches individually.

The switches on the main and extension relay boards are controlled by the operator using ProXR software from computer via Wi-Fi communication. The main interface of the ProXR software used is shown in Figure 35. As shown in the figure, the software supports up to 32 different relay banks, with each bank controlling one physical relay board. Each relay bank memory can control up to 8 relays. The software allows the operator to turn on/off each individual relay in a bank, and read the actual status of each individual relay. For our lunabotic excavator, we use four banks to control the four relay boards. Bank1 is used to control the main relay board, bank2 and bank3 are used to control the two extension relay boards with 8 relays, bank4 is used to control the extension relay board with 4 relays. Considering the data rate limit of the Wi-Fi server assigned by NASA, the Baud Rate of each relay board is chosen as 9600bps.

The relays are used to control the status of individual electromechanical components, hence driving the excavator to perform the tasks as needed. Each electromechanical component is controlled by certain relays to perform certain function. The function assignment of the bank relays for the lunabotic excavator is shown in Table 2. As we can see, certain functions may require two relays to be turned on or off simultaneously. The ProXR software allows the operator to group certain relays together so that they can be turned on or off simultaneously. The emergency switches are directly connected to the positive end of power supplies and needs to withstand large current. Thus we use two relays connected in parallel for emergency switches. Each relay has rated current of 20A, thus the total rated current is 40A. Fuses are also connected to protect the power supply and the electromechanical components.

Figure 33. Relay controller boards for lunabotic excavator

(a) Main relay controller board

(b) extension relay controller board (8 relays)

Figure 34. Connection of the main and extension relay boards

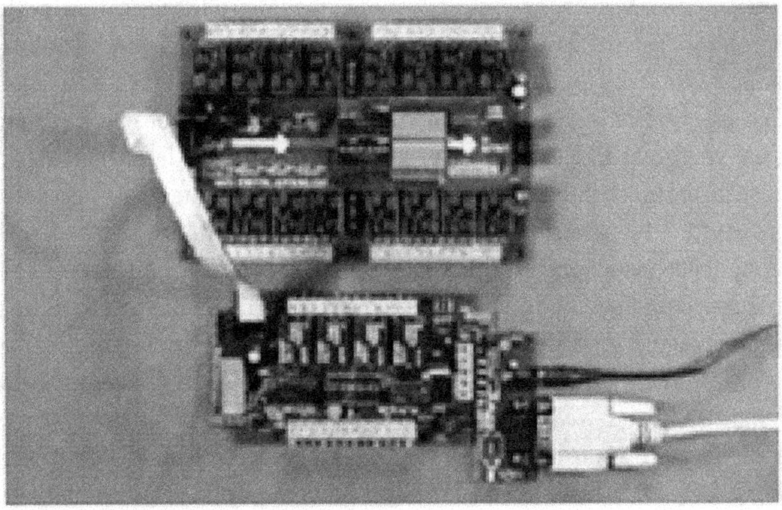

Figure 35. Main interface of NCD ProXR software used to control the relay boards

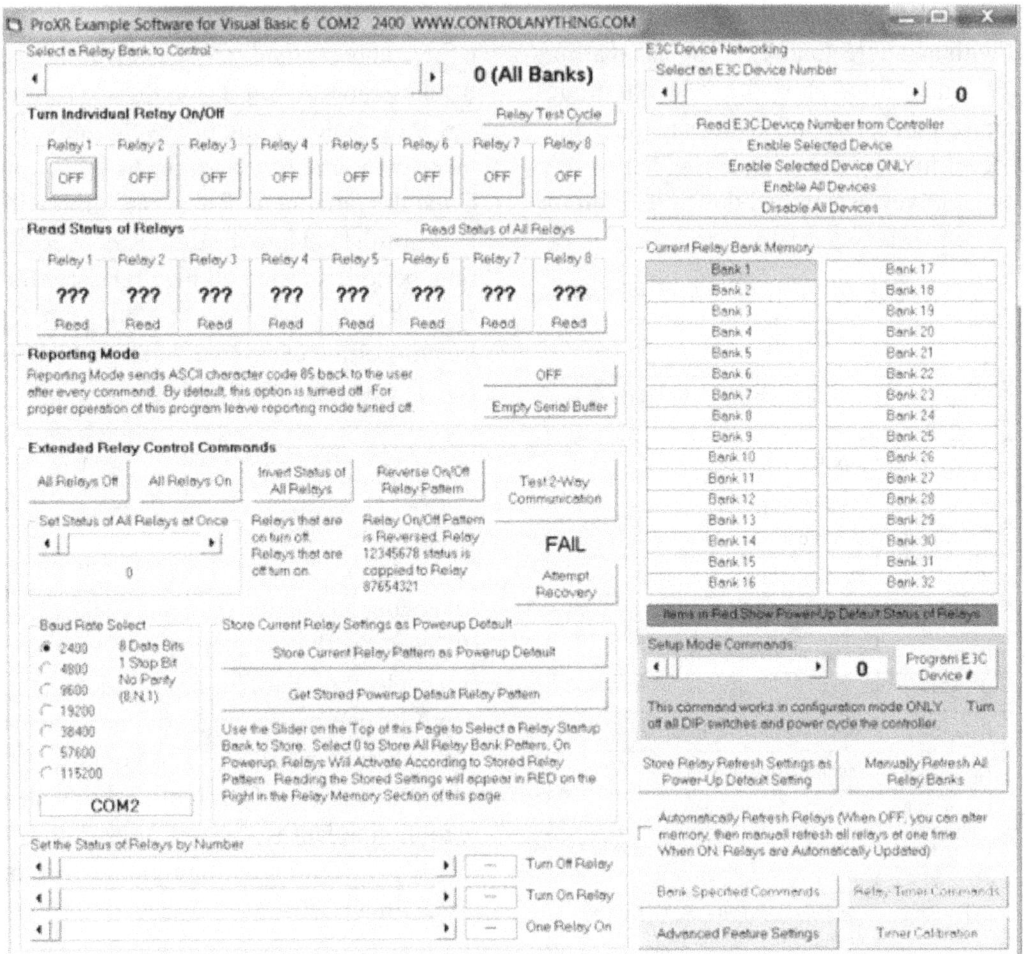

6. RESULTS AND DISCUSSIONS

The UB Excavator robotic system has been prototyped. The mechanical structure (frames, wheels and connectors) is set up first. The actuation system (DC motors, electrical linear actuators/slider) is then attached to the mechanical structure, as shown in Figure 36(a). After that, other electromechanical components are attached. With all the components integrated together, the complete lunabotic excavator is shown in Figure 36(b). The size of the excavator is 1.48m (width) × 0.749m (length) × 1.5m (height), and total weight is 71kg. The lunabotic excavator has been tested to work properly. It can move forward or backward with high speed (0.68m/s) or low speed (0.37m/s) respectively. The front wheel steering system can turn the robotic excavator to make left or right turn. The front wheels can be turned left or right with an angle up to 22°. The excavating conveyor can be lifted up in the traveling mode and dumping mode, and pulled down to touch the ground level in the excavating mode. When the excavating conveyor is pulled down, its bottom grab touches the ground. When the excavating conveyor is lifted up, its bottom grab is about 5cm above the ground. In dumping mode, once the excavator moves backward to the collector box, the dumpster can be lifted up to dump the collected regolith. When the dumpster is fully pushed up by the linear actuator, the bottom of the dumpster is lifted up by 22.8cm, and the dumpster inclines with an angle of 40°. This is good enough for the collected regolith to be dumped out from the dumpster.

Table 2. Assignment of the bank relays for lunabotic excavator

Relay bank	Relay #	Function controlled by relays
Bank 1	Relay 1&2	Steering: left turn
	Relay 3&4	Steering: right turn
Bank 2	Relay 1&2	Forward moving: high speed
	Relay 3&4	Forward moving: low speed
	Relay 5&6	Excavating conveyor lifter: up/down
	Relay 7	Excavating motor: ON/OFF
Bank 3	Relay 1&2	Backward moving: low speed
	Relay 3&4	Backward moving: high speed
	Relay 5&6	Dumpster: pushing up
	Relay 7&8	Dumpster: pulling down
Bank 4	Relay 1&2	Emergency switch for 12V power supply
	Relay 3&4	Emergency switch for 24V power supply

The excavating rate of the lunabotic excavator in excavating mode can be roughly estimated. In the excavating mode, the excavating conveyor powered by 24V DC motor rotates with constant speed. The 36 excavating grabs attached to the conveyor chain work in pipeline style. Assume the volume of the cup of one excavating grab is V_{grab}, the density of lunar regolith simulant is $\rho_{regolith}$, there are totally N_{grab} excavating grabs, and it takes time period T for the excavating conveyor to rotate for one full cycle. Initially it may take some time for the grabs to fill the pipeline. Once the pipeline is filled, one grab will dump its regolith load into the dumpster every (T/N_{grab}) time period. Assume all the grabs are filled with regolith in full, the ideal excavating rate R_{ex} (unit: kg/min) of the excavator when pipeline is filled can be calculated as

$$R_{ex} = \frac{\rho_{regolith} \cdot V_{grab}}{(T / N_{grab})} = \frac{N_{grab} \rho_{regolith} V_{grab}}{T} \tag{9}$$

For our lunabotic excavator, it takes 42 seconds (0.7min) for the excavating conveyor to rotate a full cycle, thus T=0.7min. The width, length and depth of one excavating grab are 3.2cm, 32cm and 3.2cm respectively. The volume of the cup of one grab is V_{grab}=327.7cm^3. Take the density of regolith simulant as $\rho_{regolith}$=1.3g/cm^3, and there are totally N_{grab}=36 excavating grabs, thus the ideal excavating rate is

Figure 36. Prototyping of UB Excavator robotic system

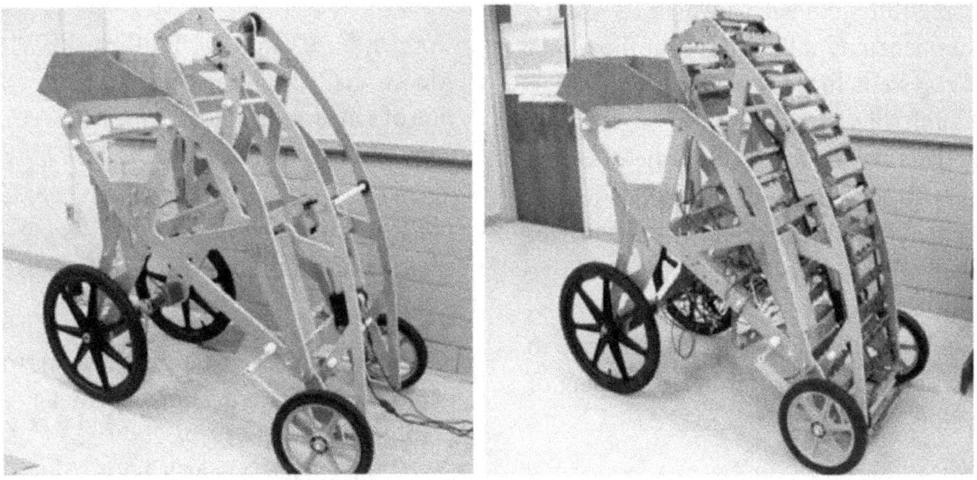

(a) excavator frames with actuators *(b) Complete UB Excavator system*

$$R_{ex} = \frac{36 \times 1.3g / cm^3 \times 327.7cm^3}{0.7\min} = 21.9kg / \min \quad (10)$$

However, this ideal excavating rate can rarely be achieved in the mining contest, because in reality the excavating cups may not be filled with regolith in full, and the regolith in the cups may not be fully passed to the dumpster due to spilling. Thus the actual excavating rate of the excavator during the contest may be much lower than the estimated ideal value. It depends on the proper gesture control of the excavator, the availability of regolith in the particular spot, and many other factors. The actually excavating rate depends on not only the effectiveness of the excavator model design, but also the proper control and manipulation of the excavator by the operator. This makes the NASA Lunabotic Mining Competition a very challenging and exciting contest.

7. CONCLUSION

In this chapter, the design and prototyping of a lunar excavator robotic system for participating in 2010 NASA Lunar Excavating Competition is introduced. Being remotely controlled by operator using a computer via Wi-Fi telecommunication, the autonomous lunabotic excavator can perform the tasks of excavating regolith stimulant, collecting it in the dumpster, and dumping it into the assigned collector box. The design considerations and implementation results of all the functional modules of the lunabotic excavator are discussed. The excavator is tested to work properly with all the designed functions. In reality, the excavating conveyor design with excavating grabs is proven to be very effective. It allows the excavator to dig and accumulate regolith continuously in a pipeline style. The proposed lunabotic excavator may offer hints to the design of new and effective robotic excavators for planetary exploration. The design and prototyping strategies used in the lunabotic excavator may be extended to ground wheeled vehicles or other autonomous robotic systems as well.

In the excavating mode, the regolith simulant collected in the moving grabs may spill off into the excavating conveyor chains and block their movement during long time operation. The routers and switch boards are mounted on the back of the control panel, hence they are less vulnerable to the spilled regolith simulant. However, the spilled regolith simulant may still contaminate the power distribution system. In the future, we will develop proper regolith shielding mechanism to overcome this challenge. During the contest, the effective excavation of regolith simulant depends on not only the excavator design, but also the proper manipulation of the excavator. In the future, we will also further improve the communication efficiency, and develop more effective strategy to control/manipulate the gesture and movement of the excavator.

ACKNOWLEDGMENT

The UB Excavator is the joint effort of UB Excavator team, which consists of multiple students and advisors from School of Engineering at the University of Bridgeport. The student members of the team include Nicolae Gari (student team leader), Fangyuan Song, Peng Lu, Yajuan Shi, Matthew Breland and Edwin Gravrand. The team is advised by Prof. Xingguo Xiong and Prof. Zheng Li. The authors are thankful to NASA for conducting the Lunabotic Mining competition. The authors are thankful to Susan Sawyer, NASA lunabotics project coordinator, for her valuable help and support in the project. The UB excavator project was funded by NASA Connecticut space grant college consortium under ESMD (Exploration Systems Mission Directorate) national design challenge project grant.

REFERENCES

Cooper, T., Cheng, A., Phillips, I., Wild, A., & Meech, J. (2010, March). Design and tele-operation of a lunar excavator for the NASA Centennial Challenge. *Proceedings of the 12th International Conference on Engineering, Science, Construction, and Operations in Challenging Environments*, Honolulu, Hawaii, USA, (pp. 1031-1038).

Craft, J., Wilson, J., Chu, P., Zacny, K., & Davis, K. (2009). Percussive digging systems for robotic exploration and excavation of planetary and lunar regolith. *Proceedings of 2009 IEEE Aerospace Conference*, Big Sky, MT, USA, (pp. 1-7).

Craft, J., Wilson, J., Chu, P., Zacny, K., & Davis, K. (2010, October). Percussive digging systems for planetary research. *IEEE Aerospace and Electronic Systems Magazine*, *25*(10), 21–26. doi:10.1109/MAES.2010.5631722

Crisp, J. A., Adler, M., Matijevic, J. R., Squyres, S. W., Arvidson, R. E., & Kass, D. M. (2003). Mars Exploration Rover mission. *Journal of Geophysical Research*, *108*(E12), 8061. doi:10.1029/2002JE002038

Digi International Inc. (n.d.). *Connect Port TS W wireless serial servers datasheet*. Retrieved September 3, 2011, from http://www.digi.com/pdf/ds_connectporttsw.pdf

Ehsani, M., Gao, Y., Gay, S. E., & Emadi, A. (2004, December). *Modern electric, hybrid electric, and fuel cell vehicles: Fundamentals, theory, and design*. Boca Raton, FL: CRC Press.

Groschopp Inc. (n.d.). *Website*. Retrieved October 2, 2011, from http://www.groschopp.com/

Heiken, G. H., Vaniman, D. T., & French, B. M. (Eds.). (1991). *Lunar sourcebook: A user's guide to the moon*. Cambridge, UK: Cambridge University Press.

Hughes, A. (2005). *Electric motors and drives: Fundamentals, types and applications*. New South Wales, Australia: Newnes.

Maniscalco, M., Lee, N., Dektor, S., Salowitz, N., Grote, E., & Twiggs, R. (2007). *A design concept for a robotic lunar regolith harvesting system*. ICRA '07 Space Robotics Workshop.

Miskin Scraper Works. (n.d.). *Website*. Retrieved September 16, 2011, from: http://www.miskin.com/

Muff, T., Johnson, L., King, R., & Duke, M. B. (2004, February). A prototype bucket wheel excavator for the Moon, Mars and Phobos. *AIP Conference Proceedings*, *699*, 967–974. doi:10.1063/1.1649662

NASA. (1973). *Lyndon B. Johnson Space Center: Apollo 17 preliminary science report*. Washington: Scientific and Technical Information Office.

NASA. (2010, January). *2010 NASA Lunabotics Mining Competition rules*.

NASA. (n.d.). *Lunar outpost*. Retrieved September 22, 2011, from http://www.nasa.gov/externalflash/LunarOutpost/index.html

NASA Lunabotic Mining Competition. (n.d.). Retrieved September 20, 2011, from http://www.nasa.gov/offices/education/centers/kennedy/technology/lunabotics.html

National Control Device. *LLC*. (n.d.). Retrieved September 5, 2011, from http://www.controlanything.com

National Fire Protection Association. (2011, November). *National electrical code 2011 handbook*.

PTC Inc. (2011) Retrieved September 22, 2011, from http://www.ptc.com/

Stieber, M. F., Trudel, C. P., & Hunter, D. G. (1997, April). Robotic systems for the International Space Station. *Proceedings of 1997 IEEE International Conference on Robotics and Automation*, Albuquerque, NM, USA, vol. 4, (pp. 3068-3073).

Thangavelautham, J., Smith, A., Samid, N. A. E., Ho, A., Boucher, D., Richard, J., & D'Eleuterio, G. M. T. (2008, February). *Multirobot lunar excavation and ISRU using artificial-neural-tissue controllers*. Space Technology and Applications International Forum—STAIF.

Volkswagen of America, Inc. (2004, December). *Electro-mechanical power steering design and function*. Self-Study Program, Course Number: 892403.

Zacny, K., Mungas, G., Parrington, L., Mungas, C., & Fisher, D. (2007, July). Pneumatic drill and excavator for planetary exploration. *Seventh International Conference on Mars*, Pasadena, California, (p. 3010).

Section 3
Robotic Systems for Medical Applications

Chapter 9
Medical Robotics

Ahmad Taher Azar
International Journal of System Dynamics Applications (IJSDA), USA

M. Sam Eljamel
The University of Dundee, UK

ABSTRACT

Medical robotics is an interdisciplinary field that focuses on developing electromechanical devices for clinical applications. The goal of this field is to enable new medical techniques by providing new capabilities to the physician or by providing assistance during surgical procedures. Medical robotics is a relatively young field, as the first recorded medical application occurred in 1985 for a brain biopsy. It has tremendous potential for improving the precision and capabilities of physicians when performing surgical procedures, and it is believed that the field will continue to grow as improved systems become available. This chapter offers a comprehensive overview about medical robotics field and its applications. It begins with an introduction to robotics, followed by a historical review of their use in medicine. Clinical applications in several different medical specialties are discusssed. The chapter concludes with a discussion of technology challenges and areas for future research.

1. INTRODUCTION

The word "robot" originated from a 1922 play called "Rossum's Universal Robots (R.U.R.)" by Karel Capek (Capek 2001). The play was about a future in which all workers are automatons, who then revolt when they acquire souls. This idea was further proved by the introduction of Maria, the first female robot on the silver screen, in Fritz Lang's silent science fiction movie 'Metropolis', which was released in 1927. A Few years later, in 1959, John McCarthy and Marvin Minsky established the Artificial Intelligence lab at the Massachusetts Institute of Technology (MIT). The first modern robotic hand was created by Heinrich Erst in 1961. The first industrial robot was invented in 1962 and named as 'Unimate'. General Motors used the robot to perform repetitive or dangerous tasks on its assembly line. The

DOI: 10.4018/978-1-4666-0176-5.ch009

year 1966 holds a special place in the history of robotics. It was during this year that the Stanford Research Institute came up with the concept and creation of 'Shakey', the first mobile robot ever created until that time, which could know and react to its own actions. Victor Scheinman, a Mechanical Engineering student working in the Stanford Artificial Intelligence Lab (SAIL) created the 'Stanford Arm' in 1969. In 1970 - researchers there built the Stanford cart, an intelligent line-follower robot. In order to assemble machines from small parts, Victor Scheinman invented the 'Silver Arm' robotic arm, in 1974. It made use of touch sensors. Robotics reached its advanced stage in the latter half of 20th century. In 1976, Shigeo Hirose came up with an innovative robot - the Soft Gripper - at the Tokyo Institute of Technology. It was designed to wrap around an object. In 1977, NASA launched its highly automated Voyager probe, in order to explore the outer solar system. In the same year, the super-hit Sci-Fi movie - Star Wars was released, which depicted the strongest image of human being's future with robots. On the academic side, robotics became an increasingly popular subject in engineering school and computer science departments in the 1980s. Several of the leading U.S. schools such as Stanford, MIT, and Carnegie Mellon developed robotics groups during this time. Carnegie Mellon formed a Robotics Institute in 1980 and began offering the first Ph.D. in robotics in 1989.

While many early students in the field started with the classic text by Paul (1981), several other textbooks began to appear (Fu et al. 1987; Craig 1989; Spong and Vidyasagar 1989). In 1986, automaker Honda started a humanoid robotics program, which went through 11 versions from 1986 to 2008. Towards the end of 20th century, robotics saw a sea of change in terms of the functionality of the robots. With passing of time, robots emerged as highly sophisticated machines, which could recognize their environment, distinguish sounds easily and observe moving objects. Some of the more remarkable inventions were 'RoboTuna', developed and built by David Barrett, 1996. 'Gastrobot', a robot that digests organic mass to produce carbon dioxide, was developed by Chris Campbell and Stuart Wilkinson at the University of South Florida, 1996 and three years later, in 1999, Sony released the AIBO robotic pet. The advent of the millennium saw further development in the field of robotics. In 2000, Honda introduced humanoid robot - ASIMO. Sony launched its third generation robotic pet in 2003. It was named 'AIBO ERS-7'. Presently, Robots can follow along with a human by holding hands. Today, with the advancement of science and technology, researchers are coming up with innovative ideas to create robots that could simplify sophisticated tasks, which are otherwise done by humans.

2. MEDICAL ROBOTICS

The field of medicine has also been invaded by robots. They are not there to replace doctors and nurses but to assist them in routine work with precision tasks. Medical robotics is a promising field that really took off in the 1990s. Since then, a wide variety of medical applications have emerged: laboratory robots, telesurgery, surgical training, remote surgery, telemedicine and teleconsultation, rehabilitation, help for the deaf and the blind, and hospital robots. Medical robots assist in operations on heart-attack victims and make possible the millimeter-fine adjustment of prostheses. There are, however, many challenges in the widespread implementation of robotics in the medical field, mainly due to issues such as safety, precision, cost and reluctance to accept this technology. Medical robotics includes a number of devices used for surgery, medical training, rehabilitation therapy, prosthetics, and assistance to people with disabilities.

2.1 Rehabilitation Robotics

Activity in the field of rehabilitation robotics began in the 1960s and has slowly evolved through the years to a point where the first commercially successful products are now available (Kim and Cook 1985).

Today, the concept of a "rehabilitation robot" may include a wide array of mechatronic devices ranging from artificial limbs to robots that support rehabilitation therapy or that provide personal assistance in hospital and residential sites. Rehabilitation robotics is often only thought of as robotic aids to assist people handicapped by a manipulative disability (Bolmsjo et al. 1995; Kassler 1993). However, recent research in rehabilitation robotics reveals more extended possibilities for the use of robot technology in rehabilitation (Hillman 2003). Hillman defined rehabilitation robotics as the application of robotic technology to the rehabilitation needs of people with disabilities as well as the growing elderly population (Hillman 1998). This extended definition includes assistive robots, prosthetics, orthotics, and therapeutic robots. The primary benefit of an assistive robot is that it reduces the need for a human attendant. The operator can use the device independently and may develop a higher level of self-esteem and functional performance, in addition to being free of the cost, loss of privacy, and personality issues of a human attendant.

Robotic prosthetics and orthotics have been developed to replace lost arms, hands, and legs and to provide assistance to weak or impaired limbs. Therapeutic robots are valuable tools for delivering neuro-rehabilitation to the limbs of individuals with disabilities following a stroke. A rehabilitation robot has many forms, which may be generally described in terms of two operating approaches, the robot as assistant and the robot as extension. The robot as assistant approach provides the user with a set of high-level commands, which are similar to those that might be given to a human attendant. The robot assistant is normally implemented in conjunction with an intelligent computer-based interface, which possesses detailed knowledge of the task environment, the command protocols, and robot control. In the robot as extension control approach, the movement of the robot is directly linked to the physical movement of the operator. This may be through a direct physical harnessing of body movement, as is the case of the control of a body-powered prosthesis, or it may be through a joystick or other adaptive interface. A number of rehabilitation robot commercial products have emerged. These products were transferred from research and development projects or developed by small companies. Table 1 contains list of some of these products, as well as information about their origin and type.

2.1.1 Robotic Prosthetics and Exoskeletons

Robotic prosthetics and exoskeletons will provide dexterity, natural mobility, and sense of touch to missing or paralyzed limbs (Kassler 1993). Individuals suffering from hip or knee conditions can use a robotically intelligent walker or wheelchair to help prevent common accidents like slipping. Robotic rehabilitation not only provides consistent and efficient therapy without tiring, it also has the potential to enhance the therapy beyond the abilities of the practitioner.

The evidence of prosthetic usage dated back as early as the ancient Egyptians. In one case, it was found that a mummy's big toe had been amputated during its life and supplanted with a carefully crafted wooden toe, which was attached by a series of wooden plates and leather strings. The oldest known leg prosthesis from 300 BCE was discovered in Capua, Italy, and was made out of copper and wood. In the 16th century, prostheses were created from iron for soldiers by the same blacksmiths who crafted their suits of arms. An iron arm had the ability to flex a fully digital hand. By the 19th century, James Potts created a leg with artificial tendons to lift the toe

Table 1. A partial list of commercial rehabilitation robotics products

Product Name	Country	Company	R&D / Support	Type
Prab Command	USA	PRAB Robotics	Boeing Company	Vocational Workstation
DeVAR	USA	Independence Works, Inc.	VA Palo Alto Rehab R&D Center, Stanford University	Vocational Workstation
Manus	Netherlands	Exact Dynamics	IRV, TPD	Wheelchair Mountable
Handy 1	UK	Rehab Robotics Limited	Undergrad thesis	Mobile Base
Helping Hand	USA	Kinetic Rehabilitation Instruments	Start-up Company	Wheelchair Mountable
Papworth Arm	UK	Papworth Group	Inventaid	Wheelchair Mountable
RAID	UK, France, Sweden	Oxford Intelligent Machines	EU TIDE Funding	Vocational Workstation
Arlyn Arm	USA	Arlyn Works	Start-up, US Dept. of Educ.	Education, Vocational Workstation
Sidekick	USA	Robotic Assistance Corporation	Start-up company	Mobile Base
Robotic Assistive Appliance	Canada	Regenesis	Neil Squire Foundation	Vocational Workstation

when bending the knee. Interest in artificial limbs increased during the American Civil War, due to the large number of amputations that occurred during this time (Dellon and Matsuoka 2007).

Advances in robotic technology that occurred are primarily due to the discovery of anesthetics. Anesthetics enable longer surgeries, so the doctor could better shape the stump, providing a better fit for the prosthetics. Newer materials and electronics have made prostheses more functional since they are lighter, more compliant, and more adaptable to specific stump shape or personal style (Dellon and Matsuoka 2007). The introduction of the C-Leg Technology in 1997 first made full intelligent control of a prosthesis system and adaptation to the individual gait pattern possible (see Figure 1). It uses microprocessor-controlled hydraulics which dynamically adapt the system to the user's walking speed in real time. At the same time, it offers real security during the stance phase. This proven regulator mechanism is made possible by a complex system of sensors, consisting of strain gauges in the tube adapter and a knee angle sensor. These sensors record stresses every 0.02 seconds, measuring both at the ankle moment above the foot

Figure 1. C-Leg (Courtesy of Ottobock Healthcare)

adapter as well as the angle and angular velocity of the knee joint. With this information, the joint continuously recognizes which walking phase the wearer is currently in.

The C-Leg compact leg prosthesis system adjusts automatically whether the user is walking down stairs, on slopes, or on uneven terrain. Physiological movement relieves the entire body. Even stumbling, darkness, or dense crowds no longer present a real problem for the patient. The knee joint is made secure with a high level of resistance in the stance phase and switches to the swing-through phase only when required. Additional advantages of C-Leg Technology include standing phase flexion (yielding), low energy expenditure while walking, possible relief of the locomotor system, and the ability to bear weight on the prosthesis during flexion. Since the introduction of the C-Leg compact, even people with a pronounced need for security can profit from the advantages of C-Leg Technology. While the swing phase is hydraulically controlled, the mechantronically controlled stance phase dampening offers the patient the highest level of security in all situations. It stabilizes the joint after heel impact and maintains that resistance until it switches over to the swing phase.

Unlike lower-limb prosthetics, upper-limb prosthetics are not yet dexterous enough to provide function comparable to healthy limbs. Commonly used upper-limb prosthetics range from a hook to a single-degree-of-freedom (opening and closing) mechanism using myoelectric control (Dellon and Matsuoka 2007). Upper-limb prosthetics with multiple degrees of freedom typically use sequential control methods, with locking mechanisms or switches used to separately activate each joint. There is a lot of room left to improve these unnatural control mechanisms.

The Utah Arm (Motion Control, Inc. U.S.A.) is a computer-controlled, above-the-elbow prosthesis developed by Jacobsen at the University of Utah in the 1980s (see Figure 2) (http://www.utaharm.com). This commercially available arm is controlled using feedback from electromyography (EMG) sensors that measure the response of a muscle to nervous stimulation (electrical activity within muscle fibers). For the last decade, the

Figure 2. Utah arm (Courtesy of Motion Control)

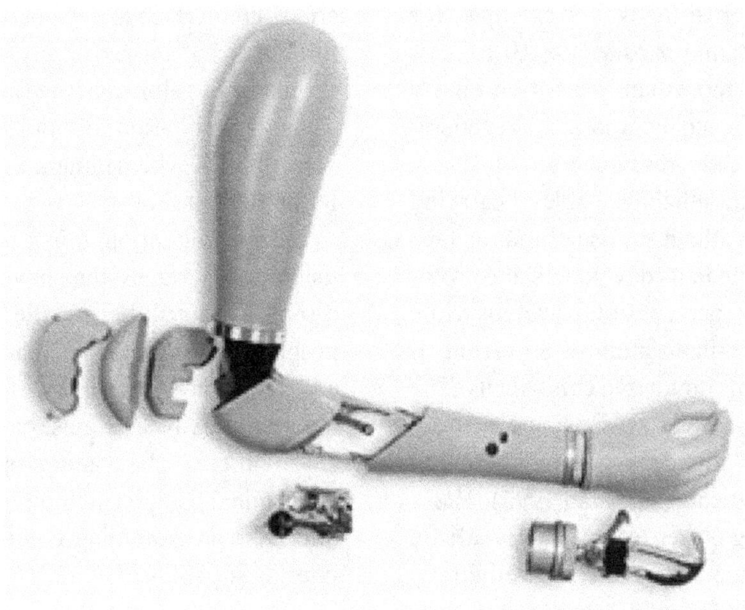

research focus in upper-limb prosthetics has been in the anthropomorphic arm and hands capable of neural interface.

Motion Control, Inc. also developed a two-fingered prosthetic hand that is controlled using myoelectric signals from the remnant limb. Another prosthetic hand is currently being developed at the Scuola Superiore Sant'Anna, Italy (Massa et al. 2002), and Rutgers University (DeLaurentis & Mavroidis, 2002) is creating a robotic prosthetic hand with five fingers and twenty Degree Of Freedom (DOF) using shape-memory alloys as artificial muscles. Robotic prosthetics can also be used to replace lower limbs. The MIT LegLab is testing an intelligent prosthetic, the Cyberhand (www.cyberhand.org), with a single degree of freedom per finger. Their goal is to interface with both afferent and efferent pathways, providing true feedback for the user.

The Anatomically-Correct Testbed (ACT) hand (Matsuoka et al, 2006) is designed around the central tenant that an anatomically correct hand can be operated by the same neural signals used to operate the muscles in the original hand (see Figure 3). Exoskeletons have links and joints that correspond to those of the human and actuators that assist the patient with moving his or her limb or lifting external loads. For example, the Berkeley lower extremity exoskeleton (BLEEX) is capable of carrying more than 75 lbs at 3 mi/h, yet it weighs only 30 lbs and incorporates a regenerative power system. Also, researchers at MIT have developed a leg exoskeleton capable of carrying an 80-pound load without the use of motors (see Figure 4). According to its developers, the prototype can support 80 percent of this weight while using less than one-thousandth of a percent of the power used by its motorized equivalents. The Wrist-Hand Orthosis (WHO) uses shape memory alloy actuators to provide a grasping function for quadriplegic patients (Makaran 1993). These systems are getting closer to providing realistic power augmentation for soldiers, firefighters, emergency personnel, and industrial workers to use for carrying heavy loads and extending their physical abilities (Kazerooni 2005). The Hybrid Assistive Limb (HAL) is an exoskeleton designed to assist the elderly and disabled individuals by using the user's myoelectric signals to control the exoskeleton movement. However, it is also strong enough to augment the user's power by 40 kg.

While these lower exoskeletons have recently made a giant leap, hand exoskeletons are still too bulky and difficult to fit into hands of different shapes and sizes. Rehabilitation robotics has many technical and non-technical problems (Tejima 2000). First, cost and maintenance are the most serious problems for popularization. Second, rehabilitation robots are used in close proximity to users and because of this proximity accidents can happen. To avoid injury to the end user, recent rehabilitation robots are designed to operate with extremely low power. This means that they can only work with light goods and can only move very slowly. Consequently, the tasks they can perform are very limited. Third, most users have physical or mental handicaps, making it difficult for them to operate a robot. An effective human interface is one of the key technologies for rehabilitation robotics. At present, the only way to adapt an interface to each user's various needs is through trial and error. Systematic adaptation should be developed in the future (Tejima 2000). Thus, rehabilitation robotics is more akin to service robotics, which integrates humans and robots in the same task.

Special attention must be paid to human-machine interfaces that have to be adapted to disabled or nonskilled people operating a specific programming device. There is a need for research and development in robotics to focus on developing more flexible systems for use in unstructured environments. The leading topics of research are mechanical design (including mobility and end-effectors), programming, control and man machine interfaces (Bolmsjö et al. 1995).

Figure 3. Anatomically correct testbed (ACT) index finger (Courtesy of Neurobotics Laboratory)

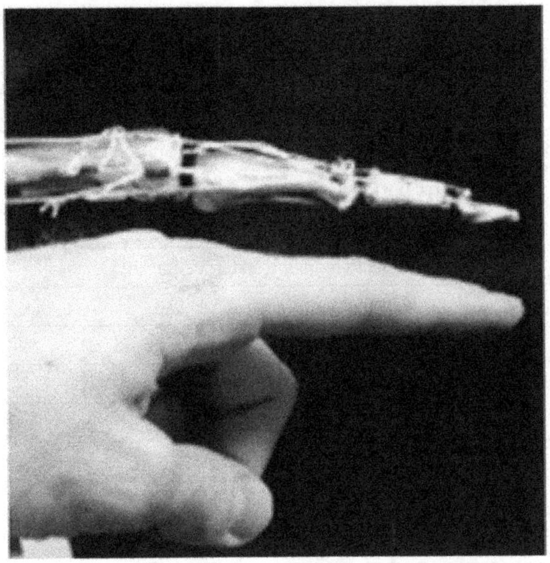

Figure 4. MIT exoskeleton (Courtesy of MIT Media Lab)

2.2 Surgical Robots

Surgical robots have begun to appear on the market in the last few years and have started to populate the operating rooms in large medical centers. These systems have already established their ability to increase a surgeon's dexterity in minimally invasive procedures and they have the potential to maximize patient outcome on a larger basis even though, for the moment, their cost is prohibitively high for widespread application (Stoianovici et al. 2005). As surgeons become increasingly aware of the clinical benefits of these systems and costs are driven down by technological advance and availability, robots will become standard operating room tools. Initial applications of a handful of robots have already demonstrated their surgical potential. As technology evolves, robots could not only improve performance in minimally invasive procedures but also enhance other existing procedures or even enable entirely new kinds of surgeries. Robots can be seen as a mechanism that has complementary capabilities to those of humans and may be used in a number of ways to increase a surgeon's ability to carry out procedures, either by making existing procedures more accurate, faster, or less invasive or by making it possible to perform otherwise infeasible interventions. In these cases, the advantages often come from exploiting the complementary strengths of human and robotic devices. Table 2 summarizes strengths and limitations of each of both, humans and robots (Howe and Matsuoka 1999). These differences in capabilities mean that current robots are restricted to simple procedures, and humans must provide detailed commands by using preoperative planning systems or by providing explicit move-by-move instructions. Even in the most sophisticated systems, robots are specialized to specific tasks within procedures; humans must prepare the patient, make many of the incisions and sutures, and perform many other functions (Howe and Matsuoka 1999).

Table 2. Strengths and limitations of humans and robots for surgery (Howe and Matsuoka (1999))

	Strength	Limitation
Humans	Strong hand-eye coordination	Limited dexterity outside natural mode
	Dexterous (at human scale)	Prone to tremor and fatigue
	Flexible and adaptable	Limited geometrical accuracy
	Can integrate extensive and diverse information	Limited ability to use quantitative information
	Able to use qualitive information	Large operation room space requirement
	Good Judgement	Limited sterility
	Easy to instruct and debrief	Susceptible to radiation and infection
Robots	Good geometric accuracy	Poor judgement
	Stable and untiririg	Limited dexterity and hand-eye coordination
	Can be designed for a wide range of scales	Limited to relative simple procedures
	May be sterilised	Expensive
	Resistant to radiation and infection	Technology in flux
	Can use diverse sensors (chemical, force, acoustic, etc.) in control	Difficult to debug

There are surgical robotics review articles with a focus on surgical procedures have been written. Troccaz and Delnondedieu (1996) give a historical review and describe passive, semiactive, and active robotic systems. Taylor (1997) discusses several taxonomies for surgical robotics and presents a different classification. Specialized reviews also exist, such as the article by Caddedu et al (Cadeddu et al, 1998a; Cadeddu et al. 1998b) on urology robotics. Howe and Matsuoka (Howe and Matsuoka 1999) overview applications in image-based procedures, orthopedic surgery, and neurosurgery, among other specialties. Davies (2000) describes the history of surgical robotics and gives one classification for the types of robot systems studied by researchers.

2.2.1 Minimally Invasive Surgery (MIS)

As long as operations exist, procedures are needed to decrease the size of the surgical incision. Smaller incisions lead to less trauma and less pain for the patient. This results in faster recovery. The patient can return home sooner, and a new spot is available for another patient. Besides the advantages to the patient's care, fast recovery has economic advantages, e.g. lower medical costs and the patient is able to return to work sooner. The search for smaller incisions has led to minimally invasive surgery (Demi et al. 2005). Small incisions of MIS are used to introduce special instruments with a long rod transmission mechanism through a cannula into the body of the patient.

By use of a small camera, the surgeon is able to follow the progress of the operation on a screen. The term minimally invasive surgery covers all surgery with small incisions and endoscopes like thorascopy (chest cavity), arthroscopy (joints), laparoscopy (abdominal cavity), pelviscopy (pelvis) and angioscopy (bloodvessels). Although MIS fastens the rehabilitation of the patients, it also has some disadvantages. In open surgery the surgeon has more space and freedom to move in a large operation room. With MIS, the operation area is smaller and the freedom to move the operating tools decreases with the use of an endoscope. Laparoscopic tools are long and have only four degrees of freedom. These four degrees of freedom include rotation around and translation along the axis perpendicular to the incision surface, and rotation around the two axes on the incision surface. The operation is less ergonomic for the surgeon. It takes more energy to use the laparoscopic tools. Furthermore, the surgeon has to concentrate on both the forward and reverse motion direction of his hands and the tip of the instrument. The surgeon has to develop more skills to perform the same operation. Another disadvantage of MIS is the interaction between the surgeon and the operative area. There is a loss of visibility for the surgeon, because he has to follow his own movements on a 2-dimensional screen with a limited angle of view. There is also a loss of tactile sensation and the contact sensation with the long endoscopic tools. This makes it hard to recognise or to distinguish different tissues or structures.

2.2.2 Microsurgery

With microsurgery the technology scales down the surgeon's motions and scales up the field of view, allowing him/her to operate on microscopic anatomy with relative ease. The technology enables the surgeon to have a more accurate view and a better control of motion and forces, than is possible with his own eyes and hands. Human perception is not lost and this enables the change for new microsurgical procedures and an improved performance. The disorder, e.g. a tumor, can be approached and removed more accurately without damaging surrounding tissue. The small forces and motions used in microsurgery cannot be achieved with conventional hand-held surgical tools under naked eye visual perception.

2.2.3 Computer Aided Surgery (CAS) and Surgical Robotics

Differences exist between Computer Aided Surgery (CAS) and Surgical Robotics. The main difference between the terms 'robotic' and 'computer aided' surgery is how they are powered. Robots are powered by a motorised system, while a CAS system generally is powered by the surgeon (Davies 2000). A robot can assist the surgeon in two ways. The robot can position the surgical tools or perform operative tasks very accurately. The position can be a predefined location or a predefined complex path. The target area has to be defined very accurately, because the robot has to be programmed for its surgical task.

The target operation area has to be defined during the pre-operative phase. With CAS systems it is only used for simulating and planning the operative procedure. Programming of instruments is not necessary. Unexpected situations can be avoided, and the operation time can be decreased. However, robots are able to provide a greater accuracy than CAS systems. Robots can also constrain the motion of the system. A robotic system has a predefined safety concept, which does not change during operation situations. The movements of a CAS system are established by the decisions of the surgeon, who can change the operation plan every moment he wants. He can neglect warnings or cut into unsafe regions.

2.2.4 Classifications of Surgical Robotics

Surgical robots can be classified based on function, technology, application, or role (Camarillo et al. 2004). The functional classification divides robots into three groups (Bann et al. 2003):

1. **Dexterity enhancement:** An input device used by the surgeon controls the robot, but by using the robot and its software the surgeons innate qualities can be improved; e.g., tremor reduction.
2. **Precision localization:** The use of preoperative investigations allows precise location of the area where surgery is required.

3. **Precision manipulation:** The generation of a preoperative plan allows the robot to operate in the predetermined area.

The technologic classification of surgical robots has three subtypes (Bann et al. 2003):

1. **Autonomous:** The preoperative plan controls the manipulator.
2. **Supervisory:** The computer to perform the surgery accurately guides the surgeon.
3. **Teleoperated:** The robot is manipulated by input devices under the surgeons control; the surgeon can be remote from the theatre.

Application-based taxonomy distinguishes robots on the basis of surgical disciplines and operative procedures. Role-based taxonomy separates robots into five discrete categories:

1. **Passive robots**, also called navigators, are global positioning and guiding systems based on the correlation between medical imaging and peroperative reality. They give mostly visual real-time sensory information on guiding and positioning surgical instruments in the operating field (Merloz at al. 1997). They need either preoperative images (CT or MRI scan navigation) or peroperative images (fluoroscopic navigation). In the first case, preoperative images, mostly coming from CT scans, are matched by image registration, by transforming the different sets of data into one coordinate system which allows 3D navigation (Peters 2000). In the second case, calibration of the image acquisition system is done before the surgical procedure by positioning referential frame few given targets from a photo- amplifier. A well-developed example is the positioning of transpedicular screws in spinal surgery (Merloz at al. 1997; Laine et al. 2000).
2. **Semi-active robots** incorporate both autonomous and surgeon driven elements. In the semi-active execution approach, the robot is used as an aiding tool during surgery for tasks such as precise guidance of the surgical cuts without actually performing them. In this mode of operation, the robot holds the surgical tool while the surgeon moves the tool. The task of the robot is to prevent the surgeon from moving the tool out of the desired region.
   Examples for this semi-active approach were presented in (Harris et al. 1997; Ho et al. 1995; Kienzel, et al. 1996) in total knee replacement surgery. Other works present systems that support both active and semi-active approaches, for example (Brandt et al. 1997; Eljamel 2009).
3. **Active robots** can autonomously achieve part of the total surgical strategy without any control from the surgeon, after specific preoperative programming. Active robots have the potential for complex interaction with the patient. Because of safety concerns these robots are used only for a specific tasks.
4. **Telemanipulators** represent a separate field in the realm of medical robotics: telesurgery. These robots supplant surgeon in the operating field by a remote-controlled robotized device. Besides hand surgery, telesurgery has found many applications in cardiothoracic (Cohn 2006) and urological (Zorn 2007) surgery. Their success is due to the outstanding features of telemanipulators with respect to the surgeon's autonomy: increase in range of motion, precision in surgical gesture (due to movement gearing and physiological tremor filtering), instrumental miniaturization, remote-controlled procedures, and reduction of hospital stay (Nelson 2007). A major disadvantage is the loss of tactile proprioception, but this can be addressed by rapidly effective visual learning compensation.

5. **Simulators** mainly have an educational and teaching role. The goal is to help surgeons practicing surgical procedures to rapidly advance their learning curves, without putting the patient's safety at stake. For this purpose, a simulator must recreate as accurately as possible the tissue behaviour under solicitation and stress something which a telemanipulator cannot do (Brouwer et al. 2001). There is a need for simulators capable of haptical feedback, i.e. perception of pressures applied to the tissues (Cotin et al. 2000) via a motorized master arm (force feedback joystick) such as ''Phantom1'' (Harja et al. 2007).

3. APPLICATIONS OF SURGICAL ROBOTS

Robots have been used in a number of clinical applications in medicine. Some of these applications are shown in Table 3 (Cleary and Nguyen 2003).

3.1 Neurosurgery

Neurosurgery is one of the first organ systems in which robotic surgery is introduced. This is due to the high precision that was required to localize and manipulate within the brain, and the relatively fixed landmarks of the cranial anatomy (Nathoo et al. 2003). Some of the neurosurgical robotic systems are discussed here.

Table 3. Clinical application of surgical robotics (Cleary and Nguyen (2003))

Clinical Area	Country	Company	System	Studies
Neurosurgery	Oxford	Advanced Research & Robotics	PUMA	Tissue samples
	USA	Integrated Surgical/ Grenoble	NeuroMate	Humans
	Switzerland	Univ. of Lausanne	Minerva	Humans
	Japan	Hospital Univ. of Tokyo	MRI compatible	Tissue samples
	Scotland	Ninewells Hospital and Medical School, Dundee	PathFinder	Humans
Orthopedic	USA	Integrated Surgical.	ROBODOC	Human
	USA	University of Tokyo/Hopkins	PAKY/RCM	Phantom
	USA	Marconi	Kawasaki	Pig
	UK	Imperial College	Acrobot	Human
Interventional	USA	Georgetown/Hopkins	PAKY/RCM	Cadaver
	Austria	ARC Seibersdorf Research/ Vienna Univ. Hospital	Custom built	Phantom
Urology	UK	Imperial College	Probot	Human
	USA	Hopkins	PAKY/RCM	Human
Maxillofacial	Germany	Charite	SurgiScope	Pig
	Germany	Karlsruhe/Heidelberg	RX 90	Pig
Radiosurgery	USA	Accuray	CyberKnife	Human
Ophthalmology	USA	Hopkins	Steady-Hand	Phantom
Cardiac	USA	Intuitive Surgical	da Vinci	Human
	USA	Computer Motion	Zeus	Human
	France	Grenoble	PADyC	Phantom

3.1.1 Programmable Universal Machine for Assembly Industrial Robot "PUMA" (1985)

PUMA is the first robot used for neurosurgery (Kwoh et al. 1988; McBeth et al. 2004). The surgeon inputs the x-y coordinates on a probe based on a preoperative image of an intracranial lesion. Programs are then used to calculate the stereotactic coordinates (frame-based), which then guide the drilling of the biopsy. This is possible with the introduction of a Cartesian robot (Compass International, Rochester, MN) which places a stereotactic head frame around the patient's head. It then uses fixed markers to record an image of the patient's brain. The device lacked safety factors, but the potential of this technology excited scientists all over the world (Davies 2000).

3.1.2 NeuroMate (1987)

NeuroMate is the first neurorobotic device to be approved by the FDA, as well as the first to be commercially available (Benabid et al. 1987; Lavallee 1989; Lavallee et al. 1995; Haidegger et al. 2008). It was developed at Grenoble University and produced by Innovative Medical Machines International (Lyon, France). In 1999, it was approved for frameless stereotactic surgery after having performed over 3000, and then had CE mark for neuro-endoscopic applications (Haidegger et al. 2008). The 5 DOF of Neuro- Mate provided an accurate and trusted assistance for supervised needle positioning for brain biopsy (Haidegger et al. 2008). The technology was bought and commercialized by Integrated Surgical Systems Inc. (Sacramento, CA, USA) in 1997 and later acquired by Shaerer Mayfield NeuroMate Sarl (Lyon, France). In the newer versions of this robot, an ultrasonic-based registration is performed using reference markers (see Figure 5). This information is used to determine the transformation matrix between the robot and the patient. The defined trajectory is used to command the robot to position a mechanical guide, which is aligned with this trajectory. The robot is then fixed in this position and the physician uses this guide to introduce the surgical tool such as a drill, probe or electrode. The robot's reported intrinsic accuracy is 0.75 mm, with a repeatability of 0.15 mm. In a real-life stereotactic surgical setup, the overall application accuracy was measured to be 1.95 ± 0.44 mm (Li et al 2002).

3.1.3 Minerva (1991 – University of Lausanne, Lausanne, Switzerland)

Minerva is the first system to provide image guidance in real-time, allowing the surgeon to change the trajectory as the brain moved, resulting in frameless stereotaxy (Burckhart et al. 1995). This was important because the structure of the brain and fiducial markers were assumed to be in the same position as before. The position of surgical tools in relative to intracranial imaging could now be seen. Intraoperative imaging can compensate for these shifts and deformations.

Figure 5. NeuroMate neurosurgical robot (Courtesy of Integrated Surgical Systems, Davis, CA)

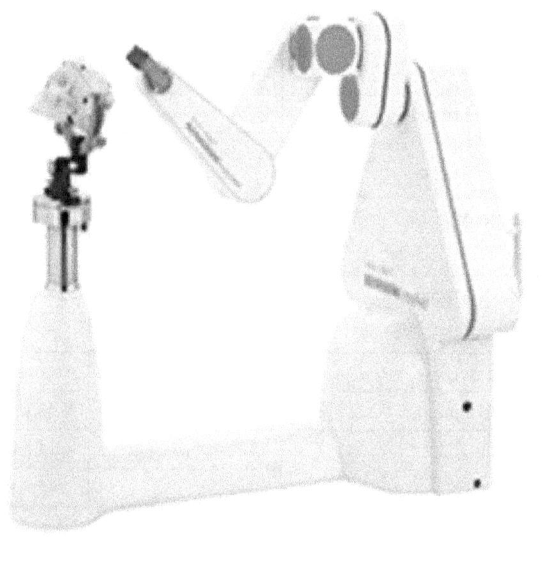

This was accomplished by placing a robotic arm inside a computed tomography (CT) scanner. The implementation of CT greatly improved 3-D localization and thus improved accuracy. The system also improved safety features which were lacking in previous models. However, the system was still limited because it could only perform single dimension incursions, and the patient had to be inside the CT system. As a result, the project was discontinued 2 years later in 1993 (Glauser et al. 1993).

3.1.4 Magnetic Resonance Imaging (MRI) Compatible Robot

This robotic system was developed by Harvard Medical School in collaboration with Mechanical Engineering Laboratory, AIST, MITI (Tsukuba, Japan) (Chinzei and Miller 2001). It has 5 degrees of freedom and is MRI compatible to provide a guide for needles and probes. MR compatibility ensures that the robot produces a minimal MR imaging (MRI) artifact and that the operation of the robot is not disturbed by the electromagnetic fields. This robot has been integrated with a software planning interface (built into the 3D Slicer) and a tracking and control system for percutaneous interventions in the prostate under MR-guidance. The surgeon interacts with the planning interface in order to specify a set of desired needle trajectories, based on anatomical structures and lesions observed in the patient's MR images. Applications included prostate biopsy and brachytherapy, but resolution and functional imaging are limited by the low magnetic field of the open magnetic resonance.

In a related development, a new MRI compatible robot has been developed to work within the interventional MRI unit at the Brigham and Women's Hospital in Boston, Massachusetts (Chinzei et al. 2000). The interventional MRI has a pair of parallel facing donut-shaped magnets with an air gap of 560 mm. The robot sits between the magnets and is mounted on top of the unit as shown in Figure 6. The benefits of MR soft-tissue

Figure 6. MRI compatible robot in an interventional MRI system (Courtesy of AIST, Tsukuba, Japan)

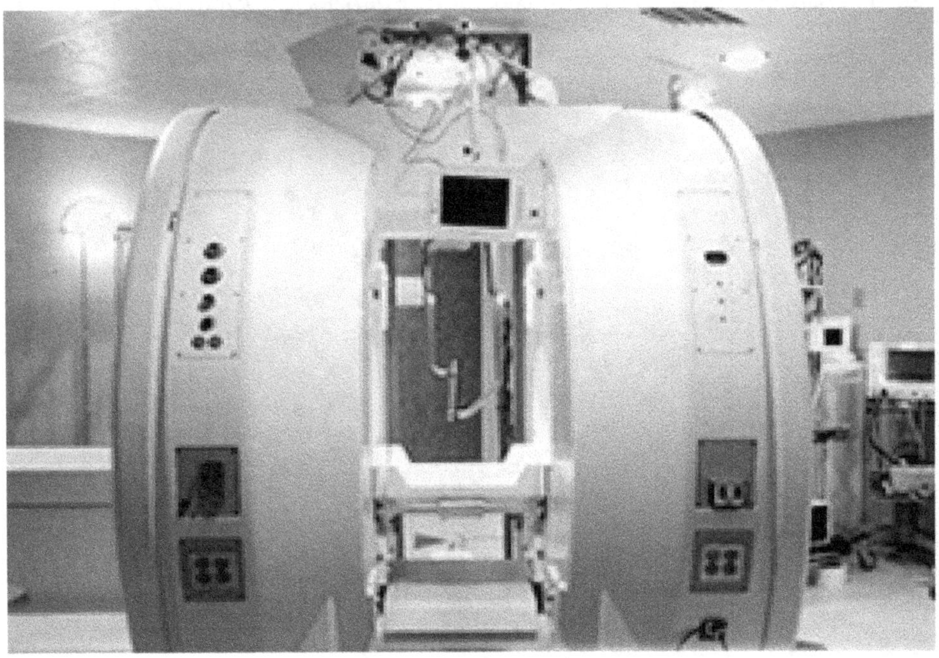

visualization nevertheless prompted investigators from Harvard University (Cambridge, MA) (Chinzei and Miller 2001), the University of Tokyo (Tokyo, Japan) (Nakamura et al, 1995), and the University of Calgary (Calgary, Alberta, Canada) (Louw et al. 2004) to develop their own MR-compatible robotic systems.

3.1.5 Robot-Assisted Microsurgery System, RAMS (1995 – NASA, Washington DC)

RAMS (see Figure 7) was developed by Jet Propulsion Laboratory (California Institute of Technology, Pasadena, CA.) to provide a dexterous platform to perform surgery with increased precision. The system is based on master-slave control; the motion of a slave arm with 6 degrees of freedom is linked to the motion of a haptic hand controller, which also has 6 degrees of freedom. RAMS is equipped with adjustable tremor filters and motion scalers to enhance dexterity (Das et al. 1999). Further studies of the feasibility of endoscopic cardiac surgery was performed by various surgical teams verifying that robotic technology could be used to accomplish a completely endoscopic anastomosis (Le Roux et al. 2001; Szymula et al. 2001; Schiff et al. 2005).

3.1.6 Steady Hand System (1995 – John Hopkins University, Baltimore, MD)

Steady-Hand System is another dexterity-enhancement system designed to augment microsurgery by filtering tremors (Das et al. 1999). It is described in detail in section 3.6. The operator's hand manipulates tools connected to and controlled by a robotic arm. The tools have strain gauges in their handles to detect force. The robot controller measures the incoming forces and filters tremor. Despite its novel design, the system has not been used in clinical applications.

Figure 7. Robot assisted microsurgery (RAMS) (Courtesy of NASA, Washington DC)

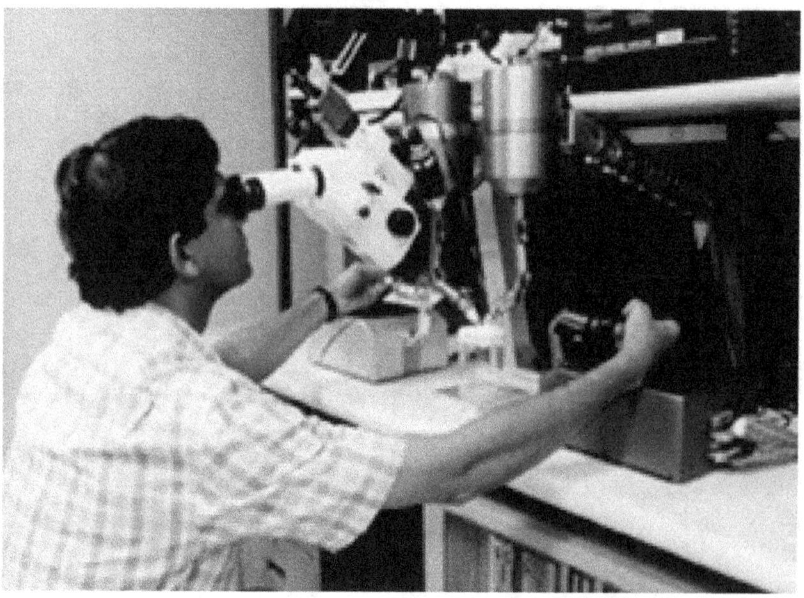

Figure 8. NeuRobot (Courtesy of Shinshu University School of Medicine, Matsumoto, Japan)

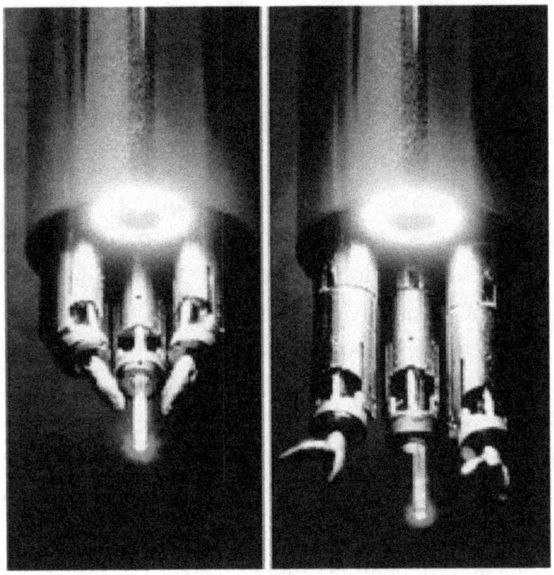

3.1.7 NeuRobot

NeuRobot (Shinshu University School of Medicine, Matsumoto, Japan) was developed for tele-controlled microneurosurgery through the portal of an endoscope as shown in Figure 8. (Hongo et al. 2002). The robot is based on a 10-mm endoscope equipped with twin tissue forceps, a camera, a light source, and a laser. Investigators performed neurosurgery on a cadaveric head and concluded that the system facilitated more accurate and less invasive surgery (Hongo et al. 2002).

Goto et al. describe a robot-assisted craniotomy in which the NeuRobot is used to resect superficial portions of an intraaxial tumor on a live human subject, citing dexterity enhancement as one of the potential advantages (Goto et al. 2003).

Figure 9. The PathFinder

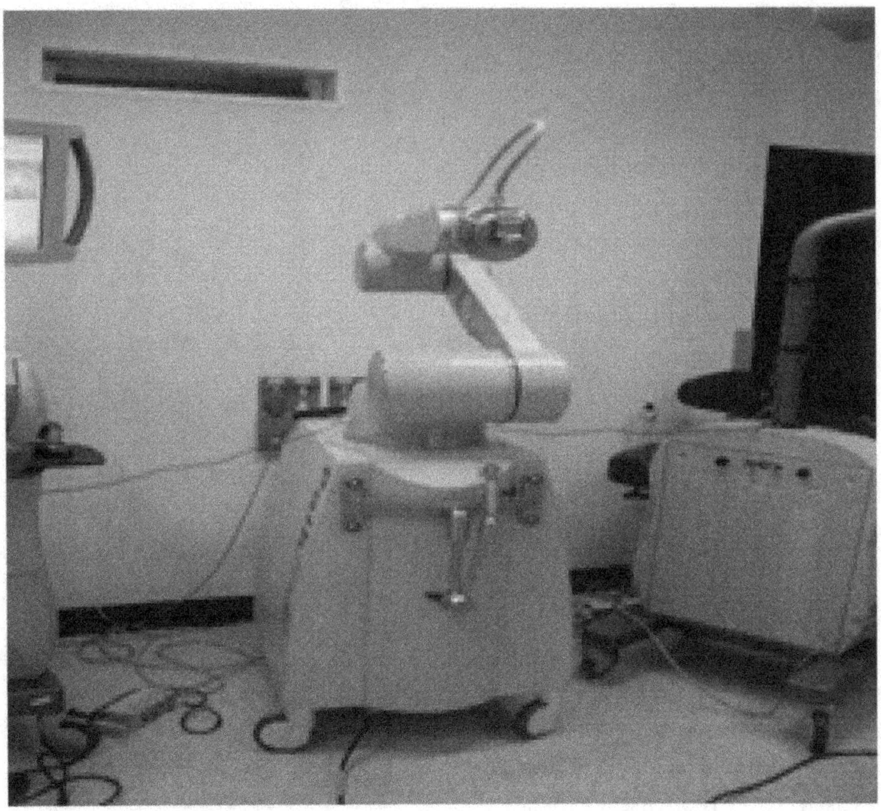

3.1.8 PathFinder

PathFinder (Prosurgics, UK, and University of Dundee, Scotland) is a collaborative robotic project which is CE marked and which is used to replace the stereotactic frame in functional neurosurgery such as lesion generation, deep brain stimulator placement, intraoperative localization, and stereotactic biopsy (Eljamel 2006, 2007, and 2009).

The PathFinder is a 6 degree of freedom robotic system mounted on a mobile base giving it a total of 8 degrees of freedom (see Figure 9). Its accuracy is better than the stereotactic frame and much better than image guided navigation system (Eljamel 2007).

3.2 Orthopedic

Robotic surgery is widely accepted by the orthopedics community for two reasons. First, the technology is well-suited to operations on bones. Since bones, unlike soft-tissues, are less-prone to deformation when being pressed on, computer navigation and mapping of bones, which use software based on the rigid-body assumption, are accurate and efficient (Bargar 2007). Furthermore, robotic surgery increases the accuracy of implantation and component placement. Current surgical techniques often result in inaccurate placing and balancing of hip replacements, knee components, or soft-tissues. With computer navigation system and robotics assisted surgery, precision and accuracy is greatly improved; this in turn leads to better long-term outcomes.

3.2.1 RoboDoc Surgical System

The Robodoc surgical system consists of the Orthodoc, a pre-operative planner, and the RoboDoc, a surgical tool. The RoboDoc (see Figure 10) was developed by Integrated Surgical Systems in the 1980s and is currently used in hospitals in Europe. The system was used in over 1000 cases at a hospital in Frankfurt, Germany from 1994 until 1998 (Borner et al. 1998). It was first used in total hip replacements, and its use is now extended to revision hip replacement and total knee replacement. The RoboDoc surgical system uses Computer Tomography to obtain structural information of the surgical object pre-operatively. The OrthoDoc allows the surgeon to construct a pre-operative bone-milling procedure. Before the bone-milling procedure, registration is conducted to match the pre-operative data with the physical location of the surgical object. In total hip replacement, RoboDoc is used to cut the femoral cavity precisely. During the surgery, the surgeon controls the robot via a hand-held terminal. The robot is manually guided to the desired location by the surgeon. Once the end-effector is brought to the designated location, the RoboDoc cuts the bone automatically (Pransky 1997).

Figure 10. RoboDoc for milling implant cavity in hip replacement surgery (Courtesy of Integrated Surgical Systems, Davis, CA.)

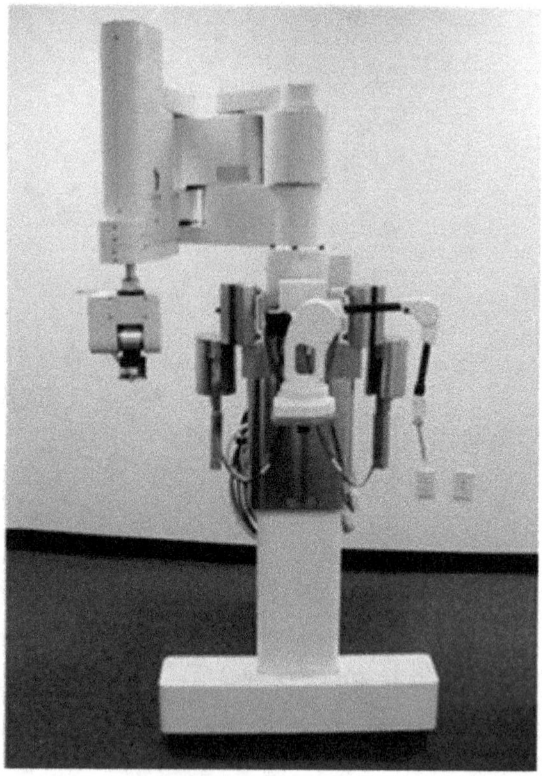

The surgeon can see inside the bone and monitor the robotic cutter's position within the bone through a "real time monitor". Once the cutting is completed, the surgeon resumes the normal manual procedures. The RoboDoc has two main advantages over manual procedure. First, the femoral pocket is more accurately formed. Second, the pre-operative planning with CT-scan data allows the surgeon to optimize the implant size and placement for each patient. Nevertheless, there are also two disadvantages inherent in the RoboDoc system. First, the registration process requires the use of registration pins on the bone surface. A traumatic pin-placing procedure and a slow pin-finding registration process are required. Second, a complex and time-consuming method is used to fix the femur to the base of the robot (Pransky 1998).

3.2.2 Acrobot

Developed by Professor Brian Davies and colleagues, Imperial College London, 1999, Acrobot (see Figure 11) is designed for Total Knee Replacement. The robotics system uses CT-scans to obtain pre-operative data for pre-operative planning. The Acrobot system employs the concept of "Active Constraint". During the operation, the surgeon stays in control, but the robot applies resistive force to the surgeon's hand based on the defined surgical region.

This system has several advantages over manual procedures (Jakopec et al. 2001). First, the use of "active constraints" confines the surgeon's actions to the safe region. Second, the Acrobot system assists the surgeon to achieve accurate cuts

Figure 11. Acrobot system (Courtesy of Imperial College London)

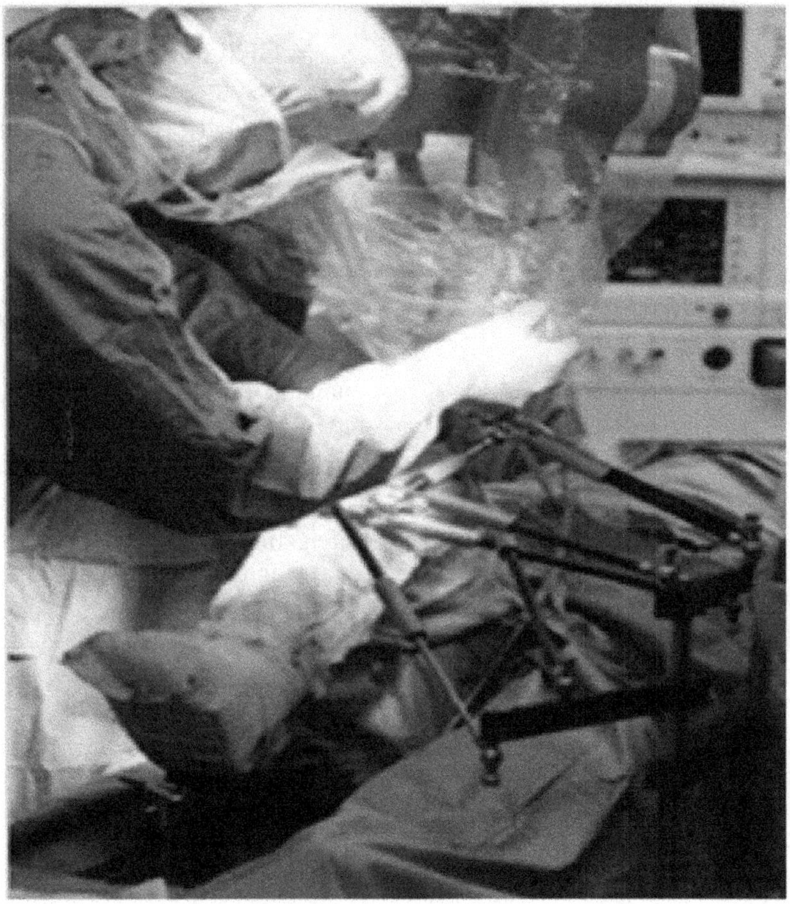

and trajectories while ensuring that the pre-operative plan is achieved. As of April, 2005, the Acrobot surgical system is not yet approved by the USFDA. No comprehensive economic analysis is available for the Acrobot system.

3.2.3 CASPAR

Computer Assisted Surgical Planning and Robotics is an integrated system developed by a German company OrtoMaquet (see Figure 12). The system has been used for implantation of hip prostheses in total-hip-replacement (THR), as well as for anterior cruciate ligament reconstruction (Petermann et al. 2000). A CASPAR procedure consists of four distinct phases. First, a surgery is performed to insert rigid bodies, commonly pins, at different locations around the knee (Grueneis et al. 1999). The procedure requires local or total anesthesia, and a tibial pin is placed 8 to 10 cm above and below the knee joint. This procedure is required for registration and referencing. Second, a CT-scan is performed to retrieve pre-operative data and images of the knee. The surgeon can then perform pre-operative planning with the data.

Figure 12. CASPAR System (Courtesy of orto-Maquet)

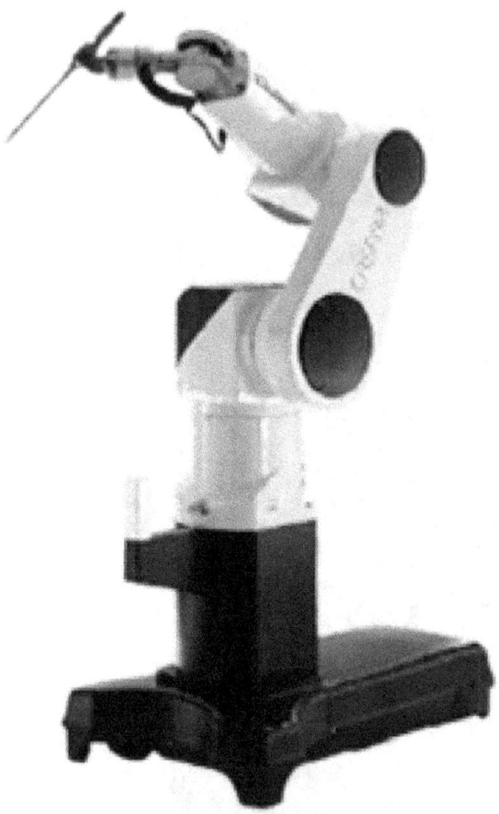

It generally involves measuring the femur and the tibia and planning the femur and tibia tunnels in a 3D environment. Finally, the robotic can do the tunnel drillings automatically under surgeon supervision. The CASPAR system offers precise tunnel drillings that cannot be done as accurately with human hands. Also, the CT-scan allows pre-operative planning and gives surgeons the chance to optimize their operation. Nevertheless, the system suffers the disadvantages of increased cost, increased surgery time, and an additional surgery for pins- implementation.

3.2.4 CRIGOS - Compact Robot System for Image-Guided Orthopedic Surgery

CRIGOS is an ongoing project by the biomedical community to build small and cost-effective robots designed specifically for surgery (Brandt et al. 1999). Current systems such as RoboDoc and CASPAR use robots based on industrial manufacturing robots. Scientists and engineers involved in the CRIGOS project aim to design specific small, cheap and sterile robots as well as the accompanying software and pre-operative navigation systems for specific surgeries (Brandt et al. 2000). Certain prototypes have been developed, but the project is still under development.

3.3 Urology

Urology robotics was slower to develop because it is more difficult to build robots to operate on soft tissue organs due to their higher deformability and mobility. Although these difficulties delayed Urology robotics development, innovative research has produced several systems either applicable to urology.

3.3.1 PROBOT

The first Urology robot was the PROBOT, introduced in 1989 by a group at the Imperial College in London for performing robot-assisted transurethral resection of the prostate (TURP) (Davies et al. 1989; Davies et al. 1991). The robot serially cored the periurethral prostatic tissue, while hemostasis was achieved manually using electrocautery after completion of the tissue resection.

The device never achieved widespread use; however, since then many minimally invasive techniques for the treatment of benign prostatic hyperplasia have been introduced. This would seem to indicate a desire to replace the standard TURP by a less invasive strategy. Transrectal ultrasound (TRUS) is used in many of these new techniques for intraoperative monitoring and image-guided robot assistance. In 1994, Potamianos et al. investigated a robotic system to assist the urologist with intraoperative percutaneous renal access (Potamianos et al. 1994). They employed a passive, encoded arm equipped with electromagnetic brakes, mounted onto the operating table. The access needle was manually positioned as prescribed by a computer, which triangulated the calyx location from multiple C-arm X-rays (Potamianos et al. 1995). In vitro experiments evaluating system performance demonstrated a targeting accuracy of less than 1.5 mm.

3.3.2 Urology Robotics Laboratory

In 1995, a research group headed by Russell Taylor at IBM developed the Remote Center of Motion (RCM) concept and implemented it on the LARS robot (Taylor et al. 1995). The RCM is a mechanism used by the surgical manipulator to enable and facilitate the pivoting motion of instruments about a fixed point in space, normally located on the instrument itself. This mechanism enables minimally invasive instruments to preserve a consistent entry point, or port, throughout the entire procedure. This technique was developed by observing the surgeon's natural motion in manual laparoscopy, but there are very few surgical systems not using this principle today. The Urology Robotics laboratory group at the Johns Hopkins School of Medicine (Baltimore, MD) has developed the Percutaneous Access of the Kidney robot (PAKY) (Stoianovici 2001).

PAKY is an active and radiolucent needle driver (Stoianovici et al. 2002). Originally held by a passive arm, the needle in the needle driver was manually positioned under C-Arm guidance. It was then locked in place and the needle inserted automatically under the surgeon's joystick control. The next step was to automate the needle orientation procedure, which was accomplished with the addition of the RCM module. The RCM supports and orients the PAKY driver while maintaining the fixed location of the needle tip. The combination of the two robotic systems enables the surgeon to automatically place the needle at a target specified on the computer screen by the urologist, based on the fluoroscopic images. The PAKY-RCM offers an unquestionable improvement in needle placement accuracy and lowers procedure time while reducing radiation exposure to patient and urologist (Cadeddu et al. 1998a; Cadeddu et al. 1998b). Clinically, the PAKY-RCM system was tested in local as well as in several transatlantic telesurgical cases (Bauer et al. 2001).

The PAKY-RCM system was also used under CT guidance with the Laser-Based CT/MR Registration. This method of registering the patient to the image makes use of the laser markers readily available on any CT scanner. Once registered in this manner, the organ of interest can be targeted precisely. The procedure was successfully used for biopsies and radio-frequency (RF) ablation of targets on the kidney and spine, as well as for nephrostomy tube placements.

3.3.3 The Automated Endoscopic System for Optimal Positioning (AESOP; Computer Motion, Goleta, CA)

The AESOP, which allows automated six DOF control of the laparoscope, was introduced in the mid-1990s (Nedas 2005). The AESOP was the first active robot used for visceral surgery. This robotic arm uses speech recognition to maneuver the endoscope, thereby eliminating the need for a member of the surgical staff to control the camera manually. The advantage of the AESOP system is that it stabilizes visualization of the internal surgical field (Guillonneau and Vallancien 2000). Laparoscopic images with the AESOP are steadier, with fewer camera changes and inadvertent instrument collisions than might occur by an inexperienced human assistant. The primary reason for AESOP's success is that it is simple to operate and, at the same time, reliable and safe. Additionally, the robot is easy to disconnect intraoperatively in the (highly unlikely) event that problems should arise. AESOP is routinely used at several institutions and in many surgical disciplines including a variety of laparoscopic urologic procedures (Partin et al. 1995). The camera is significantly steadier under robot control and neither operative set-up nor breakdown time is increased with the use of a robotic assistant (Kavoussi et al. 1995). EndoAssist (Armstrong Healthcare Limited, High Wycombe, UK) is a similar device, except that the laparoscopic camera is controlled by the surgeon's head movements. Although AESOP and EndoAssist can position an endoscope more effectively and accurately than human assistants ((Kavoussi et al. 1995), they are unable to overcome two of the major factors that limit laparoscopic surgery: the lack of 3D vision and a loss of freedom of motion. Two companies have sought to address this problem of lost dexterity by developing "master-slave" manipulators for minimally invasive surgery. These devices are composed of three components: a surgical cart, a computerized vision system, and a surgeon's console. The surgical cart, usually positioned close to the operating table, has three arms: one for the right hand, one for the left hand, and a middle arm to hold the laparoscope. The two surgical arms maneuver detachable instruments with articulated tips that allow six or seven degrees of freedom, in essence becoming the artificial hands of the surgeon.

3.3.4 da Vinci (Intuitive Surgical Inc)

The prototype of the da Vinci Surgical System was developed by the Stanford Research Institute in the 1980's and was funded by the US Army who planned to develop a robotic system to perform battlefield surgery remotely by surgeons behind the lines. The da Vinci Surgical System was first used in human surgery in 1997 in Europe. FDA approved the use of the da Vinci Surgical System in 2000 (Loulmet et al. 1999). The first robotic radical prostatectomy was performed in Europe in 2000. By the end of 2006, 559 da Vinci Surgical Systems have been installed worldwide, mostly in America.

Over 70,000 robotic surgeries have been performed in 2006. Less than 1% of the radical prostatectomies were performed by robots in 2001 in America. In 2006, it rose to over 40% or 30,000 robotic radical prostatectomies. The da Vinci Surgical System was initially designed for coronary arterial bypass surgery through small cut wounds. American Urologists quickly used the robots to perform radical prostatectomies. The daVinci system consists of a three-armed robot connected to a remote surgeon's console (see Figure 13). The surgeon operates the system while seated at the non-sterile console. The vision system is controlled using foot pedals and displays a 3D image of the surgical field similar to that seen in the case of open surgery.

The surgeon's movements are translated in real-time to movements of the pencil-sized instruments in the surgical field. These enter the patient through small ports (on the order of 5-10 mm). Two of the robotic arms are used for manipulating the surgical instruments, while the third arm manages the laparoscope. The instruments (needle holders, scissors, dissectors, scalpel, etc.) have seven DOF and are maneuvered by a robotic wrist. There are a number of safety features built into the da Vinci system.

When the camera system is being moved, the instruments stay in position, the system goes into standby mode when the surgeon takes his head away from the viewing console, and the instruments can be frozen while repositioning the robot arms. The whole system can be shut down instantly if required. The da Vinci Surgical System is expensive and has costly consumables. It is mainly used for complex laparoscopic surgery on a cost effectiveness basis. There is an intense competition between the two commercially available robots used to perform remote telerobotic heart and abdominal surgery: the da Vinci system and the ZEUS system (Computer Motion, Goleta, CA) (Sung and Gill 2001). Despite their similar appearance, these systems differ from each other in very important ways.

Figure 13. da Vinci Surgical System (Courtesy of Intuitive Surgical)

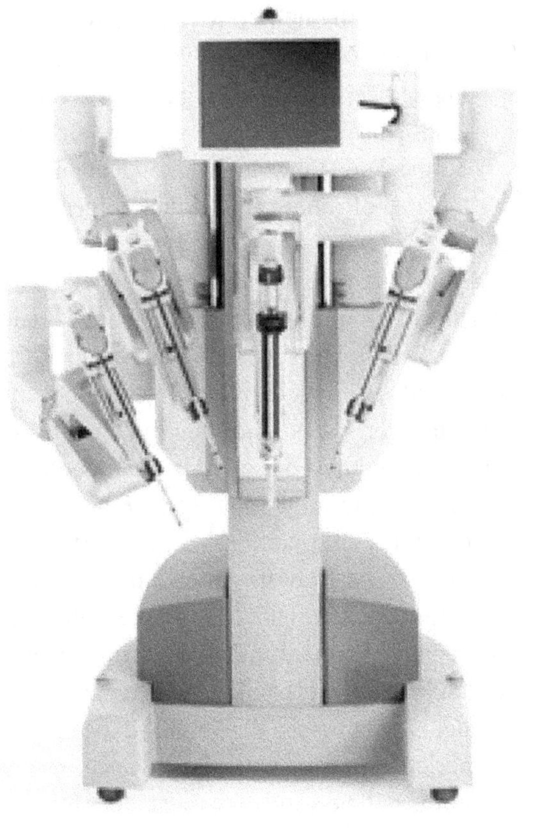

3.3.5 ZEUS™ (Computer Motion)

The Zeus system from Computer Motion (the makers of AESOP) (see Figure 14) is similar to daVinci, but Zeus consists of a combination of three robot arms and surgeon's control console. The system uses one AESOP for the laparoscope and the other two arms hold surgical instruments. Compared to daVinci, Zeus appears safer and requires significantly less preoperative setup.

Until recently, Zeus had exhibited lower dexterity for manipulating the tools within the patient. However, the company seems to have recently addressed this with the MicroWrist line of end-effector tools. Zeus received FDA clearance as a general laparoscopic tool in September, 2002. The system has been experimentally used in a number of operations including urology cases (Breda et al. 2001).

3.4 Cardiology

Robotically-assisted heart surgery is the latest advance in the move to place open heart surgical procedures into the category of minimally invasive surgery to minimize the extent and the trauma of cardiac surgery as much as possible. The entry of robotics into the field of cardiology has been a gradual scaled process that can be broken down into four levels (Kypson et al. 2003):

1. **Direct Vision and Mini-Incisions:** This level encompasses conventional minimally invasive surgery. Using, mini-sternotomies, parasternal incisions, and mini-thoracotomies, the first minimally invasive aortic valve operations were performed through this conventional tactic. However, this technique in

Figure 14. Zeus telemanipulator system operating console (Courtesy of Computer Motion Inc.)

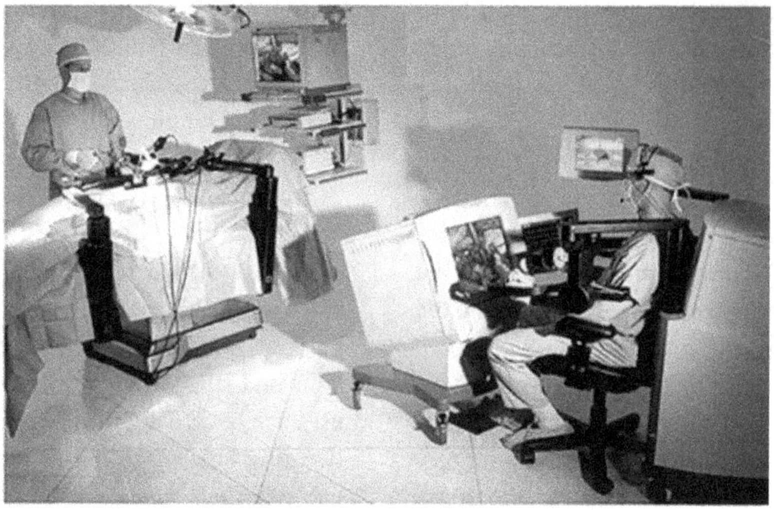

cardiac surgery proved that operations could be performed through smaller incisions and therefore, acted as a springboard to the next level.

2. **Video-assisted and Micro-Incisions:** Videoscopy could be used to enlarge the vision of the surgical field but the benefits were limited to simple repairs and replacement. Complex repairs, such as quadrangular resections and chordal replacements, were performed with few complications. Nonetheless, surgeons still awaited 3D imaging and robotic instruments.
3. **Video-Directed and Port Incisions:** Imaging modalities such as CT scans have contributed greatly to the planning and registration of procedures. This stage saw the first integration of a robotic arm in carrying out procedures. In addition to the robotic arm, surgeons use an endoscope and a conventional two-dimensional monitor to view the procedure. Devices such as AESOP enabled surgeons to increase precision and control for hand tremors through robotic actions triggered by voice-activated commands. Though the complexity of the devices cannot be compared to the Level 4 technology, these endoscopic techniques allows a wide range of minimally invasive surgery.
4. **Video-Directed and Robotic Instruments:** Current cardiac surgery involves the use of complex 3-D imaging techniques and robotic devices. Although a few years ago Computer Motion's Zeus Surgical System was the leading device in cardiac surgery, the economic landscape has changed, and now Intuitive Surgical's da Vinci Surgical Device dominates the area of robotics in cardiac surgery.

Minimally invasive surgery is carrying out using an endoscope through a very small incision (Fogoros 2011). The surgeon visualizes the surgical area and the image provided by the endoscope on a computer screen and then manipulates the surgical instruments directly via a computer console. Endoscopic surgery has some limitations in more complicated operations such as heart surgery (Fogoros 2011). Robotic procedures have been successfully performed, for instance, in mitral valve repair (Mehmanesh et al. 2002), in repairing Atrial Septal Defects (ASD) and in repairing Patent Ductus Arteriosus (PDA). It is expected that as experience and technology advance, robotic procedures will be applied to most other forms of heart surgery. Mitral valve surgeries have always been a staple of modern cardiac surgery. However, until now, the only way to either repair or replace a heart valve was to open the chest of the patient and put them on a heart-lung oxygenator.

With the advent of new robotic technologies, many centers around the world are beginning to turn to an alternative--minimally invasive mitral valve surgery using videoscopic tools like AESOP and robotic tools like the da Vinci Surgical System (Nifong et al. 2003). Rather than having to open a patient's chest, patients only have to receive small incisions in their chest roughly 2.5 inches long. Robotic arms can be manipulated inside the cavity, performing operations that may otherwise be be too difficult due to the misfortunate location of physical abnormalities or poor locations. Previously, less invasive procedures were brought into ever widening use with endoscopic instruments, but there were downsides associated with them. They only had 4 degrees of freedom, while 6 degrees of freedom are required to fully allow free function and orientation in space. Working through a fixed entry point causes problems in the standard endoscopes, both for the surgeon and for the machine.

However, the new computer-enhanced instrumentation system allows machines to overcome many of these limitations, including fixed entry points and also the deterioration of human motor skills that is commonly associated with endoscopic surgery (Mohr et al. 2001). The new systems can be classified according to the tasks they help perform. The first group is comprised of tools that hold and position various instruments, including the Automated Endoscopic System for Optimal Positioning (AESOP 3000), which is used to guide an endoscope by voice activation. The second group includes telemanipulators invented to perform fine manipulations and movements under remote conditions. With these, the operator can remotely direct the robot's movements. By these systems, the surgeon is placed remotely away from the patient, seated at an operation console with a 3D camera array. This allows the surgeon to see everything as clearly as if s/he were actually operating on an open-heart, all while removing the need to actually open the chest and expose the patient to operating room air. The surgeon is able to not only access hard-to-reach places, but also combat the limitation in vision encountered by traditional endoscopy. Recently, a new technology has allowed for a minimally invasive treatment of occluded arteries–da Vinci Cardiac Revascularization (dVCR). The dVCR is currently being used for Single-Vessel Small Thoracotomies (SVST) and Multi-Vessel Small Thoracotomies (MVST). Both procedures use the da Vinci robot from start to finish, from the harvesting the internal mammary artery, to coronary anastomosis and to revascularization. The biggest benefit of dVCR is the ability to be minimally invasive. The 3D imaging system enables surgeons to perform procedures inside a closed chest, with hand movements that parallel conventional surgery. Only a few incisions are made into the patient, ensuring a much faster recovery time.

Another upside of this new technology is the ability to use stabilizers to ensure that the patient can function without use of the lung-heart machine, something that has often been cited as being more detrimental to immediate patient post-operative health than anything else. With the ability to both avoid open heart surgery and pump use, da Vinci cardio revascularization has fast become one of the most exciting cardiac surgery innovations out there.

3.5 Radiosurgery

Radiosurgery is a medical procedure that allows non-invasive treatment of benign and malignant tumors. It is also known as stereotactic radiotherapy, (SRS) when used to target lesions in the brain, and stereotactic body radiotherpy (SBRT) when used to target lesions in the body. For more than 30 years, physicians have been using stereotactic radiosurgery to destroy tumors in the brain. While the procedure does not remove a tumor, it can destroy tumor cells or stop growth of active tissue. In stereotactic radiosurgery, high dose focused radiation beams are fired into a tumor from many angles. With older systems, such as the Gamma Knife, a metal frame is screwed into the patient's head to immobilize the patient during treatment. Wearing this head frame is uncomfortable and limits the potential application of the Gamma Knife to single treatments, used against lesions within the brain or skull base.

CyberKnife (see Figure 15) from Accuray Inc. (Sunnyvale, CA, USA) offers all of the advantages of radiosurgery (Adler et al. 1999; Chang and Adler 2001; Quinn et al. 2002; Kuo et al. 2003), but without the need for a metal head frame. Instead, patients are fitted with a more comfortable flexible mesh mask for tumors in the head or a body cradle for spinal tumors or other tumors outside the head (Gerszten et al. 2003a; Gerszten et al. 2003b; Brown et al. 2007). Because the robotic arm automatically corrects for patient movement during treatment, the CyberKnife can be used anywhere in the body where radiation is effective. CyberKnife was developed by a team of physicians, physicists and engineers at Stanford University. The CyberKnife technology was approved by the FDA for intercranial applications in August 1999 and received full-body approval in August, 2001. It has been used to treat more than 20,000 patients worldwide. The CyberKnife Radiosurgery System consists of a lightweight linear accelerator mounted on a 6-axis industrial robotic arm, two fluoroscopy systems to provide continuous stereoscopic digital imaging, and a treatment couch with electronic and manual controls for positioning the patient. Although radiosurgery has been used primarily for intracranial lesions, recent advances in robotics and image-guided therapy have extended this treatment to extracranial disorders.

Accurate irradiation of extracranial targets typically requires percutaneous implantation of 1 to 5 small gold fiducials in and around the target site, with implantation guided by images from computed tomography (CT) or fluoroscopy. A treatment plan for radiosurgery is then developed that relies on images from CT, magnetic resonance imaging (MRI), and/or CT angiography to determine the paths of the radiation beams. During treatment delivery, tracking of the fiducials with real-time x-ray imaging enables dynamic aiming of each radiation beam, which compensates for

Figure 15. CyberKnife robotic radiosurgery system. (Courtesy of Accuray Inc.)

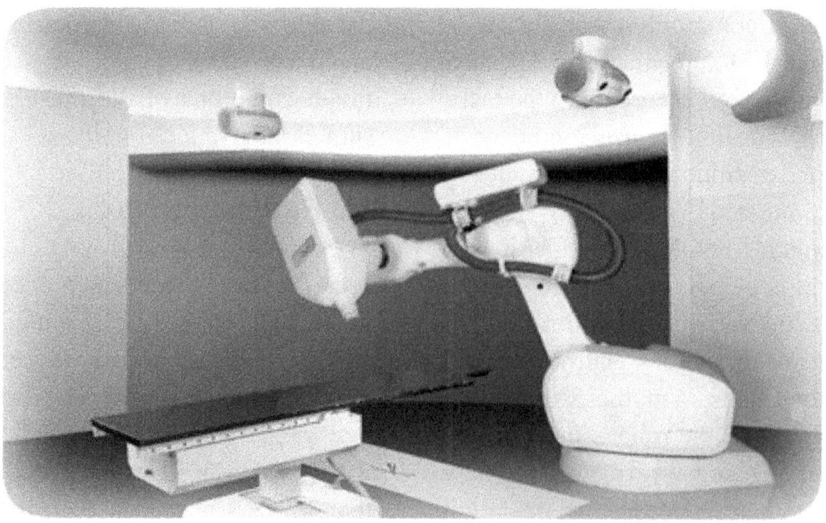

motion due to patient breathing and subtle changes in position (Chang and Adler 2001). Patients undergo CyberKnife radiosurgery in a shielded room with treatment overseen by a human operator who controls and monitors the procedure via a computer workstation and closed-circuit cameras in an adjacent shielded control room. The patient lies on the treatment couch and is positioned at the center of the imaging system with guidance from x-ray images. When treatment begins, the robotic arm moves the linear accelerator in a sequence of positions around the patient, stopping at each beam position, or node, to lock in on the treatment site. Typically, the target site is irradiated with a total of 80 to 120 beams. The CyberKnife is equipped with 12 collimators ranging from 5 to 60 mm in size, and it can fire beams from 101 specific nodes, each with 12 approach angles, offering up to 1212 possible beams for treatment flexibility. The robotic system automatically adjusts the position of the linear accelerator to correct for motions or movements as large as 10 mm of translation and/or 5 degrees of rotation. In addition, the human operator can interrupt treatment before patient motion or movement reaches maximum correction capacity or for other reasons, make corrections, and instruct the system to resume treatment.

Treatment time may range from 30 to 120 minutes and can be given in one or several fractions, depending on the condition being treated and the size of the affected area. Fiducial implantation and radiosurgery for non-neurological indications are usually performed on an outpatient basis and may involve several specialists, including thoracic surgeons, radiation oncologists, and radiologic technicians (Chang and Adler 2001).

3.6 Ophthalmology

Many surgical operations on the eye require precise microsuturing skills. In 1989, the first ocular robotic systems described by Guerrouad and Vidal was developed (Guerrouad and Jolly 1989; Guerrouad and Vidal 1989; Guerrouad and Vidal 1991; Hayat and Vidal 1995). It was called the Stereotaxical Microtelemanipulator (SMOS) and included a spherical micromanipulator mounted on a XYZ stage, which allowed 6 degrees of freedom. In 1997, Steve Charles and collaborators described

Figure 16. Steady-hand robot for microsurgical augmentation (Courtesy of Russell Taylor, Johns Hopkins University, Baltimore, MD.)

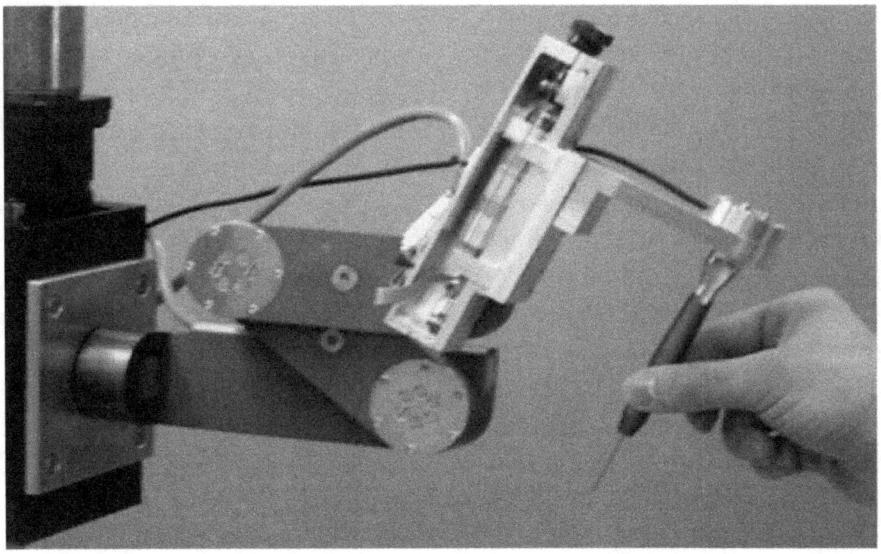

a new telerobotic platform which was called Robot Assisted MicroSurgery (RAMS) (Charles et al. 1997) (see Figure 16). Yu et al (1998) developed a patented spherical manipulator, similar to Guerrouad and Vidal, for intravascular drug delivery, implantation of microdrainage devices and the intraretinal manipulation of microelectrodes.

Taylor and colleagues at Johns Hopkins University developed the "Steady-Hand" robot (see Figure 16) for microsurgical augmentation (Taylor et al. 1999). This robotic instrument requires the physician to actually hold and manipulate the tool with the aid of the robot.

Inanimate studies evaluating the precision of suture placement have demonstrated an advantage with robotic assistance. The Steady-Hand robot offers several advantages compared to other systems. These include (Taylor et al, 1999): 1) simplicity; 2) potentially cheaper implementations; 3) a more direct coupling to the human's natural kinesthetic senses; 4) straightforward integration into the existing application environment; and 5) greater "immediacy" for the human operator. The principal drawbacks are the loss of the ability to "scale" positional motions and the loss of the ability to manipulate objects remotely.

3.7 Maxillofacial

Maxillofacial surgery is concerned primarily with operations on the jaws and surrounding soft tissues. Maxillofacial surgery often requires the manipulation of the skull bone including drilling, cutting, shaping, and repositioning operations (Burghart et al. 1998). The first experimental set up consisting of a surgical robot (RX 90), an infrared navigation system, a force torque sensor, a surgical saw, a pointer, a fixation unit and a workstation were tested at the cardiac animal laboratory of the University of Heidelberg. The complete working procedure necessary to perform robot assisted surgery was evaluated starting with data acquisition, medical image data processing, surgical planning and ending with the intraoperative support of the surgeon by an intraoperative navigation system and a surgical robot. The CT-scan and the surgery on a pig cadaver were performed at the University of Heidelberg. The surface models and the cutting trajectories of the planned osteotomies were generated at the University of Karlsruhe.

4. LIMITATIONS OF SURGICAL ROBOTICS

There are, of course, many limitations to the application of robotics in surgery. Currently, the mechanical design of manipulators limits dexterity, particularly for minimally invasive procedures with severe size constraints. There is considerable room for improved kinematic configurations, as well as more compact and efficient actuator and transmission technologies. In terms of sensing and control, robots are controlled by computers and thus share many of their all-too-familiar shortcomings, especially for autonomous operation. Robots follow instructions literally, are unable to integrate diverse sources of information, and cannot use qualitative reasoning or exercise meaningful judgment. Although complex 3-D imaging information can be pre-processed to allow execution of very precise tasks, robots have a limited ability to use information from disparate sensors to control behaviour during the course of a procedure. Increasing computational power may improve robot control capabilities, but the resulting complexity makes it increasingly difficult to program and debug these systems.

The extremely high cost required for procurement and maintenance of a robotic system is a distinct disadvantage. The current robotic systems are heavy and bulky, and the initial set-up is particularly cumbersome for an inexperienced team. Procedures like instrument exchange, aspiration and irrigation, clips and endoscopic staplers application cannot be performed by the robot. An experienced assistant is often required to carry out essential procedures on the operating table.

5. FUTURE CHALLENGES OF SURGICAL ROBOTICS

Computer-aided surgery is obviously much more economical but lacks the ability to perform actual tasks. The additional cost of the robot must be justified in improved outcomes. The use of passive robots will decline in all likelihood because of their limited roles, as the power and cost of new technologies lead to more sophisticated machines at lower prices. Robots are not intended to replace or automate the surgeon but to enhance the surgeons' own skills and this must be emphasized to both the surgical fraternity and the public (Liverneaux et al. 2009). Currently, many image-guided surgical applications use off-the-shelf industrial robot manipulators. This speeds development and reduces costs, but these devices have not been optimized for the characteristics of specific surgical tasks. For example, most industrial robots are designed for good repeatability but may lack sufficient positional accuracy. Similarly, assistive systems that share control between the robot and human surgeon would benefit from the development of low impedance manipulators in place of highly geared, stiff industrial arms. Other advantages that can accrue to specialized designs include improved sterility and compatibility with imaging systems (e.g. transparency to X-rays). In teleoperated systems, access constraints have always necessitated the development of new manipulator configurations, but kinematic structures and actuator technologies are far from being perfect.

These technologies also limit the development of microrobots for medical applications. Image-guided procedures have been an area of great success for robotic surgery, but there are many unresolved issues. Improved automatic segmentation and planning systems promise to improve efficiency and accuracy. Areas for improvement in registration include elimination of invasively placed fiducials and methods for nonrigid registration and tracking of tissue deformation in real time. The use of 2-D imaging modalities such as ultrasound in combination with 3-D tracking may lower costs and enable wider application of image-guided techniques. The interdisciplinary field of surgical robotics requires the close co-operation of medical doctors and engineers. Surgeons must be involved in the research from a very early stage to identify the targets and set the guidelines for the development.

Surgical procedures are not as easily definable as assembly tasks, and they require the integration of human and robotic skills into a common framework (Richard 2004). Physicians often work on the human-machine interface as an integrated part of the robot workcell. The control features of the system should also consider that the operator is not a robot specialist. Time constraints apply to the setup and registration procedures required before the operation and should rely minimally on the human operator to allow for fast intervention in case of emergency. Minimally invasive procedures are an advantage to the patient. Robotic systems, when combined with laparoscopy, have practical applications in surgery and could potentially benefit patients more than surgeons. However, high cost remains the main challenge of robotic surgery systems; outcome studies need to be completed to reconcile economics and clinical interest. Even if telemanipulation systems become cheap and "up to scratch", it is unlikely that they will provide assistance with increasing intelligence, much like an autopilot in an aircraft. Robotic telesurgical systems will help surgeons to extend their healing skills to places within the body that are out of reach (Liverneaux et al. 2009).

Funding for research in robotics should be divided into a number of areas to ensure continuous development in areas that are an integral part of the process. This includes—beside the obvious hardware of the robot— data fusion, haptics, image registration, and the development of end effectors. These end effectors include not only instruments but also devices such as laser and emerging technologies. Funding must also be directed toward large outcome surveys that might require the establishment of new endpoints. These endpoints will need to be good predictors of long-term outcomes; without these the introduction of new technology will be impeded. Digital patient databases will overcome the dependence of the surgeon on varied patient exposure for meeting anomalies and idiosyncratic conditions. The extra time of setting up and draping will decrease as the operating teams become familiar with instrumentation. Newer generations of robots will be smaller and consume less space, perhaps with the ability of being built into the operating theater. The ability of the surgeon to be remote from the area where the surgery is taking place or working at another console in the same theater has implications for training and minimizing learning curves. If the surgeon is to be remote then at present s/he will still require the presence of a trained surgeon in the theater for port placement, assistance, and emergencies. Although the field of medical robotics has produced much research results and many prototypes, rather few systems really entered the operating theatre and demonstrated their clinical usefulness. Indeed, one major difference with other applications of robotics is the need for the proof of clinical added-value of the medical robot.

Demonstrating that the robot can safely and accurately machine a bone surface or insert a screw is not sufficient: one must prove with quantitative data that the system has a clinical advantage over other existing techniques. The advantage may be in less complications, reduced hospitalization time and cost, higher diagnostic efficacy, etc. Such a demonstration requires careful evaluation involving a series of patients in different centres of the country, regulated by laws and ethical codes where it takes place. Developing medical robots thus appears to be a complex, expensive and long process which requires a constant collaboration of clinics, academic research labs and industrial development labs. The interaction between these different actors has to start in the very early phases of the project in order to find cost-effective, innovative technical solutions that are compatible with a clinical use and that may lead to significant breakthroughs. Safety is extremely important in automation, to prevent any kind of injury and damage. In industry, the major approach is to keep people and vulnerable structures away from the workspace of the robots. Naturally this is impossible in the case of medical applications. These robots directly contact the human body, and in many cases, the operating surgeon is also within close proximity. In every system, additional safety measures should be applied concerning both hardware and software. In surgical robotics, emergency action does not only mean halting the robot, but in certain applications immediate recovery is required in safe, further complicating safety requirements (Dohi 2005). Another great challenge is the adaptation to changing conditions. This encompasses soft tissue analysis, tissue manipulation, motion of body fluids and the radical changes in the Operating Room (OR) environment (rearrangement of the setup, lighting conditions, etc). These safety concerns prevented the approval of many automated interventional systems, and have lead the research community towards human-integrated control solutions, such as telesurgery and hands-on surgery (Kazanzides et al. 2008).

The interdisciplinary field of surgical robotics requires the close co-operation of medical doctors and engineers. Surgeons must be involved in the research from a very early stage to identify the targets and set the guidelines for development. Robots promise to become the standard modality for many common procedures, including hip replacement, heart bypass, and abdominal surgery. Surgeons, particularly researchers working to enhance and extend the field, will need to become familiar with robotic technology. The same is true for robotics researchers: Creating effective systems requires understanding the demands of surgical procedures and the culture of surgical practice

6. CONCLUSION

Medical robotics opens up new perspectives in promoting the right to physical health. Robots overcome limitations in the sensory-motor capabilities of human doctors. They may perform surgery and rehabilitation tasks faster and with greater precision than human operators. Properly used robots alleviate both post-surgery pain in patients and the fatigue of health care personnel. The most significant role of medical robots will most likely be to perform tasks that are otherwise impossible, such as enabling new microsurgery procedures by providing high-dexterity access to small anatomical structures, integrating imaging modalities into the Operating Room, providing functional replacements for lost limbs, and enabling new human–machine interfaces and techniques for delivering neuro-rehabilitation therapy. Robots are never likely to replace the highly evolutionized human hand. They will retain the benefits of the human hand along with its superlative optimization to achieve the goal of optimal precision and predictability. By combining robotic technology with human skills, robot systems will allow the performance of more precise and more dexterous operations. Present surgical training programs must include some robotic training in their curriculum, keeping in the mind its exploding potential and future use by the present and next generation of microsurgeons.

REFERENCES

Adler, J. R., Murphy, M. J., Chang, S. D., & Hancock, S. L. (1999). Image-guided robotic radiosurgery. *Neurosurgery*, *44*(6), 1299–1307.

Bann, S., Khan, M., & Hernandez, J. (2003). Robotics in surgery. *Journal of the American College of Surgeons*, *196*(5), 784–795. doi:10.1016/S1072-7515(02)01750-7

Bargar, W. L. (2007). Robots in orthopaedic surgery: past, present, and future. *Clinical Orthopaedics and Related Research*, *463*, 31–36.

Bauer, J., Lee, B. R., & Stoianovici, D. (2001). Remote percutaneous renal access using a new automated telesurgical robotic system. *Telemed Journal and Electronic Health*, *7*, 341–346. doi:10.1089/15305620152814746

Benabid, A. L., et al. (1987). Computer-driven robot for stereotactic surgery connected to CT scan and magnetic resonance imaging. Technological design and preliminary results. *Applied Neurophysiology*, *50*(1-6), 153–154.

Bolmsjo, G., Neveryd, H., & Eftring, H. (1995). Robotics in rehabilitation. *IEEE Transactions on Rehabilitation Engineering*, *3*(1), 77–83. doi:10.1109/86.372896

Borner, M. (1998). Experiences with the ROBODOC system in more than 1000 cases. In *Computer Aided Radiology and Surgery* (pp. 689–693). Amsterdam, The Netherlands: Elsevier.

Brandt, G., Radermacher, K., & Zimolong, A. (2000). CRIGOS: Development of a compact robot for image-guided orthopedic surgery. *Der Orthopade*, *29*(7), 645–649. doi:10.1007/s001320050506

Brandt, G., Zimolong, A., & Carrat, L. (1999). CRIGOS: A compact robot for image-guided orthopedic surgery. *IEEE Transactions on Information Technology in Biomedicine*, *3*(4), 252–260. doi:10.1109/4233.809169

Breda, G., Nakada, S. Y., & Rassweiler, J. J. (2001). Future developments and perspectives in laparoscopy. *European Urology*, *40*, 84–91. doi:10.1159/000049753

Brouwer, I., Ustin, J., Bentley, L., Sherman, A., Dhruv, N., & Tendick, F. (2001). Measuring in vivo animal soft tissue properties for haptic modeling in surgical simulation. In Westwood, J. D. (Eds.), *Medicine meets virtual reality*. Amsterdam, The Netherlands: IOS Press.

Brown, W. T., Wu, X., & Wen, B. C. (2007). Early results of CyberKnife image-guided robotic stereotactic radiosurgery for treatment of lung tumors. *Computer Aided Surgery*, *12*(5), 253–261.

Burckhart, C. W., Flury, P., & Glauser, D. (1995). Stereotactic brain surgery. *IEEE Engineering in Medicine and Biology*, *14*(3), 314–317. doi:10.1109/51.391771

Burghart, C. R., Muenchenberg, J. E., & Rembold, U. (1998). A system for robot assisted maxillofacial surgery. *Studies in Health Technology and Informatics*, *50*, 220–226.

Cadeddu, J. A., Stoianovici, D., Chen, R. N., Moore, R. G., & Kavoussi, L. R. (1998b). Stereotactic mechanical percutaneous renal access. *Journal of Endourology*, *12*, 121–125. doi:10.1089/end.1998.12.121

Cadeddu, J. A., Stoianovici, D., & Kavoussi, L. R. (1998a). Robotic surgery in urology. *The Urologic Clinics of North America*, *25*(1), 75–85. doi:10.1016/S0094-0143(05)70434-1

Camarillo, D. B., Krummel, T. M., & Salisbury, J. K. J. (2004). Robotic technology in surgery: past, present, and future. *American Journal of Surgery*, *188*(4A Suppl), 2-15.

Capek, K. (2001). *R.U.R. (Rossum's Universal Robots)*. Mineola, NY: Dover Publications.

Chang, S. D., & Adler, J. R. (2001). Robotics and radiosurgery--The cyberknife. *Stereotactic and Functional Neurosurgery*, *76*(3-4), 204–208. doi:10.1159/000066719

Charles, S., Das, H., Ohm, T., Boswell, C., Rodriguez, G., Steele, R., & Istrate, D. (1997). Dexterity-enhanced telerobotic microsurgery. *ICAR '97 Proceedings of IEEE, 8th International Conference on Advanced Robotics.*

Chinzei, K., et al. (2000). MR compatible surgical assist robot: system integration and preliminary feasibility study. In *Proceedings of Third International Conf. on Medical Image Computing and Computer Assisted Intervention (MICCAI),* (pp. 921–930).

Chinzei, K., & Miller, K. (2001). Towards MRI guided surgical manipulator. *Medical Science Monitor*, *7*, 153–163.

Cleary, K., & Nguyen, C. C. (2003). Medical robotics. In Zouridakis, G., & Moore, J. (Eds.), *Biomedical technology and devices handbook. CRC Press*. LLC.

Cohn, L. H. (2006). Future directions in cardiac surgery. *The American Heart Hospital Journal*, *4*, 174–178. doi:10.1111/j.1541-9215.2005.05471.x

Cotin, S., Delingette, H., & Ayache, N. (2000). A hybrid elastic model allowing realtime cutting, deformation and force-feedback for surgery training and simulation. *Visual Computing Journal*, *16*, 437–452. doi:10.1007/PL00007215

Craig, J. J. (1989). *Introduction to robotics* (2nd ed.). Reading, PA: Addison-Wesley.

Das, H., Zak, H., Johnson, J., Crouch, J., & Frambach, D. (1999). Evaluation of a telerobotic system to assist surgeons in microsurgery. *Computer Aided Surgery*, *4*, 15–25. doi:10.3109/10929089909148155

Davies, B. (2000). A review of robotics in surgery. *Proceedings of the Institute of Mechanical Engineers, Part H*, *214*, 129–140.

Davies, B. L., Hibberd, R. D., Coptcoat, M. J., & Wickham, J. E. (1989). A surgeon robot prostatectomy--a laboratory evaluation. *Journal of Medical Engineering & Technology*, *13*, 273–277. doi:10.3109/03091908909016201

Davies, B. L., Hibberd, R. D., Ng, W. S., Timoney, A. G., & Wickham, J. E. (1991). The development of a surgeon robot for prostatectomies. *Proceedings of the Institute of Mechanical Engineers Part H, 205*, 35–38. doi:10.1243/PIME_PROC_1991_205_259_02

DeLaurentis, K., & Mavroidis, C. (2002). Mechanical design of a shape memory alloy actuated prosthetic hand. *Technology and Health Care, 10*(2), 91–106.

Dellon, B., & Matsuoka, Y. (2007). Prosthetics, exoskeletons, and rehabilitation. *IEEE Robotics & Automation Magazine*, 30–34. doi:10.1109/MRA.2007.339622

Demi, B., Ortmaier, T., & Seibold, U. (2005). *The touch and feel in minimally invasive surgery*. IEEE International Workshop on Haptic Audio Visual Environments and their Applications.

Dohi, T. (2005). Surgical robots and 3D displays for computer-aided surgery. *Endourooncology, 6*, 15–26. doi:10.1007/4-431-27173-2_2

Eljamel, M. S. (2006). Robotic applications in epilepsy surgery. *International Journal of Medical Robotics and Computer Assisted Surgery, 2*(3), 233–237. doi:10.1002/rcs.97

Eljamel, M. S. (2007). Validation of a neurosurgical robot using a phantom. *International Journal of Medical Robotics and Computer Assisted Surgery, 3*(4), 372–377. doi:10.1002/rcs.153

Eljamel, M. S. (2009). Robotic neurosurgical applications: Accuracy and consistency or pure fantasy? *Stereotactic and Functional Neurosurgery, 87*, 88–93. doi:10.1159/000202974

Fogoros, R. N. (2011). *Robotic heart surgery - A status report*. Retrieved on June 28, 2011, from http://heartdisease.about.com

Fu, K. S., Lee, C. S. G., & Gonzales, R. C. (1987). *Robotics: Control, sensing, vision & intelligence*. New York, NY: McGraw- Hill.

Gerszten, P. C., Ozhasoglu, C., & Burton, S. A. (2003a). CyberKnife frameless single-fraction stereotactic radiosurgery for tumors of the sacrum. *Neurosurgical Focus, 15*(2), 1–5. doi:10.3171/foc.2003.15.2.7

Gerszten, P. C., Ozhasoglu, C., & Burton, S. A. (2003b). Evaluation of CyberKnife frameless real-time image-guided stereotactic radiosurgery for spinal lesions. *Stereotactic and Functional Neurosurgery, 81*(1-4), 84–89. doi:10.1159/000075109

Glauser, D. (1993). Mechanical concept of the neurosurgical robot Minerva. *Robotica, 11*(6), 567–575. doi:10.1017/S0263574700019421

Goto, T., Hongo, K., & Kakizawa, Y. (2003). Clinical application of robotic telemanipulation system in neurosurgery. *Journal of Neurosurgery, 99*, 1082–1084. doi:10.3171/jns.2003.99.6.1082

Grueneis, C., Richter, R., et al. (1999). Clinical introduction of the CASPAR system: Problems and initial results. 4th International Symposium of Computer Assited Orthopaedic Surgery, CAOS'99, Davos, 17-19 March 1999, Switzerland.

Guerrouad, A., & Jolly, D. (1989). Automatic analysis of weariness during a micromanipulation task by SMOS. *Proceedings of the Annual International Conference of the IEEE Engineering in Medicine and Biology Society, 3*, 906–907.

Guerrouad, A., & Vidal, P. (1989). SMOS: Stereotaxical microtelemanipulator for ocular surgery. *Proceedings of the Annual International Conference of the IEEE Engineering in Medicine and Biology Society, 3*, 879–880.

Guerrouad, A., & Vidal, P. (1991). Advantage of computer aided teleoperation (CAT) in microsurgery. *IEEE Conference Proceeding of Fifth International Conference on Advanced Robotics (ICAR),* (vol. 1, pp. 910- 914).

Guillonneau, B., & Vallancien, G. (2000). Laparoscopic radical prostatectomy: The Montsouris technique. *The Journal of Urology*, *163*, 1643–1649. doi:10.1016/S0022-5347(05)67512-X

Haidegger, T., Kovacs, L., Fordos, G., et al. (2008). Future trends in robotic neurosurgery. *14th Nordic-Baltic Conference on Biomedical Engineering and Medical Physics IFMBE Proceedings*, (vol. 20, part 4, pp. 229-233).

Harja, J., Tikkanen, J., Sorvoja, H., & Myllyla, R. (2007). Magnetic resonance imagingcompatible, three-degrees-of-freedom joystick for surgical robot. *International Journal of Medical Robotics and Computer Assisted Surgery*, *3*, 365–371. doi:10.1002/rcs.159

Hayat, S., & Vidal, P. (1995). Conception of information tools to assist the surgeon when he carries out radial keratotomy with a micromanipulator. *IEEE Conference Proceedings* (pp. 3-5).

Hillman, M. (1998). Introduction to the special issue on rehabilitation robotics. *Robotica*, *16*(5), 485–485. doi:10.1017/S0263574798000629

Hillman, M. (2003). Rehabilitation robotics from past to present. *Proceedings of the 8th International Conference on Rehabilitation Robotics.*

Hongo, K., Kobayashi, S., Kakizawa, Y., Koyama, J., Goto, T., & Okudera, H. (2002). NeuRobot: Telecontrolled micromanipulator system for minimally invasive microneurosurgery— preliminary results. *Neurosurgery*, *51*(4), 985–988.

Howe, R. D., & Matsuoka, Y. (1999). Robotics for surgery. *Annual Review of Biomedical Engineering*, *1*, 211–240. doi:10.1146/annurev.bioeng.1.1.211

Jakopec, M., Harris, S. J., Baena, F. R. Y., Gomes, P., Cobb, J., & Davies, B. L. (2001). The first clinical application of a hands-on robotic knee surgery system. *Computer Aided Surgery*, *6*, 329–339.

Kassler, M. (1993). Introduction to the Special Issue on Robotics for Health Care. *Robotica*, *11*(6), 493–494. doi:10.1017/S0263574700019330

Kavoussi, L. R., Moore, R. G., Adams, J. B., & Partin, A. W. (1995). Comparison of robotic versus human laparoscopic camera control. *The Journal of Urology*, *154*, 2134–2136. doi:10.1016/S0022-5347(01)66715-6

Kazanzides, P., Fichtinger, G., Hager, G. D., Okamura, A. M., Whitcomb, L. L., & Taylor, R. H. (2008). Surgical and interventional robotics, core concepts, technology, and design. *IEEE Robotics & Automation Magazine*, *15*(2), 122–130. doi:10.1109/MRA.2008.926390

Kazerooni, H. (2005). Exoskeletons for human power augmentation. In *Proceedings of IEEE Conference Intelligent Robots and Systems*, Edmonton, Canada, (pp. 3459–3464).

Kim, Y., & Cook, A. M. (1985). Manipulation and mobility aids. In Webster, J. G. (Eds.), *Electronic devices for rehabilitation*. London, UK: Chapman and Hall.

Kuo, J. S., Yu, C., Petrovich, Z., & Apuzzo, M. L. (2003). The CyberKnife stereotactic radiosurgery system: Description, installation, and an initial evaluation of use and functionality. *Neurosurgery*, *53*(5), 1235–1239. doi:10.1227/01.NEU.0000089485.47590.05

Kwoh, Y. S., Hou, J., Jonckheere, E. A., & Hayati, S. (1988). A robot with improved absolute positioning accuracy for CT guided stereotactic brain surgery. *IEEE Transactions on Bio-Medical Engineering*, *35*, 153–160. doi:10.1109/10.1354

Kypson, A. P., Nifong, L. W., & Chitwood, W. R. (2003). Robotic cardiac surgery. *Journal of Long-Term Effects of Medical Implants*, *13*(6), 451–464. doi:10.1615/JLongTermEffMedImplants.v13.i6.30

Laine, T., Lund, T., Ylikoski, M., Lohikoski, J., & Schlenzka, D. (2000). Accuracy of pedicle screw insertion with and without computer assistance: A randomised controlled clinical study in 100 consecutive patients. *European Spine Journal*, *9*, 235–240. doi:10.1007/s005860000146

Lavallee, S. (1989). A new system for computer assisted neurosurgery. In *Proceedings of the 11th IEEE Engineering in Medicine and Biology Conference*, (pp. 926–927).

Lavallee, S. (1995). Image-guided operating robot: A clinical application in stereotactic neurosurgery. In Taylor, R. H. (Ed.), *Computer-integrated surgery* (pp. 343–351). Cambridge, MA: MIT Press.

Le Roux, P. D., Das, H., Esquenazi, S., & Kelly, P. J. (2001). Robot-assisted microsurgery: a feasibility study in the rat. *Neurosurgery*, *48*, 584–589. doi:10.1097/00006123-200103000-00026

Li, Q. H. (2002). The application accuracy of the NeuroMate robot—A quantitative comparison with frameless and frame-based surgical localization systems. *Computer Aided Surgery*, *7*, 90–98.

Liverneaux, P., Nectoux, E., & Taleb, C. (2009). The future of robotics in hand surgery. *Chirurgie de la Main*, *28*(5), 278–785. doi:10.1016/j.main.2009.08.002

Loulmet, D., Carpentier, A., & D'Attellis, N. (1999). Endoscopic coronary artery bypass grafting with the aid of robotic assisted instruments. *The Journal of Thoracic and Cardiovascular Surgery*, *118*(1), 4–10. doi:10.1016/S0022-5223(99)70133-9

Louw, D. F., Fielding, T., McBeth, P. B., Gregoris, D., Newhook, P., & Sutherland, G. R. (2004). Surgical robotics: a review and neurosurgical prototype development. *Neurosurgery*, *54*, 525–537. doi:10.1227/01.NEU.0000108638.05274.E9

Makaran, J., Dittmer, D., Buchal, R., & MacArthur, D. (1993). The SMART(R) wrist-hand orthosis (WHO) for quadriplegic patients. *Journal of Prosthetics and Orthotics*, *5*(3), 73–76. doi:10.1097/00008526-199307000-00002

Massa, B., Roccella, S., Carrozza, M. C., & Dario, P. (2002). Design and development of an underactuated prosthetic hand. *IEEE International Conference on Robotics and Automation*, Washington, DC, (pp. 11–15).

Matsuoka, Y., Afshar, P., & Oh, M. (2006). On the design of robotic hands for brain machine interface. *Neurosurgical Focus*, *20*(5), 1–9. doi:10.3171/foc.2006.20.5.4

McBeth, P. B., Louw, D. F., Rizun, P. R., & Sutherland, G. R. (2004). *American Journal of Surgery*, *188*(4), 68–75. doi:10.1016/j.amjsurg.2004.08.004

Mehmanesh, H., Henze, R., & Lange, R. (2002). Totally endoscopic mitral valve repair. *The Journal of Thoracic and Cardiovascular Surgery*, *123*, 96–97. doi:10.1067/mtc.2002.118496

Merloz, P. H., Tonetti, J., & Eid, A. (1997). Computer-assisted spine surgery. *Clinical Orthopaedics and Related Research*, *337*, 86–96. doi:10.1097/00003086-199704000-00011

Mohr, F. W., Falk, V., Diegeler, A., Walther, T., Gummert, J. F., & Bucerius, J. (2001). Computer-enhanced "robotic" cardiac surgery: Experience in 148 patients. *The Journal of Thoracic and Cardiovascular Surgery*, *121*(5), 842–853. doi:10.1067/mtc.2001.112625

Nakamura, R., Masamune, K., & Nishikawa, Y. (1995). Development of an MRI-compatible needle insertion manipulator for stereotactic neurosurgery. *Journal of Image Guided Surgery*, *1*, 242–248. doi:10.1002/(SICI)1522-712X(1995)1:4<242::AID-IGS7>3.0.CO;2-A

Nathoo, N., Pesek, T., & Barnett, G. H. (2003). Robotics and neurosurgery. *The Surgical Clinics of North America, 83*(6), 1339–1350. doi:10.1016/S0039-6109(03)00157-9

Nedas, T. G., Challacombe, B. J., & Dasgupta, P. (2005). Robotics in urology: An update. *International Journal of Medical Robotics and Computer Assisted Surgery, 1*, 13–18. doi:10.1002/rcs.37

Nelson, B. (2007). Comparison of length of hospital stay between radical retropubic prostatectomy and robotic assisted laparoscopic prostatectomy. *The Journal of Urology, 177*, 929–931. doi:10.1016/j.juro.2006.10.070

Nifong, L. W., Chu, V. R., Bailey, B. M., Maziarz, D. M., Sorrell, V. L., Holbert, D., & Chitwood, W. R. Jr. (2003). Robotic mitral valve repair: Experience with the da Vinci system. *The Annals of Thoracic Surgery, 75*, 438–443. doi:10.1016/S0003-4975(02)04554-X

Partin, A. W., Adams, J. B., Moore, R. G., & Kavoussi, L. R. (1995). Complete robot-assisted laparoscopic urologic surgery: A preliminary report. *Journal of the American College of Surgeons, 181*, 552–557.

Paul, R. P. (1981). *Robot manipulators: Mathematics, programming, and control*. Cambridge, MA: MIT Press.

Petermann, J., Kober, R., Heinze, R., Frlich, J. J., Heeckt, P. F., & Gotzen, L. (2000). Computerassisted planning and robot-assisted surgery in anterior cruciate ligament reconstruction. *Operative Techniques in Orthopaedics, 10*, 50–55. doi:10.1016/S1048-6666(00)80042-7

Peters, T. M. (2000). Image guided surgery: From X-rays to virtual reality. *Computer Methods in Biomechanics and Biomedical Engineering, 4*, 27–57. doi:10.1080/10255840008907997

Potamianos, P., Davies, B. L., & Hibberd, R. D. (1994). *Intra-operative imaging guidance for keyhole surgery: Methodology and calibration*. International Symposium on Medical Robotics and Computer Assisted Surgery, Pittsburgh, PA.

Potamianos, P., Davies, B. L., & Hibberd, R. D. (1995). *Intra-operative registration for percutaneous surgery.* International Symposium on Medical Robotics and Computer Assisted Surgery, Baltimore, MD.

Pransky, J. (1997). ROBODOC - Surgical robot successstory. *Industrial Robot: An International Journal, 24*(3), 231–233. doi:10.1108/01439919710167444

Pransky, J. (1998). Surgeons' realizations of RoboDoc. *Industrial Robot: An International Journal, 25*(2), 105–108. doi:10.1108/01439919810205792

Quinn, A. M. (2002). CyberKnife: A robotic radiosurgery system. *Clinical Journal of Oncology Nursing, 6*(3), 149,156.

Satava, R. M. (2004). Future trends in the design and application of surgical robots. *Seminars in Laparoscopic Surgery, 11*(2), 129–135.

Schiff, J., Li, P. S., & Goldstein, M. (2005). Robotic microsurgical vasovasostomy and vasoepididymostomy in rats. *International Journal of Medical Robotics and Computing, 1*, 122–126. doi:10.1002/rcs.21

Spong, M. W., & Vidyasagar, M. (1989). *Robot dynamics and control*. New York, NY: Wiley.

Stoianovici, D. (2001). URobotics — Urology robotics at Johns Hopkins. *Computer Aided Surgery, 6*(6), 360–369.

Stoianovici, D., Kavoussi, L. R., Whitcomb, L. L., et al. (2002). *Friction transmission with axial loading and a radiolucent surgical needle driver*. United States Patent 6,400,979.

Stoianovici, D., Webster, R., & Kavoussi, L. (2005). Surgical robotic applications in minimally invasive urooncology surgery. In Moore, R., & Bishoff, J. (Eds.), *Complications of urologic laparoscopic surgery: Recognition, management and prevention* (pp. 353–363). Taylor and Francis.

Sung, G. T., & Gill, I. S. (2001). Robotic laparoscopic surgery: a comparison of the da Vinci and Zeus systems. *The Journal of Urology*, *58*, 893–898. doi:10.1016/S0090-4295(01)01423-6

Szymula, T. P., & Richter, V. (2001). Recenzie endoscopy and microsurgery. In D. Pieptu (Series Ed.), *Update in plastic surgery*. New York, NY: Spinger.

Taylor, R. H. (1997). Robots as surgical assistants: Where we are, wither we are tending, and how to get there. In *AIME 97*, Genoble, France, (pp. 3–11).

Taylor, R. H. (1999). A steady-hand robotic system for microsurgical augmentation. *The International Journal of Robotics Research*, *18*(12), 1201–1210. doi:10.1177/02783649922067807

Taylor, R. H., Funda, J., & Eldridge, B. (1995). A telerobotic assistant for laparoscopic surgery. *IEEE Engineering in Medicine and Biology Magazine*, 279–287. doi:10.1109/51.391776

Tejima, N. (2000). Rehabilitation robotics: A review. *Advanced Robotics*, *14*(7), 551–564. doi:10.1163/156855301742003

Yu, D. Y. (1998). Robotic ocular ultramicrosurgery. *Australian and New Zealand Journal of Ophthalmology*, *26*(1), 6–8. doi:10.1111/j.1442-9071.1998.tb01375.x

Zorn, K. C. (2007). Robotic radical prostatectomy learning curve of a fellowship trained laparoscopic surgeon. *Journal of Endourology*, *21*, 441–447. doi:10.1089/end.2006.0239

Chapter 10
Surgical Robots:
System Development, Assessment, and Clearance

Tamás Haidegger
Budapest University of Technology and Economics, Hungary

ABSTRACT

Information technology and robotics have been integrated into interventional medicine for over 25 years. Their primary aim has always been to provide patient benefits through increased precision, safety, and minimal invasiveness. Nevertheless, robotic devices should allow for sophisticated treatment methods that are not possible by other means. Several hundreds of different surgical robot prototypes have been developed, while only a handful passed clearance procedures, and was released to the market. This is mostly due to the difficulties associated with medical device development and approval, especially in those cases when some form of manipulation and automation is involved. This chapter is intended to present major aspects of surgical robotic prototyping and current trends through the analysis of various international projects. It spans across the phases from system planning, to development, validation, and clearance.

1. INTRODUCTION

Prototyping a new device implies creating a system for the first time, dealing with all the challenges and unforeseen pitfalls. It has an especially important role in high-risk, high-value domains, such as interventional medicine. In the past decades, numerous different robotic surgery devices have been created, and only a few reached the market. The Medical Robotic Database (Pott, 2011) lists over 450 international surgical robotic projects, of which several dozen are with the potential to become commercially available. Parallel, the number of surgical robotics related publications

DOI: 10.4018/978-1-4666-0176-5.ch010

has been steadily rising in the past years (O'Toole et al., 2010), making *Computer-Integrated Surgery* (CIS) one of the hottest areas within medical technology.

CIS refers to the entire field of technology-aided interventional medicine: theory and practical technology from image processing and augmented reality applications to automated tissue ablation. CIS means the combination of innovative algorithms, robotic devices, imaging systems, sensors and human–machine interfaces. These systems should work cooperatively with physicians in the planning and execution of surgical procedures, therefore the process involves continuous innovation and prototyping (Taylor & Kazanzides, 2008). A subfield of CIS is called *Image-Guided Surgery* (IGS), where the digital system is not necessarily involved in the physical part of the operation, but improves the quality of surgery by better visualization or guidance. IGS means the accurate registration (correlation and mapping) of the operative field to a pre-operative (typically *Magnetic Resonance Imaging*—MRI, or *Computer Tomography*—CT) imaging or intra-operative (*ultrasound*—US, fluoroscopy) data set of the patient, providing free-hand navigation, positioning accuracy of equipment, or guidance for a mechatronic system. IGS systems have been successfully prototyped and commercialized, and now being used in neurosurgery, radiotherapy, pediatrics, orthopedics and various other fields.

This chapter introduces the aims and means of surgical robot development, giving a better understanding of the difficulties the field is challenged with through examples taken from existing robots. Medical robots are mostly employed for the accuracy and reliability of their mechanics; however, it may still be hard to fully exploit their features, as surgical tasks are typically unique, involving the semi-autonomous manipulation of deformable objects in an organic, limited environment. Robot-assisted procedures offer remarkable advantages both for the patient and the surgeon. The ability to perform operations on a smaller scale makes microsurgery more accessible, and the stability of the devices. (See the *Chapter on Medical Robotics* for more details.)

Medical imaging gives the capability to navigate and position a surgical tool at the target point. Furthermore, there is the option to introduce advanced digital signal processing to control or record the spatial point-of-interests and motions (Kazanzides et al., 2010). This can be useful for surgical simulation and risk-free training. Finally, robotized equipment can greatly add to the ergonomics of the procedures. The main advantages of robotic surgery systems—based on (Karas & Chiocca, 2007) and (Lirici et al., 1997)— are the following:

- Superior 3D spatial accuracy provided by the robot,
- Specific design for maximum performance (including miniature robots),
- Stabilization of the instruments within the surgical field,
- Advanced ergonomics supporting long procedures,
- Stable performance,
- High fidelity information integration,
- Invulnerability to environmental hazards,
- Patient advantages (reduced blood loss, less trauma, shorter recovery time),
- Decreased costs (per treatment) due to shorter hospitalization and recovery,
- Possibility to provide better and more realistic training to physicians.

Further optional benefits:

- Improvement of manual dexterity, motion scaling,
- Physiological tremor filtering,
- Integrated 3D vision system with high definition (HD) resolution.

At the prototyping phase of a new system, not all of these benefits are apparent. The human–

machine interface, ergonomics and economical features are typically developed in a later phase of product development.

Robots have been introduced to the operating room primarily to provide higher accuracy and dexterity. They can support surgeons with advanced targeting, steady positioning and task execution with a precision beyond human capabilities. Therefore, the treatment delivery accuracy and objective evaluation of interventional systems is crucial, especially when they are image-guided and operating semi-autonomously (Stiehl et al., 2007). Errors and imperfections may originate in:

- CT imaging errors,
- Volume model generation errors,
- Treatment planning errors,
- Errors introduced by hardware fixturing,
- Intra-operative data noise,
- Registration errors,
- Inherent inaccuracies of surgical tools and actions.

It is important to keep in mind from the first stage of development which errors will be critical regarding the final application of the system, and how to minimize their disturbing effect.

2. BACKGROUND AND COMMUNITY

Prototyping is necessary to support the creation of innovative new systems, and to perform validation, performance assessment. Despite the fact that the advanced stage medical device assessment and especially clearance is more strictly regulated than generic industrial applications, the procedure that leads to engineering models and prototypes in CIS is very similar to other robot applications. While going through the innovation life-cycle—from idea to design, construction, evaluation and validation—many steps may result in scientifically important findings. Since surgical robot development typically begins in academic laboratories or research institutions, achievements are published in peer reviewed journals and conference proceedings. (See the *Section on Additional reading materials* for more details.)

The 24th Technical Committee of the *IEEE Robotics and Automation Society* (RAS—*ieee-ras.org*) is entirely dedicated to surgical robotics, and includes many of the most renowned peers of the field. On the medical side, probably the biggest organization promoting *Minimally Invasive Surgery* (MIS) is the *Society of American Gastrointestinal and Endoscopic Surgeon* (SAGES—www.sages.org), and in addition, the *Minimally Invasive Robotic Association* (MIRA—www.mirasurgery.org). Many other organizations also have CIS related bodies that typically organize annual conferences, where new prototypes and concepts are presented to the scientific community.

2.1 System Development Strategies

CIS systems are strongly application-oriented (ideally driven by a strong clinical need), therefore their entire architecture may be defined by the targeted treatment. It may be extremely hard to switch from one concept to another during a latter development, therefore strategic planning is a must. Different categories of surgical robots have been built for various procedures. Hand-held and directly controlled devices may serve as an incremental upgrade for existing tools, while teleoperated systems represent a whole different field. The advantage of versatility comes with the emergence of issues with control, latency handling and emergency protocols.

Robots can be involved in medical procedures with various level of autonomy (Nathoo et al., 2005), and each type requires different approach during system development. Some of them serve as a robust tool holding equipment having been directed to the desired position. Systems that are able to perform fully automated procedures—such as CT-based biopsy or cutting—are called autonomous or supervisory controlled devices. On the other hand, if the robot is entirely teleoperated or remote-controlled (robotic telesurgery system) the surgeon is absolutely in charge of its motion. The latter consists of three parts:

- One or more slave manipulators,
- A master controller
- A vision system providing visual feedback to the user.

Based on the gathered visual (and sometimes haptic) information, the surgeon guides the arm by moving the controller and closely watching its effect. In most of the cases, the slave system and camera are acting as the remote hands and eyes of the surgeon, and therefore they are key elements of the operation. (See the *Chapter Medical Robotics* for more details.)

Modifying the teleoperation control paradigm we can introduce *cooperative control* (also called shared control or hands-on surgery). It means that the surgeon is directly giving the control signals to the machine through e.g., a force sensor. It is possible to read and process these signals in real-time to create the robot's motion. The human is always in contact with the robot, as the master and the slave devices are physically identical. In this case, the robot is the extension of the doctor's hand, equipped with special features and effectors. This approach keeps the human in the loop, and still allows the surgeons to use all their senses. It is often used in the case of micro-manipulation operations, such as micro-vascular, urologic, eye or brain procedures. Cooperative control is a promising way to provide highly integrated robotic support for procedures while applying all the necessary safety standards.

Considering either a less complex needle insertion tool or a teleoperated, MR compatible robot, prototype development begins with identifying the available technology, and choosing the right architecture. Rigorous approaches to system development earlier discussed in the book (regarding technology, intellectual property, regulations, etc.) are applicable to the medical field as well. However, if it's within the academia, it might fall under the category of fundamental research, driven by scientific goals to innovate new concepts and paradigm to improve future patient treatment. Aiming for new methods, specializing on a certain disease may mean to prototype a proof-of-concept system instead of targeting the achievement of a commercial product. In cutting edge research, it may happen that during the development phase, several new results are achieved that are sometimes transforming the overall outcome of the project, making it necessary to revisit prior decision points. This means higher associated risk, something industrial companies are usually willing to avoid. It often happens that a company licenses one or more patents on prototype systems and then develops its own version, focusing on the actual market needs.

It is commonly quoted that a medical product needs 10–15 years to grow from the conception of an idea to commercialization. This extremely long time-to-market period requires wise considerations from the developers to ensure the continuous funding of the project. Developing an engineering prototype is only the first step, success in business requires various skills and sometimes entirely different approach towards R&D. The history of the first generation of surgical robots summarized in the next session provides great examples and lessons on the critical aspects of the development process.

2.2. First Prototypes

Early Concepts of Surgical Telerobots

Since the 1980s, hundreds of medical robotic research projects have been initiated, creating a set of instruments for remote and local robotic surgery. CIS and telemedicine have become widely used around the world, surgeons and engineers created systems and networks for advanced patient care, demonstrated over a hundred different procedures, transcontinental surgery and even performed procedures in weightlessness (Doarn et al., 1998).

The general idea of telerobotic health care in space was born in the early '70s, proposed in a study for the *National Aeronautics and Space Administration* (NASA) to provide surgical care for astronauts with remote controlled robots (Alexander, 1972). The concept-system presented in Figure 1. shows very well the fundamental characteristics of a complete teleoperational system, although the first prototypes were only built almost 15 years later. In the late '80s, the idea of commercial surgical robotics was born on the principle to extend the surgeon's dexterity, multiple academic centers started to develop new prototypes.

It was first proven twenty-five years ago that robotics can extend human surgeons' capabilities. The first robot used on a human patient was a Puma 200 (*Programmable Universal Machine for Assembly*), manipulating a biopsy cannulae using a *Brown–Roberts–Wells stereotactic frame* (mounted on the robot's base). The operation took place in the Memorial Medical Center (Long Beach, CA) in 1985 (Kwoh et al., 1988). In later experiments, the Puma performed complete stereotactic neurosurgical operations based on CT scan, processing the scanned images, positioning the arm and manipulating different probes.

The U.S. Army has long been interested in robotic surgery for the battlefield, and currently, the *Telemedicine & Advanced Technology Research Center* (TATRC) supports the development of prototypes to test and extend the reach of remote health care. Physicians were first able to affect distant patients with the Green telepresence system in 1991 (Bowersox et al., 1996), and the first long distance telerobotic experiment was in 1993 between NASA Jet Propulsion Laboratory (JPL) in Pasadena and Milan (Rovetta et al., 1996). The *U.S. Department of Defense* (DoD) aims to develop a system—*Trauma Pod*—by 2025 that allows combat surgeons to perform life saving operations from a safe distance (Satava, 1995) and (Garcia et al., 2009). Much of these projects created the basis of the most successful surgical robot, the da Vinci.

Figure 1. The very first design concept of a surgical robot system, aimed to support astronauts in space (Alexander, 1973)

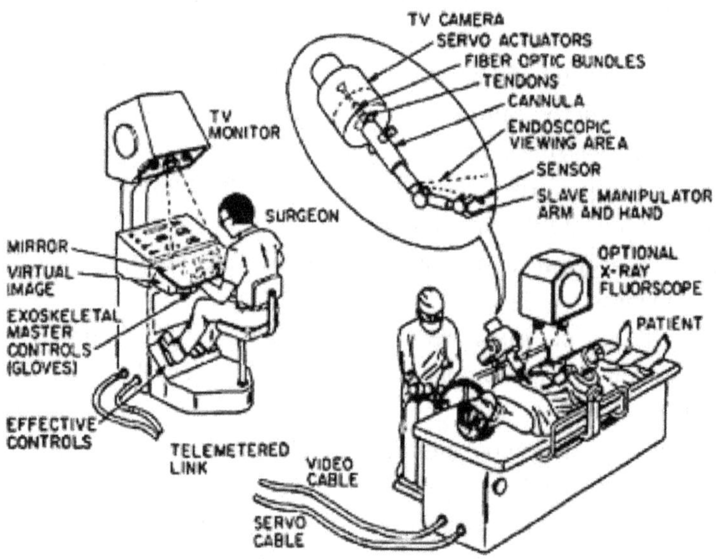

The da Vinci Surgical System

The market leader *da Vinci* robot from Intuitive Surgical Inc. (Sunnyvale, CA), is a complete teleoperated system created with roughly $500M investment (see Figure 2). The company was founded in 1995, licensing many promising technologies, and by 1997 the first functional surgical robot (*Lenny*) was ready for animal trials. Concepts from around the world and various engineering prototypes were integrated into the project (DiMaio et al., 2011):

- The *Black Falcon* teleoperated surgical robot from MIT (*Madhani et al., 1998*),
- The *Robot Assisted Microsurgery* (RAMS) workstation from NASA JPL,
- *PROBOT* prostate surgery device from Imperial College,
- Teleoperation control techniques from *University of Washington* (UW),
- *Green telepresence surgery system* developed at *Stanford Research International* (SRI).

The next prototype—*Mona*—performed the first human trials in Belgium in 1997, and the first da Vinci unit was created within a year (Guthart & Salisbury, 2000). The *U.S. Food and Drug Administration* (FDA) cleared the system for general

Figure 2. The evolution of the da Vinci robot's slave manipulators: a–c) The original Black falcon prototype, developed at MIT (Courtesy of Science Photo Library); d–f) The slave mechanisms of Leny, Mona and first da Vinci, respectively (Courtesy of Intuitive Surgical Inc.); g–h) The latest version of the slave arms (Courtesy of Intuitive Surgical Inc.)

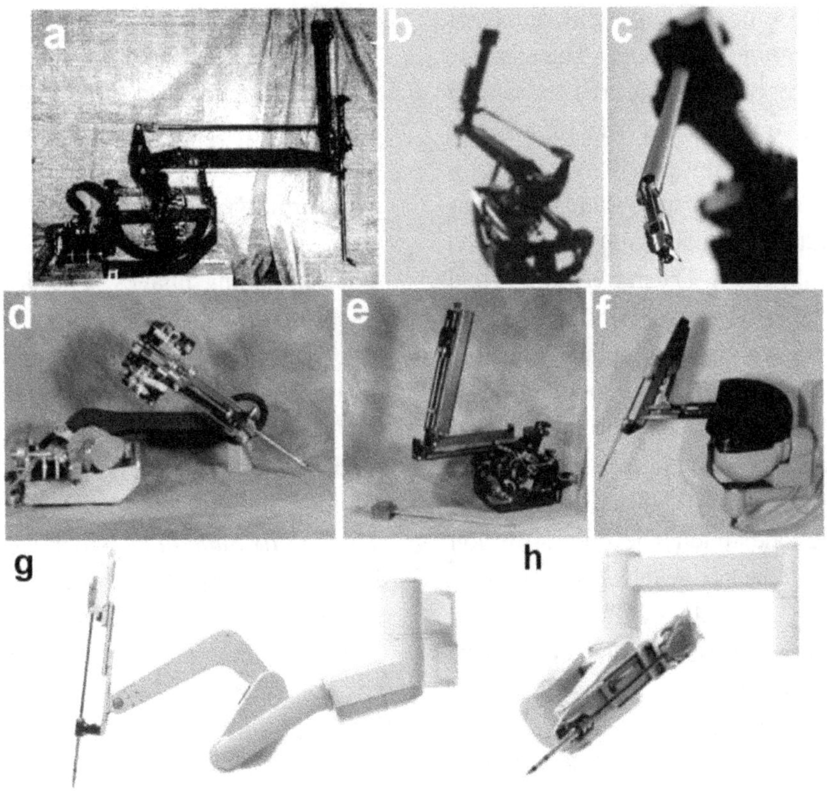

laparoscopic surgery (August 2000), thorascopic surgery (March 2001) and laparoscopic radical prostatectomy (May 2001), followed by many other approvals; most recently for transoral otolaryngologic procedures.

The da Vinci is basically a smart tool and interface between the hands of the surgeon and the laparoscopic instruments in use. The patient side consists of two/three tendon-driven, 6+1 *Degrees of Freedom* (DOF) slave manipulators. These are designed with a *Remote Center of Motion* (RCM) kinematics, resulting in an inherent safety regarding the spatial stability of the entry port. This means that the gross positioning (passive) arms can move the base point of the actuated arms, and therefore define the RCM as the fixed entry point through their semi-parallel kinematics. The camera holder arm navigates in 3 DOF, controlled with the same master console. The system provides high quality 3D vision with stereo-endoscopes, adjustable tremor filtering (6 Hz) and motion scaling (1:1–1:5). The total weight of the system is 850 kg, and the setup takes up significant floor space in the *Operating Room* (OR). The intrinsic accuracy of the robot was measured to be 1.35 mm (Target Registration Error—TRE (see next section) with the points not used in registration), with 1.02 ± 0.58 mm (mean ± standard deviation—STD) Fiducial Localization Error (FLE) (Kwartowitz et al., 2006). The newer version had similar FLE, 1.05 ± 0.24 mm (Kwartowitz et al., 2007). To compensate for any application errors (and to avoid safety hazards), the robot only follows the movements of the surgeon.

Manipulation precision comes at a price, the current version consists of 10,000 individual parts, and the operating code stretches beyond 1.4 million lines. There are 39 backdriveable *Maxon motors* (maxon motor ag, Sachseln, CH) in each robot, most of them equipped with magnetic encoders. Communications in the da Vinci is acquired via Transmission Control Protocol over Internet Protocol (TCP/IP) using NI USB-6009 data acquisition boards (National Instrument, Austin, TX). The controller computer deals with the 48 DOF at an update rate of 1.5 kHz. There are 48 encoders and 96 analog input channels supported by parallel floating point DSP architecture with a peak computational power of 384 Mflops and a sustained processing power of between 128–256 Mflops. The robot has a network of 24 microcontrollers and integer DSPs performing data transfer and health watchdog functions. Velocity control algorithms are employed mostly with 2–3 filters, the one in the forward link is used to attenuate master input commands that could cause instrument tip vibrations otherwise. Redundant sensors, hardware watchdogs and real-time error detection protocols ensure fail-safe operation of the robot. The da Vinci uses an *Insite Vision and Navigator Camera Control* with two 3-chip cameras and two separate optical channels generating two images delivered separately to each eye of the surgeon. It gives 1000 frames of the instrument position per second and filters each image through a video processor that eliminates background noise (Haidegger, 2010).

Intuitive continued perfecting the system, and the second generation—da Vinci S—was released in 2005. The setup and docking procedures were streamlined, the 4th arm became more versatile, the slave arms were redesigned and arranged in a way to avoid collision and the console was enhanced with streaming images. The latest version—da Vinci Si—became available in 2009 with improved full HD camera system, advanced ergonomic features and most importantly, the possibility to use two consoles for assisted surgery (Figure 3).

Having realized the potential in robotic surgery training, Intuitive has released its portable *da Vinci Skills Simulator*, which can be attached to a da Vinci Si system. It allows users to practice unassisted or with supervision through a set of exercises ranging from camera handling to electric coagulation. The open architecture of the software allows for the future incorporation of additional modules.

Figure 3. The da Vinci Si system, debuted in 2009 (Courtesy of Intuitive Surgical Inc.)

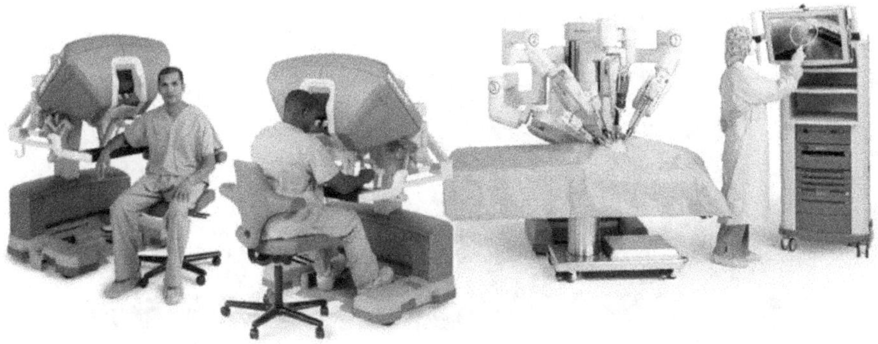

In recent years, Intuitive has been focusing on enlarging the tool inventory, developing useful and specific end effectors for various procedures (e.g. US probe, fluoroscopic imaging, etc.). Also, they put an effort to provide a generic research interface to help institutes developing compatible tools and techniques. In a joint venture with the CISST ERC they are about to release the SAW architecture (Surgical Assistant Workstation) that would further integrate the robot into any surgical information system. Current extensions allow visual overlay and augmented reality features in a prototype version (Kazanzides et al., 2010).

As of today, there are around 1950 da Vincis around the world, 3/4 of them in the U.S. The number of successful procedures performed is close to a million; the most successful applications being prostatectomy and hysterectomy. According to Intuitive, over 90% of all radical prostate removal procedures were performed robotically in the U.S. in 2010. One of the key elements in the success of the system is the universality of the platform. Despite the fact that originally it was planned to be used for beating heart surgery, it was versatile enough to shift the focus to urologic and gynecological procedures that became the most lucrative business for the company. Their extensive patenting strategy allowed them a freedom to operate in the market over the past 15 years.

The NeuroMate Neurosurgical Robot

Although the *NeuroMate* was the first robot to gain CE mark (*Conformité Européenne—European Conformity*) in Europe and then the FDA approval in 1997 for stereotactic neurosurgical procedures, it could not achieve the same success as the da Vinci. Originally prototyped based on an industrial robot at the Grenoble University and developed market-ready by *Innovative Medical Machines International* (IMMI—Lyon, France), the 5 DOF NeuroMate provides accurate and trusted assistance for supervised needle positioning for brain biopsy (Figure 4.). Combined with pre-operative images, it offers real-time visualization to give the surgeon precise location of a tumor. It has an approval for neuro-endoscopic applications and for frameless stereotactic surgery (Li et al., 2002).

The robot consists of 5 revolute joints, each mobilized by a separate, high precision servo. The joint values are read by encoders with a resolution of 1/26825 degree due to the high gearing. The Neuromate contains embedded joint controller boards that are integrated into the links of the robot, significantly reducing the required cabling. Each joint controller board contains a microprocessor and is responsible for controlling up to two axes of the robot, including the power amplification. The power supplies are placed in the triangle shaped base, eliminating the need for a separate controller rack. The ISS version of the system

Figure 4. a) IGOR, the original industrial robot-based prototype for neurosurgery (Benabid et al., 1987); b) IMMI's NeuroMate (Li et al., 2002); c–d) Current and next version of the system with newly designed interior controllers (Courtesy of Renishaw, PLC.)

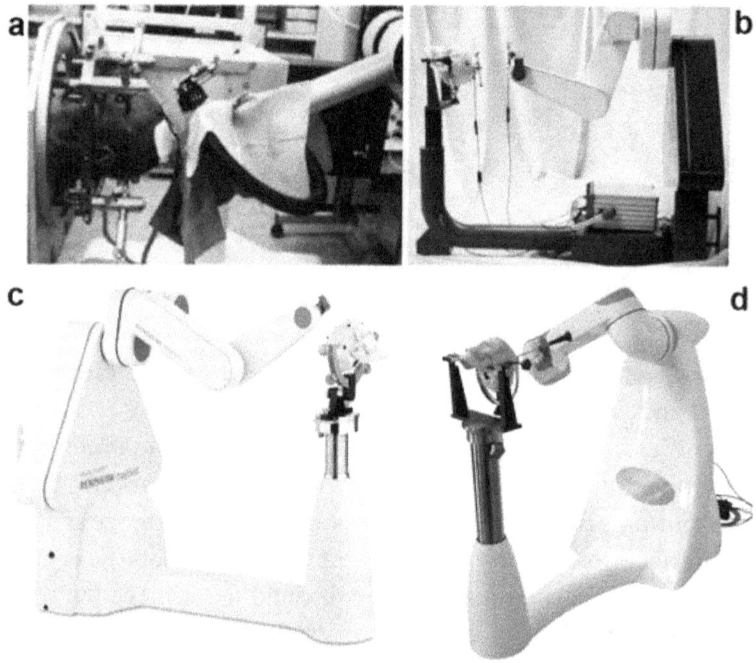

communicates with the main PC through a *Controller Area Network* (CAN) bus with a 18.2 ms communication cycle. On the lowest level, the joints are given position commands. The highest linear velocity of the robot is approximately 50 mm /sec.

The NeuroMate's reported intrinsic accuracy (i.e., the precision of the individual hardware and software components) is 0.75 mm, with a repeatability of 0.15 mm (Varma & Eldridge, 2006). In a human stereotactic surgical experiment conducted in 2002, the application accuracy (the overall precision in performing the desired task, as discussed below) was measured to be 0.86 ± 0.32 mm in frame-based configuration and 1.95 ± 0.44 mm in frameless model (Li et al., 2002). The average application accuracy of 10 different robots was measured to be 0.6 mm. In a more recent experiment, the intrinsic accuracy of the robot was 0.36 mm FRE and 0.36 ± 0.17 mm TRE (Haidegger et al., 2008).

The technology was bought by *Integrated Surgical Systems Inc.* (ISS—Sacramento, CA) in 1997. In the first couple of years of operation, the company has installed around 20 NeuroMate systems in the United States, Europe and Japan. However, due to the lack of sustainable market and investment for further innovation, the company ceased operations in the early 2000s. The NeuroMate technology was acquired first by *Siemens Finance and Leasing* (Munich, Germany), then by *Schaerer Mayfield NeuroMate AG* (Lyon, France) in 2007, and reappeared on the market in *Renishaw plc*'s (Wotton-under-Edge, UK) product line. It received a facelift, and runnning under the trademark neuro|mate (Figure 4.). More recently, the robot has been used for thousands of electrode implantation procedures for *Deep Brain Stimulation* (DBS), *Stereotactic Electroencephalography* (SEEG) and *Transcranial Magnetic Stimulation* (TMS). This also supports the importance of the ability to deploy a robotic system in various clinical fields.

Figure 5. a) The first engineering prototype of orthopedic robot at IBM T.J. Watson Center, created for canine hip surgery; b–c) Early prototype of the ROBODOC system; d) The ROBODOC, as sold by ISS until the early 2000s (Courtesy of ISS Inc.); e–f) The ROBODOC in its current form; g) Concept figure of the next generation prototype (Courtesy of Curexo Tech. Co.)

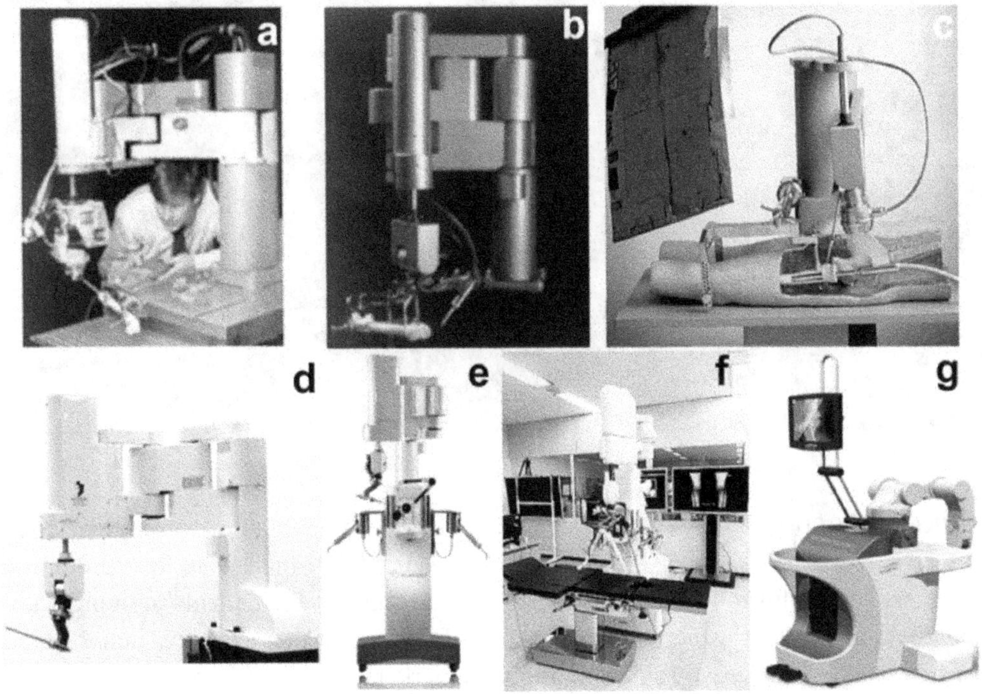

ROBODOC for Orthopaedics

ROBODOC Surgical Assistant System shares a similar story to NeuroMate, it was the first robot of ISS, when the company was founded in 1990. Development began in 1986 at *IBM T. J. Watson Center* and *U.C. Davis,* aiming to create a bone milling robot (Pransky, 1997) and (Bargar, 2007). A 5 DOF IBM SCARA robot was custom designed for Total Hip Arthroplasty (THA—surgical shaping or alteration of the joint). The 3D planning station (called *Orthodoc*) together with the ROBODOC use pre-surgical images and software to first design the surgical procedure (Figure 5). Surgeon can precisely define the desired cavity in the bone, size and position the prosthesis before the real surgery. In fact, the robot was originally aimed at canine surgery, and later changed the focus to human operations. The first-ever robotic human trial was performed in 1992 (Kazanzides, 2009), and later extended to being *Total Knee Artroplasty* (TKA) procedures al well. Overall, the company invested $80M into system development.

ROBODOC drills the bone without direct human control of the tool during the procedure; therefore the application accuracy of the system is critical. The robot has a 0.5 mm intrinsic accuracy, while the application accuracy was 1.2 mm in average, ranging from 1.0 to 3.5 mm in cadaver trials (Paul et al., 1994). The later version of the device had around 0.1 mm dimensional error and 1.0 mm placement error, providing over 95% implant-bone contact (Taylor, 2001). Since 1994, ISS has sold around 80 systems across Europe and Asia, and in 2008, it became the first FDA approved automated bone milling robot, under *Curexo Technology Corporation.*

Figure 6. a–b) First prototypes mounted on a Fanuc robot; c–d) The first and second commercially available versions of the CyberKnife radiosurgery system (Courtesy of Accuray Inc.)

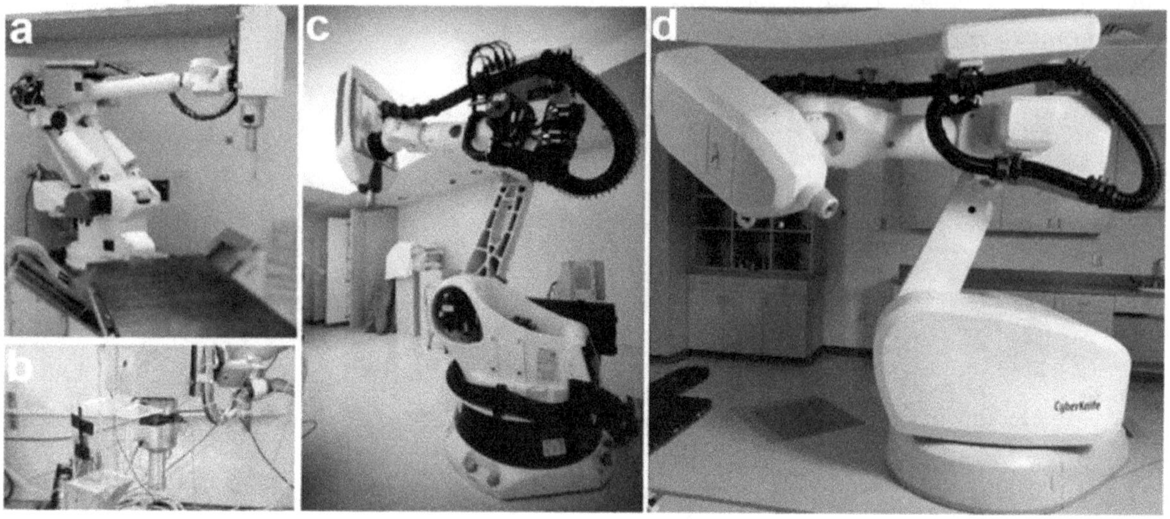

CyberKnife Radiosurgery System

One of the most successful robotic applications is the *CyberKnife* from Accuray Inc. (Sunnyvale, CA). This stereotactic radiosurgery system integrates IGS with robotic positioning. The 6 MeV *Linear Accelerator* (LINAC) relatively lightweight photon device is mounted on a KUKA KR 240, 6 DOF industrial manipulator (Figure 6). The idea was to combine existing hardware components to create an innovative device. This approach allows for faster prototyping, and shorter system development cycle. More than 200 units have been sold worldwide. The same concept has been used by SIEMENS lately with their *Artis Zeego* system for actuated X-ray imaging (SIEMENS, 2011).

CyberKnife's primary deployment is the irradiation of brain and spine tumors. X-ray cameras are used to track the spatial displacement of the patient and compensate for motion caused by e.g. breathing. The overall accuracy of the system is 0.42 ± 0.4 mm, while patient skin motions are detected with a 0.35 mm precision (Dieterich et al., 2003). The CyberKnife moves the radiation beam by physically repositioning the radiation source. It uses intra-corpuscular markers and Polaris (NDI Inc., Waterloo, ON) infrared cameras to track the patients' moving body surface. To improve the accuracy, radioopaque fiducial markers can be implanted in/near the tumor region, several days before CT scanning for treatment planning. The fiducials—which are detectable in X-ray images—are used as reference markers to locate and track tumor location during patient alignment and treatment delivery. The *Synchrony Respiratory Tracking System* builds a correlation model between the positions of periodically detected fiducials and the real-time locations of optically tracked markers placed on the chest to track tumor location. It uses 4D CT (imaging through time) to measure respiratory tissue motion and deformation and to account for the effect of displacement and deformation through the irradiation (Urshel, 2007).

The B-Rob Systems

A robotic system for CT and ultrasound-guided biopsies was developed by the robotics laboratory of *ARC Seibersdorf Research* (Austria). The first B-Rob prototype was a 7 DOF robot integrated

Figure 7. a–b) The B-ROB I; c–d) the B-ROB II prototypes (Cleary et al., 2006); e–f) The new design, created by iSYS (iSYS, 2011)

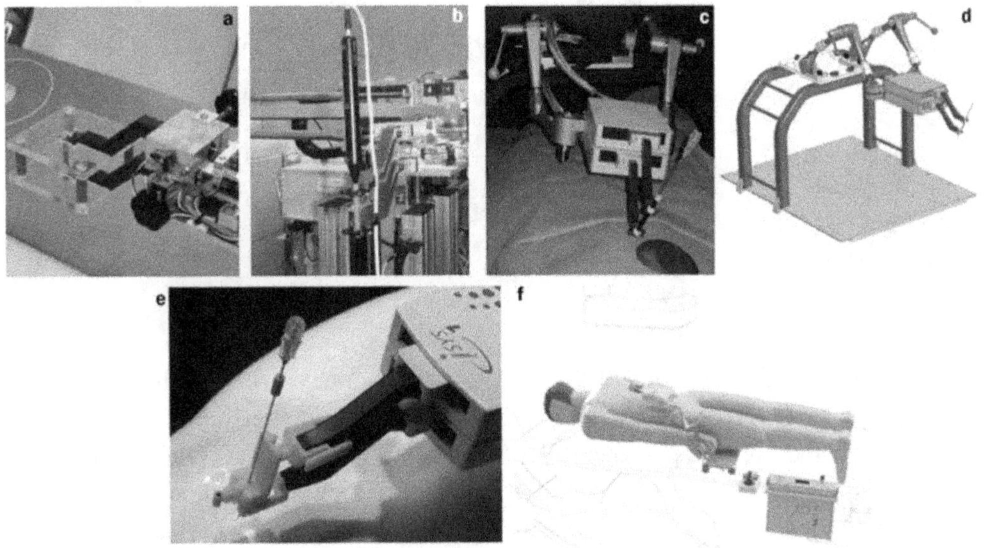

on a mobile rack. A 4 DOF positioning stage was used to direct the needle to the desired skin entry point. The complete system was thoroughly tested on needle-penetrable phantoms, where its application accuracy was 1.48 ± 0.62 mm, which was shown to be better than the traditional free-hand technique (Cleary et al., 2006). They broke with the concept of using large, universal manipulators to navigate tools, rather employed specialized hardware to suit the clinical needs of percutaneous procedures. This small-scale system better integrates with the OR.

The development of the second prototype (Figure 7c–d) was motivated by the aim to provide a modular setup for a variety of clinical applications, to allow easy integration with other systems, while reducing technical complexity and costs. The robot is equipped with a *Needle Positioning Unit* (NPU) for fine orientation. The first gel phantom tests of the B-Rob II showed 0.66 ± 0.27 mm application accuracy in IG positioning. In-vitro trials with ultrasound guided biopsy specimen harvests followed, where the mean deviation of the needle tip from the center of the target was 1.1 ± 0.8 mm. The technology was licensed by iSYS Medizintechnik GmbH, and the robot got redesigned, to better fit the market. It is currently undergoing ISO certification and CE marking.

2.3. System Capability Assessment

A major step in the evaluation of a prototype is system performance assessment, especially in terms of spatial accuracy and safety. Thorough tests are required, as the overall precision may be the highly non-linear function of the intrinsic- and registration accuracies. After 20 years of development, there is still a strong need for objective measures in medical robotics. In this section, different test methods are presented to define system characteristics, working towards their validation.

Accuracy numbers are used to convince the medical community about the improved patient outcome that robots can provide, therefore it is essential to have clear and well-founded metrics.

In CIS, registration and accuracy metrics have been commonly borrowed from industrial robotics (precision, repeatability, etc.) and image analysis (FRE, TRE, etc.), but it is important to

Figure 8. Difference between accuracy and repeatability (Based on (Stiehl et al., 2007))

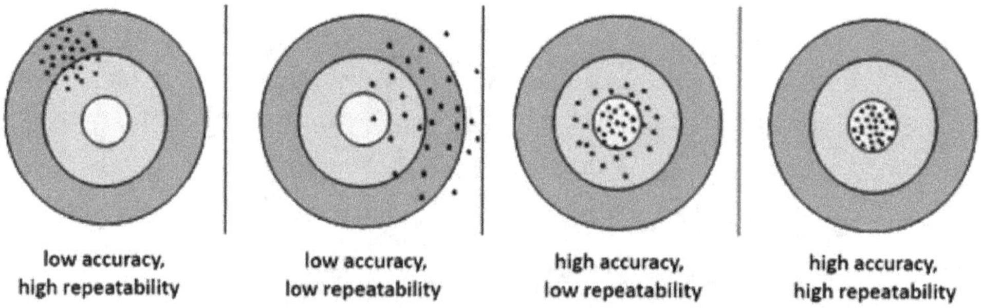

understand the validity and limitations of these concepts. Similarly, the differences between intrinsic (technical) accuracy, registration accuracy and application accuracy are to be well understood. Examples of accuracy measurement techniques are presented here, illustrating the importance of correct, consistent reporting, facilitating the comparison of different devices.

Precision of robotic systems can be represented by the accuracy and repeatability of the device to characterize the overall effect of the encoders' fineness, rigidity of the structure and the compliance of the hardware elements (the servo motors, the gears or the links). Both terms are defined for industrial robots in the *International Organization for Standardization* (ISO) 9283 standard (ISO TC-184/SC2 Robots and Robotic Devices). Accuracy refers to a robot's ability to position its end at a desired target point within the working volume. Generally, the absolute positioning accuracy shows the error of the robot when moved to a prescribed joint angle or Cartesian position. This expresses the mean difference between the actual pose (position and orientation) and the pose calculated from the mathematical model of the robot. "*Repeatability is the ability of the robot to reposition itself to a position to which it was previously commanded or trained*", as defined in (Nof, 1999). It is the standard deviation of the positioning error acquired through multiple trials to reach the same joint values (Figure 8). Repeatability is typically smaller for manipulators than accuracy, while both numbers are largely dependent on speed, payload and the range of motion (Stiehl et al., 2007). From the clinical point of view, the accuracy of treatment delivery is important, to know the task specific uncertainty. However, this may be difficult to measure routinely; it requires mock operations, cadavers or the use of pre- and post-operative imaging combined. Some manufacturers have tried to construct advanced measurement tools to facilitate system assessment. Alternatively, phantoms (artifacts) can also be used to replicate clinical conditions as much as possible.

There are three different types of accuracies (in terms of spatial errors) that can be specified with different error numbers (determined in general) according to (Grunert et al., 2003) and (Kronreif, 2011):

- Intrinsic (technical) accuracy (0.1–0.6 mm),
- Registration accuracy (0.2–3 mm),
- Application accuracy (0.6–10 mm).

Intrinsic accuracy applies to certain elements, such as the robot or the navigation system. It describes the average error of the given component in operational use. Random errors (e.g., mechanical compliance, friction, loose hardware), resolution of the imaging device, inadequate control and noise can all result in low intrinsic accuracy. On the user interface side, discretized input and modeling errors may further decrease precision.

Registration errors are also present, as computational methods involve some kind of residual errors. It is only possible to find a normalized (e.g., least squares optimized) solution for a mathematical fitting problem. In IGS, a major source of error can originate in the localization of the markers (different types, forms and materials), displacement of the fiducials and determination of the center of the fiducials. Even though the distribution of error for one fiducial in a particular set of measurements is usually Gaussian, the aggregated error for many fiducials can only be approximated to collide with it.

Application accuracy refers to the overall targeting error of the integrated system while used in a clinical procedure or a mock setup. It realistically measures the task specific effectiveness of a system and is commonly used for validation. The application accuracy depends on all other sources of errors in a complex, non-linear way, therefore typically phantom, cadaver or clinical trials are required to determine it. Further problems arise with the simple, ergonomic expression of spatial errors. Physicians may need a single number showing the precision of the system. In many applications, only the absolute distance from a desired location matters, therefore the *Root Mean Square Error* (RMSE) is given for the system:

$$\text{RMSE} = \sqrt{\frac{1}{N}\sum_{i=1}^{N}(x_i - x)^2}, \quad (1)$$

where N is the number of measurements, x is the desired point and x_i is the ith measured point. The RMSE is only an unbiased representation of isotropic and independent errors in the 3D space. For other cases, the covariance matrix of the distribution should be used.

Equation (1) does not incorporate the angular errors of the system, even though any 3D registration or tracking component with a rotational error will affect the translational accuracy. Even worse, the factor of degradation is dependent on the value of the translational vector, due to the nature of the homogeneous transformation matrix multiplication. This is especially bothersome in the case of intra-operative navigation systems that are supposed to be used with a *Dynamic Reference Base* (DRB), therefore the rotational inaccuracy's effect on the linear accuracy will depend on the distance between the camera and the patient, and will have anisotropic distribution. This model is valid for zero-mean Gaussian distributions, and RMSE gives a single value even to multi-dimensional distributions.

In IGS, typically not only robots, but tracking devices are also incorporated, providing the necessary information for navigation. In the case of IG therapies, the same metrics could be applied; however the effect of imperfect registrations and coordinate transformations has major contribution to the overall error. Performance estimation focuses on precision, noise, static/dynamic accuracy or latency. Medical device manufacturers typically provide maximum spatial error values with standard deviations, and publish limited experimental results on the distribution of errors. According to (Frantz et al., 2004), two conditions are necessary to correctly assess a positioning system's performance:

- Characteristic statistics (defining trueness and precision),
- A specifically defined assessment protocol (on which the measurement is based).

Evaluating robotic systems usually involves not only mathematical modeling and simulation, but also extensive accuracy tests. One of the difficulties in assessing an IG robot is to acquire the ground truth—the gold standard. This is feasible through the use of a significantly more precise device (e.g., laser scanner, accurate camera system), the use of a measurement phantom or other trusted method (providing the ground truth). In industrial robotics, accuracy measures and tests have been widely used, and some got straight applied to

CIS (Haidegger et al., 2010). Most commonly, the medical device is guided (directed) to different positions and orientations along a precisely known set of landmarks (fiducials) or an accuracy board. The positions can also be recorded with an independent localizer.

To evaluate the different point-based tests, certain measures had to be developed and used. Let us assume that there are $N+M$ points in total used during the experiment. These can either be artificial fiducials or anatomical landmarks $\mathbf{p}_1$, $\mathbf{p}_2$, … $\mathbf{p}_N$ points are used during the registration, and $\mathbf{p}^*_1$, $\mathbf{p}^*_2$, … $\mathbf{p}^*_M$ points are used during the procedure (and to derive the error at the target).

Specific to the tracker and the setup, the *Fiducial Localization Error* (FLE) includes the intrinsic and extrinsic sources of error, representing the accuracy of a position tracker to localize a $\mathbf{p}_i$ ($i = 1, \ldots, N$) point, consequently the centroid of the cluster of measured points (Maurer et al., 1998). FLE can be defined as the mean value of the error of all samples:

$$FLE = \frac{1}{N_{Trial} \cdot N_{Fiducial}} \sum_{i=1}^{N_{Trial}} \sum_{j=1}^{N_{Fiducial}} \varepsilon(i, j), \qquad (2)$$

where ε is the error of a single measurement at a given fiducial.

One of the most precise optical trackers available on the market is the *Optotrak Certus* from NDI. It has a 0.1–0.15 mm (RMSE) FLE according to the specifications. Typical surgical navigation systems provide less accurate measurements, a 0.2–1 mm RMSE error.

Fiducial Registration Error (FRE) is the mean square distance between homologous fiducial points; the residual error of the paired-point registration between the given subset of the known and recorded fiducial coordinates ($\mathbf{p}_i$, $i = 1, \ldots, N$) during an accuracy test (Fitzpatrick et al., 1998):

$$FRE = \sqrt{\frac{1}{N} \sum_{i=1}^{N} \left| T(p_i) - q_i \right|^2}, \qquad (3)$$

where N is the number of fiducials used during the registration, $\mathbf{q}_i$ is the position of the ith fiducial in one space (may be the robot), $\mathbf{p}_i$ is the same point in the other (patient space) and $\mathbf{T}$ is the computed homogeneous transformation.

Target Registration Error (TRE) is the deviation between points ($\mathbf{p}^*_i$, $i = 1, \ldots, M$) in the reference and the other (registered) coordinate system. FLE, FRE and TRE are presented in Figure 10. TRE is typically used for the characterization of schematic point-based registrations. Ideally, FRE and TRE both equal zero.

$$TRE = \sqrt{\frac{1}{M} \sum_{i=1}^{M} \left| T(p_i) - q_i \right|^2}, \qquad (4)$$

where M is the number of fiducials used to compute TRE (that are not identical to any of the points used during registration). In medical cases, TRE might be computed based on distinguished anatomical points.

Novel research publications show that TRE and FRE are independent for point-based registrations (Fitzpatrick, 2009) and (Danilchenko & Fitzpatrick, 2010). Many commercially available surgical navigation systems use (incorrectly) FRE as a metric for the precision of the system, while others use proprietary algorithms to define an accuracy number to display to the surgeon (see Figure 9).

2.4. Standards and Metrics to Follow

Once the prototype system steps out of the laboratory, it has to comply with various regulations and standards, therefore it's always better to keep those in mind form the beginning of the development. Safety is paramount for any surgical device, especially in the case of autonomous actuation. This issue has been addressed by many individual researchers, governmental bodies and scientific societies in the past years, pushing for a unified standardization effort to ensure patient

Figure 9. Definition of FRE and TRE used to assess point-based registration methods. The black and white circles represent corresponding point pairs in the two different spaces. FLE is the spatial deviation between the true and the recorded position of the landmark points that the registration is built on. FRE is the residual error of the applied transformation calculated over the points used to derive the **T** *transformation. TRE is the error of mapping (a set of) independent points from the original space to the registered space.*

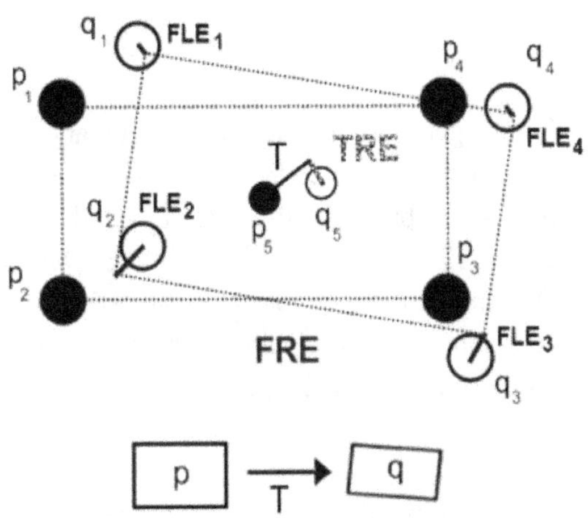

and medical staff safety in the operating room. System assessment is paramount for any prototype, and in interventional medicine, the device is most commonly in direct contact with the patient. This section collects the relevant regulations, focusing on the general approach, trends and requirements they implement.

Current Accuracy Standards

International bodies are exerting great effort to standardize medical robotics similarly to industrial robotics. However, there are no widely accepted regulations. Some of the existing robotic and medical device development standards are applicable to CIS currently. (Although these are changing periodically, the reader should be able to find the most up-to-date versions of the relevant regulations based on the ones listed here.)

In 2004, the *American Society for Testing and Materials* (ASTM) initiated a new standards committee (ASTM F04.05) under the title *Standard Practice for Measurement of Positional Accuracy of Computer Assisted Orthopaedic Surgical Systems* (CAOS). The goal was to develop an international standard for metrology, validation and performance of CAOS systems (Stiehl et al., 2007). The first draft (dating from 2007) deals with the localizer functions of navigation systems (optical, mechanical or electromagnetic). The defined generic measurement board—nicknamed *Nebraska phantom*—was machined from aluminum-alloy, and was tested with three different CIS systems (Barrera et al., 2007). It is a multi-surface object with 47 identical fiducial points (0.75 mm deep cone-shape holes) distributed on its surfaces (Figure 10a).

ASTM F04.05 plans to develop a set of standard for specific tasks (cutting, drilling, milling, reaming), distinct surgical applications (joint replacement, implant nailing, plating, osteotomy) and imaging modalities (fluoroscopy, CT, MR, ultrasound). Supporting the ASTM group, a multi-institution technical committee presented a white paper, calling for standardization in many areas of CIS (Kazanzides, 2006), (Chiao, 2008). Based on technological and economic analysis, metrology and standards should be applied especially to the following categories of medical devices:

- Computer-assisted navigation and surgery,
- Surgical robots (mostly in manual control mode),
- Surgical robots and phantom (artifact) devices,
- Stimulation devices,
- Drug-delivery and physiologic monitoring devices.

Another clinical phantom (*the Computer-Assisted Orthopaedic Hip Surgery—CAOHS Artifact*) was built at the *National Institute of Standards and Tech-*

Figure 10. a) ASTM CAOS draft standard accuracy phantom (Stiehl et al., 2007); b) NIST CAOHS Artifact for the verification of hip replacement surgery (Dagalakis et al., 2007).

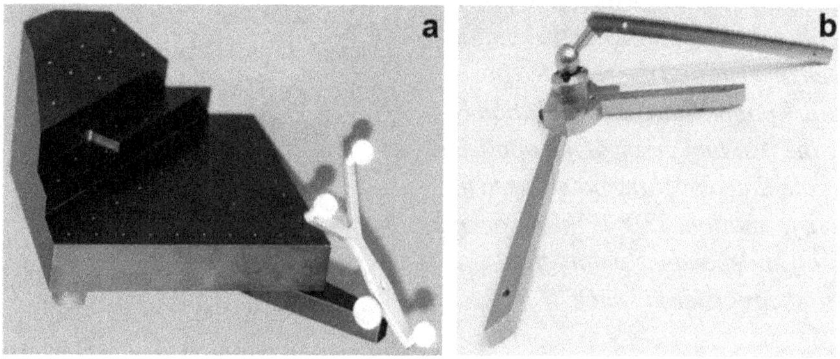

nology (NIST) to quantify task specific measurement uncertainty. It was designed to mimic hip joint using magnetic ball-and-socket to be able to simulate hip replacement procedures (Figure 10b).

Accuracy Measurement Standards Relevant to the Field

Several standards exist for medical devices in general (*International Electrotechnical Commission* (IEC) 601, ISO 14971, ISO 13485:2003), separately for industrial robotics (ISO 10219) and for safety-related systems (IEC 1508); FDA has its *Quality System Regulation* (QSR) and the *Process Validation Guidance* (GHTF) N99-10. However, none of these addresses surgical robotics directly. The ISO 10360 standard "Acceptance and reverification tests for *coordinate measuring machines* (CMM)" from 1994 (current version: 2001) can be applied to surgical robots in many cases (mostly to semi-autonomous systems). It defines the protocol for CMM based on *Volumetric Length Measuring Error* (VLME) and *Volumetric Probing Error* (VPE). To acquire VLME, a set of 5 gauges has to be measured 3 times with one probing at each end, in 7 different directions in space (Figure 11a, and all results must be within the specified error. For VPE, a precision sphere has to be measured with 25 equally distributed probing (Figure 11b). The probing error is the range of all radii.

Telesurgery require additional regulations. A recent report (Gartner, 2008) discusses technical and human issues based on the existing standards according to 1993/42/EEC, *Medical Devices Directive* (MDD) and the third edition of the IEC 601-1 (2006) for tele-neurology. According to the report, manufacturers of telemedical systems for neurology are required to:

- Determine the medical purpose §3.10 of MDD,
- Carry out risk classification and risk management (according to DIN EN ISO 14971),
- Completely meet the requirements of MDD Annex I.

Figure 11. ISO 10360-1:2001 standard for coordinate measuring machines: a) Illustrating volumetric length measuring error; b) Illustrating volumetric probing error (ISO, 2000).

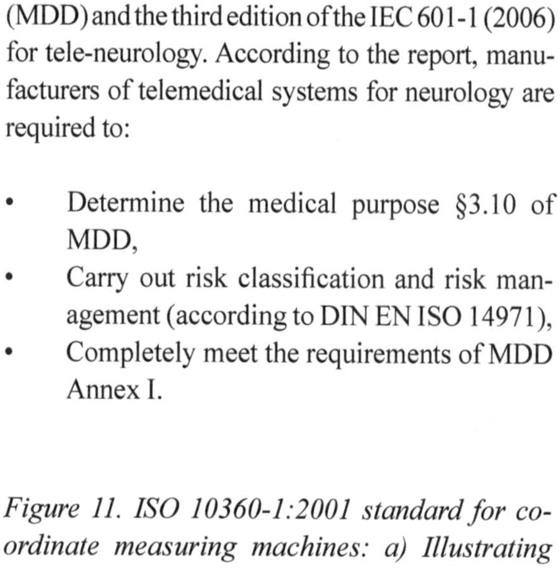

Examples of Accuracy Measurements

In indirect IGS 3–5 mm RMSE application accuracy is considered acceptable, whereas 2 mm is recommended for IG neurosurgery. However, sub-millimeter precision is recommended for robot assisted procedures—direct IGS (Grunert et al., 2003). When CIS system accuracies are reported in research papers, many times neither of the generic standards is followed. Instead, custom metrics, protocols and measurement boards are used. As a consequence, it is extremely hard to effectively evaluate and compare the results. Most often, the experiments are not described in details (e.g., application accuracy refers only to a single component of a system). While these protocols are better regulated in the industry, the assessment techniques are numerous both in the case of medical robots and surgical navigation systems.

Table 1 summarizes the published accuracy measures for well-known surgical robots. These tests were mostly conducted in engineering laboratories, following different approaches. Understandably, the control mode of the robot can greatly influence the overall accuracy of the system. It should be noted that for master–slave systems—like the da Vinci—safety originates from keeping the human in the control loop at all times through real-time sensory feedback. This means the intrinsic positioning accuracy of the robot is not crucial anymore; the surgeon uses the provided visual information to compensate for any positioning errors. However, this approach introduces the generic limitations of a human operator (such as physiological latency), and even for these systems, the intrinsic accuracy of the components has significant effect on performance.

Table 1. Different accuracies published for major surgical robot systems. Details of the experiments are not disclosed in many cases. Values are given with mean ± standard deviation. [a] Target Registration Error [b] registration performed with a stereotactic frame (head fixator) [c] registration in frameless mode [d] Fiducial Localization Error.

Robot	Company	Intrinsic accuracy	Repeatability	Application accuracy
Puma 200 [1]	Memorial Medical Center		0.05 mm	2 mm
ROBODOC [2, 3, 4]	Integrated Surgical Systems Inc. Curexo Technology Corporation	0.5–1.0 mm		1.0–2.0 mm 1.05 mm
NeuroMate [5, 6, 7]	Innovative Medical Machines Int. Integrated Surgical Systems Inc. Renishaw plc	0.75 mm 0.6 mm 0.36 ± 0.17 mm[a]	0.15 mm	0.86 ± 0.32 mm[b] 1.95 ± 0.44 mm[c]
da Vinci [8]	Intuitive Surgical Inc.	1.35 mm[d] 1.02 ± 0.58 mm[d]		
da Vinci S [9]	Intuitive Surgical Inc.	1.05 ± 0.24 mm[d]		
CyberKnife [10, 11]	Accuray Inc.			0.42 ± 0.4 mm 0.93 ± 0.29 mm
B-Rob I [12] **B-Rob II** [13]	ARC Seibersdorf Research ARC Seibersdorf Research		1.48 ± 0.62 mm 0.66 ± 0.27 mm	 1.1 ± 0.8 mm
SpineAssist [14]	Mazor Surgical Technologies			0.87 ± 0.63 mm

References: [1] (Kwoh et al., 1988); [2] (Paul et al., 1992); [3] (Taylor, 2001); [4] (Nishihara et al., 2006); [5] (Li et al., 2002); [6] (Varma et al., 2006); [7] (Haidegger et al., 2008); [8] (Kwartowitz et al., 2006); [9] (Kwartowitz et al., 2007); [10] (Dieterich et al., 2003); [11] (Jang et al., 2006); [12,13] (Cleary et al., 2006); [14] (Lieberman et al., 2006).

Application of Accuracy Measurements

The ASTM phantom and protocol were used for the assessment of the accuracy parameters of the ongoing neurosurgical research project at the *Johns Hopkins University* (JHU—Baltimore, MD). A modified NeuroMate robot was used, capable of helping and increasing the performance of human surgeons. A *StealthStation* guidance system (Medtronic Navigation, Louisville, CO) and the *3D Slicer* open source medical imaging software toolkit (www.slicer.org) were integrated to support skull base surgery. A 6 DOF force/torque sensor (JR3 Inc., Woodland, CA) and a high-speed bone drilling surgical instrument— Anspach eMax 2 (Anspach Effort Inc., Palm Beach Gardens, FL)— were attached to the last link of the NeuroMate robot (Figure 12a). Optical tracking is made possible via passive visual markers mounted on the robot and on the patient. The robot is guided in cooperative control mode for the removal of cranial bone on the skull base, while *Virtual Fixtures* (VF, i.e., safety boundaries) are applied to protect critical anatomical structures (Xia et al., 2008).

We performed accuracy tests using the ASTM draft, phantom manufactured at the University of Nebraska (Figure 12b). The accuracy was measured by directing the robot tooltip into the holes and comparing the positions measured by the joint encoders to the known positions acquired by a CMM (Figure 7). We performed five complete tests, and the collected data was evaluated based on a paired-point registration method. Every second measured point was used in the registration, resulting in an average FRE of 0.36 mm. The rest of the points were used to compute TRE; the results showed that the intrinsic accuracy of the robot is 0.34 ± 0.17 mm (TRE). The same phantom was used to test the accuracy of the StealthStation. We performed measurements tracking the *Robot Rigid Body* (the fiducial mounted at the tip of the robot) and the hand-held pointer probe (also used for CT to patient registration). Five experiments were performed with the robot and three measurements taken with the pointer probe. In the first case, the results showed that the navigation system had 0.49 ± 0.22 mm target registration error (FRE: 0.49 mm). In the second case, the TRE was 0.51 ± 0.42 mm (FRE: 0.52 mm). The ASTM phantom and protocol were used for the assessment of the accuracy parameters of the robot and the navigation system

Figure 12. a) The NeuroMate-based IGS system in action: bone cutting on the skull base in a cadaver experiment at Johns Hopkins Medical Institution (Xia et al., 2008); b) Accuracy test with the NeuroMate robot at the R. A. Swirnow Mock Operating Room at JHU CISST ERC (Haidegger et al., 2008)

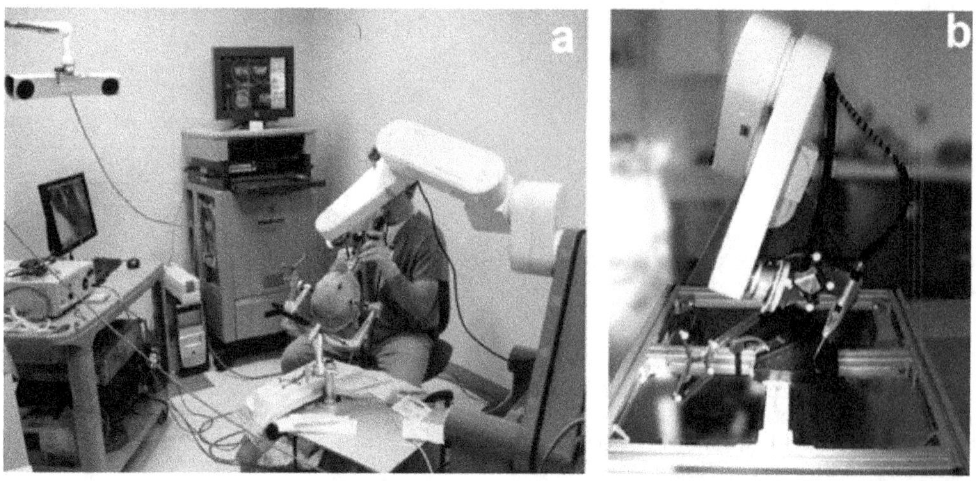

without any difficulties. The working volume of the targeted skull base surgeries is smaller than of knee joint replacement procedures, therefore the phantom was easily applicable.

Safety Standards and Methods for CIS

Safety is paramount for any surgical devices, and many kinds of errors may lead to critical conditions in the OR. According to (Satava, 2004), errors in interventional medicine can be categorized as:

- **Commission:** doing the wrong thing,
- **Omission:** not doing the right thing,
- **Execution:** doing the right thing incorrectly.

Errors can be either systematic (a series of errors resulting in an adverse event) or specific (the event itself is a form of error).

Several groups have published methodologies to support the safety of design and development of robotic surgical devices. A generic one is the *Hazard Identification and Safety Insurance Control* (HISIC) policy that has been applied to multiple surgical robotic systems so far (Fei, 2001). HISIC breaks down the issue into seven principles:

- Definitions and requirements,
- Hazard identification,
- Safety insurance control,
- Safety critical limits,
- Monitoring and control,
- Verification and validation,
- System log and documentation.

Further, a *Computational Evolution* method (Varley, 1999) and a *Unified Modeling Language* (UML) based approach have been successfully prototyped (Guiochet & Vilchis, 2002), relying on safe design, safe execution and risk assessment as cornerstones. Risk management in general is a key component of the entire medical device safety. This includes (Guiochet & Vilchis, 2002):

- Risk analysis (system definition, hazard identification and risk estimation),
- Risk evaluation (determine risk tolerance levels),
- Risk control (implementing the right action for maximum safety).

Failure mode analysis and risk assessment methods both for hardware and software in general have been addressed by many standards (Varley, 1999; Kazanzides et al., 2008; Kazanzides, 2009):

- IEC 60812 International Standard on Fault Mode and Effects Analysis (1985),
- IEC 1508 Functional Safety: Safety-Related Systems (1995),
- *European Norm* (EN) 1441 on risk management (1997),
- *American National Standards Institute* (ANSI) R15.06 standard for industrial robot safety (1999),
- GHTF/SG3/N99-10:2004 Quality Management Systems—Process Validation Guidance (2004),
- IEC 62304 on Medical Device Software—Software Life Cycle Processes (2006),
- IEC 61025 Fault Tree Analysis (Ed. 2.0, 2006),
- ISO 14971 Application of Risk Management to Medical Devices (2007),
- IEC 60601 international standard on Medical Electrical Equipment (Ed. 3.0, 2010),
- IEC 1508 draft standard on Functional Safety for software developers.

2.5. Clearance Procedures

Testing a prototype is only the first step towards commercialization, acquiring certification involves many other issues. To objectively evaluate the performance of a robot-assisted system, it is crucial to understand and apply consistent measurement methods. However, many other factors determine the success of a surgical robot beyond spatial precision. The accuracy of treatment delivery remains the baseline for applicability; a medical robotic system should still address the questions of complexity, cost and ergonomics. Different systems should be measured and assessed through the same validated experiments.

The existing industrial standards and medical robotics related drafts should be extended to all major areas of CIS. Surgical robot systems typically have an application accuracy between 1–2 mm. This means lower precision that of industrial robots, mainly due to the accumulation of different errors in IGS, preventing sub-millimeter accuracy. (Although sometimes, sub-millimeter inherent robot accuracy is wrongly claimed as overall precision.) A robot must be intuitive and require minimal maintenance and engineering skills to operate (Kazanzides et al., 2008). The user acceptance of a system will eventually determine the value of the device, therefore one of the major directives of development is to minimize the change to the existing clinical workflow. These ideas are typically represented in the medical device regulations.

International Regulations for CIS

Clearance applications (and the following continuous communication with the regulatory bodies) include discussion of electronics design, imaging systems' performance, embedded software analysis and clinical trial design and patient outcome validation. The procedures both in Europe and in the U.S. are focusing on the safety and transparency of systems (Taylor & Kazanzides, 2008). Most prototype development and testing begin with the official approval of the *Institutional Review Board* (IRB), legally taking responsibility for the primer operations.

In the EU, the CE mark must be obtained, certifying that the product complies with the essential requirements of the relevant EU health, safety and environmental protection legislation. The approval procedure is managed by independent *Notified Bodies* (NB). There are approximately 100 international, non-governmental NB for medical devices. ISO 9000 Quality Standards family is applied to verify the production management of a company (www.iso.org/iso/isocatalogue.htm). ISO 9001:2000 combines three previous standards (9001, 9002 and 9003) addressing design and development procedures under the title "Quality management systems—Requirements".

For CIS systems the ISO 13485:2003 (*Medical devices—Quality management systems—Requirements for regulatory purposes*) is in effect. It is possible for ISO 9001 complied companies to self-certify (CE mark) their products within certain limitations, and the NB would periodically audit them.

U.S. Practices

In the U.S. only the federal Food and Drug Administration can clear medical systems (www.fda.gov/MedicalDevices/ProductsandMedicalProcedures/default.htm), since the 1976 *Medical Device Amendments*. The *Safe Medical Device Act* (1990) defined the present regulatory structure requiring that all medical devices should ensure safe and effective use (DiMaio et al., 2011). "*An* Investigational Device Exemption *(IDE) allows the investigational device to be used in a clinical study in order to collect safety and effectiveness data required to support a submission to FDA. (FDA, 2011)*" For initial trials, FDA requires an IRB approval certifying that the device in question poses only insignificant risk. If an investigational use of a new system (or with a new procedure) poses significant risk to the patient, the IDE needs an FDA approval. Next, to allow further clinical investigation of a new device, the IRB and the FDA must determine the followings (Bronzino, 1990):

1. Risks to subjects are minimized,
2. Risks to subjects are reasonable in relation to the anticipated benefit and knowledge to be gained,
3. Subject selection is equitable,
4. Informed consent materials and procedures are adequate, provisions for monitoring the

study and protecting patient information are acceptable.

The *FDA review team* incorporates experts from various fields, and if needed, they may hold a Public Advisory Panel meeting, to obtain a second opinion from external professionals.

Before the whole clearance procedure is initiated, it is a good idea to ask for a pre-submission meeting from FDA, where it is possible to get informal feedback from an FDA review team on the preliminary data and concept. Clearance can be done primarily in the form of a *Pre-Market Approval* (PMA) procedure—which is for new devices—requiring extensive clinical trials and huge amount of documentation. Table 2 summarizes these requirements for different kinds of systems. A PMA submission currently costs approximately \$240K, although a reduced pricing is applied for small businesses, and it has 180-day review cycle.

On the other hand, the *Premarket Notification* pathway, the *510(k)* is for equipment that can be proved to be "*substantially equivalent*" to an existing device, already approved by FDA. Substantial equivalence meant in terms of safety and effectiveness with respect to intended use and design, compared to a predicate device. It has a 90-day-long review cycle, and may still require clinical trials, but less extensive than for PMA. All systems must comply with FDA *Quality System Regulations/Medical Device Good Manufacturing Practices* (*Code of Federal Regulations* Title 21, Part 820) (Lowery et al., 1996).

While there are 3–4000 independent 510(k) submissions annually, only 30–50 PMA arrive to FDA. The basic idea behind these regulations is to prevent failures and safety issues originating from bad design. The clinical use and patient outcome might not even be verified during the validation. At the most, the system should show the capability to perform a procedure with the same effectiveness as an existing (manual) technique. FDA can be argued that is used to rely on the selectivity of the market, which should only allow for the existence of well-sustained systems with significant added value to the surgical procedure.

Systems are classified to three main categories, depending on the risk they pose to the patient, the user and the based on the intended way of use:

Table 2. Summary of classification of CIS devices and their testing requirements in the U.S. (Janda & Buch, 2009)

Device Type	Regulatory Class	Bench Testing	Animal Testing	Software Validation	Clinical Data
Preoperative planning	II			X	
Stereotactic frames	II	X			
Computer-assisted or navigation device Computer-assisted intraoperative planning and surgical guidance	II	X		X	X
Robotic operating assistants Computer-assisted intraoperative planning and surgical guidance or action	II or III	X		X	X
Fly-by-wire	II or III	X		X	X
Robots	Unclassified	X	X	X	X

- **Class I:** low risk devices (45% of the submissions, many exempt),
- **Class II:** medium risk devices (45%, usually 510(k)),
- **Class III:** devices involving high risk to the patient, e.g., life-sustaining equipment (<10%, typically PMA).

The da Vinci robot was originally categorized as Class III, but eventually got approved as a Class II device. When performing objective system assessment, FDA looks into the following issues in the case of CIS systems (Janda & Buch, 2009):

- How accurately do the system's actions reflect the operator's intent?
- Does the device enhance surgical capabilities?
- How does the computer "intermediary" affect operative flow?
- How easily are glitches in the technology handled?
- Are there procedures and/or anatomic sites that are more or less amenable to this technology?
- How steep is the learning curve?
- What criteria are used to "optimize" a preoperative plan?
- How and when should the user override the recommendations of the system?
- How does the system define anatomic references and axes?
- How does the system dictate the surgical workflow, and how does this differ from traditional techniques?

The ROBODOC Case

The clearance procedure of the ROBODOC system was quite long and full of edifying stories (Curexo, 2009). It was the first robotic system that tried to get through the FDA clearance procedure, and it was forced into the PMA category, since no equivalent technology existed.

- October 1992, the FDA granted an IDE approval to conduct limited clinical trials (at Sutter General Hospital in Sacramento).
- November 7, 1992, the world's first robotic joint replacement surgery was performed.
- A biomedical engineer for the FDA outlined comprehensive investigative testing designed to test the safety of the robotic control software.
- 1993, ISS began the PMA clearance procedure. The ROBODOC software and hardware had to comply with robust motion control restrictions to prevent accidental tissue and bone damage.
- 1996, it received the European CE mark and approval for THA procedures.
- 2002, the FDA re-assigned the ROBODOC system to the 510(k) path. Unfortunately, by this time, ISS no longer had sufficient capital to support the procedure.
- By 2004, the significant R&D cost related to further testing and re-submission caused ISS to cease operations.
- November 2007, all assets and IP were transferred Curexo Technology Corporation, and \$12M was invested to continue operations.
- The fresh money allowed to finish trials, and to complete the robot's commercialization via FDA 510(k).
- In August 2008, the system finally received a 510(k) clearance from FDA.

Throughout these years, all the FDA-cleared surgical robots (e.g., da Vinci, NeuroMate, CyberKnife, ROBODOC, MAKO Arm, SpineAssist) went down the 510(k) procedure, proven to be substantially equivalent to existing (manual) technologies. FDA's emphasis has gradually shifted from the robotic technology itself to approving the results of the use of the robotic systems. It is now believed that the equivalence stands between the already-approved surgical techniques and the results attained through robotic interventions.

Current Trends

Regulatory bodies have realized the need for a better approach to the dynamically growing field of CIS. Since 2010, the 2007/47/EC (*European Council*) regulation extended the existing 1993/42/EEC MDD, requiring specific clinical data for new devices and companies are expected to do to ensure safety and demonstrate the performance of their product, regardless of the classification of the device. The directive also requires risk and benefit analysis, and the sustained and coordinated clinical post market surveillance of the products. It comes into effect from June 2012 in Europe and in Canada, and from June 2013 in the U.S. In addition, software developed for diagnostic or therapeutic purposes has become classified as a medical device in the EU.

More recently, a workgroup has been created to prepare an extension (the third edition) to the existing ANSI/AAMI ES60601-1:2005 standard on *Medical Electrical Equipment*, focusing on medical robotics and personal care robots. The first meeting was held in early 2010, and the new amendment is due August 2013. Working together with the subcommittee of the ISO, TC 184, dealing with robotic devices, the priorities of the new standard are:

- Defining medical software systems and associated technical requirements,
- Streamlining the application of risk management,
- Clarifying the definition of essential performance,
- Identifying essential performance and mitigating risk.

The ISO/TC 184/SC 2 technical sub-committee on *Robots and Robotic Devices* proposed the formation of a joint work group with the IEC/SC 62A subcommittee (Common Aspects of Electrical Equipment used in Medical Practice) under IEC administration to develop a new proposal for a Collateral Standard within the IEC 60601 family. The scope of the standard would be to the basic safety and essential performance requirements for medical electrical equipment and in addition, systems employing robotic technology (i.e., medical robots).

In the EU, updates to the IEC 60601-1 means that regulatory bodies decided to require supportive medical data as evidence for the safety and performance of new systems. Beginning in June 2012, risk assessment and analysis for every medical device designer and *Original Equipment Manufacturer* (OEM) will be mandatory. (Planned in the USA only from 2013.)

The FDA's 510(k) procedure is going under major revision based on the complaints filed by the device manufacturers. Two committees were created: *510(k) Working Group* and the *Task Force on the Utilization of Science in Regulatory Decision Making*. Improvements should include the clarification on clinical data requirements and the implementation of more transparent, science-based decision making structure, specifically:

- When and what type of manufacturing data to submit?
- When a pre-clearance inspection would be conducted?
- When and what types of modifications should be periodically reported in lieu of submitting a 510(k)?
- When and what type of safety and effectiveness information for the device to be reviewed that is known to the manufacturer should be submitted as a brief description?

The aim is to reduce the number of failures through more thorough tests before the release, especially for devices such as defibrillators, hip implants and hospital pumps (FDA, 2010).

The *Institute of Medicine* (IOM) concluded recently that it would be better to develop an integrated pre-market and post-market regulatory framework that provides a reasonable assurance of safety and effectiveness throughout the device life cycle (IOM, 2011). "The new framework should be:

- Based on sound science,
- Clear, predictable, straightforward, and fair,
- Risk-based,
- Be self-sustaining and self-improving,
- Facilitate innovation that improves public health by making medical devices available in a timely manner and ensuring their safety and effectiveness throughout their lifecycle,
- Use relevant and appropriate regulatory authorities and standards throughout the life cycle of devices to ensure safety and effectiveness."

3. FUTURE RESEARCH DIRECTIONS

In the past 10 years, we have seen the dynamic growth of the field of CIS, and the best surgical robots gradually conquered the hospitals. Da Vinci is dominating the market, however, several competitors are on the way, relying on the reputation surgical robots have already achieved. The success of Intuitive's system inspired many groups:

The development of the Telelap ALF-X system (*Advanced Laparoscopy through Force-RefleCT(X)ion)* began three years ago at SOFAR S.p.A., (a pharma company in Milan), started with retrofitting industrial manipulators to create a da Vinci competitor (Figure 13a). They are supported by an EU joint effort, and operating under the academic assistance of the *New European Surgical Academy* (NESA). Next, they teamed up with *Immersion Corp.* to ensure a good haptic interface (*TouchSense*) for their robot, and submitted numerous patents. Features of the robot include force control, preoperative simulation, intra-operative virtual reality overlay, automated fulcrum point identification and real-time patient monitoring. They aim to achieve 1.5 mm application accuracy and under 60 ms latency in the control. The system's main application would by gynecology. Preclinical trials with the robot

Figure 13. a) Concept of the ALF-X system (Tinelli et al., 2011); b) The Canadian Amadeus system (Courtesy of Titan Medical Inc.); c) Sofie, developed at Eindhoven University of Technology (Courtesy of EUT.); d) The Raven from the BioRobotics lab at University of Washington (Courtesy of UW.)

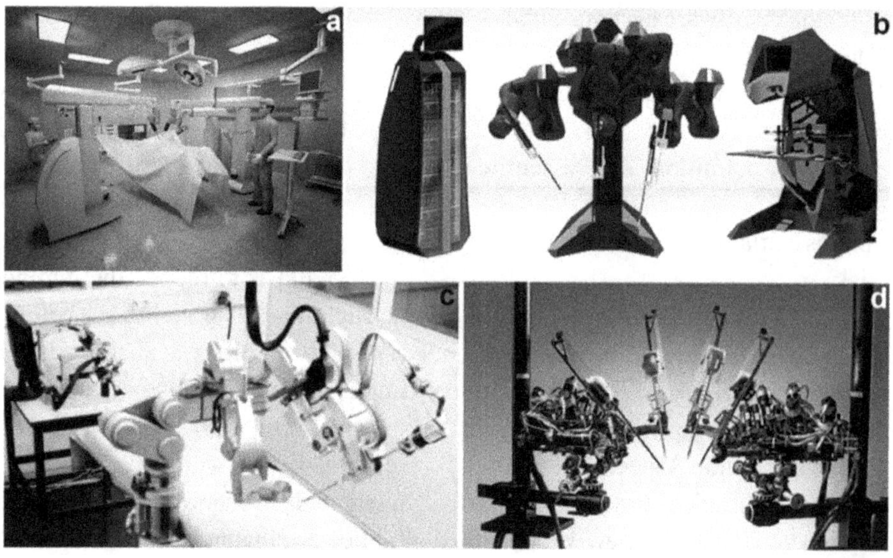

have been performed, focusing on hysterectomy, salpingo oophorectomy, myomectomy, partial and radical nephrectomy, total pelvic exenteration and cholecystectomy (Tinelli et al., 2011).

More recently, a Canadian company—*Titan Medical Inc.* (Toronto, Ontario Canada)—announced its new four-armed manipulator system, the *Amadeus* (Rayman, 2009). The robot is being built in cooperation with *Bell Canada*'s Health Division to integrate Internet Protocol based advanced telesurgery capabilities for Amadeus. Titan Inc. announced in April 2011 that they will use *KUKA Lightweight arms* (LW) for the setup (Figure 13b). This is the commercialized version of the *DLR III arm* (7 DOF, fully sensorized, compliant). The LW is preferred solution for other surgical robotic setups as well (e.g. the Active FP7 project). It is also an important feature that the KUKA robot does not use cable-drive, therefore does not collide with Intuitive's patents on the technology. Strategic agreements have been made with Canadian hospitals for the trial of the system, once they have a clinically applicable prototype (expected in 2013). Amadeus should be used in telemedicine services, including education and information sharing between sites. It is believed that the robot will be superior to da Vinci due to the redundant and compliant KUKA manipulators and to the use of more flexible tool shafts. They plan to target bowel surgery and mitral valve repairs beyond the typical abdominal and urulogical procedures.

The *Surgeon's Operating Force-feedback Interface Eindhoven* (Sofie) is a surgical robot developed at the *Eindhoven University of Technology (*van den Bedem, 2010*)*. It is a compact surgical robot, integrating force feedback to allow the surgeon to feel what he or she is doing (Figure 13c). Tactile feedback is provided in the master side joysticks. This counter pressure enables a surgeon to feel exactly what force he applies when making a suture or pushing aside a bit of tissue. Sofie is smaller, bed-mounted, hence it is less of an obstacle in the operating theater and above the patient. This avoids the otherwise necessary re-arrangement when the operating table and the patient are moved or tilted. Further, Sofie makes it possible to approach an organ from different sides thanks to its flexible tools.

The *BioRobotics Lab* at *University of Washington* has developed a portable surgical robot that can be a compromised solution to install on spacecrafts with its 22 kg overall mass (Figure 13d). The robot—called *Raven*—has two articulated arms, each holding a stainless steel shaft for different surgical tools. It can easily be assembled even by non-engineers, and its communication links have been designed for long distance remote-control. Besides the possibility of haptic feedback, additional sensors are mounted on the robot to provide more information to the surgeon and to avoid any critical failures due to communication delay. Throughout the entire development, compactness was handled as priority, the creators optimized the robot's dimensions and range-of-motion by computer, minimizing the space it occupies without compromising on manipulation capabilities (Lum, 2008).

An Italian spinoff company of the University of Verona—Surgica Robotica Ltd.—has been developing *Surgenius* since 2009. This robot is based on NASA's RAMS system that was developed in the mid-1990s. It was recognized by the euRobotics Technology Transfer Award. Currently it is undergoing pre-clinical trials for system validation, and aimed to get CE mark in 2012.

New Concepts, New Prototypes

Current research projects are trying to increase the utility of the surgical equipment along different strategies. They are mainly focusing on three areas for improvement:

- Augmenting the overall accuracy and/or efficacy of the classic stereotactic systems,

- Increasing the added-value of the equipment,
- Further enhancing the capabilities of the human surgeon, providing smarter tools.

The following ongoing research examples give insight to how these issues are addressed and what are the benefits for future patients. Safety is paramount in all cases, and should determine the way research is conducted. Patient safety is addressed differently in each discussed system.

Improvement of Stereotactic Surgery

The European Union's recent project—ROBOCAST (*Robot and Sensors Integration for Computer Assisted Surgery and Therapy*)—aimed to augment existing IGS techniques and to find new ways to perform high-precision keyhole neurosurgery (Comparetti et al., 2011). The modular system consist of two manipulators and one smaller probe, actively cooperating in a bio-mimetic sensory-motor integrated framework (Figure 14a,b). Mazor's (Caesarea, Israel) *SpineAssist* is used as a robotic end effector by itself, enhanced with a linear drive for needle insertion (Lieberman et al., 2006). The 6DOF *PathFinder* system (Prosurgics Inc., UK) forms the larger positioning robots. (The stereotactic PathFinder is already available on the European market for other IG procedures.) It works with the CT or MRI images of the patient and automatically registers the position of the probe (with at least 1.25 mm accuracy). In general practice, it is capable of aligning the surgical tools within 1 mm of the target. In 2003, human experiments showed that the application accuracy was 0.44 ± 0.02 mm using the robot, in comparison with an error of 0.98 ± 0.02 mm with the stereotactic frame and 1.96 ± 1.6 mm with a standard (frameless) navigation system. The ROBOCAST systems uses optical trackers for patient safety (to monitor and compensate for any change in the patient's position) and provide visual information of the surgical field. Given an accurate registration, the controller

Figure 14. a–b) The ROBOCAST system (Comparetti et al., 2011); c) The MRI-compatible neuroArm (Courtesy of University of Calgary); d) The Steady-Hand/Eye Robot at JHU (Courtesy of the Johns Hopkins University); e) Concept of the ACTIVE system (ACTIVE, 2011)

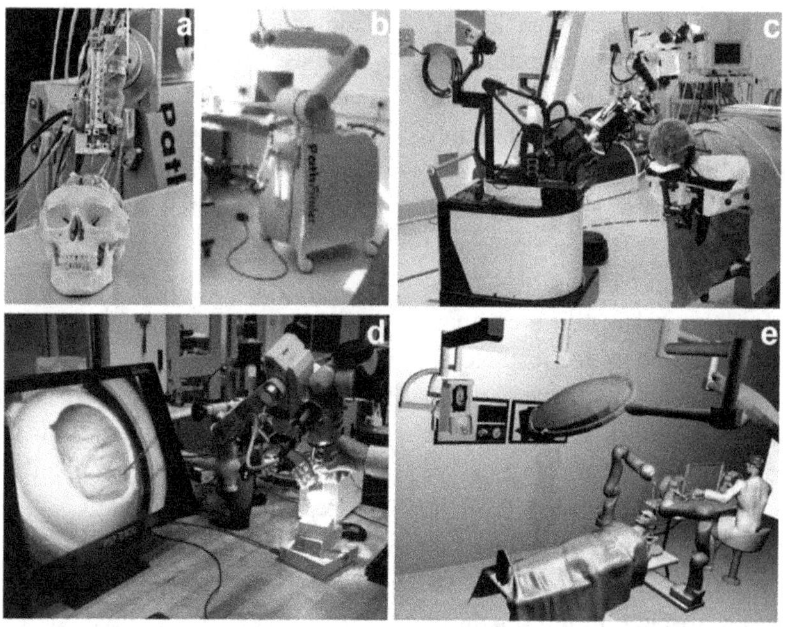

can use the preoperative diagnostic information to plan the path of the intervention.

Improving the efficiency and precision of stereotactic procedure will lead to better surgery of certain brain tumors and lesions. Deep brain stimulation electrodes could be placed very accurately with this kind of system, resulting in the routine treatment of Parkinson and similar diseases. The trajectory within the brain can be planned on the basis of a risk atlas (identifying critical structures), reproducing a fuzzy representation of statistical anatomical atlases. Construction of the atlas could be based on cognitive learning, with the possibility of simulating predicted consequences of certain surgical maneuvers. Intra-operative imaging devices (e.g. ultrasound) is planned to be used to continuously update and improve the atlas. Applications include percutaneous brachytherapy, where radioactive seeds are implanted to kill cancerous cells, and the removal of blood clots based on preoperative images. ROBOCAST was also designed to be used to inject stem cells into the brain to treat Alzheimer's and other diseases.

Integrating Imaging Devices

The other main direction of development is to integrate the robots with advanced imaging devices to increase their utility by allowing intraoperative imaging (Tsekos et al., 2007). This can be very challenging technically, as the robot has to be constructed of non-magnetic and dielectric materials such as plastics, ceramics and rubber, while being totally electricity-free. MR compatibility is considered to be one of the greatest advantages a robot may have, allowing continuous image-guidance through intra-operative imaging, and avoiding the need of image registration (Gassert et al. 2008). MRI gives a fine resolution picture of soft tissues within an acceptable timeframe, while avoiding patient and surgeon radiation exposure. MR compatible robotics has been in the focus of research interest since the mid '90s, and numerous systems have been developed.

NeuroArm (Sutherland et al., 2008) is a recent teleoperated anthropomorphic robot from a *University of Calgary* led consortium (Figure 14c). The MR safe robot (compatible up to 1.5 Tesla magnetic field) is designed for stereotaxis and microsurgery. Beyond motion scaling and high definition visual feedback, the neuroArm is able to provide very accurate 3D information of its two 7 DOF arms. It uses three displays to give complete visual coverage of the operating environment, showing in parallel the 3D stereoscopic view of the operation, the MR image of the patient and the control panel. The system has been used on a few human patients so far, and after further clinical trials, it may hit the market within a few years. A major barrier however is the increased cost of the MR compatible components.

At JHU, they have already developed several generations of MR compatible robots, mostly for prostate brachytherapy (Krieger et al., 2011). They are considered to be the leading center in brachytherapy system development, together with Queens University in Kingston, ON.

Other groups target similar procedures, such as *Deep Brain Stimulation* (DBS) with significantly smaller scale robots that can fit into the gantry of conventional, diagnostic MRI (Cole et al., 2009). Closed, high-field scanners are used for guiding DBS electrode placement interventions relying real-time guidance. It may be used for treatment of Parkinson's disease and other disorders including severe depression and Alzheimer's disease.

Hands-On Surgery with Advanced Features

Another neurosurgical research project at the JHU is a good example of the cooperative control concept (Fleming et al. 2008). The hands-on system—called formerly the Steady-Hand, and now the Eye Robot—is capable of significantly increasing the performance of human surgeons

(Figure 14d). The robot holds interchangeable tools and the force sensor that has low-threshold silicon strain gauges built in to detect forces. The force measurements are coupled to the control system after tremor filtering and smoothing. This approach has the advantage of simplicity, less expensive implementation and provides greater immediacy for the user. However, the possibility of teleoperation is completely lost. Surgeons may be more willing to accept this form of robotized equipment, as it still supports the classical way of doing the procedure. The robot is based on a Cartesian stage (allowing three orthogonal translational motions) and a RCM stage (with two orthogonal rotational DOF) that helps to keep a user-defined point (RCM) of the robot at the same position in space. This allows safe surgery through the abdomen (or other tissue layers), where the incision point limits the motion in space.

Another EU FP7 consortium has been formulated for the development of advanced robotic systems, under the name ACTIVE: *Active Constraints Technologies for Ill-defined or Volatile Environments*. A lightweight and agile 20 DOF redundant robotic cell will be created (Figure 14e). An advanced processing unit for pre- and intra-operative control will operate both autonomously and cooperatively with humans. Two cooperating KUKA LW robots will interact with the brain that prone to deform for the tool, blood pressure, breathing and deliquoration. Automated compensation techniques are to be developed. The setup is primarily aimed at neurosurgery procedures, such as awake DBS for the treatment of epilepsy (ACTIVE, 2011).

There are other concepts and approaches that seek to find ways to better serve the clinicians. Any new strategy must be developed using the same strategic principles:

- The system should pose minimal risk to the patient (compared with classical methods),
- There shall be major clinical advantages to justify its use and finally,
- The investment and maintenance costs should be reasonable for medical centers.

Minimally Invasive Surgical Techniques

Minimally invasiveness has been in the focus of clinical practice and the development of new systems, to the extreme of making no visible scars on the patient. *Natural Orifice Transluminal Endoscopic Surgery* (NOTES) was born 5 years ago to support surgical technique performed with a flexible endoscope and tools passed through a natural orifice (mouth, urethra, anus, etc.) then through an internal incision in the stomach, vagina, bladder or colon, thus avoiding any external incisions or scars. The name was invented at the Johns Hopkins Medical Institution, when first used on animals by Dr. Kalloo, as a developed version of single port laparoscopy (also called Single Incision Laparoscopic Surgery—SILS). The first human operation was in June 2007, a transgastric cholecystectomy (Dallemagne et al., 2009.)

NOTES provides numerous patient benefits, such as reduced pain, faster recovery, better cosmetic outcome (making it popular e.g. among models) and lower risk of infection. One of the great advantages is that theoretically no sterile environment is required, only sterile equipment. On the other hand, it is easier to cross-contaminate different compartment of the body within. The single port entry means serious limitations in spatial motion beyond making laparoscopy manageable in obese patients, children or burnt patients. Hyper-redundant endoscopes must be used to achieve in-body navigation, and many surgeons are needed to manipulate these devices. Research labs all around the world are trying to provide an automated tool for NOTES. One example is the *MASTER* (*Master And Slave Transluminal Endoscopic Robot*) from the Nanyang Technological Institute, Singapore, which uses a bigger, 16 mm endoscopic tube to introduce a two-armed manipulator to the surgical field. The first clinical trial (endoscopic submucosal dissection) of the system happened recently (Kok, 2011).

There are many other ways to advanced MIS this technology, such as combining it with real-time 3D visualization and augmented reality applications, such as the MUSTOF endoscope (*Multi-Sensor-Time-Of-Flight*) from *Friedrich-Alexander University*, Germany (Höller, 2010).

As of today, NOTES is still looking for a good clinical application that would best benefit from the development and justify the higher patient risk, costs and limited ergonomics of the procedures. The main goals of the *Natural Orifice Surgery Consortium for Assessment and Research* (NOSCAR) working group are to solve the question of stability and navigation in the next five years while providing more advanced tools with the capability of triangulation, and additional features such as stapling and closing.

Catheter Robots

Catheter-based procedures are wide-spread and representing an emerging field, therefore various robotic systems are aiming to improve these procedures (Antoniou et al., 2011). Hansen Medical's *Magellan* peripheral intravascular interventions robotic heart ablation system received FDA approval lately, and they also have the *Sensei X* robotic system for heart ablation on the market. Navigation of catheters is a key issue, and also in the focus of numerous research projects, such as the EU FP7 *Smart CATheterization* (SCATh) project, led by the *Katholieke Universiteit Leuven*. It is focusing on the improvement of visual and haptic tools for robust and accurate catheter guidance by fusing preoperative patient-specific anatomical and mechanical models and intra-operative data streams from in situ sensors. A remote-controlled catheter guiding robot was used in Milan in 2006 to automatically perform cardiac ablation, initiated and supervised by a group of professionals from Boston, MA. The robot used high magnetic fields to insert the catheter to the desired location, taking advantage of the pre-operative CT scans of the patient and real-time electromagnetic navigation. Initial trials were performed on 40 patients before the telesurgical experiment took place. The novelty of the system was that it could create the surgical plan on its own, relying on an anatomical atlas built on 10,000 patients' data (Pappone et al., 2006).

Microrobots

Another direction of development is toward small scale, in-body (typically capsule) robots (Toennies et al., 2010) and (Nelson et al., 2010). These offer great advantages, as they are always remote controlled, facilitating the spatial displacement of the physician from the patient. Instead of the massive rigid structure applied for human-scale robots, these are often modular and reconfigurable. *Micro-EectroMechanical Systems* (MEMS) allow for the design of structures or devices in the micro or even in the nano scale (Rebello, 2004). MEMS devices could either be (Cretu, 2010):

- **Diagnostic microsystems:** for rapid point-of-care, systems on a chip, cell and molecule sorting, DNA diagnostics,
- **Surgical microsystems:** for MIS, CAD-assisted surgery, microrobotics,
- **Therapeutic microsystems and prostheses:** drug and gene delivery, tissue augmentation/repair, biocapsules, micro/minimally invasive surgical systems.

Researchers at the University of Nebraska developed a special mobile in-vivo wheeled robot for biopsy (Rentschler et al., 2006). Equipped with a camera, the coin-sized robot can enter the abdominal cavity through one small incision and move around the organs teleoperated. The robot is able to traverse the abdomen without causing any tissue damage, therefore reducing patient trauma. More recently, the group has developed various swallowable robots, controlled with external magnets (Lehman et al., 2009).

The CRIM group at *the Scuola Superiore Sant'Anna* (Pisa, Italy) leads an FP7 funded international research collaboration to develop tethered, partially autonomous robots to perform surgery in the endolumen (*ARAKNES project*) (Menciassi & Dario, 2009) and (Harada et al., 2009). The project was originally aiming at tethered NOTES robot, but more recently, the groups has changes to SILS approach. Another EU FP7 project—*Vector*—targets the creation of effective capsule robots for local surgical procedures all along the gastro-intestinal tract (Eiriki et al., 2009).

Nanorobots

Nano-scale devices may be used for transporting, delivering and targeting drugs. This means the assembling and control of individual modules such as nano-actuators and nano-sensors through molecular *Computer-Aided Design* (CAD). System concepts exist to affect various biological procedures, e.g. repairing damaged cells or the DNA. Artificial red blood-cells could absorb a hundred times more oxygen, and nano-robot swarms could be directed around the body with external magnetic fields (Martel et al., 2008).

One of the most prominent groups in this area is at the Swiss Federal Institute of Technology Zurich (ETH). They are also involved in the NANOMA project, aiming at the development of a drug delivery microrobotic systems. They plan to use ferromagnetic microcapsules in the cardiovascular system, navigated via the induction of force from magnetic gradients generated by an MRI machine (Vartholomeos et al., 2011).

Magnetic Manipulation

Magnets are used in various ways to actuate small-size surgical robots (Menciassi, 2011). Groups around the world are focusing on permanent or uniform fields for manipulation, especially in tubular structures. The use of non-uniform fields enable the generation of more powerful magnetic fields, which is desirable for in vivo applications, and it can be used e.g. steering cochlear implants, reducing the trauma of the surgery (Kummer et al., 2010). A different approach is applied in the Italian national *Microsystems for Vascular diagnosticS and intervention* (Micro-VAST) project, to have a permanent magnet mounted on an external robotic arm to control a micro-robot (Ciuti et al., 2010). Other systems use magnets for anchoring, e.g., modular NOTES robots (Lehman et al., 2009).

4. SUMMARY

This chapter dealt with specific issues regarding the prototyping of Computer-Integrated Surgical systems and surgical robots. CIS systems are primarily developed to provide patient benefits through increased precision and minimal invasiveness. Furthermore, robotic devices may allow for refined surgical treatment unimaginable without technological support. Various system development approached have been introduced through real-life examples, also pinpointing the difficulties, such as connecting interdisciplinary technologies, validation for regulatory approvals, or entering the market, and surviving against competing treatment forms. Creating an engineering prototype is only the first necessary step toward the development of a deployable new system. The lack of standards specific to this area makes it harder to comply with applicable patient safety and process verification regulations and guidelines. The chapter gave an overview of these standards in order to facilitate future surgical robot developers. Systems currently under innovation will soon deliver great clinical advantages and improved safety features providing benefits both to the patient and the surgeon. It is believed that by time, research projects will find easier ways to advance from early phase towards more mature prototypes. Eventually, more revolutionary devices should emerge, ready for clinical trials, validation and hopefully, clearance at the end of the day.

ADDITIONAL READING MATERIALS

Surgical robotics and CIS have accumulated vast literature in the past 25 years. The most prominent articles are published in *the Institute of Electrical and Electronics Engineers* (IEEE) *Transactions on Biomedical Engineering, Transactions on Robotics* (former *Transactions on Robotics and Automation), IEEE/ASME Trans. on Mechatronics, IEEE Robotics and Automation Magazine* and in the various thematic journals, such as the *Journal of Computer Aided Surgery* (Taylor and Francis), *Interntaional Journal of Medical Robotics and Computer Assisted Surgery* (Wiley), *Interntaional Journal of Computer Assisted Radiology and Surgery* (Springer), *Journal of Robotic Surgery* (Springer), *Surgical Endoscopy* (Springer) and the *Journal of Medical Devices* (ASME). These publications are all available online.

Accordingly, the number of surgical robotics related publications has been steadily rising (O'Toole et al., 2010). Many books, tutorials and articles have been published on surgical robotics in the past 25 years, a subjective selection of the most notable ones include (Taylor et al., 2008), (Dario et al., 2003), (Taylor & Stoianovici, 2003), (Baik, 2010), (Ballantyne et al., 2004), (Pott et al., 2005), (Bozovic, 2007), (Faust, 2007) and (Kazanzides et al., 2008). The more generic field of surgical robotics is well covered by conferences, journals and periodicals.

The *Minimally Invasive Robotic Association* (MIRA), the *Society for Medical Innovation and Technology* (SMIT), the *Intl. Society and Conf. Series on Medical Image Computing and Computer-Assisted Intervention* (MICCAI) and the *Computer Assisted Orthopaedic Surgery* (CAOS) organize topical conferences every year. The *Society of Photographic Instrumentation Engineers* (SPIE) has numerous events, out of which the *SPIE Medical Imaging* conference is most attended by CIS professionals and industrial collaborators. In addition, the *Medicine Meets Virtual Reality* (MMVR) conference series, *the Congress of Computer Assisted Radiology and Surgery* (CARS), the *Annual Intl. Conf. of the IEEE Engineering in Medicine and Biology Society* (EMBC), the *IEEE Intl. Conf. on Robotics and Automation* (ICRA), the *IEEE/RSJ Intl. Conf. on Intelligent Robots and Systems* (IROS), the new *Conf. on Information Processing in Computer-Assisted Interventions* (IPCAI), *Conf. on Applied Bionics and Biomechanics* (ICABB) and the biannual *IEEE/ RAS–EMBS Intl. Conf. on Biomedical Robotics and Biomechatronics* (BioRob) all welcome original publications on surgical robot prototype development. In addition, several portals and websites are dedicated to information sharing and knowledge distribution, providing free tutorials, video materials and presentations to assist robot development in every stage.

Probably the most complete source of clinical information are *WebSurg* (www.websurg.com) virtual university run by the *European Institute of TeleSurgery* (EITS, Strasbourg, France), the *da Vinci Surgery* site (www.davincisurgery.com) and the *OR Live* (www.orlive.com). An aggregating site is the *All About Robotic Surgery* (www.allaboutroboticsurgery.com), the *MedGadget* internet journal (www.medgadget.com), the *ScienceRoll* (*scienceroll.com*) and some relevant blogs (e.g., surgrob.blogspot.com, managed by the author). For specific devices and systems, information brochures and teaching materials are available at the manufacturers' websites.

ACKNOWLEDGMENT

Tamás Haidegger's research has been supported by the US NSF EEC 9731748, Hungarian NKTH OTKA CK80316, New Hungary Development Plan (Project ID: TÁMOP-4.2.1/B-09/1/KMR-2010-0002) and the Austrian COMET-K1, EU FP7-PEOPLE-2009-IIF, StiftungAktionÖsterreich-Ungarn (AÖU) grants. Special thanks to Peter Kazanzides for his support and assistance.

REFERENCES

ACTIVE. (2011). *Project brochure*. Retrieved on August 1, 2011, from http://www.active-fp7.e
Alexander, A. D. (1972). Impacts of telemation on modern society. In *Proceedings of the 1st CISM–ITOMM Symposium*, (pp. 121–36). New York, NY: Springer.

Alexander, A. D. (1973). Impacts of telemation on modern society. In *Proceedings of the Annual Meeting of Human Factors and Ergonomics Society*, *17*(2), 299–304.

Antoniou, G. A., Riga, C. V., Mayer, E. K., Cheshire, N. J. W., & Bicknell, C. D. (2011). Clinical applications of robotic technology in vascular and endovascular surgery. *Journal of Vascular Surgery*, *53*(2), 493–499. doi:10.1016/j.jvs.2010.06.154

Baik, S. H. (Ed.). (2010). *Robot surgery*. Vienna, Austria: IN-TECH.

Ballantyne, G. H., Marescaux, J., & Giulianotti, P. C. (Eds.). (2004). *The primer of robotic and telerobotic surgery*. Philadelphia, PA: Lippincott Williams & Wilkins.

Bargar, W. L. (2007). Robots in Orthopaedic Surgery— Past, Present, and Future. *Clinical Orthopaedics and Related Research*, *463*, 31–36.

Barrera, A., Bach, J., Kazanzides, P., & Haider, H. (2007). Validation of an ASTM standard proposed to assess localizer functionality of CAOS systems: a joint effort by three laboratories. In *Proceedings of the 20th Annual Congress of Intl. Society for Technology in Arthroplasty* (ISTA), Paris, (pp. 81–81).

Benabid, A., Cinquin, P., Lavalle, S., Le Bas, J., Demongeot, J., & De Rougemont, J. (1987). Computer-driven robot for stereotactic surgery connected to CT scan and magnetic resonance imaging. Technological design and preliminary results. *AppliedNeurophysiology*, *50*(1), 153–154.

Bowersox, J., Shah, A., Jensen, G., Hill, J., Cordts, P., & Green, P. (1996). Vascular applications of telepresence surgery: initial feasibility studies in swine. *Journal of Vascular Surgery*, *23*(2), 281–287. doi:10.1016/S0741-5214(96)70272-0

Bozovic, (Ed.). (2007). *Medical robotics*. Vienna, Austria: IN-TECH.

Bronzino, J. D. (1990). Regulation of medical device innovation. In Bronzino, J. D. (Ed.), *The biomedical engineering handbook*. Hartfort, CT: CRC Press.

Chiao, J., Goldman, J., Heck, D., Kazanzides, P., Peine, W., & Stiehl, J. (2008). Metrology and standards needs for some categories of medical device. *Journal of Research of the NIST*, *113*(2), 121–129.

Ciuti, G., Valdastri, P., Menciassi, A., & Dario, P. (2010). Robotic magnetic steering and locomotion of capsule endoscope for diagnostic and surgical endoluminal procedures. *Robotica*, *28*, 199–207. doi:10.1017/S0263574709990361

Cleary, K., Melzer, A., Watson, W., Kronreif, G., & Stoianovici, D. (2006). Interventional robotic systems: Applications and technology state-of-the-art. *Minimally Invasive Therapy*, *15*(2), 101–113. doi:10.1080/13645700600674179

Cole, G., Pilitsis, J., & Fischer, G. S. (2009). Design of a robotic system for MRI-guided deep brain stimulation electrode placement. *IEEE International Conference on Robotics and Automation* (ICRA), Kobe, (pp. 4450–4456).

Comparetti, M. D., De Momi, E., Vaccarella, A., Riechmannz, M., & Ferrigno, G. (2011). Optically tracked multi-robot system for keyhole neurosurgery. In *Proceedings of the IEEE International Conference on Robotics and Automation* (ICRA), Shanghai, (pp. 661–667).

Cretu, E. (2010). *Biomedical applications of MEMS devices*. Retrieved on August 2011, from http://www.ece.ubc.ca/~leos/pdf/e331/notes/MEMS.pdf

Curexo Tech. Co. (2009). *ROBODOC®: The world's first surgical robot*. Technical report. Retrieved on August 2011, from http://pchrissanthis.com/Robodoc_investor_pitch.pdf

Dagalakis, N., Kim, Y., Sawyer, D., & Shakarji, C. (2007). Development of tools for measuring the performance of computer assisted orthopaedic hip surgery systems. In *Proceedings of the ACM Workshop on Performance Metrics for Intelligent Systems (PerMIS),* Gaithersburg, (pp. 180–187).

Dallemagne, B., Perretta, S., & Marescaux, J. (2009). Education and e-training in NOTES. *WeBSurg.com, 9*(3). Retrieved on August 1, 2011, from http://www.eats.fr/doi-ed01en0022.htm

Danilchenko, A., & Fitzpatrick, J. M. (2010). General approach to error prediction in point registration. In *Proceedings of SPIE Medical Imaging,* 7625–0F, San Diego, CA, (pp. 1–14).

Dario, P., Hannaford, B., & Menciassi, A. (2003). Smart surgical tools and augmenting devices. *IEEE Transactions on Robotics and Automation, 19*(5), 782–792. doi:10.1109/TRA.2003.817071

Dieterich, S., Tang, J., Rodgers, J., & Cleary, K. (2003). Skin respiratory motion tracking for stereotactic radiosurgery using the CyberKnife. In *Proceedings of the 17th CARS, Computer Assisted Radiology and Surgery*, (vol. 1256, pp. 130–136).

DiMaio, S., Hanuschik, M., & Kreaden, U. (2011). The da Vinci Surgical System. In Rosen, J., Hannaford, B., & Satava, R. M. (Eds.), *Surgical robotics—Systems applications and visions* (pp. 199–217). Boston, MA: Springer.

Doarn, C. R., Nicogossian, A. E., & Merrel, R. C. (1998). Applications of telemedicine in the United States Space Program. *Telemedicine Journal, 4*(1), 19–30. doi:10.1089/tmj.1.1998.4.19

Eirik, L., Johansen, B., Gjelsvik, T., & Langø, T. (2009). Ultrasound based localization of wireless microrobotic endoscopic capsule for the GI tract. In *Proceedings of the 21st Conference of the Society for Medical Innovation and Technology* (SMIT), Sinaia, (p. 77).

Faust, R. A. (Ed.). (2007). *Robotics in surgery: History, current and future applications*. Hauppauge, NY: Nova Science.

FDA. Food and Drug Administration. (2010). *510(k) and science report recommendations: Summary and overview of comments and next steps*. FDA Report. Retrieved on August 1, 2011, http://www.fda.gov/downloads/AboutFDA/CentersOffices/CDRH/CDRHReports/UCM239449.pdf

FDA. Food and Drug Administration. (2011). *Device advice: Comprehensive regulatory assistance*. Retrieved on August 1, 2011, from http://www.fda.gov/MedicalDevices/DeviceRegulationandGuidance/default.htm

Fei, B. (2001). The safety issues of medical robotics. *Reliability Engineering & System Safety, 73*(2), 183–192. doi:10.1016/S0951-8320(01)00037-0

Fitzpatrick, J. M. (2009). Fiducial registration error and target registration error are uncorrelated. *SPIE Medical Imaging*, 7261-02, Orlando, FL, (pp. 1–12).

Fitzpatrick, J. M., West, J. B., & Maurer, C. R. (1998). Predicting error in rigid-body point-based registration. *IEEE Transactions on Medical Imaging, 17*(5), 694–702. doi:10.1109/42.736021

Fleming, I., Balicki, M., Koo, J., Iordachita, I., Mitchell, B., & Handa, J. ... Taylor, R. H. (2008). Cooperative robot assistant for retinal microsurgery. *Lecture Notes in Computer Science (LNCS), Proceedings of the Annual Conf. of the Medical Image Computing and Computer Assisted Intervention Society (MICCAI)*, vol. 5242, (pp. 543–550). Palm Springs, CA: Springer.

Frantz, D. D., Kirsch, S. R., & Wiles, A. D. (2004). *Specifying 3D tracking system accuracy* (pp. 1–5). Technical report, NDI Inc.

Garcia, P., Rosen, J., Kapoor, C., Noakes, M., Elbert, G., & Treat, M. (2009). Trauma pod: A semi-automated telerobotic surgical system. *Journal of Medicine*, *5*(2), 136–146.

Gartner, A. (2008). *Teleneurology and requirements of the medical devices directive (MDD)* (pp. 1–22). Technical report, Baaske Medical GmbH & Co., Lubbecke.

Gassert, R., Burdet, E., & Chinzei, K. (2008). MRI-compatible robotics. *IEEE Engineering in Medicine and Biology Magazine*, *27*(3), 12–14. doi:10.1109/EMB.2007.910273

Grunert, P., Darabi, K., Espinosa, J., & Filippi, R. (2003). Computer-aided navigation in neurosurgery. *Neurosurgical Review*, *26*(2), 73–99. doi:10.1007/s10143-003-0262-0

Guiochet, J., & Vilchis, A. (2002). Safety analysis of a medical robot for teleechography. In *Proceedings of the 2nd IARP IEEE/RAS joint workshop on Technical Challenge for Dependable Robots in Human Environments*, Toulouse, (pp. 217–227).

Guthart, G. S., & Salisbury, J. K. (2000). The intuitive telesurgery system: Overview and application. In *Proceedings of the IEEE International Conference on Robotics and Automation (ICRA)*, San Francisco, CA, (pp. 618–621).

Haidegger, T. (2010). *Da Vinci inside—Insight*. SurgRob blog. Retrieved from http://surgrob.blogspot.com/2010/12/da-vinci-motors.html

Haidegger, T., Kazanzides, P., Rudas, I., Benyó, B., & Benyó, Z. (2010). The importance of accuracy measurement standards for computer-integrated interventional systems. In *Proceedings of the EURON GEM Sig Workshop on The Role of Experiments in Robotics Research at IEEE ICRA*, Anchorage, AK, (pp. 19–24).

Haidegger, T., Xia, T., & Kazanzides, P. (2008). Accuracy improvement of a neurosurgical robot system. In *Proceedings of the 2nd IEEE RAS/EMBS Intl. Conf. on Biomedical Robotics and Biomechatronics (BioRob),* Scottsdale, AZ, (pp. 836–841).

Harada, K., Oetomo, D., Susilo, E., Menciassi, A., Daney, D., Merlet, J.-P., & Dario, P. (2009). A reconfigurable modular robotic endoluminal surgical system: Vision and preliminary results. *Robotica, 28*(2), 171–183. doi:10.1017/S0263574709990610

Höller, K. (2010). *Novel techniques for spatial orientation in natural orifice translumenal endoscopic surgery* (NOTES), (pp. 1–190). Ph.D. thesis, Friedrich-Alexander-Universität.

IOM. Institute of Medicine. (2011). *Press release*. Retrieved on August 2011, from http://cardiobrief.org/2011/07/29/controversial-iom-report-highly-critical-of-510k-process/

ISO 10360. (2000). *ISO 10360-1:2000 standard*. Retrieved on August 1, 2011, from http://www.iso.org/iso/catalogue detail.htm?csnumber=37694

iSYS. (2011). *iSYS Medizintechnik GmbH website*. Retrieved on August 1, 2011, from http://www.isys.co.at

Janda, M., & Buch, B. (2009). The challenges of clinical validation of emerging technologies: Computer-Assisted devices for surgery. *Journal of Bone and Joint Surgery*, *91-A*(S1), 17–21. doi:10.2106/JBJS.H.01337

Jang, J., Kang, Y., Choi, B., Choi, I., Kim, M., & Shin, D. … Kwon, S. (2006). Evaluation of the accuracy of the CyberKnife. In *Proceedings of the World Congress on Medical Physics and Biomedical Engineering*, Springer, Seoul, (pp. 2024–2027).

Karas, C. S., & Chiocca, E. A. (2007). Neurosurgical robotics: A review of brain and spine applications. *Journal of Robotic Surgery*, *1*(1), 39–43. doi:10.1007/s11701-006-0006-6

Kazanzides, P. (2006). Surgical robots and phantom (artifact) devices. In *Proceedings of the ASTM Workshop on Medical Devices Metrology and Standards Needs*. ASTM, NIST, NCI, FDA and U.T. Arlington, Atlanta. Retrieved on August 1, 2011, from http://www.nist.gov/el/isd/upload/Surgical_Robots_Kazanzides.pdf

Kazanzides, P. (2009). Safety design for medical robots. In *Proceedings of the Annual International Conference of the IEEE Engineering in Medicine and Biology Society (EMBC)*, Minneapolis, MN, (pp. 7208–7211).

Kazanzides, P., DiMaio, S., Deguet, A., Vagvolgyi, B., Balicki, M., & Schneider, C. … Taylor, R. (2010). The surgical assistant workstation (SAW) in minimally-invasive surgery and microsurgery. In *Proceedings of the MICCAI International Workshop on Systems and Architectures for Computer Assisted Interventions*, Beijing. *MIDAS Journal—Computer Assisted Intervention*, June 2010.

Kazanzides, P., Fichtinger, G., Hager, G. D., Okamura, A. M., Whitcomb, L. L., & Taylor, R. H. (2008). Surgical and interventional robotics: Part I. *IEEE Robotics & Automation Magazine*, *15*(2), 122–130. doi:10.1109/MRA.2008.926390

Kok, L. (2011). *Singapore expertise pioneers quick and scarless surgery*. Press release of the Nanyang Technological University. Retrieved on August 1, 2011, from http://www.eurekalert.org/pub_releases/2011-07/ntu-sep070511.php

Krieger, A., Iordachita, I., Sang-Eun Song Cho, N. B., Guion, P., Fichtinger, G., & Whitcomb, L. L. (2011). (in press). Development and evaluation of an actuated MRI-compatible robotic system for MRI-guided prostate intervention. *IEEE/ASME Transactions on Mechatronics*. doi:10.1109/TMECH.2011.2163523

Kronreif, G. (2011). Robot systems for percutaneous needle placement. In *Proceedings of the 1st BME–MAVE International Computer-Integrated Surgery Workshop*, Budapest. Retrieved from www.eestec.hu/cis

Kummer, M. P., Abbott, J. J., Kratochvil, B. E., Borer, R., Sengul, A., & Nelson, B. J. (2010). OctoMag: An electromagnetic system for 5-DOF wireless micromanipulation. *IEEE Transactions on Robotics*, *26*(6), 1006–1018. doi:10.1109/TRO.2010.2073030

Kwartowitz, D. M., Herrell, S. D., & Galloway, R. L. (2006). Toward image-guided robotic surgery: Determining intrinsic accuracy of the da Vinci robot. *International Journal of Computer Assisted Radiology and Surgery*, *1*(3), 157–165. doi:10.1007/s11548-006-0047-3

Kwartowitz, D. M., Herrell, S. D., & Galloway, R. L. (2007). Update: Toward image-guided robotic surgery: Determining the intrinsic accuracy of the daVinci-S robot. *International Journal of Computer Assisted Radiology and Surgery*, *1*(5), 301–304. doi:10.1007/s11548-006-0064-2

Kwoh, Y., Hou, J., Jonckheere, E., & Hayati, S. (1988). A robot with improved absolute positioning accuracy for CT guided stereotactic brain surgery. *IEEE Transactions on Bio-Medical Engineering*, *35*(2), 153–160. doi:10.1109/10.1354

Lehman, A. C., Dumpert, J., Wood, N. A., Redden, L., Visty, A. Q., & Farritor, S. (2009). Natural orifice cholecystectomy using a miniature robot. *Surgical Endoscopy*, *23*(2), 260–266. doi:10.1007/s00464-008-0195-3

Li, Q., Zamorano, H. L., Pandya, A., Perez, R., Gong, F., & Diaz, F. (2002). The application accuracy of the NeuroMate robot—A quantitative comparison with frameless and frame-based surgical localization systems. *Computer Aided Surgery*, *7*(2), 90–98.

Lieberman, I. H., Togawa, D., Kayanja, M. M., Reinhardt, M. K., Friedlander, A., Knoller, N., & Benzel, E. C. (2006). Bone-mounted miniature robotic guidance for pedicle screw and translaminar facet screw placement: Part I—Technical development and a test case result. *Neurosurgery*, *59*(3), 641–650. doi:10.1227/01.NEU.0000229055.00829.5B

Lirici, M. M., Papaspyropoulos, V., & Angelini, L. (1997). Telerobotics in medicine and surgery. *Minimally Invasive Therapy & Allied Technologies*, *5*(6), 364–378. doi:10.3109/13645709709153336

Lowery, A., Strojny, J., & Puleo J. (1996). *Medical device quality systems manual: A small entity compliance guide*. HHS Publication, FDA 97-4179.

Lum, J. H. M. (2008). *Quantitative performance assessment of surgical robot systems: TeleRobotic FLS*. Ph.D. thesis, University of Washington.

Madhani, A. J., Niemeyer, G., & Salisbury, J. K., Jr. (1998). The black falcon: A teleoperated surgical instrument for minimally invasive surgery. In *Proceedings of the IEEE/RSJ International Conference on Intelligent Robots and Systems* (IROS), Victoria, BC, (pp. 936–944).

Martel, S., Felfoul, O., & Mohammadi, M. (2008). Flagellated bacterial nanorobots for medical interventions in the human body. In *Proceedings of the 2nd IEEE RAS/EMBS Intl. Conf. on Biomedical Robotics and Biomechatronics (BioRob)*, Scottsdale, AZ, (pp. 264–269).

Maurer, C. R., Fitzpatrick, J. M., Wang, M. Y., Galloway, R. L., Maciunas, R. J., & Allen, G. S. (1997). Registration of head volume images using implantable fiducial markers. *IEEE Transactions on Medical Imaging*, *16*(4), 447–462. doi:10.1109/42.611354

Menciassi, A. (2011). Single access robots for surgery. In *Proceedings of the ICRA Workshop on Mechanisms for Surgical Robotics*, Shanghai, (p. 3).

Menciassi, A., & Dario, P. (2009). Miniaturized robotic devices for endoluminal diagnosis and surgery: A single-module and a multiple-module approach. In *Proceedings of the Annual International Conference of the IEEE Engineering in Medicine and Biology Society (EMBC)*, Minneapolis, MN, (pp. 6842–6845).

Nathoo, N., Cavusoglu, M. C., Vogelbaum, M. A., & Barnett, G. H. (2005). In touch with robotics: Neurosurgery for the future. *Neurosurgery*, *56*(3), 421–433. doi:10.1227/01.NEU.0000153929.68024.CF

Nelson, B. J., Kaliakatsos, I. K., & Abbott, J. J. (2010). Microrobots for minimally invasive medicine. *Annual Review of Biomedical Engineering, 12*, 55–85. doi:10.1146/annurev-bioeng-010510-103409

Nishihara, S., Sugano, N., Nishii, T., Miki, H., Nakamura, N., & Yoshikawa, H. (2006). Comparison between hand rasping and robotic milling for stem implantation in cementless total hip arthroplasty. *The Journal of Arthroplasty, 21*(7), 957–966. doi:10.1016/j.arth.2006.01.001

Nof, S. Y. (Ed.). (1999). *Handbook of industrial robotics* (2nd ed.). USA: Wiley. doi:10.1002/9780470172506

O'Toole, M. D., Bouazza-Marouf, K., Kerr, D., Gooroochurn, M., & Vloeberghs, M. (2010). A methodology for design and appraisal of surgical robotic systems. *Robotica, 28*(2), 297–310. doi:10.1017/S0263574709990658

Pappone, C., Vicedomini, G., Manguso, F., Gugliotta, F., Mazzone, P., & Gulletta, S. (2006). Robotic magnetic navigation for atrial fibrillation ablation. *Journal of the American College of Cardiology, 47*(7), 1390–1400. doi:10.1016/j.jacc.2005.11.058

Paul, H. A., Mittlestadt, B., Bargar, W. L., Musits, B., Taylor, R. H., & Kazanzides, P. … Hanson, B. (1992). Surgical robot for total hip replacement surgery. In *Proceedings of the IEEE International Conference on Robotics and Automation*, Nice, (pp. 606–611).

Pott, P., Scharf, H.-P., & Schwarz, M. (2005). Today's state of the art in surgical robotics. *Computer Aided Surgery, 10*(2), 101–132.

Pott, P. P. (2011). *Medical robotic database: MeRoDa.* University of Heidelberg. Retrieved on August 1, 2011, from http://meroda.uni-hd.de

Pransky, J. (1997). ROBODOC—Surgical robot success story. *The Industrial Robot, 24*(3), 231–233. doi:10.1108/01439919710167444

Rayman, R. (2009). Is surgery a remote possibility? Robotic surgical system under development has telesurgery capabilities. *Health Technology Trends, 21*(7), 5–7.

Rebello, K. J. (2004). Applications of MEMS in surgery. *Proceedings of the IEEE, 92*(1), 43–55. doi:10.1109/JPROC.2003.820536

Rentschler, M., Dumpert, J., Platt, S., Oleynikov, D., Farritor, S., & Iagnemma, K. (2006). Mobile in vivo biopsy robot. In *Proceedings of the IEEE International Conference on Robotics and Automation* (ICRA), Orlando, FL, (pp. 4155–4160).

Rovetta, A., Sala, R., Cosmi, F., Wen, X., Milanesi, S., & Sabbadini, D. (1996). A new telerobotic application: Remote laparoscopic surgery using satellites and optical fiber networks for data exchange. *The International Journal of Robotics Research, 15*(3), 267–279. doi:10.1177/027836499601500303

Satava, R. M. (1995). Telesurgery, virtual reality and the new world order of medicine. *Computer Aided Surgery, 1*(1), 12–16. doi:10.3109/10929089509106821

Satava, R. M. (2005). The nature of surgical error: A cautionary tale and a call to reason. *Surgical Endoscopy, 19*(8), 1014–1016. doi:10.1007/s00464-005-8108-1

SIEMENS. (2011). *Atris Zeego product page.* Retrieved on August 1, 2011, from http://www.medical.siemens.com/

Stiehl, J. B., Bach, J., & Heck, D. A. (2007). Validation and metrology in CAOS. In *Navigation and MIS in orthopedic surgery* (pp. 68–78). Springer. doi:10.1007/978-3-540-36691-1_9

Sutherland, G. R., Latour, I., Greer, A. D., Fielding, T., Feil, G., & Newhook, P. (2008). An image-guided magnetic resonance-compatible surgical robot. *Neurosurgery, 62*(2), 286–293. doi:10.1227/01.neu.0000315996.73269.18

Taylor, R. H. (1997). Robots as surgical assistants: Where we are, wither we are tending, and how to get there. *Lecture Notes in Artificial Intelligence in Medicine*, *1211*(1), 3–11.

Taylor, R. H. (2001). *Robotic joint replacement surgery*. Course materials, Computer-Integrated Surgery, Johns Hopkins University. Retrieved on August 1, 2011, from http://www.cs.jhu.edu/~cis/cista/445/Lectures/Robodoc.pdf

Taylor, R. H., & Kazanzides, P. (2008). Medical robotics and computer-integrated interventional medicine. *Advances in Computers: Emerging Technologies*, *73*, 219–258.

Taylor, R. H., Menciassi, A., Fichtinger, G., & Dario, P. (2008). Medical robotics in computer-integrated surgery. In Siciliano, B., & Khatib, O. (Eds.), *Handbook of robotics* (2nd ed., pp. 1199–1222). Cambridge, MA: Springer. doi:10.1007/978-3-540-30301-5_53

Taylor, R. H., & Stoianovici, D. (2003). Medical robotics in computer-integrated surgery. *IEEE Transactions on Robotics and Automation*, *19*(5), 765–781. doi:10.1109/TRA.2003.817058

Tinelli, A., Malvasi, A., Buscarini, M., Morales, E. R., & Stark, M. (2011). Laparoscopy and "robotics" in cervical cancer treatment: Current insights, novel approaches and future perspectives in the new era of gynecological surgery. In E. K. Salvatti (Ed.), *Brain cancer, tumor targeting and cervical cancer*, (pp. 127–162). Nova Science Publishers.

Toennies, J. L., Tortora, G., Simi, M., Valdastri, P., & Webster, R. J. (2010). Swallowable medical devices for diagnosis and surgery: The state of the art. *Proceedings of the Institution of Mechanical Engineers. Part C, Journal of Mechanical Engineering Science*, *224*, 1397–1414. doi:10.1243/09544062JMES1879

Tsekos, N. V., Khanicheh, A., Christoforou, E., & Mavroidis, C. (2007). Magnetic resonance-compatible robotic and mechatronics systems for image-guided interventions and rehabilitation: a review study. *Annual Review of Biomedical Engineering*, *9*, 351–587. doi:10.1146/annurev.bioeng.9.121806.160642

Ummat, A., Sharma, G., Mavroidis, C., & Dubey, A. (2005). Bio-nanorobotics: state of the art and future challenges. Chapter 19 in *The Biomedical Engineering Handbook*, (pp. 19-1–19-42). CRC Press.

Urschel, H. C. (Ed.). (2007). *Robotic radiosurgery: Treating tumors that move with respiration*. Springer. doi:10.1007/978-3-540-69886-9

van den Bedem, L. (2010). *Realization of a demonstrator slave for robotic minimally invasive surgery*. Ph.D. thesis, Eindhoven University of Technology.

Varley, P. (1999). Techniques for development of safety-related software for surgical robots. *IEEE Transactions on Information Technology in Biomedicine*, *3*(4), 261–267. doi:10.1109/4233.809170

Varma, T., & Eldridge, P. (2006). Use of the NeuroMate stereotactic robot in a frameless mode for functional neurosurgery. *International Journal of Medical Robotics and Computer Assisted Surgery*, *2*(2), 107–113. doi:10.1002/rcs.88

Vartholomeos, P., Fruchard, M., Ferreira, A., & Mavroidis, C. (2011). MRI-guided robotic nanocapsules: A review study. *Annual Review of Biomedical Engineering*, *13*, 157–187. doi:10.1146/annurev-bioeng-071910-124724

Xia, T., Baird, C., Jallo, G., Hayes, K., Nakajima, N., Hata, N., & Kazanzides, P. (2008). An integrated system for planning, navigation and robotic assistance for skull base surgery. *International Journal of Medical Robotics and Computer Assisted Surgery*, *4*(4), 321–330. doi:10.1002/rcs.213

Chapter 11
Design and Evaluation of a Piezo-Driven Ultrasonic Cell Injector

Haibo Huang
Robotics and Micro-systems Center, Soochow University, China

Hao Su
Worcester Polytechnic Institute, USA

Changhai Ru
Robotics and Micro-systems Center, Soochow University, China

ABSTRACT

In this chapter, a novel piezo-driven cell injection system for automatic batch injection of suspended cells is presented; it has simplified operational procedure and better performance than the previous ones. Specifically, this new piezo-driven cell injector design has three aspects of merits: 1) by centralizing the piezo oscillation energy on the injector pipette, it eliminates the vibration amplitude of other parts of the micromanipulator; 2) meanwhile, a small piezo stack is sufficient to perform the cell injection process; and 3) detrimental lateral tip oscillations of the injector pipette are attenuated to satisfactory amount even without mercury column. The removal of mercury enables wide applications of this advanced cell injection technology in a number of cell manipulation scenarios. Furthermore, ultrasonic vibration micro-dissection (UVM) theory is utilized to analyze the piezo-driven cell injection process, and lateral oscillation of injector pipettes is investigated. Experiments on cell injection of a large amount of zebrafish embryos (n=300) indicate that the injector pipette is capable of piercing through cell membranes with low injection speed and almost no deformation of the cell wall, but with a high success rate.

DOI: 10.4018/978-1-4666-0176-5.ch011

INTRODUCTION

The year of 1995 witnessed the debut of piezo drill, a new cell injection technology that utilizes piezo-driven pipettes with a very small mercury column for cell injection, and subsequently it was successfully applied to intracytoplasmic sperm injection (ICSI) of a variety of mammal species. Although this technique significantly improves the survival rates of the ICSI process, shortcomings such as the damage to the cell membrane due to large lateral tip oscillations of the injector pipette, complexity of operation and toxicity of mercury immensely limit its application.

1 BACKGROUND

This section presents the research effort ranging from academic research to commercial products on cell injection systems. First, different gene transfer technologies are introduced and classified. Second, existing automatic cell injection systems, both adherent cell injection and suspended cell injection systems, are briefly reviewed. The automation of suspended cell injection system is the essential problem, and has emerged as the mainstream of current microinjection technology. The last section elaborates on piezo-driven cell injection technology, which is the major focus of this study.

1.1 Gene Transfer Technology

Introduction of DNA into mammalian cells is a powerful tool for two important areas of biology - assessment of gene function and gene therapy (Heiser, 2004a, 2004b). The process of introducing DNA into cells for the purpose of gene expression is called transfection or gene delivery (Heiser, 2004a). Synthetic compounds used to facilitate DNA transfer are often named synthetic vectors or transfection reagents. With the exception of microinjection, there are many new methods developed to transfer genes into target cells. The existing gene transfer methods normally include three categories: 1) virus-mediately transduction; 2) gene delivery by chemical methods and 3) gene delivery by physical methods (Ma & Chen, 2005; Watt, Sawicki, & Passaro, 1993). The gene transfer results can be transient and stable transfections.

1.1.1 Biological Vectors Methods

In the past several years, a new gene transfer technology using viruses as vectors has emerged (Walther & Stein, 2000). Viruses are naturally evolved vehicles which efficiently transfer their genes into host cells. This ability made them desirable for engineering virus vector systems for transferring genes. Nowadays this technique has broad application to mammalian cells, great efficiency and has significantly enhanced our ability to manipulate genes in humans and animals, especially in areas such as gene therapy.

The viral vectors recently used in laboratory and clinical applications are based on RNA (Ribonucleic acid) and DNA (Deoxyribonucleic acid) (Walther & Stein, 2000) viruses processed from very different genomic structures and host ranges. Particular viruses have been selected as gene delivery vehicles because of their capacities to carry foreign genes and their ability to efficiently deliver these genes associated with efficient gene expression. Among these vector systems, retrovirus vectors represent the most prominent delivery system because these vectors have high gene transfer efficiency and mediate high expression of therapeutic genes. Members of the DNA virus family such as adenovirus-, adeno-associated virus or herpesvirus have also become attractive for efficient gene delivery, as reflected by the fast growing number of clinical trials using these vectors.

The viral vectors have higher transfer efficiency and longer duration of gene expression than non-viral vectors, but the safety issues surrounding the use of viruses have been plagued.

1.1.2 Chemical Methods

These kinds of methods mainly utilize synthetic compounds to complex with DNA and transfer the gene material into cells. To accomplish DNA delivery, these complexes need to (1) bind to the cell surface, (2) cross the plasma membrane, (3) release DNA into the cytoplasm and finally, (4) transport the DNA into the nucleus. Many synthetic compounds have been developed since DEAE-dextran was first used in transfection experiments more than 40 years ago (Pagano & Vaheri, 1965). Rapid progress in developing more efficient synthetic vectors has led to successful DNA delivery into a variety of cell types in vitro and in vivo (D. Liu, Chia, Tian, & Heiser, 2004). Until now, there have been four major types of chemical reagents available to most investigators: calcium phosphate (Graham & Van Der Eb, 1973), DEAE-dextran (Pagano & Vaheri, 1965), cationic lipid (Felgner et al., 1987), and cationic polymer (Boussif et al., 1995; Tang, Redemann, & Szoka, 1996). The major advantages of synthetic vectors (or chemical methods) are their simplicity, ease of production and relatively low toxicity. Non-viral vectors could prove that they were not only safer than viruses, but also more effective in some cases. This is a newly developed technique, and more new compounds will be developed that will be able to deliver DNA to all types of cells with equally high efficiency.

1.1.3 Physical Methods

The physical methods usually use some type of physical force (pressure, sound, shock wave or electric pulse) to overcome the physical barriers presented by mammalian tissues and "force" cells or tissues to take up the genetic materials. These methods mainly include microinjection (M. Graessmann & Graessmann, 1983), particle-bombardment ("gene gun") (Armaleo et al., 1990), electroporation (Y. Sun & Nelson, 2002), ultrasound (Yanagida et al., 1999) and encapsulated microspheres (Alexakis et al., 1995). Most of the physical methods use some type of instrument, which can lead to more quantitative and standardized gene delivery. Such an "industrialized" approach is important for large-scale clinical applications and will minimize the common operator variation frequently discovered in the stage of laboratory research with nonphysical methods.

The physical gene transfer methods have higher transfection efficiency and are easier for process standardization, have fewer limits on gene size and are useful for ex vivo application, but only in a local area subjected to the physical force. However, these kinds of methods usually require specific efficiency instruments and the costs are higher.

The best results may be achieved by combining different methods of gene transfer. Physical or chemical methods have been used to deliver a biological vector-based gene transfer. For example, one research has demonstrated that inoculation of a vector encoding for the full length of proviral Simian Immunodeficiency Virus (SIV) genome, using a gene gun at the mucosal surface of nonhuman primate, was able to produce a full SIV infection in the host (S. Wang, Fuller, Manson, Wyand, & Lu, 2000). By using the same technique, a gene gun is able to deliver a plasmid that can express a live-attenuated vaccine (Chou, Biswas, Lu, & Heiser, 2004).

1.2 Robot-Assisted Microinjection

Microinjection technique could be classified as one kind of physical gene transfer method and nowadays robot-assisted microinjection is established as one of the most versatile methods of introducing genetic materials into living cells. Figure 1 shows a typical ISCI (Intracytoplasmic Sperm Injection) cell injection process of a mouse's oocyte, with a diameter of about 100*mm* (Kimura & Yanagimachi, 1995).

Microinjection - that is, the direct-pressure injection of solution into a cell through a glass

Figure 1. A typical ICSI cell injection process using piezo-driven pipette, (A, B) A sperm injection pipette was inserted into an oocyte; (C, D) When the oolemma was broken, the entire spermatozoon was injected; (E) Medium injected with the spermatozoon was retrieved as completely as possible; (F) Withdrawal of the pipette (arrow indicates sperm head), adapted from (Kimura & Yanagimachi, 1995)

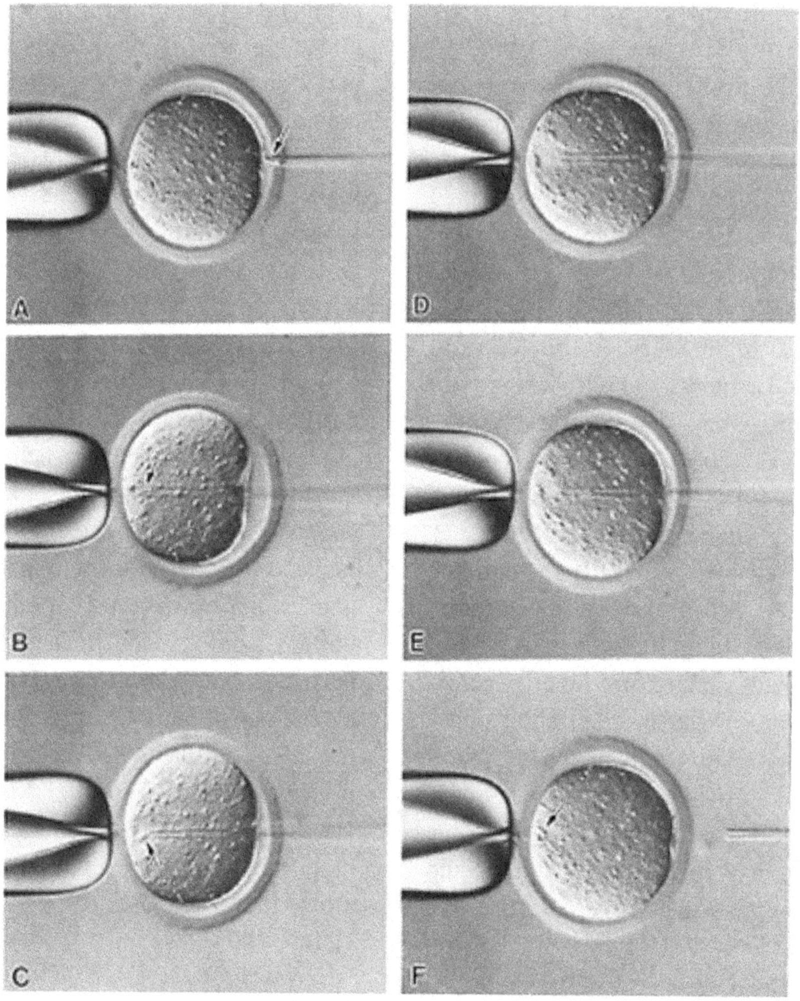

capillary - is an effective and reproducible method for introducing exogenous material into cells. Barber (Barber, 1911) first described microinjection, using a micropipette to isolate and manipulate single bacterial cells. Barber was also one of the first people to use mercury to control the movement of a small fluid volume and to use a second pipette to hold the cells (Korzh & Strähle, 2002). These techniques formed the basis for modern microinjection applications. Less than one decade later, similar techniques were applied to perform direct transfer of biological material into cells (Chambers & Chambers, 1961). A further refinement of instruments permitted intracellular transfer of solutions containing macromolecules of various origins, such as viral RNA (A. Graessmann & Graessmann, 1976), purified proteins (Mabuchi & Okuno, 1977) and viral DNA (Anderson, Killos, Sanders-Haigh, Kretchmer, & Diakumakos, 1980). Likewise, microinjection was used for DNA transfer into fertilized eggs, and a cloned gene was soon reported to be expressed into mouse somatic tissues (Gordon, Scangos, Plotkin, Barbosa, & Ruddle, 1980). Later, recombinant genes were stably introduced into the mouse germ line (Brinster et al., 1981). Since then, microinjection has been used routinely to generate transgenic animals, and applied to a wide variety of studies using living cells, such as gene expression, signal transduction or cytoskeleton studies.

The advantages of microinjection over other gene transfer methods include low cytotoxicity and the ability to deliver dosage quantitatively. Microinjection allows direct access to the two main intracellular compartments, the nucleus and the cytoplasm. This technique makes it possible to use single cells to study complicated cellular processes, structures and functions in vitro. Despite its straightforward approach, microinjection is a laborious procedure. Most cell injection operations are still performed manually. Only one cell at a time can be injected, and many injections may be needed before there is a successful DNA delivery. This is a problem if a large number of cells need to be injected in a limited time. Therefore microinjection with its current technology has little application for in vivo gene therapy.

An important improvement in this respect is introducing automation in the micromanipulation and microinjection processes, as well as the control and standardization of cell preparation or the production of injection capillaries (Ansorge, 1982). The development of computer-assisted and microprocessor-controlled injection systems makes high injection rates with optimal reproducibility feasible and allows for quantitative microinjection (Ansorge & Pepperkok, 1988).

1.3 Adherent and Suspended Cell Injection towards Automation

The cells used for microinjection in biological technology are classified as either adherent or suspended cells, corresponding to two distinct biomanipulation tasks (Pasi Kallio & Kuncova, 2003).

1.3.1 Researches on Automatic Adherent Cell Injection

Adherent cells are always much smaller in size (10-20 μm in diameter) than egg cells, which are the most typically microinjected suspended cells. These adherent cells to be injected, with a non-regular shape, always grow and get fixed on the bottom of a Petri dish. The cells are not penetrated from the side, but usually from the top. Thus, the micro capillary must go through the membrane, but it must not touch the bottom of the cultivation dish, which could break the capillary. Furthermore, the penetration movement of the capillary should be such that the injection-caused opening in the membrane is as small as possible and the injection volume should be sufficiently small so as not to harm the cell. The main challenges of automatically adherent cell injection lie in the high positioning accuracy of the micromanipulator using microscope vision feedback, especially in controlling the height of the injector pipette and the preciseness of the injection volume. As the manipulation target is fixed on the Petri dish, the injection task is more steadily controlled compared to suspended cell injection tasks. Currently, commercial devices such as those provided by Eppendorf, Narishige and Cellbiology Trading (n.d.) are available for automation of adherent cell injection tasks.

Figure 2 (a) shows a popular commercial automatic adherent cell injection system setup (computer assisted microinjection of tissue culture cells (AIS 2)) manufactured by Cellbiology Trading and Figure 2(b) is the operation interface. This automated cell injection system is mainly composed of an inverted microscope equipped with long-distance phase-contrast objectives for observation, a 3-axes-stepping motorized stage for accurate positioning of the cells, another 3-axes-stepping motorized stage controlling the capillary movement, a piezo drive axial injection movement, an Eppendorf pneumatic micro-injector with the model FemtoJet, a video system with a CCD camera and a monitor for observation of cells together with the microprocessor control unit and the required software for controlling the cell injection procedure. With the techniques of an autofocus microscope, the image identification of cells and injector pipette, as well as the image-based visual servoing for positioning control and the calibrated injection height control, this computer-assisted and microprocessor-controlled injection system can provide high injection rates with optimum reproducibility of injections, and makes quantitative adherent cell injection possible. The software reliably controls the precise injection process and allows fast axial injection at a rate of up to 1500 cells per hour. Retrieving injected cells for analysis and multiple injections into the same cells are also realized with this system.

There is still on-going research to improve the automatic adherent cell injection performance, such as the electrical detection of the contact between a microinjection capillary and cells proceeded by Lukkari et al. (M. Lukkari & Kallio, 2005; M. J. Lukkari et al., 2004), a simple computer vision microscopy based contact detection method without requiring additional sensors developed by Wang et al (W. Wang et al., 2008; W. H. Wang et al., 2008), the lab automation in cultivation of adherent cells proposed by Kallio et al. (J. K. Kallio & Kallio, 2006), the simultaneous use of several MANiPEN micromanipulators to realize the simultaneous injection of several adherent cells, as proposed by Ritala et al. (P. Kallio, Ritala, Lukkari, & Kuikka, 2006).

1.3.2 Researches on Automatic Suspended Cell Injection

On the other hand, development of methodologies for the injection of suspended cells, have been the focus of a number of research groups (T. Huang, Kimura, & Yanagimachi, 1996; X. Li et al., 2001; X. D. Li, Zong, Bi, & Zhao, 2001; Pepperkok et al., 1988). Suspended cells, such

Figure 2. Automated adherent cell injection system (AIS 2): (a) System setup with microscope, motorized stage and micromanipulator, pressure microinjector and video camera system; (b) The friendly user interface makes the microinjection as easy as possible, available: http://www.ais2.com

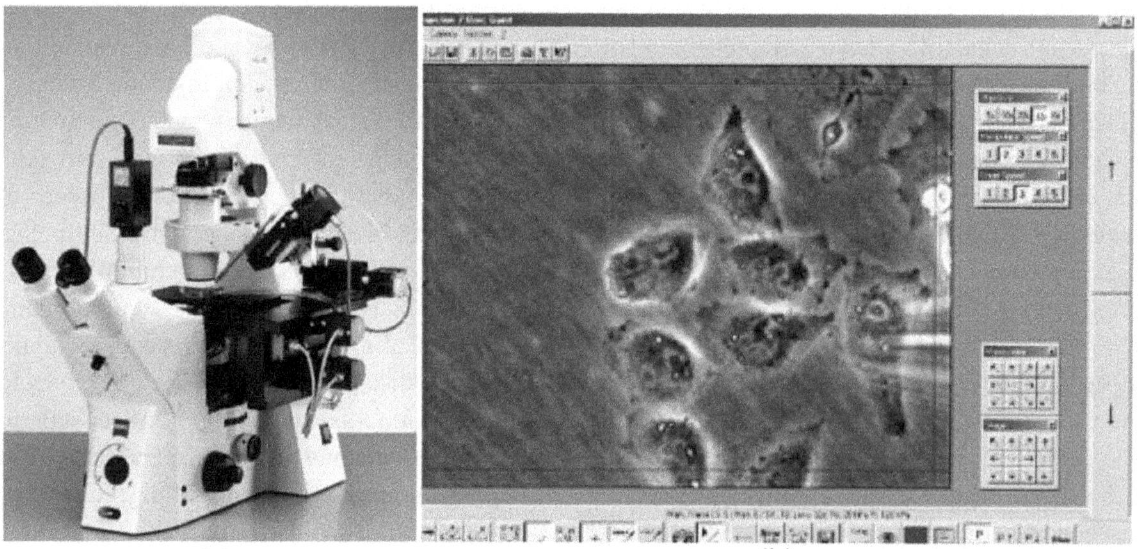

(a) (b)

as blood cells or egg cells, grow loosely in a medium. The suspended cells always have a larger size than adherent cells and float in the culture medium. Due to the freedom of movement of the cells in the solution, the automatic suspended cell injection process will be more complex and time consuming. To fix the suspended cells and facilitate the cell injection process, existing manually suspended cell injection systems all include two manipulators, as shown in Figure 3(a): a holding micro-tube with a holding pipette is mounted on the left manipulator to track and fix the suspended cells, while an injector pipette with an injector holder is mounted on the right manipulator to perform the gene injection.

The past decade has witnessed significant efforts (T. Huang et al., 1996; X. Li et al., 2001; X. D. Li et al., 2001; Pepperkok et al., 1988) performed to translate this manual process into an automated one. A representative automatic suspended cell injection system established by Li et al. (X. Li et al., 2001; X. D. Li et al., 2001) is shown in Figure 3. This kind of automatic suspended cell injection procedure mainly includes the following steps (X. Li et al., 2001; X. D. Li et al., 2001): first, it performs a visual search and recognition of cells, injecting pipette and cell holding tube through image processing; then it drives the holding tube to track and hold the cells and move them to the desired manipulation area; next it guides the injector pipette to inject material into cells and finally, it releases the injected cell in the finishing area and repeats the procedure again until all of the suspended cells have been injected. This is obviously not a reliable control process. Unpredictable situations such as the collision between cells and manipulation tools or the motion of cells themselves can easily lead to the failure of this process.

Recently, a number of potential solutions have been proposed to solve the suspended cell injection problem in a similar manner as the batch adherent cell manipulation. These solutions aim to develop schemes for managing large numbers of suspended cells in parallel, and techniques for the immobilization and positioning of suspended cells in well ordered 2-D arrays have been developed. Three promising methods have been proposed and utilized in automatic suspended cell injection (Ralph W. Bernstein, Scott, & Solgaard, 2004).

The first method is a simple and versatile for capturing suspended cells. It applies an under pressure to a chip structured with micro-machined cavities or holes. Wang et al. (W. Wang, Liu, Gelinas, Ciruna, & Sun, 2007; W. H. Wang, Liu, & Sun, 2007) firstly presents a vacuum-based cell holding device, where evenly spaced through-holes were connected to a vacuum source via a backside channel and a sucking pressure enabled each through-hole to trap a single zebrafish embryo (~1.2*mm* in diameter (including chorion)). Liu et al. (X. Liu & Sun, 2009) has extended this device to hold much tiny cells of mouse zygote (100*m*m in diameter) through the two-sided glass wet etching and PDMS-glass bonding technologies, as shown in Figure 4(a), and based on this vacuum cell holding device, an automatic suspended cell injection system was established and tested in batch ICSI process. Fujisato et al. (Fujisato et al., 1998) developed a cell holder made of micro-porous glass (MPG) with a glass or stainless coating layer, on which sand-blasted micro-pockets holes were formed, fluid containing cells are pumped into the upper chamber and across the MPG disk to the lower chamber through the holes, the pressure drop holds and fixes the cells in the micro-pockets holes. This device has been used on pig ova cell injection.

The second method is utilizing the capillary or adhesion force to immobilize suspended cells to resist the injecting and pulling force. Zhang et al. (R. W. Bernstein et al., 2003; Zhang et al., 2005) utilized fluidic self-assembly technology to position and immobilize Drosophila (fruit fly) embryos using a thin film of polychlorotrifluoroethylene oil, and realized parallel injection of genetic material in the plane with two-dimensional (2-D) Drosophila (fruit fly) embryos arrays

Figure 3. Traditional automatic suspended cell injection system: (a) sketch of the whole system; (b) operation area under the microscope, adapted from (X. Li, Zong, & Bi, 2001)

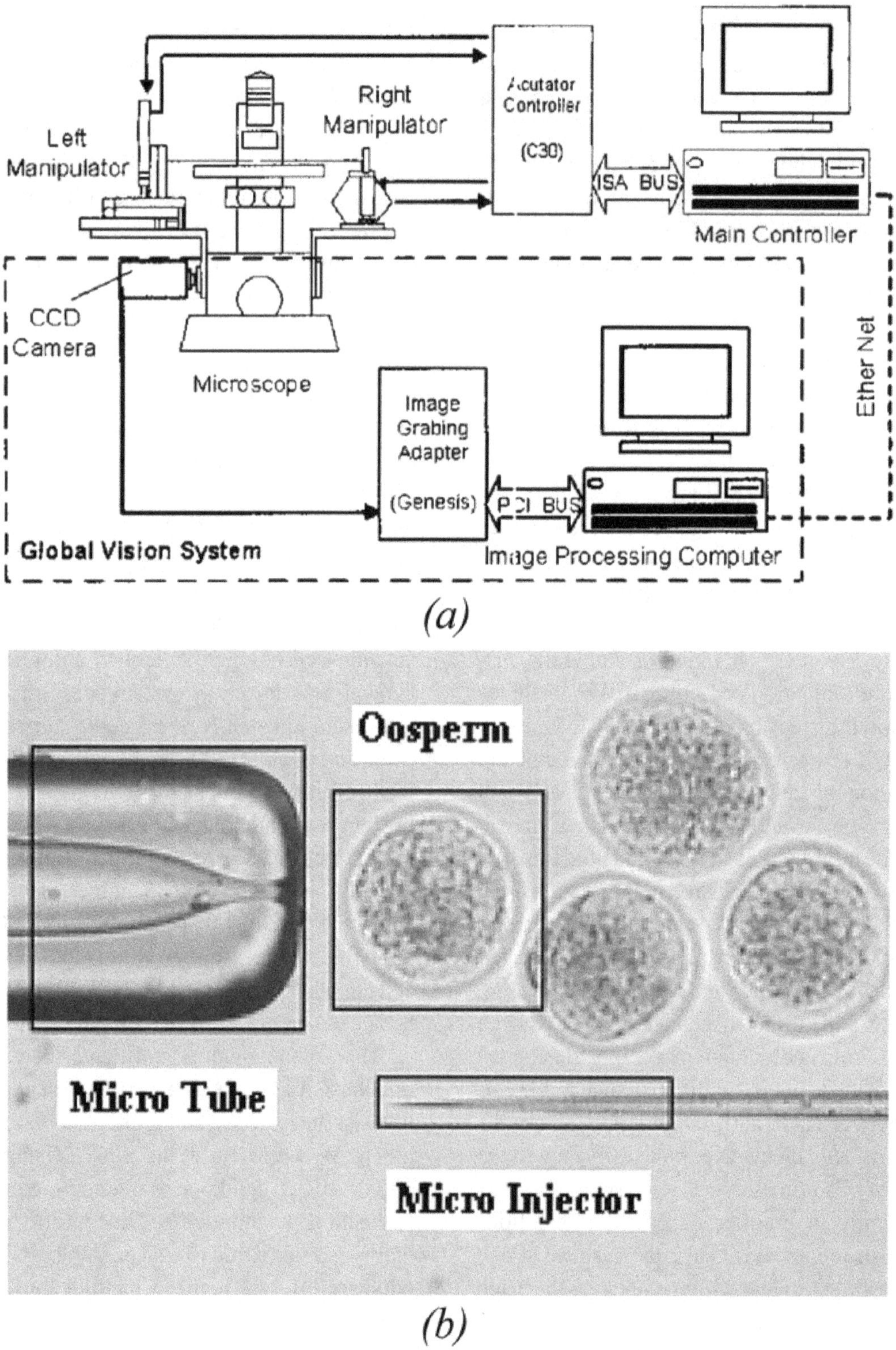

Figure 4. New solutions on automatic suspended cell injection systems

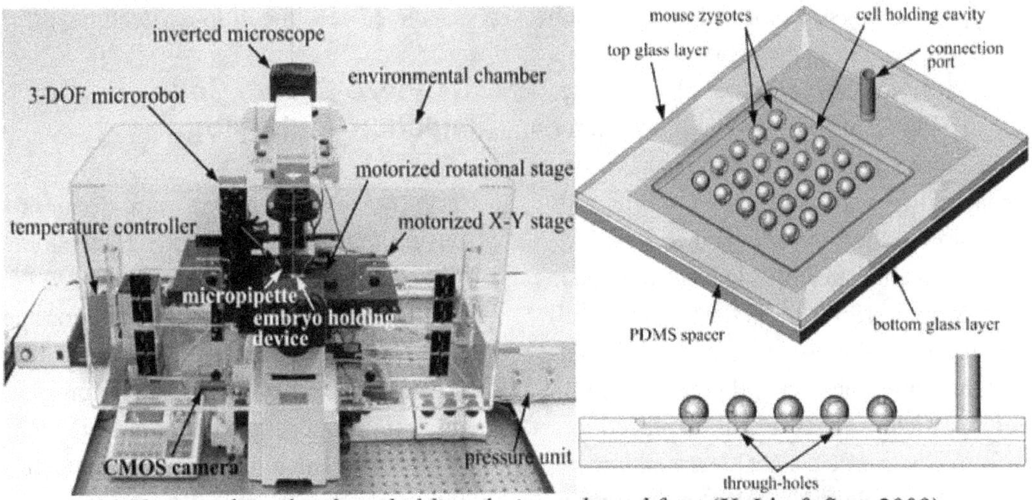

(a) Vacuum-based embryo holding device, adapted from (X. Liu & Sun, 2009).

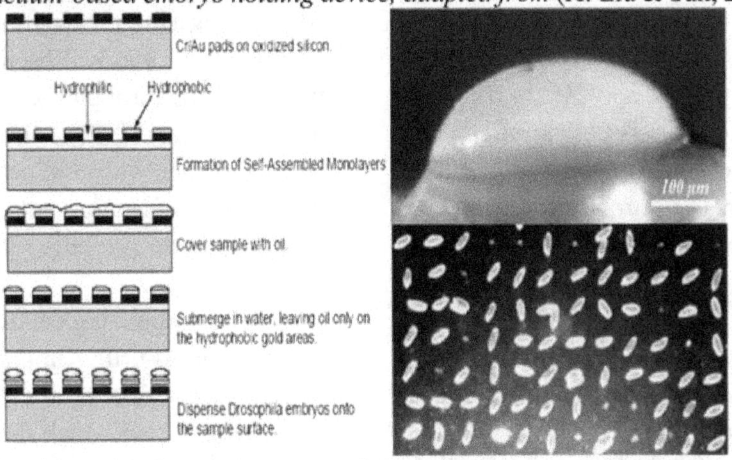

(b) Fluidic self-assembly technology to position and immobilize Drosophila (fruit fly) embryos, adapted from (R. W. Bernstein et al., 2003; Zhang et al., 2005).

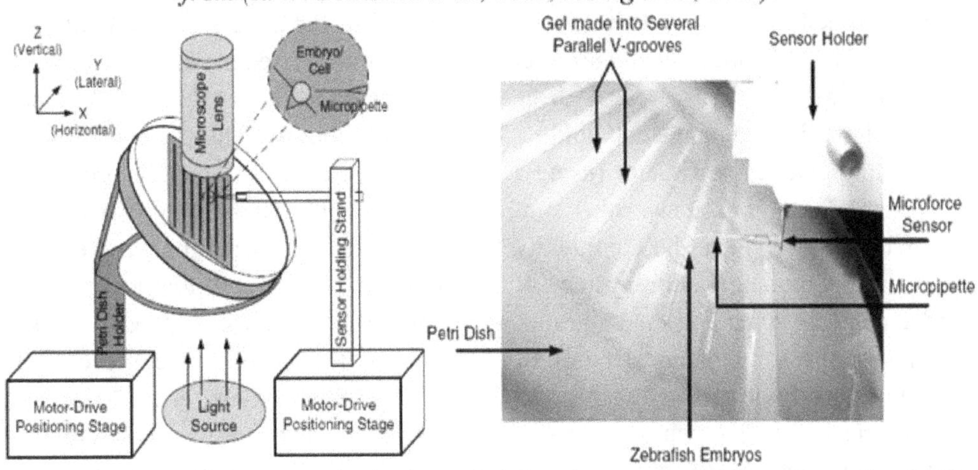

(c) Gel-made cell holder with several parallel V-grooves, adapted from (Lu et al., 2007).

(Figure 4(b)). Lu et al. (Lu, Y, Joohoo, Ge, & Lin, 2007) reported the development of a prototype micromanipulation system for automatic batch microinjection of Zebrafish embryos utilizing a gel cell holder with several parallel V-grooves (Figure 4(c)). Takeuchi et al. (Takeuchi, Nakajima, Kojima, & Fukuda) proposed a new probe device which can realize soft handling of single cells using adhesion to thermoresponsive gel.

The last promising method is Dielectrophoresis (Pethig & Markx, 1997; Pohl, 1978), is a well known method for moving, capturing, manipulating and separating bacteria and cells. When an uncharged particle, such as a cell, is placed in a non-uniform electric field a net electrical force is exerted on the particle as a result of polarization. This dielectrophoretic force is dependent on the cell size and shape, as well as on the magnitude and non-uniformity of the applied electric field. The polarity of the force is determined by the conductivity and permittivity of the cells and the suspending medium. Dielectrophoretic techniques are widely used for characterization, separation and manipulation of cells and micro-organisms (Pethig & Markx, 1997; Pohl, 1978). Demonstration of arrays of embryos realized by this method is not published in this chapter. This is a complicated structure, and since this method has great prospects for much smaller cells, it will be reserved for future research.

The suspended cell injection system derived from these methods has distinct advantages over the traditional ones, as shown in Figure 3. With the suspended cells immobilized on the cell holding device, the holding pipette manipulator (using a traditional design) does not need any other support, which will greatly reduce the complexity and costs of the micromanipulation system. If the suspended cells are spread on the cell holding device in a regular array, the localization of the manipulation target will be predictable and the searching time will be shorten greatly. Additionally, the unpredictable collision between cells and the manipulation tools, or the motion of cells themselves, can be avoided. All of these improvements will make the batch automatic suspended cell injection realizable.

1.4 Piezo-Driven Cell Injection Technology

Recently, several groups tried to improve the conventional cellular piercing technology that utilizes a pulled sharp glass pipette to penetrate the cell membrane with a high injection speed and injection force, contributing to the rupture of the cell membrane. Two novel cell piercing technologies were proposed, the Ros-Drill technique and the piezo-drill technique. Olgac (Ergenc & Olgac, 2007) developed the so-called Ros-Drill technique, which using a high speed rotation injector pipette to produce a rotational oscillation at the pipette tip as it drills through the cell membrane. With this approach, a 70-90% cleavage rate in ICSI of mouse oocytes was obtained. While successful, the whirling effect due to the eccentricity of the injector pipette presented a problem yet to be resolved.

Another promising cellular piercing technology is the piezo-drill. Kimura (Kimura & Yanagimachi, 1995) first presented the piezo-driven ICSI procedure in which the injection pipette is driven by a series of piezo-pulses with adjustable frequencies, amplitudes, and durations. With the use of the piezo-ICSI method, 80% survival rate of sperm-injected mouse oocytes were obtained, of which approximately 70% of which developed into blastocysts. The piezo-drill has been commonly applied in the ICSI process for human, mouse and bovine embryo injection and others, with satisfying performance. Researchers in the field also discovered that a very small mercury column in the injector pipette can significantly improve the success rate of piezo-ICSI (T. Huang et al., 1996; Katayose et al., 1999; Kimura et al., 1998). However, the use of mercury in the laboratory has serious potential toxicity effects, related to both personnel safety and to the development of the manipulated oocytes.

Work is being conducted to investigate the function of mercury column on the zona piercing mechanism in piezo-ICSI. Fan (Fan, Feng, Agca, & Critser, 2006), using the finite element method while investigating the effect of liquids surrounding the pipette tip, found that the use of mercury increases the amplitude of the injector pipette oscillation. Experiments conducted by Olgac (Ediz & Olgac, 2005) and in the simulations by Gan (Gan & Chen, 2008) that consider wave propagation in the pipette, support the conjecture that the presence of mercury generally reduces the amplitude of the pipette tip oscillation. Hence the role of mercury may lie in reducing the physical damage to the oocyte. As for the zona piercing mechanism in piezo-ICSI, Kimura (Kimura & Yanagimachi, 1995) and Fan (Fan et al., 2006) state that the axial pipette motion facilitates the piercing of the mouse zona pellucida. However, Olgac (Ediz & Olgac, 2005) found from their experiments that the pipette has a much larger lateral oscillation compared with axial oscillation during the piercing process, and argued that the lateral oscillation plays a critical role in the zona piercing process. Gan (Gan & Chen, 2008) modeled and simulated the mouse zona piercing process and reveals that the lateral vibration of the injection pipette plays a key role in the piercing process. But as axial and lateral vibration motion are both present in piezo-drill, the experimental validation of the function of axial and lateral motions is a difficult problem to resolve.

While there are different opinions regarding the function of mercury in the injection pipette on the zona piercing mechanism in piezo-ICSI, the main limitations of the piezo-drill technique arises from the toxic effects of mercury and the damage to the oocyte caused by the large amplitude of the pipette tip. Some researchers have attempted to reduce the vibration of the pipette tip during piezo-driven cell injection, such as the design of a partially-rotating piezoelectric based on a cylindrical piezoelectric material is presented by Tan. K. K.(Putra, Huang, Tan, Panda, & Lee, 2006; K. K. Tan & Putra, 2005a, 2005b) with application in ICSI, and simulations and experiments shows that this design can eliminate the vibration of the oocyte significantly.

In this chapter, a cell injection system for automatic batch injection of suspended cells is proposed (H. Huang, Mills, Lu, & Sun, 2011; H. Huang, Mills, & Sun, 2011). Based this system, a piezo-driven ultrasonic cell injection technique is introduced as a universal cell injection method. Comparing with the existing piezo-driven cell injector, this new designed piezo-driven cell injector centralizes the piezo oscillation power on the injector pipette through a piezo stack located near the injector pipette. Significantly, this eliminates the vibration effect on other parts of the micromanipulator. Hence, only a small piezo stack is required to actuate the pipette tip and perform the piezo-drill cell injection process. High amplitude oscillation of the injector pipette due to the system noise is substantially reduced without the use of a mercury column. Furthermore, the ultrasonic vibration micro-dissection (UVM) theory is utilized to investigate the piezo-driven cell injection process, with the main reason of the high amplitude lateral oscillation of the injector pipette investigated. From preliminary automatic cell injection experiments that were performed on zebrafish embryos, under reasonable driven frequency and amplitude, the injector pipette easily pierces through the cell membrane with much low injection speed and with virtually no deformation of the cell wall. This new technology approach shows good potential in precision special position injection and is less damaging to the injected cells.

The remaining sections of this chapter are organized as follows. In Section II, the proposed automatic batch suspended cell injection system is introduced. In Section III, the design of the new piezo-dirven ultrasonic cell injector, including the driving circuit is given. In Section IV, the ultrasonic vibration micro-dissertation theory is used to investigate the piezo-driven cell injection process. The comparative results of experiments are reported in Section V. Finally, conclusions are given in Section VI.

2 SYSTEM SETUP

Figure 5 illustrates an automatic suspended cell injection system developed in our laboratory for this research. Based on the previous research (H. Huang, Sun, Mills, Li, & Cheng, 2009; H. B. Huang, Sun, Mills, & Cheng, 2009), this newly-designed system not only considers the characteristics of the biological experiment undertaken, to make the operation more convenient, but also adds new functions such as microscope autofocus. The system is mainly comprised of executive module, sensory module and control module. The executive module includes three components: The first is the *X-Y-q* positioning table (*X-Y* axis: KR30H, THK, working distance: 60*mm* × 60*mm*, resolution: 0.1*mm*, *q* axis: MRS102, BOIF, resolution: 0.00125°) controls the motion of the cell holding device and then the injected cells. The cells to be injected are placed and fixed on the working plate through a specially designed cell holding device formed from agarose-gel in a Petri dish (H. B. Huang et al., 2009); The second part is the injection manipulator that adjusts and controls the piezo-driven cell injector, the injection pose of the pipette can be manually adjusted by the four DOF (*L-M-N-a*) MM-3 micromanipulator (Narishige)

Figure 5. Test-bed for the suspended cell injection system (H. Huang, Mills, Lu et al., 2011) ©Springer 2011

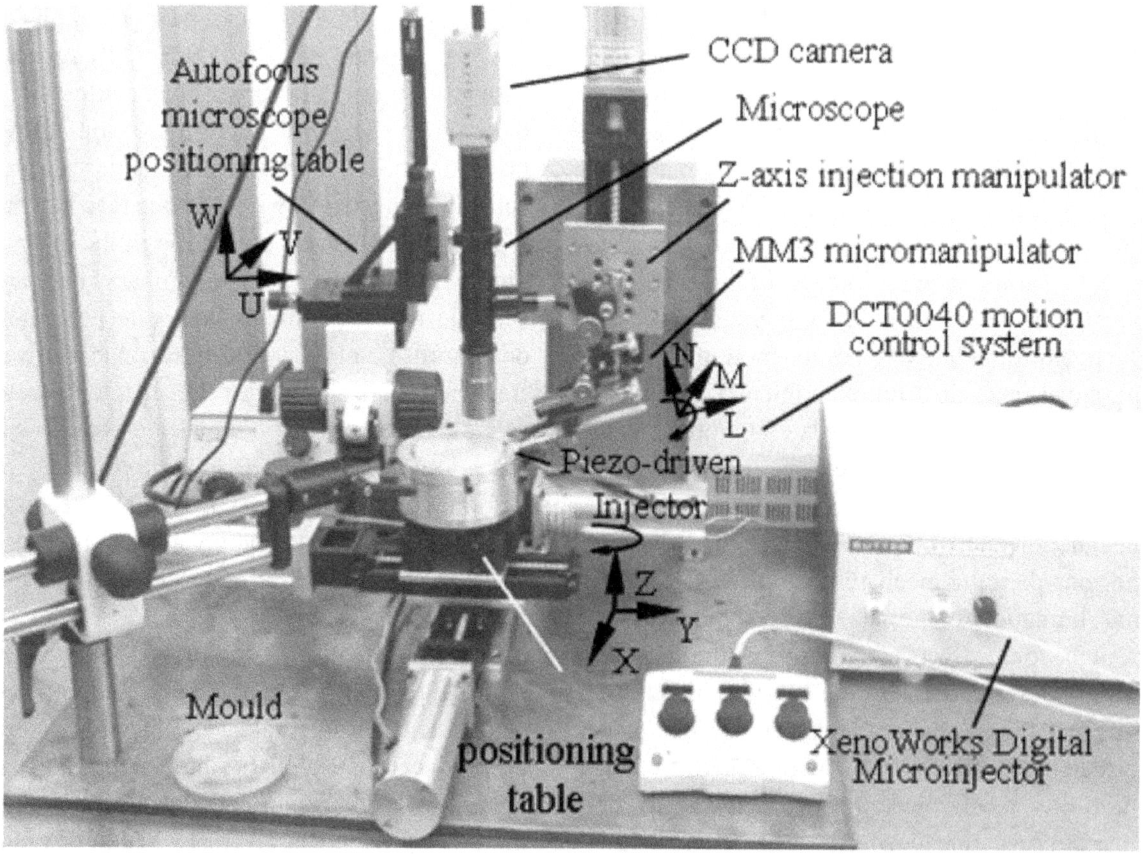

mounted on the Z-axis injection manipulator (Z axis: KR30H, THK, working distance: 60*m*m, resolution: 0.1*m*m) with a workspace 40mm × 20mm × 35mm.

The coordinated motion of the X - Y - a table and the Z-axis injection manipulator is required to perform the cell injection task. The third part is the autofocus microscope positioning table makes the adjustment of the microscope very convenient. Coarse focusing of the microscope is performed through the boomstand (SMS6B, Canimpex), then the fine-focusing of the microscope will be processed by three DOF (u - v - w) translation stage (MT3-27, Edmund, workspace 40mm × 20mm × 35mm). To realize the autofocus function, the w direction motion of the microscope is controlled by a stepping motor (MT1-Z7, Edmund, resolution: 0.1*m*m/Step). A XenoWorks Digital micro-injector (Sutter) permits a minimum volume resolution of 2nl to be delivered into the cell. The sensory module contains a vision system that includes an optical microscope, lighting system, CCD camera (PL-B776U, Pixelink, 3.1 megapixel color machine vision camera), and image processing computer. The control module consists of a 2.8 GHz host computer and a DCT0040 motion control/drive system (H. B. Huang et al., 2009). To minimize vibration, all units except the host computer and pressure units are placed on a vibration isolation table (1202, Kinetic System).

Figure 6 shows a close-up view of the injection area. Figures 5 and 6 illustrate an automatic suspended cell injection system developed in our laboratory. This system is designed to simulate automatic cell injection of large batches of suspended cells (such as fish embryos) in biological engineering processes. To achieve this purpose, the system is designed for use with cells arrays instead of holding cells individually. A specially designed cell holding device was fixed on an actuated rotary plate, permitting the cells to be held and transported one by one, into the field of view of the microscope for injection.

Figure 6. Close up view of the piezo-driven cell injector (H. Huang, Mills, Lu et al., 2011) ©Springer 2011

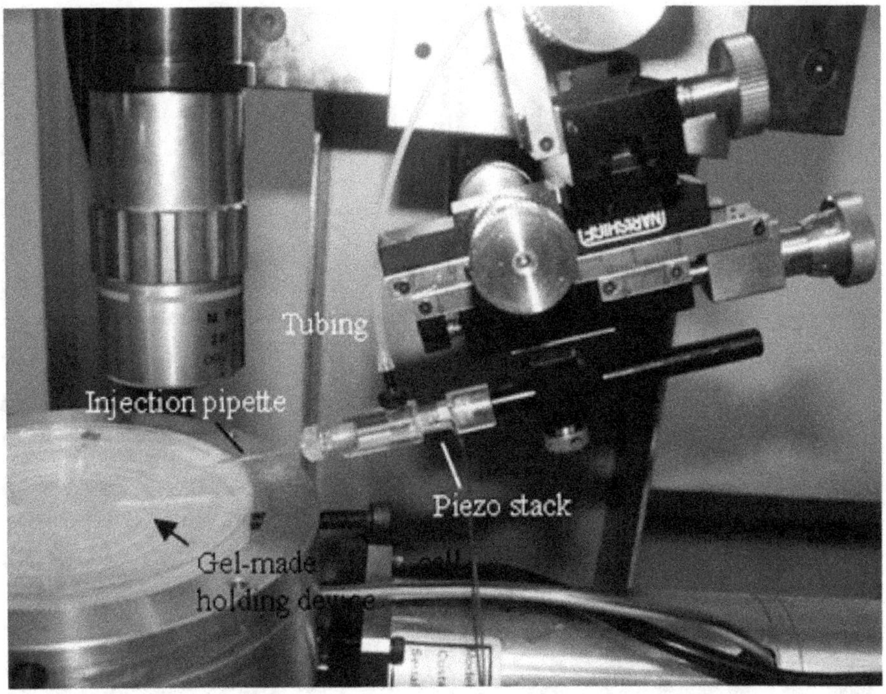

3 BATCH CELL INJECTION CONTROL DESIGN

Automatic cell injection process begins with identification of positions of the manipulated object and injector pipette by using image processing technology. The image processing procedure is as follows. First, pre-processing is conducted to obtain relatively noise-free binary images. An image is first convolved with a low-pass Gaussian filter for noise suppression. Then the gray-level image is filtered to find the edge of the cell chorion, cytoplasm and injector pipette, using the Canny edge detection algorithm, as shown in Figure 7(b). To track the movement of cell embryos, a traditional Hough transform has the limitation of long calculation time and relative low precision. A faster and more precise method called the chord midpoint Hough transform (CMHT) (H. B. Huang et al., 2009) is utilized in our research to improve the efficiency and accuracy of ellipse detection in an image. As for the injector pipette, the correlation-based pattern matching method is used to determine its position. Figure 7(c) shows the image processing results. Figure 7(d) shows the radius histograms of the zebrafish embryo with and without chorion.

Note that the image processing technology is used for detecting the cell diameter, e.g. the diameter of chorion, the diameter of yolk and outside of the cell, and then guiding the injector pipette towards the desired starting position with visual servoing. When the pipette contacts the cell and starts to inject, the position feedback obtained by the motor encoder, instead of the image feedback, will be used in the position control loop for higher processing speed and accuracy.

Figure 7. A typical image grabbed and processed by vision system, (a) Source image; (b) Edge detection; (c) Detection results; and (d) Zebrafish embryos radius histogram (H. B. Huang et al., 2009) ©IEEE 2009

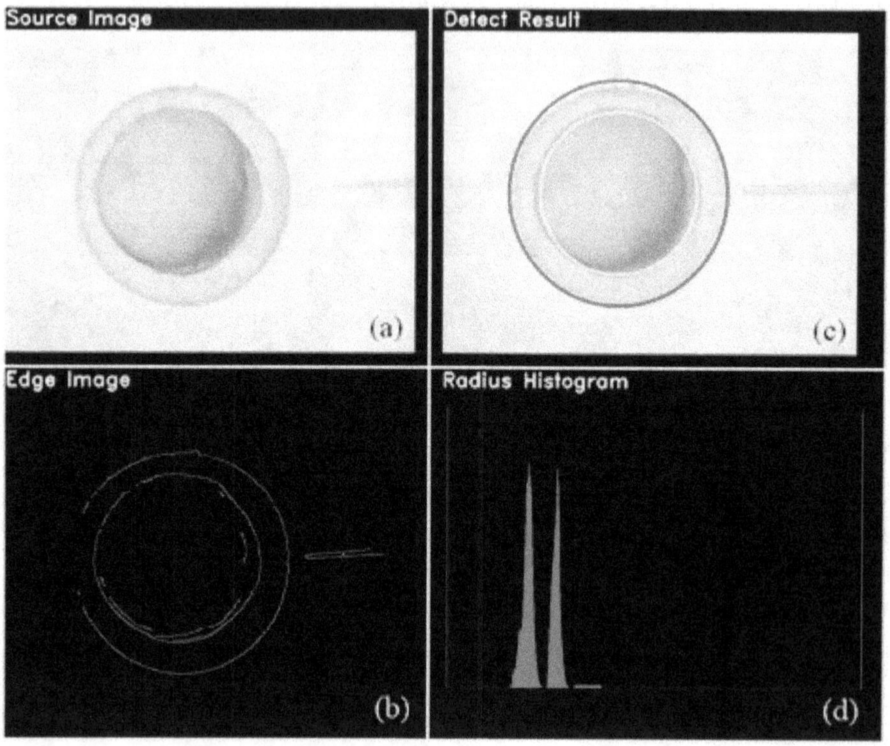

A batch of cells to be injected are placed in the hole array as shown in Figure 6. The cells are observed by the microscope and the position of the cell is adjusted to a proper injection pose by pinching a silastic tube. Figure 8 illustrates the cell injection process. Since the injector pipette may be located out of the X−Y plane such that it cannot be observed by the microscope, the Z-axis of the injector pipette must move first so that the injector pipette can be viewed by the microscope. Denote q_a as the position of the head of the injector pipette at the beginning of manipulation. The injector pipette moves to the starting injection height that is within the focal range of the microscope, and stops at the position q_b, using a vision guided method. An image-based visual servoing methodology is then used to guide the injector pipette towards the starting injection position q_s. When the injector pipette arrives in the position q_s, the control scheme switches to the more precise cell injection controller, similar

Figure 8. Batch biomanipulation process motion plan: (a) side view; (b) top view

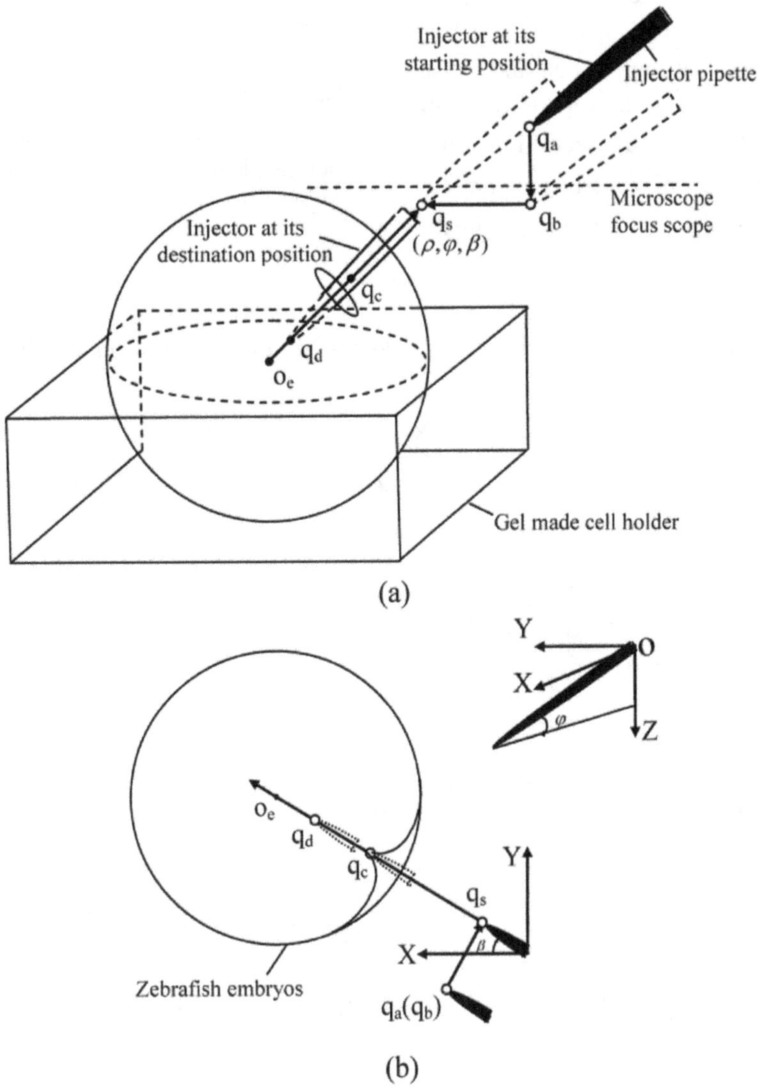

to the control scheme proposed by Sun et al. (Y. Sun & Nelson, 2002).

The precise cell injection control process is designed as follows. First, the so-called pre-piercing step is initiated by accelerating the injector pipette from the starting injection position q_s towards the cell wall. To ensure that the injector pipette is directed toward the centre of the cell, denoted by o_e, the X, Y, and Z axes must move simultaneously, with the desired trajectory of the injector pipette expressed using the spherical polar coordinates. As shown in Figure 8, when the injector pipette reaches the position q_c located on the surface of the cell membrane, the new designed piezo-driven cell injector is switched on, the cell injector is cutting the cell and cause minimal harm. Next, the injector pipette is decelerated to approach the desired injection position q_d and pierce the cell, the piezo-driven cell injector is switched off. As the piezo-driven cell injector is cutting not piercing the cell membrane, during the whole cell injection process, the velocity of the cell injector can be decreased and designed as needed, which will be much easier for control. After injecting the genetic material into the cell at q_d, the injector pipette is accelerated again in the opposite direction for removal from the cell membrane. The working plate holding the cells is then rotated to position another cell within the injection area, and the cell injection is repeated. The whole cell injection control flow is shown in Figure 9, with detail in (H. B. Huang et al., 2009).

3.1 New Piezo-Driven Cell Injection Design

The piezo-driven cell injector structure has remained the same since it was invented. A commercial piezo-driven cell injection system (PIEZO-DRILL, Burleigh) is shown in Figure 10(a). It

Figure 9. Automated suspended cell injection flow chart (H. Huang, Mills, Lu et al., 2011) ©Springer 2011

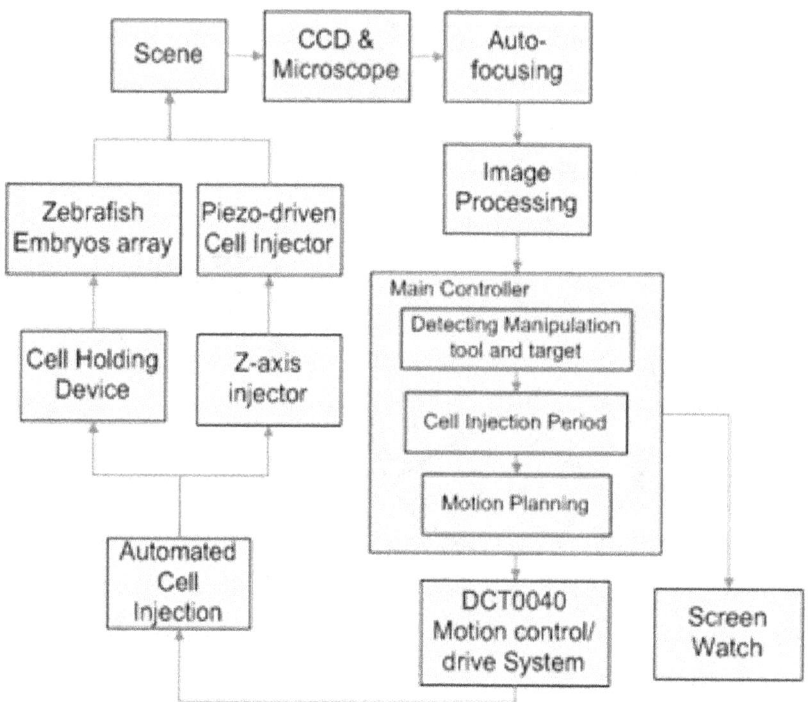

consists of the injection pipette, the pipette holder, the holding clip, and the piezo actuator, as shown in Figure 10(b). The ultrasonic mechanical pulses transfer longitudinally through pipette holder, the holding clip, the glass pipette and finally reach the pipette tip. The injection pipette, pipette holder, and the holding clip will vibrate during the cell injection process. The vibration of so many parts will unavoidably introduce unwanted mechanical vibration to the pipette tip. As discussed in the research of Olgac (Fan et al., 2006), when the ultrasonic pipette is operated in air, the largest lateral vibration of the pipette tip will reach 270*m*m. Even with a mercury column with the pipette immersed in mineral oil, the largest lateral vibration of the pipette tip still reaches 37*m*m. And the use of mineral oil and mercury not only make the cell injection process more complicated but also may lead to contamination of cells etc.

To eliminate unwanted lateral pipette vibration and centralize the piezo power on the injector pipette, a new design of this injector system was proposed, as shown in Figure 11. Figure 11(a) and (b) are respectively the exploded view and assembly view of machine parts, and Figure 12(c) is the proposed piezo-driven cell injector prototype. The piezo stack is assembled on the end of the piezo-driven cell injector and to minimize the load of piezo

Figure 10. Conventional piezo-drill system

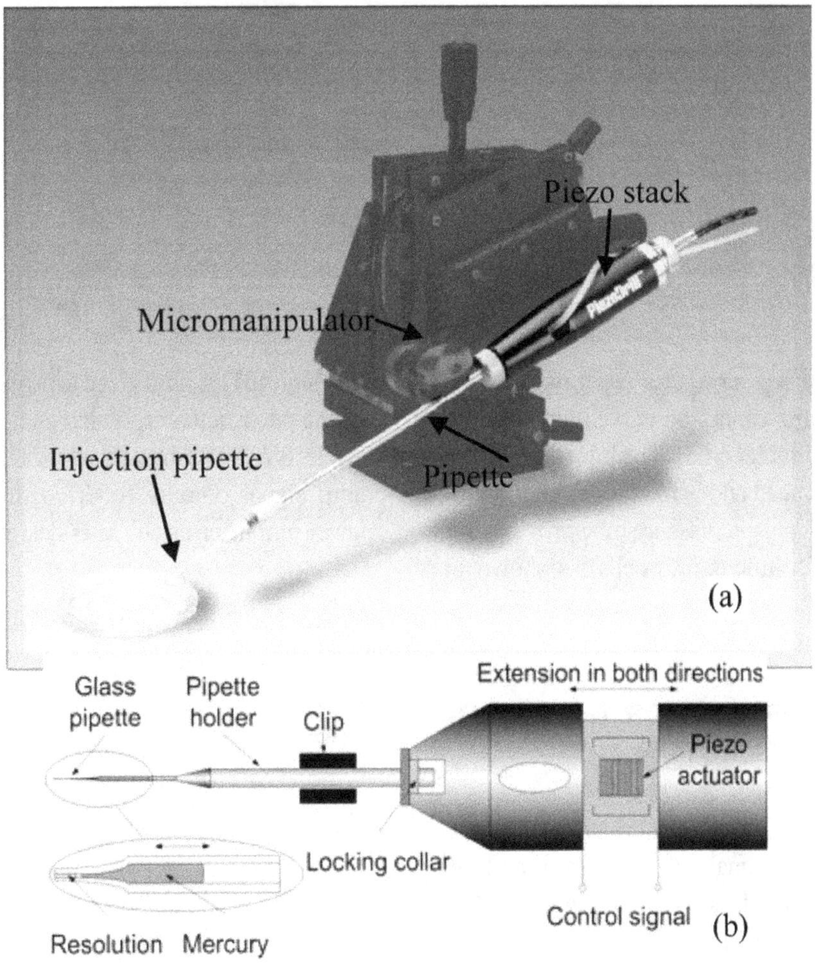

Figure 11. New design of the piezo-driven cell injector: (a) Exploded drawing; (b) Assembly drawing; (c) Proposed piezo-driven cell injector prototype

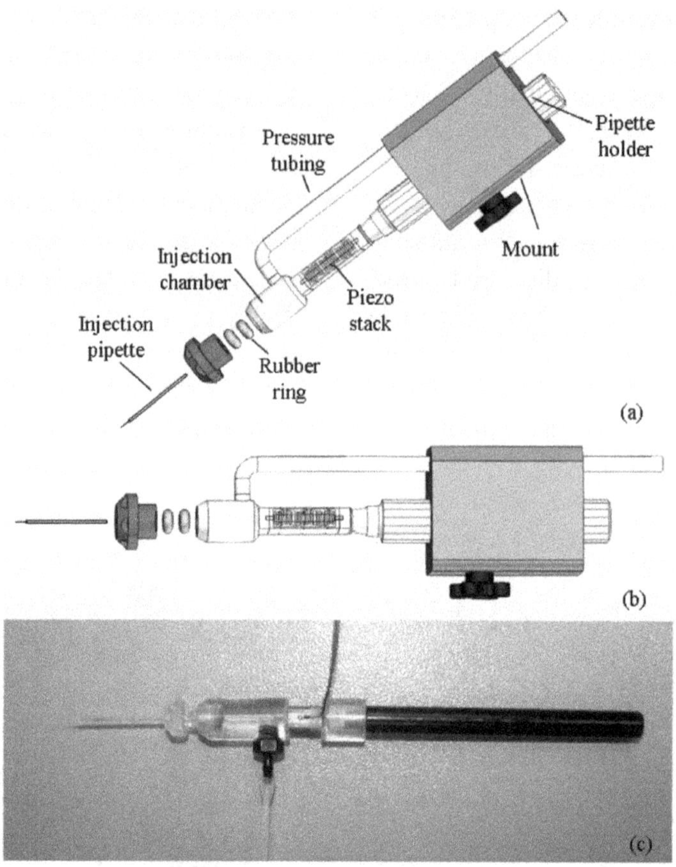

stack, all the unnecessary parts are moved to the fixed end, the parts cannot be moved are made of low density material (acrylic). The injection pipette can be easily replaced and is fixed through the rubber ring. The soft plastic pressure tube provides for the injection of solutions into cells, as shown in Figure 11(a).

4 CONTROL CIRCUIT FOR THE PIEZO-DRIVEN CELL INJECTOR

The driving system of the piezo-driven cell injector is composed of the signal wave generator and the power amplifier circuit, as shown in Figure 12. For simplicity, a commercial function generator (BK Precision 4011A) is utilized to produce a signal wave for the piezo actuator, which can produce 0.5Hz to 5MHz, 0V~20V sine, square, triangle and pulse and ramp signal output. To drive the piezo stack, a power amplifier circuit was designed and built. In

Figure 12. Block diagram of driven system for the piezo-driven cell injector

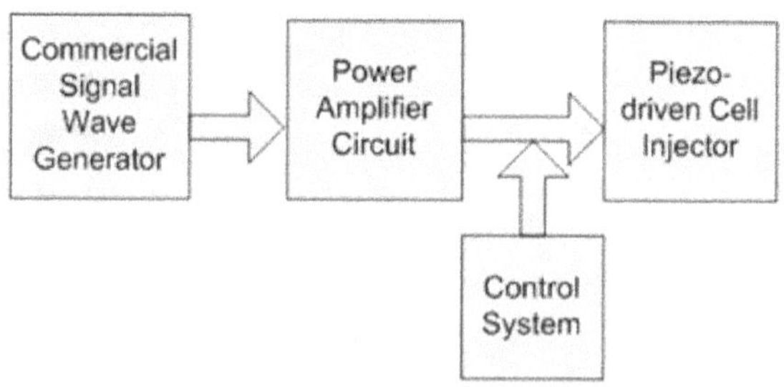

Figure 13. Schematic of the power amplifier circuit (H. Huang, Mills, Lu et al., 2011) ©Springer 2011

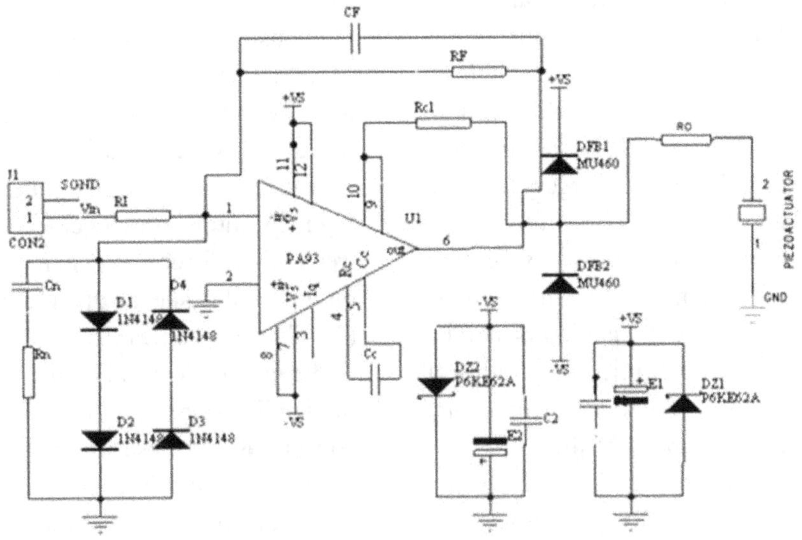

this study, a multilayer piezoelectric actuator (Model P882.10) is selected for the injector. It has many desirable characteristics: including increased lifetime and high-level dynamic performance. Its capacitance is 0.13 μF, rated displacement is 7 μm for 100V and resonance frequency is 135 kHz. To satisfy the power requirements of the piezo actuator, a high-power, low quiescent current amplifier PA93 (APEX) was used for the power amplifier circuit. The schematic of the power amplifier circuit is shown in Figure 13. The switch on or off of driven system is controlled by the I/O port of DCT0040 motion control system. Through tested, this driven circuit can supply vibrating frequency in the range of 0-40 kHz (under 1/3 resonance frequency) and amplitude from 0-3 μm to the piezo actuator according the demand of various injected cells.

5 THEORY OF ULTRASONIC VIBRATION MICRO-DISSECTION

Existing piezo-drill systems utilizes an adjustable given pulse voltage, as shown in Figure 14(b), to inspire the piezo actuator and drill the cells. The three parameters: amplitude, frequency, and duration can be adjusted and trained by the user for particular exercise. But it is found in our research that the harmful lateral tip oscillation has relationship with the frequency of the inspiring signal. Then a single frequency sinewave voltage signal, which is always used in ultrasonic cutting, was used here for piezo-driven cell injection, as shown in Figure 14(a).

Figure 14. The inspiring pulse signals for piezo-driven cell injection (H. Huang, Mills, Lu et al., 2011) ©Springer 2011

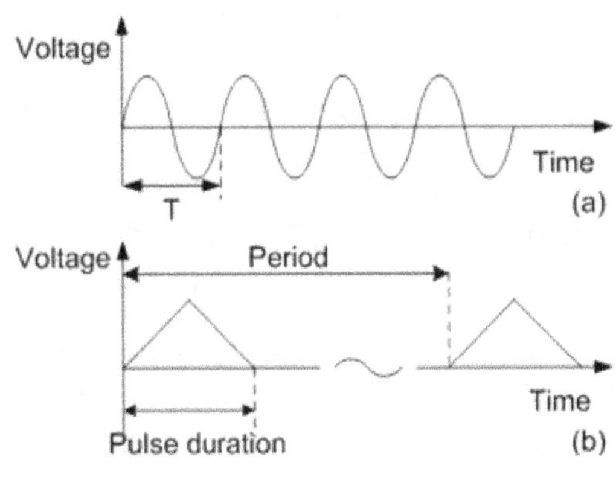

In fact, the injector pipette cuts instead of pierces through the cell membrane during the piezo-driven cell injection process. Thus the ultrasonic micro-dissertation theory is suitable to explain the course of piezo-driven cell injection. Power ultrasonics has been utilized in ultrasonic drilling machining and ultrasonic surgery (D. Sun, Zhou, Liu, & Shen, 1997) etc. These results suggest to attempt to apply power ultrasonics technology to micromanipulation of even small targets such as cells utilizing a similar cutting principle.

During piezo-driven cell injection the tip of the glass injector pipette oscillates at an ultrasonic frequency and hence propagates the ultrasonic energy to the cell membrane. Being activated by an ultrasonic wave, initially, particles of cell membrane surrounding the oscillating pipette tip vibrate at the same frequency and amplitude. As the vibrational energy propagates, inner particles of cell membrane subsequently vibrate as well.

When the piezo-driven cell injector vibrates, displacement of particles in the cell membrane are excited by ultrasonic waves created by the Peizo-actuator (L. Sun, Wang, Chen, & Liu, 2006) as given below.

Displacement: $x(t) = A\sin(2\pi ft - \theta)$ (1)

Vibration velocity: $v(t) = A2\pi f\cos(2\pi ft - \theta)$ (2)

Vibration acceleration:

$$a(t) = -A\left(2\pi f\right)^2 \sin(2\pi ft - \theta) \quad (3)$$

Force: $F = \Delta ma = \rho\Delta Va$ (4)

Sound intensity: $I = 1/2\,\rho c\left(2\pi f\right)^2 \varepsilon_m^2$ (5)

Where A is the maximal amplitude, ρ is the cell membrane density, c is the velocity of ultrasonic wave transmit in medium, f and ε_m is the vibration frequency and amplitude, I is defined as sound intensity.

Considering the stratum of the cell membrane, each stratum of particles within the cell membrane are alternately compressed and pulled by the ultrasonic wave. This is the so-called ultrasonics mechanical effect. As the frequency is very high, this ultrasonics mechanical effect generates tremendous force and instantaneous acceleration on the local tissue, as given in Equations (3) and (4). There are some micro-distortions in the inner tissue medium caused by the very large alternating applied ultrasonic force. The stress is highly concentrated in these micro-distortions. When the ultrasonic sound intensity is very high, the cell membrane ruptures as the stress exceeds its compressive elastic limit. From Equation (5), the cell membrane particle acceleration and square root of sound intensity are directly proportional to the particle vibration amplitude. Hence the mechanical effect can be discerned the tissue behaves as an elastomer and will not rupture below a specific acceleration threshold. Otherwise, the tissue can be dissected when the acceleration is greater than a specific acceleration threshold.

6 EXPERIMENTS

To evaluate the effectiveness of the proposed approach, a series of Zebrafish embryo cell based experiments were performed using the setup as shown in Figures 5 and 6. Zebrafish are commonly chosen as an animal model in biomanipulation because of its similarity in genetic makeup to higher vertebrates, easily obtained, short generation time, external fertilization and translucent embryos. The geometrical parameters to describe the cell model are as follows: the diameter of the Zebrafish egg is approximately 1.2mm (including chorion) and 600 μm (without chorion). The thickness of zebrafish embryo is 3 μm (Y. Tan, Sun, Huang, & Cheng, 2008). The radius of the injector pipette is $c = 7.5\mu m$. The injected zebrafish embryos were all placed in our custom designed cell holding device and were covered with water.

6.1 Vibration of Pipette

In order to identify cause of failure of lateral vibration of injector pipette, a large magnitude object lens (20x) was chosen to observe the vibration trajectory of the injector pipette tip. When a 0.5Hz frequency and 120Volt amplitude sinusoid wave was used to stimulate the piezo injector, the trajectory of the injector pipette tip, as shown in Figure 15(a) and (b), moves very slowly and we can see evident movement of the pipette tip under the driving of piezo actuator. During one vibration cycle, the pipette tip moves from point A to point B. However, obviously the trajectory AB does not coincide with the desired injection direction. It can be decomposed of an axial injection vibration V_I and a lateral vibration V_L. The main reason of the lateral vibration of pipette tip is due to installation error and the difference of the glass injection pipettes, as shown in Figure 15(c). In reality, the vibration direction of the piezo actuator could not exactly coincide with the axis of the pipette tip. There are some deviation angle σ exists and it can be attenuated but cannot be completely eliminated.

It is well known that overlarge lateral vibration causes cell damage during piezo-driven cell injection. A series of sinusoid wave signals (frequency 0.5Hz~40 KHz, amplitude 10Volt) were applied as the excitation signal of piezo actuator. The lateral oscillations of the pipette tip were observed and recorded by CCD camera. The video capture rate is far too low (87 Hz) to capture still images of the vibration of the pipette, but the envelope of vibration can be recorded. The images of the flexible pipette tip are captured in two separate tests: the first situation is when the tip is in the air and the second is during the piercing of zebra fish embryos, where the pipette tip is immersed in the water. The frames in

Figure 15. The vibration trajectory of the injector pipette

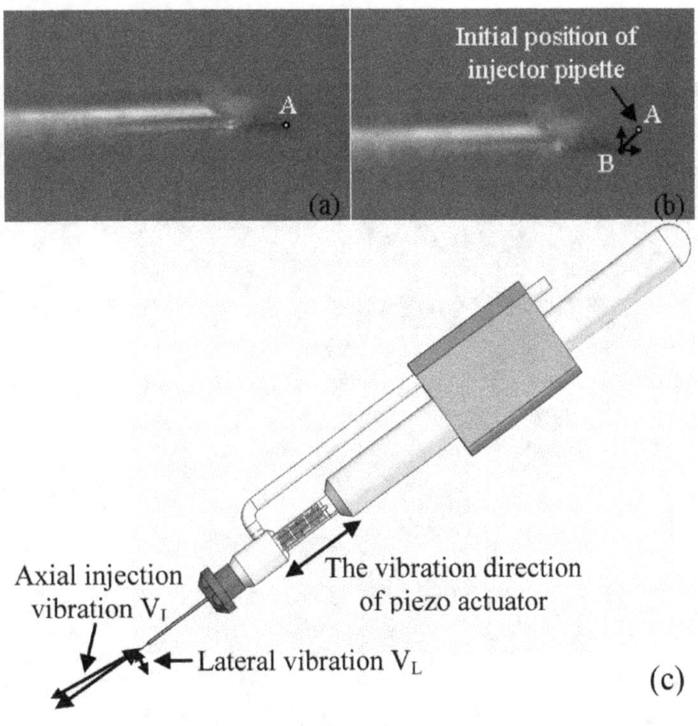

Figure16 are from the first experiment. The injector pipette is placed parallel to the cell holding device and the tip of the pipette is positioned in air. The envelope of lateral oscillations is clearly visible. When the input signal frequency is below 21 KHz, lateral vibration is marginally observable. But between 22 KHz and 23 KHz, the lateral vibration amplitudes increase significantly and the lateral oscillation amplitude reaches its maximum when the frequency of the input signal frequency is 22.5 KHz. At this time, the vibration frequency of the piezo actuator reaches the resonance frequency of the injector pipette and the amplitude of the lateral oscillation of pipette tip reaches its maximum (78 μm). When the frequency increases to values greater than 23 KHz, the lateral vibration decreased slightly.

The second experiment is conducted as during cell injection. The pipette tip is immersed in pure water and the amplitude of lateral vibration of the pipette tip is hence decreased substantially compared with the amplitude found with the pipette in air. The maximum amplitude of the lateral oscillation of pipette tip decreases from 78 μm to 30 μm, as shown in Figure 17. Hence, it is seen that the pipette lateral vibration is significantly damped by the water, which is very helpful for cell injection. When we select the frequency and amplitude of the piezo-driven cell injection, the frequency near the resonant frequency of the pipette tip should be avoided. Compared with the traditional piezo-drill system reviewed in Section III, this novel piezo-driven cell injector significantly reduces the detrimental lateral tip oscillation of the pipette tip to an acceptable range. It is necessary to note that in both experiments, the axial displacement of the tip is difficult to discern as the amplitude is smaller (less than 700nm) than the resolution of the camera.

6.2 Cell Injection Experiment

Preliminary experiments were performed in the cell injection process to demonstrate the effectiveness of piezo-driven cell injection. In all injection experiments, although each test utilized different embryo cells, the mechanical properties of all of the

Figure 16. Lateral vibration of pipette tip in the air (H. Huang, Mills, Lu et al., 2011) ©Springer 2011

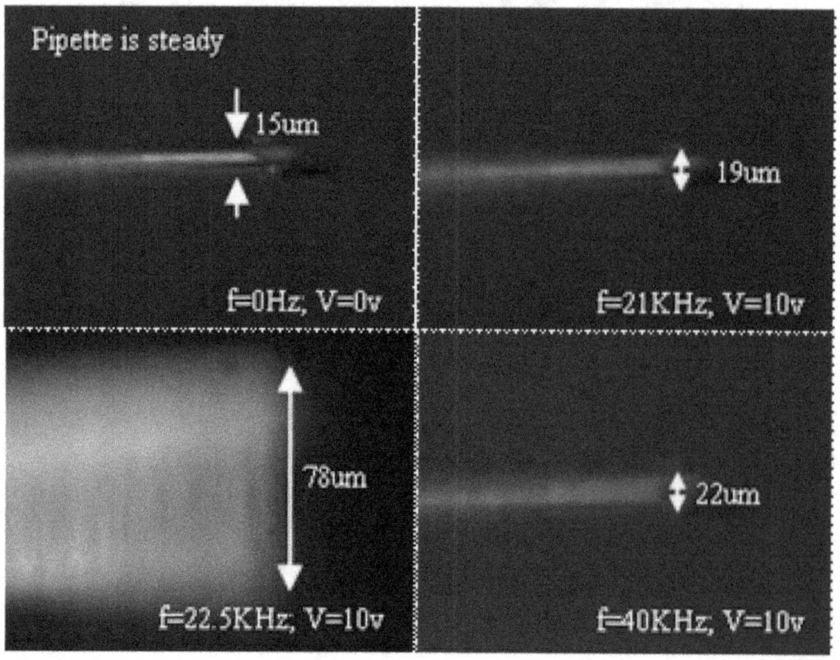

Figure 17. Lateral vibration of pipette under the water (H. Huang, Mills, Lu et al., 2011) ©Springer 2011

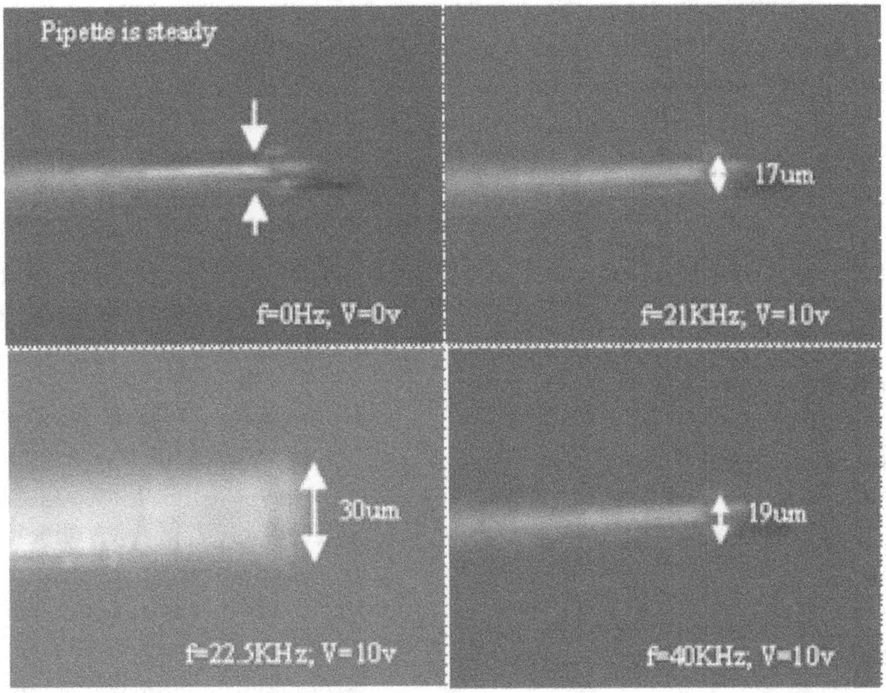

Figure 18. (a) Cell injection process under the conventional cellular piercing technology; (b) Cell injection process with the piezo-driven cell injection technology

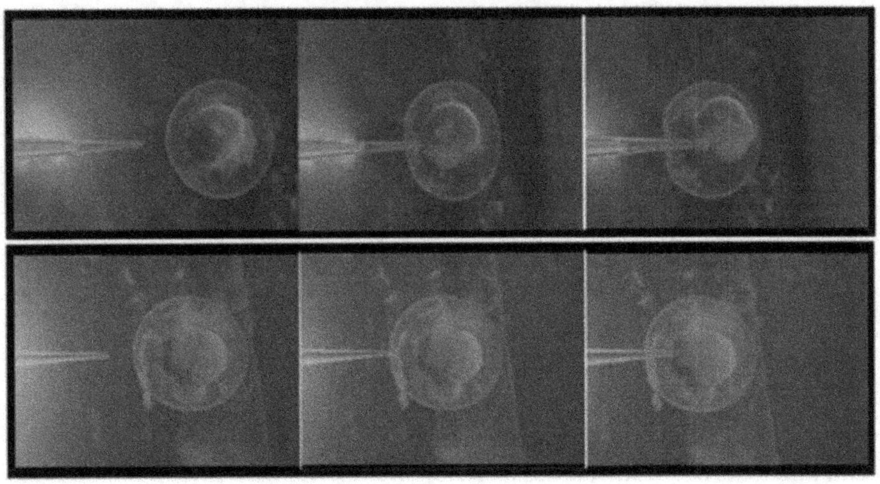

cell biomembranes are uniform, and all were injected after being placed at room temperature (22°C-24°C) for two hours after fertilization (blastula stage). According to Section IV, the trained parameter of piezo-driven cell injection of zebrafish embryo is 20.1 KHz and 10V. The injection velocity is constant with 175 $\mu m/s$.

Two entire cell injection processes with piezo-driven cell injection and conventional cellular piercing technology were shown in Figure 16 for comparison. It is easy to observe that when we inject the zebrafish embryo with piezo-driven cell injector, the injector pipette contacts with the cell membrane and cuts through it with nearly no deformation, as shown in Figure 18(b). But when the conventional cellular piercing technology is used, the cell membrane is deformed by the injection force. When injection force reaches the piercing threshold, the injector pipette pierces through the cell membrane into the cells.

When the ultrasonic energy reaches certain amount (higher than the ultrasonic cavitation threshold), the phenomena ultrasonic cavitation would be induced and can cause damage to the injected cells. Small vacuum bubbles or voids will grow in the cells and the injected cells will break.

7 CONCLUSION

In this chapter, a novel cell injection system for automatic batch injection of suspended cells is proposed. Based on the setup of cell injection system, a revised design of the piezo-driven cell injector is developed and introduced in cell injection of zebrafish embryos. Furthermore, the ultrasonic vibration micro-dissection theory is utilized to explain the piezo-driven cell injection process, and the cause of failure of the detrimental lateral oscillations of injector pipette is identified. To succinctly summarize, this piezo-driven cell injector exhibits three advantages over the conventional cellular piercing technology:

1. It utilizes the ultrasonic cutting force while not the piercing force to penetrate the cell membrane. High injecting speed of the injector pipette is not required and it helps in cell injection motion control and precision position cell injection control.
2. With nearly no deformation of the cells during the cell injection period, the damage to the cells is minimized. This technique will result in comparably high survival rate and success rate.
3. With new design of the piezo-driven cell injector, a small piezo stack is sufficient to perform the cell injection process. And with properly tuning of the stimulated frequency and amplitude, the detrimental lateral tip oscillations of injector pipette are reduced to a satisfying range even without mercury column. The removal of mercury enables a wide range of application of this advanced cell injection technology.

As commented in this chapter, without mercury column and improved performance of piezo-drill technology, this piezo-driven cell injection system have widely application areas not only in ICSI research, but also suitable for all kinds of injection researches such as adherent cell injection, PGD (preimplantation genetic diagnosis), transgenic injection. It also has promising prospect in ultrasonic cutting and needle based minimally invasive surgery, etc.

ACKNOWLEDGMENT

The authors wish to express their appreciation to Prof. Xiaoyan Wen, Miss Rui Guan and Dr. Youdong Wang (Department of Medicine, University of Toronto) for their kind assistance in biological experiments.

REFERENCES

Alexakis, T., Boadi, D., Quong, D., Groboillot, A., O'Neill, I., & Poncelet, D. (1995). Microencapsulation of DNA within alginate microspheres and crosslinked chitosan membranes for in vivo application. *Applied Biochemistry and Biotechnology*, *50*(1), 93–106. doi:10.1007/BF02788043

Anderson, W. F., Killos, L., Sanders-Haigh, L., Kretchmer, P. J., & Diakumakos, E. G. (1980). Replacation and expression of thymidine kinase and human globin genes microinjected into mouse fibroblasts. *Proceedings of the National Academy of Sciences of the United States of America, 77*, 5399–5403. doi:10.1073/pnas.77.9.5399

Ansorge, W. (1982). Improved system for capillary microinjection into living cells. *Experimental Cell Research, 140*, 31–37. doi:10.1016/0014-4827(82)90152-5

Ansorge, W., & Pepperkok, R. (1988). Performance of an automated system for capillary microinjection into living cells. *Journal of Biochemical and Biophysical Methods, 16*, 283–292. doi:10.1016/0165-022X(88)90062-0

Armaleo, D., Ye, G.-N., Klein, T. M., Shark, K. B., Sanford, J. C., & Johnston, S. A. (1990). Biolistic nuclear transformation of Saccharomyces cerevisiae and other fungi. *Current Genetics, 17*(2), 97–103. doi:10.1007/BF00312852

Barber, M. A. (1911). A technique for the inoculation of bacteria and other substances into living cells. *The Journal of Infectious Diseases, 8*, 348–352. doi:10.1093/infdis/8.3.348

Bernstein, R. W., Scott, M., & Solgaard, O. (2004). *BioMEMS for high-throughput handling and microinjection of embryos.* Paper presented at the MEMS/MOEMS Technologies and Applications II, Beijing, China.

Bernstein, R. W., Xiaojing, Z., Zappe, S., Fish, M., Scott, M., & Solgaard, O. (2003, 8-12 June 2003). *Characterization of drosophila embryos immobilized by fluidic microassembly.* Paper presented at the TRANSDUCERS, Solid-State Sensors, Actuators and Microsystems, 12th International Conference on, 2003.

Boussif, O., Lezoualc'h, F., Zanta, M. A., Mergny, M. D., Scherman, D., & Demeneix, B. (1995). A versatile vector for gene and oligonucleotide transfer into cells in culture and in vivo: polyethylenimine. *Proceedings of the National Academy of Sciences of the United States of America, 92*(16), 7297–7301. doi:10.1073/pnas.92.16.7297

Brinster, R. L., Chen, H. Y., Trumbauer, M., Senear, A. W., Warren, R., & Palmiter, R. D. (1981). Somatic expression of herpes thymidine kinase in mice following injection of a fusion gene into eggs. *Cell, 27*, 223–231. doi:10.1016/0092-8674(81)90376-7

Chambers, R., & Chambers, E. L. (1961). *Exploration into the nature of living cell.* Cambridge, MA: Harvard University. Press.

Chou, T.-H. W., Biswas, S., & Lu, S. (2004). Gene delivery using physical methods: An overview. In Heiser, W. C. (Ed.), *Gene delivery to mammalian cells: Nonviral gene transfer techniques* (pp. 147–165). Totowa, NJ: Humana Press.

Ediz, K., & Olgac, N. (2005). Effect of mercury columan on the microdynamics of the piezo-driven pipettes. *Journal of Biomechanical Engineering, 127*, 531–535. doi:10.1115/1.1894368

Eppendorf, Narishige and Cellbiology Trading. (n.d.) Retrieved August 7, 2011, from http://www.eppendorf.com, www.narishige.co.jp, www.ais2.com

Ergenc, F., & Olgac, N. (2007). New technology for cellular piercing: rotationally oscillating -injector, description and validation tests. *Biomedical Microdevices, 9*(6), 885–891. doi:10.1007/s10544-007-9102-2

Fan, M., Feng, Z. C., Agca, Y., & Critser, J. K. (2006). Vibration study of the piezodriven pipettes immersed in viscous liquids. *Journal of Applied Physics, 100*(7), 701–708. doi:10.1063/1.2353895

Felgner, P. L., Gadek, T. R., Holm, M., Roman, R., Chan, H. W., & Wenz, M. (1987). Lipofection: A highly efficient, lipid-mediated DNA-transfection procedure. *Proceedings of the National Academy of Sciences of the United States of America, 84*(21), 7413–7417. doi:10.1073/pnas.84.21.7413

Fujisato, T., Abe, S., Tsuji, T., Sada, M., Miyawaki, F., & Ohba, K. (1998, 29 Oct-1 Nov 1998). The development of an ova holding device made of microporous glass plate for genetic engineering. *Proceedings of the 20th Annual International Conference of the IEEE Engineering in Medicine and Biology Society*, 1998.

Gan, Y., & Chen, Z. (2008). A study of the zona piercing process in piezodriven intracytoplasmic sperm injection. *Journal of Applied Physics, 104*(4), 044702–044708. doi:10.1063/1.2970097

Gordon, J. W., Scangos, G. A., Plotkin, D. J., Barbosa, J. A., & Ruddle, F. H. (1980). Genetic transformation of mouse embryos by microinjection of purified DNA. *Proceedings of the National Academy of Sciences of the United States of America, 77*, 7380–7384. doi:10.1073/pnas.77.12.7380

Graessmann, A., & Graessmann, M. (1976). Early' simian-virus-40-specific RNA contains information for tumor antigen formation and chromatin replication. *Proceedings of the National Academy of Sciences of the United States of America, 73*, 366–370. doi:10.1073/pnas.73.2.366

Graessmann, M., & Graessmann, A. (1983). Microinjection of tissue culture cells. *Methods in Enzymology, 101*, 482–492. doi:10.1016/0076-6879(83)01033-2

Graham, F. L., & Van Der Eb, A. J. (1973). Transformation of rat cells by DNA of human adenovirus 5. *Virology, 54*(2), 536–539. doi:10.1016/0042-6822(73)90163-3

Heiser, W. C. (Ed.). (2004a). *Gene delivery to mammalian cells* (*Vol. 2*). Humana Press.

Heiser, W. C. (Ed.). (2004b). *Gene delivery to mammalian cells* (*Vol. 1*). Humana Press.

Huang, H., Mills, J. K., Lu, C., & Sun, D. (2011). (in press). A universal piezo-driven ultrasonic cell microinjection system. *Biomedical Microdevices*. doi:10.1007/s10544-011-9544-4

Huang, H., Mills, J. K., & Sun, D. (2011). *A new piezo-driven ultrasonic cell microinjection system.* Paper presented at the 2011 IEEE International Conference on Information and Automation.

Huang, H., Sun, D., Mills, J. K., Li, W. J., & Cheng, S. H. (2009). Visual-based impedance control of out-of-plane cell injection systems. *IEEE Transactions on Automation Science and Engineering, 6*(3), 565–571. doi:10.1109/TASE.2008.2010013

Huang, H. B., Sun, D., Mills, J. K., & Cheng, S. H. (2009). Robotic cell injection system with position and force control: Toward automatic batch biomanipulation. *IEEE Transactions on Robotics, 25*(3), 727–737. doi:10.1109/TRO.2009.2017109

Huang, T., Kimura, Y., & Yanagimachi, R. (1996). The use of piezo micromanipulation for intracytoplasmic sperm injection of human oocytes. *Journal of Assisted Reproduction and Genetics, 13*(4), 320–328. doi:10.1007/BF02070146

Kallio, J. K., & Kallio, P. J. (2006). Lab automation in cultivation of adherent cells. *IEEE Transactions on Automation Science and Engineering, 3*(2), 177–186. doi:10.1109/TASE.2006.871482

Kallio, P., & Kuncova, J. (2003). *Manipulation of living biological cells: Challenges in automation.* International Conference on Intelligent Robots and Systems. Las Vegas, USA.

Kallio, P., Ritala, T., Lukkari, M., & Kuikka, S. (2006). *Injection guidance system for cellular microinjections*. Paper presented at the Biomedical Robotics and Biomechatronics, 2006. BioRob 2006. The First IEEE/RAS-EMBS International Conference on.

Katayose, H., Yanagida, K., Shinoki, T., Kawahara, T., Horiuchi, T., & Sato, A. (1999). Efficient injection of bull spermatozoa into oocytes using a piezo-driven pipette. *Theriogenology*, *52*(7), 1215–1224. doi:10.1016/S0093-691X(99)00213-7

Kimura, Y., & Yanagimachi, R. (1995). Intracytoplasmic sperm injection in the mouse. *Biology of Reproduction*, *52*(4), 709–720. doi:10.1095/biolreprod52.4.709

Kimura, Y., Yanagimachi, R., Kuretake, S., Bortkiewicz, H., Perry, A. C., & Yanagimachi, H. (1998). Analysis of mouse oocyte activation suggests the involvement of sperm perinuclear material. *Biology of Reproduction*, *58*(6), 1407–1415. doi:10.1095/biolreprod58.6.1407

Korzh, V., & Strähle, U. (2002). Marshall Barber and the century of microinjection: From cloning of bacteria to cloning of everything. *Differentiation*, *70*(6), 221–226. doi:10.1046/j.1432-0436.2002.700601.x

Li, X., Zong, G., & Bi, S. (2001, 2001). Development of global vision system for biological automatic micro-manipulation system. *Proceedings 2001 ICRA IEEE International Conference on Robotics and Automation.*

Li, X. D., Zong, G., Bi, S., & Zhao, W. (2001). Automatic micromanipulating system for biological applications with visual servo control. *Journal of Micromechatronics*, *1*, 345–363. doi:10.1163/156856301320819534

Liu, D., Chia, E. F., Tian, H., & Heiser, W. C. (2004). Chemical methods for DNA delivery. In Walker, J. M. (Ed.), *Gene delivery to mammalian cells* (*Vol. 245*, pp. 3–23). Totowa, NJ: Humana Press. doi:10.1385/1-59259-649-5:3

Liu, X., & Sun, Y. (2009). Microfabricated glass devices for rapid single cell immobilization in mouse zygote microinjection. *Biomedical Microdevices*, *11*(6), 1169–1174. doi:10.1007/s10544-009-9333-5

Lu, Z., Y, C. P. C., Joohoo, N., Ge, R., & Lin, W. (2007). A micromanipulation system with dynamic force-feedback for automatic batch microinjection. *Journal of Micromechanics and Microengineering*, *17*(2), 314. doi:10.1088/0960-1317/17/2/018

Lukkari, M., & Kallio, P. (2005, 27-30 June 2005). Multi-purpose impedance-based measurement system to automate microinjection of adherent cells. *Proceedings IEEE International Symposium on Computational Intelligence in Robotics and Automation, CIRA 2005.*

Lukkari, M. J., Karjalainen, M. I., Sarkanen, R., Linne, M. L., Jalonen, T. O., & Kallio, P. J. (2004, 1-5 September 2004). *Electrical detection of the contact between a microinjection pipette and cells*. Paper presented at the Engineering in Medicine and Biology Society, 2004. IEMBS '04. 26th Annual International Conference of the IEEE.

Ma, H., & Chen, G. (2005). Gene transfer technique. *Nature and Science, 3*(1).

Mabuchi, I., & Okuno, M. (1977). The effect of myosin antibody on the division of starfish blastomeres. *The Journal of Cell Biology*, *74*(1), 251–263. doi:10.1083/jcb.74.1.251

Pagano, J. S., & Vaheri, A. (1965). Enhancement of infectivity of poliovirus RNA with diethylaminoethyl-dextran (DEAE-D). *Archives of Virology*, *17*(3), 456–464. doi:10.1007/BF01241201

Pepperkok, R., Zanetti, M., King, R., Delia, D., Ansorge, W., & Philipson, L. (1988). Automatic microinjection system facilitates detection of growth inhibitory mRNA. *Proceedings of the National Academy of Sciences of the United States of America*, *85*(18), 6748–6752. doi:10.1073/pnas.85.18.6748

Pethig, R., & Markx, G. H. (1997). Applications of dielectrophoresis in biotechnology. *Trends in Biotechnology*, *15*(10), 426–432. doi:10.1016/S0167-7799(97)01096-2

Pohl, H. A. (1978). *Dielectrophoresis the behavior of neutral matter in nonuniform electric fields*. Cambridge, UK: Cambridge University Press.

Putra, A. S., Huang, S., Tan, K. K., Panda, S. K., & Lee, T. H. (2006). *Partially-rotating piezoelectric actuator to assist cell penetration in intracytoplasmic sperm injection*. Paris, France.

Sun, D., Zhou, Z. Y., Liu, Y. H., & Shen, W. Z. (1997). Development and application of ultrasonic surgical instruments. *IEEE Transactions on Bio-Medical Engineering*, *44*(6), 462–467. doi:10.1109/10.581934

Sun, L., Wang, H., Chen, L., & Liu, Y. (2006). A novel ultrasonic micro-dissection technique for biomedicine. *Ultrasonics*, *44*(Supplement 1), 255–260. doi:10.1016/j.ultras.2006.06.010

Sun, Y., & Nelson, B. J. (2002). Biological cell injection using an autonomous microrobotic system. *The International Journal of Robotics Research*, *21*(10-11), 861. doi:10.1177/027836490202101083

Takeuchi, M., Nakajima, M., Kojima, M., & Fukuda, T. (7-10 November 2010). *Soft handling probe using thermal gel for single cells*. Paper presented at the Micro-NanoMechatronics and Human Science (MHS), 2010 International Symposium on.

Tan, K. K., & Putra, A. S. (2005a). *Partial-rotating piezoelectric actuators for intracytoplasmic sperm injection* (*Vol. 6048*). SPIE.

Tan, K. K., & Putra, A. S. (2005b). *Partial-rotating piezoelectric actuators for intracytoplasmic sperm injection*. Japan: Sappora.

Tan, Y., Sun, D., Huang, W., & Cheng, S. H. (2008). Mechanical modeling of biological cells in microinjection. *IEEE Transactions on Nanobioscience*, *7*(4), 257–266. doi:10.1109/TNB.2008.2011852

Tang, M. X., Redemann, C. T., & Szoka, F. C. (1996). In vitro gene delivery by degraded polyamidoamine dendrimers. *Bioconjugate Chemistry*, *7*(6), 703–714. doi:10.1021/bc9600630

Walther, W., & Stein, U. (2000). Viral vectors for gene transfer: A review of their use in the treatment of human diseases. *Drugs*, *60*(2), 249–271. doi:10.2165/00003495-200060020-00002

Wang, S., Fuller, D., Manson, K., Wyand, M., & Lu, S. (2000). A new animal model to test the AIDS vaccine efficacy in non-human primates: Gene gun-mediated mucosal infection with proviral DNA plasmids. *Antiviral Therapy*, 5.

Wang, W., Liu, X., Gelinas, D., Ciruna, B., & Sun, Y. (2007). A fully automated robotic system for microinjection of zebrafish embryos. *PLoS ONE*, *2*(9), e862. doi:10.1371/journal.pone.0000862

Wang, W., Sun, Y., Zhang, M., Anderson, R., Langille, L., & Chan, W. (2008). *A system for high-speed microinjection of adherent cells* (*Vol. 79*). AIP.

Wang, W. H., Liu, X. Y., & Sun, Y. (2007, 22-25 September). *Autonomous zebrafish embryo injection using a microrobotic system.* Paper presented at the Automation Science and Engineering, 2007. CASE 2007. IEEE International Conference on.

Wang, W. H., Sun, Y., Zhang, M., Anderson, R., Langille, L., & Chan, W. (2008, 19-23 May 2008). *A microrobotic adherent cell injection system for investigating intracellular behavior of quantum dots*. Paper presented at the Robotics and Automation, 2008. ICRA 2008. IEEE International Conference on.

Watt, P. C., Sawicki, M. P., & Passaro, E. (1993). A review of gene transfer techniques. *American Journal of Surgery*, *165*(3). doi:10.1016/S0002-9610(05)80841-4

Yanagida, K., Katayose, H., Yazawa, H., Kimura, Y., Konnai, K., & Sato, A. (1999). The usefulness of a piezo-micromanipulator in intracytoplasmic sperm injection in humans. *Human Reproduction (Oxford, England)*, *14*(2), 448–453. doi:10.1093/humrep/14.2.448

Zhang, X., Chen, C.-C., Bernstein, R. W., Zappe, S., Scott, M. P., & Solgaard, O. (2005). Microoptical characterization and modeling of positioning forces on drosophila embryos self-assembled in two-dimensional arrays. *Journal of Microelectromechanical Systems*, *14*(5), 1187–1197. doi:10.1109/JMEMS.2005.851834

Chapter 12
Prototyping of Robotic Systems in Surgical Procedures and Automated Manufacturing Processes

Zheng (Jeremy) Li
University of Bridgeport, USA

ABSTRACT

The prototyping and implementation of robotic system is a scientific and technological integrating of robotic system design, development, testing, and application. This chapter describes the recent development and applications of robotic systems to surgery procedures in biomedical engineering and automated manufacturing processes in industry. It includes the design and development, computer-aided modeling and simulation, prototype analysis, and testing of robotic systems in these two different applications.

1. INTRODUCTION

Robotic system is constructed mechanically of kinematic chains and actuators which function as muscles that transfer potential energy to body motion (Dylan, 2005, p.797 & Schreuder, 2010, p.253). The robotic mechanisms can be controlled to perform different tasks. The applications of programmable logic control to the industry bring revolution for the manufacturing techniques. It allows more sophisticated, flexible, reliable, and cost-effective manufacturing process controls (Feil-Seifer, 2007, p.425 & Kim, 2007, p.13 & Yang, 2008, p.79). Robotic system is to use control system to reduce human labor intervention during manufacturing processes and productions. It plays very important role and puts strong impact in today's industries (McComb, 2003, p.435 &

DOI: 10.4018/978-1-4666-0176-5.ch012

Pott, 2005, p.101). Computer-aided engineering design can quickly model the robotic systems and speed the design and development cycles (Kumar, 2011, p.994 & Menzel, 2000, p.180 & Tsagarakis, 2003, p.21). Computer aided manufacturing can improve the engineering integral processes of robotic design, development, engineering analysis, and production (Schiff, 2005, p.122 & Tapus, 2007, p.15 & Tholey, 2003, p.158). The current economic globalization requires significant labor cost reduction through industrial automation, applied robotic system, improved machine tools, and efficient production process (Ahmed, 2009, p.431 & Siebert, 2004, p.127). This chapter aims at the introduction of some robotic systems applied in surgical procedure and automated manufacturing processes. Two new research projects of robotic systems are introduced in this chapter through design analysis, computer modeling, computational engineering simulation, and prototype testing.

2. BACKGROUND

Robotic surgery has many advantages including minimally invasive surgical process, narrow incisions, decreased infection, reduced pains, and less hospital stays (Cadierre, 2001, p.1467 & Eirik, 2009, p.77 & Taylor, 2003). The development of surgical robots can improve the situation limited by current laparoscopic surgery and technologies (Backes, 2008, p.97 & Camarillo, 2004, p.188 & Hu, 2002 & Kazanzides, 2008). Also the robotic surgical system can be set up whereby the patients could be loaded into a vehicle by robotic surgical equipment and surgery can be performed by a surgeon remotely at a nearby mobile advanced surgical hospital (Brown, 2007, p.253 & Estey, 2009, p.488 & Ghomi, 2010, p.87 & Mataric, 2007, p.1). The robotic arms can be manipulated through surgeon's voice instructions to control the endoscopic cameras (Bargar, 2007, p.31 & Carigan, 2007, p.179 & Gerhardus, 2003, p.242 & Tapus, 2008, p.169). Figure 1 shows one robotic surgical application in gallbladder surgery. The gallbladder removal by robotic surgery is minimally invasive by way of robot technology permitting optimal viewing of the surgical field through small incisions with less pain and faster recovery time for patients (Gockley, 2006, p.150 & Hanna, 2011, p.761 & Harja, 2007, p.365 & Rosen, 2011). It closely mimics the surgery that is used in traditional 'open' procedures, but allows surgeon to perform the operation by da Vinci™ Surgical Systems through small incisions that are associated with minimally invasive and laparoscopic surgeries (Gortchev, 2010, p.153 & Kaouk, 2009, p.181 & Koh, 2011, p.1945 & Puntambekar, 2009, p.259).

Figure 2 shows the da Vinci robotics surgical system for gynecology, urogynecology, urology and cardiology procedures. It has flexible, safe, précised features that allows surgeon to operate the surgery with reduced trauma to the patients and a faster recovery time (Kwartowitz, 2006, p.157 & Li, 2002, p.90 & Melvin, 2003, p.33 & Passerotti, 2006, p.193 & Peters, 2007, p.179).

Figure 3 displays another robotic-assisted surgical system to help surgeon in surgery. It allows the surgeon to perform surgery with enhanced precision that reduces post-operative complications, less pain and discomfort after surgery, less scarring and a shorter hospital stay (Kypson, 2004, p.87 & Melvin, 2003, p.11 & Passerotti, 2006, p.193 & Patel, 2011, p.423).

The common robotic systems in manufacturing processes include robotics systems for picking and placing work pieces, assembling operations, and handling machine tools (DeLaurentis, 2002, p.91). The common type of robotic unit is the mechanical arms being used in industry and manufacturing processes (Desforges, 2011, p.963). This mechanism, like human arm, can perform many different motions, including left-and-right, up-and-down, and angular motions. Robotic arms can be either controlled through computer program or human interference. Industrial robots have a multi-linked manipulating arm

Figure 1. Robotic surgery for gallbladder removal (Gockley, 2006)

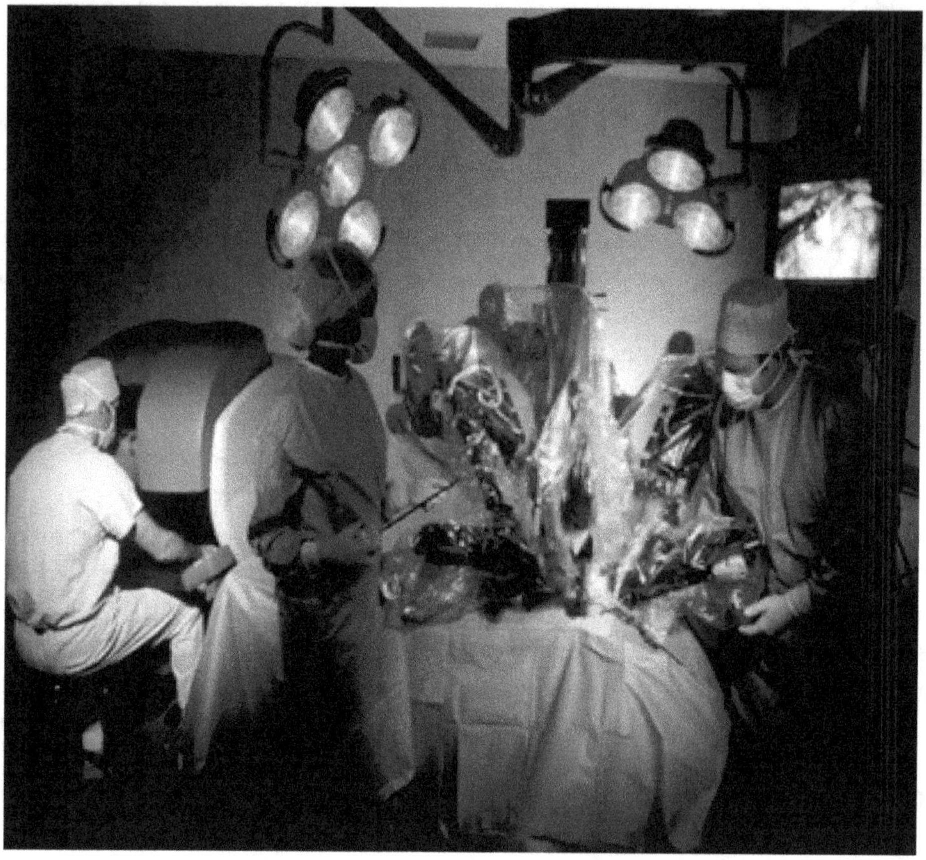

and end effecter such as a gripper mechanism. It can be applied to industrial packaging, welding, painting, printed circuit boards, hazard environment, and many other manufacturing processes (Torbjorn, 2009, p.87). Figure 4 shows one robotic system for picking and placing parts in automobile production. It can increase the efficiency at multiple working cells by automatically picking and placing product components in production (Ullman, 2002, p.178).

Figure 5 shows one heavy duty robot applied in product lifting operation. It has a payload capacity of 2200 lbs and a reach of 10 feet for extremely heavy lifting and manipulation applications in industries of automotive, foundry and building materials (Yang, 2011, p.1010).

Figure 6 displays one robotic system in material removal operations. It assists to process the material removal processes including drilling, milling, polishing, grinding, buffing and deflashing (Yun, 2011, p.977). The robot-assisted manufacturing can boost manufacturing efficiency throughout industrial applications.

3. MECHANISM AND CONTROL OF ROBOTIC SYSTEMS

The robotic systems convert the energy into motion. It may include electric motors, linear actuators, piezo motors, air wires, and electroactive polymers (Howard, 2006, p.1360). Electric motor is the most popular type of actuator (Shakernia,

Figure 2. da Vinci robotics surgical system (Passerotti, 2006)

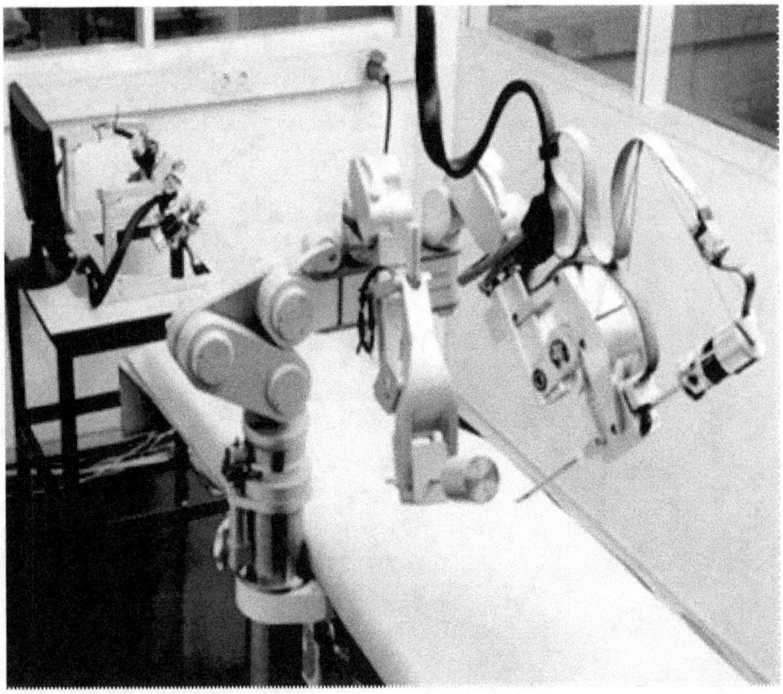

Figure 3. da Robotic-assisted surgical system (Melvin, 2003)

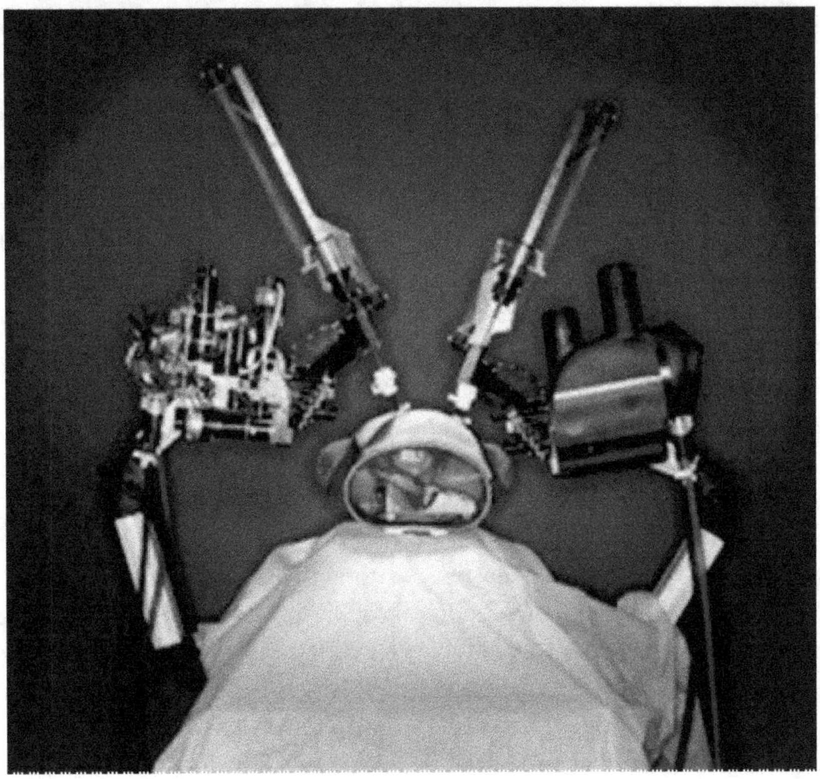

Figure 4. Robotic system for picking and placing parts in automobile production (Torbjorn, 2009)

2002, p.2793). Linear actuator is used in applications where the large forces are required in linear motion (Dagalakis, 2007, p.180, Mellinger, 2010). Piezo motor uses the vibration of the piezo elements to drive the motor. Air wire is a special tube and its length can be adjusted as internal air is exerted (Matsui, 2005, p.47). Electroactive polymer, functions as the arm of robotic system, is a special plastic material that can change its length significantly by electricity. Current robotic systems receive tactile signals with sensors simulating the mechanical properties and touch receptors of human fingertips (Gurdan, 2007, p.288). The sensor can consist of a rigid core with conductive fluid surrounded (Kazerooni, 2005, p.3459 & Kennedy, 2003, p.2106). The conductive fluid is enclosed by the elastomeric materials. Electrode is installed at the surface of the rigid core and connected to an impedance-measuring device inside the core. When the elastomeric material touches an object, the fluid volume surrounded electrode is changed, causing the impedance changes that simulate the loads received from the object. Computer-aided vision is the technology of mechanism that gets data information from images with different format including video sequences or cameras (Lerman, 2006, p.225). It uses the image sensor to detect electromagnetic radiation of visible or infra-red light.

Robots can manipulate objects by "end effectors" (hands) and "manipulators" (arms) (Eljamel, 2007, p.372 & Frutiger, 2002, p.619 & Hong, 2006, p.81). Some robotic arms equipped in exchangeable effectors with which it can perform different tasks. Mechanical grippers are the most common effectors with two fingers that can open

Figure 5. One heavy duty robot in product lifting and manipulation operation (Yang, 2011)

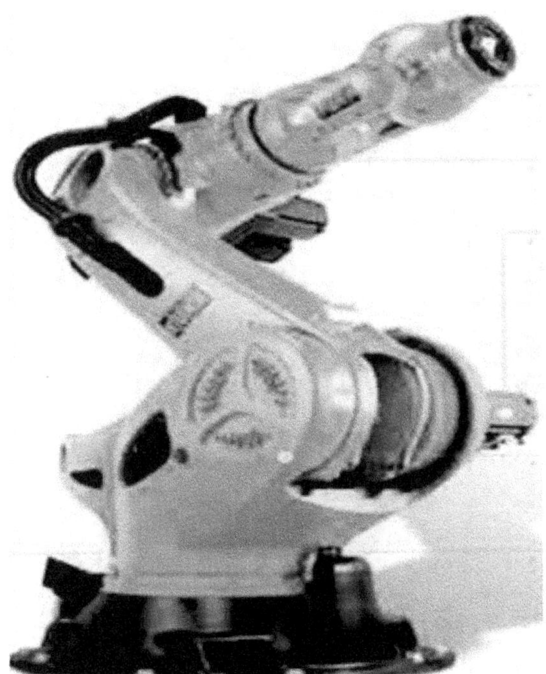

and close to pick up and place the objects. Vacuum grippers are simple and common effectors to pick and place large objects through suction. Robotic system control process is complex and dynamic related (Frisoli, 2005, p.195). Some technical methodologies include: (1). zero moment point technology—it is the algorithm used by robotic systems to keep the total forces balanced with combination of gravity, acceleration and deceleration of object body; (2). Hopping—it is the technology that keeps robotic system functioning properly and the algorithm in the control program is generated to maintain the system balance if robotic system is unbalanced; (3). dynamic balance—it is a more advanced methodology to control robotic system by dynamic balancing algorithm and more robust than zero moment point technology; (4). passive dynamics—it is applied with high efficiency in momentum of robotic system and the robotic system only requires less motor power if compared with other conventional robotic systems.

The control of robotic systems consists of three different phases: (1). Perception—sensors are applied to detect the external signal information for actuators; (2). Processing—it translates sensor information into commands to drive actuators; (3). Action—object pattern verification and computer vision can be used to track objects (Gerkey, 2004, p.939). The study of motion in robotic systems includes both kinematics and dynamics. The position, orientation, velocity, and acceleration of end effectors can be calculated through direct kinematics. The kinematics can be used to analyze the same robotic motion in different possibilities to avoid the collision and singularity in robotic systems. With all results from kinematics analysis, dynamics can be applied to study the effect of forces on motion of robotic systems. The dynamic technique is a powerful tool in computer-aided modeling and simulation of robotic systems to improve the control algorithms of a robotic system (Tinelli, 2011, p.2654).

4. NEW RESEARCHES IN SURGICAL AND INDUSTRIAL ROBOTIC SYSTEMS

Based on author's current researches, two new robotic systems have been introduced and analyzed in this chapter including robotic system design and development in surgical procedures and automated manufacturing processes. These two robotic systems are designed in such way that they can be made cheaper and simple without losing the reliability. These robotic systems have been designed and analyzed by computer-assisted modeling, simulation and analysis. The simple prototypes were being built and tested to verify its performance. The preliminary results indicated that these robotic systems function properly during the prototype testing.

Figure 6. Robotic system for material removal processes (Yun, 2011)

4.1 New Robotic System for Surgical Procedures

The surgical operations can be performed through robotic systems with more accuracy. Robotic surgery applies the surgical robots that lead direct or indirect involvement of surgeon in surgical operation. The robotic systems are computerized mechanical systems with motorized mechanisms. The basic robotic systems may consist of sensors, programmable systems, and actuators (Carigan, 2007, p.185). The important advantages of robotic surgery over conventional surgery include the accuracy of surgery, capability of repeating identical motions, and options of remote surgery (Finkelstein, 2010, p. 42). There are three categories in robotic surgery based on different surgeon involvement (Hanna, 2011, p.780): (1). supervisory—before surgical procedure, surgeon inputs the computer program into robotic system so that the surgical operation can be solely performed without surgeon direct interaction, but It requires complex program for the surgery and more advanced technologies to get images and data of patient body; (2). Telesurgery—it is a remote surgery in which surgeon directly manipulates the robotic systems during the surgical process without pre-programming the robotic system and surgeon can perform surgical operation from other locations by using real-time data information obtained from sensors in robotic system; (3). share-control—this procedure needs the most involvement of surgeon who performs the surgical operation by using robotic system that provides more accurate and stable manipulations of surgical instruments. A new surgical robotic system, shown in Figure 7, is designed and developed based on computer modeling, analysis and prototype testing.

Computer tomography scans can get cross-section images that are important to the diagnosis and surgeons can use these images to decide surgical pathways to precisely guide operation instruments with less surgical injury during pro-

Figure 7. New robotic system for surgical application

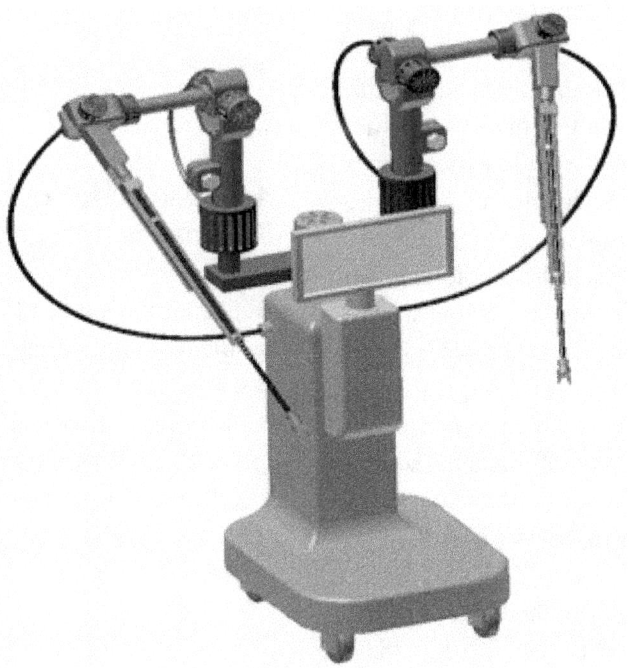

cedures. The robotic arms in this robotic system can precisely access the patient's body through tiny and narrow incisions. Surgeons can use remote manipulation device to perform complex surgical operation with accuracy and less fatigue because of proper geometrical design, wide angular motion, and 3-dimensional clear visualization. The wide view of surgical operations using this surgical robot provides surgeons with great capability to perform many surgical tasks including robotic prostate cancer surgeries for cancel removal, damage control, sexual and urinary function recovery with minimal side effects. This robotic surgery can benefit patients by quick recovery, less postoperative pain, short hospital stays, small tissue trauma, and better cosmetics. This surgical robot provides surgeons with more precise, flexible and accurate procedures than traditional surgical techniques. It also permits highly efficient suturing because of accurate placement and well controlled motion of surgical instrument in this robotic system. In summary, this surgical robot has the following features:

1. The simpler system design allows loose manufacturing tolerance control on major components in this robotic system so that it can be made cheap and maintained easily in the future;
2. The moving components in this robotic system are designed to keep its proper geometry and light weight to benefit surgeon in surgical operation with less moving vibration and higher energy efficiency.
3. Manipulated by surgeons, the robotic arms can have wider angular degrees of motion freedom with more flexible function;
4. Larger visual magnification for better surgical function;
5. Less post-operative length of stay due to the capability of performing minute surgical incisions more precisely.

4.1.1 Mathematics and Computer-Aided Simulation of Robot Arms

The robotic systems are programmed with mathematics that controls the motion of robot arms accurately and reliably. The double-link robot arm is simple, shown in Figure 8, if compared with more complicated triple-link robot arm which involves more difficult mathematics. In the double-link arm system, the first link (assistant arm) turns around the origin of an XY Cartesian coordinate system and the second link (arm to hold the surgical tools) turns around the pivot point which connects link 1 and 2. The angle between link 1 and horizontal axis is defined as A, and angle between link 1 and 2 is assumed as B. The robot arm is at the end positions (X_1, Y_1) of link 1 and (X_2, Y_2) of link 2.

There are three solution methodologies in this robotic system analysis (Carlone, 2010, p.283): (1).kinematics—determine robot arm location (X_1, Y_1) if the angles A and B are given; (2).inverse kinematics—determine angles A and B if robot arm location (X_1, Y_1) is given; (3).trajectory generation—determine robot arm new location (X_2, Y_2) by adjusting the angle A and B if current location (X_1, Y_1) is given.

4.1.2 Methodology of Kinematics (McComb, 2003)

Based on Figure 8, we can find the end location:

$$X_2 = S_1 \cdot \cos(A) + S_2 \cdot \cos(A+B) \quad (1)$$

$$Y_2 = S_1 \cdot \sin(A) + S_2 \cdot \sin(A+B) \quad (2)$$

Here,

$$X_1 = S_1 \cdot \cos(A)$$

$$Y_1 = S_1 \cdot \sin(A)$$

$$X_2 = X_1 = S_2 \cdot \cos(A+B)$$

$$Y_2 = Y_1 + S_2 \cdot \sin(A+B)$$

4.1.3 Methodology of Inverse Kinematics (McComb, 2003)

Referring kinematics Equations (1) and (2), the problem is nonlinear to determine angles A and B if end location (X_2, Y_2) is given. The algebraic methodology can be applied to solve this nonlinear problem and Equations (1) and (2) are combined to get following equations:

$$\cos(A+B) = \cos(A) \cdot \cos(B) - \sin(A) \cdot \sin(B) \quad (3)$$

$$\sin(A+B) = \cos(A) \cdot \sin(B) + \sin(A) \cdot \cos(B) \quad (4)$$

Solving these two equations can derive the new equation of finding angle B:

$$B = \arccos\left(\frac{X_2^2 + Y_2^2 - S_1^2 - S_2^2}{2S_1 \cdot S_2}\right) \quad (5)$$

Since cos (B)=cos (-B), the inverse kinematics problem has two solutions (link 2 clockwise and counter-clockwise turning), shown in Figure 9.

Angle A can be found based on solution of angle B in inverse kinematics methodology.

$$X_2 = [S_1 + S_2 \cos(B)] \cos(A) - S_2 \cdot \sin(B) \cdot \sin(A) \quad (6)$$

$$X_2 = [S_1 + S_2 \cos(B)] \cdot \cos(A) - S_2 \cdot \sin(B) \cdot \sin(A) \quad (7)$$

The angle A can be determined by solving Equations (6) and (7).

Figure 8. Double-link robot arm

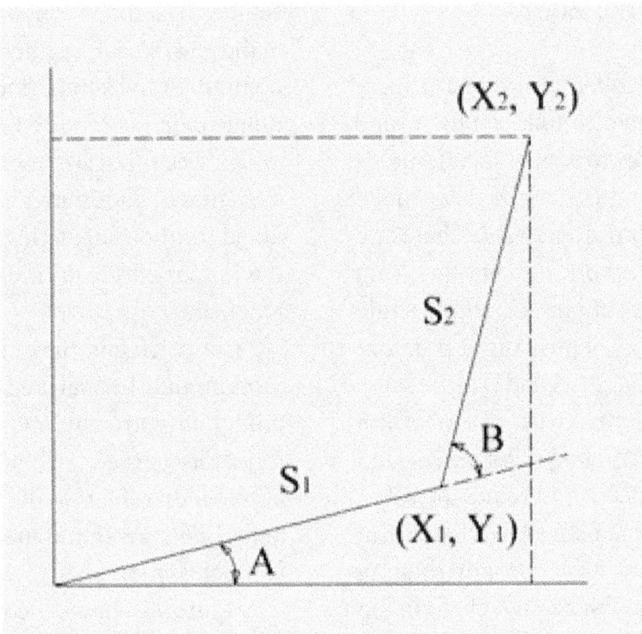

Figure 9. Inverse kinematics for robotic arm

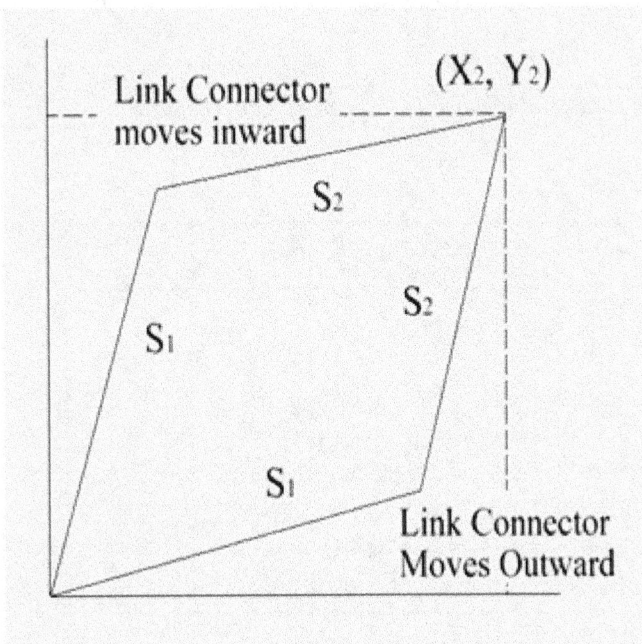

4.1.4 Methodology of Trajectory Generation (McComb, 2003)

In case of double-link robot arm, the arm location (X_1, Y_1) is associated to link angles A_1 and B_1. When rotate the arm to a new location (X_2, Y_2), the angles of A_2 and B_2 can be determined through inverse kinematics analysis. There are several ways to control robotic arm moving from old location to a new location: (1). adjust angle A before changing B; (2). adjust angle B before changing A; (3). adjust angles A and B at the same time; (4). robot arm turns clockwise; (5). robot arm turns counterclockwise. A good robot arm design can reduce the energy needed to rotate the robot arm and decrease the time needed to change arm mechanism. Figure 10 shows a robot arm rotating in trajectory generation with angle A changing in a clockwise direction.

Two moving trajectories of robot arm, shown in Figure 10, indicated the desired condition. In straight line trajectory generation methodology, a line or curve can be divided into many small segments and corresponding (X, Y) coordinates at the end of each segment are calculated through computational simulation. For each (X, Y) coordinate pair, the required angles A and B are computed. The robot arm mechanisms are adjusted as each new A and B are calculated. The computer-aided solutions of angles A and B continue, along the line or curve, until the target arm location is reached.

The more complex situation for robotic arm motion could be analyzed through computer-aided modeling and simulation. Figure 11 shows the trajectory curve of robot arm using multiple data points along the moving path. If the angles of A and B change simultaneously, the moving path is much smoother.

Figure 12 shows the robot arm moving in an alternate path with smoother curve when angles A and B are changed at the same time.

The robot arm system is a nonlinear functioning system which requires the complex mathe-

Figure 10. Robotic arm rotates clockwise in trajectory generation

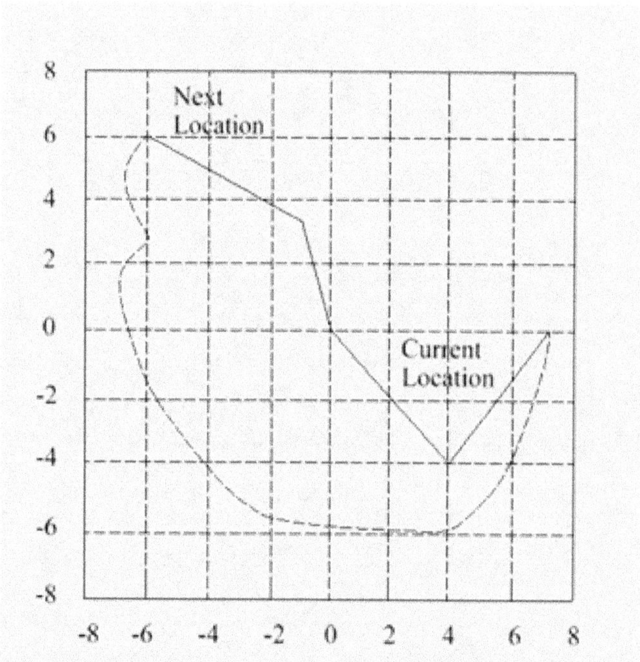

Figure 11. Simulation of robotic arm moving path with multiple data points

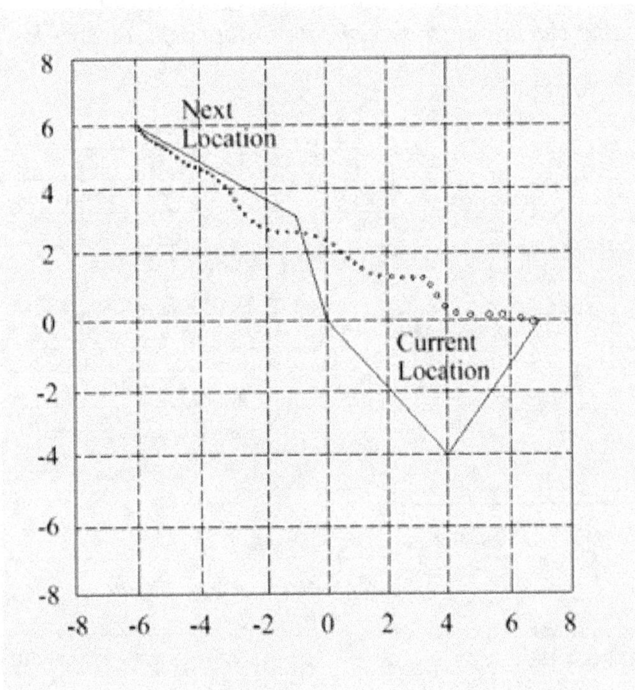

Figure 12. Simulation of robotic arm moving in an alternate path

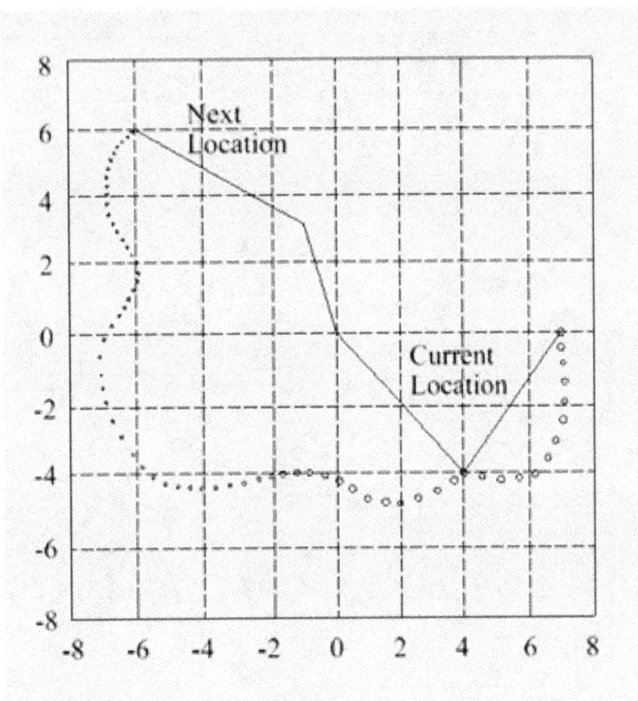

Figure 13. (a) movement of robot arm towards the front; (b) movement of robot arm towards the back; (c) movement of robot arm towards the left; (d) movement of robot arm towards the right; (e) movement of robot arm towards higher direction; (f) movement of robot arm towards lower direction

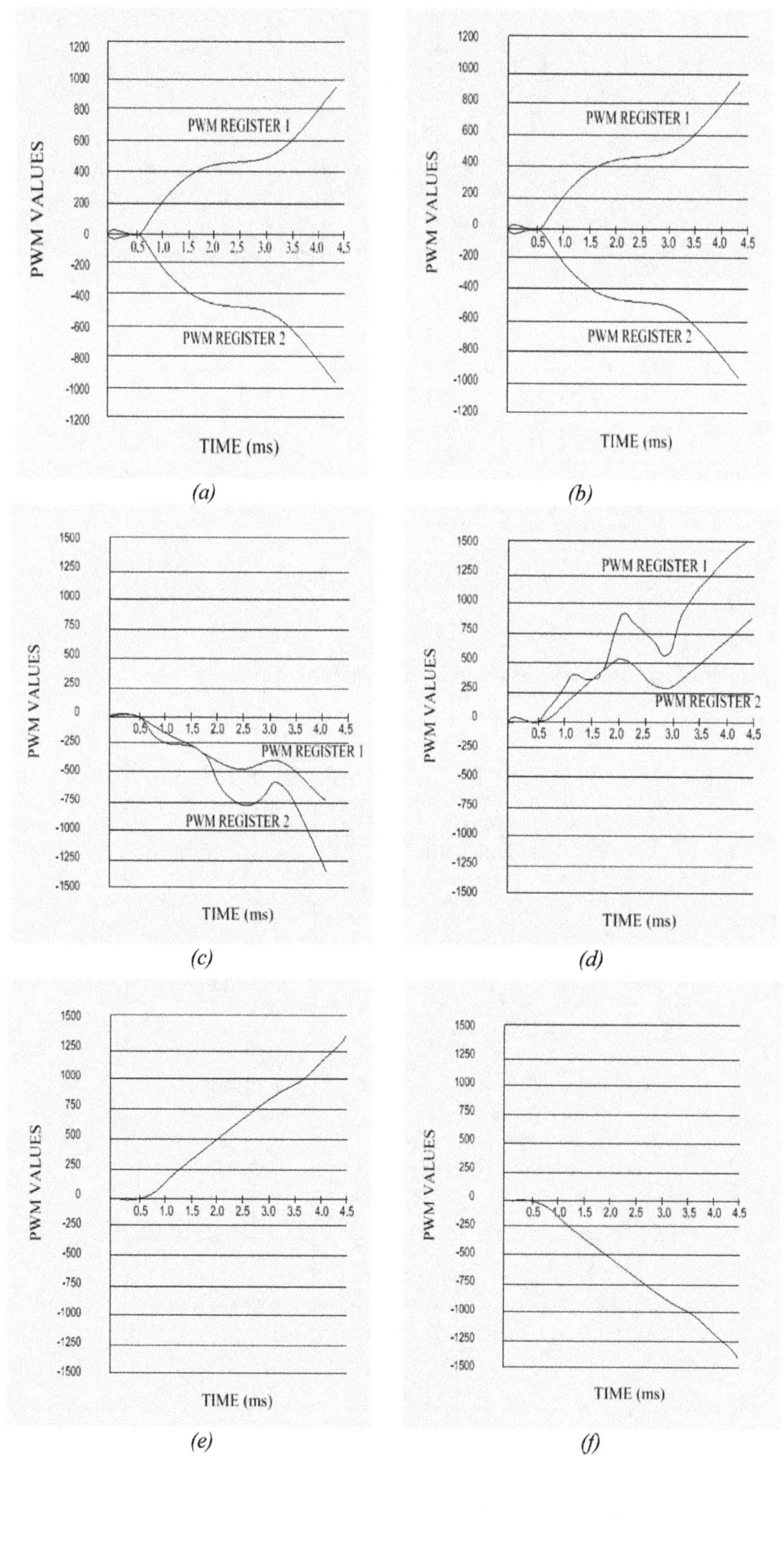

matical analysis, computer-aided modeling, and computational simulation to analyze and solve the arm motion problems. The variable length links could be used to define the potential coverage regions of robotic arms. Modifying the inverse kinematics methodology can implement the solution for trajectory generation methodology. The computational modeling and simulation are required in the analysis of robotic arm movement.

4.1.5 Prototype Testing

The kinematic motion of this robot is measured by pulse-width modulation (PWM) methodology in preliminary prototype testing. The PWM register's values are used to drive the motors of robotic system in different directions. The curves in Figs. 13 (a), (b), (c), (d), (e) and (f) show the power required to drive the movement of this surgical robot arm towards the front, back, left, right, higher and lower directions. All PWM resultant pictures corresponding to the motions of robotic arm in different directions were recorded when robotic system was manipulated by the operator. The average voltage or current values fed to the load is manipulated by controlling the switch connected between the supply and load. The switch quickly changes between on and off status. If switch stays longer in status of "on" than "off", more power will be supplied to the load which can manipulate the robotic system motions. The PWM values in these plots are related to the power that drives the robotic system in different directions.

The preliminary prototype testing indicated the maximum movement of 16 inches towards the front and back directions, 20 inches towards the left and right directions, and 24 inches towards the higher and lower directions. These preliminary results are in good agreement with design expectation since this surgical robot is expected the motion of 15 × 20 × 24 inches in 3-D design function. More tests have been planned to verify the functionality and reliability of this surgical robot. Further modification on this surgical robot will be made to determine if large moving range can be achieved.

4.2 New Robotic System for Automated and High Speed Viscous Fluid Filling

Figure 14 shows this new robotic system used in automated manufacturing processes. One of the industrial applications is to pick and place the empty cartridges from floor container to the working station in automated viscous liquids filling production line shown in Figure 15.

The precision indexing conveyor controls indexing tolerance to keep part movement within tolerance range. The empty cartridges are picked up and placed to the cartridge holders by this robotic system and securely held in indexing conveyor. The liquids filling holders are controlled to keep good dimensional and location tolerances between holders and cartridges to secure the product in automated and high speed filling process. When empty cartridges are delivered to the location below filling nozzles, the indexing conveyors stop through sequential control of PLC system that receives filling holder location signal from opposed mode sensor. Horizontal sliders mounted in gripper system drive gripper pair forward to secure cartridges in filling process. The nozzles are driven downward by vertical sliders and inserted into cartridges to begin filling process. As soon as filling process is completed, the vertical air slide drives nozzles upward. Then grippers release cartridges and move away from filling holder location for next filling cycle. The control system in this prototype can detect each filling pump to precisely determine the amount of liquids being filled. When target filling volume is reached, all pumps and nozzles shut off instantly, resulting in accurate filling process of high viscous liquid products. The control system

Figure 14. New robotic system for manufacturing applications

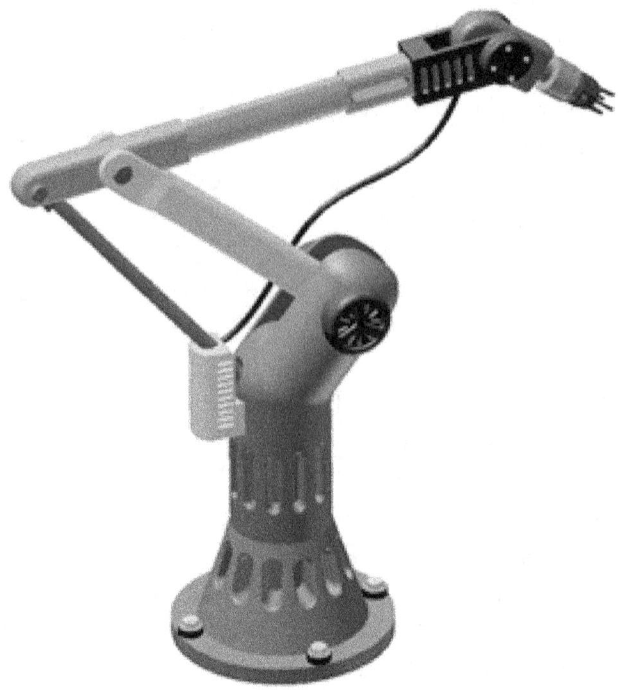

Figure 15. Robot-assisted automated and high speed viscous liquids filling production line

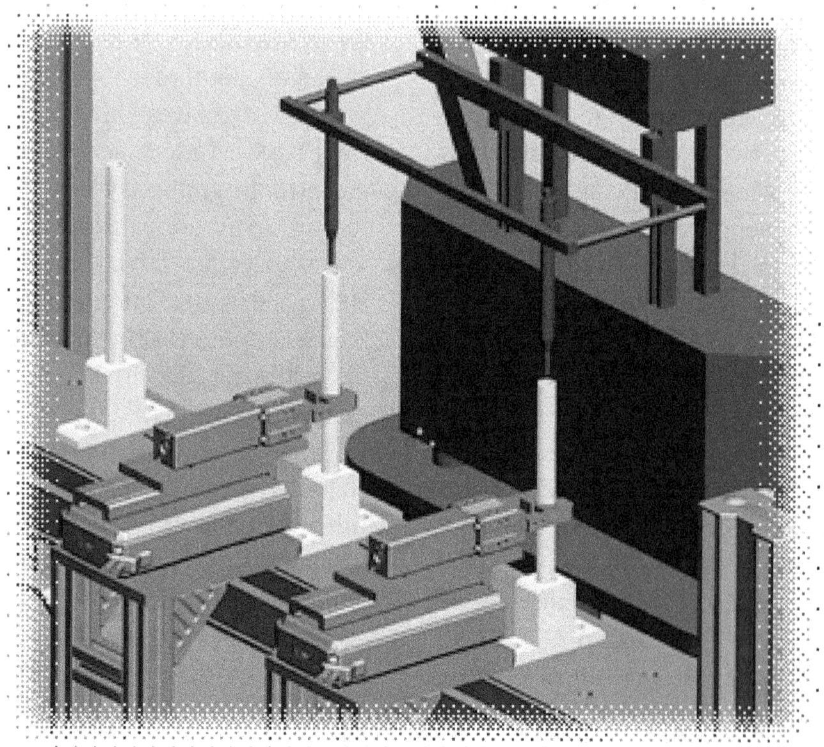

stores all filling parameters in memory for fast changeovers.

This robot-assisted automated and high speed liquid filling system is accurately and reliably controlled in filling of different high viscous liquids. It can be applied to many industries including pharmaceutical, cosmetic, dairy, chemical, food, etc. For example, some thick liquids, cosmetic creams, and chunky sauces with pasteurized temperatures can all be filled in this new robot-assisted automated filling system. The positive displacement pump will be required in high viscosity and high temperature environment. Double driving rotors are designed in this pump to make this filling system more durable in delivery of high viscosity liquids.

4.2.1 Computer Aided Modeling and Simulation

The computer-aided simulation of liquid leaking rate L during high speed liquid filling process can be determined as follow (Klopcar, 2007, p.86):

$$Y_2 = S_2 \cdot \sin(B) \cdot \cos(A) + [S_1 + S_2 \cdot \cos(B)] \cdot \sin(A) \tag{8}$$

The computer-aided simulation result, shown in Figure 16, indicated that the liquid leakage in this high speed filling process can be ignored. The computer-aided simulation shows that this new robotic-assisted automated and high speed filling system has reliable function, good sealing capability and high production rates. Also the manufacturing cost for this system is low due to less requirement of tolerance control for assembled robotic components in this new robotic-assisted manufacturing system design.

4.2.2 Prototype Testing

The prototype of this new robotic-assisted automated and high speed liquid filling system has been tested with preliminary results shown in Table 1.

Both prototype testing and computational simulation show the similar results that verify the creditability and feasibility of this new robot-

Figure 16. Viscous liquid leakages vs. piston linear speed

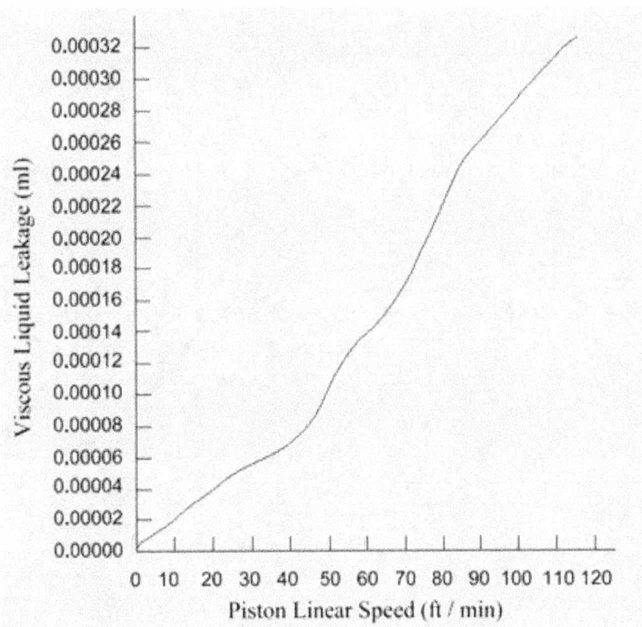

Figure 17. (a) movement of robot arm towards the front; (b) movement of robot arm towards the back; (c) movement of robot arm towards the left; (d) movement of robot arm towards the right; (e) movement of robot arm towards higher direction; (f) movement of robot arm towards lower direction

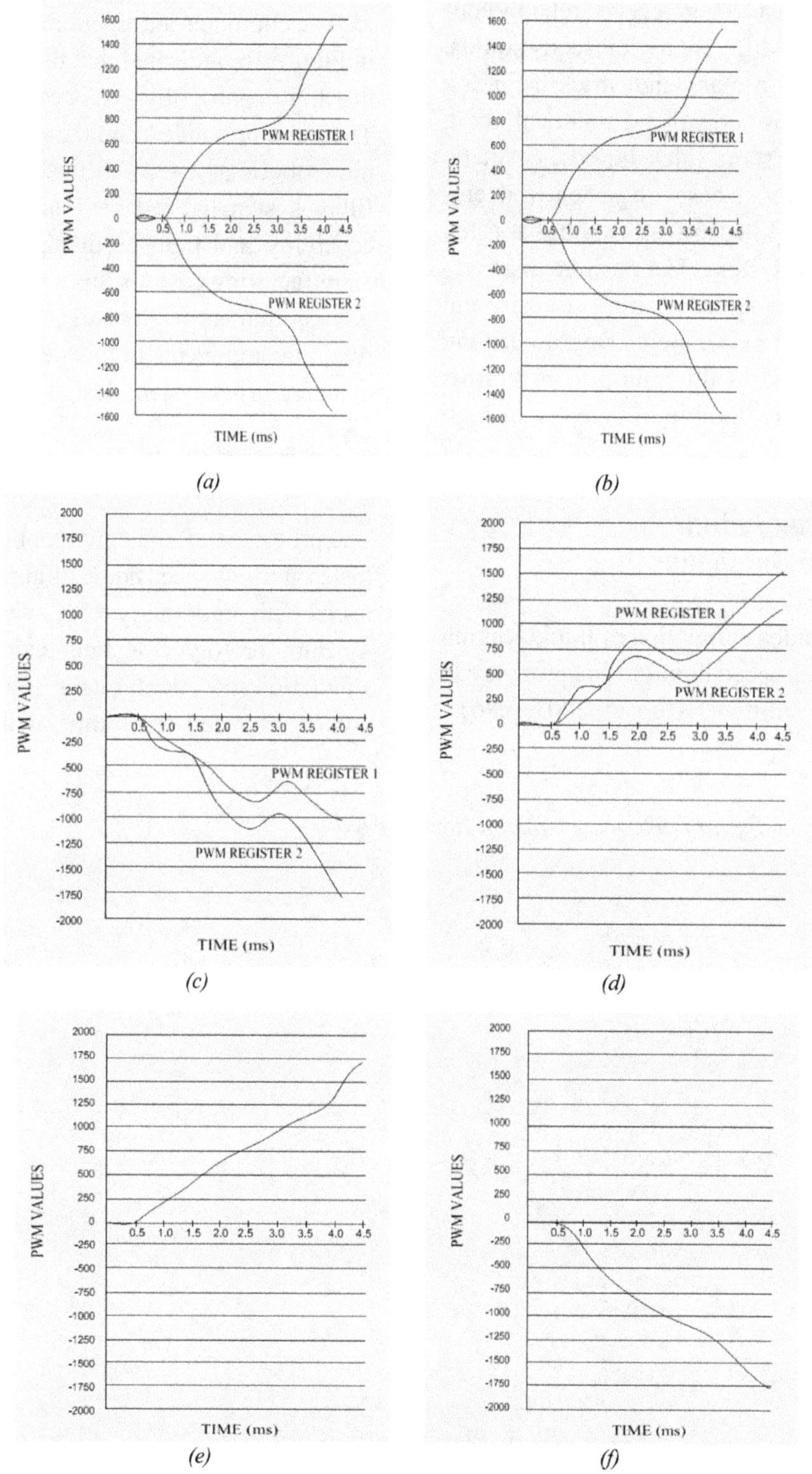

Table 1. Liquid leakage vs. piston linear speed

Piston Linear Speed (Ft/in)	Estimated Liquid Leakage (ml)	Piston Linear Speed (Ft/Min)	Estimated Liquid Leakage (ml)
10	0.000015	70	0.000145
20	0.000035	80	0.000185
30	0.000055	90	0.000205
40	0.000088	100	0.000235
50	0.000102	110	0.000265
60	0.000125	120	0.000305

assisted automated and high speed liquid filling system.

The curves in Figure 17 (a), (b), (c), (d), (e) and (f) represent the power related movement of manufacturing robot arm towards the front, back, left, right, higher and lower directions. The kinematic motion of this robot is measured by pulse-width modulation (PWM) methodology in preliminary prototype testing. The PWM register's values are used to drive the motors of robotic system in different directions. All PWM resultant pictures corresponding to the motions of robotic arm in different directions were recorded when robotic system was manipulated by the operator. The average voltage or current values fed to the load is manipulated by controlling the switch connected between the supply and load. The switch quickly changes between on and off status. If switch stays longer in status of "on" than "off", more power will be supplied to the load which can manipulate the robotic system motions. All these PWM plots represent the power needed to drive the robot arm moving in different directions.

The preliminary testing of this manufacturing robot indicated the maximum movement of 40 inches towards the front and back directions, 48 inches towards the left and right directions, and 52 inches toward the higher and lower directions. The above preliminary results are in good agreement with design expectation because this manufacturing robot is expected to move 3-D dimension in 40 × 49 × 52 inches to meet design target. To verify the functionality and reliability of this surgical robot, more prototype tests will be performed and further modifications of mechanism in this manufacturing robot will be carried out to increase moving range in different directions.

5. FUTURE IMPROVEMENT

The future improvement on these two new robotic systems include: (1). Simplify the robotic systems for easy maintenance and cost-effectiveness, including making arm opening more accessible; (2). Investigate using lighter and durable materials; (3). Modify the robotic mechanism to increase its moving range in different directions; (4). Improve system design for better function including mechanical joint connections to keep more accurate and smooth robotic motion; (5). Optimize the robotic mechanism through computer-aided simulation to verify and guide further design modification; (6). Implement the prototype and prepare to send for field evaluations.

6. CONCLUSION

Robotic system is not only significantly increasing the production speed but also more accurately controlling product quality. It can maintain consistent quality, shorten lead time, simplify material handling, optimize work flow, meet product demand for flexibility and convertibility in production, and perform some specific, precise, and dangerous tasks in industry that are very difficult using human work forces. Robotic surgery has greater levels of accuracy and precision than traditional open surgery. The robotic systems have also been widely used in many industrial fields including automated and high speed production, automobile manufacturing, hazardous duty service, and heavy facility maintenance. Two new research projects of robotic system introduced in this chapter are illustrated mathematically with verifications from computer-aided analysis and preliminary prototype testing. These new research projects can be applied to robotic surgery and automated manufacturing process. Further modifications will keep these robot-assisted systems with more reliable function, flexible features, reduced cost, and enlarged moving range.

REFERENCES

Ahmed, K., Khan, M., Vats, A., Nagpal, K., Priest, O., & Patel, V. (2009). Current status of robotic assisted pelvic surgery and future developments. *International Journal of Surgery*, *7*, 431–440. doi:10.1016/j.ijsu.2009.08.008

Albu-Schaffer, A., Haddadin, S., Ott, C., Stemmer, A., Wimbock, T., & Hirzinger, G. (2007). The DLR lightweight robot: Design and control concepts for robots in human environments. *International Journal of Industrial Robot*, *34*, 376–385. doi:10.1108/01439910710774386

Backes, F., Seamon, L., & Fowler, J. (2008). Robotic radical hysterectomy and pelvic mphadenectomy for uterine rhabdomyosarcoma. *Journal of Robotic Surgery*, *2*, 97–200. doi:10.1007/s11701-008-0095-5

Bargar, W. L. (2007). Robots in orthopaedic surgery - Past, present, and future. *Clinical Orthopaedics and Related Research*, *463*, 31–36.

Brown, W. T., Wu, X., & Wen, B. C. (2007). Early results of CyberKnife imageguided robotic stereotactic radiosurgery for treatment of lung tumors. *Journal of Computer Aided Surgery*, *12*, 253–261.

Cadierre, G. (2001). Feasibility of robotic laparoscopic surgery. *World Journal of Surgery*, *25*, 1467–1477.

Camarillo, D. (2004). Robotic technology in surgery: past, present, and future. *American Journal of Surgery*, *188*, 2–15. doi:10.1016/j.amjsurg.2004.08.025

Carigan, C., Tang, J., Roderick, S., & Naylor, M. (2007). A configuration-sapce approach to controlling a rehabilitation arm exoskeleton. *IEEE 10th International Conference Rehabilitation Robotics*, (pp. 179-187).

Carlone, L., Ng, M., Du, J., Bona, B., & Indri, M. (2010). Simultaneous localization and mapping using rao-blackwellized particle filters in multi robot systems. *Journal of Intelligent & Robotic Systems*, *63*, 283–307. doi:10.1007/s10846-010-9457-0

Chen, P., & Sheng, W. (2011). Transformative CAD based industrial robot program generation. *Journal of Robotics and Computer-Integrated Manufacturing*, *27*, 942–948. doi:10.1016/j.rcim.2011.03.006

Constantin, A. (2006). Evaluating arm movement imitation. *American Journal of Undergraduate Research*, *4*, 9–12.

Dagalakis, N., Kim, Y., Sawyer, D., & Shakarji, C. (2007). Development of tools for measuring the performance of computer assisted orthopaedic hip surgery systems. *Proceeding of the ACM Workshop on Performance Metrics for Intelligent Systems (PerMIS)*, Gaithersburg, (pp. 180–187).

DeLaurentis, K., & Mavroidis, C. (2002). Mechanical design of a shape memory alloy actuated prosthetic hand. *Journal of Health Care*, *10*, 91–106.

Desforges, X., Habbadi, A., & Archimède, B. (2011). Design methodology for smart actuator services for machine tool and machining control and monitoring. *Journal of Robotics and Computer-Integrated Manufacturing*, *27*, 963–976. doi:10.1016/j.rcim.2011.03.004

Dylan, A. S., & Matarić, M. J. (2005). Insights toward robot-assisted evacuation. *International Journal of the Robotics*, *19*, 797–818.

Eirik, L., Johansen, B., Gjelsvik, T., & Lango, T. (2009). Ultrasound based localization of wireless microrobotic endoscopic capsule for the GI tract. *Proceeding of the 21st Conference of the Society for Medical Innovation and Technology (SMIT)*, Sinaia, (pp. 77–79).

Eljamel, M. (2007). Validation of a neurosurgical robot using a phantom. *International Journal of Medical Robotics and Computer Assisted Surgery*, *3*, 372–377. doi:10.1002/rcs.153

Estey, E. (2009). Robotic prostatectomy: The new standard of care or a marketing success. *Journal of Canadian Urological Association*, *3*, 488–490.

Feil-Seifer, D., Skinner, K., & Mataric, M. J. (2007). Benchmarks for evaluating socially assistive robotics. *Journal of Interaction Science*, *8*, 423–439.

Finkelstein, J., Eckersberger, E., Sadri, H., Taneja, S. S., & Djavan, B. (2010). Open versus laparoscopic versus robot-assisted laparoscopic prostatectomy: The European and US experience. *Reviews in Urology*, *12*, 35–43.

Frisoli, A., Rocchi, F., Marcheschi, S., Dettori, A., Salsedo, F., & Bergamasco, M. (2005). A new force-feedback arm exoskeleton for haptic interaction in virtual environment. *First IEEE Joint Eurohaptics Conference and Symposium on Haptic Interfaces for Virtual Environment and Teleoperator Systems,* (pp. 195-201).

Frutiger, D. R., Bongard, J. C., & Iida, F. (2002). Iterative product engineering: evolutionary robot design. *Proceedings of the Fifth International Conference on Climbing and Walking Robots*, (pp. 619–629).

Gerhardus, D. (2003). Robot-assisted surgery: The future is here. *Journal of Healthcare Management*, *48*, 242–251.

Gerkey, B., & Matarić, M. J. (2004). A formal framework for the study of task allocation in multi-Robot systems. *The International Journal of Robotics Research*, *23*, 939–954. doi:10.1177/0278364904045564

Ghomi, A., & Askari, R. (2010). Use of a bidirectional barbed suture in robot-assisted sacrocolpopexy. *Journal of Robotic Surgery*, *4*, 87–89. doi:10.1007/s11701-010-0188-9

Gockley, R., & Matari'c, M. (2006). Encouraging physical therapy compliance with a hands-off mobile robot. *Proceedings of the 1st Annual Conference on Human-Robot Interaction,* (pp. 150–155).

Goerzen, C., Kong, Z., & Mettler, B. (2010). A survey of motion planning algorithms from the perspective of autonomous UAV guidance. *Journal of Intelligent & Robotic Systems*, *57*, 65–100. doi:10.1007/s10846-009-9383-1

Gortchev, G., Tomov, S., Tantchev, L., Velkova, A., & Radionova, Z. (2010). Da Vinci robotic surgery in the treatment of benign and malignant gynecologic tumors. *Gynecological Surgery*, *7*, 153–157. doi:10.1007/s10397-009-0535-x

Gurdan, D., Stumpf, J., Achtelik, M., Doth, K., Hirzinger, G., & Rus, D. (2007). Energy-efficient autonomous four-rotor flying robot controlled at 1 khz. *Proceeding of the IEEE International Conference on Robotics and Automation.*

Hanna, R., & Boggess, J. (2011). Applications of surgical robotics in gynecologic surgery. *Surgical Robotics, 4*, 761–789. doi:10.1007/978-1-4419-1126-1_32

Harja, J., Tikkanen, J., Sorvoja, H., & Myllyla, R. (2007). Magnetic resonance aging compatible, three-degrees-of-freedom joystick for surgical robot. *International Journal of Medical Robotics and Computer Assisted Surgery, 3*, 365–371. doi:10.1002/rcs.159

Hong, S. S., Heo, S. C., Kim, E., & Hang, Y. J. (2006). A study on the practical human robot interface design for the development of shopping support robot. *Journal of Korean Society of Design Science, 19*, 81–90.

Howard, A., & Gaurav, S. Sukhatme, S., & Matarić, M. J. (2006). Multi-robot mapping using manifold representations. *Proceedings of the IEEE – Special Issue on Multi- Robot Systems, 94,* 1360 -1369.

Hu, T., Castellanos, A., & Tholey, G. (2002). Real-time haptic feedback laparoscopic tool for use in gastro-intestinal surgery. *Fifth International Conference on Medical Image Computing and Computer Assisted Intervention, 12,* (pp. 6-16).

Kaouk, J., & Goel, R. (2009). Robotic-assisted laparoscopic bladder augmentation. *Pediatric Robotic Urology, 2*, 181–188. doi:10.1007/978-1-60327-422-7_12

Kazanzides, P., Fichtinger, G., Hager, G., Okamura, A., Whitcomb, L., & Taylor, H. (2008). Surgical and interventional robotics: core Concepts, technology, and design. *IEEE Robotics & Automation Magazine, 15*, 122–130. doi:10.1109/MRA.2008.926390

Kazerooni, H. (2005). Exoskeletons for human power augmentation, *Proceeding of IEEE Conference of Intelligent Robots and Systems,* (pp. 3459–3464).

Kennedy, C. W., & Desai, J. P. (2003). Force feedback using vision. *The 11th International Conference on Advanced Robotics,* (pp. 2106-2111).

Kim, B., Park, J., Kang, B., & Moon, C. (2011). Fabrication and property analysis of a MEMS micro-gripper for robotic micro-manipulation. *Journal of Robotics and Computer-Integrated Manufacturing, 27*, 1021–1025.

Kim, W. S. (2007). The design of humanoid robot platgorm: HUBO. *Journal of Korean Society of Design Science, 20*, 13–16.

Klopcar, N., Tomsic, M., & Lenarcic, J. (2007). A kinematic model of the shoulder complex to evaluate the arm-reachable workspace. *Journal of Biomechanics, 40*, 86–91. doi:10.1016/j.jbiomech.2005.11.010

Koh, C., Tsang, C., & Kim, S. (2011). A new application of the four-arm standard da Vinci® surgical system: Totally robotic-assisted left-sided colon or rectal resection. *Surgical Endoscopy, 25*, 1945–1952. doi:10.1007/s00464-010-1492-1

Kumar, R., Kalra, P., & Prakash, N. (2011). A virtual RV-M1 robot system. *Journal of Robotics and Computer-Integrated Manufacturing, 27*, 994–1000. doi:10.1016/j.rcim.2011.05.003

Kwartowitz, D. M., Herrell, S. D., & Galloway, R. L. (2006). Toward image-guided robotic surgery: Determining intrinsic accuracy of the da Vinci robot. *International Journal of Computer Assisted Radiology and Surgery, 1*, 157–165. doi:10.1007/s11548-006-0047-3

Kypson, P., & Chitwood, W. R. Jr. (2004). Robotic applications in cardiac surgery. *International Journal of Advanced Robotic Systems, 1*, 87–92.

Lerman, K., Jones, C., Galstyan, A., & Mataric, M. J. (2006). Analysis of dynamic task allocation in multi-robot systems. *The International Journal of Robotics Research, 25*, 225–242. doi:10.1177/0278364906063426

Li, Q. H. (2002). The application accuracy of the NeuroMate robot - A quantitative comparison with frameless and frame-based surgical localization systems. *International Journal of Computer Aided Surgery, 7*, 90–98.

Mataric, M. J., Eriksson, J., & Feil-Seifer, D. (2007). Socially assistive robotics for post-stroke rehabilitation. *International Journal of NeuroEngineering and Rehabilitation, 4*, 1–12. doi:10.1186/1743-0003-4-1

Matsui, T. (2005). *The roles and method of robot design* (pp. 47–55). Proceeding of Robot Design Forum.

McComb, G. (2003). *Robot builder's sourcebook* (pp. 435–440). McGraw-Hill.

McConnell, P., Schneeberger, E., & Michler, R. (2003). History and development of robotic cardiac surgery. *Problems in General Surgery, 20*, 20–30. doi:10.1097/01.sgs.0000081182.03671.6e

Mellinger, D., Michael, N., & Kumar, V. (2010). Trajectory generation and control for precise aggressive maneuvers with quadrotors. *Proceedings of International Symposium on Experimental Robotics,* (pp. 2520–2525).

Melvin, W., Krause, K., Needleman, B., & Ellison, E. (2003). Robotic resection of a pancreatic neuroendocrine tumor. *Journal of Laparoendoscopic & Advanced Surgical Techniques. Part A., 13*, 33–36. doi:10.1089/109264203321235449

Melvin, W., Needleman, B., Krause, K., Schneider, C., & Ellison, E. (2002). Computer–enhanced versus standard laparoscopic antireflux surgery. *Journal of Gastrointestinal Surgery, 6*, 11–16. doi:10.1016/S1091-255X(01)00032-4

Menzel, P., & D'Aluisio, F. (2000). *Robo Sapiens* (1st ed., pp. 180–189). MIT Press.

Passerotti, C., & Peters, C. (2006). Robotic-assisted laparoscopy applied to reconstructive surgeries in children. *World Journal of Urology, 24*, 193–197. doi:10.1007/s00345-006-0084-0

Patel, M., Menon, M., & Rogers, C. (2011). Robotic urologic surgery: Robot-assisted partial nephrectomy. *Robotics in Genitourinary Surgery, 4*, 423–433. doi:10.1007/978-1-84882-114-9_37

Peters, C. (2007). *Applications of robotics in pediatric urologic surgery* (pp. 179–187). Robotic Urologic Surgery. doi:10.1007/978-1-84628-704-6_24

Pettersson, P. O., & Doherty, P. (2004). Probabilistic roadmap based path planning for autonomous unmanned aerial vehicles. *Proceedings of the 14th International Workshop on Connecting Planning and Theory with Practice* (pp. 158–162).

Pott, P., Scharf, H., & Schwarz, M. (2005). Today's state of the art of surgical robotics. *Journal of Computer Aided Surgery, 10*, 101–132.

Pounds, P., Mahony, R., Gresham, J., Corke, P., & Roberts, J. (2004). Towards dynamically-favorable quad-rotor aerial robots. *Australasian Conference on Robotics and Automation* (pp. 1102 – 1106).

Puntambekar, S., Agarwai, G., Joshi, S., Rayate, N., & Sathe, R. (2009). Robotic radical hysterectomy: Applying principles of the laparoscopic Pune technique. *Journal of Robotic Surgery, 4*, 259–264. doi:10.1007/s11701-010-0224-9

Rosen, J., Hannaford, B., & Satava, R. (2011). *Surgical robotics - Systems applications and visions*. Springer. ISBN: 978-14419-4

Schiff, J., Li, P. S., & Goldstein, M. (2005). Robotic microsurgical vasovasostomy and epididymostomy in rats. *International Journal of Medical Robotics and Computer Assisted Surgery, 1*, 122–126. doi:10.1002/rcs.21

Schreuder, H., Aweemer, R., Baal, W., Lande, J., & Dijkstra, J. (2010). From open radical hysterectomy to robot-assisted laparoscopic radical hysterectomy for early stage cervical cancer: Aspects of a single institution learning curve. *Gynecological Surgery*, *7*, 253–258. doi:10.1007/s10397-010-0572-5

Senapati, S., & Advincula, A. (2011). Surgical techniques: Robot-assisted laparoscopic myomectomy with the da Vinci surgical system. *Journal of Robotic Surgery*, *1*, 69–74. doi:10.1007/s11701-007-0014-1

Shakernia, O., Vidal, R., Sharp, C. S., Ma, Y., & Sastry, S. (2002). Multiple view motion estimation and control for landing an unmanned aerial vehicle. *Proceedings of IEEE International Conference on Robotics and Automation,* (pp. 2793–2798).

Sicitiano, B. (1999). Kinematic control of redundant robot manipulators. *Journal of Intelligent & Robotic Systems*, *3*, 201–212. doi:10.1007/BF00126069

Siebert, W., Mai, S., Kober, R., & Heeckt, P. (2004). Total knee replacement: Robotic assistive technique. In DiGioia, A. M., Jaramaz, B., Picard, F., & Nolte, L.-P. (Eds.), *Computer and robotic assisted hip and knee surgery* (pp. 127–156). Oxford University.

Takahashi, T. (2007). *Emotional robot design* (pp. 32–34). Digest Book of IEEE RO-MAN.

Talamini, M., Chapman, W., Horgan, S., & Melvin, W. (2003). Evaluation of 211 robotic surgical procedures. *Surgical Endoscopy*, *17*, 1521–1524. doi:10.1007/s00464-002-8853-3

Tapus, A., Cristian, T., & Mataric, M. J. (2008). User-robot personality matching and assistive robot behavior adaptation for post-stroke rehabilitation therapy. *Intelligent Journal of Robotics*, *5*, 169–183. doi:10.1007/s11370-008-0017-4

Tapus, A., Matarić, M. J., & Scassellati, B. (2007). The grand challenges in socially assistive robotics. *IEEE Robotics & Automation Magazine*, *14*, 15–18. doi:10.1109/MRA.2007.339605

Taylor, R., & Stoianovici, D. (2003). Medical robotics in computer-integrated surgery. *IEEE Transactions on Robotics and Automation*, 19.

Tholey, G., Chanthasopeephan, T., & Hu, T. (2003). Measuring grasping and cutting forces for reality-based haptic modeling - Computer assisted radiology and surgery. *17th International Congress and Exhibition,* (pp. 158–162).

Tinelli, R., Malzoni, M., Cosentino, F., Perone, C., & Fusco, A. (2011). *Annals of Surgical Oncology*, *14*, 2654–2661.

Torbjorn, S. D., Mataric, M. J., & Gaurav, S. S. (2009). Multi-robot task allocation through vacancy chain scheduling. *Journal of Robotics and Autonomous Systems*, *57*, 87–91.

Tsagarakis, N., & Caldwell, D. (2003). Development and control of a "soft-actuated" exoskeleton for use in physiotherapy and training. *Autonomous Robots*, *15*, 21–33. doi:10.1023/A:1024484615192

Ullman, D. G. (2002). *The mechanical design process* (pp. 178–182). McGraw-Hill Science.

Yang, C., Huang, Q., & Han, J. (2011). Decoupling control for spatial six-degree-of-freedom electro-hydraulic parallel robot. *Journal of Robotics and Computer-Integrated Manufacturing*, *27*, 1010–1015.

Yang, S. H., Byun, J. H., & Seo, J. H. (2008). The design of kids robot: raBie. *Journal of Korean Society of Design Science*, *21*, 79–82.

Yun, Y., & Li, Y. (2011). Optimal design of a 3-PUPU parallel robot with compliant hinges for micromanipulation in a cubic workspace. *Journal of Robotics and Computer-Integrated Manufacturing*, *27*, 977–985. doi:10.1016/j.rcim.2011.05.001

Section 4
Prototyping of Robotic Systems for Other Applications

Chapter 13
Robotic Hardware and Software Integration for Changing Human Intentions

Akif Durdu
Middle East Technical University, Turkey

Ismet Erkmen
Middle East Technical University, Turkey

Aydan M. Erkmen
Middle East Technical University, Turkey

Alper Yilmaz
Photogrammetric Computer Vision Laboratory, The Ohio State University, USA

ABSTRACT

Estimating and reshaping human intentions are among the most significant topics of research in the field of human-robot interaction. This chapter provides an overview of intention estimation literature on human-robot interaction, and introduces an approach on how robots can voluntarily reshape estimated intentions. The reshaping of the human intention is achieved by the robots moving in certain directions that have been a priori observed from the interactions of humans with the objects in the scene. Being among the only few studies on intention reshaping, the authors of this chapter exploit spatial information by learning a Hidden Markov Model (HMM) of motion, which is tailored for intelligent robotic interaction. The algorithmic design consists of two phases. At first, the approach detects and tracks human to estimate the current intention. Later, this information is used by autonomous robots that interact with detected human to change the estimated intention. In the tracking and intention estimation phase, postures and locations of the human are monitored by applying low-level video processing methods. In the latter phase, learned HMM models are used to reshape the estimated human intention. This two-phase system is tested on video frames taken from a real human-robot environment. The results obtained using the proposed approach shows promising performance in reshaping of detected intentions.

DOI: 10.4018/978-1-4666-0176-5.ch013

1. INTRODUCTION

Human-Robot interaction, which is generally used for cooperation tasks, requires reliable and full communication capabilities. In the case when the human and robot does not communicate, the cooperation between them can be established by estimating the intention of the human and/or the robot. If the human is able to express his intention clearly for a specific task then estimating the intention reduces to how the communication is performed and the budget necessary for perform the communication (K. A. Tahboub, 2006).

As observed in the nature and studied in engineering, biomimetic interactions between two intelligent agents require estimating the intention of one another, which eventually results in "either morphing their actions to that of the other's intention" or "change their actions based on a certain strategy to adapt to other's intention". We term this strategic change the reshaping of intentions. The intention reshaping is an emerging area of research and there are only a handful of studies on plan/goal recognition and intention estimation in the recent years. The first study introduced the reshaping of human intention via robots is by Terada et al. (2007) which was limited to examining the psychological statuses of humans when they interact with robots. As described by Dennett in 1987, these psychological statuses include physical, design and intentional stance. Our problem design, however, is significantly different. We demonstrate how autonomous robots intelligently estimate the human intention in realistic scenarios, which is used to guide their motion to reshape the intention. In particular, intention estimation and recognition problems are characterized by prototypical phases of the defined problem domain. For instance, in the case when the problem is interpretation of human actions in a video, researchers use classification and learning algorithm trained from a database of known human actions. For example, Miyake et al. (2010) estimate the changes in facial expressions in a video to estimate human intentions. In similar vein, other researchers study gesture recognition and action recognition for estimating the human intentions (E. Daprati et al., 2007; F. Loula et al., 2005; V. Sevdalis et al., 2009, 2010; J.E. Cutting et al., 1977; F. Loula et al., 2005 etc.).

In this study, we adopt Aristotle's description of vision: "what is where by looking" and emphasize two most essential problems: estimation (identifying "what") and localization (identifying "where") (O.M.C. Williams, 2005). We realize this by following the conjecture that human's intention is reflected in his posture and the objects in the direction of motion, such as a person heading to a café possibly will drink coffee. In our design, a human walks in a room, which contains a set of autonomous robots and other objects, such as a bookshelf, computer, cameras and a table supporting the experiment. The motion of the human and the interactions are processed and estimated by computers connected to two cameras. Figure 1 schematically illustrates the software and hardware integration. For each captured image, the vision based inference (VIS) software detects and tracks the human and generates a feature descriptor, which is mapped to a low-dimensional vector representing the characteristics of the human intention. As will be discussed later, this reduced space provides a simple yet powerful representation.

The estimated intentions are used to guide the autonomous robots into changing intentions from one to another in the physical environment that is specifically designed to foster interactions. The guidance process is algorithmically a continuous and complex one that uses closed form mathematical models interpreted from human-human interactions. In particular, we exploit the fact that, a human tries to predict context dependent reaction of another person during their interaction by estimating intentions (T.J. Clarke et al., 2005). This observation is supported by studies on how individuals estimate intentions of others from their motion when they are in a group activity (E. Kohler et al., 2002; P. Haggard et al., 2002). We

Figure 1. The flow diagram of the prototyping system. A digital image from the camera is acquired and processed to generate a feature vector. This feature vector is then translated to a lower dimensional descriptor. With no initial models on how images are formed, the vision-based inference system mapping is learned from a training data containing example input–output pairs.

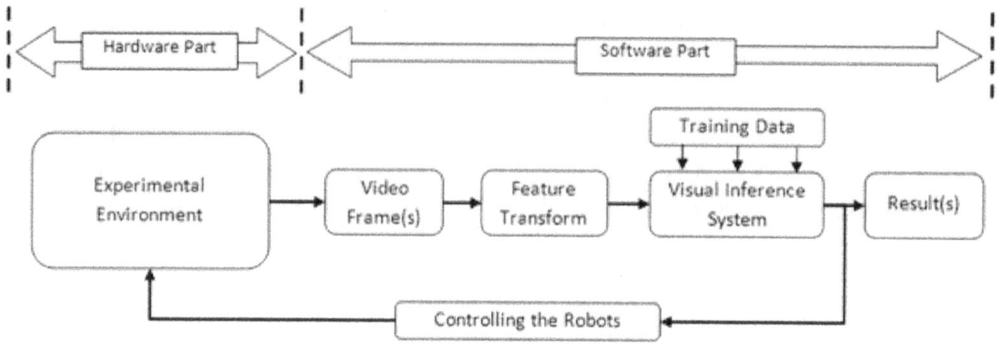

should note that the intentions in a human-human interaction are estimated subconsciously without focusing on the fundamental motives (K. A. Tahboub, 2006). This observation suggests that human-human interaction models can be adopted in non-verbal human-robot interactions.

This chapter is focused on human-robot interactions, where the interaction is guided without direct gesture-based or verbal communication. In the following text, we will first provide an overview of the current literature on intention estimation problem, its characteristics and requirements. Later, we will introduce our prototyping environment that contains a set of hardware and software components. Finally, we will discuss our approach on reshaping intentions by strategic motions of autonomous robots, and provide a discussion on future studies.

2. RELATED WORK

The focus of this chapter is twofold: Intention estimation between two agents mainly a robot predicting "what human intents to do," followed by how detected human intentions change based on moves of the robots triggering a forced interaction morphing the initial intention of the human agent into new ones. In order to reshape the intention into new ones, the robots decide the actions that need to be taken. The changes in intentions can be handled as state transitions from one to another. We therefore have to map the state transitions to a specific type of motion the robot has to undertake. The two key problems in this process are 1) estimation of the current human intention so as to identify the current state, and 2) identification of the new state which corresponds to the intention that human will be subjected. This section provides an overview of the up-to-date literature in the field of *intention estimation* and *reshaping intention.*

2.1. Intention Estimation

Within Human-Robot Interactions (HRI), the intention estimation of the human or robot agents is an option for increasing the capability of cooperation between humans and robots. If the system, which includes both agents, is able to estimate the intentions of either agent, it will foster better planning of forward actions performed by the agents, as well as adaption of agents' behaviors to better suit accompanying agents.

In order to estimate intentions of human and robots, researchers have approached the estimation problem using one or both of the two following main parameters: one being the characteristics of the problem (problem definition) itself and the other being the classification problem. For instance, if the goal is to estimate the intent of a fight between two people, the decision about the fight according to the actions of the people prior to the fight but leading to it. Classifying the actions of the human or robot agents and relating them to acceptable intentions which depend on the problem at hand is also an integral part of intention estimation. Classification of data has been extensively dealt with in the research area of Machine Learning (ML). Three closely related problems of time-series analysis, sequential supervised learning and sequence classification (T.G. Dietterich, 2002) must be necessarily be used for the intention estimation of human or agents due to the fact that the intentional action of the human or robot is a sequential process.

The characteristic defining the problem domain is an important part of intention estimation due to the fact that the actions, such as direction of gaze, body movements and facial expressions, pertaining to specific problems provide a communication channel from which a decision can be carried out for the determination of intentions. Examples of these include whole body movements and gestures, such as pointing (V. Manera et al., 2010), communicating (T.J. Clarke et al., 2005), walking and running (A. Chouchourelou et al., 2006; C.L. Roether et al., 2009), dancing (V. Sevdalis et al., 2009, 2010; W.H. Dittrich et al., 1996). Human intention, which has been discussed by philosophers for many years (M. Bratman, 1999; D.C. Dennett, 1987), can be inferred by means of one's actions and the effects of these actions to the environment they occur (K. A. Tahboub, 2006). A preliminary study to infer intentions is to recognize the actions. Researchers have used a number of different methods for investigating the tendency to recognize one's own acts and to separate them from those of others'. The most prominent paradigm in attributing actions to intentions is the visual observation of movement kinematics. It has been illustrated in the field of psychology that humans can recognize their own acts by observing kinematic displays of frames (G. Knoblich et al., 2001), hand gestures such as line tracing or finger tapping (E. Daprati et al., 2007), and body movements, such as dancing (F. Loula et al., 2005; V. Sevdalis et al., 2009, 2010), walking (J.E. Cutting et al., 1977; F. Loula et al., 2005; V. Sevdalis et al., 2009, 2010; A.N. Meltzoff, 1995) and fighting (K.K. Lee et al., 2004).

While research on communication of intention through actions is significantly outnumbered by research on other means of communication such as vocal (B. de Gelder, 2009; K. Terada et al., 2007), intentional states can be accurately recognized from actions. Particularly, neuroimaging and neurophysiology results suggest that simple grasping actions, such as eating, picking and placing in monkeys (L. Fogassi et al., 2005), and drinking and cleaning in human (F.P. de Lange et al., 2008; Iacoboni et al., 2005), depict their intentions, even when a small duration of the action is visible (M.A. Umilta et al., 2001). In similar fashion, behavioral research suggests that the intentions of deceiving or affective state, such as happiness, unhappiness can be reliably communicated to others through actions in videos (S. Runeson et al., 1981; J. Grezes et al., 2004, N. Sebanz et al., 2009).

Whole body indicative movements and gestures, such as pointing (V. Manera et al., 2010), communicating (T.J. Clarke et al., 2005), walking and running (A. Chouchourelou et al., 2006; C.L. Roether et al., 2009), dancing (V. Sevdalis et al., 2009, 2010; W.H. Dittrich et al., 1996) are also very indicative frameworks in detection of intentions. Moreover, understanding the intentional actions may not rely only on recognizing visual features. For instance, intentions based on actions can be communicated in an auditory manner by listening to an action's auditory conclusions (E. Kohler et al., 2002; P. Haggard et al., 2002). Consequently, due to the fact that the actions of robot agents or human and the relation between those actions and features of the environment where they occur form the basic characteristics of intentional states by subtle differences in movement kinematics and their affective results, human posture, position and heading angle are critical parameters used in our approach as well in estimating human intention as will be introduced in the coming section of this chapter.

Machine Learning is a research area of Computer Science that is interested in designing systems that can learn from the input data. One of the most common machine learning techniques in intention estimation is classification. Classification problem as mentioned above is also one of the two parameters of the intention estimation. For instance, the mouse is a device used for pointing purpose in the computer. However, if we want to use the human mimics instead of the mouse for pointing purposes, then the problem may be considered as an intention recognition problem achieved through classification of human actions such as winking of eyes, moving of the mouths or arms (Yıng Qi et al., 2007; B. Vlasenko et al., 2009; K. Arai et al., 2011).

The literature review provided below contains illustrative examples of classification problem applied to intention estimation and its related problems. One of the earlier studies on recognition was developed for understanding the plot of stories (R. Wilensky, 1983). H.A. Kautz et al. (1986) extended this work to a general model that recognizes general plans based on a set of observations and action taxonomy supported by environment constraints. E. Charniak et al. (1993) approached the inference estimation by measuring the uncertainty during plan recognition. In their work, decreasing the number of top-level plans provided them with computational advantage, while limiting its application to real world scenarios that inherently include uncertainty in both the perception and actuation. They performed Bayesian updating to select the most likely interpretation by using the set of observed actions. In a related work, D. Pynadath (1999) examined the Bayesian networks for monitoring traffic. This work was later extended to developing a cognitive driver assistance system to warn the driver according to the pedestrians' intention (S. Schmidt et al., 2009). S. Schmidt et al. (2009) developed a driver assistance system through capturing traffic situation and analyzing it in order to alert the driver in case a pedestrian is a potential danger. The system recognizes the intentions of the pedestrian by learning from a video database of traffic scenes.

T. Miyake et al. (2010) presented the human intention estimation from changing facial expressions. The optical flow computed using traditional computer vision techniques and neural network learning algorithm are used to recognize human intention. In their study, frame-to-frame variations in facial appearances provide a means to detect facial expressions. In another computer vision based study, K. Arai and R. Mardiyanto (2011) presented eye-based HCI system together with moving keyboard layout that reduces fatigue. The alphabetic keys were arranged into a horizontal line on the proposed key layout. When human-user is looking left, the keyboard layout shifted to the right direction or vice versa. Through the experiments with six real human-users, the moving keyboard successfully showed the effective reduction of physical fatigue for handicapped people. Aside from the systems solely depending on computer vision, B. Vlasenko et al. (2009) researched naturality in human-machine interaction by designing user behavior models in dialog systems. In their study, they used both a camera and microphone to analyze multi-model emotion source including mimics from facial images and voice from speech. Using the multi-model source, they classified intentions as cooperative, explorative and destructive.

For estimation of intentions in human-human interactions, A. Dielmann et al. (2004) analyzed meetings with multiple attendants for the classification of various actions, such as dialogs, note taking, monologues, and presenting in front of a board. In their study, they used one lapel microphone per participant of the meeting and eight microphones distributed in a circular array. Meeting participants were assumed to be either presenting in front of a board or presenting from their seats. As a result, six speech-activities were observed. They used two levels of HMM to classify and recognize meeting data using features extracted from microphones. The first level HMM performed an early integration where the different features are concatenated. The second level HMM is a multi stream approach where the different feature types are first classified separately and then integrated at the end of this level HMM. Similar to (A. Dielmann et al., 2004), D. Zhang et al. (2004) used a two-layer HMM to model individual and group actions of people during meetings using audio and visual features acquired from a camera and a microphone. The two layer HMM, corresponding to lower and higher inference, used for modeling individual and group actions.

Another body of work has been expanded to understanding of human actions from visual data, which can be used to infer intentions. For this purpose, M. Shimosaka et al. (2005) presented a method that switches linear dynamics by sorting kernel vectors for classifying human actions into walking and non-walking. They used motion modeled as motion of a human skeleton, which describes the human skeletal configuration with specific classes of standing, sitting, running, lying and translational motion from standing, for non-walking cases and varying tempo for the example of walking. Similarly, T. Mori et al. (2004) recognized human actions including sitting, standing and walking via multi-level HMMs. A tree structure was used for the occurrence of the recognition at various levels. The usage a tree structure has two advantages: simpler recognition process due to the exclusion of some irrelevant features at the detailed levels, and the possibility to give rational responses to data by coarse classification. In another study, K.K. Lee et al. (2004) used Support Vector Machines (SVM) for the classification of the motion trajectories to recognize actions of table tennis players or people fighting. For reducing dimensionality of observed trajectories, they applied Principle Component Analysis (PCA) and Independent Component Analysis (ICA). We should note that these studies are important representatives for the interpretation of the human actions and they do not include robot agents.

H. C. Lin et al. (2005) used motion classification for a different application. During MIS (Medical Imaging Solutions) tasks such as training of surgeons, motion classification was used for the estimation of the quality of the performed task. After a sequence of suturing motions, a model of the task is extracted, and linear discriminant analysis (LDA) was used to separate motions from other gestures. It is proven that the separation of motions in an expert surgeon is better than in a less experienced surgeon.

K. A. Tahboub (2005) introduced a compliant human-machine interaction architecture using a dynamic Bayesian network for probabilistic intention inference. He showed that using this system, where a human remotely commanded one of two robots in the simulation interface, the other robot tried to estimate the intention of human. Although this study is an important example of the intention recognition, simulated robots are used instead of real ones.

Oliver C. Schrempf et al. (2007) presented a different approach to create a Bayesian network for estimating intentions without a learning framework. They gave a reduced model with systematic derivation to recognize real human intentions in a virtual environment. In this study, human is real however the environment is virtual. After this point on, we will introduce works in the literature that deal with the real-time applications of intention recognition.

S. Koo et al. (2009) present a real-time experimental setup for recognizing human intention from actions using a robot equipped with Infra-Red (IR) sensors. They used the HMM model with position and velocity of the human so as to solve the action-based recognition problem. In this application, the capability of the intention recognition is limited since they only use sensory information such as human position and movement velocity. R. Kelley et al. (2008) used HMMs for autonomous mobiles robots to understand human intention. The sensor used was a camera attached on each robot enabling the system to monitor the current scene. Although this application resembles our problem, the differences lie in the aims of the two systems. Our aim includes, additionally to intention estimation, the goal of changing this current intention in a desired way.

D. Aarno et al. (2009) also exploited a layered-HMM to recognize intention of the robot-arm via its trajectories. The goal of this study is to recognize the robot (agent) intention. D. Aarno in 2007 presented another study related to intention recognition for Human-Machine Collaborative Systems (HMCS). A layered-HMM was used to model human skills in the particular applications where human and robot manipulator aims to make some tasks in a cooperative mode or tele-operative mode.

K. Noguchi et al. (2003) and Y. Nakauchi et al. (2003) designed an experimental room for detection of human intention. In their study, they used a ID4-based leaning system, complete sensors for estimation of human position and also of sounds in the environment to recognize the human intention. Their studies although constrained in their abilities are however more capable for intention recognition than the system developed by Koo et al. (2009). In this chapter, unlike the examples given above, all data is acquired from real environment using two cameras that extract and track the human as well as the robot agents. The capability of the estimation is highly dependent upon the quality of the feature vectors.

2.2. Reshaping Intention

The literature on robotics above includes many examples recognizing human intentions in a robotic environment. Reshaping of the human intention, however, has never been addressed in the field of human-robot interaction. This chapter seeks an answer to "Why do we need to study intention reshaping?" An example can explain the importance of this problem. Let's assume the environment being an area dangerous for mankind requiring certain risk taking tasks, and robots are sent into environment to cooperate with human so as to warn them for preventing putting human in danger. While textual descriptions in the form of notes may be warning signals, not everybody pay equal attention to textual information; in fact the person maybe illiterate. In this case, motion of robots may draw the attention of people and reduce the risk of getting hurt. Such systems should, however, be rapid and reliable.

In the literature, the examples of reshaping intention are typically found in psychology. Webb et al. (2006) studied whether changing behavioral intention causes behavior change. Forty-seven experimental tests of intention-behavior relations were used to satisfy this goal. In those experiments, participants were assigned randomly to a process that increases the magnitude of their own intentions relating to a control situation, and differences of the forward behavior were compared. They used meta-analysis to predict the affective ranges of intention-behavior relation. After the impact of the interventions on the intention and behavior was quantified in meta-analysis, the results indicated that the interventions could change intention also engender reshaping in behavior.

Andrew N. Meltzoff investigated whether children would change their intention towards what an adult intended to do so or shape their intentions based on what an adult currently did. The authors prepared two experiments for this purpose. In experiment 1, children are confronted with an adult who tried, but do not succeed in performing, obvious target actions. Experiment 2 matched the reaction of children to an adult towards an inanimate item. Taken together, 18-month-old children have adopted a key element of a folk psychology: although inanimate items are not understood, adults are understood within a framework which includes intentions and aims.

Kazunori Terada et al. (2007, 2010) reviewed the effect of reactive movements accomplished by a non-humanoid robot like chair and cube on human intention attribution. In their study, humans tried to understand behaviors of an artifact with respect to its goal which depends on how human can attribute intention to the artifact. The aims of this study were to observe how human behave after a movement of the robot and also were whether or not there are differences made by the appearance of the robot on human's intention attribution. Their study was related to the investigation of the human intention attribution toward the movements of the robots.

A novel contribution of this chapter is an introduction to intention reshaping in the area of HRI research. Thereby, the actual aim of this study is to realize the change in current intention via intentional robot movement within a real scenario. Our approach to this problem is to develop a hierarchical system by decoupling the low-level object detecting-tracking algorithms with the high-level intention estimation and reshaping. The system also makes decision about the changing human intention using robot movements. The low-level detection and tracking algorithm focuses on multi-robot interacting with a human. High-level capability provides a solution to the human-machine intention problem by using learning algorithms such as HMM and Principle Component Analysis (PCA).

3. EXPERIMENTAL SETUP FOR HARDWARE PROTOTYPING

An experimental application room (Figure 2) is designed to observe the reshaping of human intention by targeted motion of the robotic chair and stair shown in Figure 3c. The room, which includes worktable, bookshelf, coffee machine and mobile chair/stair robots, is monitored by two cameras mounted at oblique and orthos angles. The oblique view provides full posture of the agent, while the ortho view provides position of the agent relative to other scene components. A computer controlled stair and chair performs a series of autonomous activities for changing the intention of the human agent by interacting with that participant entering the experimental room (see Figure 3(b)-3(c)). The stair and chair robots are powered by two wheels driven by 12V DC motor and are moving in plane. A Radio Frequency (RF) communication system allows the experimenter/computer to remotely operate the robotic chair and stair. In our procedure, human agents are permitted to behave freely within the experiment area; however, damage to robots and the equipment are prohibited. Aside from these

Figure 2. The prototype of the experimental environment

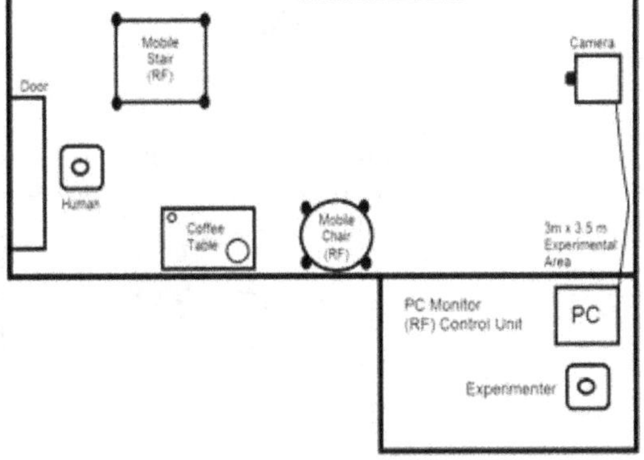

instructions no other instructions are given to the agent before the start of the experiment.

Experimental scenario: Fifteen individuals, whose ages range from 18 to 24 years, participated in the experiment. None of them were given prior knowledge about the experiment or had an experience in interacting with the mobile stair/chair robots in the experiment. First, videos are collected to estimate and learn classes of possible intentions of human agents within the context of this experimental room. Robots move according to a scenario to intentionally changing the current intentions of human into desired ones in a predictive way. The recorded video database is required to construct supervised learning based intelligent system in which the robots are controlled autonomously.

4. SOFTWARE PROTOTYPING

The algorithmic approach to change an estimated intention of the human in the scene to desired intentions by autonomous robots is achieved using three-step approach:

First step is related to low-level object detection and tracking by using background subtraction and matching techniques. In this step, people and robots are detected and tracked in the video sequence with a stationary background using the following process: 1) the first few frames of the video are used to learn the static background. 2) The moving pixels are then detected by comparing them against the learned background. 3) Moving pixels are grouped by connected component analysis, which are labeled as human or robots. 4) Finally, connected regions in the current frame are matched with those in the previous frame based on appearance and size information.

Second step deals with high-level information from low-level cues based on Bayesian inference. In this step, the state of observed activities is defined as instantaneous movement of agents that perform a set of activities that denote their current intentions. A single state observed from the video does not provide adequate information to predict the intention. We assume that the agent carries out a sequence of actions following a plan until the intention is achieved. Hence, we

Figure 3. Hardware prototyping

(a) Top view of the experimental environment

(b) Side view of the experimental environment

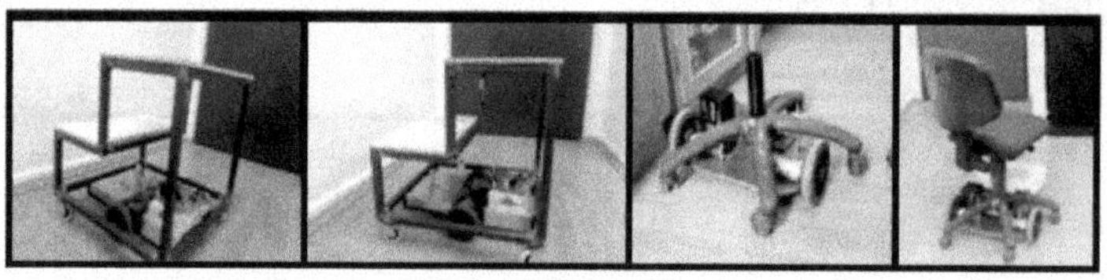

(c) Mobile stair and chair

alternatively use a sequence of human action to predict the intention of the human agent.

Once the state is constructed, we employ Bayesian inference to learn and estimate intentions. In particular, we adopt the Hidden Markov Model (HMM) that takes human posture and location information as its observation model. The proposed method classifies the posture of the human at each frame by applying the Principle Component Analysis (PCA), which is used to measure the Euclidian distance between projected data and nominal data.

The Last step in our approach is a control step to see if the robots succeeded in changing the human intention. In this step, the system compares the previous (current) intention, which is estimated when the robot starts to move, with the new intention estimated after the robot completes its movement. The flow chart of the whole system is shown in Figure 4.

4.1. Object Tracking

The flow of the tracking algorithm adopted from (W. Grimson et al., 1998) is shown in Figure 4 as a module. Tracking humans and robots in a sequence of images with a stationary background is a critical component of our experimental design. The goal of an object tracker is to generate the trajectory of an object over time by locating its position in every frame of the video. Object tracker may also provide the complete region in the image that is occupied by the object at every time instant. The tasks of detecting the object and establishing correspondence between the object instances across frames can either be performed separately or jointly. In the first case, possible object regions in every frame are obtained by means of an object detection algorithm, and then the tracker corresponds to objects across frames. In the latter case, the object region and correspondence is jointly estimated by iteratively updating object location and region information obtained from previous frames. The model selected to represent object shape limits the type of motion or deformation it can undergo. For instance, if an object is represented as a point, then only a translational model can be used. In the case where a geometric shape representation like an ellipse is used for the object, parametric motion models like affine or projective transformations are appropriate. These representations can approximate the motion of rigid objects in the scene. For a non-rigid object, silhouette or contour is the most descriptive representation and both parametric and nonparametric models can be used to specify their motion.

Tracking objects by associating detected regions is typically based on the similarity of appearance between consecutive frames. The detection can be intuitively performed by subtracting the two consecutive frames for detecting changes that manifest moving objects. A more elegant approach is to learn the background and subtract it from the images in the video. The background is generally modeled statistically by fitting parametric models to appearance observation. Most common parametric model is the Gaussian distribution, which is estimated for each pixel from a sequence of frames (A.Yilmaz et al, 2006; E. Roichman et al., 2008). In order to adapt to temporal changes in the scene, such as lighting conditions, the learned background model is updated by a moving average computed as follows:

$$B_t(x,y) = \sigma B_{t-1}(x,y) + (1-\sigma)I_t(x,y) \qquad (1)$$

where I_t is the tth input image, B_t is the tth background model, and σ is a learning element in the range [0, 1]. A recent history of each pixel is essential for the background model at each pixel. The learning factor σ for every pixel of the background model is adaptively calculated by using the correspondence value in the difference frame. A higher value of σ is obtained with larger difference between the values of current frame and background frame, therefore more weight is given

Figure 4. Flow chart of the overall system

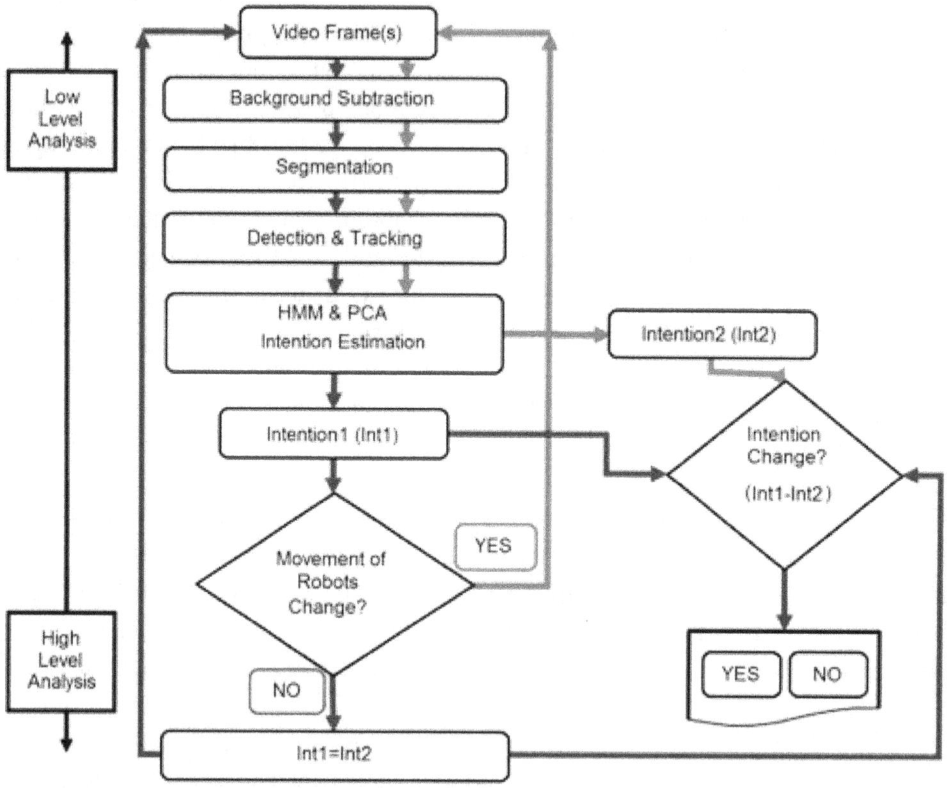

to background pixels. The background model is capable of adapting to swift changes and is robust to illumination since the tuning of σ is adaptive. Here we use a combination of the intuitive and parametric model based approach. In the case, when the moving pixels are more than 75% of the pixels in the entire image, our algorithm resets the background model and learns it again. This property allows us to handle unexpected lighting changes in the scene correctly. The average of the pixel values between background difference frame and difference frame in which movement detected is used to increase the robustness of the background subtraction system. This average calculation is formulated by:

$$D_t(x,y) = v.BD(x,y) + (1-v)SFD(x,y) \quad (2)$$

where the average difference frame is *D*, the background difference based on parametric background model is *BD*, the difference frame between successive frames based on intuitive motion estimation is *SFD* and '*v*' symbolizes the weight given to each model. We compute parameter *v* as follows:

$$v = MinW + (MaxW - MinW).ADL \quad (3)$$

where *MinW* and *MaxW* are successively the minimum weight and the maximum weight given to the background difference frame and *ADL,* which is the average difference level in *SFD*, takes values between 0 and 1. If the value of *ADL* is the highest, the weight (*v*) given to the background difference frame has the biggest value.

Figure 5 also demonstrates an example of the background subtracted image. Object detection can be achieved by building a representation of the scene called the background model and then finding deviations from the model for each incoming frame. Any significant change in an image pixel from the respective pixel in the background model signifies a moving pixel. The pixels constituting the regions undergoing change are then completed by the morphological closing which merges disconnected but proximal pixels to create moving blobs. After background subtraction, these moving blobs are subjected to a segmentation algorithm, which rejects small regions that may have been computed due to noise in observations. Finally we estimate the bounding boxes and the centers of detected regions. In the final stage of tracking, in order to determine the locations of robots and human from one frame to the next, the system compares the predicted locations of the bounding boxes with the detected locations. This enables the system to assign a unique label in the form of a color to each robot and the human (see Figure 5 for detection and Figure 6 for tracking). Figure 6 shows two trajectories corresponding to two different people in the same environment.

4.2. Learning

In this section, we explain the proposed learning mechanism to estimate human intention. Our approach exploits the changes in the human posture during an action to estimate intention using Hidden Markov Models (HMM). Let's start with an illustration to clarify how the proposed method works. Let a human with the intention of "discovering robots" (from Table 1) starts walking to the bookshelf while looking at the opposite direction from the bookshelf. In this case, we are faced with a puzzling mixture of intentions and need additional information to infer the true intention. We conjecture that the human posture defines the main posture. In order to discover the intention, we use the previous posture and location in a probabilistic setting. This postulate suggests the use of hidden Markov modeling that can learn dynamic system with hidden states.

4.2.1. Principle Component Analysis (PCA) for Human Posture

Tracking people is generally done by recognition of their actions, which is a quickly growing research field, particularly in the areas of video surveillance and human machine interaction. Several techniques such as K-Means, fuzzy C-Means, Self-Organizing Maps (M. Kiran et al., 2009) and Correlation Filter Classifier (N.M. Tahir et al., 2008) for the analyzing and recognizing of human postures have been studied in the literature. Alternatively, PCA is a popular feature mapping approach for creating new descriptors that are linear combination of the original dimensions (Jolliffe, I. T., 1986). Geometrically, the purpose of the PCA is to identify a set of orthogonal axes called the principle components (K. K. Lee et al. 2004). The new variables are formed by projected the original variables using their principle components. In this chapter, we present a Principle Component Analysis (PCA) based posture analysis approach, which identifies and classifies postures into one of the following classes "left position", "right position" and "center position".

The PCA, which is used by M. Turk et al. in 1991 to recognize faces, is based on converting *K* observations into another orthogonal axis by performing eigenspace decomposition on the K×K covariance matrix. In our case, the observations are images containing human postures. Considering the space of postures is limited, they are

Table 1. The list of intentions

Number	Intention
1	Taking a book from bookshelf
2	Sitting on the work-table
3	Taking a coffee
4	Discovering the robots

Figure 5. Detected location of the human and robots (a) the colored points show the center of the robots and human on the one real video frame. (b) The locations on background subtracted frame (c) the locations on the empty image

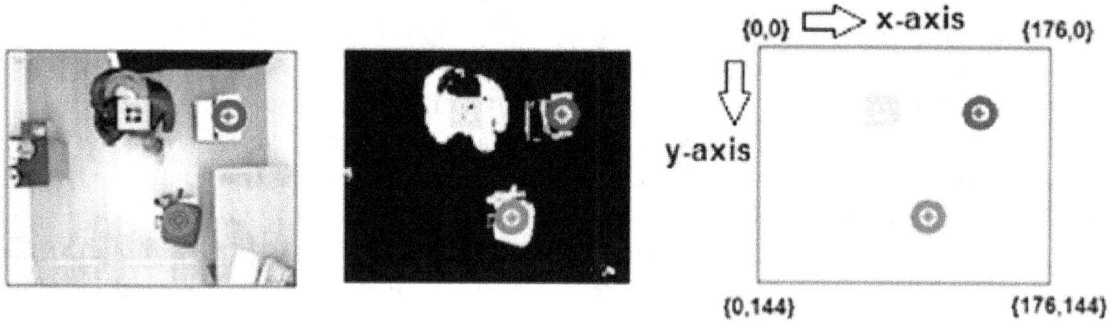

(a) (b) (c)

Figure 6. Examples frame trajectories as the location of human and robots (1500 frames) for two people

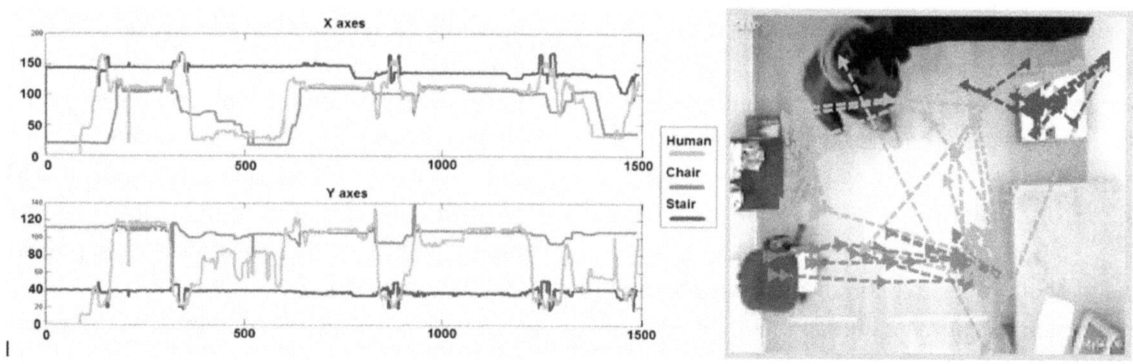

(a) Person 1

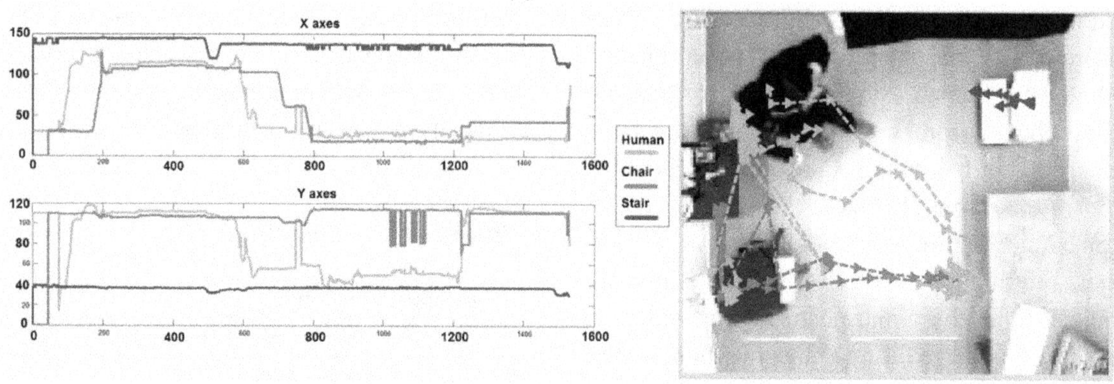

(b) Person 2

contained in a lower dimensional space. The lower dimensional space typically corresponds to a small percentage of the orthogonal dimensions computed using the PCA. The projections of K are observations to the subspace with smaller dimensionality, which represents the space of human postures. The process starts with mean normalizing the image vectors $\Gamma_1, \Gamma_2, \ldots, \Gamma_M$ containing postures: $\Phi j = \Gamma_j - \Psi$ where

$$\Psi = (1/L)\sum_{i=1}^{L}\Gamma_i. \tag{4}$$

These normalized vectors are later collected in matrix $A = [\Phi_1, \Phi_2, \ldots, \Phi_L]$, where L is the number of the images in the training set. Matrix A is then used to compute the covariance matrix:

$$C = \frac{1}{L}\sum_{i=1}^{L}\Phi_i\Phi_i^T = A.A^T \tag{5}$$

Considering that the size of C depends on the size of the image and may be significantly larger than the number of posture images. To reduce computational cost in such cases, one can estimate the subspace by multiplying both sides of $AA^Tw_i = \mu_i w_i$ with A, such that $AA^TAw_i = \mu_i Aw_i$. This decomposition suggests that Aw_i are the eigenvectors of the covariance matrix C.

Therefore, after the generation of the alternative L×L covariance matrix, $M = A^TA$, where $M_{lk} = \Phi_l^T\Phi_k$, and the eigenvectors w_l of the matrix L is calculated. The eigenvectors of the original posture space u_l are obtained by the linear combination of these eigenvectors:

$$u_m = \sum_{n=1}^{L} w_{mn}\Phi_n, m = 1, \ldots, L \tag{6}$$

Once the subspace is generated from known set of postures, recognition is achieved by projecting the input posture onto the subspace, which is a linear superposition of the most significant eigenvectors:

$$\rho_m = u_m^T(\Gamma - \Psi), m = 1, \ldots, L' \tag{7}$$

Figure 7 shows the training set and the coordinate system used in analysis and identification of postures. The weights corresponding to each eigenvector, $\Omega^T = [\rho_1, \rho_2, \ldots, \rho_{L'}]$, explain the contribution of each eigenvector related to the input image. Selection of m significant eigenvectors can be defined by selecting m which minimizes the Euclidean distance:

$$\varepsilon_m = \left\|(\Omega - \Omega_m)\right\| \tag{8}$$

We choose Ω_m as the mean of the eigenspace representations of several training images belonging to the same class of posture. When the minimum of ε_m is below a threshold θ_ε then we conjecture a known posture is observed. Otherwise, the posture is classified as unknown.

Figure 7. The example training set and the coordinate system of the image sequence (a) "left position" of the human (b) "center position" of the human (c) "right position" of the human (d) training data examples

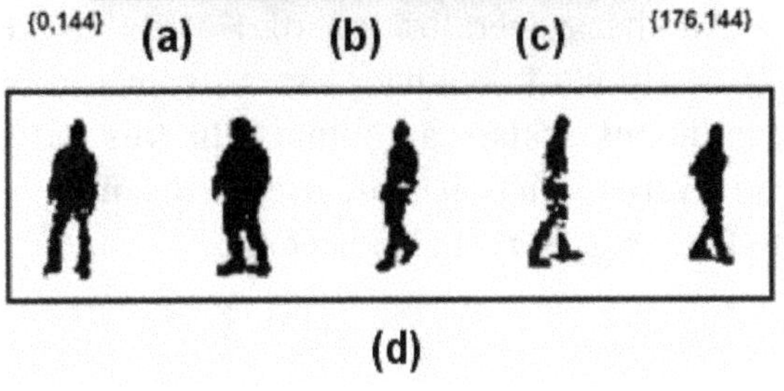

4.2.2. Hidden Markov Model (HMM) for Intention Estimation

We use the HMM to model motion patterns of the human which is used for estimating intentions. Before we detail of the estimation is achieved, we start with giving a formal explanation of HMM. A discrete HMM, $\lambda = (A, B, \pi)$ can be described N hidden states, $H_1, H_2, \ldots, H_N$, M observations (outputs/emissions), $O_1, O_2, \ldots, O_N$ transition matrix $A = \{a_{ij}\}$ which gives probability of transition from H_1 to Hj: $a_{ij} = p(H_j|H_i)$, observation matrix which encodes probability of observing O_j at H_i: $p(O_j|H_i)$, and finally, prior information, $\pi_1, \pi_2, \ldots, \pi_n$ which denotes probability of starting at state H_i: $p(H_i)$. These components of HMM are subjected to common probability rules including, $\Sigma_{i=1}^{N} a_{ij} = 1, \Sigma_{i=1}^{N} b_{ij} = 1, \Sigma_{i=1}^{N} \pi_{ij} = 1$, where, $a_{ij} \geq 0, b_{ij} \geq 0$ and $\pi_i \geq 0$. In addition to the discrete variables, the outputs of HMM can be also presented as continuous variables. In this case, probability mass function for emission probabilities a_{ij} is usually replaced by a mixture of k Gaussians, which symbolizes a probability density function (pdf) to avoid quantization and related problems (L. R. Rabiner, 1989):

$$b_i(O_j) = \sum_{k=1}^{r} w_{ik} N(O_j, \mu_k, \sigma_k), \tag{9}$$

where w_{ik} is the weight of kth component $N(x, \mu_k, \sigma_k)$, $i = 1, \ldots, N, j = 1, \ldots, M$, and

$$N(x, \mu, \sigma) = \frac{1}{\sqrt{2\pi\sigma^2}} \exp\left(\frac{-(x-\mu)^2}{2\sigma^2}\right). \tag{10}$$

Another generalization of HMMs is the introduction of continuous vectors instead of continuous scalars as outputs. In this case, multivariate Gaussians are used in the mixture so that $N(x, \mu, \sigma)$ in (10) becomes

$$N(\mathrm{x}, \mu, \Sigma) = \frac{1}{(2\pi)^{m/2} |\Sigma|^{1/2}} \exp\left(-\frac{1}{2}(\mathrm{x}-\mu)^T \Sigma^{-1} (\mathrm{x}-\mu)\right), \tag{11}$$

where m is the number of elements of vector $\mathbf{x}$ and $\boldsymbol{\Sigma}$ is an m-by-m symmetric positive semi-definite covariance matrix.

In order to model a dynamic transient system using HMMs, it is required to solve the following three problems:

1st problem: Given a model and a sequence of observations, $O = (o_1, o_2, \ldots o_T)$, how to compute the probability of sequence O belonging to this model?

2nd Problem: Given a sequence of observations, O and a model λ, how to find the most probable sequence of states $(h_1, h_2, \ldots, h_T)$ that constituted this sequence?

3rd Problem: Given a set of sequences (or a sequence of observation O), what are the parameters (A, B, π) of the model that is most likely to construct the sequence(s)?

Efficient and well-established solutions for all of the above problems have been well studied and generally use dynamic programming. The first problem is solved by the Forward algorithm (S.J. Russell et al., 2003), the solution to the second problem is given by the Viterbi algorithm (A.J. Viterbi, 1967) and the third problem is solved by the Baum-Welch algorithm (L.E. Baum et al., 1970). Since first problem and third problem are directly relevant to the method explained in the section of application statement, we supply brief overviews for the Forward and Baum-Welch algorithms in the following discussion.

Forward Algorithm: Given a model λ and a sequence O, $p(O|\lambda)$ or the probability of the sequence, which is generated by the model, can be calculated by using the Forward algorithm. It presents the sub-problem of calculating $\alpha_t(i)$, which is the probability of observing first t elements of O and end up in state H_i. Obviously,

$$\alpha_1(i) = \pi_i b_i(o_1), \tag{12}$$

for $i = 1, \ldots, N$ where b_i and π_i are as defined before. We are interested in computing $\alpha_T(i)$, where T is the number of elements (or the length) of the sequence O. It can be computed by using the iteration

$$\alpha_{t+1}(i) = \left[\sum_{j=1}^{N} \alpha_t(j) a_{ji}\right] b_i(o_{t+1}), \tag{13}$$

for $i, j = 1, \ldots, N$ so that the desired probability of a sequence belonging to a model can be calculated as

$$P(O|\lambda) = \sum_{i=1}^{N} \alpha_T(i). \tag{14}$$

Similarly, the Backward procedure includes the sub-problem of calculating $\beta_t(i)$, which is the probability of observing last $T - t$ elements starting at H_i. Again, $\beta_T(i) = 1$, for $i = 1, \ldots, N$ and

$$\beta_t(i) = \sum_{j=1}^{N} \beta_{t+1}(j) a_{ji} b_j(o_{t+1}) \tag{15}$$

for $i, j = 1, \ldots, N$.

Baum-Welch Algorithm: The Baum-Welch Algorithm is thought to train a model λ given a sequence(s) O. It is an Expectation-Maximization (EM) (A. P. Dempster et al., 1977) algorithm that finds λ^* such that

$$\lambda^* = \arg\max_{\lambda} P(O|\lambda), \tag{16}$$

where λ indicates the HMM. The algorithm entails the introduction of two new auxiliary variables:

$$\chi_t(i) = P(h_t = i \mid O, \lambda) = \frac{\alpha_t(i)\beta_t(i)}{\sum_{j=1}^{N} \alpha_t(j)\beta_t(j)}, \tag{17}$$

which is the probability of being in state H_i at time t and

$$\psi_t(i,j) = P(h_t = i, h_{t+1} = j \mid O, \lambda) = \frac{\alpha_t(i) a_{ij} \beta_{t+1}(j) b_j(o_{t+1})}{\sum_{i=1}^{N} \sum_{j=1}^{N} \alpha_t(i) a_{ij} \beta_{t+1}(j) b_j(o_{t+1})}, \tag{18}$$

is the probability of being in H_i at t at H_j at $t+1$ for $i, j = 1, \ldots, N$, and where $\alpha_t(i)$ and $\beta_t(i)$ are as defined in (13) and (16), successively. In this point, the implicit assumption is that time is sampled at discrete points so that there is a one-to-one correspondence between time and subscript t in $o_t(i)$, $h_t(i)$, $\alpha_t(i)$ and $\beta_t(i)$. In the following learning process, the parameters of HMM model λ are updated by the following rules until convergence:

$$\bar{\pi}_i = \chi_1(i), \tag{19}$$

$$\bar{a}_{ij} = \frac{\sum_{t=1}^{T-1} \psi_t(i,j)}{\sum_{t=1}^{T-1} \chi_t(i)}, \tag{20}$$

and

$$\bar{b}_i(O_k) = \frac{\sum_{t=1}^{T} \delta(o_t, O_k) \chi_t(i)}{\sum_{t=1}^{T} \chi_t(i)}, \tag{21}$$

for $i, j = 1, \ldots, N$, $k = 1, \ldots, M$, where $\bar{\pi}_i, \bar{a}_{ij}$ and $\bar{b}_i(O_k)$ are the new produced values for π_i, a_{ij} and $b_i(O_k)$ respectively.

Application Statement: The HMM model used in our framework is shown in Figure 8 (L. R. Rabiner, 1989). In this figure, each observation module (O_p), which includes a sequences of locations and posture information of the human, demonstrates the observed human action. In the transition model, a conditional probability distribution matrix represents the probability of the current observed action sequences (H_p)

Figure 8. The Hidden Markov Model (HMM)

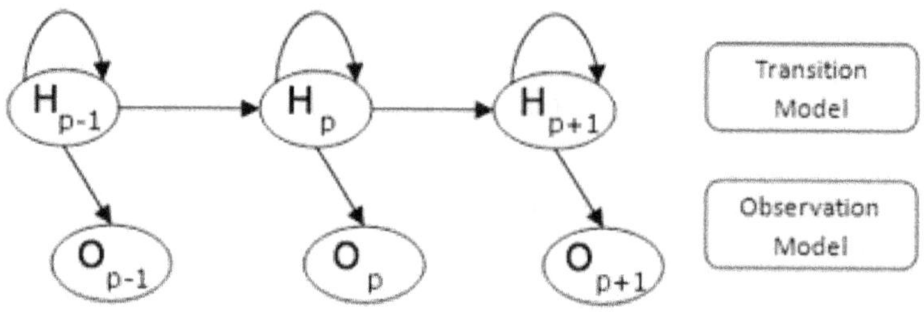

transmitted from the previous observed action sequences (H_{p-1}).

The scenarios produced by our algorithms are represented by an m-dimensional sequence of length T, where m is the number of system variables taken into account and T is the number of timestamps or the length of the evolution sequence. We choose the system variables to present the continuous multivariate observations and the underlying physical environment of the system represent the hidden states of HMM. In our setup, observable actions are defined as 'going to bookshelf', 'going to worktable', 'going to coffee-table' and 'exploring the environment'. Each of these actions, which consist of a sequence of image frames, is labeled with human intention as the hidden state of the HMM model. The labeled intention is tabulated in Table 1.

The first stage in the presented algorithm is to construct the training set. The training set comprising of sequences O can be formed by using the output trajectories of previous system simulation. We illustrate the observation model by labeling the grid cells from 144×176 to 24×25. Each cell symbolizes an observation part which depends on the position of the human agent. In Figure 9, we show a set of labeled cells in our framework. Given a set of human trajectories represented by the sequences of length T, by evaluating with human postures we can label them as intentions with respect to actions (See Figure 12). In Figure 12(a) and 12(b), intention estimation example includes a trajectory given by the human posture.

The human intention is estimated in a scenario when at least one of the variables reaches maximum value from intention list at any point of its trajectories. The scenarios whose variables do not reach maximum value are entitled as neutral. Neutral scenarios are used to generate the training set used to model parameters for the

Figure 9. Labeled observation cells

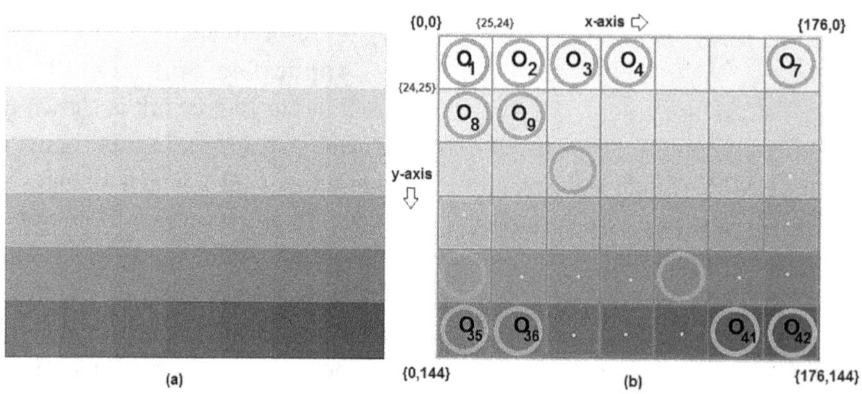

Figure 10. Example construction of the probabilistic observation matrix

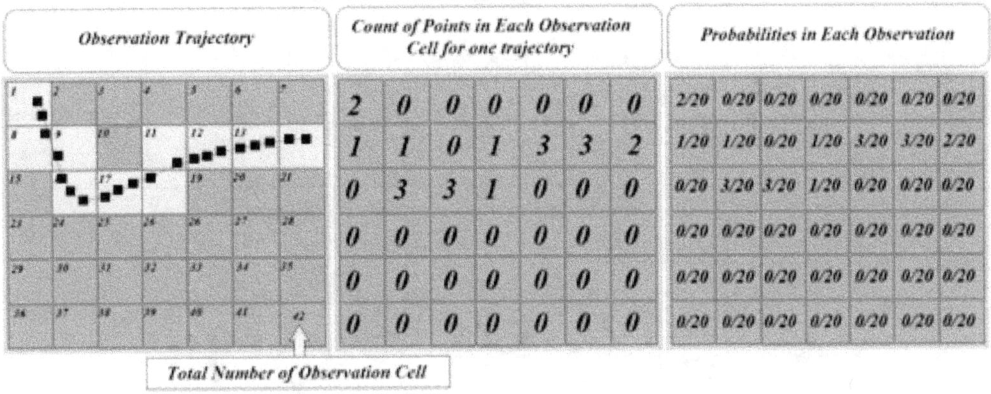

HMM. Note that rather than using all neutral scenarios, the training can be limited to a subset whose states are not similar to failed values. This will end in misclassification of some neutral scenarios but will also reduce the risk of misclassification of a failure scenario as neutral.

Once the training set is generated, we normalize it by subtracting the initial condition values from all the values in the sequence. After that point, the sequences in the HMM are trained by the aforementioned Baum-Welch algorithm. In this step, the number of hidden states N must be known in advance. Figure 10 illustrates how the probabilistic observation matrix is constructed by using the training data. The probabilistic weight of each observation cell is assigned from five different trajectories defining each intention. Figure 11 shows examples of two different training trajectories used in our experiment. On the other hand, a challenging problem with no trivial and practical solution is to estimate of number of hidden states. Hence, the number of hidden states based on the knowledge obtained from the examined system is commonly chosen. Although the number of hidden state is generally assigned by training experimentally, in this study, we propose that the number of hidden states is determined by intention list (i.e. the number of hidden states are four).

Once the model λ is constituted, given a scenario of length T or p, such that $p < T$, we can calculate the log-likelihood of this scenario being

Figure 11. Example training trajectory data's (a) "going to the work-table (b) "going to bookshelf"

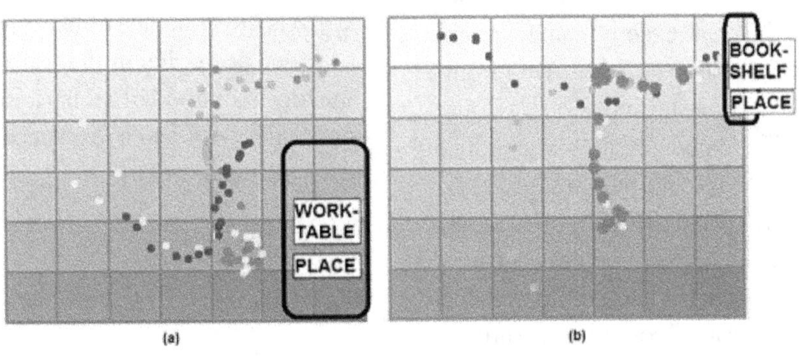

Figure 12. Example testing trajectory data and observation of the action for intention 1 (from Table 1)

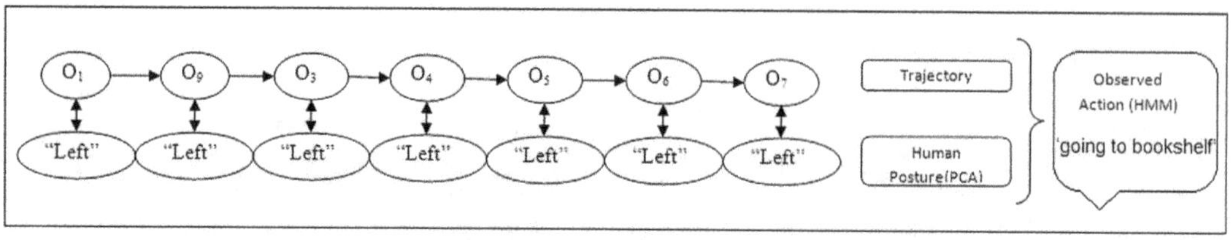

(a) example trajectory with block diagram

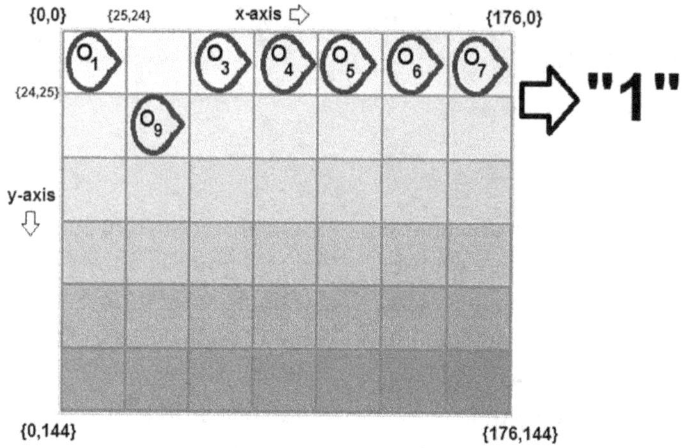

(b) example trajectory with labelled observation cells on mapping image

obtained by λ. In this case, although the training is performed for the full sequences of length T, we can compute the log-likelihood for partial sequences as well using Forward algorithm described in Section A using Equation (13). This approach allows labeling of partial scenarios, which can be called as *online* labeling. On the other hand, one important issue is the tendency to be similar between some failure scenarios and same evolution sequences as the neutral ones (at least within the chosen state variables). For example, let s_f be a failure scenario that has exactly same variable values as a neutral scenario s_n for a period from 1 to p, and starts to deviate starting at $p+1$. Consequently, the log-likelihood for both s_f and s_n of fitting the trained model will be the same at any time between 1 and p, and different after time p. Therefore, if we make decision before time $p+1$, s_f may be labeled as neutral and terminated decision. To avoid this unwanted condition, we propose using the failure scenarios that are also obtained during the training set construction. Once the model is designed, we can compute the log-likelihood of the failure scenarios being evaluated by this model and save the maximum value among all the failure scenarios at each time scale from

$$L^*(p) = \max\{L(s_{f\_i}(p)|\lambda)\}, \tag{22}$$

for $i = 1, \ldots, N_f$, $p = 1, \ldots, T$, where λ is the HMM, $s_{f\_i}(p)$ is the sequence of length p, N_f is the number of failed scenarios in the training set and $L(s)$ is the log-likelihood of scenario s. As a result, given an HMM λ, L^* and a test scenario s, the scenario ends at time p only if

$$L(s(p)|\lambda) \geq L^*(p) + \varepsilon, \tag{23}$$

where ε is the finish offset, which is the parameter that shows how consistent the ending decision is.

5. EXPERIMENTAL RESULTS

We test the scenario designed to show reshaping human intention with 15 participants. The goal in our experiments is to reshape the human intention via the movements of autonomous robots. For this purpose, all participants are asked to enter the experimental room without prior knowledge about the robots and their mobility. In Figure 13 and 14, results for two different people show that the agents changed their intention in the experimental room due to the movement of robots. In frame 760 for person 1 and in frame 510 for person 2, our system detects the intention of the agent as going and sitting at the desk, which corresponds to the second intention in Table 1. After frame 770 for person 1 and frame 530 for person 2, the stair-robot starts moving to the different side of the room. At first the human agent tried to understand what had happened. Once the robot motion caught his attention, after frame 800 for person 1 and frame 580 for person 2, he stood up from the chair and moved in the direction of the bookshelf to take a book. This motion in our set up is recognized as intention 1 for both persons. We present additional results from PCA based labeling and HMM based detection respectively in Figures 13 and 14 and Table 2.

These two and other experimental results suggest that, we have successfully changed the intention of the agents. On the other hand, this success was not true for all participants. Considering that the participants did not have knowledge about purpose of the experiment and the room, the first movement of the robots attracted the attention of the participants. Especially, we used this attraction to show that changing of the human intention was possible. Figure 13 and Figure 14 show the results for two individuals. While the person 1 is going to take a book from bookshelf after movement of the stair robot in Figure 13, the person 2 is going to take a coffee in Figure 14. Nevertheless, some participants did not respond to the robot movement. After a certain amount of time during which the participants understood the robots, we observed

Table 2. Some results for Person 1

Euclidian Distance with posture			
	Frame= 760	**Frame= 800**	**Frame= 820**
Left	2221	2938	3887
Right	4764	3669	6319
Center	5292	4002	3503
Posture Result (PCA)	Left	Left	Center
Action observation with Expectation Maximization (log-likelihood) in HMM			
	Frame Interval 750:760	***Frame Interval 790:800***	***Frame Interval 810:830***
Action 1 'going to bookshelf'	*-854.265832*	*-949.938566*	*-795.833412*
Action 2 'going to work-table'	*-744.633492*	*-662.080792*	*-708.119380*
Action 3 'going to coffee-machine'	*-826.277215*	*-840.526317*	*-956.222300*
Action 4 'exploring the environment'	*-910.068189*	*-707.836509*	*-828.892867*
Action Result (HMM)	***Action 2***	***Action 2***	***Action 1***
Intention			
Intention (from Table I)	*2*	*2*	*1*

Figure 13. The results for person 1

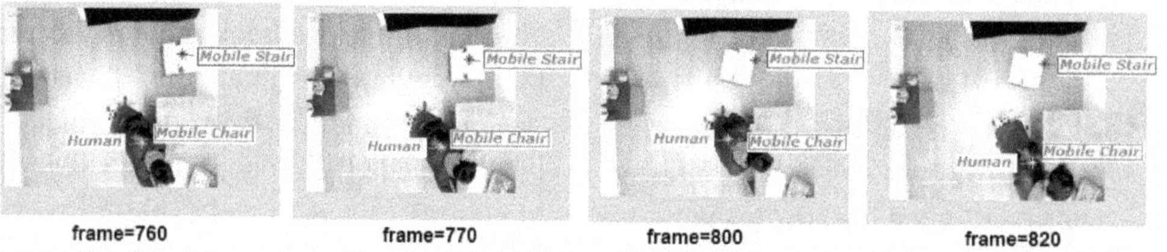

(a) Frame Captures of the experiment that show the changing the intention of the human by the movement of the mobile stair

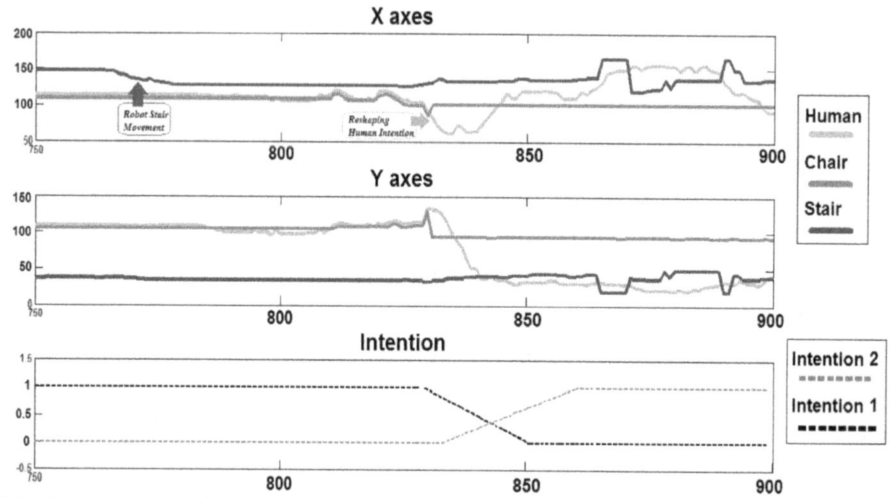

(b) (x,y) positions of the human, mobile chair and stair and the changing the intention. Intention 1 and 2 are called in Table 1.

that most of the participants not easily change the intention with movements of the robots.

6. CONCLUSION AND FUTURE DIRECTIONS

We demonstrate that changing the intention of a human agent can be achieved by interaction of autonomous robots with humans in our experimental room. Our approach analyzes a sequence of images to estimate the current intention and move the robots to change the intention to new intentions. To the best of our knowledge, is the only study that analyzes, classifies and changes intentions by learning of human motion patterns. In the future, we will increase the set of intentions from four to many that occurs in general scenarios. Our motion analysis uses a limited set of features; we will incorporate more complex feature sets as and models such as bag of words representation of optical flows. In order to understand the psychological process underlying the changes in the intentions, we will perform experiments with extended temporal length. The extended experiments will let us understand if the intentions change due to surprise of an "unknown" robot motion or change due to robots reminding the humans of other possible actions they can take. We will also look into generating a real time system that analyzes activities as they happen. We strongly believe that this study will foster new research on different aspects of human-robot interaction. In addition, the system will find applications in different problem domains including

Figure 14. The results for person 2

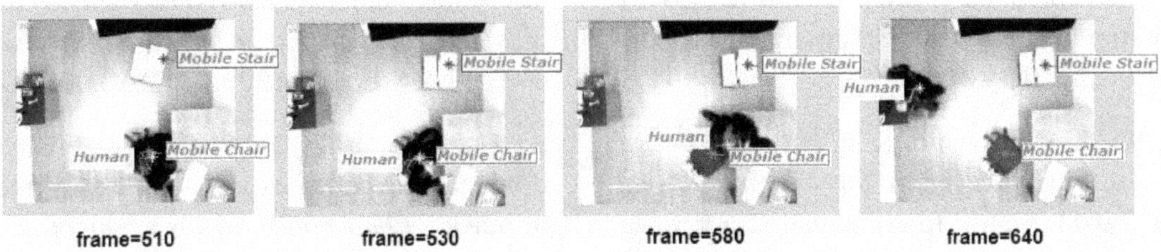

(a) Frame Captures of the experiment that show the changing the intention of the human by the movement of the mobile stair

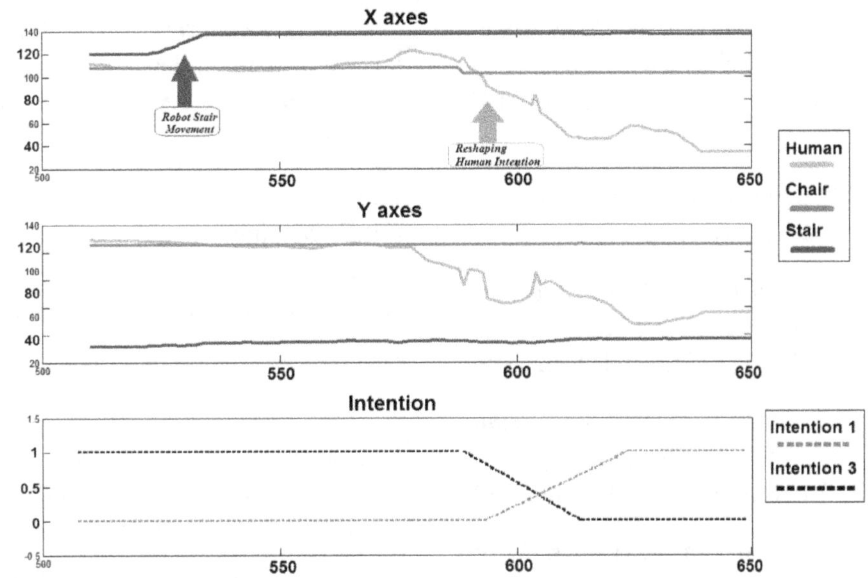

(b) (x,y) positions of the human, mobile chair and stair and the changing the intention. Intention 1 and 3 are called in Table 1.

surveillance, information services and assisted services. Especially, attracting the interest of elderly suffering from Alzheimer or infants interacting in a daycare setting may eliminate potential dangers by changing their intention, which may result in hurting themselves. Another example would be to alter intention of foul-minded individuals from performing illegal activities. In this case, robots may be modeled in different shapes and purposes may be added to security measures by using intention reshaping. Last but not least, intention reshaping can be used for training animals or chastening the inmates a prison environment. More examples can be given for this new research area.

REFERENCES

Aarno, D. (2007). *Intention recognition in human-machine collaborative systems*. Licentiate Thesis at the School of Computer Science and Communication, KTH.

Aarno, D., & Kragic, D. (2006). *Layered HMM for motion intention recognition.* In IEEE/RSJ International Conference on Intelligent Robots and Systems, IROS'06.

Arai, K., & Mardiyanto, R. (2011). *Eye based HCI with moving keyboard for reducing fatigue effects.* 2011 Eighth International Conference on Information Technology: New Generations.

Baum, L. E., Petrie, T., Soules, G., & Weiss, N. (1970). A maximization technique occurring in the statistical analysis of probabilistic functions of Markov chains. *Annals of Mathematical Statistics, 41*(1), 164–171. doi:10.1214/aoms/1177697196

Bratman, M. (1999). *Intention, plans, and practical reason.* Harvard University Press.

Bratman, M. (1999). *Faces of intention: Selected essays on intention and agency.* Cambridge University Press. doi:10.1017/CBO9780511625190

Charniak, E., & Goldman, R. (1993). A Bayesian model of plan recognition. *Artificial Intelligence, 64*, 53–79. doi:10.1016/0004-3702(93)90060-O

Chouchourelou, A., Matsuka, T., Harber, K., & Shiffrar, M. (2006). The visual analysis of emotional actions. *Social Neuroscience, 1*, 63–74. doi:10.1080/17470910600630599

Clarke, T. J., Bradshaw, M. F., Field, D. T., Hampson, S. E., & Rose, D. (2005). The perception of emotion from body movement in point-light displays of interpersonal dialogue. *Perception, 34*, 1171–1180. doi:10.1068/p5203

Cutting, J. E., & Kozlowski, L. T. (1977). Recognizing friends by their walk-gait perception without familiarity cues. *Bulletin of the Psychonomic Society, 9*, 353–356.

Daprati, E., Wriessnegger, S., & Lacquaniti, F. (2007). Kinematic cues and recognition of self-generated actions. *Experimental Brain Research, 177*, 31–44. doi:10.1007/s00221-006-0646-9

De Gelder, B. (2009). Why bodies? Twelve reasons for including bodily expressions in affective neuroscience. *Philosophical Transactions of the Royal Society B. Biological Sciences, 364*, 3475–3484. doi:10.1098/rstb.2009.0190

De Lange, F. P., Spronk, M., Willems, R. M., Toni, I., & Bekkering, H. (2008). Complementary systems for understanding action intentions. *Current Biology, 18*, 454–457. doi:10.1016/j.cub.2008.02.057

Dempster, A. P., Laird, N. M., & Rubin, D. B. (1977). Maximum likelihood from incomplete data via the EM algorithm. *Journal of the Royal Statistical Society. Series B. Methodological, 39*(1), 1–38.

Dennett, D. C. (1987). *The intentional stance.* MIT Press.

Dielmann, A., & Renals, S. (2004). Dynamic Bayesian networks for meeting structuring. In *Proceedings of the IEEE International Conference on Acoustics, Speech, and Signal Processing*, (pp. 629–632).

Dietterich, T. G. (2002). Machine learning for sequential data: A review. *Lecture Notes in Computer Science, 2396*, 15–30. doi:10.1007/3-540-70659-3_2

Dittrich, W. H., Troscianko, T., Lea, S. E. G., & Morgan, D. (1996). Perception of emotion from dynamic point-light displays represented in dance. *Perception, 25*, 727–738. doi:10.1068/p250727

Fogassi, L., Ferrari, P. F., Chersi, F., Gesierich, B., Rozzi, S., & Rizzolatti, G. (2005). Parietal lobe: From action organization to intention understanding. *Science, 308*, 662–667. doi:10.1126/science.1106138

Grezes, J., Frith, C., & Passingham, R. E. (2004). Brain mechanisms for inferring deceit in the actions of others. *The Journal of Neuroscience, 24*, 5500–5505. doi:10.1523/JNEUROSCI.0219-04.2004

Grimson, W. E. L., Stauffer, C., Romano, R., & Lee, L. (1998). Using adaptive tracking to classify and monitor activities in a site. In *IEEE Conference on Computer Vision and Pattern Recognition*, (pp. 22–29).

Haggard, P., Clark, S., & Kalogeras, J. (2002). Voluntary action and conscious awareness. *Nature Neuroscience, 5*, 382–385. doi:10.1038/nn827

Iacoboni, M., Molnar-Szakacs, I., Gallese, V., Buccino, G., Mazziotta, J. C., & Rizzolatti, G. (2005). Grasping the intentions of others with one's own mirror neuron system. *PLoS Biology, 3*, 529–535. doi:10.1371/journal.pbio.0030079

Jolliffe, I. T. (1986). *Principal component analysis* (p. 487). Springer-Verlag.

Kautz, H. A., & Allan, J. (1986). Generalized plan recognition. In *Fifth National Conference on Artificial Intelligence*, Philadelphia, PA, (pp. 32–38).

Kelley, R., Nicolescu, M., Tavakkoli, A., Nicolescu, M., King, C., & Bebis, G. (2008). *Understanding human intentions via hidden Markov Models in autonomous mobile robots.* HRI'08, March 12–15, 2008, Amsterdam, Netherlands.

Kiran, M., Kin, L. W., & Ali, K. K. H. (2009). Clustering techniques for human posture recognition: K-means, FCM and SOM. *Proceedings of the 9th WSEAS International Conference on Signal, Speech and Image Processing, and 9th WSEAS International Conference on Multimedia, Internet & Video Technologies (SSIP '09/MIV'09).*

Knoblich, G., & Prinz, W. (2001). Recognition of self-generated actions from kinematic displays of drawing. *Journal of Experimental Psychology. Human Perception and Performance, 27*, 456–465. doi:10.1037/0096-1523.27.2.456

Kohler, E., Keysers, C., Umilta, M. A., Fogassi, L., Gallese, V., & Rizzolatti, G. (2002). Hearing sounds, understanding actions: Action representation in mirror neurons. *Science, 297*, 846–848. doi:10.1126/science.1070311

Koo, S., & Kwon, D. S. (2009). Recognizing human intentional actions from the relative movements between human and robot. *The 18th IEEE International Symposium on Robot and Human Interactive Communication, RO-MAN 2009*, (pp. 939-944).

Lee, K. K., & Xu, Y. (2004). Modeling human actions from learning. In *Proceedings of the IEEE/RSJ International Conference on Intelligent Robots and Systems (IROS '04)*, (vol. 3, pp. 2787- 2792).

Lin, H. C., Shafran, I., Murphy, T. E., Okamura, A. M., Yuh, D. D., & Hager, G. D. (2005). Automatic Detection and Segmentation of Robot-Assisted Surgical Motions. In *Proceedings of the International Conference on Medical Image Computing and Computer-Assisted Intervention*, (pp. 802–810).

Loula, F., Prasad, S., Harber, K., & Shiffrar, M. (2005). Recognizing people from their movement. *Journal of Experimental Psychology. Human Perception and Performance, 31*, 210–220. doi:10.1037/0096-1523.31.1.210

Manera, V., Schouten, B., Becchio, C., Bara, B. G., & Verfaillie, K. (2010). Inferring intentions from biological motion: A stimulus set of point-light communicative interactions. *Behavior Research Methods, 42*, 168–178. doi:10.3758/BRM.42.1.168

Meltzoff, A. N. (1995). Understanding the intentions of others: Reenactment of intended acts by 18-month-old children. *Developmental Psychology, 31*(5), 1–16. doi:10.1037/0012-1649.31.5.838

Miyake, T., Matsumoto, T., Imamura, T., & Zhang, Z. E. (2011). Estimation of facial expression from its change in time. *ICIC Express Letters. Part B: Applications, 2*(3), 641–645.

Mori, T., Segawa, Y., Shimosaka, M., & Sato, T. (2004). Hierarchical recognition of daily human actions based on continuous hidden markov models. In *Proceedings of the IEEE International Conference on Automatic Face and Gesture Recognition*, (pp. 779–784).

Nakauchi, Y., Noguchi, K., Somwong, P., Matsubara, T., & Namatame, A. (2003). *Vivid room: Human intention detection and activity support environment for ubiquitous autonomy*. Intelligent Robots and Systems (IROS 2003).

Noguchi, K., Somwong, P., Matsubara, T., & Nakauchi, Y. (2003). Human intention detection and activity support system for ubiquitous autonomy. *Proceedings of IEEE International Symposium on Computational Intelligence in Robotics and Automation, 2003*, 906–911. doi:10.1109/CIRA.2003.1222300

Pynadath, D. (1999). *Probabilistic grammars for plan recognition*. Doctoral Thesis, the University of Michigan, MI.

Qi, Y., Wang, Z. L., & Huang, Y. (2007). A non-contact eye-gaze trackıng system for human computer interaction. *Proceedings of the 2007 International Conference on Wavelet Analysis and Pattern Recognition*, Beijing, China.

Rabiner, L. R. (1989). A tutorial on hidden Markov models and selected applications in speech recognition. *Proceedings of the IEEE*, *77*(2), 257–286. doi:10.1109/5.18626

Roether, C. L., Omlor, L., Christensen, A., & Giese, M. (2009). Critical features for the perception of emotion from gait. *Journal of Vision (Charlottesville, Va.)*, *9*, 1–32. doi:10.1167/9.6.15

Roichman, E., Solomon, Y., & Moshe, Y. (2008). Real-time pedestrian detection and tracking. *Proceedings of the 3rd European DSP Education and Research Symposium (EDERS 2008)*, Tel-Aviv, Israel, June 2008, (pp. 281-288).

Runeson, S., & Frykholm, G. (1981). Visual perception of lifted weight. *Journal of Experimental Psychology. Human Perception and Performance*, *7*, 733–740. doi:10.1037/0096-1523.7.4.733

Russell, S. J., & Norvig, P. (2003). *Artificial intelligence: A modern approach*, 2nd ed. Prentice Hall series in Artificial Intelligence, Prentice Hall.

Schmidt, S., & Färber, B. (2009). Pedestrians at the kerb – Recognising the action intentions of humans. *Transportation Research Part F: Traffic Psychology and Behaviour*, *12*, 300–310. doi:10.1016/j.trf.2009.02.003

Schrempf, O. C., Albrecht, D., & Hanebeck, U. D. (2007). Tractable probabilistic models for intention recognition based on expert knowledge. *Proceedings of the 2007 IEEE/RSJ International Conference on Intelligent Robots and Systems*, San Diego, CA, USA, (pp. 1429-1434).

Sebanz, N., & Shiffrar, M. (2009). Detecting deception in a bluffing body: the role of expertise. *Psychonomic Bulletin & Review*, *16*, 170–175. doi:10.3758/PBR.16.1.170

Sevdalis, V., & Keller, P. E. (2009). Self-recognition in the perception of actions performed in synchrony with music. *Annals of the New York Academy of Sciences*, *1169*, 499–502. doi:10.1111/j.1749-6632.2009.04773.x

Sevdalis, V., & Keller, P. E. (2010). Cues for self-recognition in point-light displays of actions performed in synchrony with music. *Consciousness and Cognition*, *19*(2), 617–626. doi:10.1016/j.concog.2010.03.017

Shimosaka, M., Mori, T., Harada, T., & Sato, T. (2005). Marginalized bags of vectors kernels on switching linear dynamics for online action recognition. In *Proceedings of the IEEE International Conference on Robotics and Automation*, (pp. 3072 – 3077).

Tahboub, K. A. (2005). Compliant human-robot cooperation based on intention recognition. *Proceedings of the 2005 IEEE International Symposium on Intelligent Control*, Limassol, Cyprus, June 27-29, (pp. 1417-1422).

Tahboub, K. A. (2006). Intelligent human-machine interaction based on dynamic Bayesian networks probabilistic intention recognition. *Journal of Intelligent & Robotic Systems*, *45*(1), 31–52. doi:10.1007/s10846-005-9018-0

Tahir, N. M., Hussain, A., Samad, S. A., & Husain, H. (2008). *Posture recognition using correlation filter classifier. Journal of Theoretical and Applied Information Technology (JATIT)*. EBSCO Publishing USA.

Terada, K., & Ito, A. (2010). Can a robot deceive humans? *HRI*, *2010*, 191–192. doi:10.1145/1734454.1734538

Terada, K., Shamoto, T., Mei, H., & Ito, A. (2007). Reactive movements of non-humanoid robots cause intention attribution in humans, *Proceedings of the 2007 IEEE/RSJ International Conference on Intelligent Robots and Systems*, San Diego, CA, USA, (pp. 3715-3720).

Turk, M., & Pentland, A. (1991). Face recognition using eigenfaces. *Proceedings of Computer Vision and Pattern Recognition.*

Umilta, M. A., Kohler, E., Gallese, V., Fogassi, L., Fadiga, L., & Keysers, C. (2001). I know what you are doing: A neurophysiological study. *Neuron*, *32*, 91–101.

Viterbi, A. J. (1967). Error bounds for convolutional codes and an asymptotically optimum decoding algorithm. *IEEE Transactions on Information Theory*, *13*(2), 260–269. doi:10.1109/TIT.1967.1054010

Vlasenko, B., & Wendemuth, A. (2009). *Heading toward to the natural way of human-machine interaction: The Nimitek project*. Multimedia and Expo, 2009. ICME 2009.

Webb, T. L., & Sheeran, P. (2006). Does changing behavioral intentions engender behavior change? A meta-analysis of the experimental evidence. *Psychological Bulletin*, *132*(2), 249–268. doi:10.1037/0033-2909.132.2.249

Wilensky, R. (1983). *Planning and understanding.* Reading, MA: Addison-Wesley.

Williams, O. M. C. (2005). *Bayesian learning for efficient visual inference*. The dissertation for the degree of Doctor of Philosophy.

Yilmaz, A., Javed, O., & Shah, M. (2006). Object tracking: A survey. *ACM Computing Surveys*, *38*(4), 13. doi:10.1145/1177352.1177355

Yokoyama, A., & Omori, T. (2010). *Modeling of human intention estimation process in social interaction scene*. 2010 IEEE International Conference on Fuzzy Systems (FUZZ).

Zhang, D., Gatica-Perez, D., Bengio, S., McCowan, I., & Lathoud, G. (2004). Modeling individual and group actions in meetings: A two-layer HMM framework. In *Proceedings of the IEEE Computer Society Conference on Computer Vision and Pattern Recognition Workshop*, (p. 117).

KEY TERMS AND DEFINITIONS

Autonomous Robots: Different or the same type of robots are found in the same environment for some tasks without any communication with each other.

Changing Human Intention: Depending on the human's ideas and the changes in your environment, people may change intentions at any instant.

Human-Robot Interactions: Humans and robots interact with each other for cooperation which can help each other.

Intention Estimation: Predicting intention of any agents such as robots or human according to their physical and psychological conditions.

Reshaping Human Intentions: Human may change his/her intention into another one with respect to physical environment and psychological state of mind.

Chapter 14
A Framework for Prototyping of Autonomous Multi-Robot Systems for Search, Rescue, and Reconnaissance

Sedat Dogru
Middle East Technical University, Turkey

Sebahattin Topal
Middle East Technical University, Turkey

Aydan M. Erkmen
Middle East Technical University, Turkey

Ismet Erkmen
Middle East Technical University, Turkey

ABSTRACT

Robots consistently help humans in dangerous and complex tasks by providing information about, and executing tasks in disaster areas that are highly unstructured, uncertain, possibly hostile, and sometimes not reachable to humans directly. Prototyping autonomous multi-robot systems in disaster scenarios both as hardware platforms and software can provide foundational infrastructure in comparing performance of different methodologies developed for search, rescue, monitoring and reconnaissance. In this chapter, the authors discuss prototyping modules of heterogeneous multi-robot networks and their design characteristics for two different scenarios, namely Search and Rescue in unstructured complex environments, and connectivity maintenance in Sycophant Wireless Sensor Networks which are static ecto-parasitic clandestine sensor networks mounted incognito on mobile agents using only the agent's mobility without intervention, and are cooperating with sparse mobile robot sensor networks.

DOI: 10.4018/978-1-4666-0176-5.ch014

1. INTRODUCTION

1.1 Multi-Robot Systems for Search and Rescue

Natural or manmade disasters yield catastrophe areas with many different characteristics such as variability of disaster from earthquakes, fires or terrorist attacks. Among these various characteristics, a common characteristic is that all disaster areas are highly complex, uncertain, vast in comparison to shortness of exploration time, completely unknown, and hardly measurable for search and reconnaissance tasks. During search and rescue among rubble of a disaster area, the location of buried victims must be determined as quickly as possible and the environmental conditions for reaching victims must be determined as completely as possible so that rescuers can enter the disaster area to reach and save the life of survivors. Search and rescue teams made of living beings have the important disadvantage of physical limits such as in efforts leading to exhaustion. Moreover catastrophe areas are generally hostile to humans, bearing the risk of local collapse due to unstable structures. Hence, fast and efficient search is critical for rescuing victims from their possible trapped position within the debris.

In the last decade, technological developments on robotic hardware, software, control and communication abilities caused an increase in the use of autonomous robots in search and rescue operations, although they remain still highly inefficient in real world applications. To increase the performance of the robots in the sense of time, energy and task accomplishment, a huge number of researchers have focused on the development of efficient multi-robot search and rescue systems rather than single robot ones. Because, disaster environments are generally highly complex, uncertain and unstructured, heterogeneous robot teams which consist of multiple robots with different capabilities, equipped with various sensors, actuators, computational abilities and control algorithms ranging from classical to intelligent and even bio-inspired, increase the mobility of the rescue team within debris. Hence, probability of recovering victims from rubble in a safer manner is increasing and possibility of injury during rescue is decreasing. For search and rescue in disaster environments using heterogeneous robot networks, global tasks have to be divided into small ones and subtasks should be assigned to each robot according to robots' capabilities in order to increase the efficiency in the sense of time and energy. Snake like robots for example can infiltrate among the most hectic rubble, while heavy stones and obstacles can be pulled aside by powerful robots equipped with proper end effectors. Besides, mobile robots with different sensors can generate map of debris in order to compile the inventory of the disaster and detect passages for other mobile robots to navigate further for victim search rescue and/or monitoring. On the other hand, a robot team as a whole has to be capable of decentralized decision making. In centralized control approaches, each robot unit sends collected information from the environment to the leader to carry out decision making. The one robot that works as a leader, makes all navigation plans to accomplish a given task based on data collected from each robot unit. Centralized approaches are not realistic for search and rescue operations especially because of communication burden, because all information obtained from the disaster environment has to be transmitted to the leader robot and final control decisions have to be sent back to each robot for individual execution. And also it makes response of the system very slow and inaccurate. However, decentralized heterogeneous search and rescue teams are systems least affected by communication failure or imperfect communication and computational approximation errors; and also are more robust against failure of one of the team members which can occur frequently while operating in harsh catastrophe environments.

During a search and rescue operation, communication maintenance between robot units is crucial to prevent robots from getting trapped or to convey information about the internal structure of the disaster environment and relay survivor data to a base station. Hence, connectivity maintenance algorithms have to be developed for multi robot teams.

However some challenging problems still remain unresolved because of the following source of uncertainties. Some of them are about limited and noisy sensor measurements, unpredictable environment conditions, mechanical failures and some uncertainties caused by algorithmic approximations. These limitations make search and rescue operation difficult using multi-robot systems.

Above, we mentioned open questions and challenges in the field of search and rescue using multi robots some of which are Simultaneously Localization and Mapping (SLAM), and goal oriented exploration. Physical constructions of robot team members as well as Sycophant Wireless Sensor Networks are also mentioned as requirements for hardware prototyping in this research field. Hence, we will focus on those topics as the main modules of the chapter in order to provide a spectrum of multiple prototypical tools and concept gathered from many sources in the literature that will yield rapid and efficient multi-robot systems for search and rescue.

2. HARDWARE PROTOTYPING FOR MULTI-ROBOT SEARCH AND RESCUE TEAMS

Physical construction of heterogeneous robot networks equipped with different mobility abilities are very important for the efficiency of the system, because some robots will be required to navigate through small voids and crowded rubble while others will be required to move big and heavy debris in order to open space for other robots allowing them act safely and avoid or ease cases of entrapment. Hence multi-robot teams need to be heterogeneous; some of the robots should be small and have high mobility while others should be powerful for manipulation purposes and all units should be robust to mechanical failure. Small robots could be used for debris exploration making use of their high maneuvering capabilities. Powerful robots, on the other hand, could be used for treating heavy debris, opening passages and carrying victims out of collapsed buildings.

Rescue robots are either autonomous or human operated (Guarnieri et al., 2004; Tanaka et al., 2005; Mori et al., 2006). The Helios VII arm-equipped tracked vehicle is a simple yet robust robot developed by Guarnieri et al. (2004) to explore debris in search and rescue operations. It has a mounted arm that assists motion of the robot and which can also be used for manipulation. The robot navigates through very harsh environments with the help of its tracks. Turned upside down, it is able to flip over using the mounted arm and so it gets usually out of entrapment. Tanaka et al. (2005) report a high power rescue robot that can move big obstacles and carry a victim out of debris as soon as possible using a high-pressure hydraulic actuator. After localization of victims by a small robot with high maneuvering capability, powerful robots carry the survivor out of the debris or help human beings in the transportation of heavy obstacles.

Networked heterogeneous robot teams were constructed by Sato et al. (2004) to cooperatively control each other for localization and rescue within the time critical period. These heterogeneous rescue teams are composed of three kinds of robots, each equipped with different abilities: MA-I is track type inspired from tanks and is remotely controlled by an operator over a radio signal, IGA is track type with flippers, which gives the ability of navigation in rough terrain, and KOHGA is a snake-like robot. Due to disaster variability of natural or manmade catastrophes, differently structured robots have to be used to improve team mobility for example with the help of the robots' flippers, or snake-like motion within earthquake rubble. Moreover, when a robot is trapped within rubble and cannot recover its mobility, other robot team members can physically support, and help it resume its operations. This robustness in mobility based on cooperation between robots of different capability and dynamics improves significantly safety and efficiency in navigation within challenging unstructured environments.

Task execution and penetrability can be very difficult due to structural characteristics of disaster environments. Hence, networked multi-robot systems have to handle these difficulties, not only by using information sharing but also by reshaping their morphology and passing some components to each other; this is called force cooperation in the literature. Such networked robots are called reconfigurable robots and can execute complex adaptable tasks that the information-only coordinated robot team with fixed structure cannot execute (Schenker et al., 2000; Wang et al.,2008). Reconfiguration capability provides flexibility to the robot team when accomplishing difficult tasks in urban search and rescue missions such as transportation of victims in the catastrophe, helping other failed robots in the team to recover and taking over seamlessly the task of failed units (Schenker et al., 2000). Wang et al. (2008) investigated force cooperation between networked robots to enhance mobility of the team. The proposed networked robot system can dock and adjust each team member's position and orientation according to other robots' states to easily execute a given task using force cooperation. Hard environmental conditions such as hills and being trapped in a trench are handled by using this docking skill.

Despite these advances in multi-robot hardware, one still requires ample time projected into the future for fully operational and standardized Search and Rescue (SAR) robots to appear steadily on the market. Meanwhile multi-robot networking in SAR applications has opened to emerging technologies such as wearable sensor network units. We will now treat this issue and new abilities and challenges gained in the research field. In this chapter we propose an alternative sensor subnetwork as new conceptual architecture. The proposed architecture consists of living beings or autonomous vehicles carrying wireless search and rescue sensors, like camera or other properly selected environmental sensors to acquire necessary data from a given disaster environment cooperating with mobile robots that follow these carrying agents in order to maintain a continuous communication link. This architecture had previously been developed for efficient deployment and coverage in an energy saving mode in surveillance using Wireless Sensor Network (WSN) by the authors (Dogru, Erkmen, and Erkmen, 2009). The sensor nodes deployed on the carrying agents are called sycophants and their network is called a Sycophant Sensor Network. We find necessary here to clarify the term sycophant by stating the characteristics that the sensor network possesses: a Sycophant Wireless Sensors (SWS) network is a clandestine traveler that passively uses without hostility the mobility of the carrying agent, to collect, process and communicate data harvested during the agent navigation. But it cannot in any circumstance guide or influence the motion of the carrying agent: this agent performs its navigation for its own tasks without the knowledge of the onboard "clandestine" SWS network. Depending on the number of carrying agents transporting each, SWS subnetworks, various problems of communication between SWS subnets may arise; and difficulties in seamless coverage of a certain area may occur since no guidance can be provided to the carrying agents by SWS subnets. We realize connectivity in communication between SWS nets using mobile units of a sparse mobile robot sensor networks.

In a search operation within a collapsed building the carrying agent could be a network of educated mice, each equipped with a small wireless camera, traveling easily in the labyrinths of the rubble, passing through small holes or swimming whenever needed and reaching into larger spaces where the needy can be found. Since the SWS network cannot guide its carrying agents, upon whom it is clandestinely hooked, the SWS can convey its collected data only to a mobile tracking unit, following the SWS carrying agent. Thus, the SWS convey its collected information to the search and rescue base station through the tracking unit of heterogeneous mobile robot network. This robot network's job is not only to act as relays to a base station, but also actively participate in search and rescue through environment inspection and mapping. Note that the transmitters on the mice will probably be small and not have the chance to transmit the collected information directly to a base station both due to distance and due to irregularities of rubble in the environment.

Such a hybrid network consisting of sycophant wireless networks and mobile robot networks could also be deployed on firemen as they try to rescue in a building, or on search and rescue team members rescuing trapped mountain climbers under difficult environmental conditions like after avalanche and landslides.

Wearable WSN are now appearing in the literature (Aylward, Lovell, & Paradiso, 2006; Aylward & Paradiso, 2007; Ye, Xu, & Lee, 2005; Mizuno et al., 2007; Kroc & Delic, 2003) but contrary to our conceptual approach all are there to get human centered data from the human himself/herself and his/her environment under his/her knowledge.

Moreover, in wearable WSN, the assumptions made do not fit the SAR requirements: in many research approaches, it is assumed that the sinks, which are the final destinations for the data collected, are within the wireless communication range so that they are connected (Volgyesi, Balogh, Nadas, Nash, & Ledeczi, 2007; Aylward et al., 2006; Aylward & Paradiso, 2007), or it is assumed that in systems such as on-transporter deployment architectures, the static network has an infinite energy source and the wireless sensor network is able to connect to far nodes like satellites (Mahlknecht & Madani, 2007). Consequently such assumptions cannot be made for our SWS subnets due to their nature of being placed incognito on different carrying agents: aerial, underwater or land vehicles as well as a human being. Here SWS subnets are sycophants, thus needing to use abilities of the carrier such as mobility. Therefore they do not posses infinite energy and since they cannot guide the carrier, communication with a station can be frequently and easily broken. We will deal with these issues in the forthcoming paragraphs.

Volgyesi et al. (2007) present a WSN based mobile counter-sniper system consisting of sensor nodes placed on the helmet(s) of solder(s) and tripods getting human centered data rather than from the environment. The sensor node consists of a microphone array, a custom FPGA based sensor-board for acoustic processing of time and angle of arrivals, and a MICAz ("Crossbow Technology: IRIS 2.4GHz," n.d.) mote for inter-node communication. During the test, soldiers wearing the wireless sensor nodes did not move and the system was able to estimate the trajectories of the shots ranging from 50-300m with a 96% success rate using 10 sensor nodes. However the success rate for single sensor nodes was 42%.

Aylward et al. (2006) and Aylward & Paradiso (2007) describe a design of custom made high communication rate wearable WSN capturing expressive gestures of dance ensembles in real time, thus, gathering dancer centric data. In their design, a sensor node consisted of three orthogonal gyroscopes and accelerometers used for capturing load dynamics together with capacitive sensors used to measure node to node distances. The sensor nodes can be placed on various parts of the body, like arms or shoes or an accessory worn. Sensors are able to communicate with each other, and as a result extract information about the correlation of gesture patterns of a group of dancers. A group of three persons each having a single sensor node on their right arm was tested and simple correlations for simple arm movements were calculated.

There are also other studies that use wearable wireless sensor networks, but this time to monitor person health, identify fall detection (J. Chen, Kwong, Chang, Luk, & Bajcsy, 2005), and even to learn context-dependent personal preferences (Krause, Smailagic, & Siewiorek, 2006).

A tracking mobile WSN example which is most relevant to the SAR case, can be found in (Mahlknecht & Madani, 2007) where the authors propose deploying a WSN on containers to track containers used in transportation, based on an architecture consisting of internal monitors, container monitors and prime monitors. Internal monitors are sensor nodes monitoring the inside of the containers for temperature, light, humidity, etc. Container monitors are sensor nodes that send data from the internal monitors to prime monitors, and are placed one per container. Prime monitors are sensor nodes having global location awareness and an infinite amount of energy. They have global connectivity and are placed as a single monitor per ship/train.

As our literature survey shows, the vast majority of research works concentrate on networks collecting data related to the agent itself, where the agent is aware of the network it is carrying, and the sink is somehow reachable, either by being in the vicinity or by letting some of network nodes have a large energy source. Our seminal contribution in our research work is a SWS network where the agent (carrier) is not aware of the ecto-parasitic SWS network it carries and the SWS network collects information about the environment of the agent, in a way which is independent of the agent task.

In our work we assume that the sink is far enough from the sycophant WSN, so that no direct communication is possible, and the nodes of the sycophant network have scarce energy resources: just enough for periodic data transmission to the tracking units of the mobile robot network. Therefore, a connectivity problem may occur between SWS subnetwork and mobile unit. To solve the connectivity problem in this case, we assign dedicated units of a deployed mobile robot sensor network to follow the sycophant network and cooperate with it at an acceptable distance. This of course introduces new problems, such as assignment of tracking task to a unit or units, navigation, path planning, localization and mapping, for which solutions exist individually in the literature (Sebastian Thrun, Wolfram Burgard, & Dieter Fox, 2005; Bar-Shalom & Fortmann, 1987; Lavalle, 2006). However the hybrid network that includes mobile robot subnets working cooperatively with SWS subnets has to handle all problems at once. This means that it has to properly attack each problem, applying solutions to each, correctly without generating any interruptions of the communication and of the navigation of mobile units. Since direct line of sight may not always be possible, the mobile network has to be able to locate the sycophant WSN using estimation techniques and be able to process information received from the SWS subnets according to parameters, such as received signal strength indicator, time or angle of arrival. It also has to be able to estimate the trajectory of the sycophant network based on previous knowledge about its motion, which is actually a tracking problem. However, it should be kept in mind that depending on the host carrying the sycophant WSN, the problem may simplify if the host is a slowly moving one, but may also become more difficult if the host has high maneuvering capabilities. The hybrid network also has to successfully map its own environment which means the fusion of two different mapping abilities with two different navigation capabilities: that of SWS subnets and that of the mobile robot sensor network. This also implies modeling such collaboration of SWS nets and mobile robot network within a "hybrid" SLAM. Therefore, SWS subnets cooperating with mobile multi-robot subnetworks, forming a hybrid network provides a good hardware prototyping for SAR applications.

Hence, the mobile robot network should track a target (a SWS in our case) based on a decentralized strategy: all its nodes moving disorderly on the same line-of-sight route to the target within a complex environment may mean a quick and sad end for the mission, since the same heading strategy with which all nodes are competently moving, may lead units to get stuck in a narrow passage between obstacles of an environment. Backtracking of units after being stuck by choosing another route could mean losing the connectivity with the target which is the sycophant network. Another problem worth mentioning, which will make the tracking problem for seamless communication more difficult, is the following: the sycophant network is only able to send periodic signals due to the power scarce nature of its nodes. So localization and tracking should be solved using this sparse information, only sent periodically to any mobile unit within range of communication.

In the remainder of this chapter we will provide the control prototyping environment of sycophant subnetworks by covering the issues of SLAM for mobile robot networks together with hybrid SLAM over heterogeneous (SWS + mobile subnet) robot networks, the map merging and map growing issues over such networks, exploitation versus exploration approaches, and the connectivity maintenance between nodes or group of nodes of heterogeneous networks such as those of sycophant and mobile subnetworks, while their nodes asynchronously track each sycophant subnetwork without breaking connectivity. This chapter will conclude by prototyping issues remaining unanswered and dwell with requirements of future prototyping networks of heterogeneous robot subnetworks and their connectivity maintenance for search, rescue and reconnaissance.

3. PROTOTYPING CONTROL APPROACHES FOR SEARCH AND RESCUE MULTI-ROBOT TEAMS

3.1 Simultaneously Localization and Mapping

Map generation and exploration are important tasks for real world applications where global positioning data (GPS) is not available such as environments where satellite data transmission is occluded during forest explorations or within rubble of collapsed buildings. In simultaneously localization and mapping (SLAM), and exploration missions robots try to answer the questions of "*What does the mission environment look like?*", "*Where am I in the disaster area?*" and "*Where do I have to navigate next, to localize and rescue a buried victim?*" by using noisy range sensor measurements obtained from completely unknown environments and noisy travelled distance measurements coming from encoders. SLAM is a chicken–and–egg like problem, which makes the problem very difficult to solve, because each robot needs an accurate environment map to localize itself accurately in the work space, however robots also need their exact position and orientation to generate an accurate environment map.

A huge number of researchers have focused on the SLAM problem, but their work bears different points of view. Smith, Self and Cheeseman (1990) proposed an Extended Kalman Filter (EKF) based method for estimating the posterior distribution of robot position along with landmark positions in the work environment. It is an initial work that established the fact that there is a statistical relationship between each landmark location and observations. Assumptions made by a Kalman Filter (KF) are that noise in the system is Gaussian distributed with zero mean and the process is linear. This assumption is far from applicable to complex real world applications. Linearity constraint is relaxed by the extended version of KF, namely Extended Kalman Filter, which is used considerably for the solution of SLAM problems in the literature (Dissanayake et al., 2001; Durrant-Whyle and Bailey, 2006; Se, 2002). Although EKF based SLAM methods are historically the earliest used approaches that work well for environments in which there are limited number of features, computational complexity increases with an increasing amount of features and the number of landmarks has to be smaller than 900. Hence, computational burden dramatically increases in proportion to the number of landmarks in the environment. This characteristic together with the Gaussian assumption in the probability density function form the main shortcomings of EKF based SLAM algorithms (Montemerlo et al., 2002; Guivant and Nebot, 2001). Moreover, in EKF based SLAM approaches, the amount of uncertainty should be small, otherwise huge errors may result especially due to the linearization issue.

An alternative Particle Filter (PF) based method for SLAM problem was introduced by Montemerlo et al. (2002). Particle filter based approaches are approximations of Bayesian filters such that probability distributions are approximately quantized by a finite set of particles (Montemerlo et al., 2003). Thus, arbitrary multimodal distributions can be approximated by using a sufficient number of particles. The most advantageous characteristics of PF for SLAM problems are the abilities of handling nonlinearities and relaxation of Gaussian noise constraint. However computational complexity increases with the number of particles, which makes the real-time implementation of PF difficult in complex and hard real world applications. Hence, the FastSLAM method, which is a modified version of PF, has been proposed to handle the computational burden in classical particle filter methods making implementation on real robots possible (Montemerlo et al., 2002). This proposed method utilizes a Rao-Blackwellized representation which integrates PF and KF representations. FastSLAM uses PF to estimate robot path and each particle runs EKF to estimate map feature locations. Since each particle generates an individual map representation of the work environment, the FastSLAM algorithm suffers from memory constraints. Feng et al. (2008) improve memory efficiency of the Rao-Blackwellized FastSLAM approach using genetic algorithms.

3.2 Map Merging in Decentralized SLAM

In order to increase efficiency of SLAM algorithms, putting emphasis to better map building of the unknown work environments, multi-robot teams have been equipped with map merging capabilities (Fenwick et al., 2002). Each robot extracts a partial map of a different portion of the large environment and those partial maps are merged into a single joint map to speed up the mapping process. Decentralized SLAM problems become complex in the phase of data fusion which requires integration of different partial maps into a global one based on the cooperation of robot units.

Map merging in general, assumes that the robots know or can measure their relative poses with respect to each other at some time instants, called "rendezvous", during the mapping process. Then this knowledge is used to find either the basic map transformation or an initial guess for that transformation.

Thrun, Burgard and, Fox (2000) assume that a leader robot has the global map, and although robots initial poses are not known they start within the map of their leader, and use Monte Carlo localization to localize themselves in this map and join the mapping process. Konolige (2006) use a methodology they call zippering, where common locations between the past trajectories of robot pairs are found using a sequential algorithm, and then all scans of one map are fused to the other as if all scans were collected by a single robot. Williams (2002) let the robots create local maps which are merged into the global one at regular intervals. The transformation between the maps is found by relative robot positions if available, if not it is found by feature matching. Howard

(2006) uses multi agent SLAM where each robot starts its own SLAM succeeded by interchanging measurements. They are first able to measure each others positions and then subsequently extend their particle filter to include states of other robots as well as those of virtual pairs which are assumed to move backward in time, providing the previous measurements. The author assumes that the uncertainty in relative pose measurements is small. Although the method is a direct extension of single robot SLAM, it has some drawbacks: it requires the robots to be always within communication range during mapping; it has latency and requires the robots to move at comparable speeds. Zhou and Roumeliotis (2006) use the relative robot pose only as a first estimate of the transformation matrix and use matching landmarks as constraints to further improve the transformation. León et al. (2009) require very precise inter-robot pose measurements for occupancy grids, then use scan matching to complete the merging process. Andersson and Nygards (2009) use a set of synchronized inter-robot measurements at rendezvous points as an additional constraint to the non-linear least squares minimization problem of finding the optimum map and trajectory.

Other papers on map merging do not require any knowledge of robot poses and make use of various methods that will maximize the match between two maps. Dedeoglu and Sukhatme (2000) use heuristics to select a subset of the available landmarks and then compare properties of same type of landmarks to find the matching ones. They subsequently calculate the transformation for each pair of landmarks and choose the one maximizing the number of landmark matches. Birk and Carpin (2006) find the transformation that maximizes the match between occupancy grid maps to be an adaptive Gaussian random walk that generates random samples at each iteration, and measure the fitness of these transformations by an image similarity measure. However the method is computationally expensive. Carpin (2008) uses sampled cross correlation of Hough spectrum of the occupancy grids to find the rotation component of the transformation, then utilizes cross correlation of the X-, Y- spectra to find the x, y displacements of the transformation. Although the method is brittle when X-, Y- projections of the map are not distinctive, the method is very fast compared to the previous ones. Adluru, Latecki, Sobel, and Lakaemper (2008) use a virtual robot with local maps and the shapes of real robots as range and odometry sensors respectively. Odometry is found by structure registration.

3.3 Hybrid SLAM and Multi-agent Map Growing

Our setup is slightly different than the ones in the literature; one of the mobile agents, namely the SWS subnetwork, does not have any odometry information since it uses mobility of the carrier and thus, its pose needs to be estimated by other agents of the mobile subnetworks using only range measurements. Therefore our proposed solution needs to comply with this slightly different setup by relying upon a slightly different methodology than those existent in the literature.

In the hybrid SLAM problem that we propose as a prototyping control framework, we assume that the major map merging is done on a mobile node of the robot sensor network, receiving environmental sensory data from the sycophant sensor network (SWS) creating the very local, partial map of the SWS from the received data. The mobile tracker then fuses this map with its own and grows it by added exploration when necessary, without interrupting its tracking mission. The critical problem in such an incomplete map merging and a subsequent completion arises from the fact that the exact location of the sycophant sensor network is not known and needs to be estimated at all times through the tracker. Therefore the map alignment process has to take this uncertainty into account by using both the generated map and the range measurements. In our recent work we extend the posterior probability for multi-agent SLAM to include range measurements as well as the state and map of the SWS. So in order to join the maps created by the sycophant as well as the mobile robot we develop the following approach.

The single robot full SLAM problem is defined as finding the conditional posterior probability $p(x_{1:t}, m \mid z_{1:t}, u_{1:t})$ where *x* is the robot state, *m* is the map, *z* is the vector of sensor measurements, and *u* is the input. This posterior probability is decomposed into:

$$p(x_{1:t}, m \mid z_{1:t}, u_{1:t}) = \eta \cdot p(z_t \mid x_t, m) \cdot p(m \mid x_{1:t}, z_{1:t-1}) \cdot p(x_t \mid x_{t-1}, u_{t-1}) \cdot p(x_{1:t-1}, m \mid z_{1:t-1}, u_{1:t-1})$$

where $p(m \mid x_{1:t}, z_{1:t-1})$ and $p(x_t \mid x_{t-1}, u_{1:t-1})$ are state updates, and $p(z_t \mid x_t, m)$ is the measurement update.

The posterior probability to be used when solving a two robot SLAM problem is $p({}^1x_{1:t}, {}^2x_{1:t}, m \mid {}^1z_{1:t}, {}^2z_{1:t}, {}^1u_{1:t}, {}^2u_{1:t})$.

However, in our own scenario, there is another measurement *r*, the distance between the tracker and the SWS, which is measured using Time Difference of Arrival (TDoA) and is the only measurement between the tracker and the SWS. And also, the SWS does not have any odometry. So the posterior probability for this problem becomes

$$p({}^1x_{1:t}, {}^2x_{1:t}, m \mid {}^1z_{1:t}, {}^2z_{1:t}, {}^1u_{1:t}, r_{1:t})$$

where ${}^1x$ and ${}^2x$ belong to the mobile robot and the SWS respectively. Decomposing this posterior probability we get:

$$\begin{aligned} &p({}^1x_{1:t}, {}^2x_{1:t}, m \mid {}^1z_{1:t}, {}^2z_{1:t}, {}^1u_{1:t}, r_{1:t}) \\ &= \eta \cdot p({}^1z_t \mid {}^1x_t, m) \cdot p({}^2z_t \mid {}^2x_t, m) \\ &\cdot p(r_t \mid {}^1x_t, {}^2x_t) \cdot p(m \mid {}^1x_{1:t}, {}^2x_{1:t}, {}^1z_{1:t-1}, {}^2z_{1:t-1}) \\ &\cdot p({}^1x_t \mid {}^1x_{t-1}, {}^1u_{t-1}) \cdot p({}^2x_t \mid {}^2x_{t-1}) \\ &\cdot p({}^1x_{1:t-1}, {}^2x_{1:t-1}, m \mid {}^1z_{1:t-1}, {}^2z_{1:t-1}, {}^1u_{1:t-1}, r_{1:t-1}) \end{aligned}$$

We implement in our work, this expression by extending the grid based SLAM (Sebastian Thrun et al., 2005) algorithm to include the new terms; such as the state and measurement updates for the SWS, which are $p({}^2x_t \mid {}^2x_{t-1})$ and $p(r_t \mid {}^1x_t, {}^2x_t)$ respectively. Since in this work we do not possess any heading information due to the fact that the SWS does not have any ranger/camera attached on the Sycophant, we assume that its mapping consists of marking a rectangular area around its pose as empty. Our hybrid SLAM approach will be illustrated within the context of connectivity maintenance in the subsequent section.

3.4 Connectivity Maintenance

Connectivity maintenance is critical for search & rescue robotics as well as for Sycophant Wireless Sensor Networks. In search & rescue, the robots may lose connectivity with the base station and even with each other due to cluttered highly irregular environment and thus may not be able to convey information to the rescue station, or may even be trapped without being able to ask for help. To prevent such situations, and increase collaboration between robots, it is necessary that the robots maintain connectivity with each other, as well as with the base station. In Sycophant WSN deployments, the sycophant nodes usually do not have strong RF transceivers and it is also possible that the agent carrying the sycophant subnet wander in a difficult terrain, for example for search & rescue, penetrating of a collapsed buildings. At that point the communication with the SWS will be lost consequently. We had to propose in our research work, hybrid architecture to supplement the sycophant subnet with mobile robots, acting as relays to maintain connectivity with a base station.

Connectivity related issues have been discussed widely both in Wireless Sensor Networks, as well as in Mobile Ad Hoc Networks (MANET). In this subsection, we overview works on connectivity maintenance in mobile robot networks, and give detailed descriptions of works on leader-follower networks. Search & rescue operations that include some leading robots or sycophant subnets doing the exploration could be classified as leader-follower networks and therefore connectivity maintenance in such networks requires special attention. Connectivity maintenance in leader-follower networks could also be seen as a mobile relaying problem, due to the fact that non-leading robots act as relays.

Nguyen, Everett, Manouk, and Verma (2002) use a convoying strategy where all the relay robots convoy the leading robot as it advances in the environment until the need for a relay shows up, which is detected as the signal level dropping below a predefined threshold. At this point the last robot in the convoy stays (is deployed) at that point, and the rest of the convoy goes on following the leading robot. As the leading robot moves, RF shortcuts may be detected causing a re-localization of the robots that do not relay anymore, which happens by mobbing the robots through the path of the leading robot till they join the convoy again. The authors show that this strategy, although consuming more power compared to the case when robots are called from a base station as needed, lets the leading robot reach its destination faster, with fewer pauses.

Dynia, Kutyłowski, Lorek, and Heide (2006) propose a strategy where mobile relays are placed equidistant to each other along the line connecting the leader and the base station. The authors assume that the terrain is obstacle free. In this strategy the line is formed by each relay moving to the middle of its neighbors using a look, compute, move strategy. Connectivity is maintained by attaching a partner to each mobile relay and the leader. As soon as the leader moves too far away from its neighbors, its partner moves to the middle point between the leader and it neighbors. This strategy requires 2n robots where n is the number of relays + the leader.

Dynia, Kutylowski, Heide, and Schrieb (2007) improve their previous strategy (Dynia et al., 2006) by injecting new relays directly between their base station and the leader, removing the need for partner robots. This strategy is called Chase Explorer. In this strategy in addition to keeping close to the line connecting the leader and the base, a relay tries to maintain a predefined distance to its neighbor and to the leader. Authors also handle obstacles, by taking the relay at the obstacle point as a base station for the relaying between this relaying and the leader.

Chen, Zhuang, and Tan (2009) use a potential field consisting of repulsive and propulsive parts and received signal strength as the distance measure between the robots to place the relays. They also handle obstacles by a virtual potential field between the relays and the obstacles.

Correll, Bachrach, Vickery, and Rus (2009) propose a probabilistic approach that does not need robot localization. Assuming that the relays are able to estimate the number of their neighbors and can avoid obstacles in a bounded environment, a relay node makes Brownian motion if either the robot is not connected to the gateway or its neighbors are too much or too few. So the robots mobbing randomly and avoiding each other in a bounded environment are able to uniformly cover the environment.

Dixon and Frew (2007) propose using the signal to noise ratio of the communication links instead of the distance between the relays and find the optimum location for the relays by moving the robots, autonomous aircrafts, on a globally stable limit cycle produced by a Lyapunov Guidance Vector Field.

Stump, Jadbabaie, and Kumar (2008) solve mobile relaying in a leader-follower network indoors by trying to maximize the second largest eigenvalue (Fiedler value) of the weighted Laplacian describing the connectivity of the robot network. During this process, they also aim at keeping the hop count between any two robots below a desired value, which is done using a centralized algorithm.

Stump et al. (2008) utilize (Kim & Mesbahi, 2006), where connectivity constraints are expressed in algebraic terms as a function of the adjacency matrix, in order to maintain connectivity despite obstacles and combine both the connectivity and the obstacle constraint in a convex optimization problem solved at each step.

Tekdas and Isler (2008) solve the same problem by providing wireless network connectivity to a mobile user in mind. They utilize dynamic programming to generate relay trajectories that establish a continuous communication link between the base station and the user. The running time of the proposed algorithm is exponential, and processing is central.

Nguyen, Farrington, & Pezeshkian (2004), Pezeshkian, Nguyen, and Burmeister, (2006) and Souryal, Geissbuehler, Miller, and Moayeri, (2007) propose using relay bricks that are deployed by the leading robot (actually the only robot in their scenario), instead of mobile relays, as soon as detecting that signal strength is below a certain threshold. Nguyen et al take the idea of bricks one step further and propose a system called Automatically Deployed Communication Relays (ADCR) system, where the relay bricks are intelligent and are capable of correcting their position and opening their antenna after being dropped from the leading robot. Though relaying completely using such a deployment of relays is usually infeasible, mixing mobile deployment of static relays with mobile relays provides new opportunities.

Although not directly related to connectivity maintenance in leader-follower networks perspectives of works on other multi-agent systems such as consensus, rendezvous (Cortes, Martinez, & Bullo, 2006; Dimarogonas & Kyriakopoulos, 2007; Ji & Egerstedt, 2007), leader following (Zavlanos & Pappas, 2008), flocking and formation control (Ji & Egerstedt, 2007) also provide good insight for connectivity. Therefore a brief overview of the methods used in the literature for the aforementioned usage is appropriate. In such systems, decisions and actions are generally based on local observations and actions of a connected network, therefore research works focus upon connectivity of the whole robot network. Most research on connectivity maintenance use potential fields assuming large values for the potentials at the boundaries. Dimarogonas and Johansson (2008) take into account agent constraints and propose a solution based on bounded potential functions. Schuresko and Cortes (2008) take into account delays in communication when robots determine actions that would prevent loss of connectivity. Esposito and Dunbar (2006) still use potential functions in formation control but take also into account visibility problems caused by obstacles. Kim and Mesbahi (2006) instead of using potential functions, maximizes the second largest eigenvalue of the graph Laplacian to maintain connectivity. In Zavlanos and Pappas (2008) connections between nodes are formed and deleted dynamically using an auction algorithm, whereas other studies assume that any two nodes are connected as soon as one node is within the sensing range of another.

Another distinct class of methods uses virtual springs between robot nodes. Virtual springs approach has been previously used for connectivity management in formation control by Aizawa & Kubota (2009), Esin, Unel, and Yildiz (2008), MacArthur and Crane (2007), Reif and Wang (1995), Shucker and Bennett (2005), Urcola, Riazuelo, Lazaro, and Montano (2008). Reif and Wang (1995) use a spring based graph to reach a predefined structure. They initially adjust each spring's equilibrium length such that the desired structure is the only equilibrium point. Under local disturbances, robot nodes are found to reach the equilibrium state and the graph achieves a final desired structure. Esin et al. (2008) use virtual springs between robots to prevent collisions and keep a desired distance between them during shape formation. MacArthur and Crane (2007) use a virtual spring damper system between robots in order to maintain a wedge or vee formation when

Figure 1. Block diagram showing algorithm for motion control block ("zk" stands for measurements)

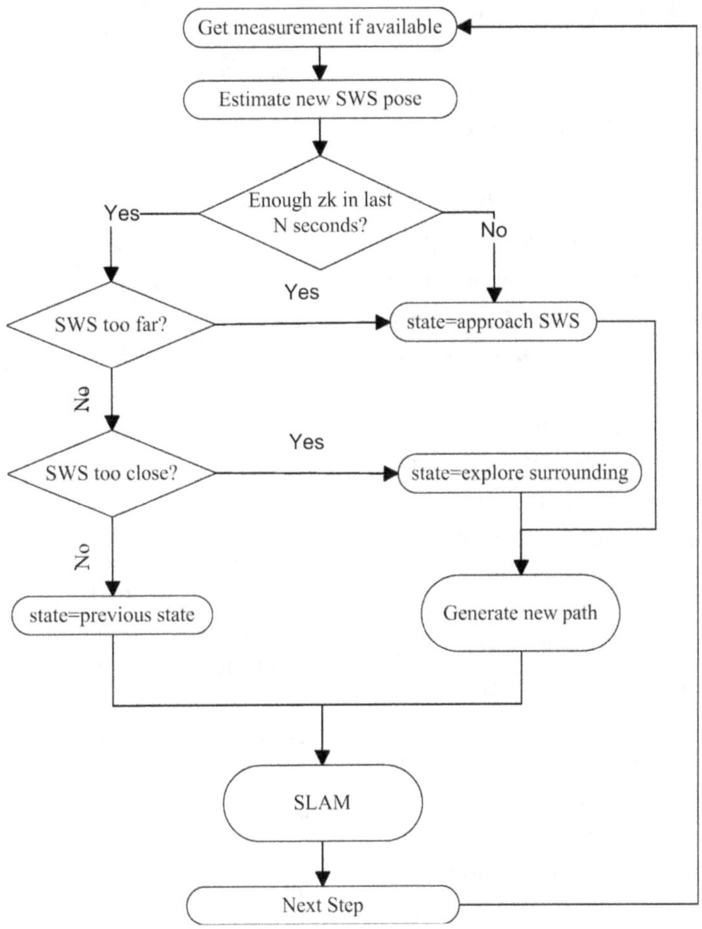

the formation is following a trajectory. Urcola et al. (2008) use linear spring-damper to maintain distance between the nodes, and torsional springs to maintain a given angle between them during leader following, allowing a deformable structure which can comply with the environment. Aizawa and Kubota (2009) use directional springs to avoid some equilibrium points during leader following in a triangular formation. Shucker and Bennett (2005) use a mesh made of spring edges between selected vertices and the robots are allowed move under the influence of forces due to these springs that provide damping for motion control during target tracking.

We end this subsection by providing our connectivity maintenance algorithm for a single leader single follower type network in a crowded environment. The leading node consists of four sycophant nodes forming a SWS mounted on a vest that a person wears. The sensors on the vest are placed at different quadrants of the vest front and back and this way we make sure that ultrasound signals that will be used for TDoA are not blocked by the human body itself and distance measurements in all directions are possible. The follower node is a MagellanPro mobile robot with rangers and odometry for mapping and wireless communication modules. TDoA is used to localize the SWS through distance measurements.

The connectivity maintenance algorithm integrates hybrid SLAM and path generation and is best explained using the flowchart of Figure 1.

In obstacle free regions, the solution is straightforward: as the signal strength decreases, the mobile node moves closer to the SWS network. Getting closer is allowed when the distance is above a predefined threshold and this allowance stops when the distance is below another predefined threshold. When the mobile robot is too close, instead of standing still, the mobile robot begins to explore its surrounding without losing contact with the SWS. This exploration increases tracking performance as well as mapping performance. Keeping the interval between thresholds large enough helps avoid high oscillations in the movement and prevents unnecessary quick maneuvers. In our implementation the lower and higher threshold values were assigned differently for different scenarios but were generally between 0.5 to 2 meters and 2 to 4 meters respectively.

Characteristic of the path followed during exploration and leader following is critical: The SWS network is localized through Multi-Agent SLAM but using only range measurements. This makes the system unobservable unless the follower robot's degree of motion equation is at least one degree higher than the targets', as is shown by Song (1999). For example, if the target is following a constant velocity trajectory, the tracker has to have acceleration. In our implementation, the follower robot dynamics was considered jerky (3rd order) in order to follow any type of acceleration/deceleration of the human wearing the SWS built-in vest.

When obstacles are present in the medium, the behavior of the tracker becomes a bit different. Assuming that range measurements are interrupted due to an obstacle in the communication path, the mobile robot does not have to go on with a complex motion to preserve observability of the system, instead it goes to the estimated position of the SWS using the shortest path in order not to lose the SWS that would cause missing measurement updates. When communication is reestablished the mobile agent resumes the standard tracking and map building operations.

Figure 2. The simulation environment used by the SWS, denoted by the triangle, and the mobile robot, denoted by a square, for the presented connectivity analysis

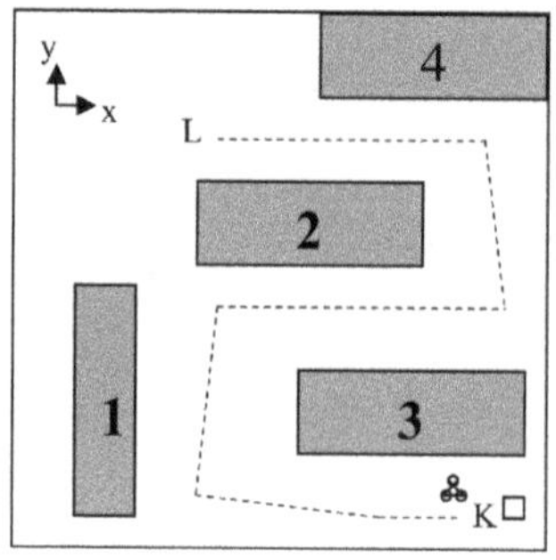

So the main motion characteristic of the connectivity algorithm is to keep the SWS always within a predefined distance range and be visible to the SWS maintaining an uninterrupted communication link, to make jerky motions for everlasting observability, and explore the immediate neighborhood when possible.

Note that, the sycophant sensor network broadcast over a large period of time in order to economize an energy, so getting a distance measurement by the tracker for every time step is not possible. This decreases the quality of target SWS state estimation.

The required jerky motion could easily be provided using Rapidly-exploring Random Trees (RRT) (Lavalle, 1998) for path generation. RRT is a randomized data structure that explores the free space by randomly selecting points and adding them to the tree and is designed for a broad class of path planning problems. The exploration is biased towards unexplored regions of the space and is able to quickly cover the empty space. RRTs are able to handle nonholonomic constraints, including dynamics, and high degrees of freedom

Here we also provide an example simulation scenario to show the effectiveness of the algorithm presented. The simulation environment for the hybrid WSN imitates our somehow cluttered lab environment (Figure 2). During the simulation the

Figure 3. Trajectories of the mobile robot, the SWS network and its estimated pose as it follows the trajectory KL. Measurement update is done every 0.64 seconds, and the state update is executed every 0.064 seconds

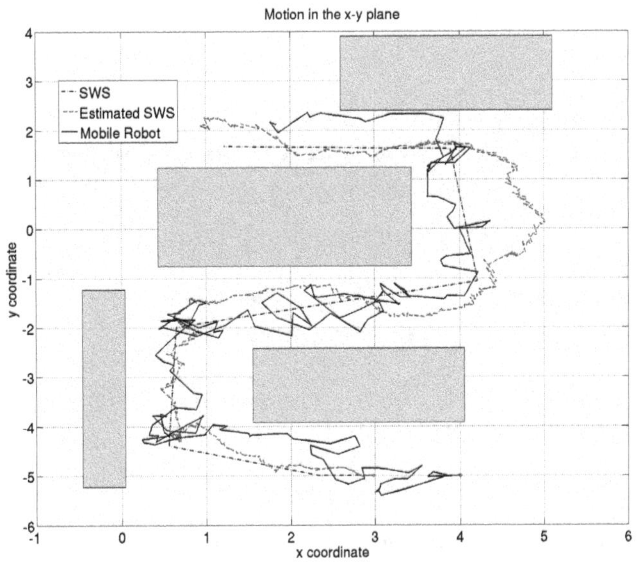

agent carrying the SWS moves along the dashed line segments KL and the mobile robot tries to keep a distance to this agent by following the given algorithm.

For a pictorial description of the scenario and the results, we provide the x-y trajectories of the mobile node and SWS subnet as well as its estimated position as obtained by one of our simulation runs (Figure 3). The time evolution of trajectories is given in Figure 4. As can be seen in the figures, the tracker follows a random trajectory which is the outcome of the RRT algorithm that guides the robot. Path of the tracker was intentionally not smoothed in order to provide the jerky motion required to make the system observable. The estimated trajectory is close to the original one in form, and the average error is 0.33 meters.

Since wireless sensor nodes do not continuously broadcast RF signals, it is critical that we run the experiments using different broadcast intervals. Therefore, this scenario was also run using different measurement update intervals in order to find a correlation between the error and measurement update intervals. The results are presented on Table 1. The first column of the table gives the measurement update interval in seconds; the second column shows the average error of 10 standard runs in meters. As the table shows, having a measurement update interval 10 times the state update interval causes larger errors as seen comparatively between first and third rows.

The results are satisfactory: the robot can properly track and maintain connectivity with the SWS using range measurements only, despite a non-continuous communication link.

Here we give preliminary results showing how hybrid SLAM improves mapping in the context of connectivity maintenance within the environment depicted in Figure 5. Here the tracker is allowed a minimum proximity with SWS of 2 meters. Although the mobile robot has the ability to fully map the environment, getting passive help from the SWS can help the mobile robot conserve energy. SWS enable mapping areas not reachable by the mobile robot but easy to infiltrate by a human (carrier of the SWS).

Figure 4. Time evolution of the trajectories given in Figure 3

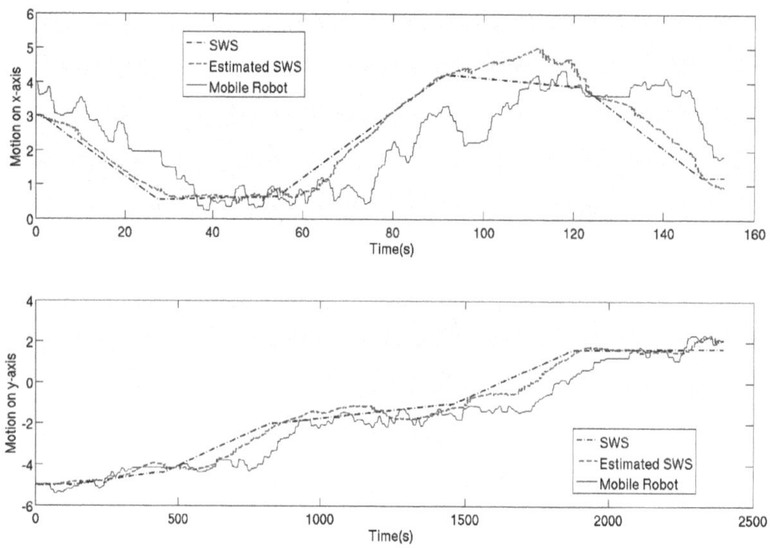

Table 1. Average error for different measurement update intervals, where the state update interval was chosen to be 0.064 seconds

Measurement Update Interval (seconds)	Average error after 20 runs (meters), Standard Deviation
0.064	0.33 0.07
0.320	0.51 0.11
0.640	0.68 0.11

Figure 5. The simulation environment used by the SWS, denoted by the triangle, and the mobile robot, denoted by a square, while perform hybrid mapping as connectivity is maintained

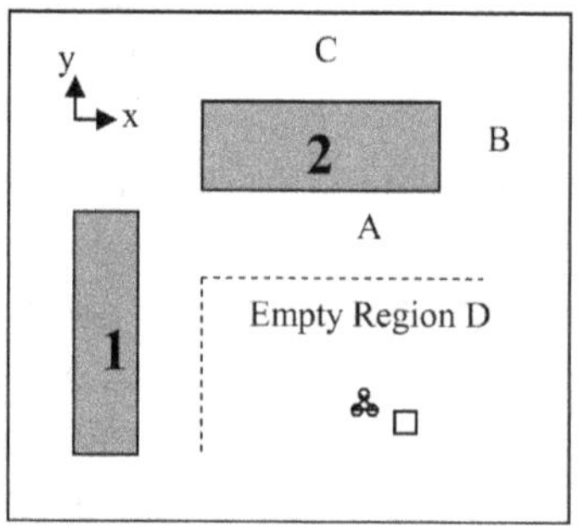

In Figure 6 the top left region is not yet explored by the mobile robot, but is already traversed and partially mapped by the SWS. This partial map is merged to the mobile robot's map, and a little later the mobile robot explores the small region left between the SWS and itself (Figure 7).

Results from another run of hybrid mapping are given in Figure 8 and Figure 9, illustrating another advantage of hybrid mapping, namely its contribution to path planning. In Figure 8, the SWS is in region B of Figure 5 and the mobile robot is still exploring the empty region due to the large proximity constraint. A few steps later, the SWS enters region C, and due to proper target tracking, its path in region C is also merged to the final map. Knowing this extra map patch, the mobile robot can easily generate a path that will lead it to the SWS, and that is obstacle free and bearing high likelihood of pointing possible directions of growth for RRTs during path planning.

Having a large hybrid WSN consisting of many mobile robots as well as many SWS networks will make advantages of hybrid mapping more obvious.

3.5 Goal Oriented Exploration

The map growing made by mobile robots tracking the SWS in our hybrid SLAM is based on aim oriented exploration. The aim is to grow the map merged from individual SWS and mobile subnetwork maps so mobile agents explore the surrounding in order to grow a given incomplete map. Aim oriented exploration is also an important task for search and rescue multi robot teams: it is the navigation of robots on the disaster environment to execute predetermined task and to obtain necessary information from the work space. Unfortunately, numerous researches have focused on single robot systems to simultaneously explore and create maps of an unknown environment in search and rescue operations (Thrun, 1993; Choset, 1996; Hahnel, 2003). The most important disadvantage of single robot usage is that it takes much more time than networked robot systems and the task execution fails in the case of damages or failures occurring in robot units. Consequently such systems are not robust. Hence coordinating and cooperating multi-robot teams have to be developed for the effectiveness of such systems. Time efficiency and robustness in distributed multi-robot system can be achieved by partitioning tasks and coordination between robot units. Coordinating multi-robot systems win over single robot approaches, dealing with limited communication range, map merging into a better global map of the workspace and better time constraints (Burgard et al., 2005; Fox et al., 2006).

To date, the major portion of coordinated exploration works has focused upon coverage of an entire area in minimum time (Amigoni, 2008; Sheng et al., 2006). These techniques are based on the selection of the best next move for each robot in order to reduce the uncertainty in the mapping of the environment. Frontier cells, which are the cells on the borders between explored and unknown space, are taken as possible next observation points. Most existing methodologies only dwell on SLAM solutions that minimize the exploration time by navigating towards nearest unexplored space by using frontier cells (Burgard et al., 2002; Koenig et al. 2001). Entropy based methodologies, which use the cost of reaching the target observation point and expected information gain, are geared towards optimal data acquisition for localization but with very limited exploration (Baglietto et al., 2002).

Very few tasks focus on guided navigation such as localizing odor source (Ferri et al., 2009; Farrell et al., 2003). However these are not suitable in search and rescue in complex disaster environments where strong odor source like carbon dioxide does not exist. Hence for SAR applications, guidance can be provided based on prior information about the environment and extracted growing map until the frontier of an explored region during navigation of an autonomous robot (Topal et al., 2009). Regions with the highest likelihood for finding victims can give a priority to the exploration, called prioritized exploration. We have developed an active and goal oriented exploration approach based on percolation guidance, where a percolator estimates the existence of connected voids in the upcoming yet unexplored region ahead of the robot, so as to increase the efficiency of the reconnaissance operation by the superior ability of the percolator guidance for speedy coverage of the area (Topal et al., 2009). There, space coverage and time optimization are mainly aimed through guidance, based on either prior knowledge, or on extra information reflecting the characteristics of the environments that leads to the estimation of connected voids in the regions to be explored. These techniques have high map coverage accuracy, while being adversely affected by errors in robot localization that would be minimized if exploitation was done. The next section will bring more details on guided exploration based on percolation.

Figure 6. Starting from left, map of the SWS, map of the robot and both maps merged, as host of the SWS and the mobile robot travel in the environment given in Figure 5

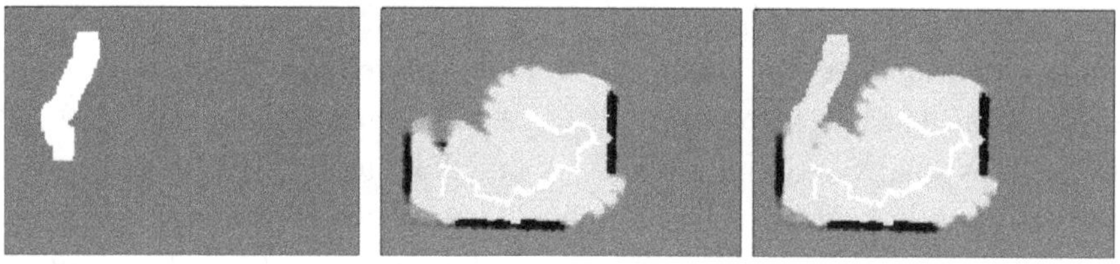

Figure 7. Starting from left, map of the SWS, map of the robot and both maps merged, after some time has elapsed on the state given in Figure 6

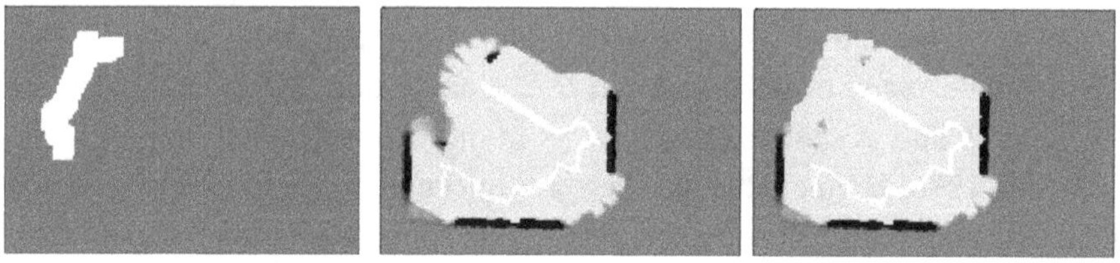

Figure 8. Starting from left, map of the SWS, map of the robot and both maps merged, as host of the SWS and the mobile robot travel in the environment given in Figure 5

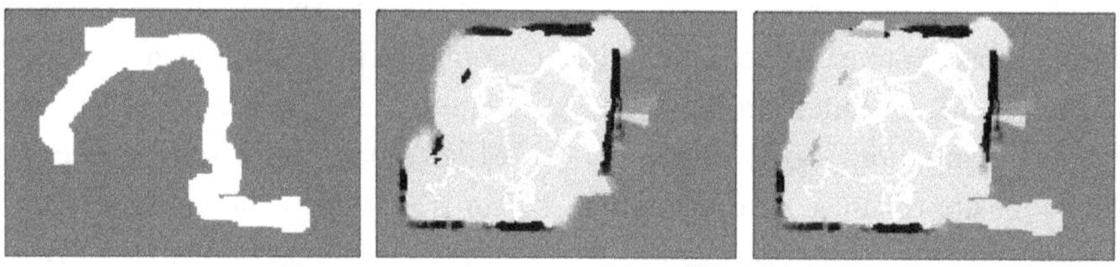

Figure 9. Starting from left, map of the SWS, map of the robot and both maps merged, as host of the SWS and the mobile robot travel in the environment given in Figure 5

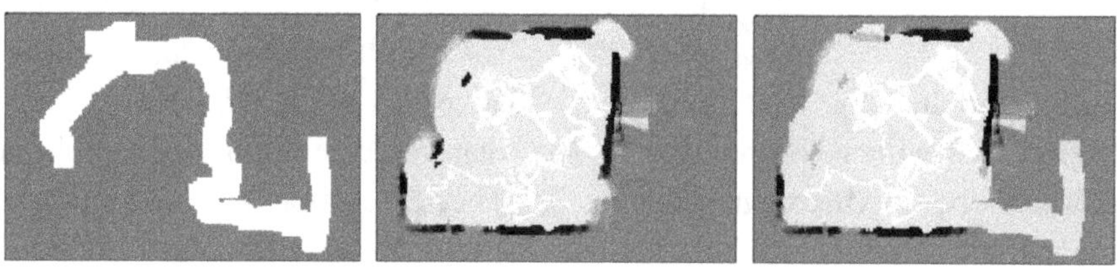

Figure 10. Bond (left) and site (right) percolation

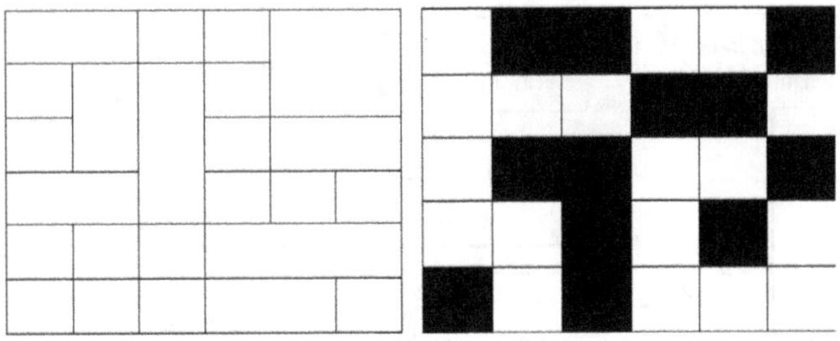

3.6 Percolation Inspired Goal Oriented Exploration

In our proposed method based on percolation inspired guidance, we consider disaster area traversability without being caught in a dead end and where no prior knowledge of victim location exists: the guidance of exploration is therefore, towards avoiding dead ends and following the most possible connected voids. Those connected voids, also called spanning clusters, and dead ends are the two local characteristics of the environment, among which our priority goes towards estimating and exploring connected voids. Percolation theory is found to be a fluidics approach to best fit our objective of robot penetrability. Percolation theory has been introduced by John M. Hammersley and Simon R. Broadbent who have focused on the behavior of fluid through a random medium such as pumice stone. They considered the following problem: if a large porous pumice stone is immersed in a bucket of water, would the water reach the centre of the pumice stone? They showed that there is a value of the density of porosity in the stone enabling a fluid to reach the pumice stone centre (Grimmert, 1999; Hunt and Ewing, 2009). In the literature, percolation model is applied to describe the dynamical behavior of the spreading epidemics and forest fires (Loehle et al., 1996), air traffic control simulations (Amor et al., 2006) and image processing (Yamaguchi, 2006), but extremely few applications exist in robotics. Generally, in percolation theory, media is described based on two types of lattices where a navigating element percolates as shown in Figure 10.

Two types of percolation that have been developed and applied in the literature are: site and bond (Grimmet, 1999). Site percolation considers lattices sites (vertices), in which the vertices are declared to be open with probability p (white cells) and closed with probability $1 - p$ (black cells). Bond percolation considers edges instead of vertices and are asserted to be open with probability p or closed with probability $1 - p$. There exist a critical value of p, termed as pc such that there exists an open cluster between the top of the lattice and the bottom of the lattice when $p > pc$. The prioritized exploration that we develop and introduce in this subsection is a percolator guided exploration that aims at continuous navigation through the estimation of connected voids to cover the disaster area as much as possible without getting trapped in dead ends. Site percolation model is very close to our environment model. There, while the probability of a cell being an obstacle, represented in black is approximately one, the white cells' occupancy probabilities are equal to zero and correspond to void cells.

In our percolation-based approach, we consider the search environment as a large porous medium, each robot trying to find connected clusters with good porosity (voids) to cover, through exploration, all accessible points of the environment in quest of survivors. Robots quit the local area being explored and switch to a new area of the environment when the obstacle density in the area being explored exceeds a predetermined threshold

Figure 11. Flow chart of multi-robot exploration strategy

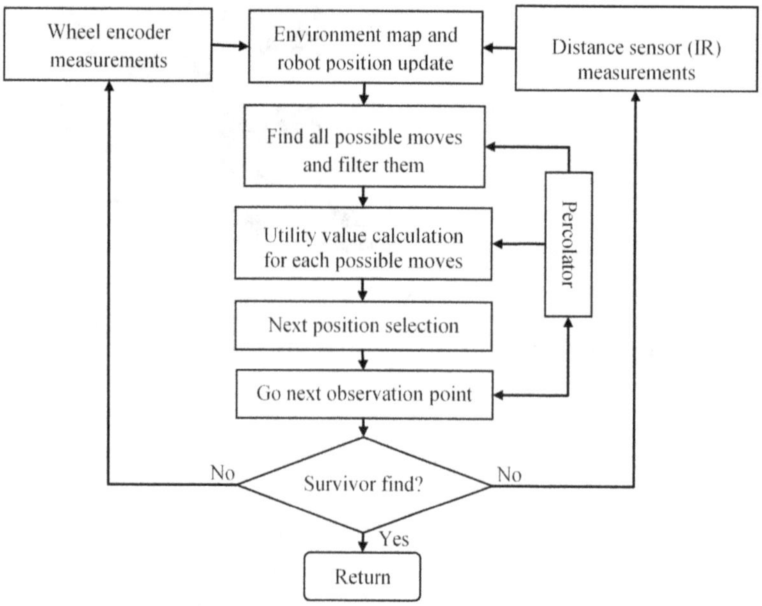

value. Motion of the robot can be interpreted as the propagation of water in the porous medium of a pumice stone, which is a labyrinth of connected voids and dead ends of a disaster environment. If robot exploration is guided into connected big voids (safe spaces), efficiency and coverage of the area increase. Our approach is motivated by a prioritized exploration that is required when search is done within rubble with the aim of maximum possible coverage without getting trapped to dead ends, based on the structural characteristics of the disaster area. The main idea of coordinated multi-robot exploration is to create possible moves for each robot where each robot selects the best one for itself, to cover as much as possible, the area cooperatively. Our multi-robot exploration strategy proceeds as given in the following flowchart of Figure 11.

At the first step of the control algorithm, each robot unit of the team gets odometry data by measurements from their individual wheel encoders and from their IR distance sensors which are generally corrupted by noise due to environmental and hardware uncertainties. There is no initial knowledge about the environment map which makes the methodology more realistic for real world applications. By using noisy sensor measurements, position, orientation and occupancy grid environment map of each robot unit are generated by using particle filter based SLAM. Secondly, each robot finds all frontier cells which represent the boundary between explored and unexplored areas in the common environment map. These frontier cells are the possible next position state for each robot. Frontier based exploration forces robots to dive into the unexplored regions by moving cells that are stemming cells of connected voids at the frontier. Small passages on the frontier, through which a robot cannot infiltrate, are then eliminated. At the next step of the process each robot calculates utility values for each of its possible moves and navigates in the direction of maximizing this utility value. Utility values are calculated by using the utility function of Equation (1), which consists of two parts. First one is the amount of expected information gain available for each possible frontier cell

Figure 12. Percolation model based robot search area controller, percolator

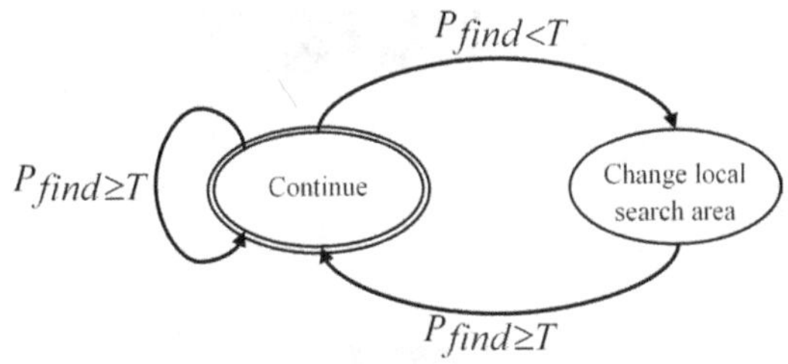

mi, (Equation (2)) and second part is the cost of reaching target frontier cell.

$$^{i}U_{Total} = ^{i}U_{I}(m_i) - ^{i}U_{C}(m_i) \quad (1)$$

$$^{i}U_{I}(m_i) = \sum_{p \in R^i} H(p) \quad (2)$$

Here R^i is the region around the frontier cell m_i. $H(p)$ is the entropy value for each cell p in the R^i region and calculated as follows;

$$H(p) = -p \log_2(p) - (1-p)\log_2(1-p) \quad (3)$$

where p is the probability of the cell being empty.

Second term $^{i}U_{C}(m_i)$ is the cost of reaching frontier cells in the sense of Euclidean distance. Manhattan distance is used to compute the distance from the robots current position to the selected frontier cell.

Every robot chooses its next observation position from candidate frontier cells by maximizing its expected utility and navigates towards this next position. Multiple area coverage is handled by reducing utility of target locations whenever they are expected to be covered by another robot.

Percolator, as seen in Figure 11, checks the probability of finding any survivor in the local search area when the robot navigates towards the next position. P_{find}, given in Equation (4), triggers the controller that toggles to other state according to the structural characteristic of the local area. If that likelihood decreases under a predefined threshold that is given to the robot according to the priori information about mission space, then the robot navigates towards other parts of the environment to gain more information about new unexplored parts of the environment and robot decides that there is no free big connected cluster in the current exploration area. Equation (4) gives the probability of finding any victim (P_{find}) in a local region and is calculated as 1 minus the ratio of obstacle area to the total swept area in the Δt time interval.

$$P_{find} = 1 - \frac{A_{obstacle}}{2v(r_{robot} + S_{\max})\Delta t} \quad (4)$$

Here v is the robot average speed for the time interval Δt, $S_{\max}$ is the maximum sensing range of the robot in question and r_{robot} is the robot radius. If P_{find} is smaller than the threshold, the robot decides that the likelihood of finding victim is low in the local search region and selects the next observation point from another part of the work space by using only the expected information gain, and omits the cost value to reach frontier cells, which is irrelevant at this step.

The derivation of Equation (4) is given with the help of Figure 13. The figure represents motion of a robot between X_{t-1} (previous position) and X_t (current position) with the average speed v in the Δt time interval. Swept area between robot's previous and current position is equals to

Figure 13. Movement plot of the search and rescue robot

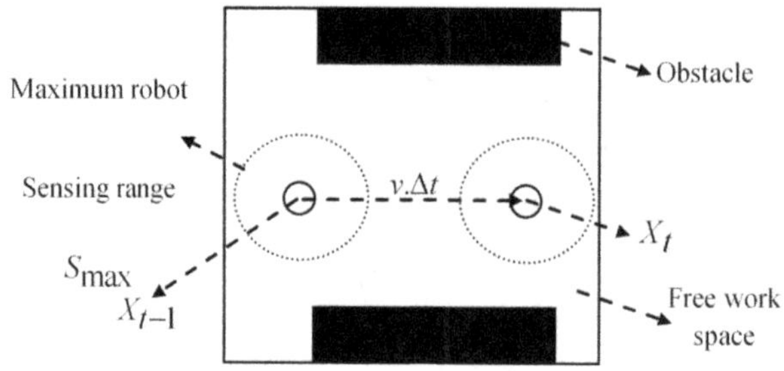

Figure 14. (a) Simulation environment, (b) Resultant environment map

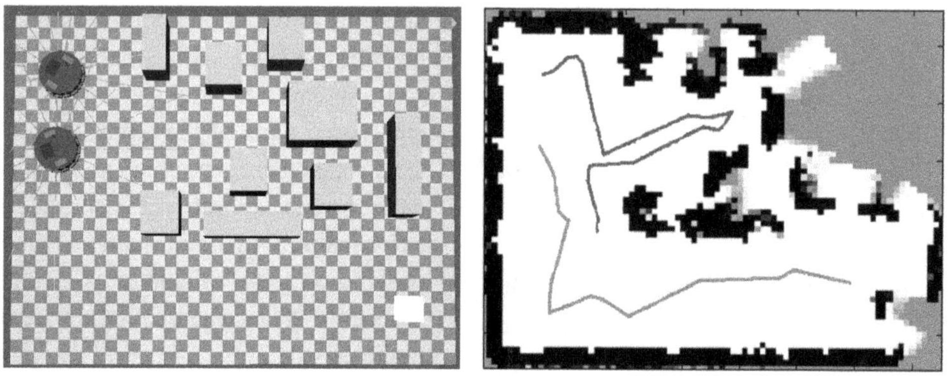

$2v(r_{robot} + S_{\max})\Delta t$. The ratio of obstacle related area in the swept area to the total swept area is equal to the complementary of probability of finding survivors in the local search environment.

In order to evaluate our proposed multi-robot coordination control strategy presented above, a victim search scenario illustrated in Figure 14 (a) was run on a computer with a 2.67GHz Pentium™ Core 2 processor and 2GB of memory. Two Magellan robots equipped with autonomous navigation and simultaneously localization and mapping ability enter the disaster environment and start to explore mission space to localize buried victim who is shown by the white box. Figure 14 (b) shows us robots' common environment map beliefs and individual navigation paths for each robot at the end of the given task which was victim localization. Colored lines show us the navigation path for each robot. We map the environment using an occupancy grid based mapping algorithm (Thrun, 2003). The algorithm decomposes the high dimensional continuum mapping problem into a one-dimensional one, where the occupancy of each grid cell is estimated independently. Occupancy of a grid is a numerical value representing the posterior probability of a grid cell being occupied. This provides a simple spatial representation of the environment. In our demonstrative examples of map grids, as seen in Figure 14 b, the different cell states are represented by different colors. Black cells are related with obstacles which can not be visited by the robots, gray ones show the unknown part of the environment and free space, where robots can navigate safely, represented with white color. We use the threshold value as 0.53.

We observe from Figure 14 (b) that the robot following the blue path changed its search direction because of the high obstacle density and thus, the probability of finding survivor decreasing

Figure 15. Average exploration times with different number of robots and control strategies; uncoordinated and Percolator-based controller with different environments; (a) for completely unstructured environment (b) for semi-unstructured one

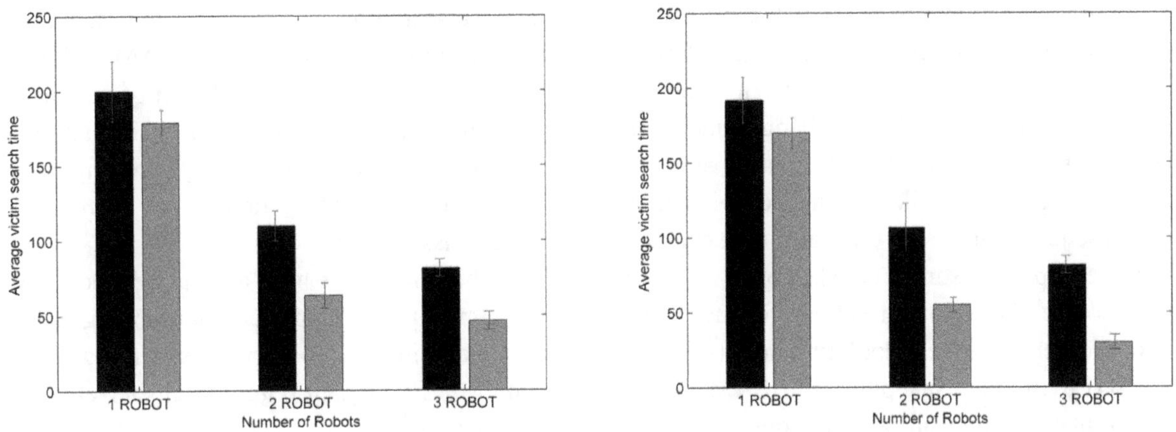

under the threshold (T). We also see that there is no multiple area coverage by the two robots: the overlapped of coverage is minimized, thus increasing the efficiency of the exploration. Multiple area coverage is prevented by the utility function and a heuristic method, reducing utility of target locations for each robot whenever they are expected to be covered by another robot.

In order to make detailed analysis about the performance of the proposed multi-robot coordination control strategy presented above, victim search scenarios were run 20 times for completely unstructured and semi unstructured environments. Semi-unstructured environments are building rubble where room walls remain erected. Whereas in completely unstructured rubble, all walls also have collapsed leading to no structured reference point. However, due to walls that still remain erected, semi-unstructured environments contain more dead ends than completely unstructured rubble. Robots' initial positions and victim locations were selected randomly in each experiment. In these simulations, our proposed method's performance is compared to the classical uncoordinated exploration approaches in literature by using one, two and three robots. In the uncoordinated methodology, each robot selects the next observation point from closest frontier cell, which is to say that the robot always navigates towards the closest cell of the unexplored area as a first come closest to observed target.

Figure 15 shows that the average exploration times with different number of robots by using uncoordinated and percolator coordinated exploration approaches in a completely unstructured environment (Figure 15 a) and a semi unstructured one (Figure 15 b). Our goal-oriented exploration strategy is seen to decrease the task accomplishment time because of non-existent redundancy due to reduced multiple area coverage. And also it is observed that, multi-robot systems are more efficient than single robot systems in the sense of time which is the most important issue in search and rescue operation. We observe during the experiments that percolator guided exploration can easily detect dead ends. Hence, this method is more efficient in semi-unstructured environments. For example, average exploration time using three robots is 25 seconds in a semi-unstructured environment and 47 seconds in a completely unstructured environment.

4. FUTURE RESEARCH DIRECTIONS

Although computation abilities of software and technologies for modular, scalable hardware increased considerably and many and efficient theoretical developments have been made in the last decade, significant challenges and open questions still exist for future research areas of prototyping multi-mobile robot networks in search and rescue operations. In this part of the chapter, we briefly present some of the key open research questions and develop a roadmap for the multi-robot research area, extrapolating into its future.

3D Environment mapping with SLAM done at different elevation using autonomous unmanned air or underwater vehicles gains much more importance these days for high level and complex task executions within non planar environments. Major parts of the existing works have focused on planar approaches using 2D environment models. Navigation by heterogeneous robots within air, water and/or surfaces of different orientations requires 3D SLAM and 3D map merging capabilities. Guided 3D explorations are also issues that will be challenged in the very near future. For example, 3D SLAM in underwater environments is a very hard problem because of uncertainties such as unpredictable water currents resulting in unpredictable motion of the robot body. Fish, other marine species together with currents cause a very dynamic environment which adversely affects distance measurements. Visual devices are used to extract landmarks, however lightening conditions and the effect of water flow makes the problem much more complex.

Existing networked mobile robot systems are not yet fully operable during real and hard catastrophe management activities due to limitations of the hardware, software, sensing, navigation and decision making capabilities. Hence, physical construction of robust robot units and development of advanced software and sensor abilities which have the ability to solve ambiguities in real time, are gaining importance for real life search and rescue applications.

Each robot unit encounters many challenges during a search and rescue operation, including but not limited to exploration, SLAM and path planning. These challenges should be solved in real time. Hence computational efficiency and fast response in emergency cases are important for the accuracy and efficiency of search and rescue missions.

In this chapter we have also introduced a new prototyping platform of energy scavenging sensor networks using the mobility of its carrying agent, namely the Sycophant Wireless Sensor Networks (SWS). In order to improve the usability of SWS networks we proposed using them in combination with sparse Mobile Robot Sensor Network. Moreover this cooperative decentralized architecture can be enriched by cooperation with static sensor networks or networks made up of different agents. Cooperative acting and communication of heterogeneous subnetworks yielding the necessity of hybrid SLAM, multi sensorial multi resolution sensor fusion, multitasking provide a richness of challenges to which novelties can be added especially like relaying: architecture based on elastic networks, multi layered or 3D SLAM, multilayer to 3D map merging and map growing, network to network exploration guidance, fractional control based on network complexity that may lead to fractional exploration. Although in this chapter we have mentioned their usage for reconnaissance in search & rescue, Sycophant Wireless Sensor Networks could be utilized in military applications as well as space exploration applications. Each application has its own constraints requiring adaptation of the proposed concept and the underlying architecture.

5. CONCLUSION

In this chapter we have presented a general hardware and software framework for prototyping a search and rescue multi robot system made of different types of sensor subnetworks. After discussing the necessity for such networks, we gave a detailed literature on and our implementations of SLAM, Map Merging, Goal Oriented Exploration, Connectivity Maintenance in Robot Networks and Robot Construction as well as Sycophant Wireless Sensor Networks. We introduced a new goal oriented exploration approach based on percolation theory as well as a new hybrid sensor robot network architecture, namely the Sycophant Wireless Sensor Network, which is followed by a mobile robot network.

The proposed hybrid (SWS + mobile robots) architecture provides new perspectives into robotic applications of search and rescue, possibly increasing their efficiency and usability as well as opening the way to new hybrid methodologies that promise a boost in 3D control approaches. The proposed percolation guidance yields a predictive exploration based on fluidics.

REFERENCES

Adluru, N., Latecki, L. J., Sobel, M., & Lakaemper, R. (2008). Merging maps of multiple robots. In *Pattern Recognition ICPR 2008* (pp. 1-4).

Aizawa, N., & Kubota, N. (2009). Intelligent formation control based on directionality of multi-agent system. In *Robotic Intelligence in Informationally Structured Space, RIISS '09* (pp. 87-92).

Amigoni, F. (2008). Experimental evaluation of some exploration strategies for mobile robots. *In Proceedings of the IEEE International Conference on Robotics and Automation* (pp. 2818–2823).

Amor, S. B., Huy, T. D., & Bui, M. (2006). A percolation based model for ATC simulation. I*EEE International Conference on Research, Innovation and Vision for the Future* (pp. 17-22).

Andersson, L. A., & Nygards, J. (2009). On multi-robot map fusion by inter-robot observations. In *12th International Conference on Information Fusion, FUSION '09* (pp. 1712 -1721).

Aylward, R., Lovell, S. D., & Paradiso, J. A. (2006). A compact, wireless, wearable sensor network for interactive dance ensembles. *International Workshop on Wearable and Implantable Body Sensor Networks, BSN 2006*, (pp. 4-70).

Aylward, R., & Paradiso, J. A. (2007). A compact, high-speed, wearable sensor network for biomotion capture and interactive media. *6th International Symposium on Information Processing in Sensor Networks,* (pp. 380-389).

Baglietto, M., Paolucci, M., Scardovi, L., & Zoppoli, R. (2002). Information based multi-agent exploration. *IEEE International Workshop on Robot Motion and Control* (pp. 173-179).

Bar-Shalom, Y., & Fortmann, T. E. (1987). *Tracking and data association. Mathematics in Science and Engineering* (*Vol. 179*). San Diego, CA: Academic Press Professional.

Birk, A., & Carpin, S. (2006). Merging occupancy grid maps from multiple robots. *Proceedings of the IEEE, 94*(7), 1384–1397. doi:10.1109/JPROC.2006.876965

Burgard, W., Fox, D., Moors, M., Simmons, R., & Thrun, S. (2000). Collaborative multi-robot exploration. In *Proceedings of the IEEE International Conference on Robotics and Automation* (pp. 476–481).

Burgard, W., Moors, M., & Schneider, F. (2002). Collaborative exploration of unknown environments with teams of mobile robots. *Dagstuhl Seminar on Plan-based Control of Robotic Agents* (pp. 187-215).

Burgard, W., Moors, M., Stachniss, C., & Schneider, F. E. (2005). Coordinated multi-robot exploration. *IEEE Transactions on Robotics*, *21*(3), 376–386. doi:10.1109/TRO.2004.839232

Carpin, S. (2008). Merging maps via Hough transform. In *IEEE/RSJ International Conference on Intelligent Robots and Systems, IROS 2008* (pp. 1878 -1883).

Carpin, S. (2008). Fast and accurate map merging for multi-robot systems. *Autonomous Robots*, *25*(3), 305–316. doi:10.1007/s10514-008-9097-4

Carpin, S., Birk, A., & Jucikas, V. (2005). On map merging. *Robotics and Autonomous Systems*, *53*(1), 1–14. doi:10.1016/j.robot.2005.07.001

Chen, J., Kwong, K., Chang, D., Luk, J., & Bajcsy, R. (2005). Wearable sensors for reliable fall detection. *27th Annual International Conference of the Engineering in Medicine and Biology Society, IEEE-EMBS 2005*, (pp. 3551-3554).

Chen, X., Zhuang, W., & Tan, J. (2009). The maintaining of communication links quality in unknown environment. In *International Conference on Information and Automation, ICIA '09* (pp. 218-223).

Choset, H. (1996). *Sensor based motion planning: The hierarchical generalized Voronoi graph.* Doctoral dissertation, California Institute of Technology, Pasadena, CA, USA.

Correll, N., Bachrach, J., Vickery, D., & Rus, D. (2009). Ad-hoc wireless network coverage with networked robots that cannot localize. In *IEEE International Conference on Robotics and Automation, ICRA '09* (pp. 3878-3885).

Cortes, J., Martinez, S., & Bullo, F. (2006). Robust rendezvous for mobile autonomous agents via proximity graphs in arbitrary dimensions. *Automatic Control. IEEE Transactions on*, *51*(8), 1289–1298. doi:10.1109/TAC.2006.878713

Dedeoglu, G., & Sukhatme, G. (2000). Landmark-based matching algorithm for cooperative mapping by autonomous robots. In *Distributed Autonomous Robotics Systems* (pp. 251–260). Springer-Verlag.

Dimarogonas, D. V., & Johansson, K. H. (2008). Decentralized connectivity maintenance in mobile networks with bounded inputs. In *IEEE International Conference on Robotics and Automation, ICRA 2008* (pp. 1507-1512).

Dimarogonas, D. V., & Kyriakopoulos, K. J. (2007). On the rendezvous problem for multiple nonholonomic agents. *IEEE Transactions on Automatic Control*, *52*(5), 916–922. doi:10.1109/TAC.2007.895897

Dissanayake, A., Newman, P., Clark, S., Durrant-Whyte, H. F., & Csorba, M. (2001). A solution to the simultaneous localization and map building problem. *IEEE Transactions on Robotics and Automation*, *17*, 229–241. doi:10.1109/70.938381

Dixon, C., & Frew, E. W. (2007). Maintaining optimal communication chains in robotic sensor networks using mobility control. In *Proceedings of the 1st international conference on Robot communication and coordination* (pp. 1-8). Athens, Greece: IEEE Press.

Dogru, S., Erkmen, A. M., & Erkmen, I. (2009). Tracking a sycophant wireless sensor network for its seamless integration to mobile wireless sensor networks. In *2nd International Conference on Adaptive Science & Technology, ICAST, 2009* (pp. 110-116).

Durrant-Whyle, H. F., & Bailey, T. (2006). Simultaneous localization and mapping: Part I. *IEEE Robotics & Automation Magazine*, *13*(2), 99–108. doi:10.1109/MRA.2006.1638022

Dynia, M., Kutylowski, J., Heide, F. M. A. D., & Schrieb, J. (2007). Local strategies for maintaining a chain of relay stations between an explorer and a base station. In *Proceedings of the Nineteenth Annual ACM Symposium on Parallel Algorithms and Architectures* (pp. 260-269).

Dynia, M., Kutyłowski, J., Lorek, P., & auf der Heide, F. (2006). *Maintaining communication between an explorer and a base station*. In Biologically Inspired Cooperative Computing.

Esin, Y. H., Unel, M., & Yildiz, M. (2008). Formation control of multiple robots using parametric and implicit representations. In *ICIC '08: Proceedings of the 4th International Conference on Intelligent Computing* (pp. 558–565). Berlin, Germany: Springer-Verlag.

Esposito, J. M., & Dunbar, T. W. (2006). Maintaining wireless connectivity constraints for swarms in the presence of obstacles. In *Proceedings 2006 IEEE International Conference on Robotics and Automation, ICRA 2006* (pp. 946-951).

Farrell, J. A., Pang, S., & Li, W. (2003). Plume mapping via hidden Markov methods. *IEEE Transactions on Systems, Man, and Cybernetics. Part B, Cybernetics, 33*(6), 850–863. doi:10.1109/TSMCB.2003.810873

Feng, D. J., & Wijesoma, W. S. (2008). Improving rao-blackwellised genetic algorithmic filter SLAM through genetic learning. In *Proceedings of the 10th International Conference on Control, Automation, Robotics and Vision*, Hanoi, Vietnam, (pp. 1200–1205).

Fenwick, J. W., Newman, P. M., & Leonard, J. J. (2002). Cooperative concurrent mapping and localization. *IEEE International Conference on Robotics and Automation*, Washington DC, USA, (pp. 1810–1817).

Ferri, G., Caselli, E., Mattoli, V., Mondini, A., & Mazzolai, B. (2009). SPRIAL: A novel biologicallyinspired algorithm for gas/odor source localization in an indoor environment with no strong airflow. *Robotics and Autonomous Systems, 57*(4), 393–402. doi:10.1016/j.robot.2008.07.004

Fox, D., Ko, J., Konolige, K., Limketkai, B., & Stewart, B. (2006). Distributed multi-robot exploration and mapping. *Proceeding of the IEEE Special Issue on Multi-Robot Systems, 94*(7), 1325–1339.

Grimmet, G. (1999). *Percolation*. Berlin, Germany: Springer-Verlag.

Guarnieri, M., Debenest, P., Inoh, T., Fukushima, E. F., & Hirose, S. (2004). Development of Helios VII: An arm equipped tracked vehicle for search and rescue operations. In *Proceedings of the International Conference Intelligent Robots and Systems*, Sendai, Japan, (pp. 39–45).

Guivant, J., & Nebot, E. (2001). Optimization of the simultaneous localization and map building algorithm for real time implementation. *IEEE Transactions on Robotics and Automation, 17*(3), 242–257. doi:10.1109/70.938382

Hähnel, D., Burgard, W., Fox, D., & Thrun, S. (2003). An efficient FastSLAM algorithm for generating maps of large-scale cyclic environments from raw laser range measurements. In *Proceedings of the IEEE International Conference on Intelligent Robots and Systems*, Las Vegas, NV, USA, (pp. 206–211).

Howard, A. (2006). Multi-robot simultaneous localization and mapping using particle filters. *The International Journal of Robotics Research, 25*(12), 1243–1256. doi:10.1177/0278364906072250

Hunt, A., & Ewing, R. (2009). *Percolation theory for flow in porous media*. Springer.

Ji, M., & Egerstedt, M. (2007). Distributed coordination control of multiagent systems while preserving connectedness. *IEEE Transactions on Robotics*, *23*(4), 693–703. doi:10.1109/TRO.2007.900638

Kim, Y., & Mesbahi, M. (2006). On maximizing the second smallest eigenvalue of a state-dependent graph Laplacian. *IEEE Transactions on Automatic Control*, *51*(1), 116–120. doi:10.1109/TAC.2005.861710

Koenig, S., Tovey, C., & Halliburton, W. (2001). Greedy mapping of terrain. *IEEE International Conference on Robotics & Automation* (pp. 3534-3539).

Konolige, K. A. (2006). Centibots: Very large scale distributed robotic teams. *Springer Tracts in Advanced Robotics*, *21*, 131–140. doi:10.1007/11552246_13

Krause, A., Smailagic, A., & Siewiorek, D. P. (2006). Context-aware mobile computing: learning context- dependent personal preferences from a wearable sensor array. *IEEE Transactions on Mobile Computing*, *5*(2), 113–127. doi:10.1109/TMC.2006.18

Kroc, S., & Delic, V. (2003). Personal wireless sensor network for mobile health care monitoring. *6th International Conference on Telecommunications in Modern Satellite, Cable and Broadcasting Service, TELSIKS 2003*, (vol. 2, 471-474).

Lavalle, S. M. (1998). *Rapidly-exploring random trees: A new tool for path planning.*

Lavalle, S. M. (2006). *Planning algorithms.* Cambridge University Press. León, A., Barea, R., Bergasa, L., López, E., Ocaña, M., & Schleicher, D. (2009). SLAM and map merging. *Journal of Physical Agents*, *3*(1), 13–23.

Loehle, C., Li, B., & Sundell, R. C. (1996). Forest spread and phase transitions at forest-prairie ecotones in Kansas. *Journal of Landscape Ecology*, *11*(4), 225–235. doi:10.1007/BF02071813

MacArthur, E. Z., & Crane, C. D. (2007). Compliant formation control of a multi-vehicle system. In *International Symposium on Computational Intelligence in Robotics and Automation, CIRA 2007* (pp. 479-484).

Mahlknecht, S., & Madani, S. A. (2007). On architecture of low power wireless sensor networks for container tracking and monitoring applications. *2007 5th IEEE International Conference on Industrial Informatics,* (vol. 1, 353-358).

Mizuno, H., Nagai, H., Sasaki, K., Hosaka, H., Sugimoto, C., Khalil, K., & Tatsuta, S. (2007). Wearable sensor system for human behavior recognition (first report: basic architecture and behavior prediction method). *Solid-State Sensors, Actuators and Microsystems Conference, TRANSDUCERS 2007,* (pp. 435-438).

Montemerlo, M., Thrun, S., Koller, D., & Wegbreit, B. (2002). FastSLAM: a factored solution to the simultaneous localization and mapping problem. In *Proceedings of the National Conference on Artificial Intelligence,* Edmonton, Canada, (pp. 593–598).

Montemerlo, M., Thrun, S., Koller, D., & Wegbreit, B. (2003). FastSLAM 2.0: An improved particle filtering algorithm for simultaneous localization and mapping that provably converges. In *Proceedings of the International Conference on Artificial Intelligence (IJCAI)*, Acapulco, Mexico, (pp. 1151–1156).

Mori, M., Tanaka, J., Suzumori, K., & Kanda, T. (2006). Field test for verifying the capability of two high-powered hydraulic small robots for rescue operations. In *the Proceedings of the IEEE International Conference on Intelligent Robots and Systems*, Beijing, China, (pp. 3492-3497).

Nguyen, H. G., Everett, H. R., Manouk, N., & Verma, A. (2002). *Autonomous mobile communication relays.* Presented at the SPIE AeroSense/ Unmanned Ground Vehicle Technology IV, Orlando, FL.

Nguyen, H. G., Farrington, N., & Pezeshkian, N. (2004). *Maintaining communication link for tactical ground robots.* Presented at the AUVSI Unmanned Systems North America, Anaheim, CA, Pezeshkian, N., Nguyen, H. G., & Burmeister, A. (2006). Unmanned ground vehicle non-line-of-sight operations using relaying radios. In *Proceedings of the IASTED International Conference on Robotics and Applications* (pp. 1-6).

Reif, J. H., & Wang, H. (1995). Social potential fields: a distributed behavioral control for autonomous robots. In *WAFR: Proceedings of the workshop on Algorithmic foundations of robotics* (pp. 331–345). Natick, MA, USA: A. K. Peters, Ltd.

Sato, N., Matsuno, F., Yamasaki, T., Kamegawa, T., Shiroma, N., & Igarashi, H. (2004). Cooperative task execution by a multiple robot team and its operators in search and rescue operations. *In Proceedings of the International Conference on Intelligent Robots and Systems*, Sendai, Japan, (pp. 1083–1088).

Schenker, P. S., Pirjanian, P., Balaram, B., Ali, K. S., Trebi-Ollennu, A., & Huntsberger, T. L. (2000). Reconfigurable robots for all terrain exploration. In *Proceedings of the SPIE Symposium on Sensor Fusion and Decentralized Control in Robotic Systems III, 4196*, 454–468.

Schuresko, M., & Cortes, J. (2008). Distributed motion constraints for algebraic connectivity of robotic networks. In *47th IEEE Conference on Decision and Control, CDC 2008* (pp. 5482-5487).

Se, S., Lowe, D., & Little, J. (2002). Mobile robot localization and mapping with uncertainty using scale invariant visual landmarks. *The International Journal of Robotics Research, 21*(8), 735–758. doi:10.1177/027836402761412467

Sheng, W., Yang, Q., & Tan, J. (2006). Distributed multi-robot coordination in area exploration. *Robotics and Autonomous Systems, 54*(12), 945–955. doi:10.1016/j.robot.2006.06.003

Shucker, B., & Bennett, J. K. (2005). Target tracking with distributed robotic macrosensors. In *Military Communications Conference, MILCOM 2005 IEEE* (vol. 4, pp. 2617-2623).

Smith, R., Self, M., & Cheeseman, P. (1990). *Estimating uncertain spatial relationships in robotics* (pp. 167–193). Autonomous Robot Vehicles.

Song, T. L. (1999). Observability of target tracking with range-only measurements. *IEEE Journal of Oceanic Engineering, 24*(3), 383–387. doi:10.1109/48.775299

Souryal, M. R., Geissbuehler, J., Miller, L. E., & Moayeri, N. (2007). Real-time deployment of multihop relays for range extension. In *Proceedings of the 5th International Conference on Mobile Systems, Applications and Services* (pp. 85-98). San Juan, Puerto Rico: ACM.

Stump, E., Jadbabaie, A., & Kumar, V. (2008). Connectivity management in mobile robot teams. In *IEEE International Conference on Robotics and Automation, ICRA 2008* (pp. 1525-1530).

Tanaka, J., Suzumori, K., Takata, M., Kanda, T., & Mori, M. (2005). A mobile jack robot for rescue operation. In *the Proceedings of the IEEE International Workshop on Safety, Security and Rescue Robotics*, Kobe, Japan, (pp. 99–104).

Technology, C. (n.d.). *IRIS 2.4GHz*. Retrieved February 27, 2009, from http://www.xbow.com/Products/productdetails.aspx?sid=264

Tekdas, O., & Isler, V. (2008). Robotic routers. In *IEEE International Conference on Robotics and Automation, ICRA 2008* (pp. 1513-1518).

Thrun, S. (1993). Exploration and model building in mobile robot domains. In *Proceeding of the IEEE International Conference on Neural Networks*, San Francisco, CA, USA, (pp. 175–180).

Thrun, S. (2003). Learning occupancy grids with forward sensor models. *Autonomous Robots, 15*(2), 111–127. doi:10.1023/A:1025584807625

Thrun, S., Burgard, W., & Fox, D. (2000). A real-time algorithm for mobile robot mapping with applications to multi-robot and 3D mapping. In *Proceedings IEEE International Conference on Robotics and Automation, CRA '00* (vol. 1, pp. 321 -328).

Thrun, S., Burgard, W., & Fox, D. (2005). *Probabilistic robotics (Intelligent robotics and autonomous agents)*. The MIT Press.

Topal, S., Erkmen, İ., & Erkmen, A. M. (2009). New prioritized multi-robot exploration strategy based on percolation model in unknown environment. In *Proceedings of the International Conference on Climbing and Walking Robots and the Support Technologies for Mobile Machines*, Istanbul, Turkey, (pp. 967–976).

Urcola, P., Riazuelo, L., Lazaro, M. T., & Montano, L. (2008). Cooperative navigation using environment compliant robot formations. In *IEEE/RSJ International Conference on Intelligent Robots and Systems, IROS 2008* (pp. 2789-2794).

Volgyesi, Balogh, Nadas, Nash, & Ledeczi. (2007). Shooter localization and weapon classification with soldier-wearable networked sensors. In *MobiSys '07: Proceedings of the 5th International Conference on Mobile Systems, Applications and Services* (pp. 113–126). New York, NY: ACM.

Wang, W., Zhang, H., Zong, G., & Zhang, J. (2008). Force cooperation in a reconfigurable field multi-robot system. *Journal of Field Robotics, 25*(11-12), 922–938. doi:10.1002/rob.20261

Williams, S. B. (2002). Towards multi-vehicle simultaneous localisation and mapping. In *Proceedings - IEEE International Conference on Robotics and Automation* (vol. 3, pp. 2743-2748).

Yamaguchi, T. (2006). Image processing based on percolation model. *Journal of IEICE Transactions on Information and Systems. E (Norwalk, Conn.), 89-D*(7), 2044–2052.

Ye, W., Xu, Y., & Lee, K. K. (2005). Shoe-Mouse: An integrated intelligent shoe. *2005 IEEE/RSJ International Conference on Intelligent Robots and Systems, (IROS 2005)*, (pp. 1163-1167).

Zavlanos, M. M., & Pappas, G. J. (2008). Distributed connectivity control of mobile networks. *IEEE Transactions on Robotics, 24*(6), 1416–1428. doi:10.1109/TRO.2008.2006233

Zhou, X. S., & Roumeliotis, S. I. (2006). Multi-robot SLAM with unknown initial correspondence: The robot rendezvous case. In *IEEE/RSJ International Conference on Intelligent Robots and Systems* (pp. 1785 -1792).

ADDITIONAL READING

Bar-Shalom, Y., & Fortmann, T. E. (1987). *Tracking and data association. Mathematics in Science and Engineering* (*Vol. 179*). San Diego, CA, USA: Academic Press Professional.

Carpin, S. (2008). Fast and accurate map merging for multi-robot systems. *Autonomous Robots, 25*(3), 305–316. doi:10.1007/s10514-008-9097-4

Durrant-Whyle, H. F., & Bailey, T. (2006). Simultaneous localization and mapping: Part I. *IEEE Robotics & Automation Magazine, 13*(2), 99–108. doi:10.1109/MRA.2006.1638022

Grimmet, G. (1999). *Percolation*. Berlin: Springer-Verlag.

Hunt, A., & Ewing, R. (2009). *Percolation Theory for Flow in Porous Media*. Springer.

Konolige, K. A. (2006). Centibots: Very large scale distributed robotic teams. *Springer Tracts in Advanced Robotics, 21*, 131–140. doi:10.1007/11552246_13

Lavalle, S. M. (2006). *Planning Algorithms*. Cambridge University Press. doi:10.1017/CBO9780511546877

Montemerlo, M., Thrun, S., Koller, D., & Wegbreit, B. (2003). FastSLAM 2.0: an improved particle filtering algorithm for simultaneous localization and mapping that provably converges. *In Proceedings of the International Conference on Artificial Intelligence (IJCAI)*, Acapulco, Mexico, (pp. 1151–1156).

Murphy, R. R., Tadokoro, S., Nardi, D., Jacoff, A., Fiorini, P., Choset, H., & Erkmen, A. M. (2008). Search and rescue robotics. In Siciliano, B., & Khatib, O. (Eds.), *Springer Handbook of Robotics* (pp. 1151–1173). Springer Berlin Heidelberg. doi:10.1007/978-3-540-30301-5_51

Reif, J. H., & Wang, H. (1995). Social potential fields: a distributed behavioral control for autonomous robots. In *WAFR: Proceedings of the workshop on Algorithmic foundations of robotics* (pp. 331–345). Natick, MA, USA: A. K. Peters, Ltd.

Song, T. L. (1999). Observability of target tracking with range-only measurements. *Oceanic Engineering. IEEE Journal of, 24*(3), 383–387.

Stachniss, C. (2006). *Exploration and Mapping with Mobile Robots*. Doctoral Dissertation, University of Freiburg, Department of Computer Science, Germany

Thrun, S., Burgard, W., & Fox, D. (2005). *Probabilistic Robotics (Intelligent Robotics and Autonomous Agents)*. The MIT Press.

KEY TERMS AND DEFINITIONS

Connectivity Maintenance: Ensuring a direct communication path between any two robots or any robot and a predefined base station in a multi robot system.

Leader Follower Robot Networks: A robot network with some of the nodes designated as leader and the others as follower, where the leaders work towards accomplishing own missions, not necessarily cooperating with the rest of the network, and the followers act to maintain connectivity in the robot network and a base station.

Multi Robot Systems: Systems of many robots collaborating and cooperating to accomplish a given task.

Percolation: Penetration of autonomous robots in connected voids of catastrophe environments to accomplish a given task.

Robot Exploration: Autonomous navigation of robots in an unknown environment in order to construct map of the environment.

Search and Rescue: Quick localization of a victim in a catastrophe environment and his/her transportation to a safe place.

SLAM: Mapping of an unknown environment by a robot (or many robots) while localizing itself (or themselves) in the constructed map.

Sycophant Wireless Sensor Networks: Static ecto-parasitic clandestine wireless sensor networks mounted incognito on mobile agents using only the agent's mobility without intervention, and are cooperating with sparse mobile robot sensor networks.

Chapter 15
A Heuristic Approach for Disassembly Sequencing Problem for Robotic Disassembly Operations

Ahmed ElSayed
University of Bridgeport, USA

Elif Kongar
University of Bridgeport, USA

Surendra M. Gupta
Laboratory for Responsible Manufacturing, Northeastern University, USA

ABSTRACT

Electronic products enter the waste stream rapidly due to technological enhancements. Their parts and material recovery involve significant economic and environmental gain. To regain the value added to such products a certain level of disassembly may be required. Disassembly operations are often expensive and the complexity of determining the best disassembly sequence increases as the number of parts in a product grows. Therefore, it is necessary to develop methodologies for obtaining optimal or near optimal disassembly sequences to ensure efficient recovery process. To that end, this chapter introduces a Genetic Algorithm based methodology to develop disassembly sequencing for end-of-life products. A numerical example is presented to provide and demonstrate better understating and functionality of the algorithm.

1. INTRODUCTION

Industrial robots have become one of the main components in products assembly and industry automation, due to their positive impact on efficiency and productivity by reducing time and increasing operational performance. RV-M1 Movemaster EX Mitsubishi manipulator is one of the many industrial robots, which is an industrial micro-manipulator used widely in academic and industry. Proposed study also uses this robot manipulator for the automated disassembly of end-of-life (EOL) products.

DOI: 10.4018/978-1-4666-0176-5.ch015

Studies seeking sustainable solutions for accumulating waste have been growing in recent years. Electronic wastes constitute a major part in these studies since their recovery involves significant economic and environmental gain. Furthermore, electronic products enter the waste stream much faster than many of the products due to the rapid pace of technological innovation. These products are classified as end-of-life (EOL) products and contain valuable materials as well as reusable components in their bills-of-materials (BOMs). In many cases reused electronic components may be more reliable than their newly manufactured counterparts since they have already survived the "burn-out" period. On the other hand, electronic products may also contain hazardous materials which require special handling. Therefore, EOL products recovery mainly involves one or more of the following processing options: (1) recycling operations to regain the material value, (2) reuse operations with additional refurbishing or remanufacturing to regain the technological value and (3) proper disposal operations to prevent environmental hazard. Storage could also be included as yet another alternative for components that are not reusable at the moment due to lack of demand or financial considerations.

In majority of EOL processing, a certain level of disassembly may be required. Disassembly can be *partial* or *complete* and may use a methodology that is *destructive* or *non-destructive*. In this chapter, the case of complete disassembly is considered where the components are retrieved by either destructive or non-destructive methodology.

For electronic products, modeling of EOL processing can be complex due to the large number of components in their product structures. In such cases, exhaustive search algorithms are computationally expensive rendering heuristic methods more appealing for obtaining optimal or near-optimal solutions. With this motivation, we present a genetic algorithm for disassembly sequencing in the presence of constraints and precedence relationships.

2. LITERATURE REVIEW

This section first introduces the manipulator used in the process. Following the introduction a literature review listing similar studies in the field is provided.

RV-M1 Movemaster EX Mitsubishi manipulator by Mitsubishi Co. (n.d.) is a five degree of freedom (5DOF) manipulator arm as shown in Figure 1.

The robotic manipulator includes a waist motor to affect rotations about the Z-axis. In the X-Y plane; a shoulder, elbow and wrist joints enable navigation in 3D. The arm is connected to a hand griper that has the capability of rotation about the X-axis for the handling of parts. The top speed of the arm is 1,000 m/s according to Mitsubishi co. (n.d.), thus enabling the applications within the mass production arena; including computers disassembly which is the main focus of this chapter.

As mentioned before, EOL products disassembly is one of new applications that can be used to reduce waste and improve recycling processes. The

Figure 1. RV-M1 Movemaster-EX Mitsubishi manipulator (courtesy of Mitsubishi Co.)

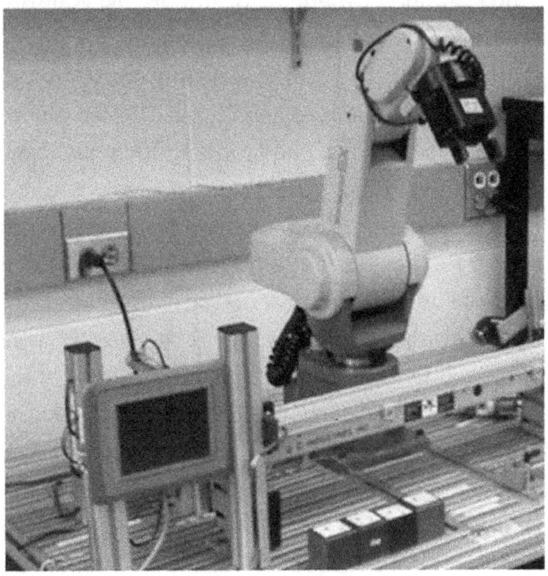

main design challenges in this process include the assessment of the usefulness of parts to be disassembled and the reduction of the utilized time. A robotic manipulator can be utilized to save time and improve the performance of the process. One of the main issues in using the robotic arm in disassembly is process planning, through which we need to compute the path required to remove items in the EOL product with minimum waste. This can be achieved by feeding an optimum disassembly sequence into the robotic arm. The optimal sequence is the path that would generate the maximum gain while minimizing the overall disassembly time. In this chapter, we implement Genetic Algorithm (GA) in determining the optimal/near optimal disassembly sequence.

In recent years, genetic algorithm (GA) has been gaining popularity for solving combinatorial and *NP*-complete problems. One of the multi-objective optimization applications was proposed by Valenzuela-Rendón, Uresti-Charre & Monterrey (1997). Keung *et al.* (2001) also applied a multi-objective GA approach to a tool selection model. Lazzerini and Marcelloni (2000) utilized genetic algorithm in assembly scheduling problem. In the area of disassembly, Kongar and Gupta (2006) proposed a GA for disassembly sequencing problem, while McGovern and Gupta (2007) applied genetic algorithm to disassembly line balancing. For further literature on disassembly scheduling and processing, see Ilgin and Gupta (2010), Gungor and Gupta (1999), Lee *et al.* (2001), Lambert and Gupta (2005), and McGovern and Gupta (2011).

The complexity of sequencing problems is one of the factors that increases the precedence relationships in the product structure. A study that considered precedence relationships was proposed by Sanderson *et al.* (1990). Seo *et al.* (2001) proposed a genetic algorithm for generating optimal disassembly sequences considering both economical and environmental factors. Bierwirth *et al.* (1996) and Bierwirth and Mattfeld (1999) proposed a methodology to overcome this problem by introducing the precedence preservative crossover (PPX) technique for scheduling problems which preserves the precedence relationships during the crossover function of GA. Shimizu *et al.* (2007) developed a prototype system for strategic decision-making on disassembly for recycling at the design stage of the product life cycle. Hui *et al.* (2008) utilized a genetic algorithm to determine feasible and optimal disassembly solutions.

In the area of automated disassembly, Torres *et al.* (2004, 2003) presented a personal computer disassembly cell that is able to handle a certain degree of automatism for the non-destructive disassembly process. This work is then followed by Pomares *et al.* (2004) who generated an object-oriented model required for developing a disassembly process. Gil *et al.* (2007) proposed a flexible multi-sensorial system for automatic disassembly using cooperative robots. As a follow-up work, Torres *et al.* (2009) presented a task planner for disassembly process based on decision trees.

3. GENETIC ALGORITHM FOR DISASSEMBLY SEQUENCING

Genetic algorithm starts with a set of randomly selected potential solutions called the *population*. Each member of the population is encoded as an artificial *chromosome* which contains information about the solution mapping. Each chromosome is assigned a score based on a predefined *fitness function*. A new population of chromosomes is iteratively created in the hope of finding a chromosome with a better score. At each step of creation, a *mutation* may occur in a chromosome, and/or two chromosomes may mate to produce a child; this process is known as *crossover*. The selection of parent chromosomes is biased towards fitter members of the population. The process is iterated until some predetermined termination conditions are satisfied.

Figure 2. Original bill-of-materials of the end-of-life PC

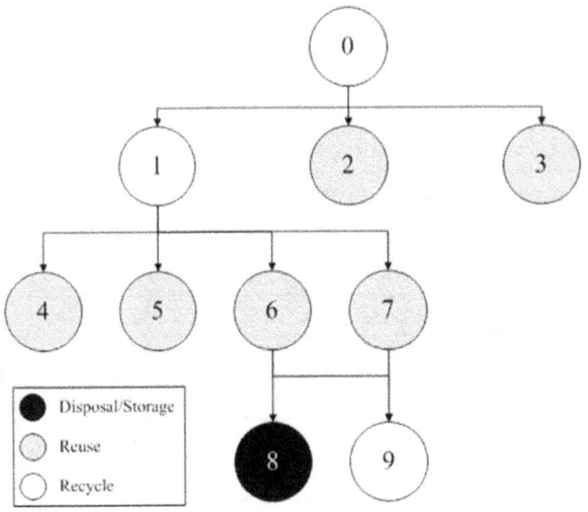

3.1. Elements of Genetic Algorithm for Disassembly Sequencing

The structure of the example EOL product and related data are provided in Figure 1. The given product consists of ten components indexed by integers from 0 to 9. Therefore, $j \in \{0,1, .., n\text{-}1\}$. Location of each component is defined by its 3-D coordinates. The methodology used for the disassembly of each component may be destructive (*D*) or non-destructive (*N*). A component may or may not be demanded. If it is not demanded, it is represented by *s*. If it is demanded, it may be demanded for reuse (*u*) or recycling (*r*). A component has one of three types of joints, viz., latch (*L*), screw (*S*) or slot entry (*E*). In addition, the precedence relationships are given as follows: component 0 must be disassembled prior to any other component; component 1 must be disassembled prior to components 4, 5, 6, and 7; and components 6 and 7 must be disassembled prior to components 8 and 9.

Chromosomes: In every GA problem, the solution and parameters must be coded into chromosomes, represented by a combination of numbers, alphabets and/or other characters, before they can be processed. In this study, chromosomes are codified in the form of a string consisting of four ordered sections of equal length, representing the disassembly sequence, the disassembly method, the demand type of each component, and the material type of each component respectively. Coordinate data are kept separately from the chromosome structure to provide ease in calculations.

Initial Population: The initial population consists of *ncr* random chromosomes that satisfy the precedence relationships and any other constraints imposed by the product structure. For the example provided in Figure 1, hundred chromosomes (feasible solutions) are randomly created to form the initial population (*ncr* = 100).

The chromosomes in the initial random population is provided in Table 1 (20 out of 100 are shown).

Crossover: The crossover methodology employed in this study is the precedence preservative crossover (PPX). PPX, in addition to the two strings representing the chromosomes of the parents (Parent_1 and Parent_2), creates two additional strings that pass on the precedence relationship based on the two parental permutations to two new offspring. This operation eliminates the possibility of new precedence relationships being introduced. A vector that represents the number of operations involved in the problem is then randomly filled with elements of the set. This vector defines the order in which the operations are successively drawn from Parent_1 and Parent_2.

The algorithm is initialized with an empty offspring. The leftmost operation in one of the two parents is selected in accordance with the order of parents given in the vector. After an operation is selected it is deleted in both parents. Finally, the selected operation is appended to the offspring. This step is repeated until both parents are empty and the offspring contains all operations involved.

Mutation: The population is subject to mutation operation with a given probability. If the probability holds, the mutation operator selects a random number of genes (*rnd* = 0, …, 9), and

Table 1. Partial initial population for the genetic algorithm example

Sequence	Method	Demand	Material	F(ch,gn)
0165872394	DDNNNNNNDN	rruusuuuru	ACAAPAPPAP	148.8409
0147632958	DDNNNNNDNN	rruuuuurus	ACPAAPPAAP	156.9266
0321657948	DNNDNNNDNN	ruuruuurus	APPCAAAAPP	105.9101
0231657489	DNNDNNNNND	ruuruuuusr	APPCAAAPPA	125.7518
0231756948	DNNDNNNDNN	ruuruuurus	APPCAAAAPP	108.8020
0137625849	DDNNNNNNND	rruuuuusur	ACPAAPAPPA	160.7229
0176532498	DDNNNNNNDN	rruuuuuurs	ACAAAPPPAP	96.2360
0127648539	DDNNNNNNND	rruuuusuur	ACPAAPPAPA	160.3261
0176385492	DDNNNNNNDN	rruuusuuru	ACAAPPAPAP	176.2894
0126537984	DDNNNNNDNN	rruuuuursu	ACPAAPAAPP	142.3334
0167385249	DDNNNNNNND	rruuusuuur	ACAAPPAPPA	139.1801
0176489325	DDNNNNDNNN	rruuusruuu	ACAAPPAPPA	129.6855
0317268594	DNDNNNNNDN	ruruuusuru	APCAPAPAAP	199.9146
0213674859	DNDNNNNNND	ruruuuusur	APCPAAPPAA	174.1672
0137425689	DDNNNNNNND	rruuuuuusr	ACPAPPAAPA	158.2748
0213768954	DNDNNNNDNN	ruruuusruu	APCPAAPAAP	187.9381
0314627598	DNDNNNNNDN	ruruuuuurs	APCPAPAAAP	188.4772
0167452983	DDNNNNNDNN	rruuuuursu	ACAAPAPAPP	166.9929
0176289543	DDNNNNDNNN	rruuusruuu	ACAAPPAAPP	137.4591
0127354698	DDNNNNNNDN	rruuuuuurs	ACPAPAPAAP	183.7432

exchanges them in such a way that the same precedence relationships are preserved. Otherwise the population remains unchanged and carried to the next generation. The mutation operator proposed in this chapter exchanges components 0 and 1. The rest of the strings remain the same.

Fitness Evaluation: The fitness function is dependent on the increment in disassembly time. There are three factors, which add up to the disassembly time of a component. The first one is basic disassembly time for component j in sequence seq ($dt_{j,seq}$).

The second function ($ct_{j,seq}$) is the penalty (in seconds) for each travel time to disassemble component j in sequence seq, which includes a function of the distance traveled between the (seq-1)th and seq^{th} sequences and the robot arm speed factor (sf):

$$ct_{j,seq} = \sqrt{(x_{j'(seq-1)} - x_{j'seq})^2 + (y_{j'(seq-1)} - y_{j'seq})^2 + (z_{j'(seq-1)} - z_{j'seq})^2} \Big/ sf$$

The last criterion in fitness function is the penalty for disassembly method change ($mt_{j,seq}$). For each disassembly method change, the sequence is penalized by 1 second:

0, if no method change is required, (*e.g. N* to *N*)

$mt_{j,seq}$ =1, if method change is required, (*e.g. N* to *D*)

In addition, the algorithm searches for a "recycling pair" and does not penalize the sequence if the two adjacent components are made of the same material and if they are both demanded for recycling.

Let T_{seq} denote the cumulative disassembly time after the disassembly operation in sequence *seq* is completed for component *j*:

$$T_{seq} = T_{seq\text{-}1} + dt_{j,seq} + ct_{j,seq} + mt_{j,seq}, \text{ for } seq = 0, \ldots, n\text{-}2$$

$$T_{seq} = T_{seq\text{-}1} + dt_{j,seq}, \text{ for } seq = n\text{-}1 \quad (1)$$

In the proposed GA model, the objective is to minimize the total fitness function (F) by minimizing (i) the traveled distance, (ii) the number of disassembly method changes, and (iii) by combining the identical-material components together, eliminating unnecessary disassembly operations. Let $F(ch,gn)$ denote the total fitness for chromosome *ch* in generation *gn*. Hence, total time to disassemble all the components can be calculated as follows:

$$F(ch, gn) = \sum_{seq=0}^{n-1} dt_{j,seq} + \sum_{seq=0}^{n-2} ct_{j,seq} + \sum_{seq=0}^{n-2} mt_{j,seq}, \quad \forall j, j = 0, \ldots, n\text{-}1. \quad (2)$$

Selection and Regeneration Procedure: After every generation, the chromosomes obtain a certain expectation depending on their fitness values. A roulette wheel is then implemented to select the sequence of parents that will be included in the next generation (the higher the fitness value the higher the chance to be selected). This method aims at allowing the parents in the current generation to be selected for the next generation without getting trapped in the local optima. In addition, a new population is generated eliminating the weak chromosomes.

Table 2. EOL product data for the GA model example

Item #	Description	Material**	Disassembly		Coordination (x,y,z)
			Time (sec.)	Method*	
0	Side Cover	A	3	D	(83.6, 78.6,100)
1	Power Supply	C	6	D	(42, 80.5, 110)
2	Sound Card	P	3	N	(127, 89, 210)
3	Modem Card	P	3	N	(146.3, 90.3, 210)
4	CPU	P	5	N	(57.5, 88, 300)
5	Hard Drive	A	4	N	(86, 31,150)
6	CD Drive	A	4	N	(20, 50,150)
7	Zip Drive	A	4	N	(64, 35,150)
8	RAM	P	3	N	(63, 54, 320)
9	Driver Slot	A	2	D	(43, 33, 170)

*D: Destructive N: Non-destructive
**A: Aluminum C: Copper P: Plastic

Table 3. Final population for the numerical example

Chromosome Sequence	*F(ch,gn)*
0 1 6 5 7 9 2 3 4 8 DDNNNDNNNN rruuuruuus ACAAAAPPPP	63.3167 secs

Figure 3. Flowchart of the proposed genetic algorithm with crossover and mutation operators

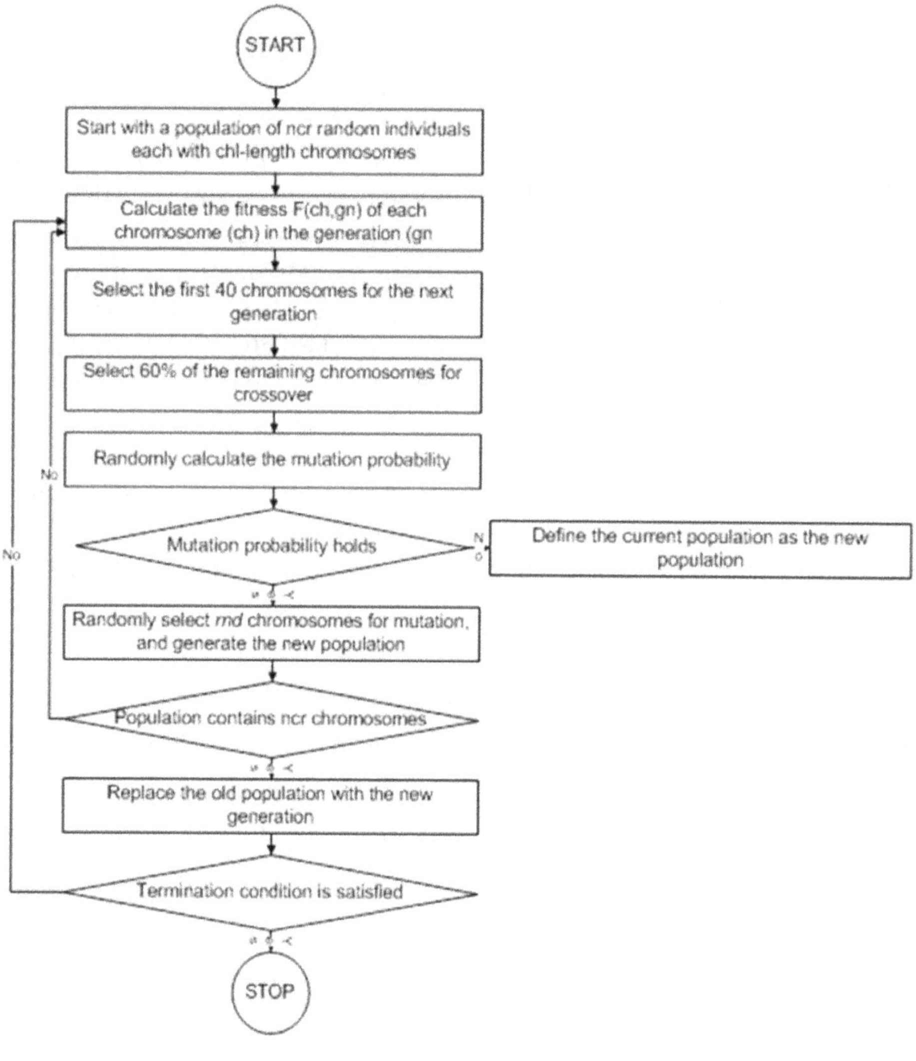

Termination: The execution of GA terminates if the number of generations reaches up to a maximum value (100 in our example).

The steps of the proposed GA are depicted in Figure 3.

3.2. Genetic Algorithm Model Assumptions

Proposed model assumes that the end effector speed for the robot arm is a constant value of 25 cm/sec. In addition, the time spent for robot arm angle change (for all three angles) is assumed to be embedded in the disassembly time for each component. In addition, every component is assumed to have one joint that connects the component with the rest of the product structure.

4. Numerical Example

The bill-of-materials provided in Figure 1 with the crossover and mutation probabilities 0.60 and 0.005 respectively is fed into the proposed genetic

Figure 4. Best and average fitness values of the proposed genetic algorithm

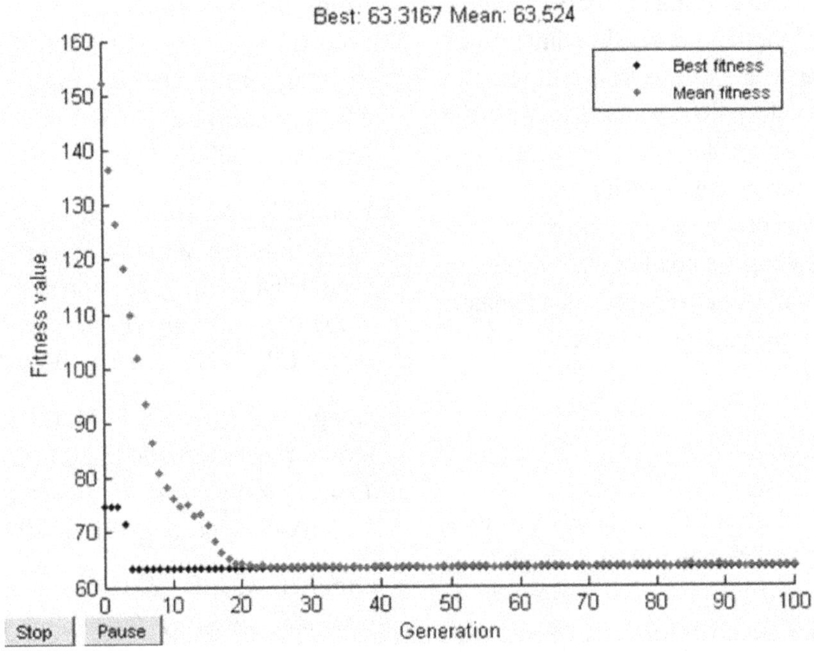

algorithm model. The initial population partially provided in Table 1 consists of 100 chromosomes (*ncr* = 100) included four duplicate chromosomes, where as the rest was unique sequences. The parameters for the EOL product are provided in Table 2.

The GA model provided single optimal solution for the given EOL product with a fitness function value of 63.3167 seconds. The average fitness value throughout the whole 100 generations is found to be 63.524 seconds (Figure 4).

The best chromosome appeared in the fourth generation even though the algorithm is run till the maximum number of generations are reached. The result is obtained in 3.6348 seconds (Table 3). It is important to note that the sequence is found in a few iterations even though it did not take place in the initial random population.

The algorithm is then run hundred times in order to obtain the average calculation time and average number of generations to obtain the best chromosome. The runs have depicted that at the average it took 1.4251 seconds to obtain the best chromosomes with a mean of 9.2 generations. In order to further test the proposed algorithm, the best chromosome is searched via an exhaustive search algorithm. That is, every possible combination and their corresponding fitness values are obtained in order to compare with the heuristics search results. The exhaustive search took 37.8926 seconds with the identical best chromosome and the fitness value.

The code was written in Matlab (version 7.7.0.471(R2008b)) using Genetic Algorithm and Direct Search Toolbox. The code was run on a computer with a Processor Intel Core 2 Duo CPU P8400 2.26 GHz, 4.00 GB RAM.

5. CONCLUSION

In this chapter, a genetic algorithm model was presented in order to determine the optimal disassembly sequence of a given product. The model provides quick and reliable input to the disassembly scheduling environments. As also stated by Keung *et al.* (2001), GAs do not make unrealistic assumptions such as linearity, convexity and/or differentiability.

This adds further importance to the proposed model and makes it even more desirable. For the example considered, the algorithm provided optimal disassembly sequence in a short execution time. The algorithm is practical, as it is easy to use, considers the precedence relationships and additional constraints in the product structure and is easily applicable to problems with multiple objectives. Future research will include embedding the proposed algorithm into a disassembly cell that contains a robotic manipulator, camera and visual recognition software.

REFERENCES

Bierwirth, C., & Mattfeld, D. C. (1999). Production scheduling and rescheduling with genetic algorithms. *Evolutionary Computation*, *7*(1), 1–18. doi:10.1162/evco.1999.7.1.1

Bierwirth, C., Mattfeld, D. C., & Kopfer, H. (1996). On permutation representations for scheduling problems. In Voigt, H. M. (Eds.), *Parallel Problem Solving from Nature -- PPSN IV, Lecture Notes in Computer Science* (pp. 310–318). Berlin, Germany: Springer-Verlag. doi:10.1007/3-540-61723-X_995

Gil, P., Pomares, J., Puente, S. V., Diaz, C., Candelas, F., & Torres, F. (2007). Flexible multisensorial system for automatic disassembly using cooperative robots. *International Journal of Computer Integrated Manufacturing*, *20*(8), 757–772. doi:10.1080/09511920601143169

Gungor, A., & Gupta, S. M. (1999). Issues in environmentally conscious manufacturing and product recovery: A survey. *Computers & Industrial Engineering*, *36*(4), 811–853. doi:10.1016/S0360-8352(99)00167-9

Hui, W., Dong, X., & Guanghong, D. (2008). A genetic algorithm for product disassembly sequence planning. *Neurocomputing*, *71*, 2720–2726. doi:10.1016/j.neucom.2007.11.042

Ilgin, M. A., & Gupta, S. M. (2010). Environmentally conscious manufacturing and product recovery (ECMPRO): A review of the state of the art. *Journal of Environmental Management*, *91*(3), 563–591. doi:10.1016/j.jenvman.2009.09.037

Keung, K. W., Ip, W. H., & Lee, T. C. (2001). The solution of a multi-objective tool selection model using the GA approach. *International Journal of Advanced Manufacturing Technology*, *18*, 771–777. doi:10.1007/s001700170001

Kongar, E., & Gupta, S. M. (2006). Disassembly sequencing using genetic algorithm. *International Journal of Advanced Manufacturing Technology*, *30*(5-6), 497–506. doi:10.1007/s00170-005-0041-x

Lambert, A. J. D., & Gupta, S. M. (2005). *Disassembly modeling for assembly, maintenance, reuse, and recycling*. Boca Raton, FL: CRC Press.

Lazzerini, B., & Marcelloni, F. (2000). A genetic algorithm for generating optimal assembly plans. *Artificial Intelligence in Engineering*, *14*, 319–329. doi:10.1016/S0954-1810(00)00011-X

Lee, D.-H., Kang, J.-G., & Xirouchakis, P. (2001). Disassembly planning and scheduling: Review and further research. *Journal of Engineering Manufacture*, *215*(B5), 695–709. doi:10.1243/0954405011518629

McGovern, S. M., & Gupta, S. M. (2007). A balancing method and genetic algorithm for disassembly line balancing. *European Journal of Operational Research*, *179*(3), 692–708. doi:10.1016/j.ejor.2005.03.055

McGovern, S. M., & Gupta, S. M. (2011). *The disassembly line*. New York, NY: McGraw Hill Inc.

Mitsubishi Co. (n.d.). *Movemaster-EX industrial micro-robot system model RV-M1 user manual.*

Pomares, J., Puente, S. T., Torres, F., Candelas, F. A., & Gil, P. (2004). Virtual disassembly of products based on geometric models. *Computers in Industry*, *55*(1), 1–14. doi:10.1016/j.compind.2004.03.001

Sanderson, A. C., Homem de Mello, L. S., & Zhang, H. (1990). Assembly sequence planning. *AI Magazine*, 62–81.

Seo, K.-K., Park, J.-H., & Jang, D.-S. (2001). Optimal disassembly sequence using genetic algorithms considering economic and environmental aspects. *International Journal of Advanced Manufacturing Technology*, *18*, 371–380. doi:10.1007/s001700170061

Shimizu, Y., Tsuji, K., & Nomura, M. (2007). Optimal disassembly sequence generation using a genetic programming. *International Journal of Production Research*, *45*(18-19), 4537–4554. doi:10.1080/00207540701440261

Torres, F., Gil, P., Puente, S. T., Pomares, J., & Aracil, R. (2004). Automatic PC disassembly for component recovery. *International Journal of Advanced Manufacturing Technology*, *23*(1-2), 39–46. doi:10.1007/s00170-003-1590-5

Torres, F., Puente, S., & Diaz, C. (2009). Automatic cooperative disassembly robotic system: task planner to distribute tasks among robots. *Control Engineering Practice*, *17*(1), 112–121. doi:10.1016/j.conengprac.2008.05.013

Torres, F., Puente, S. T., & Aracil, R. (2003). Disassembly planning based on precedence relations among assemblies. *International Journal of Advanced Manufacturing Technology*, *21*(5), 317–327. doi:10.1007/s001700300037

Valenzuela-Rendon, M., Uresti-Charre, E., & Monterrey, I. (1997). *A non-generational genetic algorithm for multiobjective optimization*. In Seventh International Conference on Genetic Algorithms.

KEY TERMS AND DEFINITIONS

Disassembly: *Disassembly* is the process of the systematic removal of desirable constituents (components and/or materials) from the original assembly so that there is no impairment to any useful constituent.

Disassembly Sequencing: The order in which the parts will be removed from the end-of-life product.

End-of-Life (EOL) Processing Options: EOL processing options, e.g., reuse, recycle, and remanufacturing, are effective ways to regain the materials and the components in electronic EOL products.

Genetic Algorithm: A heuristic approach that optimize rules by mimicking the Darwinian Law of survival of the fittest.

Product Recovery: Operations that aim at regaining the part and material value added to the product via reuse, remanufacturing and recycling.

Robotics: The study of machines that resemble a human and conduct mechanical, routine tasks on command.

Selective Disassembly: Disassembly operations limited to recyclable materials and reusable components in the EOL product.

Compilation of References

Aarno, D. (2007). *Intention recognition in human-machine collaborative systems*. Licentiate Thesis at the School of Computer Science and Communication, KTH.

Aarno, D., & Kragic, D. (2006). *Layered HMM for motion intention recognition*. In IEEE/RSJ International Conference on Intelligent Robots and Systems, IROS'06.

Achtelik, M., Bachrach, A., He, R., Prentice, S., & Roy, N. (2009). Stereo vision and laser odometry for autonomous helicopters in GPS-denied indoor environments. In *Proceedings of SPIE Conference on Unmanned Systems Technology XI*, Orlando, FL, USA. doi:10.1117/12.819082

ACTIVE. (2011). *Project brochure*. Retrieved on August 1, 2011, from http://www.active-fp7.e Alexander, A. D. (1972). Impacts of telemation on modern society. In *Proceedings of the 1st CISM–ITOMM Symposium*, (pp. 121–36). New York, NY: Springer.

Adler, J. R., Murphy, M. J., Chang, S. D., & Hancock, S. L. (1999). Image-guided robotic radiosurgery. *Neurosurgery, 44*(6), 1299–1307.

Adluru, N., Latecki, L. J., Sobel, M., & Lakaemper, R. (2008). Merging maps of multiple robots. In *Pattern Recognition ICPR 2008* (pp. 1-4).

Aggarwal, P., Syed, Z., Noureldin, A., & El-Sheimy, N. (2010). *MEMS-based integrated navigation*. Artech House.

AggieAir Flying Circus. (2011, March 3). *Home*. Retrieved March 25, 2011, from http://aggieair.usu.edu/

Ahmed, K., Khan, M., Vats, A., Nagpal, K., Priest, O., & Patel, V. (2009). Current status of robotic assisted pelvic surgery and future developments. *International Journal of Surgery, 7*, 431–440. doi:10.1016/j.ijsu.2009.08.008

Aizawa, N., & Kubota, N. (2009). Intelligent formation control based on directionality of multi-agent system. In *Robotic Intelligence in Informationally Structured Space, RIISS '09* (pp. 87-92).

Albu-Schaffer, A., Haddadin, S., Ott, C., Stemmer, A., Wimbock, T., & Hirzinger, G. (2007). The DLR lightweight robot: Design and control concepts for robots in human environments. *International Journal of Industrial Robot, 34*, 376–385. doi:10.1108/01439910710774386

Alexakis, T., Boadi, D., Quong, D., Groboillot, A., O'Neill, I., & Poncelet, D. (1995). Microencapsulation of DNA within alginate microspheres and crosslinked chitosan membranes for in vivo application. *Applied Biochemistry and Biotechnology, 50*(1), 93–106. doi:10.1007/BF02788043

Alexander, A. D. (1973). Impacts of telemation on modern society. In *Proceedings of the Annual Meeting of Human Factors and Ergonomics Society, 17*(2), 299–304.

Altug, E., Ostrowski, J. P., & Mahony, R. (2002). Control of a quadrotor helicopter using visual feedback. In *Proceedings of IEEE International Conference on Robotics and Automation 2002,* (vol. 1, 72–77). doi:10.1109/ROBOT.2003.1242264

Amigoni, F. (2008). Experimental evaluation of some exploration strategies for mobile robots. *In Proceedings of the IEEE International Conference on Robotics and Automation* (pp. 2818–2823).

Amor, S. B., Huy, T. D., & Bui, M. (2006). A percolation based model for ATC simulation. I*EEE International Conference on Research, Innovation and Vision for the Future* (pp. 17-22).

Anderson, W. F., Killos, L., Sanders-Haigh, L., Kretchmer, P. J., & Diakumakos, E. G. (1980). Replacation and expression of thymidine kinase and human globin genes microinjected into mouse fibroblasts. *Proceedings of the National Academy of Sciences of the United States of America*, *77*, 5399–5403. doi:10.1073/pnas.77.9.5399

Andersson, L. A., & Nygards, J. (2009). On multi-robot map fusion by inter-robot observations. In *12th International Conference on Information Fusion, FUSION '09* (pp. 1712 -1721).

Ansorge, W. (1982). Improved system for capillary microinjection into living cells. *Experimental Cell Research*, *140*, 31–37. doi:10.1016/0014-4827(82)90152-5

Ansorge, W., & Pepperkok, R. (1988). Performance of an automated system for capillary microinjection into living cells. *Journal of Biochemical and Biophysical Methods*, *16*, 283–292. doi:10.1016/0165-022X(88)90062-0

Antoniou, G. A., Riga, C. V., Mayer, E. K., Cheshire, N. J. W., & Bicknell, C. D. (2011). Clinical applications of robotic technology in vascular and endovascular surgery. *Journal of Vascular Surgery*, *53*(2), 493–499. doi:10.1016/j.jvs.2010.06.154

Arai, K., & Mardiyanto, R. (2011). *Eye based HCI with moving keyboard for reducing fatigue effects.* 2011 Eighth International Conference on Information Technology: New Generations.

Arducopter. (n.d.). *Project hosting on Google Code*. Retrieved March 11, 2011, from http://code.google.com/p/arducopter/

Armaleo, D., Ye, G.-N., Klein, T. M., Shark, K. B., Sanford, J. C., & Johnston, S. A. (1990). Biolistic nuclear transformation of Saccharomyces cerevisiae and other fungi. *Current Genetics*, *17*(2), 97–103. doi:10.1007/BF00312852

Avraam, M., Horodinca, M., Romanescu, I., & Preumont, A. (2010). Computer controlled rotational MR-brake for wrist rehabilitation device. *Journal of Intelligent Material Systems and Structures*, *21*(15), 1543–1557. doi:10.1177/1045389X10362274

Aylward, R., & Paradiso, J. A. (2007). A compact, high-speed, wearable sensor network for biomotion capture and interactive media. *6th International Symposium on Information Processing in Sensor Networks,* (pp. 380-389).

Aylward, R., Lovell, S. D., & Paradiso, J. A. (2006). A compact, wireless, wearable sensor network for interactive dance ensembles. *International Workshop on Wearable and Implantable Body Sensor Networks, BSN 2006*, (pp. 4-70).

Backes, F., Seamon, L., & Fowler, J. (2008). Robotic radical hysterectomy and pelvic mphadenectomy for uterine rhabdomyosarcoma. *Journal of Robotic Surgery*, *2*, 97–200. doi:10.1007/s11701-008-0095-5

Baglietto, M., Paolucci, M., Scardovi, L., & Zoppoli, R. (2002). Information based multi-agent exploration. *IEEE International Workshop on Robot Motion and Control* (pp. 173-179).

Baik, S. H. (Ed.). (2010). *Robot surgery*. Vienna, Austria: IN-TECH.

Bailey, T., & Durrant-Whyte, H. F. (2006). Simultaneous localisation and mapping (SLAM): Part II state of the art. *Robotics and Automation Magazine*, *13*(3), 108–117. doi:10.1109/MRA.2006.1678144

Balafoutis, C. A., & Patel, R. V. (1991a). *Robot dynamic analysis of robot manipulators: A Cartesian tensor approach.* Dordrecht, The Netherlands: Kluwer Academic Publishers Group.

Balafoutis, C. A., & Patel, R. V. (1991b). Efficient computation of manipulator inertia matrices and direct dynamics problem. *IEEE Transactions on Systems, Man, and Cybernetics*, *19*(5), 1313–1320. doi:10.1109/21.44054

Ballantyne, G. H., Marescaux, J., & Giulianotti, P. C. (Eds.). (2004). *The primer of robotic and telerobotic surgery*. Philadelphia, PA: Lippincott Williams & Wilkins.

Bann, S., Khan, M., & Hernandez, J. (2003). Robotics in surgery. *Journal of the American College of Surgeons*, *196*(5), 784–795. doi:10.1016/S1072-7515(02)01750-7

Barber, M. A. (1911). A technique for the inoculation of bacteria and other substances into living cells. *The Journal of Infectious Diseases*, *8*, 348–352. doi:10.1093/infdis/8.3.348

Bargar, W. L. (2007). Robots in orthopaedic surgery - Past, present, and future. *Clinical Orthopaedics and Related Research, 463*, 31–36.

Barker, C. R. (1985). A complete classification of planar four-bar linkages. *Mechanism and Machine Theory, 20*(6), 535–554. doi:10.1016/0094-114X(85)90071-0

Barrera, A., Bach, J., Kazanzides, P., & Haider, H. (2007). Validation of an ASTM standard proposed to assess localizer functionality of CAOS systems: a joint effort by three laboratories. In *Proceedings of the 20th Annual Congress of Intl. Society for Technology in Arthroplasty* (ISTA), Paris, (pp. 81–81).

Bar-Shalom, Y., & Fortmann, T. E. (1987). *Tracking and data association. Mathematics in Science and Engineering* (*Vol. 179*). San Diego, CA: Academic Press Professional.

Bauer, J., Lee, B. R., & Stoianovici, D. (2001). Remote percutaneous renal access using a new automated telesurgical robotic system. *Telemed Journal and Electronic Health, 7*, 341–346. doi:10.1089/15305620152814746

Baum, L. E., Petrie, T., Soules, G., & Weiss, N. (1970). A maximization technique occurring in the statistical analysis of probabilistic functions of Markov chains. *Annals of Mathematical Statistics, 41*(1), 164–171. doi:10.1214/aoms/1177697196

Benabid, A. L., et al. (1987). Computer-driven robot for stereotactic surgery connected to CT scan and magnetic resonance imaging. Technological design and preliminary results. *Applied Neurophysiology, 50*(1-6), 153–154.

Benabid, A., Cinquin, P., Lavalle, S., Le Bas, J., Demongeot, J., & De Rougemont, J. (1987). Computer-driven robot for stereotactic surgery connected to CT scan and magnetic resonance imaging. Technological design and preliminary results. *Applied Neurophysiology, 50*(1), 153–154.

Bernstein, N. L., Lawrence, D. A., & Pao, L. Y. (2005). Friction modeling and compensation for haptic interfaces. *First IEEE Joint Eurohaptics Conference and Symposium on Haptic Interfaces for Virtual Environment and Teleoperator Systems,* (pp. 290-295).

Bernstein, R. W., Scott, M., & Solgaard, O. (2004). *BioMEMS for high-throughput handling and microinjection of embryos.* Paper presented at the MEMS/MOEMS Technologies and Applications II, Beijing, China.

Bernstein, R. W., Xiaojing, Z., Zappe, S., Fish, M., Scott, M., & Solgaard, O. (2003, 8-12 June 2003). *Characterization of drosophila embryos immobilized by fluidic microassembly*. Paper presented at the TRANSDUCERS, Solid-State Sensors, Actuators and Microsystems, 12th International Conference on, 2003.

Bicker, R., Burn, K., Hu, Z., Pongean, W., & Bashir, A. (2004). The early development of remote tele-manipulation systems. *International Symposium on History of Machines and Mechanims*, (pp. 391-404).

Bierwirth, C., & Mattfeld, D. C. (1999). Production scheduling and rescheduling with genetic algorithms. *Evolutionary Computation, 7*(1), 1–18. doi:10.1162/evco.1999.7.1.1

Bierwirth, C., Mattfeld, D. C., & Kopfer, H. (1996). On permutation representations for scheduling problems. In Voigt, H. M. (Eds.), *Parallel Problem Solving from Nature -- PPSN IV, Lecture Notes in Computer Science* (pp. 310–318). Berlin, Germany: Springer-Verlag. doi:10.1007/3-540-61723-X_995

Bin, A., Massie, T., & Vayner, V. (1995). *Sensory feedback exoskeleton Armmaster.* (US Patent 1995/006376).

Birk, A., & Carpin, S. (2006). Merging occupancy grid maps from multiple robots. *Proceedings of the IEEE, 94*(7), 1384–1397. doi:10.1109/JPROC.2006.876965

Bolmsjo, G., Neveryd, H., & Eftring, H. (1995). Robotics in rehabilitation. *IEEE Transactions on Rehabilitation Engineering, 3*(1), 77–83. doi:10.1109/86.372896

Bolopion, A., Cagneau, B., Redon, S., & Régnier, S. (2010). Haptic teleoperation for 3D microassembly of spherical objects. *IEEE Transaction on Mechatronics, 99*, 1–12.

Borenstein, J. (1995). Control and kinematic design for multi-degree-of-freedom mobile robots with compliant linkage. *IEEE Transactions on Robotics and Automation, 11*(1), 21–35. doi:10.1109/70.345935

Borenstein, J., & Koren, Y. (1989). Real-time obstacle avoidance for fast mobile robots. *IEEE Transactions on Systems, Man, and Cybernetics*, *19*(5), 1179–1187. doi:10.1109/21.44033

Borenstein, J., & Koren, Y. (1991). The vector field histogram – Fast obstacle avoidance for mobile robots. *IEEE Journal on Robotics and Automation*, *7*(3), 278–288. doi:10.1109/70.88137

Borner, M. (1998). Experiences with the ROBODOC system in more than 1000 cases. In *Computer Aided Radiology and Surgery* (pp. 689–693). Amsterdam, The Netherlands: Elsevier.

Bouabdallah, S., & Siegwart, R. (2007a). Full control of a quadrotor. In *Proceedings of IEEE/RSJ International Conference on Intelligent Robots and Systems, 2007,* (pp. 153-158). doi:10.1109/IROS.2007.4399042

Bouabdallah, S., Murrieri, P., & Siegwart, R. (2004). Design and control of an indoor micro quadrotor. In *Proceedings of IEEE International Conference on Robotics and Automation, 2004,* (vol. 5, 4393-4398). doi:10.1109/ROBOT.2004.1302409

Bouabdallah, S., Noth, A., & Siegwart, R. (2004). PID vs LQ control techniques applied to an indoor micro quadrotor. In *Proceedings of IEEE/RSJ International Conference on Intelligent Robots and Systems 2004*, (vol. 3, pp. 2451-2456). doi:10.1109/IROS.2004.1389776

Bouabdallah, S., Becker, M., & Siegwart, R. (2007). Autonomous miniature flying robots: Coming soon! Research, development, and results. *IEEE Robotics & Automation Magazine*, *14*(3), 88–98. doi:10.1109/MRA.2007.901323

Bouabdallah, S., & Siegwart, R. (2005). Backstepping and sliding-mode techniques applied to an indoor micro quadrotor. In *Proceedings of IEEE International Conference on Robotics and Automation, 2005*, 2247–2252. doi:10.1109/ROBOT.2005.1570447

Bouabdallah, S., & Siegwart, R. (2007b). Design and control of a miniature quadrotor. In Valavanis, K. (Ed.), *Design and control of a miniature quadrotor*. Springer.

Boussif, O., Lezoualc'h, F., Zanta, M. A., Mergny, M. D., Scherman, D., & Demeneix, B. (1995). A versatile vector for gene and oligonucleotide transfer into cells in culture and in vivo: polyethylenimine. *Proceedings of the National Academy of Sciences of the United States of America*, *92*(16), 7297–7301. doi:10.1073/pnas.92.16.7297

Bowersox, J., Shah, A., Jensen, G., Hill, J., Cordts, P., & Green, P. (1996). Vascular applications of telepresence surgery: initial feasibility studies in swine. *Journal of Vascular Surgery*, *23*(2), 281–287. doi:10.1016/S0741-5214(96)70272-0

Bozovic, (Ed.). (2007). *Medical robotics*. Vienna, Austria: IN-TECH.

Brandt, G., Radermacher, K., & Zimolong, A. (2000). CRIGOS: Development of a compact robot for image-guided orthopedic surgery. *Der Orthopade*, *29*(7), 645–649. doi:10.1007/s001320050506

Brandt, G., Zimolong, A., & Carrat, L. (1999). CRIGOS: A compact robot for image-guided orthopedic surgery. *IEEE Transactions on Information Technology in Biomedicine*, *3*(4), 252–260. doi:10.1109/4233.809169

Bratman, M. (1999). *Faces of intention: Selected essays on intention and agency*. Cambridge University Press. doi:10.1017/CBO9780511625190

Bratman, M. (1999). *Intention, plans, and practical reason*. Harvard University Press.

Breant, F., & Paviot-Adet, E. (1992). OCCAM prototyping from hierarchical Petri nets. In Becker, M., Litzler, L., & Trehel, M. (Eds.), *Transputers '92 Advanced Research and Industrial Applications* (pp. 189–209). Amsterdam, Netherlands: IOS Press.

Breda, G., Nakada, S. Y., & Rassweiler, J. J. (2001). Future developments and perspectives in laparoscopy. *European Urology*, *40*, 84–91. doi:10.1159/000049753

Brinster, R. L., Chen, H. Y., Trumbauer, M., Senear, A. W., Warren, R., & Palmiter, R. D. (1981). Somatic expression of herpes thymidine kinase in mice following injection of a fusion gene into eggs. *Cell*, *27*, 223–231. doi:10.1016/0092-8674(81)90376-7

Bronzino, J. D. (1990). Regulation of medical device innovation. In Bronzino, J. D. (Ed.), *The biomedical engineering handbook*. Hartfort, CT: CRC Press.

Brouwer, I., Ustin, J., Bentley, L., Sherman, A., Dhruv, N., & Tendick, F. (2001). Measuring in vivo animal soft tissue properties for haptic modeling in surgical simulation. In Westwood, J. D. (Eds.), *Medicine meets virtual reality*. Amsterdam, The Netherlands: IOS Press.

Brown, W. T., Wu, X., & Wen, B. C. (2007). Early results of CyberKnife imageguided robotic stereotactic radiosurgery for treatment of lung tumors. *Journal of Computer Aided Surgery, 12*, 253–261.

Burckhart, C. W., Flury, P., & Glauser, D. (1995). Stereotactic brain surgery. *IEEE Engineering in Medicine and Biology, 14*(3), 314–317. doi:10.1109/51.391771

Burgard, W., Cremers, A. B., Fox, D., Hähnel, D., Lakemeyer, G., & Schulz, D. ... Thrun, S. (1998). The interactive museum tour-guide robot. In *Proceedings of the AAAI Fifteenth National Conference on Artificial Intelligence.*

Burgard, W., Fox, D., Moors, M., Simmons, R., & Thrun, S. (2000). Collaborative multi-robot exploration. In *Proceedings of the IEEE International Conference on Robotics and Automation* (pp. 476–481).

Burgard, W., Moors, M., & Schneider, F. (2002). Collaborative exploration of unknown environments with teams of mobile robots. *Dagstuhl Seminar on Plan-based Control of Robotic Agents* (pp. 187-215).

Burgard, W., Moors, M., Stachniss, C., & Schneider, F. E. (2005). Coordinated multi-robot exploration. *IEEE Transactions on Robotics, 21*(3), 376–386. doi:10.1109/TRO.2004.839232

Burghart, C. R., Muenchenberg, J. E., & Rembold, U. (1998). A system for robot assisted maxillofacial surgery. *Studies in Health Technology and Informatics, 50*, 220–226.

Burks, B. L., de Saussure, G., Weisbin, C. R., Jones, J. P., & Hamel, W. R. (1987). Autonomous navigation, exploration, and recognition using the HERMIES-IIB robot. *IEEE Expert, 2*(4), 18–27. doi:10.1109/MEX.1987.5006527

Butler, B., Esser, R., & Mattmann, R. (1991). A distributed simulator for high order Petri nets. In Rozenberg, G. (Ed.), *Advances in Petri Nets LNCS 483* (pp. 47–63). Berlin, Germany: Springer-Verlag.

Cadeddu, J. A., Stoianovici, D., Chen, R. N., Moore, R. G., & Kavoussi, L. R. (1998b). Stereotactic mechanical percutaneous renal access. *Journal of Endourology, 12*, 121–125. doi:10.1089/end.1998.12.121

Cadeddu, J. A., Stoianovici, D., & Kavoussi, L. R. (1998a). Robotic surgery in urology. *The Urologic Clinics of North America, 25*(1), 75–85. doi:10.1016/S0094-0143(05)70434-1

Cadierre, G. (2001). Feasibility of robotic laparoscopic surgery. *World Journal of Surgery, 25*, 1467–1477.

Caflisch, R. E. (1998). Monte Carlo and quasi-Monte Carlo methods. *Acta Numerica, 7*, 1–49. doi:10.1017/S0962492900002804

Caloini, A., Magnani, G., & Pezze, M. (1998). A technique for designing robotic control systems based on Petri nets. *IEEE Transactions on Control Systems Technology, 6*(1), 72–87. doi:10.1109/87.654878

Camarillo, D. B., Krummel, T. M., & Salisbury, J. K. J. (2004). Robotic technology in surgery: past, present, and future. *American Journal of Surgery, 188*(4A Suppl), 2-15.

Camarillo, D. (2004). Robotic technology in surgery: past, present, and future. *American Journal of Surgery, 188*, 2–15. doi:10.1016/j.amjsurg.2004.08.025

Capek, K. (2001). *R.U.R. (Rossum's Universal Robots)*. Mineola, NY: Dover Publications.

Caprari, G., Breitenmoser, A., Fischer, W., Hurzeler, C., Tache, F., & Siegwart, R. ... Moser, R. (2010, October). *Highly Compact Robots for Inspection of Power Plants*, Paper presented at *2010 1st International Conference on Applied Robotics for the Power Industry (CARPI)*, (pp. 1-6), Montreal, QC.

Carigan, C., & Liszka, M. S. (2007). *Portable arm exoskeleton for shoulder rehabilitation*. US Patent 2007/0225620.

Carigan, C., Liszka, M., & Roderick, S. (2005). Design of an arm exoskeleton with scapula motion for shoulder rehabilitation. *International Conference on Advanced Robotics ICAR05* (pp. 524-531).

Carigan, C., Tang, J., Roderick, S., & Naylor, M. (2007) A configuration-sapce approach to controlling a rehabilitation arm exoskeleton. *IEEE 10th International Conference on Rehabilitation Robotics*, (pp. 179-187).

Carlone, L., Ng, M., Du, J., Bona, B., & Indri, M. (2010). Simultaneous localization and mapping using rao-blackwellized particle filters in multi robot systems. *Journal of Intelligent & Robotic Systems*, *63*, 283–307. doi:10.1007/s10846-010-9457-0

Carpin, S. (2008). Merging maps via Hough transform. In *IEEE/RSJ International Conference on Intelligent Robots and Systems, IROS 2008* (pp. 1878 -1883).

Carpin, S. (2008). Fast and accurate map merging for multi-robot systems. *Autonomous Robots*, *25*(3), 305–316. doi:10.1007/s10514-008-9097-4

Carpin, S., Birk, A., & Jucikas, V. (2005). On map merging. *Robotics and Autonomous Systems*, *53*(1), 1–14. doi:10.1016/j.robot.2005.07.001

Chambers, R., & Chambers, E. L. (1961). *Exploration into the nature of living cell*. Cambridge, MA: Harvard University. Press.

Chang, S. D., & Adler, J. R. (2001). Robotics and radiosurgery--The cyberknife. *Stereotactic and Functional Neurosurgery*, *76*(3-4), 204–208. doi:10.1159/000066719

Chang, W., Lin, C., & Lin, L. (2005). A note on Grashof's theorem. *Journal of Marine Science and Technology*, *13*(4), 239–248.

Chao, H. (2010). *Cooperative remote sensing and actuation using networked unmanned vehicles*. Doctoral Dissertation, Utah State University. Retrieved from http://digitalcommons. usu.edu/ctd/597

Chao, H., Baumann, M., Jensen, A., Chen, Y. Q., Cao, Y., Ren, W., & McKee, M. (2008). Band-reconfigurable multi-UAV-based cooperative remote sensing for real-time water management and distributed irrigation control. In *Proceedings of International Federation of Automatic Control World Congress.* Seoul, Korea.

Chao, H., Cao, Y., & Chen, Y. Q. (2010). Autopilots for small unmanned aerial vehicles: A survey. *International Journal of Control Automation and Systems*, *8*(1), 36–44. doi:10.1007/s12555-010-0105-z

Chao, H., Coopmans, C., Long, D., & Chen, Y. Q. (2010). A comparative evaluation of low-cost IMUs for unmanned autonomous systems. In *Proceedings of IEEE Conference on Multisensor Fusion and Integration for Intelligent Systems, 2010*, 211–216. doi:.doi:10.1109/MFI.2010.5604460

Charles, S., Das, H., Ohm, T., Boswell, C., Rodriguez, G., Steele, R., & Istrate, D. (1997). Dexterity-enhanced telerobotic microsurgery. *ICAR '97 Proceedings of IEEE, 8th International Conference on Advanced Robotics*.

Charniak, E., & Goldman, R. (1993). A Bayesian model of plan recognition. *Artificial Intelligence*, *64*, 53–79. doi:10.1016/0004-3702(93)90060-O

Chen, J., Kwong, K., Chang, D., Luk, J., & Bajcsy, R. (2005). Wearable sensors for reliable fall detection. *27th Annual International Conference of the Engineering in Medicine and Biology Society, IEEE-EMBS 2005*, (pp. 3551-3554).

Chen, X., Zhuang, W., & Tan, J. (2009). The maintaining of communication links quality in unknown environment. In *International Conference on Information and Automation, ICIA '09* (pp. 218-223).

Chen, P., & Sheng, W. (2011). Transformative CAD based industrial robot program generation. *Journal of Robotics and Computer-Integrated Manufacturing*, *27*, 942–948. doi:10.1016/j.rcim.2011.03.006

Chiao, J., Goldman, J., Heck, D., Kazanzides, P., Peine, W., & Stiehl, J. (2008). Metrology and standards needs for some categories of medical device. *Journal of Research of the NIST*, *113*(2), 121–129.

Chinzei, K., et al. (2000). MR compatible surgical assist robot: system integration and preliminary feasibility study. In *Proceedings of Third International Conf. on Medical Image Computing and Computer Assisted Intervention (MICCAI)*, (pp. 921–930).

Chinzei, K., & Miller, K. (2001). Towards MRI guided surgical manipulator. *Medical Science Monitor*, *7*, 153–163.

Choset, H. (1996). *Sensor based motion planning: The hierarchical generalized Voronoi graph.* Doctoral dissertation, California Institute of Technology, Pasadena, CA, USA.

Chouchourelou, A., Matsuka, T., Harber, K., & Shiffrar, M. (2006). The visual analysis of emotional actions. *Social Neuroscience, 1*, 63–74. doi:10.1080/17470910600630599

Chou, T.-H. W., Biswas, S., & Lu, S. (2004). Gene delivery using physical methods: An overview. In Heiser, W. C. (Ed.), *Gene delivery to mammalian cells: Nonviral gene transfer techniques* (pp. 147–165). Totowa, NJ: Humana Press.

Ciuti, G., Valdastri, P., Menciassi, A., & Dario, P. (2010). Robotic magnetic steering and locomotion of capsule endoscope for diagnostic and surgical endoluminal procedures. *Robotica, 28*, 199–207. doi:10.1017/S0263574709990361

Clarke, T. J., Bradshaw, M. F., Field, D. T., Hampson, S. E., & Rose, D. (2005). The perception of emotion from body movement in point-light displays of interpersonal dialogue. *Perception, 34*, 1171–1180. doi:10.1068/p5203

Cleary, K., Melzer, A., Watson, W., Kronreif, G., & Stoianovici, D. (2006). Interventional robotic systems: Applications and technology state-of-the-art. *Minimally Invasive Therapy, 15*(2), 101–113. doi:10.1080/13645700600674179

Cleary, K., & Nguyen, C. C. (2003). Medical robotics. In Zouridakis, G., & Moore, J. (Eds.), *Biomedical technology and devices handbook. CRC Press*. LLC.

Cloud Cap Technology. (n.d.). *Piccolo II Autopilot for unmanned systems*. Retrieved March 11, 2011, from http://www.cloudcaptech.com/piccolo_II.shtm

Cohn, L. H. (2006). Future directions in cardiac surgery. *The American Heart Hospital Journal, 4*, 174–178. doi:10.1111/j.1541-9215.2005.05471.x

Cole, G., Pilitsis, J., & Fischer, G. S. (2009). Design of a robotic system for MRI-guided deep brain stimulation electrode placement. *IEEE International Conference on Robotics and Automation* (ICRA), Kobe, (pp. 4450–4456).

Comparetti, M. D., De Momi, E., Vaccarella, A., Riechmannz, M., & Ferrigno, G. (2011). Optically tracked multi-robot system for keyhole neurosurgery. In *Proceedings of the IEEE International Conference on Robotics and Automation* (ICRA), Shanghai, (pp. 661–667).

Constantin, A. (2006). Evaluating arm movement imitation. *American Journal of Undergraduate Research, 4*, 9–12.

Cooper, T., Cheng, A., Phillips, I., Wild, A., & Meech, J. (2010, March). Design and tele-operation of a lunar excavator for the NASA Centennial Challenge. *Proceedings of the 12th International Conference on Engineering, Science, Construction, and Operations in Challenging Environments*, Honolulu, Hawaii, USA, (pp. 1031-1038).

Correll, N., Bachrach, J., Vickery, D., & Rus, D. (2009). Ad-hoc wireless network coverage with networked robots that cannot localize. In *IEEE International Conference on Robotics and Automation, ICRA '09* (pp. 3878-3885).

Cortes, J., Martinez, S., & Bullo, F. (2006). Robust rendezvous for mobile autonomous agents via proximity graphs in arbitrary dimensions. *Automatic Control. IEEE Transactions on, 51*(8), 1289–1298. doi:10.1109/TAC.2006.878713

Cotin, S., Delingette, H., & Ayache, N. (2000). A hybrid elastic model allowing realtime cutting, deformation and force-feedback for surgery training and simulation. *Visual Computing Journal, 16*, 437–452. doi:10.1007/PL00007215

Craft, J., Wilson, J., Chu, P., Zacny, K., & Davis, K. (2009). Percussive digging systems for robotic exploration and excavation of planetary and lunar regolith. *Proceedings of 2009 IEEE Aerospace Conference*, Big Sky, MT, USA, (pp. 1-7).

Craft, J., Wilson, J., Chu, P., Zacny, K., & Davis, K. (2010, October). Percussive digging systems for planetary research. *IEEE Aerospace and Electronic Systems Magazine, 25*(10), 21–26. doi:10.1109/MAES.2010.5631722

Craig, J. J. (1989). *Introduction to robotics* (2nd ed.). Reading, PA: Addison-Wesley.

Cretu, E. (2010). *Biomedical applications of MEMS devices*. Retrieved on August 2011, from http://www.ece.ubc.ca/~leos/pdf/e331/notes/MEMS.pdf

Crisp, J. A., Adler, M., Matijevic, J. R., Squyres, S. W., Arvidson, R. E., & Kass, D. M. (2003). Mars Exploration Rover mission. *Journal of Geophysical Research, 108*(E12), 8061. doi:10.1029/2002JE002038

Crowley, J. (1985). Navigation for an intelligent mobile robot. *IEEE Journal on Robotics and Automation*, *1*(1), 31–41. doi:10.1109/JRA.1985.1087002

CSOIS - Center for Self-Organizing Intelligent Systems. (n.d.). Retrieved February 21, 2011, from http://www.csois.usu.edu/

Curexo Tech. Co. (2009). *ROBODOC®: The world's first surgical robot*. Technical report. Retrieved on August 2011, from http://pchrissanthis.com/Robodoc_investor_pitch.pdf

Cutting, J. E., & Kozlowski, L. T. (1977). Recognizing friends by their walk-gait perception without familiarity cues. *Bulletin of the Psychonomic Society*, *9*, 353–356.

Dagalakis, N., Kim, Y., Sawyer, D., & Shakarji, C. (2007). Development of tools for measuring the performance of computer assisted orthopaedic hip surgery systems. *Proceeding of the ACM Workshop on Performance Metrics for Intelligent Systems (PerMIS)*, Gaithersburg, (pp. 180–187).

Dai, J. S., & Shah, P. (2002). Orientation capability of planar serial manipulators using rotatability analysis based on workspace decomposition. *Proceedings of the Institution of Mechanical Engineers. Part C, Journal of Mechanical Engineering Science*, *216*(C4), 275–288. doi:10.1243/0954406021524990

Dai, J. S., & Shah, P. (2003). Orientation capability of planar manipulators using virtual joint angle analysis. *Mechanism and Machine Theory*, *28*(3), 241–252. doi:10.1016/S0094-114X(02)00118-0

Dallemagne, B., Perretta, S., & Marescaux, J. (2009). Education and e-training in NOTES. *WeBSurg.com*, *9*(3). Retrieved on August 1, 2011, from http://www.eats.fr/doi-ed01en0022.htm

Danilchenko, A., & Fitzpatrick, J. M. (2010). General approach to error prediction in point registration. In *Proceedings of SPIE Medical Imaging,* 7625–0F, San Diego, CA, (pp. 1–14).

Daprati, E., Wriessnegger, S., & Lacquaniti, F. (2007). Kinematic cues and recognition of self-generated actions. *Experimental Brain Research*, *177*, 31–44. doi:10.1007/s00221-006-0646-9

Dario, P., Hannaford, B., & Menciassi, A. (2003). Smart surgical tools and augmenting devices. *IEEE Transactions on Robotics and Automation*, *19*(5), 782–792. doi:10.1109/TRA.2003.817071

Das, H., Zak, H., Johnson, J., Crouch, J., & Frambach, D. (1999). Evaluation of a telerobotic system to assist surgeons in microsurgery. *Computer Aided Surgery*, *4*, 15–25. doi:10.3109/10929089909148155

Dautenhahn, K., Woods, S., Kaouri, C., Walters, M. L., Koay, K. L., & Werry, I. (2005). What is a robot companion – Friend, assistant or butler? In *Proceedings of IEEE/RSJ International Conference on Intelligent Robots and Systems (IROS)*, (pp. 1488-1493).

David, R., & Alla, H. (1992). *Petri Nets and Grafcet: Tools for modelling discrete events systems*. UK: Prentice-Hall International.

Davies, B. (2000). A review of robotics in surgery. *Proceedings of the Institute of Mechanical Engineers, Part H, 214*, 129–140.

Davies, B. L., Hibberd, R. D., Coptcoat, M. J., & Wickham, J. E. (1989). A surgeon robot prostatectomy--a laboratory evaluation. *Journal of Medical Engineering & Technology*, *13*, 273–277. doi:10.3109/03091908909016201

Davies, B. L., Hibberd, R. D., Ng, W. S., Timoney, A. G., & Wickham, J. E. (1991). The development of a surgeon robot for prostatectomies. *Proceedings of the Institute of Mechanical Engineers Part H*, *205*, 35–38. doi:10.1243/PIME_PROC_1991_205_259_02

De Gelder, B. (2009). Why bodies? Twelve reasons for including bodily expressions in affective neuroscience. *Philosophical Transactions of the Royal Society B. Biological Sciences*, *364*, 3475–3484. doi:10.1098/rstb.2009.0190

De Lange, F. P., Spronk, M., Willems, R. M., Toni, I., & Bekkering, H. (2008). Complementary systems for understanding action intentions. *Current Biology*, *18*, 454–457. doi:10.1016/j.cub.2008.02.057

Dedeoglu, G., & Sukhatme, G. (2000). Landmark-based matching algorithm for cooperative mapping by autonomous robots. In *Distributed Autonomous Robotics Systems* (pp. 251–260). Springer-Verlag.

DeLaurentis, K., & Mavroidis, C. (2002). Mechanical design of a shape memory alloy actuated prosthetic hand. *Technology and Health Care*, *10*(2), 91–106.

Dellaert, F., Fox, D., Burgard, W., & Thrun, S. (1999). Monte Carlo localization for mobile robots. In *Proceedings of IEEE International Conference on Robotics and Automation (ICRA)*, (pp. 1322-1328).

Dellon, B., & Matsuoka, Y. (2007). Prosthetics, exoskeletons, and rehabilitation. *IEEE Robotics & Automation Magazine*, 30–34. doi:10.1109/MRA.2007.339622

Demi, B., Ortmaier, T., & Seibold, U. (2005). *The touch and feel in minimally invasive surgery*. IEEE International Workshop on Haptic Audio Visual Environments and their Applications.

Dempster, A. P., Laird, N. M., & Rubin, D. B. (1977). Maximum likelihood from incomplete data via the EM algorithm. *Journal of the Royal Statistical Society. Series B. Methodological*, *39*(1), 1–38.

Dennett, D. C. (1987). *The intentional stance*. MIT Press.

Department of Defense. (2005). *Unmanned aircraft systems (UAS) roadmap, 2005-2030*. Washington, DC: Office of the Secretary of Defense.

Desforges, X., Habbadi, A., & Archimède, B. (2011). Design methodology for smart actuator services for machine tool and machining control and monitoring. *Journal of Robotics and Computer-Integrated Manufacturing*, *27*, 963–976. doi:10.1016/j.rcim.2011.03.004

Desrochers, A. D., & Al-Jaar, R. Y. (1995). *Applications of Petri Nets in manufacturing systems: Modeling, control and performance analysis*. New York, NY: IEEE Press.

Dielmann, A., & Renals, S. (2004). Dynamic Bayesian networks for meeting structuring. In *Proceedings of the IEEE International Conference on Acoustics, Speech, and Signal Processing*, (pp. 629–632).

Dieterich, S., Tang, J., Rodgers, J., & Cleary, K. (2003). Skin respiratory motion tracking for stereotactic radiosurgery using the CyberKnife. In *Proceedings of the 17th CARS, Computer Assisted Radiology and Surgery*, (vol. 1256, pp. 130–136).

Dietterich, T. G. (2002). Machine learning for sequential data: A review. *Lecture Notes in Computer Science*, *2396*, 15–30. doi:10.1007/3-540-70659-3_2

Digi International Inc. (n.d.). *ConnectPort TS W wireless serial servers datasheet*. Retrieved September 3, 2011, from http://www.digi.com/pdf/ds_connectporttsw.pdf

DiMaio, S., Hanuschik, M., & Kreaden, U. (2011). The da Vinci Surgical System. In Rosen, J., Hannaford, B., & Satava, R. M. (Eds.), *Surgical robotics—Systems applications and visions* (pp. 199–217). Boston, MA: Springer.

Dimarogonas, D. V., & Johansson, K. H. (2008). Decentralized connectivity maintenance in mobile networks with bounded inputs. In *IEEE International Conference on Robotics and Automation, ICRA 2008* (pp. 1507-1512).

Dimarogonas, D. V., & Kyriakopoulos, K. J. (2007). On the rendezvous problem for multiple nonholonomic agents. *IEEE Transactions on Automatic Control*, *52*(5), 916–922. doi:10.1109/TAC.2007.895897

Dissanayake, A., Newman, P., Clark, S., Durrant-Whyte, H. F., & Csorba, M. (2001). A solution to the simultaneous localization and map building problem. *IEEE Transactions on Robotics and Automation*, *17*, 229–241. doi:10.1109/70.938381

Dissanayake, G., Newman, P., Durrant-Whyte, H. F., Clark, S., & Csobra, M. (2001). A solution to the simultaneous localisation and map building (SLAM) problem. *IEEE Transactions on Robotics and Automation*, *17*(3), 229–241. doi:10.1109/70.938381

Dittrich, W. H., Troscianko, T., Lea, S. E. G., & Morgan, D. (1996). Perception of emotion from dynamic point-light displays represented in dance. *Perception*, *25*, 727–738. doi:10.1068/p250727

Dixon, C., & Frew, E. W. (2007). Maintaining optimal communication chains in robotic sensor networks using mobility control. In *Proceedings of the 1st international conference on Robot communication and coordination* (pp. 1-8). Athens, Greece: IEEE Press.

Doarn, C. R., Nicogossian, A. E., & Merrel, R. C. (1998). Applications of telemedicine in the United States Space Program. *Telemedicine Journal*, *4*(1), 19–30. doi:10.1089/tmj.1.1998.4.19

Dogru, S., Erkmen, A. M., & Erkmen, I. (2009). Tracking a sycophant wireless sensor network for its seamless integration to mobile wireless sensor networks. In *2nd International Conference on Adaptive Science & Technology, ICAST, 2009* (pp. 110-116).

Dohi, T. (2005). Surgical robots and 3D displays for computer-aided surgery. *Endourooncology*, *6*, 15–26. doi:10.1007/4-431-27173-2_2

Draganfly.com. (n.d.). *Industrial aerial video systems & UAVs*. Retrieved February 21, 2011, from http://www.draganfly.com/

Drone, A. R. com USA. (n.d.). *Parrot Wi-Fi quadricopter- Augmented reality games on iPhone, iPod touch and iPad.* Retrieved February 21, 2011, from http://ardrone.parrot.com/parrot-ar-drone/usa/

Durrant-Whyte, H. F., & Bailey, T. (2006). Simultaneous localisation and mapping (SLAM): Part I the essential algorithms. *Robotics and Automation Magazine*, *13*(2), 99–110. doi:10.1109/MRA.2006.1638022

Dylan, A. S., & Matarić, M. J. (2005). Insights toward robot-assisted evacuation. *International Journal of the Robotics*, *19*, 797–818.

Dynia, M., Kutylowski, J., Heide, F. M. A. D., & Schrieb, J. (2007). Local strategies for maintaining a chain of relay stations between an explorer and a base station. In *Proceedings of the Nineteenth Annual ACM Symposium on Parallel Algorithms and Architectures* (pp. 260-269).

Dynia, M., Kutyłowski, J., Lorek, P., & auf der Heide, F. (2006). *Maintaining communication between an explorer and a base station.* In Biologically Inspired Cooperative Computing.

Ediz, K., & Olgac, N. (2005). Effect of mercury columan on the microdynamics of the piezo-driven pipettes. *Journal of Biomechanical Engineering*, *127*, 531–535. doi:10.1115/1.1894368

Ehsani, M., Gao, Y., Gay, S. E., & Emadi, A. (2004, December). *Modern electric, hybrid electric, and fuel cell vehicles: Fundamentals, theory, and design.* Boca Raton, FL: CRC Press.

Eirik, L., Johansen, B., Gjelsvik, T., & Langø, T. (2009). Ultrasound based localization of wireless microrobotic endoscopic capsule for the GI tract. In *Proceedings of the 21st Conference of the Society for Medical Innovation and Technology* (SMIT), Sinaia, (p. 77).

Eirik, L., Johansen, B., Gjelsvik, T., & Lango, T. (2009). Ultrasound based localization of wireless microrobotic endoscopic capsule for the GI tract. *Proceeding of the 21st Conference of the Society for Medical Innovation and Technology (SMIT),* Sinaia, (pp. 77–79).

Eljamel, M. S. (2006). Robotic applications in epilepsy surgery. *International Journal of Medical Robotics and Computer Assisted Surgery*, *2*(3), 233–237. doi:10.1002/rcs.97

Eljamel, M. S. (2007). Validation of a neurosurgical robot using a phantom. *International Journal of Medical Robotics and Computer Assisted Surgery*, *3*(4), 372–377. doi:10.1002/rcs.153

Eljamel, M. S. (2009). Robotic neurosurgical applications: Accuracy and consistency or pure fantasy? *Stereotactic and Functional Neurosurgery*, *87*, 88–93. doi:10.1159/000202974

Engineering Services Inc. (n.d.). Retrieved September 2, 2011, from http://www.esit.com

Eppendorf, Narishige and Cellbiology Trading. (n.d.) Retrieved August 7, 2011, from http://www.eppendorf.com, www.narishige.co.jp, www.ais2.com

Erbatur, K., & Kurt, O. (2009). Natural ZMP trajectories for biped robot reference generation. *IEEE Transactions on Industrial Electronics*, *56*(3), 835–845. doi:10.1109/TIE.2008.2005150

Ergenc, F., & Olgac, N. (2007). New technology for cellular piercing: rotationally oscillating -injector, description and validation tests. *Biomedical Microdevices*, *9*(6), 885–891. doi:10.1007/s10544-007-9102-2

Esin, Y. H., Unel, M., & Yildiz, M. (2008). Formation control of multiple robots using parametric and implicit representations. In *ICIC '08: Proceedings of the 4th International Conference on Intelligent Computing* (pp. 558–565). Berlin, Germany: Springer-Verlag.

Esposito, J. M., & Dunbar, T. W. (2006). Maintaining wireless connectivity constraints for swarms in the presence of obstacles. In *Proceedings 2006 IEEE International Conference on Robotics and Automation, ICRA 2006* (pp. 946-951).

Estey, E. (2009). Robotic prostatectomy: The new standard of care or a marketing success. *Journal of Canadian Urological Association, 3*, 488–490.

Fan, M., Feng, Z. C., Agca, Y., & Critser, J. K. (2006). Vibration study of the piezodriven pipettes immersed in viscous liquids. *Journal of Applied Physics, 100*(7), 701–708. doi:10.1063/1.2353895

Farrell, J. A., Pang, S., & Li, W. (2003). Plume mapping via hidden Markov methods. *IEEE Transactions on Systems, Man, and Cybernetics. Part B, Cybernetics, 33*(6), 850–863. doi:10.1109/TSMCB.2003.810873

Faust, R. A. (Ed.). (2007). *Robotics in surgery: History, current and future applications*. Hauppauge, NY: Nova Science.

FDA. Food and Drug Administration. (2010). *510(k) and science report recommendations: Summary and overview of comments and next steps*. FDA Report. Retrieved on August 1, 2011, http://www.fda.gov/downloads/AboutFDA/CentersOffices/CDRH/CDRHReports/UCM239449.pdf

FDA. Food and Drug Administration. (2011). *Device advice: Comprehensive regulatory assistance*. Retrieved on August 1, 2011, from http://www.fda.gov/MedicalDevices/DeviceRegulationandGuidance/default.htm

Featherstone, R. (2008). *Rigid body dynamics algorithms*. New York, NY: Springer Publishing. doi:10.1007/978-0-387-74315-8

Fei, B. (2001). The safety issues of medical robotics. *Reliability Engineering & System Safety, 73*(2), 183–192. doi:10.1016/S0951-8320(01)00037-0

Feil-Seifer, D., Skinner, K., & Mataric, M. J. (2007). Benchmarks for evaluating socially assistive robotics. *Journal of Interaction Science, 8*, 423–439.

Felgner, P. L., Gadek, T. R., Holm, M., Roman, R., Chan, H. W., & Wenz, M. (1987). Lipofection: A highly efficient, lipid-mediated DNA-transfection procedure. *Proceedings of the National Academy of Sciences of the United States of America, 84*(21), 7413–7417. doi:10.1073/pnas.84.21.7413

Feng, D. J., & Wijesoma, W. S. (2008). Improving rao-blackwellised genetic algorithmic filter SLAM through genetic learning. In *Proceedings of the 10th International Conference on Control, Automation, Robotics and Vision*, Hanoi, Vietnam, (pp. 1200–1205).

Fenwick, J. W., Newman, P. M., & Leonard, J. J. (2002). Cooperative concurrent mapping and localization. *IEEE International Conference on Robotics and Automation*, Washington DC, USA, (pp. 1810–1817).

Ferri, G., Caselli, E., Mattoli, V., Mondini, A., & Mazzolai, B. (2009). SPRIAL: A novel biologicallyinspired algorithm for gas/odor source localization in an indoor environment with no strong airflow. *Robotics and Autonomous Systems, 57*(4), 393–402. doi:10.1016/j.robot.2008.07.004

Finkelstein, J., Eckersberger, E., Sadri, H., Taneja, S. S., & Djavan, B. (2010). Open versus laparoscopic versus robot-assisted laparoscopic prostatectomy: The European and US experience. *Reviews in Urology, 12*, 35–43.

Fitzpatrick, J. M. (2009). Fiducial registration error and target registration error are uncorrelated. *SPIE Medical Imaging*, 7261-02, Orlando, FL, (pp. 1–12).

Fitzpatrick, J. M., West, J. B., & Maurer, C. R. (1998). Predicting error in rigid-body point-based registration. *IEEE Transactions on Medical Imaging, 17*(5), 694–702. doi:10.1109/42.736021

Fleming, I., Balicki, M., Koo, J., Iordachita, I., Mitchell, B., & Handa, J. … Taylor, R. H. (2008). Cooperative robot assistant for retinal microsurgery. *Lecture Notes in Computer Science (LNCS), Proceedings of the Annual Conf. of the Medical Image Computing and Computer Assisted Intervention Society (MICCAI)*, vol. 5242, (pp. 543–550). Palm Springs, CA: Springer.

Fogassi, L., Ferrari, P. F., Chersi, F., Gesierich, B., Rozzi, S., & Rizzolatti, G. (2005). Parietal lobe: From action organization to intention understanding. *Science, 308*, 662–667. doi:10.1126/science.1106138

Fogoros, R. N. (2011). *Robotic heart surgery - A status report*. Retrieved on June 28, 2011, from http://heartdisease.about.com

Fox, D., Ko, J., Konolige, K., Limketkai, B., & Stewart, B. (2006). Distributed multi-robot exploration and mapping. *Proceeding of the IEEE Special Issue on Multi-Robot Systems, 94*(7), 1325–1339.

Fox, D., Burgard, W., & Thrun, S. (1997). The dynamic window approach to collision avoidance. *IEEE Robotics & Automation Magazine*, *4*(1), 23–33. doi:10.1109/100.580977

Frantz, D. D., Kirsch, S. R., & Wiles, A. D. (2004). *Specifying 3D tracking system accuracy* (pp. 1–5). Technical report, NDI Inc.

Frisoli, A., Rocchi, F., Marcheschi, S., Dettori, A., Salsedo, F., & Bergamasco, M. (2005). A new force-feedback arm exoskeleton for haptic interaction in virtual environment. *First IEEE Joint Eurohaptics Conference and Symposium on Haptic Interfaces for Virtual Environment and Teleoperator Systems,* (pp. 195-201).

Frisoli, A., Montagner, A., Borelli, L., Salsedo, F., & Bergamasco, M. (2009). A force-feedback exoskeleton for upper limb rehabilitation in Virtual Reality. *Applied Bionics and Biomechanics*, *6*(2), 115–126. doi:10.1080/11762320902959250

Frutiger, D. R., Bongard, J. C., & Iida, F. (2002). Iterative product engineering: evolutionary robot design. *Proceedings of the Fifth International Conference on Climbing and Walking Robots*, (pp. 619–629).

Fujimoto, Y., & Kawamura, A. (1998). Simulation of an autonomous biped walking robot including environmental force interaction. *IEEE Robotics & Automation Magazine*, *5*(2), 33–42. doi:10.1109/100.692339

Fujisato, T., Abe, S., Tsuji, T., Sada, M., Miyawaki, F., & Ohba, K. (1998, 29 Oct-1 Nov 1998). The development of an ova holding device made of microporous glass plate for genetic engineering. *Proceedings of the 20th Annual International Conference of the IEEE Engineering in Medicine and Biology Society*, 1998.

Fu, K. S., Lee, C. S. G., & Gonzales, R. C. (1987). *Robotics: Control, sensing, vision & intelligence*. New York, NY: McGraw- Hill.

Gan, Y., & Chen, Z. (2008). A study of the zona piercing process in piezodriven intracytoplasmic sperm injection. *Journal of Applied Physics*, *104*(4), 044702–044708. doi:10.1063/1.2970097

Garcia, F. J., & Villarroel, J. L. (1998). Decentralized implementation of real-time systems using time Petri nets. Application to mobile robot control. In *Proceedings of the 5th IFAC Workshop on Algorithms and Architectures for Real-time Control*, (pp. 11-16).

Garcia, P., Rosen, J., Kapoor, C., Noakes, M., Elbert, G., & Treat, M. (2009). Trauma pod: A semi-automated telerobotic surgical system. *Journal of Medicine*, *5*(2), 136–146.

Gartner, A. (2008). *Teleneurology and requirements of the medical devices directive (MDD)* (pp. 1–22). Technical report, Baaske Medical GmbH & Co., Lubbecke.

Gassert, R., Burdet, E., & Chinzei, K. (2008). MRI-compatible robotics. *IEEE Engineering in Medicine and Biology Magazine*, *27*(3), 12–14. doi:10.1109/EMB.2007.910273

Gerhardus, D. (2003). Robot-assisted surgery: The future is here. *Journal of Healthcare Management*, *48*, 242–251.

Gerkey, B., & Matarić, M. J. (2004). A formal framework for the study of task allocation in multi-Robot systems. *The International Journal of Robotics Research*, *23*, 939–954. doi:10.1177/0278364904045564

Gerszten, P. C., Ozhasoglu, C., & Burton, S. A. (2003a). CyberKnife frameless single-fraction stereotactic radiosurgery for tumors of the sacrum. *Neurosurgical Focus*, *15*(2), 1–5. doi:10.3171/foc.2003.15.2.7

Gerszten, P. C., Ozhasoglu, C., & Burton, S. A. (2003b). Evaluation of CyberKnife frameless real-time image-guided stereotactic radiosurgery for spinal lesions. *Stereotactic and Functional Neurosurgery*, *81*(1-4), 84–89. doi:10.1159/000075109

Ghomi, A., & Askari, R. (2010). Use of a bidirectional barbed suture in robot-assisted sacrocolpopexy. *Journal of Robotic Surgery*, *4*, 87–89. doi:10.1007/s11701-010-0188-9

Gil, P., Pomares, J., Puente, S. V., Diaz, C., Candelas, F., & Torres, F. (2007). Flexible multi-sensorial system for automatic disassembly using cooperative robots. *International Journal of Computer Integrated Manufacturing*, *20*(8), 757–772. doi:10.1080/09511920601143169

Girault, C., & Valk, R. (2003). *Petri Nets for systems engineering*. Berlin, Germany: Springer-Verlag.

Glauser, D. (1993). Mechanical concept of the neurosurgical robot Minerva. *Robotica*, *11*(6), 567–575. doi:10.1017/S0263574700019421

Gockley, R., & Matari'c, M. (2006). Encouraging physical therapy compliance with a hands-off mobile robot. *Proceedings of the 1st Annual Conference on Human-Robot Interaction,* (pp. 150–155).

Goerzen, C., Kong, Z., & Mettler, B. (2010). A survey of motion planning algorithms from the perspective of autonomous UAV guidance. *Journal of Intelligent & Robotic Systems*, *57*, 65–100. doi:10.1007/s10846-009-9383-1

Gogoussis, A., & Donath, M. (1987). Coulomb friction joint and drive effects in robot mechanisms. *IEEE International Conference on Robotics and Automation*, (vol. 4, pp. 828-836).

Goldenberg, A., Grinewsky, M., & Campbell, T. (2010, October). AARM: A robot for internal operations in nuclear reactors. *2010 1st International Conference on Applied Robotics for the Power Industry (CARPI)*, (pp. 1-6). Montreal, QC.

Goldenberg, A., Trachtenberg, J., Kucharczyk, W., Yang, Y., Haider, M., Sussman, M., … Weersink, R. (2010, March). Robot-assisted MRI-guided prostatic interventions. *Robotica – Special Issue on Surgical Robotics*, *28*(2), 215-234.

Goldenberg, A., Yang, Y., Ma, L., Trachtenberg, J., Kucharczyk, W., & Haider, M. … Weersink, R. (2010, October). *MRI compatible robot with calibration phantom & phantom*, PCT Patent Application No. PCT/CA2009/000846, filed in Jun 2009, 35 claims accepted Oct 2010.

Goldenberg, A. (2007). Robotics, mechatronics, and intelligent automation. In Avallone, E. A., Baumeister, T., & Sadegh, A. (Eds.), *Marks' standard handbook for mechanical engineers*. New York, NY: McGraw Hill.

Goldenberg, A., Trachtenberg, J., Kucharczyk, W., Yang, Y., Haider, M., & Ma, L. (2008, June). Robotic system for closed-bore MRI-guided prostatic interventions. *IEEE/ASME Transactions on Mechatronics*, *13*(3), 374–379. doi:10.1109/TMECH.2008.924122

Gordon, J. W., Scangos, G. A., Plotkin, D. J., Barbosa, J. A., & Ruddle, F. H. (1980). Genetic transformation of mouse embryos by microinjection of purified DNA. *Proceedings of the National Academy of Sciences of the United States of America*, *77*, 7380–7384. doi:10.1073/pnas.77.12.7380

Gortchev, G., Tomov, S., Tantchev, L., Velkova, A., & Radionova, Z. (2010). Da Vinci robotic surgery in the treatment of benign and malignant gynecologic tumors. *Gynecological Surgery*, *7*, 153–157. doi:10.1007/s10397-009-0535-x

Gosselin, C. (1992). The optimum design of robotic manipulators using dexterity indices. *Robotics and Autonomous Systems*, *9*(4), 213–226. doi:10.1016/0921-8890(92)90039-2

Goswami, A. (2008, December). Kinematic and dynamic analogies between planar biped robots and the reaction mass pendulum (RMP) model. In *IEEE International Conference on Humanoid Robotics* (pp. 182-188).

Goto, T., Hongo, K., & Kakizawa, Y. (2003). Clinical application of robotic telemanipulation system in neurosurgery. *Journal of Neurosurgery*, *99*, 1082–1084. doi:10.3171/jns.2003.99.6.1082

Graessmann, A., & Graessmann, M. (1976). Early' simian-virus-40-specific RNA contains information for tumor antigen formation and chromatin replication. *Proceedings of the National Academy of Sciences of the United States of America*, *73*, 366–370. doi:10.1073/pnas.73.2.366

Graessmann, M., & Graessmann, A. (1983). Microinjection of tissue culture cells. *Methods in Enzymology*, *101*, 482–492. doi:10.1016/0076-6879(83)01033-2

Graham, F. L., & Van Der Eb, A. J. (1973). Transformation of rat cells by DNA of human adenovirus 5. *Virology*, *54*(2), 536–539. doi:10.1016/0042-6822(73)90163-3

Greenwood, D. T. (2003). *Advanced dynamics*. New York, NY: Cambridge University Press.

Grehan, R., Moote, R., & Cyliax, I. (1998). *Real-time programming: A guide to 32-bit embedded development*. Reading, MA: Addison-Wesley.

Grezes, J., Frith, C., & Passingham, R. E. (2004). Brain mechanisms for inferring deceit in the actions of others. *The Journal of Neuroscience*, *24*, 5500–5505. doi:10.1523/JNEUROSCI.0219-04.2004

Grimmet, G. (1999). *Percolation*. Berlin, Germany: Springer-Verlag.

Grimson, W. E. L., Stauffer, C., Romano, R., & Lee, L. (1998). Using adaptive tracking to classify and monitor activities in a site. In *IEEE Conference on Computer Vision and Pattern Recognition*, (pp. 22–29).

Groschopp Inc. (n.d.). *Website*. Retrieved October 2, 2011, from http://www.groschopp.com/

Grueneis, C., Richter, R., et al. (1999). Clinical introduction of the CASPAR system: Problems and initial results. 4th International Symposium of Computer Assited Orthopaedic Surgery, CAOS'99, Davos, 17-19 March 1999, Switzerland.

Grunert, P., Darabi, K., Espinosa, J., & Filippi, R. (2003). Computer-aided navigation in neurosurgery. *Neurosurgical Review*, *26*(2), 73–99. doi:10.1007/s10143-003-0262-0

Grzonka, S., Grisetti, G., & Burgard, W. (2009). Towards a navigation system for autonomous indoor flying. In *Proceedings of IEEE International Conference on Robotics and Automation*, *2009*, 2878–2883. doi:.doi:10.1109/ROBOT.2009.5152446

Guarnieri, M., Debenest, P., Inoh, T., Fukushima, E. F., & Hirose, S. (2004). Development of Helios VII: An arm equipped tracked vehicle for search and rescue operations. In *Proceedings of the International Conference Intelligent Robots and Systems*, Sendai, Japan, (pp. 39–45).

Guerrouad, A., & Vidal, P. (1991). Advantage of computer aided teleoperation (CAT) in microsurgery. *IEEE Conference Proceeding of Fifth International Conference on Advanced Robotics (ICAR)*, (vol. 1, pp. 910- 914).

Guerrouad, A., & Jolly, D. (1989). Automatic analysis of weariness during a micromanipulation task by SMOS. *Proceedings of the Annual International Conference of the IEEE Engineering in Medicine and Biology Society*, *3*, 906–907.

Guerrouad, A., & Vidal, P. (1989). SMOS: Stereotaxical microtelemanipulator for ocular surgery. *Proceedings of the Annual International Conference of the IEEE Engineering in Medicine and Biology Society*, *3*, 879–880.

Guillonneau, B., & Vallancien, G. (2000). Laparoscopic radical prostatectomy: The Montsouris technique. *The Journal of Urology*, *163*, 1643–1649. doi:10.1016/S0022-5347(05)67512-X

Guiochet, J., & Vilchis, A. (2002). Safety analysis of a medical robot for teleechography. In *Proceedings of the 2nd IARP IEEE/RAS joint workshop on Technical Challenge for Dependable Robots in Human Environments*, Toulouse, (pp. 217–227).

Guivant, J., & Nebot, E. (2001). Optimization of the simultaneous localization and map building algorithm for real time implementation. *IEEE Transactions on Robotics and Automation*, *17*(3), 242–257. doi:10.1109/70.938382

Guizzo, E., & Goldstein, H. (2005). The rise of the body bots. *IEEE Spectrum*, *42*(10), 50–56. doi:10.1109/MSPEC.2005.1515961

Gungor, A., & Gupta, S. M. (1999). Issues in environmentally conscious manufacturing and product recovery: A survey. *Computers & Industrial Engineering*, *36*(4), 811–853. doi:10.1016/S0360-8352(99)00167-9

Gurdan, D., Stumpf, J., Achtelik, M., Doth, K., Hirzinger, G., & Rus, D. (2007). Energy-efficient autonomous four-rotor flying robot controlled at 1 khz. *Proceeding of the IEEE International Conference on Robotics and Automation.*

Guthart, G. S., & Salisbury, J. K. (2000). The intuitive telesurgery system: Overview and application. In *Proceedings of the IEEE International Conference on Robotics and Automation (ICRA)*, San Francisco, CA, (pp. 618–621).

Haggard, P., Clark, S., & Kalogeras, J. (2002). Voluntary action and conscious awareness. *Nature Neuroscience*, *5*, 382–385. doi:10.1038/nn827

Hähnel, D., Burgard, W., Fox, D., & Thrun, S. (2003). An efficient FastSLAM algorithm for generating maps of large-scale cyclic environments from raw laser range measurements. In *Proceedings of the IEEE International Conference on Intelligent Robots and Systems*, Las Vegas, NV, USA, (pp. 206–211).

Haidegger, T. (2010). *Da Vinci inside—Insight*. SurgRob blog. Retrieved from http://surgrob.blogspot.com/2010/12/da-vinci-motors.html

Haidegger, T., Kazanzides, P., Rudas, I., Benyó, B., & Benyó, Z. (2010). The importance of accuracy measurement standards for computer-integrated interventional systems. In *Proceedings of the EURON GEM Sig Workshop on The Role of Experiments in Robotics Research at IEEE ICRA*, Anchorage, AK, (pp. 19–24).

Haidegger, T., Kovacs, L., Fordos, G., et al. (2008). Future trends in robotic neurosurgery. *14th Nordic-Baltic Conference on Biomedical Engineering and Medical Physics IFMBE Proceedings*, (vol. 20, part 4, pp. 229-233).

Haidegger, T., Xia, T., & Kazanzides, P. (2008). Accuracy improvement of a neurosurgical robot system. In *Proceedings of the 2nd IEEE RAS/EMBS Intl. Conf. on Biomedical Robotics and Biomechatronics (BioRob)*, Scottsdale, AZ, (pp. 836–841).

Hamel, T., & Mahony, R. (2006). Attitude estimation on SO[3] based on direct inertial measurements. In *Proceedings of IEEE International Conference on Robotics and Automation, 2006*, 2170–2175. doi:.doi:10.1109/ROBOT.2006.1642025

Hamilton, A., Burany, S. J., Peralta, S. B., & Greenland, L. (2010, October). Robotic removal of high-activity debris from a nuclear primary heat transfer system. *2010 1st International Conference on Applied Robotics for the Power Industry (CARPI)*, (pp. 1-6). Montreal, QC.

Hanna, R., & Boggess, J. (2011). Applications of surgical robotics in gynecologic surgery. *Surgical Robotics, 4*, 761–789. doi:10.1007/978-1-4419-1126-1_32

Harada, K., Oetomo, D., Susilo, E., Menciassi, A., Daney, D., Merlet, J.-P., & Dario, P. (2009). A reconfigurable modular robotic endoluminal surgical system: Vision and preliminary results. *Robotica, 28*(2), 171–183. doi:10.1017/S0263574709990610

Harja, J., Tikkanen, J., Sorvoja, H., & Myllyla, R. (2007). Magnetic resonance imagingcompatible, three-degrees-of-freedom joystick for surgical robot. *International Journal of Medical Robotics and Computer Assisted Surgery, 3*, 365–371. doi:10.1002/rcs.159

Hasegawa, K., Takahashi, K., Masuda, R., & Ohno, H. (1984). Proposal of mark flow graph for discrete system control. *Transactions of SICE, 20*(2), 122–129.

Hayat, S., & Vidal, P. (1995). Conception of information tools to assist the surgeon when he carries out radial keratotomy with a micromanipulator. *IEEE Conference Proceedings* (pp. 3-5).

Heiken, G. H., Vaniman, D. T., & French, B. M. (Eds.). (1991). *Lunar sourcebook: A user's guide to the moon*. Cambridge, UK: Cambridge University Press.

Heiser, W. C. (Ed.). (2004a). *Gene delivery to mammalian cells* (*Vol. 2*). Humana Press.

Heiser, W. C. (Ed.). (2004b). *Gene delivery to mammalian cells* (*Vol. 1*). Humana Press.

He, R., Prentice, S., & Roy, N. (2008). Planning in information space for a quadrotor helicopter in a GPS-denied environment. In *Proceedings of IEEE International Conference on Robotics and Automation, 2008*, 1814–1820. doi:.doi:10.1109/ROBOT.2008.4543471

Herisse, B., Russotto, F. X., Hamel, T., & Mahony, R. (2008). Hovering flight and vertical landing control of a VTOL unmanned aerial vehicle using optical flow. In *Proceedings of IEEE/RSJ International Conference on Intelligent Robots and Systems, 2008*, (pp. 801–806.) doi:10.1109/IROS.2008.4650731

Herisse, B., Oustrieres, S., Hamel, T., Mahony, R., & Russotto, F. (2010). A general optical flow based terrain-following strategy for a VTOL UAV using multiple views. In *Proceedings of IEEE International Conference on Robotics and Automation, 2010*, 3341–3348. doi:.doi:10.1109/ROBOT.2010.5509536

Higgins, W. T. (1975). A comparison of complementary and Kalman filtering. *IEEE Transactions on Aerospace and Electronic Systems, 3*, 321–325. doi:10.1109/TAES.1975.308081

Hillman, M. (2003). Rehabilitation robotics from past to present. *Proceedings of the 8th International Conference on Rehabilitation Robotics.*

Hillman, M. (1998). Introduction to the special issue on rehabilitation robotics. *Robotica, 16*(5), 485–485. doi:10.1017/S0263574798000629

Hirukawa, H., Hattori, S., Harada, K., Kajita, S., Kaneko, K., & Kanehiro, K. ... Morisawa, M. (2006, October). A universal stability criterion of the foot contact of legged robots – Adios ZMP. In *IEEE International Conference on Robotics and Automation* (pp. 1976-1983).

Hoffmann, G., Huang, H., Waslander, S. L., & Tomlin, C. J. (2007). Quadrotor helicopter flight dynamics and control: Theory and experiment. In *Proceedings of American Institute of Aeronautics and Astronautics Guidance, Navigation, and Control Conference 2007*. Hilton Head, South Carolina.

Hoffmann, G., Rajnarayan, D. G., Waslander, S. L., Dostal, D., Jang, J. S., & Tomlin, C. J. (2005). The Stanford testbed of autonomous rotorcraft for multi agent control (STARMAC). In *Proceedings of Digital Avionics Systems Conference, 2004.* Salt Lake City, Utah.

Hofmann, A., Popovic, M., & Herr, H. (2009, April). Exploiting angular momentum to enhance bipedal center-of-mass control. In *IEEE International Conference on Robotics and Automation* (pp. 4423-4429).

Holding, D. J., & Sagoo, J. S. (1992). A formal approach to the software control of high-speed machinery. In G. W. Irwin & P. J. Fleming (Eds.), *Transputers in real-time control,* (pp. 239-282). Taunton, Somerset, UK: Research Studies Press.

Höller, K. (2010). *Novel techniques for spatial orientation in natural orifice translumenal endoscopic surgery* (NOTES), (pp. 1–190). Ph.D. thesis, Friedrich-Alexander-Universität.

Hongo, K., Kobayashi, S., Kakizawa, Y., Koyama, J., Goto, T., & Okudera, H. (2002). NeuRobot: Telecontrolled micromanipulator system for minimally invasive microneurosurgery— preliminary results. *Neurosurgery, 51*(4), 985–988.

Hong, S. S., Heo, S. C., Kim, E., & Hang, Y. J. (2006). A study on the practical human robot interface design for the development of shopping support robot. *Journal of Korean Society of Design Science, 19*, 81–90.

Howard, A., & Gaurav, S. Sukhatme, S., & Matarić, M. J. (2006). Multi-robot mapping using manifold representations. *Proceedings of the IEEE – Special Issue on Multi- Robot Systems, 94,* 1360 -1369.

Howard, A. (2006). Multi-robot simultaneous localization and mapping using particle filters. *The International Journal of Robotics Research, 25*(12), 1243–1256. doi:10.1177/0278364906072250

Howe, R. D., & Matsuoka, Y. (1999). Robotics for surgery. *Annual Review of Biomedical Engineering, 1*, 211–240. doi:10.1146/annurev.bioeng.1.1.211

Hu, T., Castellanos, A., & Tholey, G. (2002). Real-time haptic feedback laparoscopic tool for use in gastro-intestinal surgery. *Fifth International Conference on Medical Image Computing and Computer Assisted Intervention, 12,* (pp. 6-16).

Huang, H., Mills, J. K., & Sun, D. (2011). *A new piezo-driven ultrasonic cell microinjection system.* Paper presented at the 2011 IEEE International Conference on Information and Automation.

Huang, H. B., Sun, D., Mills, J. K., & Cheng, S. H. (2009). Robotic cell injection system with position and force control: Toward automatic batch biomanipulation. *IEEE Transactions on Robotics, 25*(3), 727–737. doi:10.1109/TRO.2009.2017109

Huang, H., Hoffmann, G. M., Waslander, S. L., & Tomlin, C. J. (2009). Aerodynamics and control of autonomous quadrotor helicopters in aggressive maneuvering. In *Proceedings of IEEE International Conference on Robotics and Automation, 2009*, 3277–3282. doi:.doi:10.1109/ROBOT.2009.5152561

Huang, H., Mills, J. K., Lu, C., & Sun, D. (2011). (in press). A universal piezo-driven ultrasonic cell microinjection system. *Biomedical Microdevices.* doi:10.1007/s10544-011-9544-4

Huang, H., Sun, D., Mills, J. K., Li, W. J., & Cheng, S. H. (2009). Visual-based impedance control of out-of-plane cell injection systems. *IEEE Transactions on Automation Science and Engineering*, *6*(3), 565–571. doi:10.1109/TASE.2008.2010013

Huang, Q., Yokoi, K., Kajita, S., Kaneko, K., Arai, H., Koyachi, N., & Tanie, K. (2001). Planning walking patterns for a biped robot. *IEEE Transactions on Robotics and Automation*, *17*(3), 280–289. doi:10.1109/70.938385

Huang, T., Kimura, Y., & Yanagimachi, R. (1996). The use of piezo micromanipulation for intracytoplasmic sperm injection of human oocytes. *Journal of Assisted Reproduction and Genetics*, *13*(4), 320–328. doi:10.1007/BF02070146

Hughes, A. (2005). *Electric motors and drives: Fundamentals, types and applications*. New South Wales, Australia: Newnes.

Hui, W., Dong, X., & Guanghong, D. (2008). A genetic algorithm for product disassembly sequence planning. *Neurocomputing*, *71*, 2720–2726. doi:10.1016/j.neucom.2007.11.042

Hunt, A., & Ewing, R. (2009). *Percolation theory for flow in porous media*. Springer.

Hyon, S., & Cheng, G. (2007, April). Disturbance rejection for biped humanoid robots. In *IEEE International Conference on Robotics and Automation* (pp. 2668-2673).

Iacoboni, M., Molnar-Szakacs, I., Gallese, V., Buccino, G., Mazziotta, J. C., & Rizzolatti, G. (2005). Grasping the intentions of others with one's own mirror neuron system. *PLoS Biology*, *3*, 529–535. doi:10.1371/journal.pbio.0030079

Ilgin, M. A., & Gupta, S. M. (2010). Environmentally conscious manufacturing and product recovery (ECMPRO): A review of the state of the art. *Journal of Environmental Management*, *91*(3), 563–591. doi:10.1016/j.jenvman.2009.09.037

IOM. Institute of Medicine. (2011). *Press release*. Retrieved on August 2011, from http://cardiobrief.org/2011/07/29/controversial-iom-report-highly-critical-of-510k-process/

ISO 10360. (2000). *ISO 10360-1:2000 standard*. Retrieved on August 1, 2011, from http://www.iso.org/iso/catalogue detail.htm?csnumber=37694

iSYS. (2011). *iSYS Medizintechnik GmbH website*. Retrieved on August 1, 2011, from http://www.isys.co.at

Iwakura, D., Wang, W., Nonami, K., & Haley, M. (2011). Movable range-finding sensor system and precise automated landing of quad-rotor MAV. *Journal of System Design and Dynamics*, *5*(1), 17–29. doi:10.1299/jsdd.5.17

Jafari, A., Tsagarakis, N. G., Vanderborght, B., & Caldwell, D. G. (2010, October). A novel actuator with adjustable stiffness (AwAS). In *IEEE International Conference on Intelligent Robots and Systems* (pp. 4210-4215).

Jakopec, M., Harris, S. J., Baena, F. R. Y., Gomes, P., Cobb, J., & Davies, B. L. (2001). The first clinical application of a hands-on robotic knee surgery system. *Computer Aided Surgery*, *6*, 329–339.

Janda, M., & Buch, B. (2009). The challenges of clinical validation of emerging technologies: Computer-Assisted devices for surgery. *Journal of Bone and Joint Surgery*, *91-A*(S1), 17–21. doi:10.2106/JBJS.H.01337

Jang, J., Kang, Y., Choi, B., Choi, I., Kim, M., & Shin, D. … Kwon, S. (2006). Evaluation of the accuracy of the CyberKnife. In *Proceedings of the World Congress on Medical Physics and Biomedical Engineering*, Springer, Seoul, (pp. 2024–2027).

Jensen, A. M., Baumann, M., & Chen, Y. Q. (2009). Low-cost multispectral aerial imaging using autonomous runway-free small flying wing vehicles. In *Proceedings of IEEE International Geoscience and Remote Sensing Symposium, 2008,* (vol. 5, pp. 506-509). doi:10.1109/IGARSS.2008.4780140

Ji, M., & Egerstedt, M. (2007). Distributed coordination control of multiagent systems while preserving connectedness. *IEEE Transactions on Robotics*, *23*(4), 693–703. doi:10.1109/TRO.2007.900638

Jolliffe, I. T. (1986). *Principal component analysis* (p. 487). Springer-Verlag.

Kajita, S., & Tani, K. (1991, April). Study of dynamic biped locomotion on rugged terrain. In *IEEE International Conference on Robotics and Automation* (pp. 1405-1410).

Kajita, S., Kanehiro, F., Kaneko, K., Yokoi, K., & Hirukawa, H. (2001, October). The 3D linear inverted pendulum mode: A simple modeling for a biped walking pattern generation. In *IEEE International Conference on Intelligent Robots and Systems* (pp. 239-246).

Kajita, S., Kaneko, K., Morisawa, M., Nakaoka, S., & Hirukawa, H. (2007, April). ZMP-based biped running enhanced by toe springs. In *IEEE International Conference on Robotics and Automation* (pp. 3963-3969).

Kajita, S., Morisawa, M., Harada, K., Kaneko, K., Kanehiro, F., Fujiwara, K., & Hirukawa, H. (2006, October). Biped walking generator allowing auxiliary ZMP control. In *IEEE International Conference on Intelligent Robots and Systems* (pp. 2993-2999).

Kajita, S., & Espiau, B. (2008). Legged robots. In Siciliano, B., & Khatib, O. (Eds.), *Springer handbook of robotics*. Berlin, Germany: Springer Publishing. doi:10.1007/978-3-540-30301-5_17

Kajita, S., & Sugihara, T. (2009). Humanoid robots in the future. *Advanced Robotics*, *23*(11), 1527–1531. doi:10.1163/016918609X12469692711804

Kallio, P., & Kuncova, J. (2003). *Manipulation of living biological cells: Challenges in automation.* International Conference on Intelligent Robots and Systems. Las Vegas, USA.

Kallio, P., Ritala, T., Lukkari, M., & Kuikka, S. (2006). *Injection guidance system for cellular microinjections.* Paper presented at the Biomedical Robotics and Biomechatronics, 2006. BioRob 2006. The First IEEE/RAS-EMBS International Conference on.

Kallio, J. K., & Kallio, P. J. (2006). Lab automation in cultivation of adherent cells. *IEEE Transactions on Automation Science and Engineering*, *3*(2), 177–186. doi:10.1109/TASE.2006.871482

Kanehiro, F., Hirukawa, H., & Kajita, S. (2004). OpenHRP: Open architecture humanoid robotics platform. *The International Journal of Robotics Research*, *23*(2), 155–165. doi:10.1177/0278364904041324

Kaouk, J., & Goel, R. (2009). Robotic-assisted laparoscopic bladder augmentation. *Pediatric Robotic Urology*, *2*, 181–188. doi:10.1007/978-1-60327-422-7_12

Kapandji, I. A. (1997). *Physiologie articulaire: Schemas commentés de mécanique humaine (5ème edition), Tome 1: Members supérieurs*. Maloine.

Karas, C. S., & Chiocca, E. A. (2007). Neurosurgical robotics: A review of brain and spine applications. *Journal of Robotic Surgery*, *1*(1), 39–43. doi:10.1007/s11701-006-0006-6

Kassler, M. (1993). Introduction to the Special Issue on Robotics for Health Care. *Robotica*, *11*(6), 493–494. doi:10.1017/S0263574700019330

Katayose, H., Yanagida, K., Shinoki, T., Kawahara, T., Horiuchi, T., & Sato, A. (1999). Efficient injection of bull spermatozoa into oocytes using a piezo-driven pipette. *Theriogenology*, *52*(7), 1215–1224. doi:10.1016/S0093-691X(99)00213-7

Kautz, H. A., & Allan, J. (1986). Generalized plan recognition. In *Fifth National Conference on Artificial Intelligence*, Philadelphia, PA, (pp. 32–38).

Kavoussi, L. R., Moore, R. G., Adams, J. B., & Partin, A. W. (1995). Comparison of robotic versus human laparoscopic camera control. *The Journal of Urology*, *154*, 2134–2136. doi:10.1016/S0022-5347(01)66715-6

Kazanzides, P. (2006). Surgical robots and phantom (artifact) devices. In *Proceedings of the ASTM Workshop on Medical Devices Metrology and Standards Needs*. ASTM, NIST, NCI, FDA and U.T. Arlington, Atlanta. Retrieved on August 1, 2011, from http://www.nist.gov/el/isd/upload/Surgical_Robots_Kazanzides.pdf

Kazanzides, P. (2009). Safety design for medical robots. In *Proceedings of the Annual International Conference of the IEEE Engineering in Medicine and Biology Society (EMBC)*, Minneapolis, MN, (pp. 7208–7211).

Kazanzides, P., DiMaio, S., Deguet, A., Vagvolgyi, B., Balicki, M., & Schneider, C. ... Taylor, R. (2010). The surgical assistant workstation (SAW) in minimally-invasive surgery and microsurgery. In *Proceedings of the MICCAI International Workshop on Systems and Architectures for Computer Assisted Interventions*, Beijing. *MIDAS Journal—Computer Assisted Intervention*, June 2010.

Kazanzides, P., Fichtinger, G., Hager, G. D., Okamura, A. M., Whitcomb, L. L., & Taylor, R. H. (2008). Surgical and interventional robotics, core concepts, technology, and design. *IEEE Robotics & Automation Magazine*, *15*(2), 122–130. doi:10.1109/MRA.2008.926390

Kazerooni, H. (2005). Exoskeletons for human power augmentation. In *Proceedings of IEEE Conference Intelligent Robots and Systems*, Edmonton, Canada, (pp. 3459–3464).

Kelley, R., Nicolescu, M., Tavakkoli, A., Nicolescu, M., King, C., & Bebis, G. (2008). *Understanding human intentions via hidden Markov Models in autonomous mobile robots.* HRI'08, March 12–15, 2008, Amsterdam, Netherlands.

Kennedy, C. W., & Desai, J. P. (2003). Force feedback using vision. *The 11th International Conference on Advanced Robotics,* (pp. 2106-2111).

Keung, K. W., Ip, W. H., & Lee, T. C. (2001). The solution of a multi-objective tool selection model using the GA approach. *International Journal of Advanced Manufacturing Technology*, *18*, 771–777. doi:10.1007/s001700170001

Khatib, O. (1986). Real-time obstacle avoidance for manipulators and mobile robots. *The International Journal of Robotics Research*, *5*(1), 90–98. doi:10.1177/027836498600500106

Kim, B., Park, J., Kang, B., & Moon, C. (2011). Fabrication and property analysis of a MEMS micro-gripper for robotic micro-manipulation. *Journal of Robotics and Computer-Integrated Manufacturing*, *27*, 1021–1025.

Kimura, Y., & Yanagimachi, R. (1995). Intracytoplasmic sperm injection in the mouse. *Biology of Reproduction*, *52*(4), 709–720. doi:10.1095/biolreprod52.4.709

Kimura, Y., Yanagimachi, R., Kuretake, S., Bortkiewicz, H., Perry, A. C., & Yanagimachi, H. (1998). Analysis of mouse oocyte activation suggests the involvement of sperm perinuclear material. *Biology of Reproduction*, *58*(6), 1407–1415. doi:10.1095/biolreprod58.6.1407

Kim, W. S. (2007). The design of humanoid robot platgorm: HUBO. *Journal of Korean Society of Design Science*, *20*, 13–16.

Kim, Y., & Cook, A. M. (1985). Manipulation and mobility aids. In Webster, J. G. (Eds.), *Electronic devices for rehabilitation*. London, UK: Chapman and Hall.

Kim, Y., Lee, J., Lee, S., & Kim, M. (2005). A force reflected exoskeleton-type masterarm for human-robot interaction. *IEEE Transactions on Systems, Man and Cybernetics. Part A*, *35*(2), 198–212.

Kim, Y., & Mesbahi, M. (2006). On maximizing the second smallest eigenvalue of a state-dependent graph Laplacian. *IEEE Transactions on Automatic Control*, *51*(1), 116–120. doi:10.1109/TAC.2005.861710

Kingston, D. B., & Beard, R. W. (2004). Real-time attitude and position estimation for small UAVs using low-cost sensors. In *Proceedings of American Institute of Aeronautics and Astronautics 3rd Unmanned Unlimited Systems Conference and Workshop*. Chicago, Illinois.

Kiran, M., Kin, L. W., & Ali, K. K. H. (2009). Clustering techniques for human posture recognition: K-means, FCM and SOM. *Proceedings of the 9th WSEAS International Conference on Signal, Speech and Image Processing, and 9th WSEAS International Conference on Multimedia, Internet & Video Technologies (SSIP '09/MIV'09).*

Klopcar, N., Tomsic, M., & Lenarcic, J. (2007). A kinematic model of the shoulder complex to evaluate the arm-reachable workspace. *Journal of Biomechanics*, *40*, 86–91. doi:10.1016/j.jbiomech.2005.11.010

Knoblich, G., & Prinz, W. (2001). Recognition of self-generated actions from kinematic displays of drawing. *Journal of Experimental Psychology. Human Perception and Performance*, *27*, 456–465. doi:10.1037/0096-1523.27.2.456

Koenig, S., Tovey, C., & Halliburton, W. (2001). Greedy mapping of terrain. *IEEE International Conference on Robotics & Automation* (pp. 3534-3539).

Koh, C., Tsang, C., & Kim, S. (2011). A new application of the four-arm standard da Vinci® surgical system: Totally robotic-assisted left-sided colon or rectal resection. *Surgical Endoscopy*, *25*, 1945–1952. doi:10.1007/s00464-010-1492-1

Kohler, E., Keysers, C., Umilta, M. A., Fogassi, L., Gallese, V., & Rizzolatti, G. (2002). Hearing sounds, understanding actions: Action representation in mirror neurons. *Science*, *297*, 846–848. doi:10.1126/science.1070311

Kok, L. (2011). *Singapore expertise pioneers quick and scarless surgery*. Press release of the Nanyang Technological University. Retrieved on August 1, 2011, from http://www.eurekalert.org/pub_releases/2011-07/ntu-sep070511.php

Kongar, E., & Gupta, S. M. (2006). Disassembly sequencing using genetic algorithm. *International Journal of Advanced Manufacturing Technology*, *30*(5-6), 497–506. doi:10.1007/s00170-005-0041-x

Konolige, K. A. (2006). Centibots: Very large scale distributed robotic teams. *Springer Tracts in Advanced Robotics*, *21*, 131–140. doi:10.1007/11552246_13

Konvalin, C. (2009). Compensating for tilt, hard-iron, and soft-iron effects. *Sensors Magazine*. Retrieved February 22, 2011, from http://www.sensorsmag.com/sensors/motion-velocity-displacement/compensating-tilt-hard-iron-and-soft-iron-effects-6475

Koo, S., & Kwon, D. S. (2009). Recognizing human intentional actions from the relative movements between human and robot. *The 18th IEEE International Symposium on Robot and Human Interactive Communication, RO-MAN 2009*, (pp. 939-944).

Kormushev, P., Ugurlu, B., Calinon, S., Tsagarakis, N. G., & Caldwell, D. G. (September, 2011). Bipedal walking energy minimization by reinforcement learning with evolving policy parameterization. *Proceedings of the IEEE International Conference on Intelligent Robots and Systems.*

Korzh, V., & Strähle, U. (2002). Marshall Barber and the century of microinjection: From cloning of bacteria to cloning of everything. *Differentiation*, *70*(6), 221–226. doi:10.1046/j.1432-0436.2002.700601.x

Krause, A., Smailagic, A., & Siewiorek, D. P. (2006). Context-aware mobile computing: learning context-dependent personal preferences from a wearable sensor array. *IEEE Transactions on Mobile Computing*, *5*(2), 113–127. doi:10.1109/TMC.2006.18

Krieger, A., Iordachita, I., Sang-Eun Song Cho, N. B., Guion, P., Fichtinger, G., & Whitcomb, L. L. (2011). (in press). Development and evaluation of an actuated MRI-compatible robotic system for MRI-guided prostate intervention. *IEEE/ASME Transactions on Mechatronics*. doi:10.1109/TMECH.2011.2163523

Kroc, S., & Delic, V. (2003). Personal wireless sensor network for mobile health care monitoring. *6th International Conference on Telecommunications in Modern Satellite, Cable and Broadcasting Service, TELSIKS 2003*, (vol. 2, 471-474).

Kronreif, G. (2011). Robot systems for percutaneous needle placement. In *Proceedings of the 1st BME–MAVE International Computer-Integrated Surgery Workshop*, Budapest. Retrieved from www.eestec.hu/cis

Kumar, A., & Waldron, K. (1981). The workspace of mechanical manipulator. *Journal of Mechanical Design. Transactions of the ASME*, *103*, 665–672. doi:10.1115/1.3254968

Kumar, R., Kalra, P., & Prakash, N. (2011). A virtual RV-M1 robot system. *Journal of Robotics and Computer-Integrated Manufacturing*, *27*, 994–1000. doi:10.1016/j.rcim.2011.05.003

Kummer, M. P., Abbott, J. J., Kratochvil, B. E., Borer, R., Sengul, A., & Nelson, B. J. (2010). OctoMag: An electromagnetic system for 5-DOF wireless micromanipulation. *IEEE Transactions on Robotics*, *26*(6), 1006–1018. doi:10.1109/TRO.2010.2073030

Kuo, J. S., Yu, C., Petrovich, Z., & Apuzzo, M. L. (2003). The CyberKnife stereotactic radiosurgery system: Description, installation, and an initial evaluation of use and functionality. *Neurosurgery*, *53*(5), 1235–1239. doi:10.1227/01.NEU.0000089485.47590.05

Kwartowitz, D. M., Herrell, S. D., & Galloway, R. L. (2006). Toward image-guided robotic surgery: Determining intrinsic accuracy of the da Vinci robot. *International Journal of Computer Assisted Radiology and Surgery*, *1*, 157–165. doi:10.1007/s11548-006-0047-3

Kwartowitz, D. M., Herrell, S. D., & Galloway, R. L. (2007). Update: Toward image-guided robotic surgery: Determining the intrinsic accuracy of the daVinci-S robot. *International Journal of Computer Assisted Radiology and Surgery*, *1*(5), 301–304. doi:10.1007/s11548-006-0064-2

Kwoh, Y. S., Hou, J., Jonckheere, E. A., & Hayati, S. (1988). A robot with improved absolute positioning accuracy for CT guided stereotactic brain surgery. *IEEE Transactions on Bio-Medical Engineering*, *35*, 153–160. doi:10.1109/10.1354

Kwon, T. B., Song, J. B., & Joo, J. H. (2010). Elevation moment of inertia: A new feature for Monte Carlo localization in outdoor environment with elevation map. *Journal of Field Robotics*, *27*(3), 371–386.

Kypson, A. P., Nifong, L. W., & Chitwood, W. R. (2003). Robotic cardiac surgery. *Journal of Long-Term Effects of Medical Implants*, *13*(6), 451–464. doi:10.1615/JLong-TermEffMedImplants.v13.i6.30

Kypson, P., & Chitwood, W. R. Jr. (2004). Robotic applications in cardiac surgery. *International Journal of Advanced Robotic Systems*, *1*, 87–92.

Laine, T., Lund, T., Ylikoski, M., Lohikoski, J., & Schlenzka, D. (2000). Accuracy of pedicle screw insertion with and without computer assistance: A randomised controlled clinical study in 100 consecutive patients. *European Spine Journal*, *9*, 235–240. doi:10.1007/s005860000146

Lambert, A. J. D., & Gupta, S. M. (2005). *Disassembly modeling for assembly, maintenance, reuse, and recycling*. Boca Raton, FL: CRC Press.

Lavalle, S. M. (1998). *Rapidly-exploring random trees: A new tool for path planning*.

Lavallee, S. (1989). A new system for computer assisted neurosurgery. In *Proceedings of the 11th IEEE Engineering in Medicine and Biology Conference*, (pp. 926–927).

Lavallee, S. (1995). Image-guided operating robot: A clinical application in stereotactic neurosurgery. In Taylor, R. H. (Ed.), *Computer-integrated surgery* (pp. 343–351). Cambridge, MA: MIT Press.

LaValle, S. M. (2006). *Planning algorithms*. Cambridge, UK: Cambridge University Press. doi:10.1017/CBO9780511546877

Lawrence, D. A. (1993). Stability and Transparency in bilateral teleoperation. *IEEE Transactions on Robotics and Automation*, *9*(5), 624–637. doi:10.1109/70.258054

Lazzerini, B., & Marcelloni, F. (2000). A genetic algorithm for generating optimal assembly plans. *Artificial Intelligence in Engineering*, *14*, 319–329. doi:10.1016/S0954-1810(00)00011-X

Le Roux, P. D., Das, H., Esquenazi, S., & Kelly, P. J. (2001). Robot-assisted microsurgery: a feasibility study in the rat. *Neurosurgery*, *48*, 584–589. doi:10.1097/00006123-200103000-00026

Lee, E. J., Togueni, A., & Dangoumau, N. (2006). A Petri net based decentralized synthesis approach for the control of flexible manufacturing systems. In *Proceedings of the 2006 IMACS Multiconference Computational Engineering in Systems Applications*.

Lee, K. K., & Xu, Y. (2004). Modeling human actions from learning. In *Proceedings of the IEEE/RSJ International Conference on Intelligent Robots and Systems (IROS '04)*, (vol. 3, pp. 2787- 2792).

Lee, D.-H., Kang, J.-G., & Xirouchakis, P. (2001). Disassembly planning and scheduling: Review and further research. *Journal of Engineering Manufacture*, *215*(B5), 695–709. doi:10.1243/0954405011518629

Lee, J.-S., & Chung, W. K. (2010). Robust mobile robot localization in highly non-static environments. *Autonomous Robots*, *29*, 1–16. doi:10.1007/s10514-010-9184-1

Lehman, A. C., Dumpert, J., Wood, N. A., Redden, L., Visty, A. Q., & Farritor, S. (2009). Natural orifice cholecystectomy using a miniature robot. *Surgical Endoscopy*, *23*(2), 260–266. doi:10.1007/s00464-008-0195-3

Leishman, J. G. (2002). The breguet-richet quad-rotor helicopter of 1907. *Vertiflite*, *47*(3), 58–60.

Leonard, J. J., & Durrant-Whyte, H. F. (1991). Mobile robot localization by tracking geometric beacons. *IEEE Transactions on Robotics and Automation*, *7*(3), 376–382. doi:10.1109/70.88147

Lerman, K., Jones, C., Galstyan, A., & Mataric, M. J. (2006). Analysis of dynamic task allocation in multi-robot systems. *The International Journal of Robotics Research*, *25*, 225–242. doi:10.1177/0278364906063426

Letier, P. (2010). *Bras Exosquelette Haptique: Conception et Contrôle*. PhD Thesis, Active Structures Laboratory, Université Libre de Bruxelles, July 2010.

Letier, P., Avraam, M., Horodinca, M., Schiele, A., & Preumont, A. (2006). Survey of actuation technologies for body-grounded exoskeletons. *Eurohaptics Conference*, (pp. 497-500).

Levine, W. S. (2010). *The control handbook: Control system fundamentals* (2nd ed.). Taylor & Francis US.

Li, X., Zong, G., & Bi, S. (2001, 2001). Development of global vision system for biological automatic micromanipulation system. *Proceedings 2001 ICRA IEEE International Conference on Robotics and Automation.*

Li, Y., Wang, J., Rizos, C., Mumford, P., & Ding, W. (2006). Low-cost tightly coupled GPS/INS integration based on a nonlinear Kalman filtering design. In *Proceedings of ION National Technical Meeting*, (pp. 18–20).

Lieberman, I. H., Togawa, D., Kayanja, M. M., Reinhardt, M. K., Friedlander, A., Knoller, N., & Benzel, E. C. (2006). Bone-mounted miniature robotic guidance for pedicle screw and translaminar facet screw placement: Part I—Technical development and a test case result. *Neurosurgery*, *59*(3), 641–650. doi:10.1227/01.NEU.0000229055.00829.5B

Lim, C. W., & Li, Y. P. (2004). *Torque distribution and slip minimisation in an Omni-directional mobile base*. Technical Report, Singapore Institute of Manufacturing Technology.

Lin, H. C., Shafran, I., Murphy, T. E., Okamura, A. M., Yuh, D. D., & Hager, G. D. (2005). Automatic Detection and Segmentation of Robot-Assisted Surgical Motions. In *Proceedings of the International Conference on Medical Image Computing and Computer-Assisted Intervention*, (pp. 802–810).

Lin, J., & Goldenberg, A. (2009, February). *Variable configuration articulated tracked vehicle*. U.S. Patent #7,493,976.

Li, Q. H. (2002). The application accuracy of the NeuroMate robot - A quantitative comparison with frameless and frame-based surgical localization systems. *International Journal of Computer Aided Surgery*, *7*, 90–98.

Li, Q., Zamorano, H. L., Pandya, A., Perez, R., Gong, F., & Diaz, F. (2002). The application accuracy of the NeuroMate robot—A quantitative comparison with frameless and frame-based surgical localization systems. *Computer Aided Surgery*, *7*(2), 90–98.

Li, R. Q., & Dai, J. S. (2009). Orientation angle workspace of planar serial three-link manipulators. *Science in China Series E: Technological Sciences*, *52*(4), 975–985. doi:10.1007/s11431-009-0083-7

Lirici, M. M., Papaspyropoulos, V., & Angelini, L. (1997). Telerobotics in medicine and surgery. *Minimally Invasive Therapy & Allied Technologies*, *5*(6), 364–378. doi:10.3109/13645709709153336

Liu, B. (2011). *5 actual rounds localization by QMCL*. Retrieved August 10th, 2011, from http://www.youtube.com/watch?v=-pCzy9oFZL8

Liu, D., Chia, E. F., Tian, H., & Heiser, W. C. (2004). Chemical methods for DNA delivery. In Walker, J. M. (Ed.), *Gene delivery to mammalian cells* (*Vol. 245*, pp. 3–23). Totowa, NJ: Humana Press. doi:10.1385/1-59259-649-5:3

Liu, X., & Sun, Y. (2009). Microfabricated glass devices for rapid single cell immobilization in mouse zygote microinjection. *Biomedical Microdevices*, *11*(6), 1169–1174. doi:10.1007/s10544-009-9333-5

Liverneaux, P., Nectoux, E., & Taleb, C. (2009). The future of robotics in hand surgery. *Chirurgie de la Main*, *28*(5), 278–785. doi:10.1016/j.main.2009.08.002

Li, X. D., Zong, G., Bi, S., & Zhao, W. (2001). Automatic micromanipulating system for biological applications with visual servo control. *Journal of Micromechatronics*, *1*, 345–363. doi:10.1163/156856301320819534

Loehle, C., Li, B., & Sundell, R. C. (1996). Forest spread and phase transitions at forest-prairie ecotones in Kansas. *Journal of Landscape Ecology*, *11*(4), 225–235. doi:10.1007/BF02071813

Loula, F., Prasad, S., Harber, K., & Shiffrar, M. (2005). Recognizing people from their movement. *Journal of Experimental Psychology. Human Perception and Performance*, *31*, 210–220. doi:10.1037/0096-1523.31.1.210

Loulmet, D., Carpentier, A., & D'Attellis, N. (1999). Endoscopic coronary artery bypass grafting with the aid of robotic assisted instruments. *The Journal of Thoracic and Cardiovascular Surgery*, *118*(1), 4–10. doi:10.1016/S0022-5223(99)70133-9

Louw, D. F., Fielding, T., McBeth, P. B., Gregoris, D., Newhook, P., & Sutherland, G. R. (2004). Surgical robotics: a review and neurosurgical prototype development. *Neurosurgery*, *54*, 525–537. doi:10.1227/01.NEU.0000108638.05274.E9

Lowery, A., Strojny, J., & Puleo J. (1996). *Medical device quality systems manual: A small entity compliance guide*. HHS Publication, FDA 97-4179.

Lukkari, M. J., Karjalainen, M. I., Sarkanen, R., Linne, M. L., Jalonen, T. O., & Kallio, P. J. (2004, 1-5 September 2004). *Electrical detection of the contact between a microinjection pipette and cells*. Paper presented at the Engineering in Medicine and Biology Society, 2004. IEMBS '04. 26th Annual International Conference of the IEEE.

Lukkari, M., & Kallio, P. (2005, 27-30 June 2005). Multi-purpose impedance-based measurement system to automate microinjection of adherent cells. *Proceedings IEEE International Symposium on Computational Intelligence in Robotics and Automation, CIRA 2005*.

Lum, J. H. M. (2008). *Quantitative performance assessment of surgical robot systems: TeleRobotic FLS*. Ph.D. thesis, University of Washington.

Lumelski, V. J., & Stepanov, A. A. (1986). Dynamic path planning for a mobile automaton with limited information on the environment. *IEEE Transactions on Automatic Control*, *31*(11).

Luo, Y., Di, L., Han, J., Chao, H., & Chen, Y. Q. (2010). VTOL UAV altitude flight control using fractional order controllers. In *Proceedings of 4th IFAC Workshop Fractional Differentiation and its Applications*. Badajoz, Spain.

Lu, Z., Y, C. P. C., Joohoo, N., Ge, R., & Lin, W. (2007). A micromanipulation system with dynamic force-feedback for automatic batch microinjection. *Journal of Micromechanics and Microengineering*, *17*(2), 314. doi:10.1088/0960-1317/17/2/018

Ma, H., & Chen, G. (2005). Gene transfer technique. *Nature and Science*, *3*(1).

Mabuchi, I., & Okuno, M. (1977). The effect of myosin antibody on the division of starfish blastomeres. *The Journal of Cell Biology*, *74*(1), 251–263. doi:10.1083/jcb.74.1.251

MacArthur, E. Z., & Crane, C. D. (2007). Compliant formation control of a multi-vehicle system. In *International Symposium on Computational Intelligence in Robotics and Automation, CIRA 2007* (pp. 479-484).

Madhani, A. J., Niemeyer, G., & Salisbury, J. K., Jr. (1998). The black falcon: A teleoperated surgical instrument for minimally invasive surgery. In *Proceedings of the IEEE/RSJ International Conference on Intelligent Robots and Systems* (IROS), Victoria, BC, (pp. 936–944).

Maeda, T., Ugurlu, B., & Kawamura, A. (2008, March). Two legged jumping simulation and experiment on biped robot MARI-3. In *IEEE International Workshop on Advanced Motion Control* (pp. 301-305).

Mahlknecht, S., & Madani, S. A. (2007). On architecture of low power wireless sensor networks for container tracking and monitoring applications. *2007 5th IEEE International Conference on Industrial Informatics*, (vol. 1, 353-358).

Mahony, R., Cha, S., & Hamel, T. (2006). A coupled estimation and control analysis for attitude stabilisation of mini aerial vehicles. In *Proceedings of the Australasian Conference on Robotics and Automation*. Auckland, New Zealand.

Makaran, J., Dittmer, D., Buchal, R., & MacArthur, D. (1993). The SMART(R) wrist-hand orthosis (WHO) for quadriplegic patients. *Journal of Prosthetics and Orthotics*, *5*(3), 73–76. doi:10.1097/00008526-199307000-00002

Manera, V., Schouten, B., Becchio, C., Bara, B. G., & Verfaillie, K. (2010). Inferring intentions from biological motion: A stimulus set of point-light communicative interactions. *Behavior Research Methods*, *42*, 168–178. doi:10.3758/BRM.42.1.168

Maniscalco, M., Lee, N., Dektor, S., Salowitz, N., Grote, E., & Twiggs, R. (2007). *A design concept for a robotic lunar regolith harvesting system*. ICRA '07 Space Robotics Workshop.

Marder-Eppstein, E., Berger, E., Foote, T., Gerkey, B., & Konolige, K. (2010). The office marathon: Robust navigation in an indoor office environment. In *Proceedings of IEEE International Conference on Robotics and Automation (ICRA)*.

Martel, S., Felfoul, O., & Mohammadi, M. (2008). Flagellated bacterial nanorobots for medical interventions in the human body. In *Proceedings of the 2nd IEEE RAS/EMBS Intl. Conf. on Biomedical Robotics and Biomechatronics (BioRob),* Scottsdale, AZ, (pp. 264–269).

Massa, B., Roccella, S., Carrozza, M. C., & Dario, P. (2002). Design and development of an underactuated prosthetic hand. *IEEE International Conference on Robotics and Automation*, Washington, DC, (pp. 11–15).

Mataric, M. J., Eriksson, J., & Feil-Seifer, D. (2007). Socially assistive robotics for post-stroke rehabilitation. *International Journal of NeuroEngineering and Rehabilitation, 4*, 1–12. doi:10.1186/1743-0003-4-1

Matsui, T. (2005). *The roles and method of robot design* (pp. 47–55). Proceeding of Robot Design Forum.

Matsuoka, Y., Afshar, P., & Oh, M. (2006). On the design of robotic hands for brain machine interface. *Neurosurgical Focus, 20*(5), 1–9. doi:10.3171/foc.2006.20.5.4

Maurer, C. R., Fitzpatrick, J. M., Wang, M. Y., Galloway, R. L., Maciunas, R. J., & Allen, G. S. (1997). Registration of head volume images using implantable fiducial markers. *IEEE Transactions on Medical Imaging, 16*(4), 447–462. doi:10.1109/42.611354

McBeth, P. B., Louw, D. F., Rizun, P. R., & Sutherland, G. R. (2004). *American Journal of Surgery, 188*(4), 68–75. doi:10.1016/j.amjsurg.2004.08.004

McComb, G. (2003). *Robot builder's sourcebook* (pp. 435–440). McGraw-Hill.

McConnell, P., Schneeberger, E., & Michler, R. (2003). History and development of robotic cardiac surgery. *Problems in General Surgery, 20*, 20–30. doi:10.1097/01.sgs.0000081182.03671.6e

McGovern, S. M., & Gupta, S. M. (2007). A balancing method and genetic algorithm for disassembly line balancing. *European Journal of Operational Research, 179*(3), 692–708. doi:10.1016/j.ejor.2005.03.055

McGovern, S. M., & Gupta, S. M. (2011). *The disassembly line*. New York, NY: McGraw Hill Inc.

Mehmanesh, H., Henze, R., & Lange, R. (2002). Totally endoscopic mitral valve repair. *The Journal of Thoracic and Cardiovascular Surgery, 123*, 96–97. doi:10.1067/mtc.2002.118496

Mellinger, D., Michael, N., & Kumar, V. (2010). Trajectory generation and control for precise aggressive maneuvers with quadrotors. In *Proceedings of International Symposium on Experimental Robotics 2010*. Delhi, India.

Meltzoff, A. N. (1995). Understanding the intentions of others: Reenactment of intended acts by 18-month-old children. *Developmental Psychology, 31*(5), 1–16. doi:10.1037/0012-1649.31.5.838

Melvin, W., Krause, K., Needleman, B., & Ellison, E. (2003). Robotic resection of a pancreatic neuroendocrine tumor. *Journal of Laparoendoscopic & Advanced Surgical Techniques. Part A., 13*, 33–36. doi:10.1089/109264203321235449

Melvin, W., Needleman, B., Krause, K., Schneider, C., & Ellison, E. (2002). Computer–enhanced versus standard laparoscopic antireflux surgery. *Journal of Gastrointestinal Surgery, 6*, 11–16. doi:10.1016/S1091-255X(01)00032-4

Menciassi, A. (2011). Single access robots for surgery. In *Proceedings of the ICRA Workshop on Mechanisms for Surgical Robotics*, Shanghai, (p. 3).

Menciassi, A., & Dario, P. (2009). Miniaturized robotic devices for endoluminal diagnosis and surgery: A single-module and a multiple-module approach. In *Proceedings of the Annual International Conference of the IEEE Engineering in Medicine and Biology Society (EMBC)*, Minneapolis, MN, (pp. 6842–6845).

Menzel, P., & D'Aluisio, F. (2000). *Robo Sapiens* (1st ed., pp. 180–189). MIT Press.

Merloz, P. H., Tonetti, J., & Eid, A. (1997). Computer-assisted spine surgery. *Clinical Orthopaedics and Related Research, 337*, 86–96. doi:10.1097/00003086-199704000-00011

Meyer, J.-A., & Filliat, D. (2003). Map-based navigation in mobile robotics: II. A review of map-learning and path-planning strategies. *Cognitive Systems Research*, *4*(4), 283–317. doi:10.1016/S1389-0417(03)00007-X

Michaud, F., Brosseau, Y., Cote, C., Letourneau, D., Moisan, P., & Ponchon, A. … Kabanza, F. (2005). Modularity and integration in the design of a socially interactive robot. In *Proceedings of IEEE International Workshop on Robot and Human Interactive Communication (RO-MAN)*, (pp. 172-177).

MicroPilot. (n.d.). *World leader in small UAV autopilots*. Retrieved March 11, 2011, from http://www.micropilot.com/

MikroKopter. (n.d.). *Wiki*. Retrieved March 15, 2011, from http://www.mikrokopter.de/

Miskin Scraper Works. (n.d.). *Website*. Retrieved September 16, 2011, from: http://www.miskin.com/

Mitchell, J. S. B. (1996). Shortest paths among obstacles in the plane. *International Journal of Computational Geometry & Applications*, *6*(3), 309–332. doi:10.1142/S0218195996000216

Mitobe, K., Capi, G., & Nasu, Y. (2004). A new control method for walking robots based on angular momentum. *Mechatronics*, *18*(2), 163–174. doi:10.1016/S0957-4158(03)00028-X

Mitsubishi Co. (n.d.). *Movemaster-EX industrial micro-robot system model RV-M1 user manual.*

Miyake, T., Matsumoto, T., Imamura, T., & Zhang, Z. E. (2011). Estimation of facial expression from its change in time. *ICIC Express Letters. Part B: Applications*, *2*(3), 641–645.

Mizuno, H., Nagai, H., Sasaki, K., Hosaka, H., Sugimoto, C., Khalil, K., & Tatsuta, S. (2007). Wearable sensor system for human behavior recognition (first report: basic architecture and behavior prediction method). *Solid-State Sensors, Actuators and Microsystems Conference, TRANSDUCERS 2007*, (pp. 435-438).

MobileRobots. (2011). *MobileRobots Pioneer docking station.* Retrieved March 30th, 2011 from http://mobilerobots.com/ResearchRobots/Accessories/ChargingSolutions.aspx

Mohr, F. W., Falk, V., Diegeler, A., Walther, T., Gummert, J. F., & Bucerius, J. (2001). Computer-enhanced "robotic" cardiac surgery: Experience in 148 patients. *The Journal of Thoracic and Cardiovascular Surgery*, *121*(5), 842–853. doi:10.1067/mtc.2001.112625

Montemerlo, M., Thrun, S., Koller, D., & Wegbreit, B. (2002). FastSLAM: a factored solution to the simultaneous localization and mapping problem. In *Proceedings of the National Conference on Artificial Intelligence*, Edmonton, Canada, (pp. 593–598).

Montemerlo, M., Thrun, S., Koller, D., & Wegbreit, B. (2003). FastSLAM 2.0: An improved particle filtering algorithm for simultaneous localization and mapping that provably converges. In *Proceedings of the International Conference on Artificial Intelligence (IJCAI)*, Acapulco, Mexico, (pp. 1151–1156).

Mori, M., Tanaka, J., Suzumori, K., & Kanda, T. (2006). Field test for verifying the capability of two high-powered hydraulic small robots for rescue operations. In *the Proceedings of the IEEE International Conference on Intelligent Robots and Systems*, Beijing, China, (pp. 3492-3497).

Mori, T., Segawa, Y., Shimosaka, M., & Sato, T. (2004). Hierarchical recognition of daily human actions based on continuous hidden markov models. In *Proceedings of the IEEE International Conference on Automatic Face and Gesture Recognition*, (pp. 779–784).

Motoi, N., Suzuki, T., & Ohnishi, K. (2009). A bipedal locomotion planning based on virtual linear inverted pendulum mode. *IEEE Transactions on Industrial Electronics*, *56*(1), 835–845. doi:10.1109/TIE.2008.2004663

Muff, T., Johnson, L., King, R., & Duke, M. B. (2004, February). A prototype bucket wheel excavator for the Moon, Mars and Phobos. *AIP Conference Proceedings*, *699*, 967–974. doi:10.1063/1.1649662

Muir, P., & Neuman, C. (1987). Kinematic modeling for feedback control of an omnidirectional wheeled mobile robot. In *Proceedings of IEEE International Conference on Robotics and Automation (ICRA)*, (pp. 1772-1778).

Murata, T. (1989). Petri nets: Properties, analysis and applications. *Proceedings of the IEEE*, *77*(4), 541–580. doi:10.1109/5.24143

Murata, T., Komoda, N., Matsumoto, K., & Haruna, K. (1986). A Petri net based controller for flexible and maintainable sequence control and its application in factory automation. *IEEE Transactions on Industrial Electronics*, *33*(1), 1–8. doi:10.1109/TIE.1986.351700

Nakamura, R., Masamune, K., & Nishikawa, Y. (1995). Development of an MRI-compatible needle insertion manipulator for stereotactic neurosurgery. *Journal of Image Guided Surgery*, *1*, 242–248. doi:10.1002/(SICI)1522-712X(1995)1:4<242::AID-IGS7>3.0.CO;2-A

Nakauchi, Y., Noguchi, K., Somwong, P., Matsubara, T., & Namatame, A. (2003). *Vivid room: Human intention detection and activity support environment for ubiquitous autonomy*. Intelligent Robots and Systems (IROS 2003).

NASA Lunabotic Mining Competition. (n.d.). Retrieved September 20, 2011, from http://www.nasa.gov/offices/education/centers/kennedy/technology/lunabotics.html

NASA. (1973). *Lyndon B. Johnson Space Center: Apollo 17 preliminary science report*. Washington: Scientific and Technical Information Office.

NASA. (2010, January). *2010 NASA Lunabotics Mining Competition rules*.

NASA. (n.d.). *Lunar outpost*. Retrieved September 22, 2011, from http://www.nasa.gov/externalflash/LunarOutpost/index.html

Nathoo, N., Cavusoglu, M. C., Vogelbaum, M. A., & Barnett, G. H. (2005). In touch with robotics: Neurosurgery for the future. *Neurosurgery*, *56*(3), 421–433. doi:10.1227/01.NEU.0000153929.68024.CF

Nathoo, N., Pesek, T., & Barnett, G. H. (2003). Robotics and neurosurgery. *The Surgical Clinics of North America*, *83*(6), 1339–1350. doi:10.1016/S0039-6109(03)00157-9

National Control Device. *LLC*. (n.d.). Retrieved September 5, 2011, from http://www.controlanything.com

National Fire Protection Association. (2011, November). *National electrical code 2011 handbook*.

Navarro-Serment, L. E., Mertz, C., & Hebert, M. (2010). Pedestrian detection and tracking using three-dimensional LADAR data. *The International Journal of Robotics Research*, *29*, 516–1528. doi:10.1177/0278364910370216

Nedas, T. G., Challacombe, B. J., & Dasgupta, P. (2005). Robotics in urology: An update. *International Journal of Medical Robotics and Computer Assisted Surgery*, *1*, 13–18. doi:10.1002/rcs.37

Nef, T., Guidali, M., & Riener, R. (2009). ARMin III - Arm therapy exoskeleton with an ergonomic shoulder actuation. *Applied Bionics and Biomechanics*, *6*(2), 127. doi:10.1080/11762320902840179

Nelson, B. (2007). Comparison of length of hospital stay between radical retropubic prostatectomy and robotic assisted laparoscopic prostatectomy. *The Journal of Urology*, *177*, 929–931. doi:10.1016/j.juro.2006.10.070

Nelson, B. J., Kaliakatsos, I. K., & Abbott, J. J. (2010). Microrobots for minimally invasive medicine. *Annual Review of Biomedical Engineering*, *12*, 55–85. doi:10.1146/annurev-bioeng-010510-103409

Neumann, P. (2007). Communication in industrial automation – What is going on? *Control Engineering Practice*, *15*(11), 1332–1347. doi:10.1016/j.conengprac.2006.10.004

Nguyen, H. G., Everett, H. R., Manouk, N., & Verma, A. (2002). *Autonomous mobile communication relays*. Presented at the SPIE AeroSense/Unmanned Ground Vehicle Technology IV, Orlando, FL.

Nguyen, H. G., Farrington, N., & Pezeshkian, N. (2004). *Maintaining communication link for tactical ground robots*. Presented at the AUVSI Unmanned Systems North America, Anaheim, CA, Pezeshkian, N., Nguyen, H. G., & Burmeister, A. (2006). Unmanned ground vehicle non-line-of-sight operations using relaying radios. In *Proceedings of the IASTED International Conference on Robotics and Applications* (pp. 1-6).

Nifong, L. W., Chu, V. R., Bailey, B. M., Maziarz, D. M., Sorrell, V. L., Holbert, D., & Chitwood, W. R. Jr. (2003). Robotic mitral valve repair: Experience with the da Vinci system. *The Annals of Thoracic Surgery*, *75*, 438–443. doi:10.1016/S0003-4975(02)04554-X

Nishihara, S., Sugano, N., Nishii, T., Miki, H., Nakamura, N., & Yoshikawa, H. (2006). Comparison between hand rasping and robotic milling for stem implantation in cementless total hip arthroplasty. *The Journal of Arthroplasty*, *21*(7), 957–966. doi:10.1016/j.arth.2006.01.001

Nof, S. Y. (Ed.). (1999). *Handbook of industrial robotics* (2nd ed.). USA: Wiley. doi:10.1002/9780470172506

Noguchi, K., Somwong, P., Matsubara, T., & Nakauchi, Y. (2003). Human intention detection and activity support system for ubiquitous autonomy. *Proceedings of IEEE International Symposium on Computational Intelligence in Robotics and Automation, 2003*, 906–911. doi:10.1109/CIRA.2003.1222300

Nokleby, S. B., & Podhorodeski, R. P. (2001). Optimization-based synthesis of Grashof geared five-bar mechanisms. *Journal of Mechanical Design*, *123*(4), 529–534. doi:10.1115/1.1401736

Nonami, K., Kendoul, F., Suzuki, S., Wang, W., & Nakazami, D. (2010). *Autonomous flying robots*. Springer. doi:10.1007/978-4-431-53856-1

O'Malley, M., & Goldfarb, M. (2002). The effect of force saturation on the haptic perception of detail. *IEEE/ASME Transactions on Mechatronics*, *7*(3), 280–288. doi:10.1109/TMECH.2002.802725

O'Toole, M. D., Bouazza-Marouf, K., Kerr, D., Gooroochurn, M., & Vloeberghs, M. (2010). A methodology for design and appraisal of surgical robotic systems. *Robotica*, *28*(2), 297–310. doi:10.1017/S0263574709990658

Okamura, A. M. (2009). Haptic feedback in robot-assisted minimally invasive surgery. *Current Opinion in Urology*, *19*(1), 102–107. doi:10.1097/MOU.0b013e32831a478c

Oliveira, H. P., Sousa, A. J., Moreira, P., & Costa, P. J. (2008). Dynamical models for Omni directional robots with 3 and 4 wheels. In *Proceedings of International Conference on Informatics in Control, Automation and Robotics*.

Oliveira, H. P., Sousa, A. J., Moreira, A. P., & Costa, P. J. (2009). Modeling and assessing of Omni-directional robots with three and four wheels. In Rodić, A. D. (Ed.), *Contemporary robotics - Challenges and solutions*. InTech.

Orin, D. E., & Goswami, A. (2008, October). Centroidal momentum matrix of a humanoid robot: Structure and properties. In *IEEE International Conference on Intelligent Robots and Systems* (pp. 653-659).

Paden, B., & Sastry, S. (1988). Optimal kinematic design of 6R manipulators. *The International Journal of Robotics Research*, *7*(2), 43–61. doi:10.1177/027836498800700204

Pagano, J. S., & Vaheri, A. (1965). Enhancement of infectivity of poliovirus RNA with diethylaminoethyl-dextran (DEAE-D). *Archives of Virology*, *17*(3), 456–464. doi:10.1007/BF01241201

Paparazzi. (n.d.). *Main page*. Retrieved March 11, 2011, from http://paparazzi.enac.fr/wiki/Main_Page

Pappone, C., Vicedomini, G., Manguso, F., Gugliotta, F., Mazzone, P., & Gulletta, S. (2006). Robotic magnetic navigation for atrial fibrillation ablation. *Journal of the American College of Cardiology*, *47*(7), 1390–1400. doi:10.1016/j.jacc.2005.11.058

Park, J. H. (2001). Impedance control for biped robot locomotion. *IEEE Transactions on Robotics and Automation*, *17*(6), 870–882. doi:10.1109/70.976014

Partin, A. W., Adams, J. B., Moore, R. G., & Kavoussi, L. R. (1995). Complete robot-assisted laparoscopic urologic surgery: A preliminary report. *Journal of the American College of Surgeons*, *181*, 552–557.

Passenberg, C., Peer, A., & Buss, M. (2010). A survey of environment-, operator-, and task-adapted controllers for teleoperation systems. *Mechatronics*, *20*(7), 787–801. doi:10.1016/j.mechatronics.2010.04.005

Passerotti, C., & Peters, C. (2006). Robotic-assisted laparoscopy applied to reconstructive surgeries in children. *World Journal of Urology*, *24*, 193–197. doi:10.1007/s00345-006-0084-0

Patel, M., Menon, M., & Rogers, C. (2011). Robotic urologic surgery: Robot-assisted partial nephrectomy. *Robotics in Genitourinary Surgery*, *4*, 423–433. doi:10.1007/978-1-84882-114-9_37

Paul, H. A., Mittlestadt, B., Bargar, W. L., Musits, B., Taylor, R. H., & Kazanzides, P. … Hanson, B. (1992). Surgical robot for total hip replacement surgery. In *Proceedings of the IEEE International Conference on Robotics and Automation*, Nice, (pp. 606–611).

Paul, B. (1979). A reassessment of Grashof's criterion. *Journal of Mechanical Design. Transactions of the ASME*, *101*(3), 515–518. doi:10.1115/1.3454089

Paul, R. P. (1981). *Robot manipulators: Mathematics, programming, and control*. Cambridge, MA: MIT Press.

Pepperkok, R., Zanetti, M., King, R., Delia, D., Ansorge, W., & Philipson, L. (1988). Automatic microinjection system facilitates detection of growth inhibitory mRNA. *Proceedings of the National Academy of Sciences of the United States of America*, *85*(18), 6748–6752. doi:10.1073/pnas.85.18.6748

Perrot, Y., Gargiulo, L., Houry, M., Kammerer, N., Keller, D., & Measson, Y. … Verney, A. (2010, October). Long reach articulated robots for inspection in hazardous environments, recent developments on robotics and embedded diagnostics. *2010 1st International Conference on Applied Robotics for the Power Industry (CARPI)*, (pp. 1-5). Montreal, QC.

Perry, J. C., Rosen, J., & Burns, S. (2007). Upper-limb powered exoskeleton design. *IEEE/ASME Transactions on Mechatronics*, *12*(4), 408–417. doi:10.1109/TMECH.2007.901934

Petermann, J., Kober, R., Heinze, R., Frlich, J. J., Heeckt, P. F., & Gotzen, L. (2000). Computerassisted planning and robot-assisted surgery in anterior cruciate ligament reconstruction. *Operative Techniques in Orthopaedics*, *10*, 50–55. doi:10.1016/S1048-6666(00)80042-7

Peters, C. (2007). *Applications of robotics in pediatric urologic surgery* (pp. 179–187). Robotic Urologic Surgery. doi:10.1007/978-1-84628-704-6_24

Peters, T. M. (2000). Image guided surgery: From X-rays to virtual reality. *Computer Methods in Biomechanics and Biomedical Engineering*, *4*, 27–57. doi:10.1080/10255840008907997

Pethig, R., & Markx, G. H. (1997). Applications of dielectrophoresis in biotechnology. *Trends in Biotechnology*, *15*(10), 426–432. doi:10.1016/S0167-7799(97)01096-2

Pettersson, P. O., & Doherty, P. (2004). Probabilistic roadmap based path planning for autonomous unmanned aerial vehicles. *Proceedings of the 14th International Workshop on Connecting Planning and Theory with Practice* (pp. 158–162).

Piedrafita, R., Tardioli, D., & Villarroel, J. L. (2008). Distributed implementation of discrete event control systems based on Petri nets. In *Proceedings of the 2008 IEEE International Symposium on Industrial Electronics*, (pp. 1738-1745).

Pohl, H. A. (1978). *Dielectrophoresis the behavior of neutral matter in nonuniform electric fields*. Cambridge, UK: Cambridge University Press.

Pomares, J., Puente, S. T., Torres, F., Candelas, F. A., & Gil, P. (2004). Virtual disassembly of products based on geometric models. *Computers in Industry*, *55*(1), 1–14. doi:10.1016/j.compind.2004.03.001

Popovic, M., Goswami, A., & Herr, H. (2005). Ground reference points in legged locomotion: Definitions, biological trajectories and control implications. *The International Journal of Robotics Research*, *24*(12), 1013–1032. doi:10.1177/0278364905058363

Potamianos, P., Davies, B. L., & Hibberd, R. D. (1994). *Intra-operative imaging guidance for keyhole surgery: Methodology and calibration*. International Symposium on Medical Robotics and Computer Assisted Surgery, Pittsburgh, PA.

Potamianos, P., Davies, B. L., & Hibberd, R. D. (1995). *Intra-operative registration for percutaneous surgery.* International Symposium on Medical Robotics and Computer Assisted Surgery, Baltimore, MD.

Pott, P. P. (2011). *Medical robotic database: MeRoDa*. University of Heidelberg. Retrieved on August 1, 2011, from http://meroda.uni-hd.de

Pott, P., Scharf, H., & Schwarz, M. (2005). Today's state of the art of surgical robotics. *Journal of Computer Aided Surgery*, *10*, 101–132.

Poulakakis, I., & Grizzle, J. W. (2009). The spring loaded inverted pendulum as the hybrid zero dynamics of an asymmetric hopper. *IEEE Transactions on Automatic Control*, *54*(8), 1779–1793. doi:10.1109/TAC.2009.2024565

Pounds, P., Mahony, R., & Corke, P. (2006). Modelling and control of a quad-rotor robot. In *Proceedings of Australasian Conference on Robotics and Automation.* Auckland, New Zealand.

Pounds, P., Mahony, R., Gresham, J., Corke, P., & Roberts, J. (2004). *Towards dynamically-favourable quad-rotor aerial robots*. In Australasian Conference on Robotics and Automation. Canberra, Australia.

Pounds, P., Mahony, R., Hynes, P., & Roberts, J. F. (2002). Design of a four-rotor aerial robot. In *Proceedings of Australasian Conference on Robotics and Automation*. Auckland, New Zealand.

Pounds, P., Hamel, T., & Mahony, R. (2007). Attitude control of rigid body dynamics from biased IMU measurements. In *Proceedings of IEEE Conference on Decision and Control, 2007*, 4620–4625. doi:.doi:10.1109/CDC.2007.4434173

Pounds, P., Mahony, R., & Corke, P. (2009). Design of a static thruster for microair vehicle rotorcraft. *Journal of Aerospace Engineering, 22*(1), 85–94. doi:10.1061/(ASCE)0893-1321(2009)22:1(85)

Pounds, P., Mahony, R., & Corke, P. (2010). Modelling and control of a large quadrotor robot. *Control Engineering Practice, 18*(7), 691–699. doi:10.1016/j.conengprac.2010.02.008

Pransky, J. (1997). ROBODOC - Surgical robot success story. *Industrial Robot: An International Journal, 24*(3), 231–233. doi:10.1108/01439919710167444

Pransky, J. (1998). Surgeons' realizations of RoboDoc. *Industrial Robot: An International Journal, 25*(2), 105–108. doi:10.1108/01439919810205792

Pratt, J., Carff, J., Drakunov, S., & Goswami, A. (2006, December). Capture point: A step toward humanoid push recovery. In *IEEE International Conference on Humanoid Robotics* (pp. 200-207).

Press, W. H., Tekolsky, S. A., Vetterling, W. T., & Flannery, B. P. (1992). *Numerical recipes in C: The art of scientific computing* (2nd ed.). New York, NY: Cambridge University Press.

Procerus Technologies. (n.d.). *Fly light with world's smallest UAV autopilot*. Retrieved March 11, 2011, from http://www.procerusuav.com/index.php

PTC Inc. (2011) Retrieved September 22, 2011, from http://www.ptc.com/

Puntambekar, S., Agarwai, G., Joshi, S., Rayate, N., & Sathe, R. (2009). Robotic radical hysterectomy: Applying principles of the laparoscopic Pune technique. *Journal of Robotic Surgery, 4*, 259–264. doi:10.1007/s11701-010-0224-9

Putra, A. S., Huang, S., Tan, K. K., Panda, S. K., & Lee, T. H. (2006). *Partially-rotating piezoelectric actuator to assist cell penetration in intra-cytoplasmic sperm injection*. Paris, France.

Pynadath, D. (1999). *Probabilistic grammars for plan recognition*. Doctoral Thesis, the University of Michigan, MI.

Qi, Y., Wang, Z. L., & Huang, Y. (2007). A non-contact eye-gaze tracking system for human computer interaction. *Proceedings of the 2007 International Conference on Wavelet Analysis and Pattern Recognition*, Beijing, China.

Quinn, A. M. (2002). CyberKnife: A robotic radiosurgery system. *Clinical Journal of Oncology Nursing, 6*(3), 149,156.

Rabiner, L. R. (1989). A tutorial on hidden Markov models and selected applications in speech recognition. *Proceedings of the IEEE, 77*(2), 257–286. doi:10.1109/5.18626

Rayman, R. (2009). Is surgery a remote possibility? Robotic surgical system under development has telesurgery capabilities. *Health Technology Trends, 21*(7), 5–7.

Rebello, K. J. (2004). Applications of MEMS in surgery. *Proceedings of the IEEE, 92*(1), 43–55. doi:10.1109/JPROC.2003.820536

Reif, J. H., & Wang, H. (1995). Social potential fields: a distributed behavioral control for autonomous robots. In *WAFR: Proceedings of the workshop on Algorithmic foundations of robotics* (pp. 331–345). Natick, MA, USA: A. K. Peters, Ltd.

Reisig, W. (1985). *Petri Nets*. Berlin, Germany: Springer-Verlag.

Rentschler, M., Dumpert, J., Platt, S., Oleynikov, D., Farritor, S., & Iagnemma, K. (2006). Mobile in vivo biopsy robot. In *Proceedings of the IEEE International Conference on Robotics and Automation* (ICRA), Orlando, FL, (pp. 4155–4160).

Rocon, E., & Pons, J. L. (2011). *Exoskeletons in rehabilitation robotics: Tremor suppression*. Springer Tracts in Advanced Robotics.

Roether, C. L., Omlor, L., Christensen, A., & Giese, M. (2009). Critical features for the perception of emotion from gait. *Journal of Vision (Charlottesville, Va.), 9*, 1–32. doi:10.1167/9.6.15

Roichman, E., Solomon, Y., & Moshe, Y. (2008). Real-time pedestrian detection and tracking. *Proceedings of the 3rd European DSP Education and Research Symposium (EDERS 2008),* Tel-Aviv, Israel, June 2008, (pp. 281-288).

Rosen, J., Hannaford, B., & Satava, R. (2011). *Surgical robotics - Systems applications and visions*. Springer. ISBN: 978-14419-4

Rosen, J., Perry, J. C., Manning, N., Burns, S., & Hannaford, B. (2005). The human arm kinematics and dynamics during daily activities – Toward a 7 DOF upper limb powered exoskeleton. *International Conference on Advanced Robotics* (pp. 532-539).

Ross, P. (2011). Top 11 technologies of the decade. *Spectrum, IEEE, 48*(1), 27-63. doi:10.1109/MSPEC.2011.5676379

Roufas, K. D., Zhang, Y., Duff, D. G., & Yim, M. H. (2001). Six degree of freedom sensing for docking using IR LED emitters and receivers. *Experimental Robotics, VII*, 91–200.

Rovetta, A., Sala, R., Cosmi, F., Wen, X., Milanesi, S., & Sabbadini, D. (1996). A new telerobotic application: Remote laparoscopic surgery using satellites and optical fiber networks for data exchange. *The International Journal of Robotics Research, 15*(3), 267–279. doi:10.1177/027836499601500303

Runeson, S., & Frykholm, G. (1981). Visual perception of lifted weight. *Journal of Experimental Psychology. Human Perception and Performance, 7*, 733–740. doi:10.1037/0096-1523.7.4.733

Russell, S. J., & Norvig, P. (2003). *Artificial intelligence: A modern approach*, 2nd ed. Prentice Hall series in Artificial Intelligence, Prentice Hall.

Ryu, J. H., Artigas, J., & Preusche, C. (2010). A passive bilateral control scheme for a teleoperator with time-varying communication delay. *Mechatronics, 20*(7), 812–823. doi:10.1016/j.mechatronics.2010.07.006

Sanderson, A. C., Homem de Mello, L. S., & Zhang, H. (1990). Assembly sequence planning. *AI Magazine*, 62–81.

Santos, R. R., Steffen, V., & Saramago, S. (2010). Optimal task placement of a serial robot manipulator for manipulability and mechanical power optimization. *Intelligent Information Management, 2*(9), 512–525. doi:10.4236/iim.2010.29061

Satava, R. M. (1995). Telesurgery, virtual reality and the new world order of medicine. *Computer Aided Surgery, 1*(1), 12–16. doi:10.3109/10929089509106821

Satava, R. M. (2004). Future trends in the design and application of surgical robots. *Seminars in Laparoscopic Surgery, 11*(2), 129–135.

Satava, R. M. (2005). The nature of surgical error: A cautionary tale and a call to reason. *Surgical Endoscopy, 19*(8), 1014–1016. doi:10.1007/s00464-005-8108-1

Sato, N., Matsuno, F., Yamasaki, T., Kamegawa, T., Shiroma, N., & Igarashi, H. (2004). Cooperative task execution by a multiple robot team and its operators in search and rescue operations. *In Proceedings of the International Conference on Intelligent Robots and Systems*, Sendai, Japan, (pp. 1083–1088).

Schenker, P. S., Pirjanian, P., Balaram, B., Ali, K. S., Trebi-Ollennu, A., & Huntsberger, T. L. (2000). Reconfigurable robots for all terrain exploration. In *Proceedings of the SPIE Symposium on Sensor Fusion and Decentralized Control in Robotic Systems III, 4196*, 454–468.

Schiele, A. (2008). Case study: The ergonomic EXARM exoskeleton. In *Wearable robots: Biomechatronic exoskeleton* (pp. 248–255). John Wiley & Sons Ltd.

Schiff, J., Li, P. S., & Goldstein, M. (2005). Robotic microsurgical vasovasostomy and epididymostomy in rats. *International Journal of Medical Robotics and Computer Assisted Surgery, 1*, 122–126. doi:10.1002/rcs.21

Schmidt, S., & Färber, B. (2009). Pedestrians at the kerb – Recognising the action intentions of humans. *Transportation Research Part F: Traffic Psychology and Behaviour, 12*, 300–310. doi:10.1016/j.trf.2009.02.003

Schrempf, O. C., Albrecht, D., & Hanebeck, U. D. (2007). Tractable probabilistic models for intention recognition based on expert knowledge. *Proceedings of the 2007 IEEE/RSJ International Conference on Intelligent Robots and Systems*, San Diego, CA, USA, (pp. 1429-1434).

Schreuder, H., Aweemer, R., Baal, W., Lande, J., & Dijkstra, J. (2010). From open radical hysterectomy to robot-assisted laparoscopic radical hysterectomy for early stage cervical cancer: Aspects of a single institution learning curve. *Gynecological Surgery*, *7*, 253–258. doi:10.1007/s10397-010-0572-5

Schur, C. (2011). *Chris & Dawn Schur's docking logic.* Retrieved March 30th, 2011 from http://www.schursastrophotography.com/robotics/dockinglogic.html

Schuresko, M., & Cortes, J. (2008). Distributed motion constraints for algebraic connectivity of robotic networks. In *47th IEEE Conference on Decision and Control, CDC 2008* (pp. 5482-5487).

Sebanz, N., & Shiffrar, M. (2009). Detecting deception in a bluffing body: the role of expertise. *Psychonomic Bulletin & Review*, *16*, 170–175. doi:10.3758/PBR.16.1.170

Senapati, S., & Advincula, A. (2011). Surgical techniques: Robot-assisted laparoscopic myomectomy with the da Vinci surgical system. *Journal of Robotic Surgery*, *1*, 69–74. doi:10.1007/s11701-007-0014-1

Seo, K.-K., Park, J.-H., & Jang, D.-S. (2001). Optimal disassembly sequence using genetic algorithms considering economic and environmental aspects. *International Journal of Advanced Manufacturing Technology*, *18*, 371–380. doi:10.1007/s001700170061

Se, S., Lowe, D., & Little, J. (2002). Mobile robot localization and mapping with uncertainty using scale invariant visual landmarks. *The International Journal of Robotics Research*, *21*(8), 735–758. doi:10.1177/027836402761412467

Sevdalis, V., & Keller, P. E. (2009). Self-recognition in the perception of actions performed in synchrony with music. *Annals of the New York Academy of Sciences*, *1169*, 499–502. doi:10.1111/j.1749-6632.2009.04773.x

Sevdalis, V., & Keller, P. E. (2010). Cues for self-recognition in point-light displays of actions performed in synchrony with music. *Consciousness and Cognition*, *19*(2), 617–626. doi:10.1016/j.concog.2010.03.017

Shakernia, O., Vidal, R., Sharp, C. S., Ma, Y., & Sastry, S. (2002). Multiple view motion estimation and control for landing an unmanned aerial vehicle. *Proceedings of IEEE International Conference on Robotics and Automation,* (pp. 2793–2798).

Sheng, W., Yang, Q., & Tan, J. (2006). Distributed multi-robot coordination in area exploration. *Robotics and Autonomous Systems*, *54*(12), 945–955. doi:10.1016/j.robot.2006.06.003

Shimizu, Y., Tsuji, K., & Nomura, M. (2007). Optimal disassembly sequence generation using a genetic programming. *International Journal of Production Research*, *45*(18-19), 4537–4554. doi:10.1080/00207540701440261

Shimosaka, M., Mori, T., Harada, T., & Sato, T. (2005). Marginalized bags of vectors kernels on switching linear dynamics for online action recognition. In *Proceedings of the IEEE International Conference on Robotics and Automation*, (pp. 3072 – 3077).

Shucker, B., & Bennett, J. K. (2005). Target tracking with distributed robotic macrosensors. In *Military Communications Conference, MILCOM 2005 IEEE* (vol. 4, pp. 2617-2623).

Sicitiano, B. (1999). Kinematic control of redundant robot manipulators. *Journal of Intelligent & Robotic Systems*, *3*, 201–212. doi:10.1007/BF00126069

Siebert, W., Mai, S., Kober, R., & Heeckt, P. (2004). Total knee replacement: Robotic assistive technique. In DiGioia, A. M., Jaramaz, B., Picard, F., & Nolte, L.-P. (Eds.), *Computer and robotic assisted hip and knee surgery* (pp. 127–156). Oxford University.

Siegwart, R., & Nourbakhsh, I. R. (2004). *Introduction to autonomous mobile robots*. The MIT Press.

SIEMENS. (2011). *Atris Zeego product page*. Retrieved on August 1, 2011, from http://www.medical.siemens.com/

Silva, C. W. D. (2005). *Mechatronics: An integrated approach.* CRC Press.

Silva, M. (1990). Petri nets and flexible manufacturing. In Rozenberg, G. (Ed.), *Advances in Petri Nets LNCS 424* (pp. 374–417). Berlin, Germany: Springer-Verlag.

Silva, M., & Velilla, S. (1982). Programmable logic controllers and Petri nets: A comparative study. In Ferrate, G., & Puente, E. A. (Eds.), *In IFAC Software for Computer Control 1982* (pp. 83–88). Oxford, UK: Pergamon.

Silverman, M., Nies, D. M., Jung, B., & Sukhatme, G. S. (2002). Stay alive: A docking station for autonomous robot recharging. In *Proceedings of IEEE International Conference on Robotics and Automation (ICRA)*.

Sitti, M., & Hashimoto, H. (2003). Teleoperated touch feedback from the surfaces at the nanoscale: Modeling and experiments. *IEEE/ASME Transactions on Mechatronics*, *8*(2), 287–298. doi:10.1109/TMECH.2003.812828

Smith, R., Self, M., & Cheeseman, P. (1990). Estimating uncertain spatial relationships in robotics. In Cox, I. J., & Wilfon, G. T. (Eds.), *Autonomous robot vehicles* (pp. 167–193). Springer-Verlag. doi:10.1007/978-1-4613-8997-2_14

Sobh, T., & Toundykov, D. (2004). Kinematic synthesis of robotic manipulators from task descriptions. *IEEE Magazine on Robotics and Automation*, *11*(2), 78–85. doi:10.1109/MRA.2004.1310944

Song, J.-B., & Byun, K.-S. (2004). Design and control of a four-wheeled Omnidirectional mobile robot with steerable Omnidirectional wheels. *Journal of Robotic Systems*, *21*, 193–208. doi:10.1002/rob.20009

Song, T. L. (1999). Observability of target tracking with range-only measurements. *IEEE Journal of Oceanic Engineering*, *24*(3), 383–387. doi:10.1109/48.775299

Souryal, M. R., Geissbuehler, J., Miller, L. E., & Moayeri, N. (2007). Real-time deployment of multihop relays for range extension. In *Proceedings of the 5th International Conference on Mobile Systems, Applications and Services* (pp. 85-98). San Juan, Puerto Rico: ACM.

Spong, M. W., & Vidyasagar, M. (1989). *Robot dynamics and control*. New York, NY: John Wiley & Sons.

Srinivasa, S. S., Ferguson, D., Helfrich, C. J., Berenson, D., Collet, A., & Diankov, R. (2010). HERB: A home exploring robotic butler. *Autonomous Robots*, *28*(1), 5–20. doi:10.1007/s10514-009-9160-9

Stark, B., Jensen, A., Chen, Y. Q., & McKee, M. (2012) A personal remote sensing system with VTOL UAV for natural resource management applications. (submitted) In *Proceedings of IEEE Aerospace Conference 2012*. Big Sky, MT, USA.

Stieber, M. F., Trudel, C. P., & Hunter, D. G. (1997, April). Robotic systems for the International Space Station. *Proceedings of 1997 IEEE International Conference on Robotics and Automation*, Albuquerque, NM, USA, vol. 4, (pp. 3068-3073).

Stiehl, J. B., Bach, J., & Heck, D. A. (2007). Validation and metrology in CAOS. In *Navigation and MIS in orthopedic surgery* (pp. 68–78). Springer. doi:10.1007/978-3-540-36691-1_9

Stoianovici, D., Kavoussi, L. R., Whitcomb, L. L., et al. (2002). *Friction transmission with axial loading and a radiolucent surgical needle driver*. United States Patent 6,400,979.

Stoianovici, D. (2001). URobotics—Urology robotics at Johns Hopkins. *Computer Aided Surgery*, *6*(6), 360–369.

Stoianovici, D., Webster, R., & Kavoussi, L. (2005). Surgical robotic applications in minimally invasive urooncology surgery. In Moore, R., & Bishoff, J. (Eds.), *Complications of urologic laparoscopic surgery: Recognition, management and prevention* (pp. 353–363). Taylor and Francis.

Strikwerda, J. C. (1989). *Finite difference schemes and partial differential equations*. Wadsworth & Brooks/Cole Advanced Books & Software.

Stump, E., Jadbabaie, A., & Kumar, V. (2008). Connectivity management in mobile robot teams. In *IEEE International Conference on Robotics and Automation, ICRA 2008* (pp. 1525-1530).

Sugihara, T., & Nakamura, Y. (2007, October). Enhancement of boundary condition relaxation method for 3D hopping motion of biped robots. In *IEEE International Conference on Intelligent Robots and Systems* (pp. 444-449).

Sun, D., Zhou, Z. Y., Liu, Y. H., & Shen, W. Z. (1997). Development and application of ultrasonic surgical instruments. *IEEE Transactions on Bio-Medical Engineering*, *44*(6), 462–467. doi:10.1109/10.581934

Sung, G. T., & Gill, I. S. (2001). Robotic laparoscopic surgery: a comparison of the da Vinci and Zeus systems. *The Journal of Urology*, *58*, 893–898. doi:10.1016/S0090-4295(01)01423-6

Sun, L., Wang, H., Chen, L., & Liu, Y. (2006). A novel ultrasonic micro-dissection technique for biomedicine. *Ultrasonics*, *44*(Supplement 1), 255–260. doi:10.1016/j.ultras.2006.06.010

Sun, Y., & Nelson, B. J. (2002). Biological cell injection using an autonomous microrobotic system. *The International Journal of Robotics Research*, *21*(10-11), 861. doi:10.1177/027836490202101 0833

Sutherland, G. R., Latour, I., Greer, A. D., Fielding, T., Feil, G., & Newhook, P. (2008). An image-guided magnetic resonance-compatible surgical robot. *Neurosurgery*, *62*(2), 286–293. doi:10.1227/01.neu.0000315996.73269.18

Szymula, T. P., & Richter, V. (2001). Recenzie endoscopy and microsurgery. In D. Pieptu (Series Ed.), *Update in plastic surgery*. New York, NY: Spinger.

Tahboub, K. A. (2005). Compliant human-robot cooperation based on intention recognition. *Proceedings of the 2005 IEEE International Symposium on Intelligent Control*, Limassol, Cyprus, June 27-29, (pp. 1417-1422).

Tahboub, K. A. (2006). Intelligent human-machine interaction based on dynamic Bayesian networks probabilistic intention recognition. *Journal of Intelligent & Robotic Systems*, *45*(1), 31–52. doi:10.1007/s10846-005-9018-0

Tahir, N. M., Hussain, A., Samad, S. A., & Husain, H. (2008). *Posture recognition using correlation filter classifier. Journal of Theoretical and Applied Information Technology (JATIT)*. EBSCO Publishing USA.

Tajima, R., Honda, D., & Suga, K. (2009, April). Fast running experiments involving a humanoid robot. In *IEEE International Conference on Robotics and Automation* (pp. 1571-1576).

Takahashi, T. (2007). *Emotional robot design* (pp. 32–34). Digest Book of IEEE RO-MAN.

Takenaka, T., Matsumoto, T., & Yoshiike, T. (2009a, October). Real time motion generation and control for biped robot – 3rd report: Dynamics error compensation. In *IEEE International Conference on Intelligent Robots and Systems.* (pp. 1594-1600).

Takenaka, T., Matsumoto, T., Yoshiike, T., & Shirokura, S. (2009b, October). Real time motion generation and control for biped robot –2ndreport: Running gait pattern generation. In *IEEE International Conference on Intelligent Robots and Systems* (pp. 1092-1099).

Takeuchi, M., Nakajima, M., Kojima, M., & Fukuda, T. (7-10 November 2010). *Soft handling probe using thermal gel for single cells*. Paper presented at the Micro-NanoMechatronics and Human Science (MHS), 2010 International Symposium on.

Talamini, M., Chapman, W., Horgan, S., & Melvin, W. (2003). Evaluation of 211 robotic surgical procedures. *Surgical Endoscopy*, *17*, 1521–1524. doi:10.1007/s00464-002-8853-3

Tanaka, J., Suzumori, K., Takata, M., Kanda, T., & Mori, M. (2005). A mobile jack robot for rescue operation. In *the Proceedings of the IEEE International Workshop on Safety, Security and Rescue Robotics*, Kobe, Japan, (pp. 99–104).

Tanev, T., & Stoyanov, B. (2000). On the performance indexes for robot manipulators. In *Problems of Engineering* (pp. 64–70). Cybernetics and Robotics.

Tang, M. X., Redemann, C. T., & Szoka, F. C. (1996). In vitro gene delivery by degraded polyamidoamine dendrimers. *Bioconjugate Chemistry*, *7*(6), 703–714. doi:10.1021/bc9600630

Tanigawa, M., Luinge, H., Schipper, L., & Slycke, P. (2008). Drift-free dynamic height sensor using MEMS IMU aided by MEMS pressure sensor. In *Proceedings of 5th Workshop on Positioning, Navigation and Communication, 2008*, (pp. 191–196).

Tan, K. K., & Putra, A. S. (2005a). *Partial-rotating piezoelectric actuators for intracytoplasmic sperm injection (Vol. 6048)*. SPIE.

Tan, Y., Sun, D., Huang, W., & Cheng, S. H. (2008). Mechanical modeling of biological cells in microinjection. *IEEE Transactions on Nanobioscience*, *7*(4), 257–266. doi:10.1109/TNB.2008.2011852

Tapus, A., Cristian, T., & Mataric, M. J. (2008). User-robot personality matching and assistive robot behavior adaptation for post-stroke rehabilitation therapy. *Intelligent Journal of Robotics*, *5*, 169–183. doi:10.1007/s11370-008-0017-4

Tapus, A., Matarić, M. J., & Scassellati, B. (2007). The grand challenges in socially assistive robotics. *IEEE Robotics & Automation Magazine*, *14*, 15–18. doi:10.1109/MRA.2007.339605

Tayebi, A., & McGilvray, S. (2004). Attitude stabilization of a four-rotor aerial robot. In *Proceedings of IEEE Conference on Decision and Control, 2004,* (vol. 2, pp. 1216–1221). doi:10.1109/CDC.2004.1430207

Tayebi, A., & McGilvray, S. (2006). Attitude stabilization of a VTOL quadrotor aircraft. *IEEE Transactions on Control Systems Technology*, *14*(3), 562–571. doi:10.1109/TCST.2006.872519

Taylor, R. H. (1997). Robots as surgical assistants: Where we are, wither we are tending, and how to get there. In *AIME 97*, Genoble, France, (pp. 3–11).

Taylor, R. H. (2001). *Robotic joint replacement surgery*. Course materials, Computer-Integrated Surgery, Johns Hopkins University. Retrieved on August 1, 2011, from http://www.cs.jhu.edu/~cis/cista/445/Lectures/Robodoc.pdf

Taylor, R. H. (1999). A steady-hand robotic system for microsurgical augmentation. *The International Journal of Robotics Research*, *18*(12), 1201–1210. doi:10.1177/02783649922067807

Taylor, R. H., Funda, J., & Eldridge, B. (1995). A telerobotic assistant for laparoscopic surgery. *IEEE Engineering in Medicine and Biology Magazine*, 279–287. doi:10.1109/51.391776

Taylor, R. H., & Kazanzides, P. (2008). Medical robotics and computer-integrated interventional medicine. *Advances in Computers: Emerging Technologies*, *73*, 219–258.

Taylor, R. H., Menciassi, A., Fichtinger, G., & Dario, P. (2008). Medical robotics in computer-integrated surgery. In Siciliano, B., & Khatib, O. (Eds.), *Handbook of robotics* (2nd ed., pp. 1199–1222). Cambridge, MA: Springer. doi:10.1007/978-3-540-30301-5_53

Taylor, R. H., & Stoianovici, D. (2003). Medical robotics in computer-integrated surgery. *IEEE Transactions on Robotics and Automation*, *19*(5), 765–781. doi:10.1109/TRA.2003.817058

Teal Group. (2011). *Teal Group predicts worldwide UAV market will total over $80 billion in its just released 2010 UAV market profile and forecast*. Retrieved February 21, 2011, from www.tealgroup.com

Technology, C. (n.d.). *IRIS 2.4GHz*. Retrieved February 27, 2009, from http://www.xbow.com/Products/product-details.aspx?sid=264

Tejima, N. (2000). Rehabilitation robotics: A review. *Advanced Robotics*, *14*(7), 551–564. doi:10.1163/156855301742003

Tekdas, O., & Isler, V. (2008). Robotic routers. In *IEEE International Conference on Robotics and Automation, ICRA 2008* (pp. 1513-1518).

Terada, K., Shamoto, T., Mei, H., & Ito, A. (2007). Reactive movements of non-humanoid robots cause intention attribution in humans, *Proceedings of the 2007 IEEE/RSJ International Conference on Intelligent Robots and Systems*, San Diego, CA, USA, (pp. 3715-3720).

Terada, K., & Ito, A. (2010). Can a robot deceive humans? *HRI, 2010*, 191–192. doi:10.1145/1734454.1734538

Thangavelautham, J., Smith, A., Samid, N. A. E., Ho, A., Boucher, D., Richard, J., & D'Eleuterio, G. M. T. (2008, February). *Multirobot lunar excavation and ISRU using artificial-neural-tissue controllers*. Space Technology and Applications International Forum—STAIF.

Tholey, G., Chanthasopeephan, T., & Hu, T. (2003). Measuring grasping and cutting forces for reality-based haptic modeling - Computer assisted radiology and surgery. *17th International Congress and Exhibition,* (pp. 158–162).

Thrun, S. (1993). Exploration and model building in mobile robot domains. In *Proceeding of the IEEE International Conference on Neural Networks*, San Francisco, CA, USA, (pp. 175–180).

Thrun, S., Burgard, W., & Fox, D. (2000). A real-time algorithm for mobile robot mapping with applications to multi-robot and 3D mapping. In *Proceedings IEEE International Conference on Robotics and Automation, CRA '00* (vol. 1, pp. 321 -328).

Thrun, S. (2003). Learning occupancy grids with forward sensor models. *Autonomous Robots*, *15*(2), 111–127. doi:10.1023/A:1025584807625

Thrun, S., Burgard, W., & Fox, D. (2005). *Probabilistic robotics (Intelligent robotics and autonomous agents)*. The MIT Press.

Timoshenko, S., & Young, D. H. (1948). *Advanced dynamics*. US: McGraw-Hill Book Company.

Tinelli, A., Malvasi, A., Buscarini, M., Morales, E. R., & Stark, M. (2011). Laparoscopy and "robotics" in cervical cancer treatment: Current insights, novel approaches and future perspectives in the new era of gynecological surgery. In E. K. Salvatti (Ed.), *Brain cancer, tumor targeting and cervical cancer*, (pp. 127–162). Nova Science Publishers.

Tinelli, R., Malzoni, M., Cosentino, F., Perone, C., & Fusco, A. (2011). *Annals of Surgical Oncology*, *14*, 2654–2661.

Ting, K. L. (1986). Five-bar Grashof criteria. *Journal of Mechanisms Transmissions and Automation in Design*, *108*(4), 533–537. doi:10.1115/1.3258765

Ting, K. L., & Liu, Y. W. (1991). Rotatability laws for N-bar kinematic chains and their proofs. *ASME Journal of Mechanical Design*, *113*(3), 32–39. doi:10.1115/1.2912747

Toennies, J. L., Tortora, G., Simi, M., Valdastri, P., & Webster, R. J. (2010). Swallowable medical devices for diagnosis and surgery: The state of the art. *Proceedings of the Institution of Mechanical Engineers. Part C, Journal of Mechanical Engineering Science*, *224*, 1397–1414. doi:10.1243/09544062JMES1879

Topal, S., Erkmen, İ., & Erkmen, A. M. (2009). New prioritized multi-robot exploration strategy based on percolation model in unknown environment. In *Proceedings of the International Conference on Climbing and Walking Robots and the Support Technologies for Mobile Machines*, Istanbul, Turkey, (pp. 967–976).

Torbjorn, S. D., Mataric, M. J., & Gaurav, S. S. (2009). Multi-robot task allocation through vacancy chain scheduling. *Journal of Robotics and Autonomous Systems*, *57*, 87–91.

Torres, F., Gil, P., Puente, S. T., Pomares, J., & Aracil, R. (2004). Automatic PC disassembly for component recovery. *International Journal of Advanced Manufacturing Technology*, *23*(1-2), 39–46. doi:10.1007/s00170-003-1590-5

Torres, F., Puente, S. T., & Aracil, R. (2003). Disassembly planning based on precedence relations among assemblies. *International Journal of Advanced Manufacturing Technology*, *21*(5), 317–327. doi:10.1007/s001700300037

Torres, F., Puente, S., & Diaz, C. (2009). Automatic cooperative disassembly robotic system: task planner to distribute tasks among robots. *Control Engineering Practice*, *17*(1), 112–121. doi:10.1016/j.conengprac.2008.05.013

Tournier, G. P., Valenti, M., How, J. P., & Feron, E. (2006). Estimation and control of a quadrotor vehicle using monocular vision and Moire patterns. In *Proceedings of American Institute of Aeronautics and Astronautics Guidance, Navigation and Control Conference and Exhibit*, (pp. 21–24). Keystone, Colorado.

Tricaud, C. (2010). *Optimal sensing & actuation policies for networked mobile agents in a class of cyber-physical systems*. Doctoral Dissertation, Utah State University. Retrieved from http://www.digitalcommons.usu.edu/ctd/673

Tsagarakis, N., & Caldwell, D. (2003). Development and control of a "soft-actuated" exoskeleton for use in physiotherapy and training. *Autonomous Robots*, *15*, 21–33. doi:10.1023/A:1024484615192

Tsai, C.-C., Wu, Z.-R., Wang, Z.-C., & Hisu, M.-F. (2010). Adaptive dynamic motion controller design for a four-wheeled omnidirectional mobile robot. In *Proceedings of International Conference on System Science and Engineering.*

Tsekos, N. V., Khanicheh, A., Christoforou, E., & Mavroidis, C. (2007). Magnetic resonance-compatible robotic and mechatronics systems for image-guided interventions and rehabilitation: a review study. *Annual Review of Biomedical Engineering*, *9*, 351–587. doi:10.1146/annurev.bioeng.9.121806.160642

Turk, M., & Pentland, A. (1991). Face recognition using eigenfaces. *Proceedings of Computer Vision and Pattern Recognition.*

Udengaard, M., & Iagnemma, K. (2007). Kinematic analysis and control of an omnidirectional mobile robot in rough terrain. In *Proceedings of IEEE/RSJ International Conference on Intelligent Robots and Systems (IROS)*, (pp. 795-800).

Ugurlu, B. (2010). *Bipedal motion planning based on composite rigid body angular momentum resolution.* Doctoral dissertation, Yokohama National University, Yokohama, Japan.

Ugurlu, B., & Kawamura, A. (2009, October). Real-time running and jumping pattern generation for bipedal robots based on ZMP and Euler's equations. In *IEEE International Conference on Intelligent Robots and Systems.* (pp. 1100-1105).

Ugurlu, B., & Kawamura, A. (2010b, May). Eulerian ZMP resolution based bipedal walking: Discussions on the intrinsic angular momentum rate change about center of mass. In *IEEE International Conference on Robotics and Automation* (pp. 4218-4223).

Ugurlu, B., & Kawamura, A. (2010c, December). Bipedal walking trajectory generation based on ZMP and Euler's equations. In *IEEE International Conference on Humanoid Robotics* (pp. 468-473).

Ugurlu, B., Hirabayashi, T., & Kawamura, A. (2009, November). A unified control frame for stable bipedal walking. In *IEEE International Conference on Industrial Electronics and Control* (pp. 4162-4172).

Ugurlu, B., Saglia, J. A., Tsagarakis, N. G., & Caldwell, D. G. (2012, May). Yaw moment compensation for bipedal walking robots based on Euler's equations. In *IEEE International Conference on Robotics and Automation* (Under Review).

Ugurlu, B., Tsagarakis, N. G., Spyrakos-Papastravridis, E., & Caldwell, D. G. (2011, April). Compliant joint modification and real-time dynamic walking implementation on bipedal robot cCub. In *IEEE International Conference on Mechatronics* (to be presented).

Ugurlu, B., & Kawamura, A. (2010a). ZMP-based online jumping pattern generation for a one-legged robot. *IEEE Transactions on Industrial Electronics*, *57*(5), 1701–1709. doi:10.1109/TIE.2009.2032439

Ullman, D. G. (2002). *The mechanical design process* (pp. 178–182). McGraw-Hill Science.

Umilta, M. A., Kohler, E., Gallese, V., Fogassi, L., Fadiga, L., & Keysers, C. (2001). I know what you are doing: A neurophysiological study. *Neuron*, *32*, 91–101.

Ummat, A., Sharma, G., Mavroidis, C., & Dubey, A. (2005). Bio-nanorobotics: state of the art and future challenges. Chapter 19 in *The Biomedical Engineering Handbook*, (pp. 19-1–19-42). CRC Press.

Urcola, P., Riazuelo, L., Lazaro, M. T., & Montano, L. (2008). Cooperative navigation using environment compliant robot formations. In *IEEE/RSJ International Conference on Intelligent Robots and Systems, IROS 2008* (pp. 2789-2794).

Urschel, H. C. (Ed.). (2007). *Robotic radiosurgery: Treating tumors that move with respiration.* Springer. doi:10.1007/978-3-540-69886-9

Valavanis, K. (2007). *Advances in unmanned aerial vehicles: State of the art and the road to autonomy*. Springer. doi:10.1007/978-1-4020-6114-1

Valenzuela-Rendon, M., Uresti-Charre, E., & Monterrey, I. (1997). *A non-generational genetic algorithm for multiobjective optimization.* In Seventh International Conference on Genetic Algorithms.

van den Bedem, L. (2010). *Realization of a demonstrator slave for robotic minimally invasive surgery*. Ph.D. thesis, Eindhoven University of Technology.

Varley, P. (1999). Techniques for development of safety-related software for surgical robots. *IEEE Transactions on Information Technology in Biomedicine*, *3*(4), 261–267. doi:10.1109/4233.809170

Varma, T., & Eldridge, P. (2006). Use of the NeuroMate stereotactic robot in a frameless mode for functional neurosurgery. *International Journal of Medical Robotics and Computer Assisted Surgery*, *2*(2), 107–113. doi:10.1002/rcs.88

Vartholomeos, P., Fruchard, M., Ferreira, A., & Mavroidis, C. (2011). MRI-guided robotic nanocapsules: A review study. *Annual Review of Biomedical Engineering*, *13*, 157–187. doi:10.1146/annurev-bioeng-071910-124724

Viterbi, A. J. (1967). Error bounds for convolutional codes and an asymptotically optimum decoding algorithm. *IEEE Transactions on Information Theory*, *13*(2), 260–269. doi:10.1109/TIT.1967.1054010

Vlasenko, B., & Wendemuth, A. (2009). *Heading toward to the natural way of human-machine interaction: The Nimitek project*. Multimedia and Expo, 2009. ICME 2009.

Volgyesi, Balogh, Nadas, Nash, & Ledeczi. (2007). Shooter localization and weapon classification with soldier-wearable networked sensors. In *MobiSys '07: Proceedings of the 5th International Conference on Mobile Systems, Applications and Services* (pp. 113–126). New York, NY: ACM.

Volkswagen of America, Inc. (2004, December). *Electro-mechanical power steering design and function*. Self-Study Program, Course Number: 892403.

Vukobratovic, M., & Borovac, B. (2004). Zero moment point – Thirty five years of its life. *International Journal of Humanoid Robotics*, *1*(1), 157–173. doi:10.1142/S0219843604000083

Walther, W., & Stein, U. (2000). Viral vectors for gene transfer: A review of their use in the treatment of human diseases. *Drugs*, *60*(2), 249–271. doi:10.2165/00003495-200060020-00002

Wang, W. H., Liu, X. Y., & Sun, Y. (2007, 22-25 September). *Autonomous zebrafish embryo injection using a microrobotic system.* Paper presented at the Automation Science and Engineering, 2007. CASE 2007. IEEE International Conference on.

Wang, W. H., Sun, Y., Zhang, M., Anderson, R., Langille, L., & Chan, W. (2008, 19-23 May 2008). *A microrobotic adherent cell injection system for investigating intracellular behavior of quantum dots*. Paper presented at the Robotics and Automation, 2008. ICRA 2008. IEEE International Conference on.

Wang, S., Fuller, D., Manson, K., Wyand, M., & Lu, S. (2000). A new animal model to test the AIDS vaccine efficacy in non-human primates: Gene gun-mediated mucosal infection with proviral DNA plasmids. *Antiviral Therapy*, 5.

Wang, W., Liu, X., Gelinas, D., Ciruna, B., & Sun, Y. (2007). A fully automated robotic system for microinjection of zebrafish embryos. *PLoS ONE*, *2*(9), e862. doi:10.1371/journal.pone.0000862

Wang, W., Sun, Y., Zhang, M., Anderson, R., Langille, L., & Chan, W. (2008). *A system for high-speed microinjection of adherent cells* (*Vol. 79*). AIP.

Wang, W., Zhang, H., Zong, G., & Zhang, J. (2008). Force cooperation in a reconfigurable field multi-robot system. *Journal of Field Robotics*, *25*(11-12), 922–938. doi:10.1002/rob.20261

Watt, P. C., Sawicki, M. P., & Passaro, E. (1993). A review of gene transfer techniques. *American Journal of Surgery*, *165*(3). doi:10.1016/S0002-9610(05)80841-4

Webb, T. L., & Sheeran, P. (2006). Does changing behavioral intentions engender behavior change? A meta-analysis of the experimental evidence. *Psychological Bulletin*, *132*(2), 249–268. doi:10.1037/0033-2909.132.2.249

Wilensky, R. (1983). *Planning and understanding*. Reading, MA: Addison-Wesley.

Williams, O. M. C. (2005). *Bayesian learning for efficient visual inference*. The dissertation for the degree of Doctor of Philosophy.

Williams, S. B. (2002). Towards multi-vehicle simultaneous localisation and mapping. In *Proceedings - IEEE International Conference on Robotics and Automation* (vol. 3, pp. 2743-2748).

Willow Garage. (2011a). *PR2 users manual*. Retrieved August 10th, 2011, from http://pr2support.willowgarage.com/wiki/PR2%20Manual

Willow Garage. (2011b). *Docking station and autonomous parking for the Texas robot*. Retrieved August 10th, 2011 from http://www.willowgarage.com/blog/2010/01/26/docking-station-and-autonomous-parking-texas

Xia, T., Baird, C., Jallo, G., Hayes, K., Nakajima, N., Hata, N., & Kazanzides, P. (2008). An integrated system for planning, navigation and robotic assistance for skull base surgery. *International Journal of Medical Robotics and Computer Assisted Surgery*, *4*(4), 321–330. doi:10.1002/rcs.213

Yamaguchi, T. (2006). Image processing based on percolation model. *Journal of IEICE Transactions on Information and Systems. E (Norwalk, Conn.)*, *89-D*(7), 2044–2052.

Yanagida, K., Katayose, H., Yazawa, H., Kimura, Y., Konnai, K., & Sato, A. (1999). The usefulness of a piezo-micromanipulator in intracytoplasmic sperm injection in humans. *Human Reproduction (Oxford, England)*, *14*(2), 448–453. doi:10.1093/humrep/14.2.448

Yang, C., Huang, Q., & Han, J. (2011). Decoupling control for spatial six-degree-of-freedom electro-hydraulic parallel robot. *Journal of Robotics and Computer-Integrated Manufacturing*, *27*, 1010–1015.

Yang, S. H., Byun, J. H., & Seo, J. H. (2008). The design of kids robot: raBie. *Journal of Korean Society of Design Science*, *21*, 79–82.

Yasuda, G. (2010). Distributed cooperative control of industrial robotic systems using Petri net based multitask processing. In H. Liu, H. Ding, Z. Xiong, & X. Zhu (Eds.), *Proceedings of the 3rd International Conference on Intelligent Robotics and Applications LNAI 6425*, (pp. 32-43). Berlin, Germany: Springer-Verlag.

Yasuda, G., & Ge, B. (2010). Petri net model based specification and distributed control of robotic manufacturing systems. In *Proceedings of the 2010 IEEE International Conference on Information and Automation*, (pp. 699-705).

Yasuda, G., & Tachibana, K. (1992). Implementation of real-time control schemes on a parallel processing architecture using transputers. In *Proceedings of IEEE Singapore International Conference on Intelligent Control and Instrumentation*, (pp. 760-765).

Yasuda, G. (2000). A distributed control and communication structure for multiple cooperating robot agents. In *IFAC Artificial Intelligence in Real Time Control 2000* (pp. 283–288). Oxford, UK: Pergamon.

Ye, W., Xu, Y., & Lee, K. K. (2005). Shoe-Mouse: An integrated intelligent shoe. *2005 IEEE/RSJ International Conference on Intelligent Robots and Systems, (IROS 2005)*, (pp. 1163-1167).

Yilmaz, A., Javed, O., & Shah, M. (2006). Object tracking: A survey. *ACM Computing Surveys*, *38*(4), 13. doi:10.1145/1177352.1177355

Yinga, C., Jia-Fan, Z., Can-Jun, Y., & Bin, N. (2007). Design and hybrid control of the pneumatic force-feedback systems for Arm-Exoskeleton by using on/off valve. *Mechatronics*, *17*(6), 325–335. doi:10.1016/j.mechatronics.2007.04.001

Yokokohji, Y., & Yoshikawa, T. (1994). Bilateral control of master-slave manipulators for ideal kinesthetic coupling - Formulation and experiment. *IEEE Transactions on Robotics and Automation*, *10*(5), 605–620. doi:10.1109/70.326566

Yokoyama, A., & Omori, T. (2010). *Modeling of human intention estimation process in social interaction scene*. 2010 IEEE International Conference on Fuzzy Systems (FUZZ).

Yu, D. Y. (1998). Robotic ocular ultramicrosurgery. *Australian and New Zealand Journal of Ophthalmology*, *26*(1), 6–8. doi:10.1111/j.1442-9071.1998.tb01375.x

Yun, Y., & Li, Y. (2011). Optimal design of a 3-PUPU parallel robot with compliant hinges for micromanipulation in a cubic workspace. *Journal of Robotics and Computer-Integrated Manufacturing*, *27*, 977–985. doi:10.1016/j.rcim.2011.05.001

Zacny, K., Mungas, G., Parrington, L., Mungas, C., & Fisher, D. (2007, July). Pneumatic drill and excavator for planetary exploration. *Seventh International Conference on Mars*, Pasadena, California, (p. 3010).

Zaman, B., McKee, M., & Jenson, A. (2011) Use of high-resolution multispectral imagery acquired with an autonomous unmanned aerial vehicle to quantify the spread of an invasive wetland species (to appear) In *Proceedings of IEEE International Geoscience and Remote Sensing Symposium 2011*.

Zavlanos, M. M., & Pappas, G. J. (2008). Distributed connectivity control of mobile networks. *IEEE Transactions on Robotics*, *24*(6), 1416–1428. doi:10.1109/TRO.2008.2006233

Zhang, D., Gatica-Perez, D., Bengio, S., McCowan, I., & Lathoud, G. (2004). Modeling individual and group actions in meetings: A two-layer HMM framework. In *Proceedings of the IEEE Computer Society Conference on Computer Vision and Pattern Recognition Workshop*, (p. 117).

Zhang, P., Gu, J., Milios, E. E., & Huynh, P. (2006). Navigation with IMU/GPS/digital compass with unscented Kalman filter. In *Proceedings of IEEE International Conference Mechatronics and Automation, 2005,* (vol. 3, pp. 1497–1502). doi: 10.1109/ICMA.2005.1626777

Zhang, X., Chen, C.-C., Bernstein, R. W., Zappe, S., Scott, M. P., & Solgaard, O. (2005). Microoptical characterization and modeling of positioning forces on drosophila embryos self-assembled in two-dimensional arrays. *Journal of Microelectromechanical Systems*, *14*(5), 1187–1197. doi:10.1109/JMEMS.2005.851834

Zhou, X. S., & Roumeliotis, S. I. (2006). Multi-robot SLAM with unknown initial correspondence: The robot rendezvous case. In *IEEE/RSJ International Conference on Intelligent Robots and Systems* (pp. 1785 -1792).

Zhou, M., DiCesare, F., & Desrochers, A. A. (1993). *Petri Net synthesis for discrete event control of manufacturing systems*. London, UK: Kluwer.

Zhou, M., & Venkatesh, K. (1999). *Modeling, simulation, and control of flexible manufacturing systems: A Petri Net approach*. Singapore: World Scientific.

Zhu, C., & Kawamura, A. (2010). Development of biped robot MARI-3 for jumping. *Advanced Robotics*, *24*(11), 1661–1675. doi:10.1163/016918610X512640

Zorn, K. C. (2007). Robotic radical prostatectomy learning curve of a fellowshiptrained laparoscopic surgeon. *Journal of Endourology*, *21*, 441–447. doi:10.1089/end.2006.0239

Zwicker, E., Zesch, W., & Moser, R. (2010, October). A modular inspection robot platform for power plant applications. *2010 1st International Conference on Applied Robotics for the Power Industry (CARPI)*, (pp. 1-6). Montreal, QC.

About the Contributors

Tarek M. Sobh received the B.Sc. in Engineering degree with honors in Computer Science and Automatic Control from the Faculty of Engineering, Alexandria University, Egypt in 1988, and M.S. and Ph.D. degrees in Computer and Information Science from the School of Engineering, University of Pennsylvania in 1989 and 1991, respectively. He is currently the Vice President for Graduate Studies and Research, Dean of the School of Engineering and Distinguished Professor of Engineering and Computer Science at the University of Bridgeport (UB), Connecticut; the Founding Director of the Interdisciplinary Robotics, Intelligent Sensing, and Control (RISC) laboratory; the Founder of the High-Tech Business Incubator at UB (CTech IncUBator), and a Professor of Computer Engineering, Computer Science, Electrical Engineering, and Mechanical Engineering.

Xingguo Xiong is an Associate Professor in Department of Electrical and Computer Engineering, University of Bridgeport, Connecticut, USA. He obtained his B.S. degree in Physics in Wuhan University, Wuhan, China in 1994. He obtained his first PhD degree in Electrical Engineering from Shanghai Institute of Microsystem and Information Technology, Chinese Academy of Sciences, China in 1999. He obtained his second PhD degree in Computer Engineering in University of Cincinnati, OH, USA in 2005. He joined University of Bridgeport as faculty since Fall 2005. Prof. Xiong's research areas include Microelectromechanical Systems (MEMS), nanotechnology, as well as VLSI design and testing. He is also interested in robotic system design and prototyping, microrobots, and nanorobots. He has two patents, and published one book chapter and numerous conference/journal papers. He is the recipient of 2009 Northeast ASEE (American Society of Engineering Education) Outstanding Teacher Award.

* * *

Ahmad Azar has received the M.Sc. degree (2006) in System Dynamics and Ph.D degree (2009) in Adaptive Neuro-Fuzzy Systems from Faculty of Engineering, Cairo University (Egypt). Dr. Ahmad Azar has worked in the areas of system dynamics, intelligent control, and modelling in biomedicine and is the author of more than 35 papers in these subjects. He is an editor of two books in the field of fuzzy logic systems and biomedical engineering. Dr. Ahmad Azar is closely associated with several international journals as a reviewer. He serves as international programme committee member in many international and peer-reviewed conferences. He is the Editor in Chief of *International Journal of System Dynamics Applications* (IJSDA), and he currently serves as the editor of a lot of international journals. His biography was selected to appear in the 27th Edition of Who's Who in the World, Marquis Who's Who, USA, 2010.

YangQuan Chen received the B.S. degree in Industrial Automation from the University of Science and Technology of Beijing, Beijing, China, in 1985, the M.S. degree in Automatic Control from the Beijing Institute of Technology, Beijing, in 1989, and the Ph.D. degree in Advanced Control and Instrumentation from the Nanyang Technological University, Singapore, Singapore, in 1998. He is currently an Associate Professor of Electrical Engineering at Utah State University, Logan, and the Director of the Center for Self-Organizing and Intelligent Systems (CSOIS). His current areas of research interests include: applied fractional calculus in controls, signal processing and energy informatics; distributed measurement and distributed control of parameter systems using mobile actuator and sensor networks; mechatronics; multi-UAV based cooperative multi-spectral "personal remote sensing," and real-time water management and irrigation control.

Sedat Dogru received from Middle East Technical University Turkey his B.Sc. degree in Electrical and Electronics Engineering and his B.Sc. degree in Physics (double major) in 2000 and 2001, respectively. He has served in the industry as software developer and project leader for almost 4 years. He is currently a Ph.D. student at the Department of Electrical and Electronics Engineering of Middle East Technical University Turkey. His research interests are mainly Wireless Sensor Networks, SLAM, path and motion planning, and connectivity maintenance.

Akif Durdu received his B.Sc. degree in Electrical and Electronics Engineering from the Selcuk University, (SU), Turkey, in 2001. He had worked as a researcher at Selcuk University for one year. He has been a PhD candidate and a research assistant at the Department of Electrical and Electronics Engineering from the Middle East Technical University in Ankara since 2002. His research interests are controlling systems, making robots, intention estimation and machine learning algorithms in human-robot interaction. He has been a visiting scholar in order to make research at the Ohio State University since January, 2011. He is a member of IEEE.

Sam Eljamel, Consultant Neurosurgeon, graduated with honours MBBCh-1982, fellow of Royal College of Surgeons-Ireland and Royal College of Surgeons of Edinburgh. His qualifications include intercollegiate fellowship in Neurosurgery and M.D with distinction in neurosciences Liverpool University. The higher surgical training committee in Neurosurgery-UK awarded him a certificate of completion of specialist training in Neurosurgery-1993 and had completed clinical fellowship in Neurosurgery-University of Connecticut & Hartford Hospital, Connecticut-1995. He has been a National Panelist in Neurosurgery for 8 years and Head of the Service at Ninewells Hospital & Medical School. His practice concentrates on skull base surgery, keyhole neurosurgery, and functional neurosurgery. 1996, he was appointed Head of Department, Surgical Neurology, University of Dundee, 2006 he took the Chairmanship of Clinical Governance Committee. He was elected secretary of British Stereotactic & Functional Neurosurgery 2006, joined Executive Committee of AANS/CNS-Tumor Section in 2007, elected board member of International Photodynamic Association 2008, and Vice President of BIDA 2008. His main research interests are surgical robotics, stereotaxy, photodynamic therapy and neuro-oncology. He had published more than 80 papers, edited two books, and coauthored more than 6 chapters.

Ahmed ElSayed received his B.Sc. in Electrical Engineering in 2003 from the Department of Electrical Engineering, Alexandria University, Egypt, and M.Sc. in Engineering Mathematics in 2006 from the Department of Engineering Mathematics and Physics, Alexandria University, Egypt. Currently, he is a Ph.D. candidate and holds a Teaching and Research Assistant position in the Department of Computer Science and Engineering, University of Bridgeport, Bridgeport, CT. He was awarded a Teaching Assistantship by the Department of Engineering Mathematics and Physics, Alexandria University, Egypt. He published and presented papers in the fields of Fuzzy Control systems and Soft Computing methods and its applications. His research interests include AI, fuzzy systems, soft computing, hybrid system, computer vision, and signal and image processing.

Vikas Reddy Enti, currently at Kiva Systems Inc., was a Research Engineer at the A*STAR Data Storage Institute. He started the DAvinCi project to integrate Cloud Computing and Robotics and is interested in systems engineering of robots and large systems. He specializes in the mechanical design of mobile robots. Vikas has a B.S in Mechanical Engineering (2008) from Cornell University.

Aydan M Erkmen received her B.Sc. from Bogazici Univ. Turkey in Electrical Eng double majoring in Math, her M.Sc. in ECE from Drexel University, USA, Ph.D. in Info Tech & Eng from George Mason Univ, USA in 1989. She is full Professor in Electrical Eng. Dept Middle East Tech Univ Turkey. Her research interests include mechatronic design, search & rescue robotics, robot hands, sensor fusion, biologically inspired control, multi-robots networks, and sensor networks. She served 2 terms in IEEE Robotics and Automation Society (RAS) Admin Committee; was IEEE RAS vice president 2000 to 2003; was till 2010 associate Editor *Int. Journal of Robotics & Automation*. She is since 2009 Editor in Springer Verlag-Paladyn *Journal of Behavioral Robotics*. She served 3 years on the editorial board of *IEEE Robotics and Automation Magazine*. She was 4 years at Turkish Scientific and Technological Research Council (TUBITAK) as Vice President for industrial funds (TIDEB), the FP6 national coordinator till 2002, IST NCP till 2003. Between November 2007-Dec 2009, she was the Vice Dean of the Faculty of Engineering in Middle East Technical University. Since 2009 she serves in EURON as an Award Jury member for George Giralt PhD Theses Awards.

Ismet Erkmen received his B.Sc. and M.Sc. degrees in EE from the Middle East Technical University, (METU), Turkey, in 1970 and 1972 respectively, and Ph.D. degree from Drexel University, Philadelphia, in 1984. His employment experience included the Turkish Electricity Authority; Brown Boveri, Switzerland; Drexel University; Carlsen and Fink Associates, Fairfax; and Turkish Scientific and Technological Research Council. He is currently a full Professor and chairman of the Electrical and Electronics Engineering Dept. of METU. He is the founder of the Mechatronics Lab. in the dept. He also serves as the executive board member of METU-BILTER CAD/CAM, Robotics and Automation Research Center. He has been awarded as the excellency in education in 93-94 at METU. He is a member of IEEE.

Nicolae Gari received his Bachelor of Science degree in Electronics and Digital Telecommunications from Polytechnic University of Bucharest, Romania in 2002. He obtained his Master of Science degree in Electrical Engineering from University of Bridgeport, Connecticut, USA in 2010. His research interests include robotic system design and prototyping, automations, controls, telecommunications, renewable and sustainable energy, smart grid technology, et cetera.

Andrew Goldenberg, Ph.D., is the founder and President of Engineering Services Inc. (ESI) (www.esit.com), established in 1982. ESI is a high-technology company involved in the development of technology for robotics-based automation. In addition, Dr. Goldenberg is President of Anviv Mechatronics Inc. (AMI), which he founded in 2006. Anviv is involved in the development of mechatronics products. From 2000-2001 Dr. Goldenberg was also the President of Virtek Engineering Science Inc. (VESI) a company formed after the acquisition of part of ESI by Virtek Vision International Ltd. Since 1982 Dr. Goldenberg has been a Professor of Mechanical and Industrial Engineering at the University of Toronto (www.mie.utoronto.ca/labs/ral). He is cross-appointed in the Department of Electrical and Computer Engineering, and the Institute of Biomaterials and Biomedical Engineering, both of University of Toronto. Dr. Goldenberg is also an Adjunct Professor at Ryerson University in Toronto, Ontario, and a Guest Professor at Nanjing University of Science and Technology in P.R. China. From 1975 – 1981, Dr. Goldenberg was an employee of SPAR Aerospace Ltd. in Toronto. There he worked on the first Space Shuttle Remote Manipulator System - the Canadarm. Dr. Goldenberg is a former Editor of the archival international journal *IEEE Transactions on Robotics and Automation*. He also serves on the editorial boards of *Journal of Robotics, Robotica*, and *Robotics in Japan*. He is a Fellow of IEEE, a Fellow of ASME, a Member of Professional Engineers of Ontario, a Designated Consulting Engineer in Ontario, a Fellow of Engineering Institute of Canada, and a Fellow of Canadian Academy of Engineering. Dr. Goldenberg has authored and co-authored over 33 patents, more than 135 peer-reviewed journal papers, over 360 conference proceedings papers, 12 chapters in books, and has made numerous presentations at scientific and engineering conferences. He has supervised over 100 graduate students to date (38 PhD).

Surendra M. Gupta, P.E., is a Professor of Mechanical and Industrial Engineering and the director of the Laboratory for Responsible Manufacturing, Northeastern University. He has authored or coauthored well over 400 technical papers published in books, journals, and international conference proceedings. Dr. Gupta has taught over 100 courses in such areas as operations research, inventory theory, queuing theory, engineering economy, supply chain management, and production planning and control. Among the many recognitions received, he is the recipient of outstanding research award and outstanding Industrial Engineering Professor award (in recognition of teaching excellence) from Northeastern University as well as a national outstanding doctoral dissertation advisor award.

Tamás Haidegger received his M.Sc. degrees from the Budapest University of Technology and Economics (BME) in Electrical Engineering and Biomedical Engineering in 2006 and 2008, respectively, then Ph.D. in Electrical Engineering in 2011. His main field of research is control/teleoperation of surgical robots, image-guided therapy and supportive medical technologies. Tamás spent the 2007/08 academic year with the CISST ERC at the Johns Hopkins University, working on a skull base surgery setup. Currently, he is a research fellow at BME, and in the mean time, he is in post-doc position at the recently founded Austrian Center of Medical Innovation and Technology (ACMIT), working on minimally invasive surgical simulation and training, medical education, and the technological improvement of gynecological brachytherapy. Tamás is the CEO of a university start-up, focusing on objective hand hygiene control. He has been an active member of various professional organizations since 2002. Currently, he is the student chair of the IEEE Robotics and Automation Society and the secretary of the

Hungarian MATE Control Engineering Chapter. He is the author of more than 50 papers published in various scientific meeting and conference proceedings, refereed journals and books in the field of biomedical, control engineering, and computer-integrated surgery. He has been maintaining a professional blog on CIS for over 4 years: http://surgrob.blogspot.com/.

Lee Jun Hong, currently at Dyson Operations, was a Research Engineer at Data Storage Institute. He is interested and specializes in embedded systems design of mobile robots. Jun Hong has an MSc in Embedded Systems (2008) from Nanyang Technological University and a BEng (Hons) in Electronics Engineering (2003) from University of Surrey.

Haibo Huang received the Ph.D. degree in Manufacturing Engineering and Engineering Management from the City University of Hong Kong, Kowloon, Hong Kong. He was a Postdoctoral Research Fellow at the Department of Mechanical and Industrial Engineering, University of Toronto, ON, Canada. He is now an Associate Professor with the Soochow University, Suzhou, China. His current research interests include microrobotic cell manipulation, microforce sensing/controlling, bio-device and instrumentation, and micro/nano manipulation.

Atsuo Kawamura was born in Yamaguchi, Japan, in 1953. He received the B.E., M.E., and Ph.D. degrees in Electrical Engineering from the University of Tokyo, Tokyo, Japan, in 1976, 1978, and 1981, respectively. Upon receiving the PhD degree, he joined Department of Electrical and Computer Engineering at the University of Missouri-Columbia as a post-doctoral fellow, and was an Assistant Professor there from 1983 to 1986. In 1986, he joined Department of Electrical and Computer Engineering at Yokohama National University, Yokohama, Japan as an Associate Professor, and became a Professor in 1996. His research interests are power electronics, electric vehicles, Shinkansen traction control, bipedal robotics, and mechatronic system control. He is a recipient of IEEE IAS Transactions Prize Paper Award in 1988, IEE of Japan Transactions Prize Paper Award in 1996, and IEEE IES Transactions Best paper Awards in 20001 and 2002, and also IEEE Fellow in 2002. Prof. Kawamura is a member of the IEE of Japan, IEEE (Fellow), Robotics Society of Japan, the Institute of Electronics, Information and Communication Engineers, the Society of Instrument and Control Engineering.

Elif Kongar is an Assistant Professor at Bridgeport University and a Visiting Researcher in the Center for Industrial Ecology at Yale University. Her research interests are in sustainability, energy and environment, environmentally conscious manufacturing, product recovery, disassembly systems, and multiple criteria decision making. Dr. Kongar is the author of numerous journal and conference papers, and has presented her work at various national and international conferences. She is a member of the ASEE, SWE, Scientific Research Society, Sigma Xi, the Industrial Engineering Honor Society, Alpha Pi Mu, the Phi Beta Delta Honor Society, and the Phi Kappa Phi Honor Society. Dr. Kongar received her BS and MS degrees in industrial engineering from Yildiz Technical University, and PhD degree in Industrial Engineering from Northeastern University. Before joining Bridgeport University, Dr. Kongar was an Assistant Professor of Industrial Engineering at Yildiz Technical University where she was also the Coordinator and Lecturer of the Logistics certificate program.

Pierre Letier received his Ph.D. in Mechanical Engineering from the Université Libre de Bruxelles in 2010. His Ph.D. was dedicated to the development and control of force-feedback haptic interfaces. He joined the Company Space Application Services in 2010, as robotics and systems engineer. He is involved in several ESA and European Research projects related to human-machine interfaces, haptic, rehabilitation, and virtual reality applications.

Zheng (Jeremy) Li is an Associate Professor in University of Bridgeport. He has many years of research, engineering, and industrial experiences before working as a full time faculty in university. He has two book chapters published in the past and two more book chapters pending for publication. He also has 18 journal papers and 40 conference papers published. His research projects include robotic system design and development in surgical and industrial applications, biomedical instrument design and development, automated and high speed machinery design and development, sustainable energy system design and development including solar panel and wind power, automated manufacturing and process control, computer-aided modeling and simulation in industrial applications, and structure / stress (FEA) analysis.

Bingbing Liu is a Research Scientist with Institute of Infocomm, A*STAR, Singapore. He was a Senior Research Fellow with Data Storage Institute, A*STAR, Singapore. He received the B.E. degree in control engineering from Harbin Institute of Technology, China in 2000 and the Ph.D. Degree from Nanyang Technological University, Singapore in 2007. His current research interests include inertial navigation system, simultaneous localization and mapping for mobile robots, multi-sensor data fusion, estimation theory and computer vision for robotic navigation. He has published more than 20 international journal/conference papers and book chapters in these areas. Bingbing Liu has served as a program committee member or associate editor for multiple international conferences, such as WiSARN 2010 & 2011, ICIRA 2011, and CIS-RAM 2011. He is an IEEE member.

Mac McKee is Professor of Civil and Environmental Engineering and the Director of the Utah Water Research Laboratory (UWRL), a world renowned water research institution. Dr. McKee's famous water management slogan: "*If you do not measure it, you cannot manage it*" motivated the UAV research program at Utah State University in 2005. Dr. McKee started a unique UAV service center at the UWRL called the AggieAir Flying Circus, (http://aggieair.usu.edu) which now offers commercial flights under official FAA CoAs in support of water and environmental research projects. His recent interests are in using UAV based "personal remote sensing" platforms for various real world missions such as the invasive species monitoring and control in wetlands, multiscale information fusion techniques using various learning vector machines, and water resource management.

Sarosh Patel received the B.E. degree in Electrical and Electronics Engineering with Distinction from the Faculty of Engineering Osmania University, India and M.S. degrees in Electrical Engineering and Technology Management from the School of Engineering, University of Bridgeport (UB). He is currently pursuing Ph.D. in Computer Engineering at U.B. He currently works as a Research Assistant at the Interdisciplinary RISC (Robotics and Intelligent Systems Control) Lab. He had been nominated for inclusion in 2005 & 2006 edition of Who's Who Among Students in American Universities and has been elected to the Phi Kappa Phi honor society.

André Preumont has been a Professor of Mechanical Engineering and Robotics at ULB since 1987. Before that, he was a consultant for Belgonucléaire and Visiting Professor at Virginia Tech. (USA). He also held visiting appointments at UTC Compiègne and INSA de Lyon (France). He is a member of the Belgian Royal Academy and was the recipient of the Five-year prize in Applied Exact Sciences of FNRS in 2000 and the Alexander von Humboldt Research Award in 2011. He is the author of several books, including: Random Vibrations and Spectral Analysis, Kluwer 1994; Vibration Control of Active Structures, An Introduction, Kluwer, 2nd Edition in 2002; Mechatronics, Dynamics of Electromechanical and Piezoelectric Systems, Springer 2006; Active Control of Structures (with Prof. K.Seto), Wiley, 2008.

Changhai Ru received the Ph.D. degree in Mechatronics Engineering from Harbin Institute of Technology, Harbin, China, in 2005. He is currently a Professor at Robotics and Microsystems Center, Soochow University, China. He was a Postdoctoral Fellow at the Advanced Micro and Nanosystems Laboratory, Department of Mechanical and Industrial Engineering, University of Toronto, ON, Canada. His research interests include micro–nanomanipulation, nano–positioning technology, and solid-state actuators' driving and control methods.

Brandon Stark is a Ph.D. student at the Center for Self Organizing and Intelligent Systems (CSOIS) in the Department of Electrical & Computer Engineering at Utah State University. He received the B.S. degree in Computer Engineering at University of California, Irvine in 2007 and the M.S. degrees in Computer Engineering and Electrical Engineering at the University of Bridgeport in 2010. His research interests include the control of Unmanned Aerial Vehicles (UAVs), inertial navigation, and image processing.

Hao Su received the Bachelor's degree in Control Science and Engineering from the Harbin Institute of Technology and Master's degree from the Mechanical and Aerospace Engineering from State University of New York University at Buffalo, and he is currently a Ph.D. candidate in the department of mechanical engineering of Worcester Polytechnic Institute. His current research interests include surgical robotics, image-guided therapy, and cell manipulation.

Sebahattin Topal received his B.Sc. degree in Electronics and Communication Engineering from Süleyman Demirel University, (SDU), Turkey, in 2002. He is currently a Ph.D. student at the Department of Electrical and Electronics Engineering from the Middle East Technical University. His research interests are in the coordination of multi-robots, simultaneously localization, and mapping algorithms and particle filter.

Barkan Ugurlu was born in Cankiri, Turkey in 1981. He received his B.E. in Electrical Engineering with high honor from Yildiz Technical University, Istanbul, Turkey, and his M.E. in Mechatronics Engineering from Istanbul Technical University, Istanbul, Turkey, in 2004 and 2006, respectively. Subsequently, he obtained his Ph.D. degree from the Department of Electrical and Computer Engineering, Yokohama National University, Yokohama, Japan. He is a recipient of Japan, Ministry of Education, Culture, Sports and Technology (MEXT) Monbukagakusho governmental scholarship, from 2006 to 2010. He was also a research assistant fellow of Global Center of Excellence (GCOE) between 2008 and 2010. Upon receiving the PhD degree, he served as a post-doctoral researcher in the Department of

Advanced Robotics, Italian Institute of Technology, Genoa, Italy. In April 2011, he joined Department of Advanced Science and Technology, Toyota Technological Institute, Nagoya, Japan as a post-doctoral associate. His research interests are bipedal robot dynamics and control, humanoid motion planning, rigid body dynamics, space robotics, and power assist systems. Dr. Ugurlu is a member of IEEE.

Xiaojun Wu, currently is a Research Scientist at Data Storage Institute, A*STAR, Singapore. He was the co-Project leader of the A*STAR Social Robotics Program during year 2008 - 2011, focused on the development of indoor mobile robots. He specializes in robot motion planning, navigation, and sensor integration. His other research interests include human-robot interaction and cloud-computing for group of robots etc. Xiaojun has a Ph.D. degree in the field of robot motion planning from Nanyang Technological University, a M.Sc. in Vehicle Engineering from Huazhong University of Science and Technology and a B.E. in Mechanical Engineering from Wuhan University of Technology, respectively.

Gen'ichi Yasuda was a Professor in Department of Human and Computer Intelligence, Faculty of Informatics, Nagasaki Institute of Applied Science, Japan, from 1996 to 2010. He was also a Professor in Division of Instrumentation and Control, Department of Electronics and Information Technology, Graduate School of Engineering, Nagasaki Institute of Applied Science, from 1996 to 2010. He is now a guest Professor of Nagasaki Institute of Applied Science. His teaching and research interests include industrial intelligent robotics, advanced intelligent mechatronics, and distributed autonomous robotic systems.

Alper Yilmaz is an Assistant Professor with The Civil, Environmental and Geodetic Engineering Department at The Ohio State University since 2006. He earned Ph.D. degree in Computer Science from the University of Central Florida in 2004. He was awarded the Turkish Informatics Foundation Academic Award (1999), two merit fellowships from UCF (2000, 2001), the Hillman Award for research excellence from UCF (2003), and the Duane Brown Photogrammetry award from OSU (2008). Dr. Yilmaz's area of expertise is computer vision and image understanding. He has served as an associate editor for the Machine Vision and Applications journal from 2006 to 2011. Dr. Yilmaz is a program committee member of leading Computer Vision Conferences. He has served as an area chair for the IEEE Conf. on Computer Vision and Pattern Recognition (2008) and the International Conference on Pattern Recognition (2010). He acted as treasurer for 2010 ACM International Conference on Health Informatics.

Xi Zheng was educated at University of Washington and Nanyang Technological University (NTU). He graduated in 2010 and received his B.E. degree in Electrical & Electronic Engineering from NTU. He then founded the software consultancy firm, Thinking Dots, in the following year. He has worked on varying fields such as design of microwave circuits to fiber optics communication. He also has keen interest in autonomous robots and has worked on various platforms. His area of research focuses on localization and probabilistic robotics, in particular, the effect of different models on the convergence of particles.

Index

E

F

G

H

I

J

K

L

M

S

T

U

V

W

Z

www.ingramcontent.com/pod-product-compliance
Lightning Source LLC
LaVergne TN
LVHW060351300926
849162LV00006B/1244
* 9 7 8 1 4 6 6 6 0 1 7 6 5 *